T0304641

UNIVERSAL THEMES OF BOSE-EINSTEIN CONDENSATION

Following an explosion of research on Bose-Einstein condensation (BEC) ignited by demonstration of the effect by 2001 Nobel Prize winners Cornell, Wieman, and Ketterle, this book surveys the field of BEC studies. Written by experts in the field, it focuses on BEC as a universal phenomenon, covering topics such as cold atoms, magnetic and optical condensates in solids, liquid helium, and field theory. Summarising general theoretical concepts and the research to date – including novel experimental realisations in previously inaccessible systems and their theoretical interpretation – it is an excellent resource for researchers and students in theoretical and experimental physics who wish to learn of the general themes of BEC in different subfields.

NICK P. PROUKAKIS is Professor of Quantum Physics at Newcastle University and the Associate Director of the Joint Quantum Centre Durham-Newcastle. He is an expert on finite temperature non-equilibrium modelling of quantum gases; his research interests span from atomic physics to optical condensates, being mainly focused on universal features, non-equilibrium dynamics, superfluid turbulence, and multi-component quantum matter.

DAVID W. SNOKE is Professor of Physics at the University of Pittsburgh and a Fellow of the American Physical Society. He is author of over 130 articles and 4 books, including *Bose-Einstein Condensation* (Cambridge University Press, 1996), one of the first books to survey the phenomenon. His research focuses on nonequilibrium dynamics, semiconductor optics, and Bose-Einstein condensation.

PETER B. LITTLEWOOD is a professor of physics at the University of Chicago, former Head of the Cavendish Laboratory, Cambridge University, and former Director of Argonne National Laboratory. Author of more than 200 articles, he has played an important role in the theory of Bose-Einstein condensation in cold atoms, polaritons, and excitonic systems.

UNIVERSAL THEMES OF BOSE-EINSTEIN CONDENSATION

NICK P. PROUKAKIS
Newcastle University

DAVID W. SNOKE
University of Pittsburgh

PETER B. LITTLEWOOD
University of Chicago

Shaftesbury Road, Cambridge CB2 8EA, United Kingdom

One Liberty Plaza, 20th Floor, New York, NY 10006, USA

477 Williamstown Road, Port Melbourne, VIC 3207, Australia

314–321, 3rd Floor, Plot 3, Splendor Forum, Jasola District Centre, New Delhi – 110025, India

103 Penang Road, #05–06/07, Visioncrest Commercial, Singapore 238467

Cambridge University Press is part of Cambridge University Press & Assessment, a department of the University of Cambridge.

We share the University's mission to contribute to society through the pursuit of education, learning and research at the highest international levels of excellence.

www.cambridge.org
Information on this title: www.cambridge.org/9781107085695

10.1017/9781316084366

First published 2017

A catalogue record for this publication is available from the British Library

Library of Congress Cataloging-in-Publication data
Names: Proukakis, Nick, editor. | Snoke, David W., editor. | Littlewood, Peter B., editor.
Title: Universal themes of Bose-Einstein condensation / edited by
Nick P. Proukakis (Newcastle University), David W. Snoke (University of Pittsburgh),
Peter B. Littlewood (University of Chicago).
Description: Cambridge, United Kingdom ; New York, NY : Cambridge University Press, 2017. |
Includes bibliographical references and index.
Identifiers: LCCN 2016039245| ISBN 9781107085695
(Hardback ; alk. paper) | ISBN 1107085691 (Hardback ; alk. paper)
Subjects: LCSH: Bose-Einstein condensation.
Classification: LCC QC175.47.B65 U55 2017 | DDC 530.4/2–dc23
LC record available at https://lccn.loc.gov/2016039245

ISBN 978-1-107-08569-5 Hardback

Contents

Foreword

At the time of the first workshop in this series in 1993, the only experimentally realized Bose condensate (at least in the simple sense conjectured by Einstein) was liquid ^{4}He. In the intervening twenty-plus years, much has happened in the world of Bose-Einstein condensation (BEC). Probably the most exciting development has been the attainment of condensation in ultracold bosonic atomic gases such as ^{87}Rb and ^{23}Na in 1995, followed a few years later by the achievement of degeneracy and eventually Bardeen-Cooper-Schrieffer (BCS) pairing in their fermionic counterparts, and the experimental realization of the theoretically long-anticipated "BEC-BCS crossover" by using the magnetic field degree of freedom to tune the system through a Feshbach resonance. One particularly fascinating aspect of the latter has been the realization of a "unitary gas" at the resonance itself – a system which prima facie has no characteristic length scale other than the interparticle separation, and is therefore a major challenge to theorists. Other systems in which BEC has been realized, sometimes transiently, include exciton-polariton complexes in semiconducting microcavities and, at least in a formal sense, the magnons in a magnetic insulator, as well as ultracold gases with a nontrivial and sometimes large "spin" degree of freedom.

As compared with our "traditional" Bose condensate, liquid ^{4}He, these new systems typically have many more (and more rapidly adjustable) control parameters, and have therefore permitted qualitatively new types of experiment. One particularly fascinating development has been the use of optical techniques to generate "synthetic gauge fields" and thus mimic some of the topologically nontrivial systems which have recently been of such intense interest in a condensed-matter setting. At the same time, there remain long-standing issues from helium physics, such as the nature and consequences of "spontaneously broken U(1) symmetry," the "Kibble-Zurek" mechanism, and more generally the relaxation of strongly non-equilibrium states to equilibrium; in some cases, the new systems have been used to address these more quantitatively than was possible with ^{4}He. The chapters in this volume address all of these questions and more, and should be of intense interest to both the experimental and the theoretical sides of the BEC community.

Tony Leggett
University of Illinois at Urbana-Champaign, USA

Preface

This book marks the twentieth anniversary of the publication of the book *Bose-Einstein Condensation* by Cambridge University Press. The book was the result of the 1993 meeting in Levico-Terme, Italy, organized by Allan Griffin, David Snoke, and Sandro Stringari, with significant help from Andre Mysyrowicz. That meeting grew out of a desire by many theorists and experimentalists to discuss the general themes of Bose-Einstein condensation, to draw connections between different physical systems.

One of the major driving forces for that meeting was the desire to have another example of Bose-Einstein condensation besides liquid helium. There was serious discussion at the time of whether nature abhorred a condensate, and liquid helium was a special, anomalous case. Experiments on spin-polarized hydrogen, excitons in semiconductors, and optically trapped atoms had been going on for more than a decade, without success. To move the field forward, the organizers of the 1993 meeting brought together world experts on the general theory, and experimentalists of all types, to discuss the universal themes of Bose-Einstein condensation generally. There were fascinating and heated debates about such topics as the time scale for condensation (could it be possible that condensation will not occur in a system with finite lifetime?), the concept of spontaneous symmetry breaking (could there ever be a universal "phase standard" for condensates?), and how superfluidity and condensation are related.

The situation is quite changed now. We now have many experimental examples of Bose-Einstein condensation, most notably atoms at very low temperature in optical traps, which led to the Nobel Prize in Physics in 2001. This work of Eric Cornell and Carl Wieman was first announced at the second general meeting on Bose-Einstein condensation in 1995, in Mt. Ste. Odile, France. This led to the successful conference series on atomic Bose-Einstein condensation, which now takes place regularly in San Feliu de Guixols, Spain.

Because of this changed situation, the present book does not have the same form as the 1995 book. At that time, it was possible to survey a good fraction of all the experimental and theoretical efforts in the field. The field is now so large that no book can do that comprehensively, and this book leaves out a good many significant topics. But the meeting[1] which led to this book, held at the Lorentz Center[2] in Leiden, Netherlands, in 2013, had much the same spirit as the original 1993 meeting, namely to bring together many of the world's experts on the general theory and diverse experiments on Bose-Einstein condensation, with the aim of discussing universal questions, some of which are still debated. This book aims to have that spirit of looking at the larger questions, while also surveying many of the particular experimental systems. Several of the people at the 1993 and/or 1995 meetings gave impetus to the 2013 Leiden meeting, such as Gordon Baym, Wolfgang Ketterle, Tony Leggett, David Snoke, Henk Stoof, and Sandro Stringari.

It is with great sadness that we note that the chair of the original 1993 meeting, Allan Griffin, who was a driving force of nonequilibrium and condensate physics and the San Feliu de Guixols conference series for many years, passed away in 2011, before the meeting in Leiden. In fact, the first discussions for organizing a twenty-year anniversary meeting took place between David Snoke and Nick Proukakis during "Griffinfest," a research symposium held in Toronto in May of 2011, attended by colleagues, friends, and family of Allan Griffin, just a few days before he passed away.[3] His energy and zeal would surely have made a significant contribution to this book.

Nick P. Proukakis
Joint Quantum Centre Durham-Newcastle, Newcastle University, UK

David W. Snoke
University of Pittsburgh, USA

Peter B. Littlewood
University of Chicago, USA

[1] "Universal themes of Bose-Einstein Condensation" workshop, organized by K. Burnett, P. B. Littlewood, N. P. Proukakis, D. W. Snoke, and H. T. C. Stoof, 11–15 March 2013. Details can be found at www.lorentzcenter.nl/lc/web/2013/546/info.php3?wsid=546.

[2] We gratefully acknowledge the wonderful support received by all the staff at the Lorentz Center, and in particular Corrie Kuster and Mieke Schutte, whose constant support ensured we could focus on the "science," thus indirectly assisting us in the early stages of planning of this book.

[3] Details can be found at ultracold.physics.utoronto.ca/GriffinFest.html.

Part I

Introduction

1

Universality and Bose-Einstein Condensation: Perspectives on Recent Work

DAVID W. SNOKE

Department of Physics and Astronomy, University of Pittsburgh, Pennsylvania, USA

NICK P. PROUKAKIS

Joint Quantum Centre (JQC) Durham–Newcastle, Newcastle University, UK

THIERRY GIAMARCHI

Department of Quantum Matter Physics, University of Geneva, Switzerland

PETER B. LITTLEWOOD

Argonne National Laboratory, Lemont, Illinois, USA
James Franck Institute and Department of Physics, University of Chicago, Illinois, USA

The study of Bose-Einstein condensation has undergone a remarkable expansion during the last twenty years. Observations of this phenomenon have been reported in a number of diverse atomic, optical, and condensed matter systems, facilitated by remarkable experimental advances. The synergy of experimental and theoretical work in this broad research area is unique, leading to the establishment of Bose-Einstein condensation as a universal interdisciplinary area of modern physics. This chapter reviews the broad expansion of Bose-Einstein condensation physics in the past two decades.

1.1 Introduction

The field of Bose-Einstein condensation (BEC) has undergone an explosive expansion during the past twenty years. Newcomers to this field are now often introduced to this as a universal phenomenon, which nonetheless exhibits diverse (and sometimes strikingly different) manifestations. Despite such differences, the common underlying theme creates a unique identity across many different energy and length scales.

The study of BEC as a universal phenomenon was highlighted in a focused conference[1] in 1993, leading to the publication of the well-known "green book" [1], which surveyed the breadth of the field of condensate physics at that time. The success of the conference led to a second meeting[2] in 1995, at which Eric Cornell and Carl Wieman announced the achievement of Bose-Einstein condensation of ultracold ^{87}Rb atoms in a harmonic trap. That work began an explosion of new research in the field of cold atoms, which has continued to this day, and this very success inevitably led many of those studying cold atoms to pay less attention to other types of condensates. Wolfgang Ketterle gives a historical overview of this exciting period of time in Chapter 3. Recently, various scientific meetings have worked to re-establish the physical connections across different BEC systems, and in 2013 a workshop[3] was held with the focused goal of improving communications across disciplines. This present book is an outgrowth of that meeting.

1.2 The Situation Before the Revolution

Because of the great success of the cold atom BEC and other BEC systems in the past twenty years, it may be hard for young scientists to understand the climate of BEC research in the early 1990s. At that time, there was only one known example of BEC, namely liquid helium-4, and there was a small but vocal minority of scientists who questioned whether BEC was established even in that system. One reason for these objections was that there was no clear-cut demonstration in liquid helium of the canonical property of a condensate, namely macroscopic occupation of the ground state. Neutron scattering data (reviewed by Paul Sokol in the 1995 green book [1]) were consistent with macroscopic occupation of the ground state, but because liquid helium is strongly interacting, there was no obvious condensate peak at zero momentum.

At the same time, up to the early 1990s, some scientists still argued that BEC had little or nothing to do with Bardeen–Cooper–Schrieffer (BCS) superconductivity or the BCS state of liquid helium-3, following John Bardeen's own original objections to such a connection, although J. M. Blatt, [2] Tony Leggett, [3] and others had argued for a deep connection between the two very early on; Leggett's recent book, *Quantum Fluids* [4], elegantly draws the connection between the two. Mohit Randeria's chapter on BEC-BCS crossover was considered groundbreaking

[1] First International Conference on Bose-Einstein Condensation, Levico Terme, Italy, 1993; organized by D. Snoke, A. Griffin, S. Stringari, and A. Mysyrowicz.

[2] Second International Conference on Bose-Einstein Condensation, Mt. Ste. Odile, France, 1995; organized by J. T. M. Walraven, J. Treiner, and M. W. Reynolds.

[3] "Universal Themes of Bose-Einstein Condensation," Lorentz Center, Leiden, The Netherlands; organized by K. Burnett, P. B. Littlewood, N. P. Proukakis, D. W. Snoke, and H. T. C. Stoof.

to include in the green book [1]. Indeed, this was one of the universal themes explored at the 1993 meeting, with the general theory of BEC-BCS crossover originally explored in the context of excitons in semiconductors [5, 6] (see also the chapter by L. V. Keldysh in the green book [7]), and later extended to superconductors [8]). BEC-BCS crossover has been an active field of cold atom research in the past decade.

The state of the field in 1993, therefore, was that of having one physical example of Bose-Einstein condensation, liquid helium-4, and a large number of proposed systems, with various degrees of theoretical and experimental work. The main candidates at that time were cold atoms, excitons in semiconductors, and spin-polarized hydrogen. (BEC of positronium was also proposed at that time, as discussed in a short chapter by P. M. Platzmann and A. P. Mills in the green book [9], and still remains a possibility.) Condensation of spin-polarized hydrogen was obtained in 1998 [10]; Thomas Greytak and Daniel Kleppner review the history of that work in Chapter 2. The remarkable early work on spin-polarized hydrogen contributed to the eventual success of BEC in cold atoms, through the development of evaporative cooling methods, a feat recognized by the scientific community through the Senior BEC Award in 2015 to Thomas Greytak, Harald Hess, and Daniel Kleppner for "inventing the technique of evaporative cooling, and for the observation of Bose-Einstein condensation in a dilute gas of atomic hydrogen."

1.3 Early Work on Excitons in Semiconductors

In the early 1990s, excitons in semiconductors (especially in bulk Cu_2O and GaAs coupled quantum wells) seemed to be the next most promising system. Good fits of the Bose-Einstein distribution to the experimental exciton energy distribution in Cu_2O had been observed over a wide range of density and temperature. However, questions about this interpretation arose already in 1996 [11], and by 2000 it was known [12] that the experiments on excitons in Cu_2O did not show BEC. The spatial distribution was found to be much more inhomogeneous than initially believed. The fits of the exciton kinetic-energy distribution to the Bose-Einstein distribution were surprisingly good given the accidental nature of how they arose.

A review of work on Cu_2O in the past two decades is given in Ref. [13]. Until very recently, the behavior of the excitons at high density in Cu_2O was believed to be dominated by a density-dependent, nonradiative recombination process known as Auger recombination. Recently, indications have emerged [14] that there may be a strongly bound biexciton state in Cu_2O. (The biexciton, or excitonic molecule, is comprised of two excitons, and is well known to exist in other semiconductors, but was widely believed to be nonexistent or very weakly bound in Cu_2O, based on theoretical calculations [15, 16].) Weak biexciton light emission could be masked by a nearby, brighter single-exciton emission line. If a stable biexciton state exists, then it would be the ground state which would undergo condensation, not the single exciton states. It may also play a role in assisting nonradiative recombination [17].

Similarly, reports of evidence for BEC of dipolar excitons in coupled semiconductor quantum wells, reported at the 1995 meeting, did not play out as hoped. Most of the anomalous results at low temperature in this system have found other explanations (for a review, see Ref. [18]). Recently, two groups [19, 20] have reported evidence for increased coherence in the light emission from this system at low temperature. Objections have been raised [21, 22] that this coherence is an artifact of the way the experiments were done at the limit of the spatial resolution of the imaging lenses – a reduction in the spatial extent of the exciton cloud could give a similar effect. Although these objections have been addressed in part [23], the interpretation of the dipolar exciton experiments as BEC remains controversial. Spectral narrowing with increase of particle density, considered a key test of coherence in optical systems, and a tell-tale for excitonic BEC, has not been observed for Cu_2O or coupled quantum well excitons.

Although the early work on Cu_2O had to be reinterpreted, experiments pursuing BEC of excitons in this unique semiconductor continue to this day. Two groups [24, 25, 26], in particular, have pursued studies of excitons in this material at millikelvin temperatures. The low temperature allows the possibility of BEC at very low exciton density, so that nonradiative and biexciton effects become much less important.

1.4 The "Atomic Revolution"

As mentioned above, the year 1995 was a turning point for ultracold atoms and for the broader field of weakly interacting Bose-Einstein condensation. Within a period of only a few weeks, three different US groups reported the observation of Bose-Einstein condensation in trapped, dilute weakly interacting atomic gases of ^{87}Rb [27], ^{7}Li [28], and ^{23}Na [29]. The observations and subsequent investigations led to the 2001 Nobel Prize in Physics being jointly awarded to Eric Cornell, Wolfgang Ketterle, and Carl Wieman "for the achievement of Bose-Einstein condensation in dilute gases of alkali atoms, and for early fundamental studies of the properties of the condensates." Randy Hulet's groundbreaking work with effectively attractive atoms of ^{7}Li turned out to have some issues with imaging resolution, with the definitive realization of BEC in ^{7}Li reported by the same group (as an *erratum* to the original manuscript) in 1997 [30].

A unique attraction of the cold atom experiments was that they provided a "smoking gun" of BEC: although BEC refers to condensation in *momentum* space (for a homogeneous gas), the harmonic traps used in cold atom experiments also enabled the phase transition to be observed in *position* space.[4] The traps used in liquid helium experiments had nominally flat potential energy profile, and therefore the density inhomogeneity in the cold atoms experiments added both excitement and a challenge to identify (both experimentally and theoretically) the precise nature of the observable effects and novel features present in such systems due to the zero-point motion and the modified density of states. Atomic BECs were realized at remarkably low temperatures of the order of 100 nK. The low temperatures reflect the required condition to reach quantum degeneracy for sufficiently dilute atomic gases. At higher densities, which nominally would allow higher critical temperature for BEC, three-body collisions severely limit the lifetime of the atoms in the traps. The extremely low temperatures are impressive experimental achievements, but pose a challenge to any technological applications, as they require an extremely good vacuum.

The experimental observation of BEC in cold atoms sparked a significant interest in the broader public domain, with coverage reaching mainstream newspaper headlines, nonscientific journals,[5] and national news sources. Bose-Einstein condensation, recently identified as one of the top five physics achievements of the past twenty-five years,[6] was no longer an obscure physical phenomenon.

Atomic BEC experiments during the first few years focused primarily on equilibrium properties, elementary and nonlinear excitations (vortices, solitons), low-order correlation functions, and the effects of thermal fluctuations (see, e.g., the 1999 review by Dalfovo et al. [31]). Soon atomic BEC experiments, which were by now starting to be done in numerous labs worldwide, moved toward enhanced control (e.g., of interaction strength [32] or dimensionality [33]), characterization of dynamics, and manifestation of distinct regimes mimicking the solid state. These latter include periodic "optical lattice" potentials (reviewed in Chapter 13 by Immanuel Bloch) and spinor condensates/quantum magnetism (reviewed in Chapter 18 by Masahito Ueda). The achievement of quantum degeneracy of fermionic gases [34] has paved the way for studies of Fermi superfluids and a controlled characterization of the BEC-BCS crossover.

By the late 1990s, then, we had three good experimental examples of BEC: liquid helium-4, trapped cold alkali atoms, and spin-polarized hydrogen experiments, with cold atoms already demonstrating an immense potential for characterizing and harnessing the properties of quantum gases. Superconductors and helium-3 in

[4] Actually, the initial experiments measured the density distribution after expansion, thus corresponding to information on the velocity distributions, with subsequent experiments able to directly image density profiles "in situ."

[5] See, e.g., "Bose knows," in the Science & Technology section, *The Economist*, 1 July 1995, pages 93–94.

[6] *Physics World*, October 2013 Special Issue (Institute of Physics); see also www.bbc.co.uk/news/science-environment-24282059.

the BCS limit provided alternative, more involved, manifestations of macroscopic quantum coherence.

1.5 BEC Physics in the New Millennium: Excitons, Polaritons, Photons, Magnons, and Triplons

The decade of the 2000s turned another important corner for controlled quantum gas studies, with numerous credible examples of BEC or quasi-BEC in different systems in the optical and condensed matter regimes.

Two variations of the excitonic systems made it much easier to observe BEC than in the early types of experiments discussed above. In one variation, magnetic field and electronic gating was used to control the density of electrons and holes in a coupled quantum well system. The pairing of electrons and holes in this system can be described as thermodynamically stable excitons. (For a review, see Ref. [35].) These experiments have much in common with earlier experiments on the integer and fractional quantum Hall effects, but show a different physics, with transport through the bulk of the two-dimensional (2D) gas [36] instead of just along the edges.

The other variation which exploded on the scene in the early to mid 2000s was to place the excitons in a high-Q optical cavity, creating strong coupling between an optical cavity mode and the exciton state. The eigenstates of this system are known as polaritons. This system can be visualized in either of two ways. Starting with excitons, one can think of the polariton effect as giving the exciton much lighter mass (typically by three orders of magnitude), so that the excitons have much longer wavelength, and therefore much greater quantum effects, at typical densities and temperatures. The light mass also means that disorder plays much less of a role, since the polaritons average over disorder on length scales less than their wavelength. On the other hand, starting with photons, one can think of the polariton mixing as giving the photons much stronger interactions than they would normally have in a solid, so that they can thermalize. (Recent work [37] indicates that these interactions may even be much stronger than theoretically expected.)

Following promising early work [38, 39], the canonical effects expected for BEC, namely onset of coherence, a bimodal momentum distribution with a peak in the ground state, and spatial condensation in the ground state of a trap (effects previously demonstrated for cold atoms), were also shown in the polariton system in 2006–2007 [40, 41]. This book contains a general review of polariton condensation (Chapter 4), along with several chapters reviewing the broad diversity of work now done in such systems (Chapters 10–11 and 20–24). The early work was all done with structures with fairly short polariton lifetime, as they could turn into photons which leaked out of the optical cavity. In the initial experiments, the lifetime was about 1–2 ps, comparable to the thermalization time. For this reason, polariton condensates have often been referred to as "nonequilibrium condensates." However, the lifetime of polaritons in cavities has been gradually increasing, so that lifetimes of 20–30 ps are now routine and lifetimes up to 300 ps have been obtained [42]. In many cases, it is now more appropriate to say that the polaritons are "nearly equilibrated," and true equilibrium results have recently been reported [43]. In general, there are now many examples of polariton systems that "look like" condensates, such as the ring condensate shown in Fig. 4.4 of Chapter 4 (and on the front cover of this book).

Because the early polariton experimental work arose in systems that were to some extent not in equilibrium, theory in the field addressed at length the universal questions of how dissipation and nonequilibrium affects BEC [44, 45]. Jonathan Keeling and colleagues review some of this work in Chapter 11. Note that the equilibrium of cold atomic BEC is not perfect: three-body recombination gradually converts the gaseous BEC into a solid (for the typical experimental densities this happens on a time scale of many seconds). The decay of a polariton into an external photon by tunneling through a cavity mirror is more analogous to tunneling of an atom through a barrier. While the destruction of BEC by molecular formation is physically distinct from the out-tunneling of a particle, it is still debatable how sharp a distinction can be made between equilibrium and nonequilibrium condensates. Even in the experiments with paired electrons and holes in the quantum-Hall-type experiments discussed above, the population is not truly permanent – the average number of electrons in the two-dimensional plane depends on the balance of electrons scattering into and out of the trapped state from the rest of the semiconductor structure, with rates that depend on the temperature and the applied gate voltages. Only in liquid helium can we say that the number of particles in the system can be truly conserved. Nonequilibrium effects have become a major topic of BEC studies, and are also addressed in the studies of quenches and the onset of coherence, discussed below.

The broad range of work with polariton condensates brings out another general theme in condensate theory, namely BEC-laser crossover, analogous to BEC-BCS crossover (the latter reviewed in Chapter 12 of this book). In general, lasing occurs in a system of photons and electronic transitions when the interaction between the photons is weak or nonexistent, while BEC of polaritons occurs in the strong coupling regime, when the interaction between the photons dominates their behavior. Both emit coherent light and have a coherent medium [46]. In practice, there is not a smooth transition between these two limits; in the experiments, the system "pops" between lasing and polariton condensation [47, 48] at different excitation densities. Alessio Chiocchetta et al. review the connection of lasing and polariton BEC in their chapter in this book (Chapter 20). There is also a crossover between lasing and photon BEC. The latter was achieved in 2010 [49] by trapping photons in a high-Q cavity containing a solution of dye molecules; Jan Klaers and Martin Weitz review this work in Chapter 19. In this case, the photons are essentially noninteracting, but they can thermalize by repeated absorption into the dye and re-emission, because the high Q of the cavity gives a photon lifetime long compared with the time scale for re-emission. Although the photons have a finite lifetime to leak out of the cavity, a condensate can be established in steady state with an optical pump to replenish the particles. Similar thermalization and condensation behavior has been observed for lasers with a one-dimensional continuum of states [50, 51].

Besides the photonic/excitonic condensates, in the 2000s two more varieties of condensate were demonstrated in magnetic systems, namely magnons created as a nonequilibrium population in a ferromagnet, and "triplons" in antiferromagnets, in which the condensate is the stable ground state at certain values of the magnetic field and temperature. Salman et al. review the former in this book (Chapter 25), while Kollath et al. review the latter (Chapter 28), with the related topic of "spintronics" discussed by Duine et al. in Chapter 26.

Three of these systems, namely photons in a high-Q cavity, magnons in ferromagnets, and triplons in antiferromagnets, can undergo BEC at room temperature. Some polariton systems also show characteristics of BEC at room temperature [52, 53, 54, 55]. Polaritons in organic semiconductors may soon become a ubiquitous room temperature condensate used for optical devices.

All of these examples can be called BEC of quasiparticles, but "quasiparticle" here should not be read as "not real." Quasiparticles are simply particles which arise when a simple, diagonal Hamiltonian is constructed out of a more complicated, underlying Hamiltonian. A universal question arising out of the observation of BEC in bosonic quasiparticles, however, is the degree to which we can make a clear demarcation between "quantum" and "classical" effects. It is standard in BEC theory to describe a condensate by a Gross-Pitaevskii equation [56, 57], which is essentially a classical nonlinear wave equation. This is because a coherent macroscopic condensate acts just the same as a classical wave – this macroscopic coherent behavior is one of the attributes which makes condensates so interesting. But any classical nonlinear system will behave the same way in the low-frequency limit. Jason Fleischer and coworkers [58] have shown that classical optical waves with random phases in a self-defocusing refractive crystal condense in a way fully analogous to that of a BEC. Of course, from the quantum perspective, classical waves exist because the bosonic quantum field relations allow macroscopic occupation and coherence, so that we cannot say that any such classical wave behavior is not at all quantum.

1.6 Recent Experimental Trends

In the meantime, experiments on cold atoms have continued apace, opening up in even more diverse directions, from the very fundamental to the technological. For the purposes of this chapter, we only mention in passing some of the most notable advances over the past two decades. Those include detailed studies in guided, or controlled, geometries (e.g., atom chips [59], ring-traps [60], and box-like potentials [61]), multicomponent and spinor BECs [62] (see also Chapter 18), studies of superfluidity and sound propagation (see also Chapter 16) and nonlinear effects including dark and bright solitons and vortices [63] extending to the realm of quantum turbulence (see Chapter 17), Dirac monopoles [64], observations of Josephson effects [65, 66], and the use of atoms in optical lattices to create strongly correlated systems and as quantum simulators of condensed matter systems [67] (see also Chapter 13). Controlled experiments have further enabled the study of different phase transitions, including the superfluid-Mott insulator transition [68], the BEC-BCS crossover (reviewed in Chapter 12), the BEC phase transition itself (both statically and through dynamic quenches obeying the Kibble-Zurek scaling law – see Chapter 7) and the effect of interactions on it (Chapter 6), the related study of the Berezinskii-Kosterlitz-Thouless (BKT) phase transition in two-dimensional settings and scale invariance (see Chapter 9), studies of integrability in one-dimensional settings [69], and prethermalization (see Chapter 8). Other topics of interest include condensates with controllable [70] or long-range interactions [71], the controlled study of disorder [72] and quasiperiodic potentials [73], creation of artificial magnetic fields (see Chapter 15) and topological band structures (see Chapter 14), and studies of analogue gravity and Hawking radiation [74]. Moreover, the unprecedented experimental control in such systems is setting the scene for technological applications, such as in precision measurements, quantum simulations, or atomtronics.

Most of the above interesting features are now also being routinely studied in polariton condensates, reviewed in Chapter 4 by Peter Littlewood and Alexander Edelman (see also the review article by Carusotto and Ciuti [44]). While by no means a complete survey, this book highlights the study of vortex dynamics (Chapter 21) and other topological excitations (Chapter 24), the role of disorder and phase-locking (Chapter 23), and the versatility of optical control for such systems (Chapter 22). As mentioned above, there is a strong trend toward studies of BEC of polaritons and other quasiparticles at room temperature.

In addition to the laboratory examples discussed above, there are several proposals that BEC may occur at the astronomical and cosmological scales. The theory of BEC of neutron matter is on sound footing; possible condensates in neutron stars are reviewed by Chris Pethick et al. in Chapter 29. There is a long history of proposals that the universe itself may already have some particles in a BEC state. Gerry Brown proposed kaon condensation in the whole universe in the 1995 green book [75]; in this book, Nilanjan Banik and Pierre Sikivie propose axion condensation as a candidate for cold dark matter (Chapter 31), and Gia Dvali and Cesar Gomez discuss graviton BEC as a new approach to quantum gravity (Chapter 32).

The state of the field in 2015 is thus that we have a cornucopia of experimental systems with different, and often complementary, properties. We can indeed talk of seeing universal aspects of the theory of condensation across many different systems. A survey of the different types of condensates now being explored and their key properties and identifying features, initially generated with input from the participants of the "Universal Themes of Bose-Einstein Condensation" workshop in 2013, and subsequently updated accordingly by authors of this volume, is shown in Table 1.1. Topics not covered in this volume are noted with citations to other reviews. The majority of those systems (but not all of them) are discussed by some of the leading researchers in the respective fields in the remainder of this book.

Thus, cumulatively, we see that BEC is a truly universal phenomenon which is independent of time scale (from picoseconds for polaritons, to fractions of minutes for cold atoms, to many permanent systems) and independent of length scale (from the microscopic to the cosmic).

1.7 Ongoing Theoretical Questions

The general theory of condensates has also progressed enormously in the past two decades. At the 1993 meeting, there was significant debate about the time scale for condensation. Since there were so few experimental examples of condensation, there was serious argument that perhaps the time scale for condensation could be very long, effectively preventing BEC from appearing in all but an infinite-lifetime system such as liquid helium. The two chapters in the 1995 green book by Henk Stoof [76] and Yuri Kagan [77] reflected this debate. On one hand, it could be argued that the natural time scale for equilibration is of the order of the local scattering time, independent of the size of the system; this seemed to be the implication of quantum Boltzmann-type approaches. But a hallmark of condensation is long-range phase order, and it is natural to expect that the time to establish long-range order would scale with the size of the system. These two different perspectives were put together nicely in work by Berloff and Svistunov in 2000 [78]. They showed that if the system reached local phase coherence on the time scale of the microscopic scattering time, it would obtain long-range phase coherence on a longer time scale, which scales with the size of the system, as random vortices disappear.

The topic of critical dynamics and the physics of the phase transition itself remains one of the most interesting areas of active study within ultracold atoms, with broader consequences for other condensation effects. Already by the early 1990s, numerical simulations could describe the overall features of the early condensate growth experiments fairly well, but the actual dynamics of crossing the phase transition through controlled quenches has been studied in detail only during the past few years in ultracold atoms, with numerous open questions. Our current understanding of this most fundamental question is reviewed at length in Chapter 7 (see also related discussions in Chapters 6 and 8). A central feature of the growth of coherence is the Kibble-Zurek mechanism, first proposed in a cosmological setting [79], and subsequently broadly analyzed in the condensed matter regime [80]. Chapter 30 presents a related discussion on using a superfluid to simulate cosmological processes, such as cosmic string formation and brane annihilation in the early universe.

This question becomes even more interesting when changing the dimensionality of the system: two-dimensional systems exhibit the related BKT phase transition, where there is no true long-range order, but only algebraic long-range order. Although experimentally observed in helium, superconductors, and ultracold atoms (see Chapter 10), interesting questions remain as to the precise nature of the BKT transition in a driven-dissipative system, such as polariton systems, with the latter question discussed extensively in Chapter 11.

Related to all these studies is the justification and applicability of the Gross-Pitaevskii equation, which is so ubiquitous in BEC theory. As mentioned above, the Gross-Pitaevskii equation is essentially a nonlinear classical wave equation, presuming local phase coherence. Its microscopic justification, limitations and appropriate generalizations are reviewed in Refs. [81, 82]. One can think of Bose-Einstein condensation as the process by which an initially incoherent system becomes a classical wave. In a very good analogy, one can view a BEC as a bell which begins to ring spontaneously when it is cooled below a certain temperature! In BEC and superconductors, the classical wave is known as the "order parameter" – a time- and space-dependent single-valued (but complex) wave function. The onset of the coherence can be seen as arising from a thermodynamic "enphasing" intrinsic to a many-particle bosonic system [83], the opposite of the dephasing seen in fermionic or low-density many-particle systems. As mentioned above, because condensates are effectively classical waves, a purely classical system can be designed to simulate a condensate [58].

The use of the Gross-Pitaevskii equation seems to presume the effect of "spontaneous symmetry breaking," in which the condensate acquires a definite phase,

Table 1.1 *Various Physical Examples of Condensates*

	Dimensionality	Homogeneity	Typical probes	Typical control parameters
cold atoms	3D, 2D, 1D, 0D	harmonic and box-like traps, periodic and tailored potentials	light imaging (*in situ*, absorption, time of flight), Bragg scattering, $g^{(n)}$ interferences	interaction strength/ sign, trap profile, density, dispersion, temperature
liquid helium-4	3D, quasi-2D	homogeneous	transport, heat capacity, moment of inertia, neutron scattering	temperature, pressure, dilution
liquid helium-3[a]	3D, quasi-2D	homogeneous	transport, heat capacity, moment of inertia	temperature, pressure
spin-polarized hydrogen	3D	harmonic traps	two-photon absorption	temperature, density
polaritons	2D, 1D, 0D	homogeneous pumping, periodic and tailored potentials	real-space / k-space imaging, $g^{(1)}$ and $g^{(2)}$ interferences	density, potential energy profile, photon fraction
photons	2D	2D harmonic traps	same as polaritons	density, temperature, reservoir size
classical waves (e.g., nonlinear optics)	2D, 1D	homogeneous	direct measure of complex amplitude	intensity, quasi temperature, nonlinear term
bilayer excitons	2D	homogeneous in puddles	transport	density, temperature
superconductors[a]	3D, quasi-2D	homogeneous, with surface proximity effects	transport, particle scattering	temperature, density (doping) magnetic field
magnons in ferromagnets	3D	homogeneous	Brillouin scattering, real-space imaging, coherent radiation	density, quasi temperature
antiferromagnets	3D, 2D, 1D	homogeneous	heat capacity, magnetization, NMR	density, temperature
nuclei[b]	3D	up to several hundred nucleons per nucleus	odd/even staggering in masses, excitation spectrum	neutron and proton number
neutron stars	3D	homogeneous locally	glitches (sudden rotational speedups)	mass
hot quark matter				
-heavy ion collisions[c]	3D	-varies during collision	-energy spectra and correlations	-beam energies, target nuclei
-early universe[d]		-homogeneous locally	-$b\bar{b}$ asymmetry	-no control
axions	3D	homogeneous on Mpsec scale	axion to microwave conversion	no control—may control the universe

[a] See, e.g., Leggett, *A. J., Quantum Liquids*, 2006, Oxford University Press.
[b] See, e.g., Brink, D. M. and Broglia, R. A., 2010, *Nuclear Superfluidity: Pairing in Finite Systems*, Cambridge University Press.

Table 1.1 *(cont.)*

	Type	Superfluidity seen in	Condensate seen in	Equilibration
cold atoms	weakly to strongly interacting, BCS	quantized vortices, critical velocity, persistent flow, second sound	real-space and k-space profiles, suppression of density/phase fluctuations	equilibrium (slow loss), quenches, no steady state
liquid helium-4	strongly interacting	same as above, plus moment of inertia and fountain effect	indirectly deduced from neutron scattering	equilibrium
liquid helium-3[a]	BCS	moment of inertia	—	equilibrium
spin-polarized hydrogen	weakly interacting	—	narrow spectral feature	equilibrium
polaritons	weakly to strongly interacting	quantized vortices, critical velocity, suppressed scattering	real-space and k-space profiles, long-range coherence	pulse generation, steady state, quasi-equilibrium to equilibrium
photons	non-interacting	—	real-space and k-space profiles	quasi-equilibrium, non-equilibrium (lasing)
classical waves (e.g. nonlinear optics)	weakly interacting	—	k-space profiles	reversible near-equilibrium
bilayer excitons	BCS	coherent transport	—	equilibrium
superconductors[a]	BCS, high-T_c	quantized vortices, persistent flow, Meissner effect	—	equilibrium
magnons in ferromagnets	weakly interacting	—	k-space profiles, superradiance	quasi-equilibrium
antiferromagnets	strongly interacting	—	magnetization phase diagram	equilibrium
nuclei[b]	BCS	moment of inertia	two-particle transfer reactions	equilibrium
neutron stars	BCS (BEC of mesons)	possible vortex pinning in crust	effects on cooling	equilibrium
hot quark matter				
-heavy ion collisions[c]	-chiral quark condensate	—	-particle masses	-freeze out
-early universe[d]	-Higgs, other scalar fields		-particle masses	-freeze out
axions	weakly interacting	—	possibly seen in caustics of galactic halos	quasicollisionless, equilibrates in ~ age of universe

[c] See, e.g., Braun-Munzinger, P. and Wambach, J., 2009, *Rev. Mod. Phys.* **81**, 1031.
[d] See, e.g., Dine, M., and Kusenko, A., 2003, *Rev. Mod. Phys.* **76**, 1.

i.e., a definite complex amplitude, out of a continuum of possibilities. In the 1993 green book, Tony Leggett addressed some of the debated questions about spontaneous symmetry breaking, and has called this notion into question, especially in the case of the Bogoliubov-de Gennes equations as applied to superconductors. Debate still continues, and is reviewed by Snoke and Daley in Chapter 5 of this book. The effectiveness of the Gross-Pitaevskii equation in all the work discussed above shows that a condensate can be well described by a macroscopic coherent wave function, but nothing in that theory shows exactly what its absolute phase must be. In solid-state physics, it is common to see spontaneous symmetry breaking as the amplification of some small fluctuation. In atomic physics, however, it is not clear what could create such a fluctuation in a number-conserving system. The choice of one value of phase over all others is instead commonly attributed to the unpredictable effect of quantum measurement.

All of the above questions are part of the larger field of nonequilibrium thermodynamics, which has become quite active. Condensates make good systems to study nonequilibrium physics; cold atom systems can be quenched rapidly to out-of-equilibrium states, while polaritons and magnons can be generated far from equilibrium.

Other general topics also cross between different types of condensate. One is the crossover from 2D to 3D physics – technically, BEC cannot occur in 2D, but finite 2D and 3D systems can have nearly the same properties. Another is the effect of disorder: the crossover from a "Bose glass" phase [84, 85, 86, 87], with local, uncorrelated condensates, to a true condensate with long-range, or quasi-long-range, order (this transition is discussed by Eastham and Rosenow in Chapter 23). Both of these topics deal with the basic question: if we define BEC as having infinite long-range phase coherence (a.k.a. "off-diagonal order"), but no system we study is truly infinite, can we define how far the correlations must be for a system to be "effectively" condensed? Another general topic is the effect of the types of interactions, e.g., the crossover from weakly to strongly interacting gases, and the crossover from short-range to long-range interactions, both of which can be probed in cold gases, including the "unitary" gas regime of infinitely strong interactions.

The theory of universal themes of condensates, as opposed to modeling the details of specific condensates, thus has grown to a field in its own right. We hope that this book will stimulate further progress in this advancing field.

Acknowledgments: We gratefully acknowledge the Lorentz Centre in Leiden, and in particular Corrie Kuster and Mieke Schutte, for helping organize the "Universal Themes of Bose-Einstein Condensation" workshop in March 2013, which was an essential step toward the realization of this book. Many participants and others contributed to the table appearing in this chapter.

References

[1] Griffin, A., Snoke, D. W., and Stringari, S. (eds). 1995. *Bose-Einstein Condensation.* Cambridge University Press.
[2] Blatt, J. M. 1964. *Theory of Superconductivity: Pure and Applied Physics.* Academic Press.
[3] Leggett, A. J. 1980. Diatomic molecules and Cooper pairs. In: Pekalski, A., and Przystawa, J. A. (eds), *Modern Trends in the Theory of Condensed Matter, Proceedings of the XVI Karpacz Winter School.* Springer.
[4] Leggett, A. J. 2006. *Quantum Liquids: Bose Condensation and Cooper Pairing in Condensed-Matter Systems.* Oxford Graduate Texts in Mathematics. Oxford University Press.
[5] Comte, C., and Nozieres, P. 1982. Exciton Bose condensation – the ground-state of an electron-hole gas. *J. Physique*, **43**, 1069.
[6] Comte, C., and Nozieres, P. 1982. Spin states, screening and band-structure effects. *J. Physique*, **43**, 1083.
[7] Keldysh, L. V. 1995. Macroscopic coherent states of excitons in semiconductors. Page 558 of: Griffin, A., Snoke, D. W., and Stringari, S. (eds.), *Bose-Einstein Condensation.* Cambridge University Press.
[8] Nozières, P., and Schmitt-Rink, S. 1985. Bose condensation in an attractive fermion gas: from weak to strong coupling superconductivity. *J Low Temp Phys.* **59**, 195.
[9] Platzmann, P. M., and Mills, A. P. 1995. Possibilities for BEC of positronium. Page 558 of: Griffin, A., Snoke, D. W., and Stringari, S. (eds.), *Bose-Einstein Condensation.* Cambridge University Press.
[10] Fried, D. G., Killian, T. C., Willmann, L., Landhuis, D., Moss, S. C., Kleppner, D., and Greytak, T. J. 1998. Bose-Einstein condensation of atomic hydrogen. *Phys. Rev. Lett.*, **81**, 3811.
[11] Snoke, D. W. 1996. Bose-Einstein condensation of excitons: II. Results from Cu_2O at high density. *Comments on Condensed Matter Physics*, **17**, 325.
[12] O'Hara, K. E., and Wolfe, J. P. 2000. Relaxation kinetics of excitons in cuprous oxide. *Phys. Rev. B*, **62**, 12909.
[13] Snoke, D. W., and Kavoulakis, G. M. 2014. Bose–Einstein condensation of excitons in Cu2O: progress over 30 years. *Reports on Progress in Physics*, **77**, 116501.
[14] Nelson, K., private communication.
[15] Bobrysheva, A. I., and Moskalenko, S. A. 1983. Biexcitons in crystals with large ortho-para-exciton splitting. *Phys. Status Solidi B*, **119**, 141.
[16] Bobrysheva A. I., Moskalenko, S. A., and Russu, S. S. 1991. The biexciton influence on Bose-Einstein condensation of orthoexcitons in Cu_2O. *Phys. Status Solidi B*, **167**, 625.
[17] Wolfe, J. P., and Jang, J. I. 2014. The search for Bose-Einstein condensation of excitons in Cu_2O: exciton Auger recombination versus biexciton formation. *New J. Phys.*, **16**, 123048.
[18] Snoke, D. W. 2013. Dipole excitons in coupled quantum wells: toward an equilibrium exciton condensate. Pages 419–432 of: Proukakis, N. P., Gardiner, S. A., Davis, M. J., and Szymanska, M. H. (eds), *Quantum Gases: Finite Temperature and Non-Equilibrium Dynamics*. Imperial College Press (London).
[19] High, A., Leonard, J., Remeika, M., Butov, L., and Hanson, M. 2012. Condensation of excitons in a trap. *Nano Lett.*, **12**, 2605.
[20] Alloing, M., Beian, M., Lewenstein, M., Fuster, D., Gonzalez, Y., Gonzalez, L., Combescot, R., Combescot, M., and Dubin, F. 2014. Evidence for a Bose-Einstein condensate of excitons. *Europhysics Lett.*, **107**, 10012.

[21] Semkat, D., Sobkowiak, S., Manzke, G., and Stolz, H. 2012. Comment on "Condensation of Excitons in a Trap." *Nano Lett.*, **12**, 5055.

[22] Repp, J., Schinner, G. J., Schubert, E., Rai, A. K., Reuter, D., Wieck, A. D., Wurstbauer, U., Kotthaus, J. P., and Holleitner, A. W. 2014. Confocal shift interferometry of coherent emission from trapped dipolar excitons. *Applied Physics Lett.*, **105**, 241101.

[23] High, A., Leonard, J., Remeika, M., Butov, L., and Hanson, M. 2012. Reply to "Comment on Condensation of Excitons in a Trap." *Nano Lett.*, **12**, 5422.

[24] Yoshioka, K., Morita, Y., Fukuoka, K., and Kuwata-Gonokami, M. 2013. Generation of ultracold paraexcitons in cuprous oxide: a path toward a stable Bose-Einstein condensate. *Phys. Rev. B*, **88**, 041201.

[25] Sandfort, C., Brandt, J., Finke, C., Fröhlich, D, and Bayer, M. 2011. Paraexcitons of Cu_2O confined by a strain trap and high magnetic fields. *Phys. Rev. B*, **84**, 165215.

[26] Sobkowiak, S., Semkat, D., and Stolz, H. 2014. Modeling of the thermalization of trapped paraexcitons in Cu_2O at ultralow temperatures. *Phys. Rev. B*, **82**, 064505.

[27] Anderson, M. H., Ensher, J. R., Matthews, M. R., Wieman, C. E., and Cornell, E. A. 1995. Observation of Bose–Einstein condensation in a dilute atomic vapor. *Science*, **269**, 198.

[28] Bradley, C. C., Sackett, C. A., Tollett, J. J., and Hulet, R. G. 1995. Evidence of Bose–Einstein condensation in an atomic gas with attractive interactions. *Phys. Rev. Lett.*, **75**, 1687.

[29] Davis, K. B., Mewes, M. O., Andrews, M. R., van Druten, N. J., Durfee, D. S., Kurn, D. M., and Ketterle, W. 1995. Bose–Einstein condensation in a gas of sodium atoms. *Phys. Rev. Lett.*, **75**, 3969.

[30] Bradley, C. C., Sackett, C. A., and Hulet, R. G. 1997. Bose–Einstein condensation of lithium: observation of limited condensate number. *Phys. Rev. Lett.*, **78**, 985.

[31] Dalfovo, Franco, Giorgini, Stefano, Pitaevskii, Lev P., and Stringari, Sandro. 1999. Theory of Bose–Einstein condensation in trapped gases. *Rev. Mod. Phys.*, **71**, 463.

[32] Inouye, S., Andrews, M. R., Stenger, J., J., Miesner H., Stamper-Kurn, D. M., and Ketterle, W. 1998. Observation of Feshbach resonances in a Bose-Einstein condensate. *Nature*, **392**, 151.

[33] Görlitz, A., Vogels, J. M., Leanhardt, A. E., Raman, C., Gustavson, T. L., Abo-Shaeer, J. R., Chikkatur, A. P., Gupta, S., Inouye, S., Rosenband, T., and Ketterle, W. 2001. Realization of Bose–Einstein condensates in lower dimensions. *Phys. Rev. Lett.*, **87**, 130402.

[34] DeMarco, B., and Jin, D. S. 1999. Onset of Fermi degeneracy in a trapped atomic gas. *Science*, **285**, 1703.

[35] Eisenstein, J. P. 2014. Exciton condensation in bilayer quantum hall systems. Pages 159–181 of: Langer, J. S. (ed), *Annual Review of Condensed Matter Physics*, vol. 5. Palo Alto: Annual Reviews.

[36] Finck, A. D. K., Eisenstein, J. P., Pfeiffer, L. N., and West, K. W. 2011. Exciton transport and Andreev reflection in a bilayer quantum hall system. *Phys. Rev. Lett.*, **106**, 236807.

[37] Sun, Y., Yoon, Y., Steger, M., Liu, G., Pfeiffer, L. N., West, K., Snoke, D. W., and Nelson, K. A. 2015. Direct Measurement of Polariton-Polariton Interaction Strength, *Nature Physics*, in press *(arXiv:1508.06698)*.

[38] Deng, H., Weihs, G., Santori, C., Bloch, J., and Yamamoto, Y. 2002. Condensation of semiconductor microcavity exciton polaritons. *Science*, **298**, 199.

[39] Deng, H., Press, D., Götzinger, S., Solomon, G. S., Hey, R., Ploog, K. H., and Yamamoto, Y. 2006. Quantum degenerate exciton-polaritons in thermal equilibrium. *Phys. Rev. Lett.*, **97**, 146402.

[40] Kasprzak, J., Richard, M., Kundermann, S., Baas, A., Jeambrun, P., Keeling, J. M. J., Marchetti, F. M, Szymanska, M. H., Andre, R., Staehli, J. L., Savona, V., Littlewood, P. B., Deveaud, B., and Dang, L. S. 2006. Bose-Einstein condensation of exciton polaritons. *Nature*, **443**, 409.

[41] Balili, R., Hartwell, V., Snoke, D. W., Pfeiffer, L., and West, K. 2007. Bose-Einstein condensation of microcavity polaritons in a trap. *Science*, **316**, 1007.

[42] Steger, M., Gautham, C., Snoke, D. W., Pfeiffer, L., and West, K. 2015. Slow reflection and two-photon generation of microcavity exciton-polaritons. *Optica*, **2**, 1.

[43] Sun, Y., Wen, P., Yoon, Y., Liu, G., Steger, M., Pfeiffer, L. N., West, K., Snoke, D. W., and Nelson, K. A. 2017. Bose-Einstein condensation of long-lifetime polaritons in thermal equilibrium. *Phys. Rev. Lett.* 118, 016602 (2017).

[44] Carusotto, I., and Ciuti, C. 2013. Quantum fluids of light. *Rev. Mod. Phys.*, **85**, 299–366.

[45] Keeling, J. 2011. Superfluid density of an open dissipative condensate. *Phys. Rev. Lett.*, **107**, 080402.

[46] Snoke, D. W. 2012. Polariton condensation and lasing. Pages 307–328 of: Sanvitto, D., and Timofeev, V. (eds), *Exciton Polaritons in Microcavities*. Springer Series in Solid State Sciences, vol. 172. Springer.

[47] Nelsen, B., Balili, R., Snoke, D. W., Pfeiffer, L., and West, K. 2009. Polariton condensation: two distinct transitions in GaAs microcavities with stress traps. *Journal of Applied Physics*, **105**, 122414.

[48] Yamaguchi, M., Kamide, K., Nii, R., Ogawa, T., and Yamamoto, Y. 2013. Second thresholds in BEC-BCS-laser crossover of exciton-polariton systems. *Phys. Rev. Lett.*, **111**, 026404.

[49] Klaers, J., Schmitt, J. Vewinger, F., and Weitz, M. 2010. Bose-Einstein condensation of photons in an optical microcavity. *Nature*, **468**, 545.

[50] Gordon, A., and Fischer, B. 2002. Phase transition theory of many-mode ordering and pulse formation in lasers. *Phys. Rev. Lett.*, **89**, 103901.

[51] Weill, R., Levit, B., Bekker, A., Gat, O., and Fischer, B. 2010. Laser light condensate: experimental demonstration of light-mode condensation in actively mode locked laser. *Optics Express*, **18**, 16520.

[52] Baumberg, J. J., Kavokin, A. V., Christopoulos, S., Grundy, A. J. D., Butté, R., Christmann, G, Solnyshkov, D. D, Malpuech, G., Baldassarri Höger von Högersthal, G., Feltin, E., Carlin, J.-F., and Grandjean, N. 2008. Spontaneous polarization buildup in a room-temperature polariton laser. *Phys. Rev. Lett.*, **101**, 136409.

[53] Kéna-Cohen, S., and Forrest, S. R. 2010. Room-temperature polariton lasing in an organic single-crystal microcavity. *Nature Photonics*, **4**, 371.

[54] Plumhof, J. D., Stoferle, T., Mai, L. J., Scherf, U., and Mahrt, R. F. 2014. Room-temperature Bose-Einstein condensation of cavity exciton-polaritons in a polymer. *Nature Materials*, **13**, 248.

[55] Bhattacharya, P., Frost, T., Deshpande, S., Baten, M. Z., Hazari, A., and Das, A. 2014. Room temperature electrically injected polariton laser. *Phys. Rev. Lett.*, **112**, 236802.

[56] Pethick, C. J., and Smith, H. 2002. *Bose–Einstein Condensation in Dilute Gases*. Cambridge University Press.

[57] Pitaevskii, L. P., and Stringari, S. 2003. *Bose–Einstein Condensation*. Clarendon Press.

[58] Sun, C., Jia, S., Barsi, C., Rica, S., Picozzi, A., and Fleischer, J. W. 2012. Observation of the kinetic condensation of classical waves. *Nature Phys.*, **8**, 470.

[59] Reichel, J., and Vuletic, V. (eds). 2011. *Atom Chips*. Wiley-VCH.

[60] Ryu, C., Andersen, M. F., Cladé, P., Natarajan, Vasant, Helmerson, K., and Phillips, W. D. 2007. Observation of persistent flow of a Bose-Einstein condensate in a toroidal trap. *Phys. Rev. Lett.*, **99**, 260401.
[61] Gaunt, A. L., Schmidutz, T. F., Gotlibovych, I., Smith, R. P., and Hadzibabic, Z. 2013. Bose-Einstein condensation of atoms in a uniform potential. *Phys. Rev. Lett.*, **110**, 200406.
[62] Stamper-Kurn, D. M., and Ueda, M. 2013. Spinor Bose gases: symmetries, magnetism, and quantum dynamics. *Rev. Mod. Phys.*, **85**, 1191–1244.
[63] Kevrekidis, P. G., Frantzeskakis, D. J., and Carretero-Gonzalez, R. (eds). 2008. *Emergent Nonlinear Phenomena in Bose-Einstein Condensates: Theory and Experiment*. Springer.
[64] Ray, M. W., Ruokoski, E., Kandel, S., M., Mottonen, and Hall, D. S. 2014. Observation of Dirac monopoles in a synthetic magnetic field. *Nature*, **505**, 657.
[65] Albiez, Michael, Gati, Rudolf, Fölling, Jonas, Hunsmann, Stefan, Cristiani, Matteo, and Oberthaler, Markus K. 2005. Direct observation of tunneling and nonlinear self-trapping in a single bosonic Josephson junction. *Phys. Rev. Lett.*, **95**, 010402.
[66] Levy, S., Lahoud, E., Shomroni, I., and Steinhauer, J. 2007. The a.c. and d.c. Josephson effects in a Bose-Einstein condensate. *Nature*, **449**, 579.
[67] Lewenstein, M., Sanpera, A., and Ahufinger, V. 2012. *Ultracold Atoms in Optical Lattices: Simulating Quantum Many-Body Systems*. Oxford University Press: Oxford.
[68] Greiner, M., Mandel, O., Esslinger, T., Hänsch, Th. W., and Bloch, I. 2002. Quantum phase transition from a superfluid to a Mott insulator in a gas of ultracold atoms. *Nature*, **415**, 39.
[69] Kinoshita, T., Wenger, T., and Weiss, D. S. 2006. A quantum Newton's cradle. *Nature*, **440**, 900.
[70] Chin, C., Grimm, R., Julienne, P., and Tiesinga, E. 2010. Feshbach resonances in ultracold gases. *Rev. Mod. Phys.*, **82**, 1225–1286.
[71] Lahaye, T., Menotti, C., Santos, L., Lewenstein, M., and Pfau, T. 2009. The physics of dipolar bosonic quantum gases. *Reports on Progress in Physics*, **72**, 126401.
[72] Billy, J., Josse, V., Zuo, Z., Bernard, A., Hambrecht, B., Lugan, P., Clement, D., Sanchez-Palencia, L., Bouyer, P., and Aspect, A. 2008. Direct observation of Anderson localization of matter waves in a controlled disorder. *Nature*, **453**, 891.
[73] Roati, G., D'Errico, C., Fallani, L., Fattori, M., Fort, C., Zaccanti, M., Modugno, G., Modugno, M., and Inguscio, M. 2008. Anderson localization of a non-interacting Bose-Einstein condensate. *Nature*, **453**, 895.
[74] Steinhauer, J. 2014. Observation of self-amplifying Hawking radiation in an analogue black-hole laser. *Nature Physics*, **10**, 864.
[75] Brown, G. E. 1995. Kaon condensation in dense matter. Page 438 of: Griffin, A., Snoke, D. W., and Stringari, S. (eds), *Bose-Einstein Condensation*. Cambridge University Press.
[76] Stoof, H. 1995. Condensate formation in a Bose gas. Page 226 of: Griffin, A., Snoke, D. W., and Stringari, S. (eds), *Bose-Einstein Condensation*. Cambridge University Press.
[77] Kagan, Yu. 1995. Kinetics of Bose-Einstein condensation formation in an interacting Bose gas. Page 202 of: Griffin, A., Snoke, D. W., and Stringari, S. (eds), *Bose-Einstein Condensation*. Cambridge University Press.
[78] Berloff, N. G., and Svistunov, B. V. 2002. Scenario of strongly nonequilibrated Bose-Einstein condensation. *Phys. Rev. A*, **66**, 013603.
[79] Kibble, T. W. B. 1976. Topology of cosmic domains and strings. *J. Phys. A*, **9**, 1387.
[80] Zurek, W. H. 1985. Cosmological experiments in superfluid helium. *Nature*, **317**, 505.

[81] Proukakis, N. P., and Jackson, B. 2008. Finite temperature models of Bose–Einstein condensation. *J. Phys. B: At. Mol. Opt.*, **41**, 203002.

[82] Proukakis, N. P., Gardiner, S. A., Davis, M. J., and Szymańska, M. H. 2013. *Quantum Gases: Finite Temperature and Non-Equilibrium Dynamics*. Cold Atoms. World Scientific Publishing Company.

[83] Snoke, D. W., and Girvin, S. M. 2013. Dynamics of phase coherence onset in Bose condensates of photons by incoherent phonon emission. *Journal of Low Temperature Physics*, **171**, 1.

[84] Fisher, M. P. A., Weichman, P. B., Grinstein, G., and Fisher, D. S. 1989. Boson localization and the superfluid-insulator transition. *Phys. Rev. B*, **40**, 546–570.

[85] Giamarchi, T., and Schulz, H. J. 1988. Anderson localization and interactions in one-dimensional metals. *Phys. Rev. B*, **37**, 325–340.

[86] Scalettar, R. T., Batrouni, G. G., and Zimanyi, G. T. 1991. Localization in interacting, disordered, Bose systems. *Physical Review Letters*, **66**, 3144–3147.

[87] D'Errico, C., Lucioni, E., Tanzi, L., Gori, L., Roux, G., McCulloch, I. P., Giamarchi, T., Inguscio, M., and Modugno, G. 2014. Observation of a disordered Bosonic insulator from weak to strong interactions. *Phys. Rev. Lett.*, **113**, 095301.

2

A History of Bose-Einstein Condensation of Atomic Hydrogen

THOMAS GREYTAK AND DANIEL KLEPPNER

Department of Physics, Massachusetts Institute of Technology, Cambridge, USA

We describe the principal events that occurred in the journey from the first realization that atomic hydrogen could be a candidate for observing Bose-Einstein Condensation in a gas to the realization of that goal forty years later.

2.1 Introduction

Phase transitions are ubiquitous in the world around us. Examples include the gas-to-liquid transition, the liquid-to-solid transition, the transition from a paramagnet to a ferromagnet or antiferromagnet, and the separation of a solution into concentrated and dilute phases. As the temperature is lowered, the disruptive effects of thermal energy are diminished relative to the attractive interactions between the molecules or spins. As a result, the system makes a transition to a state of higher spatial order and lower entropy. An active subfield of condensed matter physics is devoted to understanding just how this occurs.

2.1.1 Bose-Einstein Condensation

In 1924 Einstein [1], expanding on the work of Bose [2], pointed out that a new type of phase transition could occur in a gas of particles with integer spin, a transition that was intimately associated with quantum statistics and that could take place in the absence of any physical interaction between the particles. Furthermore, for a gas in a uniform potential, the order below the transition appears not in real space, but in momentum space. The transition, referred to as Bose-Einstein condensation (BEC), occurs when the mean spacing between the particles $n^{-1/3}$ decreases to a value comparable to the thermal de Broglie wavelength,

$$\Lambda(T) \equiv h/\sqrt{2\pi M k_B T}. \tag{2.1}$$

Specifically, the transition temperature T_C for particles with zero spin is

$$T_C = \frac{h^2}{2\pi M k_B}\left(\frac{n}{\zeta(3/2)}\right)^{2/3}. \tag{2.2}$$

Here h and k_B are the Planck and Boltzmann constants, M is the particle mass, n is the number density, and $\zeta(3/2) \approx 2.61238$ is the Riemann zeta function of $3/2$.

BEC became a familiar topic in statistical mechanics textbooks because of its intimate connection to basic quantum theory. The effect, however, could not be observed because all elements heavier than helium become solid under their saturated vapor pressure before temperatures low enough for BEC to occur can be reached. In helium, the second lightest element, there is a competition between the attractive interactions and the disruptive effects of the large zero point motion associated with the very light mass. The result is a draw: the attractive interactions do not win because the system does not form a solid under its own vapor pressure, the zero point motion does not win because the system remains a liquid near absolute zero. Hydrogen, the lightest element, forms molecules, H_2, which are lighter than the helium atoms. However, due to low-lying molecular states the interaction between the molecules is much stronger than the interaction between the helium atoms. As a result molecular hydrogen is a solid below 14 K.

2.1.2 Spin-Polarized Hydrogen

The interaction between two hydrogen atoms is either dominantly attractive or dominantly repulsive depending on the relative orientation of their electronic spins. If the spins are antiparallel, the spin part of the electronic wavefunction satisfies the anti-symmetry requirement for fermions. Consequently, the spatial wavefunction must be symmetric, permitting the lowest energy eigenstate to be occupied. The electronic state of such a system is the $^1\Sigma_g^+$ state, a spin singlet. If the spins are parallel, the spin wavefunction is symmetric, thus the spatial wavefunction must be antisymmetric, and a higher energy spatial state is required. The lowest lying electronic state in this situation is the $^3\Sigma_u^+$ state, a spin triplet. Accurate pair potentials for hydrogen atoms in these two states have been calculated by Kolos and Wolniewicz [3]. In both cases, there is strong repulsion at small separations. The potential energy for the $^1\Sigma_g^+$ state, however, has a deep attractive minimum of –4.75 eV at a separation of 0.74 angstroms. The potential for the $^3\Sigma_u^+$ state is basically repulsive with only a shallow van der Waals minimum of -5.3×10^{-4} eV (corresponding to –6.2 K) at a separation of 4.15 angstroms.

As early as 1959, Hecht [4] used the quantum theory of corresponding states [5] to point out that if the electronic spins of atomic hydrogen were polarized by a strong magnetic field, the system would remain a gas at temperatures down to absolute zero. Such a gas would be a candidate for displaying BEC. In 1975, Etters, Dugan, and Palmer [6, 7] published the results of a more accurate many-body calculation. They confirmed that spin-polarized hydrogen would be a gas at absolute zero and calculated the ground-state energy, pressure, and compressibility as a function of molar volume. They also found that in the triplet state deuterium would be a permanent gas, but that tritium would be a liquid at $T = 0$.

In 1976, Stwalley and Nosanow [8] published a letter that brought these issues to the attention of a wider community and sparked the experimental search for BEC. They returned to the quantum theory of corresponding states, now analyzed in more depth. Most importantly, they discussed the quantum statistical mechanical opportunities that would be opened by the creation of these low-temperature gases. Spin-polarized atomic hydrogen appeared to be the best candidate for observing BEC. It was the only such system that would remain gas at absolute zero. Furthermore, because of its low mass, for any given number density spin-polarized hydrogen would have a higher transition temperature than any other element.[1]

2.1.3 The Role of Nuclear Spin

The discussion above has included only the electron spin. However, the spin of the proton couples to the electron spin through the hyperfine interaction and thus plays an important role in the stability of a gas of atomic hydrogen. Fig. 2.1 shows the energy and composition of the four eigenstates of the hyperfine interaction. The **a** and **b** states are referred to as high-field-seeking states because they are attracted to the highest magnetic field region available to the gas. The **c** and **d** states are low-field-seeking states. Two-body collisions in the gas can cause transitions between the various hyperfine states, a process called relaxation. In particular, collisions can convert high-field-seeking states into low-field states and vice versa. This turned out to be a problem when trying to confine the hydrogen in a magnetic trap.

A two-body collision in the gas cannot cause the atoms to recombine to form a hydrogen molecule because it is not possible to conserve both energy and momentum in such an event. Recombination can occur, however, in a three-body collision. In this case, the molecule leaves with a high momentum and the remaining atom leaves with a nearly equal and opposite momentum. A two-body collision on a surface can also cause recombination because the surface acts as the third body in taking up the recoil energy and momentum. Surface recombination limits the performance of atomic hydrogen masers. In these devices, the surfaces are often coated with Teflon©, a material that has a low binding energy for hydrogen. When confining atomic hydrogen at cryogenic temperatures, the surface coating of choice

[1] As it turned out, both of these advantages were illusory. With the development of laser cooling during the following decade, and techniques for trapping atoms with magnetic or optical fields, it became possible to form metastable gases of alkali metal atoms at ultra-low temperatures. BEC was first observed in such systems [9, 10, 11].

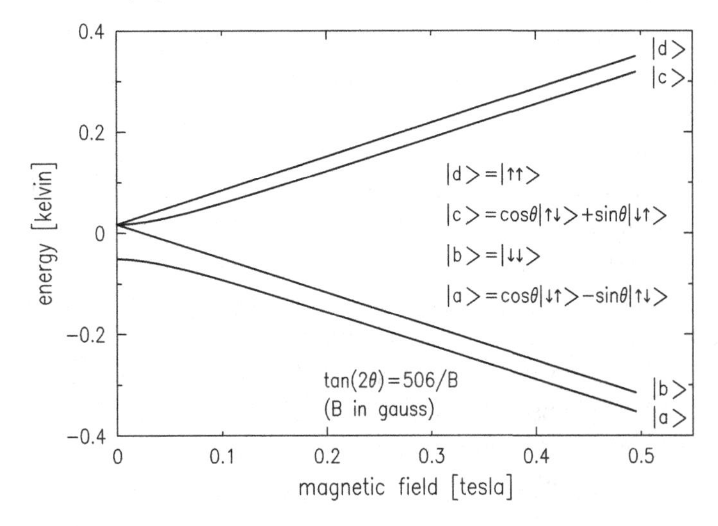

Figure 2.1 Hyperfine diagram for the ground state of atomic hydrogen. Reprinted with permission from T. J. Greytak (1995), Prospects for Bose-Einstein condensation in magnetically trapped atomic hydrogen, in *Bose-Einstein Condensation*, A. Griffin, D. W. Snoke, and S. Stringari (eds.) (Cambridge University Press, Cambridge, England).

is superfluid ^{4}He. Not only does it have a low binding energy for hydrogen atoms, ~1 K, but it flows to cover any bare spots on the underlying substrate.

2.2 Early Studies of Hydrogen at Cryogenic Temperatures

In 1979, two groups [12, 13] managed to accumulate enough hydrogen atoms at cryogenic temperatures to study the properties of the gas using hyperfine resonance. Relaxation times, the adsorption energy of H on H_2, the recombination rate on the surface, and the shift of the hyperfine frequency due to surface collisions were all measured. However, the interpretation of these numbers was unclear due to uncertainty in the temperature and physical state of the molecular hydrogen coating the surface of the confinement cell.

The initial approach to creating spin-polarized hydrogen involved confining the high-field seekers. A gas of cold atomic hydrogen was directed toward a cell in the center of a superconducting magnet. The high-field seekers were accelerated into the cell by the field gradient. There they lost kinetic energy through collisions with the walls and became trapped. Silvera and Walraven [14] in 1980 were the first to demonstrate this approach. Their cell was precoated with solid H_2 and then covered with a saturated film of superfluid ^{4}He. In their geometry, the refluxing action of ^{4}He gas helped to confine the hydrogen atoms in the cell, giving rise to a longer hold time than provided by the magnetic forces alone. The number of trapped H atoms was measured by forcing them to recombine and measuring the energy released. That energy was found from the temperature rise of the cell. Silvera and Walraven were able to accumulate a density of approximately 2×10^{14} atoms/cm^3 at 270 mK and 7 tesla. Under these conditions, they saw no measurable decay after 9 minutes, their longest wait time.

Later that year, Cline et al. [15] achieved higher densities and longer confinement times by confining the high-field seekers only by the magnetic field and the cell walls.[2] The number of trapped H atoms was measured by the triggered recombination technique developed by Silvera and Walraven [14]. Cline et al. achieved a density of 0.8×10^{17} atoms/cm^3 at 0.3 K and 8 tesla.

In the absence of recombination or relaxation mechanisms, high-field-seeking H atoms escape from the cell with a time constant

$$\tau_B = t_0 \exp(\mu B_0/k_B T), \tag{2.3}$$

where μ is the magnetic moment of the hydrogen atom, B_0 is the maximum field in the cell, and T is the temperature at the entrance to the cell. t_0 is the escape time in the absence of a field. The experimental results were in excellent agreement with Eq. (2.3) up to a magnetic field of 8 tesla. Above 8 tesla (where the decay time was longer than 10^3 seconds), the measured times began to fall below Eq. (2.3), presumably due to relaxation and three-body recombination, processes which depend on the density. For one sample with a density of 10^{16} atoms/cm^3 at 0.3 K and 8 tesla, the measured decay time was 4 hours.

2.3 Relaxation and Recombination

After the problems associated with accumulating useful densities of high-field seekers were solved, attention turned to understanding the processes that ultimately limited the density and lifetime of the atoms: relaxation and recombination.

The high-field seekers consist of two hyperfine states (see Fig. 2.1), the "pure" **b** state, $|b\rangle = |-\frac{1}{2}, -\frac{1}{2}\rangle$ (in the notation $|m_e, m_p\rangle$) and the "mixed" **a** state, $|a\rangle = \cos\theta|-\frac{1}{2}, \frac{1}{2}\rangle - \sin\theta|\frac{1}{2}, -\frac{1}{2}\rangle$, where $\tan 2\theta = A_{hf}/[h(\gamma_e + \gamma_p)B]$. Here A_{hf} is the hyperfine constant, γ_e and γ_p are the electron and proton gyromagnetic ratios, and B is the magnetic field. In a 10 tesla field, $\sin\theta \simeq 2 \times 10^{-3}$.

There are two requirements for recombination of the high-field seekers to occur: at least one atom must be in the **a** state because of its admixture of the "wrong" direction of the electron spin and a third body is required, either another atom or

[2] Confinement of the high-field seekers by a magnetic field alone is not possible because the magnitude of the magnetic field cannot have a maximum in a source-free region.

a surface. Studies of the temperature dependence of the reaction rate show that recombination occurs predominantly on the surface. At low surface densities, the surface density σ is proportional to the volume density

$$\sigma = n\Lambda(T)\exp(E_B/k_BT), \tag{2.4}$$

where E_B is the binding energy of the atomic hydrogen on the liquid helium surface. The relaxation proceeds as if it were described by a two-body gas collision process with an effective two-body recombination rate constant K [16].

If recombination were the only process causing changes in the **a** and **b** populations, after the source was turned off all the **a** states would eventually disappear, leaving a stable gas of **b** states with total electronic and nuclear polarization. However, nuclear relaxation also plays a role [16]. The nuclear relaxation rate is proportional to the total density, $n_t \equiv n_b + n_a$, and causes the population difference $n_b - n_a$ to decay to zero. $T_1^{-1} \equiv Gn_t$ is the effective nuclear-spin relaxation rate which can have contributions from collisions in the gas phase or on the surface. In this expression, G is the effective two-body relaxation rate constant. In the situations usually encountered in the laboratory, $K \gg G$ and the cell is filled in a time short compared with $(Kn_t)^{-1}$. Initially, $n_a \approx n_b$ and n_t begins to decay rapidly according to the equation

$$\dot{n}_t \approx -\gamma_0 n_t - Kn_t^2, \tag{2.5}$$

where γ_0 is the rate of one-body decay processes such as escape from the magnetic potential well.

After a time that is long compared with $(Kn_t)^{-1}$, most **a**-state atoms have disappeared and the decay of n_t becomes much slower, limited by the rate at which the **b** state is converted to the **a** state:

$$\dot{n}_t \approx -\gamma_0 n_t - 2Gn_t^2. \tag{2.6}$$

In this limit, the remaining gas has a high nuclear polarization

$$n_b/n_t \approx 1 - G/K. \tag{2.7}$$

A significant feature of this model of relaxation and recombination is the prediction that the decay rate $\dot{n}_t$ can have different values at equal values of n_t, depending on the history of the sample.

Experimental studies of recombination and relaxation were facilitated by using nondestructive methods to monitor the atomic density in the traps: hyperfine resonance [17] and high-sensitivity pressure transducers [18, 19]. These studies confirmed all the features of the decay model outlined above.

2.3.1 The Problem with High-Field Seekers

Once the effective recombination rates K and relaxation rates G were known, it was possible to realistically address the problem of achieving Bose-Einstein condensation. BEC requires a situation where the mean separation between the particles $n^{-1/3}$ is comparable to the thermal de Broglie wavelength $\Lambda(T)$. This can only be achieved by an appropriate combination of high density and low temperature. From Eq. (2.2), the density and temperature at the BEC transition for atomic hydrogen are related by the expression $T_C^3 = 4.05 \times 10^{-42} n^2$, where the particle density n is expressed in cm^{-3}. At a temperature of 300 mK, the density required is 8.2×10^{19} cm^{-3}, which is much higher than could realistically be achieved. However, if one tries to take a reasonably stable gas of **b**-state atoms at 300 mK and cool it to temperatures below about 100 mK, the equilibrium density of atoms on the wall rises quickly (see Eq. (2.4)) and the sample is lost due to recombination on the wall.

An alternative approach is to take that same gas in equilibrium with the wall at 300 mK and compress it rapidly to a higher density. Our MIT group carried out such a compression using a piston [20]. The Amsterdam group compressed a bubble of H in liquid ^{4}He [21]. However, the compression approach turned out to be limited by three-body recombination. Kagan et al. [22] had predicted earlier that three-body recombination would occur in spin-polarized hydrogen by a dipolar interaction between the electronic magnetic moments of the atoms. Two of the atoms recombine to form a molecule which speeds off in one direction, while the third atom exits at high speed in the opposite direction, conserving energy and momentum. This process, which occurs both in the bulk and on the surface, is not suppressed by nuclear polarization or high magnetic fields. Bulk densities achieved by compression were limited to a few times 10^{18} cm^{-3} by spontaneous destruction of the sample due to three-body recombination.

Although these experiments ended hopes of attaining BEC in traps requiring physical surfaces for confinement, they did allow detailed measurements of two- and three-body rate constants in the cold atomic hydrogen gas and their comparison with theoretical calculations. The article by Bell et al. [23] presents an extensive study of decay mechanisms in spin-polarized atomic hydrogen: physical mechanisms, experimental techniques, comparison with the theories, and references to work by other groups.

2.3.2 A New Path to BEC

In 1986, Harald Hess [24] suggested a new strategy for approaching BEC in atomic hydrogen. Although Maxwell's equations preclude creating a maximum in the magnitude of the magnetic field in a source-free region, they do allow creating isolated minima or even zeros in the magnitude of the field. Thus it could be possible to trap the upper two hyperfine states, the low-field seekers, without the use of physical walls. This would eliminate the most important source of recombination.

Hess suggested using a trap of the Ioffe-Pritchard form [25]. He showed that by carefully manipulating the shape and depth of the trapping potential, the hottest atoms in the trap would evaporate, cooling the remaining atoms well below their initial temperature. Spin-exchange collisions of the type **c+c** $\rightarrow$ **b+d** would lead to the creation of a doubly polarized gas of **d** atoms as the **b** atoms are expelled. Finally, the lifetime of the **d** atoms would be limited by dipolar electron-spin relaxation **d+d** $\rightarrow$ **a+a**, where the **a** atoms are promptly expelled from the trap. Once the rate constant for this dipolar relaxation is known, the measured relaxation rate can be used to determine the density of the atoms in the trap.

By pursuing Hess's strategy, our MIT group evaporatively cooled a doubly polarized gas of **d** atoms to a temperature of 3.0 mK with a central density of 7.6×10^{12} cm^{-3} [26]. At this density, the transition to BEC is still a factor of 50 lower in temperature. Though this represented a major step forward, two problems remained: the method used for evaporation became increasingly inefficient at low temperatures and there was no direct means of studying the gas while confined in the trap.

2.4 Achieving BEC in Atomic Hydrogen

Based on our understanding of recombination, relaxation, and thermalizing collisions in hydrogen, we were able to set out once again to observe BEC in hydrogen. The critical elements are a suitable magnetic trap with a strategy for evaporative cooling, a technique for observing the gas *in situ*, and techniques for confirming that BEC has taken place.

2.4.1 The Magnetic Trap

Near the bottom of a Ioffe-Pritchard trap, the potential rises parabolically from a nonzero minimum; thus surfaces of constant magnetic field are ellipsoids of revolution about the primary axis. A unique feature of our trap is its high aspect ratio of 400:1, the ratio of the major (vertical) axis to the minor axis. Details of the trap and the loading process are given elsewhere [27]. After loading, the trap contains about 10^{14} atoms at a temperature of 40 mK.

Initially we allow the atoms to escape by lowering the confining potential at one end of the trap. This method becomes inefficient below about 100 μK because atoms promoted to high-energy states by collisions near the center of the trap are likely to lose that energy in a collision before reaching the end of the trap and escaping. To solve this problem, below a temperature of 120 μK we use RF ejection of the atoms [28]. RF ejection was first applied to the evaporative cooling of alkali-metal vapors [29]. An RF magnetic field flips the spins of atoms in that region of the trap having a particular value of the trapping magnetic field. Atoms whose spins are reversed are no longer confined. By starting with an RF field resonant with the highest fields in the trap, then slowly lowering the frequency, successively lower energy atoms are expelled and fly out to recombine on the nearest surface. Since all the atoms on a specific energy surface in the trap are excited simultaneously, the process is more efficient than evaporation through one end, particularly in a trap as long and thin as ours. This technique finally allowed us to achieve BEC in hydrogen.

2.4.2 Two-Photon Spectroscopy

Spectroscopy is the most attractive, and possibly the only, technique for monitoring a gas trapped in space. Unfortunately, the principal transition for hydrogen, the Lyman-α transition, is in the far UV (121 nm), which is difficult to generate. Furthermore, the recoil energy, $(1/2M)/(h\nu/c)^2$, corresponds to a prohibitively high temperature of $\approx$20 mK. The alternative, which we adopted, is two-photon excitation of the $1S$-$2S$ transition. Because the $2S$ state is metastable, the intrinsic resolution of the transition is high. Furthermore, by exciting the transition in a standing wave, the atom can be excited by oppositely directed photons, yielding a Doppler-free signal with zero recoil. To observe the metastable atoms, we apply an electric field which mixes the long-lived $2S$ state with the short-lived $2P$ state. The atom then returns to the ground state by the emission of a Lyman-α photon. We detect the resonance by recording the Lyman-α production as a function of the frequency of the illuminating beam.

In our experiments [30], a laser beam at 243 nm is reflected back on itself by a mirror at the bottom of the cell creating a standing wave in the trap. Atoms that absorb two co-propagating photons produce a spectral line that is recoil shifted by 6.7 MHz. At a temperature of 40 μK, the normal component of the gas is Doppler broadened with a half width of $\sim$ 2 MHz. The shape of the Doppler line gives the momentum distribution in the gas and thus provides a reliable measure of the temperature. Atoms that absorb two counterpropagating photons transfer no momentum to the atoms and thus give rise to a feature which is centered near a zero frequency shift. In our final experiments, the width of the normal component of this feature was limited to 1 kHz (or one part in 10^{12}) by laser jitter [32].

In 1996, Jamieson, Dalgarno, and Doyle [31] pointed out that the interactions between the atoms cause a density-dependent mean field shift of each of the hydrogen energy levels. They calculated that the resulting shift in the $1S$-$2S$ transition frequency would be negative and proportional to the density with a magnitude that would be observable in our experiments. We measured the shift [32] by studying the Doppler-free component in normal hydrogen and found $\Delta\nu_{1S-2S} = 2\Delta\nu_{243\text{nm}} = \chi n$, where n is the atomic density and $\chi = -3.8\pm0.8\times10^{-16}$ MHz cm^3. Since there is a distribution of densities in the trap, the Doppler-free component is broadened as well as shifted by the interaction. The shape of this feature in the spectrum allows us to determine the density distribution in the trap. The Doppler-sensitive component is broadened and shifted by similar amounts, but these effects are small compared with the Doppler broadening and recoil shift.

2.4.3 Observing the Transition

The improved evaporative cooling and the use of two-photon spectroscopy enabled our MIT group to achieve BEC in hydrogen in 1998 [33]. We confirmed the nature of the transition by observing condensation in real space and also in momentum space, and by measuring the phase diagram. We outline these first measurements below. The most detailed and up-to-date discussion of the apparatus and techniques, and a comparison between theory and experiment (including studies of the growth and decay of the condensate), can be found in the PhD thesis of Stephen Moss [34].

Condensation in Real Space

For a non-interacting Bose gas in a uniform potential, there is no condensation in real space. For a harmonic trap, however, the lowest energy eigenstate is that of a three-dimensional harmonic oscillator. This gives rise to a condensation in real space because the spatial extent of the ground-state wavefunction is much smaller than that of the normal gas. Consequently, the density of atoms becomes very high over a narrow region at the bottom of the trap. This effect is evident in Fig. 2.2, which shows the Doppler-free portion of the two-photon spectrum of a gas cooled just below its transition at a temperature of 50 μK and a normal component density of 2×10^{14} cm^{-3}. The strong sharp peak on the right is due to the noncondensed (normal) component, which is spread throughout the trap. It is shifted slightly to the red of the free atom resonance by an amount determined by its relatively low density. The weak broad feature shifted farther to the red, which appears only below the transition, is due to the condensate concentrated near the center of the trap. The dashed line is a fit of the condensate feature to a Thomas-Fermi density profile in a parabolic trap. Based on this fit, the maximum condensate density is 4.8×10^{15} cm^{-3}. Integration of both densities over the trap volume gives the total number of atoms in the trap as 2×10^{10} and the number of atoms in the condensate as 10^9. The resulting condensate fraction is 5%.

Although the peak condensate density we measure is 20 times higher than the density of the normal gas, it is about 200 times smaller than would be the case

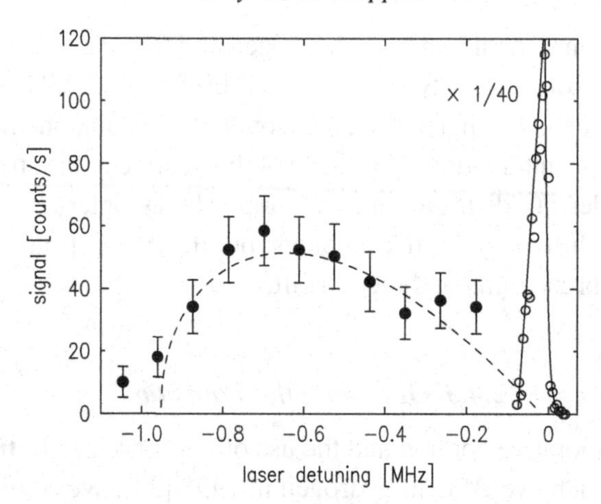

Figure 2.2 Doppler-free spectra of the condensate (broad feature) and the normal component (narrow feature). A red shift of 1.0 MHz corresponds to a density of 5.2×10^{15} cm^{-3}. Reprinted with permission from D. G. Fried et al. (1998), *Phys. Rev. Lett.*, **81**, 3811 [33]. Copyright (1998) by the American Physical Society.

if the gas were noninteracting. The observed density is the result of a mean field repulsion or spreading pressure due to the interactions between the atoms. The standard measure of the interaction strength in an ultra-low temperature gas is the *s*-wave scattering length, which, for hydrogen, has been calculated theoretically to be $a_{1S-1S} = 0.0648$ nm [35]. Using this value in a Thomas-Fermi calculation, together with the measured maximum condensate density, the temperature of the normal component, and the trap geometry, results in a condensate fraction of 6%, in good agreement with the value determined by spectral weights.

Condensation in Momentum Space

A drastic reduction in momentum occurs for condensate atoms in a trap because for them the momentum spread is determined only by the uncertainty principle and the spatial confinement of the condensate. In our case, where the condensate has a length of about 5 mm along the propagation direction of the photons, the Doppler width would be about 100 Hz. This intrinsic width is obscured by the distribution of density shifts in the condensate. The resulting width, similar to that in the condensate contribution to the Doppler-free portion of the spectrum, is still much narrower than the Doppler width of the normal component. Were there no condensation in momentum space this narrow, rapidly decaying feature would be absent. Fig. 2.3 shows the condensate feature in the Doppler-sensitive part of the two-photon spectrum, plotted relative to the recoil shifted resonance and fit to a Thomas-Fermi density profile.

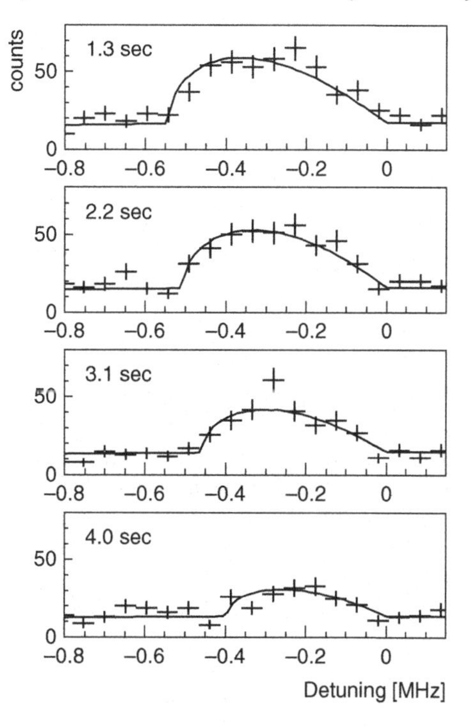

Figure 2.3 Condensate feature sitting on top of the much broader contribution (full width at half height of about 4 MHz) from the normal component in the Doppler-sensitive portion of the two-photon spectrum. Zero detuning corresponds to the recoil shift of a single stationary atom. Times are measured from the end of forced evaporative cooling. The number of atoms in the condensate is about the same as in Fig. 2.2, but the trap volume is larger resulting in a lower condensate density. Reprinted with permission from Elsevier from T. J. Greytak et al. (2000), Bose-Einstein condensation in atomic hydrogen, *Physica B*, **280**, 20 [36].

Phase Diagram

If one were to add atoms to a non-interacting Bose gas held at a fixed temperature, the density would increase until it reached a critical value $n_c(T) = 2.612(2\pi M k_B T)^{3/2}/h^3$. If more atoms were added, they would condense into the lowest single particle energy eigenstate and density of the normal component would remain constant. Thus $n_c(T)$, which is virtually unchanged by the presence of a weak interaction, can be regarded as a phase transition line in a plot of the density of the normal component versus temperature. Two-photon spectroscopy allows us to determine the density of the normal component directly. The temperature during evaporative cooling is an almost constant fraction of the trap depth. By plotting the normal density verses the trap depth, as in Fig. 2.4, one can both determine this fraction and display the BEC transition line.

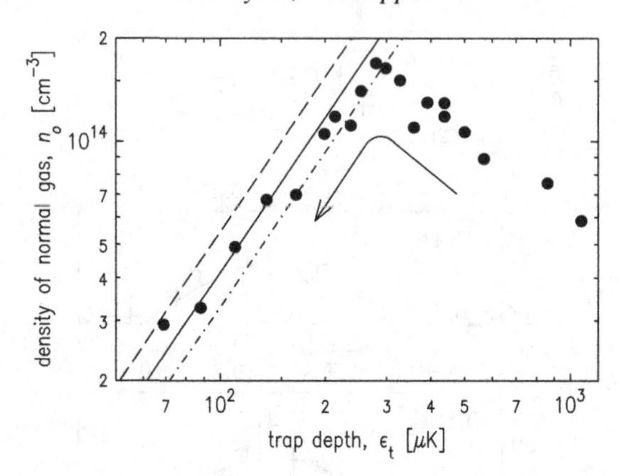

Figure 2.4 Density of the normal component of the gas as the trap depth is reduced. The lines (dashed, solid, dot-dashed) indicate the BEC transition line assuming sample temperatures of $\frac{1}{5}$th, $\frac{1}{6}$th, $\frac{1}{7}$th the trap depth. Reprinted with permission from D. G. Fried et al. (1998), *Phys. Rev. Lett.* **81**, 3811 [33]. Copyright (1998) by the American Physical Society.

2.5 Epilogue

The creation of atomic quantum gases is among the success stories of physics in recent decades. The field has expanded beyond anyone's expectations, providing a new arena for many-body physics and a consilience between atomic and condensed matter physics. The search for BEC in hydrogen helped to precipitate interest in this development, and the early years of the research were full of promise. Hydrogen was soon cooled into the cryogenic regime and produced at unprecedented densities. Even after it was understood that three-body recombination would force new strategies, those strategies – magnetic confinement and evaporative cooling – were close at hand. At the same time, laser cooling had been developed. This technique, which worked even better than theory predicted, was crucial not only to the first observations of BEC in atomic gases but to the entire development of ultracold atomic physics, discussed in the next chapter by Wolfgang Ketterle (see also Chapters 6, 8, 9, and 13–18 for a subset of current activity in this area, addressing both fundamental questions and selected research topics).

In contrast, it can now be seen that we were lucky to achieve Bose-Einstein condensation at all. Hydrogen's short scattering length, which we initially viewed as an advantage, turned out to be a near-fatal disadvantage. Evaporative cooling requires collisions, and the collision cross section is proportional to the square of the scattering length. As a result, cooling in the alkali metal atoms proceeds typically ten thousand times more rapidly than in hydrogen. Moreover, the experimental techniques for hydrogen are so complex (and expensive) compared with the laser-cooling approach that there is little temptation to pursue hydrogen further. Nevertheless, the search for BEC in hydrogen helped to precipitate interest in this field, and the field's success is a great satisfaction to all involved.

Acknowledgments: We wish to thank our colleagues who participated in the journey to observe BEC in hydrogen and generously shared the results of their labors. Space constraints made it impossible for us to do justice to their achievements.

References

[1] A. Einstein. 1924. Quantum theory of the monatomic ideal gas. *Proc. Prussian Acad. of Sciences* **22**, 261–277.

[2] S. N. Bose. 1924. Planck's law and the light quantum hypothesis. *Z. Physik* **26**, 178.

[3] W. Kolos and L. Wolniewicz. 1965. Potential-energy curves for the $X\,^1\Sigma_g^+$, $b\,^3\Sigma_u^+$, and $C\,^1\Pi_u$ states of the hydrogen molecule. *J. Chem. Phys.* **43**, 2429.

[4] C. E. Hecht. 1959. The possible superfluid behaviour of hydrogen atom gases and liquids. *Physica* **25**, 1159;

[5] J. De Boer. 1948. Quantum theory of condensed permanent gases: I. The law of corresponding states. *Physica* **14**, 139; J. De Boer and B. S. Blaisse. 1948. Quantum theory of condensed permanent gases: II. The solid state and the melting line. *Physica* **14**, 149; J. De Boer and R. J. Lunbeck, Quantum theory of condensed permanent gases: III. The equation of state of liquids. *Physica* **14**, 520.

[6] R. D. Etters, J. V. Dugan Jr., and R. W. Palmer. 1975. The ground state properties of spin-aligned atomic hydrogen, deuterium, and tritium. *J. Chem. Phys.* **62**, 313.

[7] J. V. Dugan Jr. and R. D. Etters. 1973. Ground state properties of spin-aligned atomic hydrogen. *J. Chem. Phys.* **59**, 6171.

[8] W. C. Stwalley and L. H. Nosanow. 1976. Possible "new" quantum systems. *Phys. Rev. Lett.* **36**, 910.

[9] M. H. Anderson, J. R. Ensher, M. R. Matthews, C. E. Wieman, and E. A. Cornell. 1995. Observation of Bose-Einstein condensation in a dilute atomic vapor. *Science* **269**, 198.

[10] K. B. Davis, M.-O. Mewes, M. R. Andrews, N. J. van Druten, D. S. Durfee, D. M. Kurn, and W. Ketterle. 1995. Bose-Einstein condensation in a gas of sodium atoms. *Phys. Rev. Lett.* **75**, 3969.

[11] C. C. Bradley, C. A. Sackett, J. J. Tollett, and R. G. Hulet. 1995. Evidence of Bose-Einstein condensation in an atomic gas with attractive interactions. *Phys. Rev. Lett.* **75**, 1687; C. C. Bradley, C. A. Sackett, and R. G. Hulet. 1997. Bose-Einstein condensation of lithium: Observation of limited condensate number. *Phys. Rev. Lett.* **78**, 985.

[12] S. B. Crampton, T. J. Greytak, D. Kleppner, W. D. Phillips, D. A. Smith, and A. Weinrib. 1979. Hyperfine resonance of gaseous atomic hydrogen at 4.2 K. *Phys. Rev. Lett.* **42**, 1039.

[13] N. W. Hardy, A. J. Berlinsky, and L. A. Whitehead. 1979. Magnetic resonance studies of gaseous atomic hydrogen at low temperatures. *Phys. Rev. Lett.* **42**, 1042.

[14] I. F. Silvera and J. T. M. Walraven. 1980. Stabilization of atomic hydrogen at low temperature. *Phys. Rev. Lett.* **44**, 164; see also J. T. M. Walraven, I. F. Silvera, and

A.P. M. Matthey. 1980. Magnetic equation of state of a gas of spin-polarized atomic hydrogen. *Phys. Rev. Lett.* **45**, 449.

[15] R. W. Cline, D. A. Smith, T. J. Greytak, and D. Klepper. 1980. Magnetic confinement of spin-polarized atomic hydrogen. *Phys. Rev. Lett.* **45**, 2117.

[16] B. W. Statt and A. J. Berlinsky. 1980. Theory of spin relaxation and recombination in spin-polarized atomic hydrogen. *Phys. Rev. Lett.* **45**, 2105.

[17] M. Morrow, R. Jochemsen, A. J. Berlinsky, and W. N. Hardy. 1981. Zero-field hyperfine resonance of atomic hydrogen for 0.18<~T<~1 K: The binding energy of H on liquid He4. *Phys. Rev. Lett.* **46**, 195. Erratum published as *Phys Rev. Lett.* **47**, 455.

[18] A. P. M. Matthey, J. T. M. Walraven, and I. F. Silvera. 1981. Measurement of pressure of gaseous H↓: Adsorption energies and surface recombination rates on helium. *Phys. Rev. Lett.* **46**, 668.

[19] R. W. Cline, T. J. Greytak, and D. Kleppner. 1981. Nuclear polarization of spin-polarized hydrogen. *Phys. Rev. Lett.* **47**, 1195.

[20] H. F. Hess, D. A. Bell, G. P. Kochanski, R. W. Cline, D. Kleppner, and T. J. Greytak. 1983. Observation of three-body recombination in spin-polarized hydrogen. *Phys. Rev. Lett.* **51**, 483; H. F. Hess, D. A. Bell, G. P. Kochanski, D. Kleppner, and T. J. Greytak. 1984. Temperature and magnetic field dependence of three-body recombination in spin-polarized hydrogen. *Phys. Rev. Lett.* **52**, 1520.

[21] R. Sprik, J. T. M. Walraven, and I. F. Silvera. 1983. Compression of spin-polarized hydrogen to high density. *Phys. Rev. Lett.* **51**, 479. Erratum published as *Phys. Rev. Lett.* **51**, 942.

[22] Yu. Kagan, G. V. Shlyapnikov, I. V. Vartanyantz, and N. A. Glukhov. 1982. Quasi-2-dimensional spin-polarized atomic-hydrogen. *JETP Letters* **35**(9), 477–481.

[23] D. A. Bell, H. F. Hess, G. P. Kochaski, S. Buchman, L. Pollack, Y. M. Xiao, D. Kleppner, and T. J. Greyta. 1986. Relaxation and recombination in spin-polarized atomic hydrogen. *Phys. Rev. B* **34**, 7670.

[24] H. F. Hess. 1986. Evaporative cooling of magnetically trapped and compressed spin-polarized hydrogen. *Phys. Rev. B* **34**, 3476.

[25] D. E. Pritchard. 1983. Cooling neutral atoms in a magnetic trap for precision spectroscopy. *Phys. Rev. Lett.* **51**, 1336.

[26] H. F. Hess, G. P. Kochanski, J. M. Doyle, N. Masuhara, D. Kleppner, and T. J. Greytak. 1987. Magnetic trapping of spin-polarized atomic hydrogen. *Phys. Rev. Lett.* **59**, 672; N. Masuhara, J. M. Doyle, J. C. Sandberg, D. Kleppner, T. J. Greytak, H. F. Hess, and G. P. Kochanski. 1988. Evaporative cooling of spin-polarized atomic hydrogen. *Phys. Rev. Lett.* **61**, 935.

[27] T. J. Greytak. 1995. Prospects for Bose-Einstein condensation in magnetically trapped atomic hydrogen, in A. Griffin, D. W. Snoke, and S. Stringari (eds.), *Bose-Einstein Condensation* (Cambridge University Press, Cambridge, England), p. 131.

[28] D. E. Pritchard, K. Helmerson, and A. G. Martin. 1989. Atom traps, in S. Haroche, J. C. Gay, and G. Grynberg (eds.), *Atomic Physics 11* (World Scientific, Singapore), p. 179.

[29] W. Petrich, M. H. Anderson, J. R. Ensher, and E. A. Cornell. 1995. Stable, tightly confining magnetic trap for evaporative cooling of neutral atoms. *Phys. Rev. Lett.* **74**, 3352; K. B. Davis, M.-O. Mewes, M. A. Joffe, M. R. Andrews, and W. Ketterle, 1995. Stable, tightly confining magnetic trap for evaporative cooling of neutral atoms. *Phys. Rev. Lett.* **74**, 5202.

[30] C. L. Cesar, D. G. Fried, T. C. Killian, A. D. Polcyn, J. C. Sandberg, I. A. Yu, T. J. Greytak, D. Kleppner, and J. M. Doyle. 1996. Two-photon spectroscopy of trapped atomic hydrogen. *Phys. Rev. Lett.* **77**, 255.

[31] M. J. Jamieson, A. Dalgarno, and J. M. Doyle. 1996. Scattering lengths for collisions of ground state and metastable state hydrogen atoms. *Mol. Phys.*, **87**, 817.

[32] T. C. Killian, D. G. Fried, L. Willmann, D. Landhuis, S. C. Moss, T. J. Greytak, and D. Kleppner. 1998. Cold collision frequency shift of the 1S- 2S transition in hydrogen *Phys. Rev. Lett.* **81**, 3807.

[33] D. G. Fried, T. C. Killian, L. Willmann, D. Landhuis, S. C. Moss, D. Kleppner, and T. J. Greytak. 1998. Bose-Einstein condensation of atomic hydrogen. *Phys. Rev. Lett.* **81**, 3811.

[34] S. C. Moss. 2002. Formation and decay of a Bose-Einstein condensate in atomic hydrogen. Ph.D. thesis, Massachusetts Institute of Technology.

[35] M. J. Jamieson, A. Dalgarno, and M. Kimura. 1995. Scattering lengths and effective ranges for He-He and spin-polarized H-H and D-D scattering. *Phys. Rev. A* **51**, 2626.

[36] T. J. Greytak, D. Kleppner, D. G. Fried, T. C. Killian, L. Willmann, D. Landhuis, and S. C. Moss. 2000. Bose-Einstein condensation in atomic hydrogen. *Physica B* **280**, 20.

[37] Yu. Kagan, I. A. Vartanyantz, and G. Shlyapnikov. 1982. Kinetics of decay of metastable gas phase of polarized atomic hydrogen at low temperatures. *JETP* **54**, 590.

3

Twenty Years of Atomic Quantum Gases: 1995–2015

WOLFGANG KETTERLE
Department of Physics, Massachusetts Institute of Technology, Cambridge, USA

The field of atomic quantum gases has seen rapid and sometimes surprising developments since its beginnings in 1995. In this chapter, I summarize, highlight, and comment on selected topics.

3.1 Introduction

Bose-Einstein condensation (BEC) in atomic gases was first observed in 1995, and has changed the face of atomic physics. It is for atoms or matter waves what the laser is for photons: a macroscopically occupied quantum state. It was regarded as an elusive goal until it was discovered in 1995. Although BEC was immediately viewed as a major accomplishment, its impact has far exceeded expectations. Now, twenty years on, there is no question that the field remains exciting.

It is impossible to give a review over the developments during those twenty years. Instead, in this chapter, I want to illustrate how often predictions or expectations changed in the pursuit of Bose-Einstein condensation. There were many surprises, and some advances and breakthroughs happened although they were predicted to be impossible. A lesson we can learn from this is that we should always carefully read the fine print when something is proven or assumed impossible, and try to figure out if the assumptions can be circumvented!

Since the early history of Bose-Einstein condensation provides several examples for such "impossibility theorems," I digress into those earlier developments in the first part of this chapter.

3.2 Early Theoretical Questions

3.2.1 Validity of the Prediction of Bose-Einstein Condensation

Einstein predicted Bose-Einstein condensation in 1924 in the second of two papers where he generalized Bose's treatment from photons to massive particles [1]. Using statistical arguments introduced by Bose, he found that below a critical temperature, bosonic particles condense in the lowest energy state of the system. In contrast, at very low temperature photons can simply disappear.

However, for about a decade, this prediction was not taken at face value. Einstein himself wrote to Ehrenfest, "From a certain temperature on, the molecules 'condense' without attractive forces; that is, they accumulate at zero velocity. The theory is pretty, but is there some truth in it?" [2].

In 1927, George E. Uhlenbeck concluded that the predicted BEC phase transition was an artifact of the replacement of the summation over states by an integral and wrote that no "splitting into two phases" would occur [3]. It was understood only in 1937 that discontinuities of thermodynamic quantities require an infinite number of particles, and true phase transitions occur only in the thermodynamic limit [4].

Kahn and Uhlenbeck showed that Bose-Einstein condensation is analogous to condensation of real gases, and that the quantization of the translational motion can be neglected in the thermodynamic limit. With these conclusions, the previous objections against BEC were no longer valid [5]. A fascinating account of how our current understanding of phase transitions emerged and Einstein's predictions were vindicated is given in Ref. [4].

3.2.2 Can Quantum Gases Be Realized in Nature?

Even if the theory was consistent, the question arose whether quantum gases would exist in nature. Already in his 1925 paper Einstein pointed out that real gases would not reach values of densities required for an ideal gas to be saturated, although helium and hydrogen come close. Since their critical densities are five and twenty-six times lower, respectively, than the saturation density of the ideal gas, Einstein speculated that quantum statistics could affect their equation of state [1].

Fritz London wrote in 1938 about Bose-Einstein condensation: "Since, from the very first, the mechanism appeared to be devoid of any practical significance, all real gases being condensed at the temperature in question, the matter has never been examined in detail; and it has been generally supposed that there is no such condensation phenomenon" [6]. He reiterates this view in his famous *Nature* paper: "but in the course of time, the degeneracy of the Bose-Einstein gas has rather got the reputation of having only a purely imaginary existence" [7]. But then he connects the observed lambda phenomenon in the specific heat of liquid helium to the discontinuity of the specific heat for an ideal Bose gas. This paper made the first close connection between Bose-Einstein condensation and an observable phenomenon.

However, superfluid helium is a liquid, and it was generally assumed that Bose-Einstein condensation of a gas would never be observed. Schrödinger wrote in a textbook in 1946: "The densities are so high and the temperatures so low that the van der Waals corrections are bound to coalesce with the possible effects of degeneration, and there is little prospect of ever being able to separate the two kinds of effect" [8]. What Schrödinger didn't consider were dilute systems in a metastable gaseous phase or quasiparticles!

For completeness, I want to mention the work on liquid helium in vycor. This example also illustrates that at least in principle, there is another way around the prediction that interactions dominate in all real material at relevant densities and don't allow the observation of a BEC in a weakly interacting gas. In the 1980s, experimental [9, 10] and theoretical [11] work showed that the onset of superfluidity for liquid helium in vycor has features of dilute-gas Bose-Einstein condensation. At sufficiently low coverage, the helium adsorbed on the porous spongelike glass behaved like a dilute three-dimensional gas. However, the interpretation of these results is not unambiguous [12].

3.3 Why Hydrogen Is Not the Only Quantum Gas

Bose-Einstein condensation in gases appeared on the agenda when Hecht [13] and Stwalley and Nosanow [14] used the quantum theory of corresponding states to conclude that spin-polarized hydrogen would remain gaseous down to zero temperature and should be a good candidate to realize Bose-Einstein condensation in a dilute atomic gas. These suggestions triggered several experimental efforts, most notably by Greytak and Kleppner at the Massachusetts Institute of Technology (MIT) and Silvera and Walraven in Amsterdam. The stabilization of a spin-polarized hydrogen gas [15, 16] created great excitement about the prospects of exploring quantum-degenerate gases. Experiments were first done by filling cryogenic cells with the spin-polarized gas, then by compressing the gas, and since 1985 by magnetic trapping and evaporative cooling. BEC was finally accomplished in 1998 by Kleppner, Greytak, and collaborators [17]. See Refs. [18, 19, 20, 21] and the contribution of Kleppner and Greytak to this volume (Chapter 2) for a full account of the pursuit of Bose-Einstein condensation in atomic hydrogen. Major efforts have also been directed toward reaching quantum degeneracy and superfluidity in a two-dimensional gas of spin-polarized hydrogen, at Harvard [22], in Amsterdam [23], in Kyoto [24], and at the University of Turku, where evidence for the two-dimensional (2D) phase transition was reported in 1998 [25].

It is interesting to look at the unique role which was given to spin-polarized atomic hydrogen in the early suggestions [13, 14, 21]. In the quantum theory of corresponding states, one defines a dimensionless parameter η, which is related to the ratio of the zero-point energy to the molecular binding energy. This parameter determines whether the system will be gaseous down to zero temperature. For large η, the zero-point motion dominates and the system is gaseous; for small

η, it condenses into a liquid or solid. The critical η value is 0.46, and only spin-polarized hydrogen with $\eta = 0.55$ exceeds this value [21]; alkali vapors have η values in the range 10^{-5} to 10^{-3}.

In reality, all spin-polarized gases are only *metastable* at $T = 0$ due to depolarization processes. The lifetime of the gas is limited by three-body recombination. Since the triplet potential of molecular hydrogen supports no bound states, spin-polarized hydrogen can only recombine into the singlet state with a spin-flip. In contrast, alkali atoms have both bound singlet and triplet molecular states, and their three-body recombination coefficient is ten orders of magnitude larger than for spin-polarized hydrogen. However, the rate of three-body processes depends on the square of the atomic density, is suppressed at sufficiently low density, and is almost negligible during the cooling to BEC. For magnetically trapped atoms, dipolar relaxation is an additional loss process, and hydrogen and the alkalis have comparable rate coefficients. So in hindsight, the unique benefits of hydrogen over other gases were not crucial for gaseous BEC. Although spin-polarized hydrogen has been called the only "true quantum gas," the difference from alkali vapors is just the range of densities and lifetimes of the metastable gaseous phase.

The work in alkali atoms is based on the work in spin-polarized hydrogen in several respects:

- Studies of spin-polarized hydrogen showed that systems can remain in a metastable gaseous state close to BEC conditions. The challenge was then to find the window in density and temperature where this metastability is sufficient to realize BEC.
- Many aspects of BEC in an inhomogeneous potential [26, 27, 28] and the theory of cold collision processes (see e.g. [29]) developed in the 1980s for hydrogen could be applied directly to the alkali systems.
- The technique of evaporative cooling was developed first for hydrogen [30, 31] and then used for alkali atoms.

The work on spin-polarized hydrogen paved the way for the alkali atoms and provided inspiration and guidance to the author and the whole field.

3.4 Ultracold Atomic Gases Were Not the Top Contender for BEC Until 1995

The first BEC conference in 1993 in Levico featured Bose-Einstein condensation as a shared goal of different communities. The review book based on this conference summarized work on different systems, from helium to excitons and positrons [32]. The articles in this book and the participants of this conference demonstrate that laser-cooled alkali atoms were not regarded as a top contender, with phase space densities a million times lower than required for BEC. A major focus was on spin-polarized hydrogen, where the phase space density was only a factor of 6 away from BEC, and exciton and polariton systems. Right before the conference, evidence for BEC in excitons in cuprous oxide was reported. However, the initial evidence [33] was later retracted [34] when it turned out that the non-Maxwellian distribution function of the exciton gas was not caused by quantum degeneracy, but by nonequilibrium effects due to inelastic collisions. I made a deliberate decision not to attend the 1993 BEC conference since I felt that BEC was so far away that I should better spend my time working in the lab than learning about the possible science with Bose-Einstein condensates.

Indeed, many people doubted that BEC could ever be achieved, and it was regarded as an elusive goal. Many believed that pursuing BEC would result in new and interesting physics, but whenever one would come close, some new phenomenon or technical limitation would show up. A news article in 1994 quoted Steve Chu: "I am betting on nature to hide Bose condensation from us. The last 15 years she's been doing a great job" [35]. It was not clear if laser cooling could be combined with evaporative cooling. Laser cooling works best at low densities where light absorption and light-induced collisions are avoided, whereas evaporative cooling requires a high collision rate and high density. The problem is the much higher cross section for light scattering of 10^{-9} cm^{-2}, while the cross section for elastic scattering of atoms is a thousand times smaller. In hindsight, it would have been sufficient to provide tight magnetic compression after laser cooling and an extremely good vacuum to obtain a lifetime of the sample that is much longer than the time between collisions. However, many researchers assumed that a major improvement had to be done to laser cooling to bridge the gap in density between the two cooling schemes and pursued research into such directions. Examples include the dark Spontaneous Force Optical Trap (SPOT) magneto-optical trap (MOT) [36] (to boost densities), an AC magnetic trap [37] (to avoid dipolar relaxation), sub-recoil cooling [38] (to achieve lower temperatures), and a cryogenic MOT [39] (to obtain extremely high vacuum and use superconducting coils). Another reason for doubts was the unknown collisional properties at very low temperatures. Would three-body recombination or dipolar relaxation lead to an unfavorable ratio of good to bad collisions when BEC was approached? This actually happened when cesium was cooled in magnetic traps [40], but turned out not to be a problem for sodium and rubidium. Would the condensate have such high densities that it would self-destruct by rapid inelastic losses [41]? This actually almost happened for atomic hydrogen where only a small condensate fraction was achieved, since the condensate suffered from rapid loss [17]. In contrast, alkali condensates were limited in density by their much larger scattering length and the resulting repulsive interactions. In hindsight, the Joint Institute for Laboratory Astrophysics (JILA) and MIT groups were very lucky when they picked rubidium and sodium. Even now, twenty years later, these are still the condensates with the most favorable properties for cooling and stability.

It is interesting to recall some doubtful comments from reviewers of my National Science Foundation (NSF) proposal in early 1995: "It seems that vast improvements are required (in order to reach BEC) ... the current techniques are so far from striking range for BEC that it is not yet possible to make an assessment ..."; "The scientific payoffs, other than the importance of producing a BEC itself, are unclear." And a third reviewer: "... there have been few specific (or realistic) proposals of interesting experiments that could be done with a condensate." Despite the skepticism, all reviewers concluded that the proposed "experiments are valuable and worth pursuing," and I received the funding I urgently needed.

The 1995 BEC conference took place in June in St. Odile, France, and this time, I attended it after having seen evaporative cooling the year before [42]. There was much more representation of the laser cooling approach, with six invited talks, but twice as many talks covered excitons, helium, and hydrogen. However, since the Boulder group observed BEC a few weeks earlier, BEC in atomic gases became the hot topic and the central subject for discussions. Since then, ultracold atoms have dominated the agenda of the BEC meetings, and the field has been exciting and rapidly developing for more than twenty years.

3.5 Quasiparticle Condensates

Metastability is key to the observation of Bose-Einstein condensation. Ultracold atoms have finite lifetimes due to inelastic collisions and background gas collisions of seconds to minutes, much longer than equilibration times and other dynamical time scales of the system. It is this huge separation of time scales which distinguishes ultracold atoms from most quasiparticle systems for which BEC has been studied, including excitons [43, 44], polaritons, magnons [45], and photons [46] (the latter three reviewed in this book in Chapters 4, 25–28, and 19, respectively). The polariton system turned out to be very rich [47], and was used to study quantized vortices [48, 49], half-vortices [50], superfluid flow [51], and Josephson oscillations [52], going even beyond what has been done with ultracold atoms. (See Chapters 4, 19–24 for a nonexhaustive selection of topics in polariton research.) Very recently, polariton lifetimes have been increased to be much longer than equilibration times [53]. For magnons, room temperature BEC has been demonstrated [45].

The condensation of magnons in the B phase of superfluid helium-3 deserves a special note. In 1984, the spontaneous phase coherent precession of spins was discovered [54], but only in 2007 was it interpreted as BEC of magnons [55]. One can argue that this system was the first realization of a phenomenon in nature described by the equations for a weakly interacting BEC.

Magnons provide even systems where the BEC is stable, not metastable [56, 57]. In certain magnetic insulators, a phase transition between different magnetic phases can be described as Bose-Einstein condensation of magnons. The magnon BEC is a magnetically ordered equilibrium state featuring a transverse magnetization which spontaneously breaks the rotational symmetry of the spin Hamiltonian [58]. Such an ordered state has diagonal long-range order, in contrast to superfluids, which have off-diagonal long-range order [59]. Ref. [60] distinguishes these two kinds of Bose-Einstein condensates. (See also related discussion on magnon BEC in Chapters 25–27.)

The very different physical regimes for quasiparticle condensates have expanded the research agenda for Bose-Einstein condensation and led to a deeper understanding of the underlying concepts.

3.6 Early Surprises

When BEC was realized in 1995 [61, 62], the robustness of the condensate was a positive surprise. Inelastic collision rates were favorably low and resulted in long lifetimes of the condensed state, which immediately enabled detailed studies of the properties of BECs. The Rice group showed in 1996 that even for attractive interactions, small condensates of about 10^3 atoms were stable due to the zero-point energy of the trapping potential [63]. This corrected their earlier evidence for reaching the quantum-degenerate regime with 100 times larger condensates [64].

Even the phase of the condensate turned out to be robust, as shown in the early interference experiments of two Bose-Einstein condensates [65]. I was personally very surprised when I learned about the experiment in Boulder, where the phase coherence between the two components of a condensate (using two hyperfine states of rubidium) survived even during phase separation and damping of the relative motion between the two components [66].

Creating BECs was conceptually simple, but it required adaption of technology that was not available in most atomic physics labs at this time, including sensitive cameras, ultrahigh vacuum technology, and high current control for magnetic traps. This explains why it took almost two years after the first demonstrations of BEC in 1995 before the next groups succeeded in reaching BEC (in 1997, the groups of Dan Heinzen [67], Lene Hau [68], Mark Kasevich [69], and Gerhard Rempe [70] followed).

3.7 Feshbach Resonances

A game-changing element for atomic quantum gases has been the use of Feshbach resonances. Fano-Feshbach type resonances have been studied for a long time in nuclear and atomic physics as features (usually enhancement of inelastic scattering) showing up when the collision energy is varied. Ultracold atoms are always close to zero collision energy; therefore, the resonances have to be tuned (e.g., via external magnetic fields) through zero energy [71]. It was only two years before BEC was observed that B. Verhaar's group pointed out that external fields could also change the elastic scattering properties and therefore the strength and sign of interactions of quantum gases [72].

Now Feshbach resonances are an indispensable tool in many quantum gases experiments [73]. However, it took a few steps before that happened. Initially, only magnetic traps were used to confine BECs, and the search for Feshbach resonances focused on Rb-85. However, very soon after transferring BECs into optical traps, Feshbach resonances were discovered (in sodium [74], almost simultaneously with enhanced photoassociation in rubidium [75]). Feshbach resonances in condensates were accompanied by fast losses – the three-body rate increased by three orders of magnitude [76]. It looked like the "gain is not worth the pain." These first experiments were carried out with only a few mW of infrared laser power for optical trapping requiring a tightly focused beam and therefore a very small trap volume. Now, much longer lifetimes are achieved with higher-power lasers and much lower densities of atoms. Still, for very strong interactions, near the unitarity limit, the lifetime of BECs is extremely short, preventing the clouds from reaching equilibrium [77, 78]. This is in contrast to fermions [79], where near the Feshbach resonance Pauli blocking strongly suppresses inelastic three-body collisions, but not the two-body collisions which are responsible for interactions and equilibration [80].

Finally, Feshbach resonances are used to create extremely weakly interacting quantum gases (by tuning the scattering length to zero). Depending on the properties of the Feshbach resonance, losses can be very small in these regions. For lithium-7 it was possible to vary the scattering length over seven decades [81]. Feshbach resonances are almost ubiquitous. In 1994, in rubidium-87, a negative search was conducted to find a resonance over a range of 500 G [82] (this atom has several Feshbach resonances but not for the lowest hyperfine state in this range of magnetic field). In contrast, for the lanthanides (erbium and dysprosium), about two hundred resonances are found over a 50 G range [83].

It is hard to imagine where the field would be without Feshbach resonances. They were crucial to obtain BEC in some atomic species, prominently for Rb-85, where a detailed study of the collapse of BECs with attractive interactions was performed [84], and for cesium [85]. They enabled the study of the BEC-BCS crossover [86, 87] and of Fermi gases with unitarity limited interactions [88, 89], two major highlights in our field. They are the preferred technique to create ultracold molecules by combining ultracold atoms [73], bypassing the need to cool the molecules directly (which is still a challenge and frontier of the field). These developments have created a new subfield, molecular collisions in the ultracold regime, and ultracold chemistry with full control over the initial quantum states. Another new subfield enabled by Feshbach resonances is three-body physics and Efimov resonances [73].

None of these developments based on Feshbach resonances was anticipated in 1995! It took until 2002, when research (and number of publications; see figure in Ref. [73]) on Feshbach resonances took off and transformed the field.

3.8 New Directions

3.8.1 From Weak to Strong Interactions

When BEC was realized in ultracold atoms, the new features (compared with liquid helium) were the weak interactions and the exciting prospect to observe and verify Bogoliubov and mean field theories in a quantum gas. For ultracold Fermi gases, I expected that the main direction would be demonstrations of Pauli blocking and other concepts, assuming one would be limited to weak interactions, e.g., smaller than the Fermi energy.

Now, in 2015, some of the most impressive accomplishments for quantum gases have been studies of strongly interacting systems. This became possible by increasing the strength of interactions via Feshbach resonances (see above), by reducing the kinetic energy in optical lattices, or by confining the system to lower dimensions. A comprehensive overview summarizes the various frontiers in many-body physics with ultracold atoms [90].

3.8.2 Many New Species

The first years focused on alkali atoms, the atoms for which laser cooling was well established. By now, many more atomic species have been condensed. This includes atoms that cannot be confined in magnetic traps [91], one of the key technologies used in earlier BEC work, and thought to be indispensable for cooling to BEC. These atoms have to be confined in optical traps. Using high-power infrared lasers, such traps can now accommodate large atom clouds, and have even replaced the magnetic trap in many experiments with alkali atoms [92].

The list of atoms includes all stable alkali isotopes; some alkaline earth atoms (calcium and strontium, which are used as atomic clocks and offer a richer electronic structure with metastable states and very narrow transitions), chromium; the lanthanides ytterbium, erbium and dysprosium (the last two and chromium have large magnetic dipole moments due to unpaired inner-shell electrons); and hydrogen and metastable helium.

Mixtures of different atomic species are being studied, including spinor condensates, which are mixtures of atoms in different internal states that interconvert depending on external parameters, resulting in a rich phase diagram. See Ref. [93] and Chapter 18 for recent reviews.

Still, I continue to be surprised at how big a market share of research the alkali atoms still command.

3.8.3 Techniques and Technology

Evaporative cooling is the crucial technology in all quantum gas experiments. Before BEC was accomplished, there was a lot of research on new laser-cooling schemes, including sub-recoil schemes with the goal of achieving BEC with laser cooling. Within a very short time, most of the groups working on laser cooling transitioned to studies of BECs. Evaporative cooling of atoms was just so much more robust and applicable to higher densities than any sub-recoil laser-cooling scheme. First demonstrated in 1988 by Thomas Greytak, Harald Hess, and Dan Kleppner [31] (see also their account in Chapter 2), evaporative cooling was the enabling technology for the first atomic BECs and is still the *only* cooling method that works for cooling atoms to quantum degeneracy. A paper on purely optical cooling of strontium to degeneracy also relied on elastic collisions when a "dimple" trap was used to increase the phase space density [94] – this can be regarded as evaporative cooling of atoms in the dimple, but the "evaporated" atoms are now not discarded, but kept in a shallow trap.

For species that are not favorable for evaporative cooling, sympathetic cooling can be used. This may be necessary for many molecular species, which can have large rate coefficients for inelastic collisions, and was required for fermions for which elastic collisions freeze out due to the Pauli exclusion principle. The latter does not apply for fermions with dipolar interactions, which are long range, and therefore elastic collisions don't freeze out at zero temperature [95].

Laser cooling has been used as a precooling stage to cool atoms and load them into optical or magnetic traps. Only two BEC experiments loaded magnetic traps directly via collisions at cryogenic temperatures, using helium-covered walls [17] or helium buffer gas [96]. Fifteen years ago, when I started to collaborate with John Doyle on buffer gas cooling, I expected cryogenic cooling of atoms to become more prevalent for two reasons: to cool other species and to overcome the limits of laser cooling in density (due to light-assisted collisions) and atom number (caused by absorption of laser light in large clouds). Now it has become clear that laser cooling can be used for many atoms even when they have a complicated level structure, and only very few experiments use condensates larger than a few million atoms [17, 97, 98]. In contrast, cryogenic buffer gas cooling is frequently used as a first cooling stage for molecules since direct laser cooling of them is difficult.

New techniques are constantly developed, and become indispensable tools. Optical lattices have opened broad new research directions [90]. More recently, quantum microscopes combined lattices with single-site resolution, and now allow observations of single atoms and their correlations [99]. (These features are further discussed by Bloch in Chapter 13.)

New techniques are often enabled by new technology. Laser and optical technology has rapidly advanced. Laser sources are now much cheaper, more powerful, and more reliable, and it is possible to do experiments with many different lasers of different colors. This is a big step forward from my own labs in the mid- and late 1990s, which fully depended on dye lasers that are now almost universally thought of as giant extinct species.

3.8.4 Quantum Simulations

BEC was immediately recognized as a new quantum liquid, extending the tradition from the superfluid quantum liquids helium-3 and helium-4 to gaseous systems at a billion times lower density. This immediately created a shopping list of scientific directions, including the study of phase transitions, sound and other collective excitations, vortices, and superfluidity. (See also Chapters 6–7 for a discussion of our current understanding of the critical region and condensate formation, Chapter 16 for the use of second sound to study superfluidity, and Chapter 17 for an overview of the emerging field of controlled superfluid turbulence studies.) Ultracold quantum gases have now realized the simplest case of bosonic and fermionic superfluids (as reviewed in Refs. [100, 87]).

In the spirit of a quantum simulator, one can now add bells and whistles to those systems and realize different Hamiltonians (as also discussed in Chapter 13). A toolbox of various atomic species, multiple laser beams, and radio-frequency or microwave radiation is used to "quantum engineer" interesting, often paradigmatic, Hamiltonians and to study their properties. Prime examples are the BEC-BCS crossover (reviewed in Chapter 12), fermions with infinitely strong interactions (see also Chapters 9 and 16), population-imbalanced fermion systems, Bose-Hubbard and Fermi-Hubbard models, and Anderson localization [101]. Temporal modulation of lattices can break time-reversal symmetry, realizing systems with nontrivial topological properties [102]. Twenty years ago, I would have never imagined that Bose-Einstein condensates can now be used in a quantum simulator for charged particles in strong magnetic fields [103]!

3.9 Outlook

Many new developments were featured at the biannual international conference on Bose-Einstein condensation held in September 2015 in Spain (see Fig. 3.1). These conferences started in 1993 when progress toward achieving BEC intensified. Despite keeping the name, the meeting now covers all frontiers in quantum gases. The 2015 conference opened with a celebratory session "BEC 20 years." Bill Phillips, 1997 Nobel Laureate for laser cooling, captured the spirit of the meeting in a talk entitled "40 years of laser cooling, 20 years of BEC: still surprises." It seemed apt that the 2015 senior BEC prize was given to Greytak, Hess, and Kleppner for their early demonstration of evaporative cooling of atoms. The award talk highlighted the major obstacles during the early days of research toward Bose-Einstein condensates, and the solutions that many younger researchers now take for granted.

Some of the current frontiers discussed at the meeting include new systems such as ultracold atomic gases with Rydberg excitations to obtain strong interactions and correlations, highly magnetic atoms that show strong dipolar interactions, and quantum fluids of photons. New techniques are being developed, such as single-atom microscopy, and shaping quantum gases into two reservoirs connected by a thin channel for transport measurements. And new scientific avenues have emerged,

Figure 3.1 This photo, taken at the 2015 BEC conference in Spain, shows past, current, and future chairmen of this biannual conference (1993–2017), together with three Nobel laureates and Dan Kleppner, an early pioneer in BEC research. Front row (left to right): Christophe Salomon (2015), Ehud Altman (2013), Eric Cornell (NP 2001), Thierry Giamarchi (2017); back row: Ignacio Cirac (2001), Tilman Esslinger (2011), Jook Walraven (1995), Massimo Inguscio (2007), Wolfgang Ketterle (NP 2001), Sandro Stringari (1993), Dan Kleppner, Gora Shlyapnikov (2005), Yvan Castin (2003), Peter Zoller (2009), Maciej Lewenstein (1999), and Bill Phillips (NP 1997). The chairman of the 1997 conference, Martin Wilkens, is missing. Photo credit: Sébastien Laurent.

including spin-orbit coupling and artificial gauge fields (see also Chapters 14–15), the creation of topological defects (Kibble-Zurek physics) during quenches across the BEC phase transition (see also Chapters 6–7), disorder and many-body localization, ergodicity and prethermalization (see also Chapter 8), the entanglement of few atoms, and polarons in BECs and Fermi gases. Although the community of researchers in this field has grown rapidly, it still feels like a big family marked by a friendly atmosphere with a collaborative spirit.

One major goal for the future is to obtain a deeper understanding of entanglement, strong interactions, and correlations in few and many-body systems. This can be realized by using ultracold atoms and molecules to assemble interesting quantum systems. Materials with topological properties, including the fractional quantum Hall effect, topological insulators, and Majorana fermions, new forms of superfluidity (including p-wave and d-wave pairing, Fulde-Ferrell-Larkin-Ovchinnikov [FFLO] states, models for high-temperature superconductors), and frustrated spin systems all rank high on this list. A challenge is to use fermions as particles and bosons as fields to simulate dynamic gauge fields, such as toy models for quantum chromodynamics (QCD).

Atomic physics can go beyond the realizations available to an electron system by using bosonic and fermionic atoms in various spin states – finding the bosonic version of fractional quantum Hall states, for example, or superfluidity in a three-component Fermi system. But almost certainly, there will be surprises and unexpected breakthroughs. In the long term, the hope is that insight into new quantum phases of matter will pave the way toward fundamentally new materials and new devices.

Acknowledgments: The author thanks Alan Jamison for a critical reading of the manuscript, Nick Proukakis and his team for organizing the memorable 2013 BEC workshop in Leiden and editing this book, and the NSF and Department of Defense (DoD) for funding this research for more than twenty years.

References

[1] Einstein, A. 1925. Quantentheorie des einatomigen idealen Gases. II. *Sitzungsber. Preuss. Akad. Wiss.*, **Bericht 1**, 3–14.

[2] Einstein, A. 1924. Letter to Ehrenfest (Dec.). Quoted in Abraham Pais, *Subtle Is the Lord: The Science and the Life of Albert Einstein*. Oxford: Oxford University Press, 2005, p. 432.

[3] Uhlenbeck, G. E. 1927. Over Statistische Methoden in de Theorie der Quanta. 's Gravenhage: Martinus Nijhoff, pp. 69–71, cited and translated in [4].

[4] Monaldi, D. 2008. First steps (and stumbles) of Bose-Einstein condensation. In Christian Joas, Christoph Lehner, and Jürgen Renn (eds), *HQ-1: Conference on the History of Quantum Physics*, vols. I and II. Berlin: Max-Planck-Institut

fr Wissenschaftsgeschichte, p. 135; www.mpiwg-berlin.mpg.de/en/file/26683/download?token=kdl9Wa0y.
[5] Kahn, B., and Uhlenbeck, G. E. 1937. On the theory of condensation. *Physica*, **4**, 1155.
[6] London, F. 1938. On the Bose-Einstein condensation. *Phys. Rev.*, **54**, 947.
[7] London, F. 1938. The λ-phenomenon of liquid helium and the Bose-Einstein degeneracy. *Nature*, **141**, 643.
[8] Schrödinger, E. 1952. *Statistical Thermodynamics*. Cambridge: Cambridge University Press, reprinted by Dover Publications (New York, 1989).
[9] Crooker, B. C., Hebral, B., Smith, E. N., Takano, Y., and Reppy, J. D. 1983. Superfluidity in a dilute Bose gas. *Phys. Rev. Lett.*, **51**, 666–669.
[10] Reppy, J. D. 1984. ^{4}He as a dilute Bose gas. *Physica B*, **126**, 335–341.
[11] Rasolt, M., Stephen, M. H., Fisher, M. E., and Weichman, P. B. 1984. Critical behavior of a dilute interacting Bose fluid. *Phys. Rev. Lett.*, **53**, 798.
[12] Cho, H., and Williams, G. A. 1995. Vortex core size in submonolayer superfluid ^{4}He films. *Phys. Rev. Lett.*, **75**, 1562.
[13] Hecht, C. E. 1959. The possible superfluid behaviour of hydrogen atom gases and liquids. *Physica*, **25**, 1159.
[14] Stwalley, W. C., and Nosanow, L. H. 1976. Possible "new" quantum systems. *Phys. Rev. Lett.*, **36**, 910.
[15] Silvera, I. F., and Walraven, J. T. M. 1980. Stabilization of atomic hydrogen at low temperature. *Phys. Rev. Lett.*, **44**, 164–168.
[16] Cline, R. W., Smith, D. A., Greytak, T. J., and Kleppner, D. 1980. Magnetic confinement of spin-polarized atomic hydrogen. *Phys. Rev. Lett.*, **45**, 2117.
[17] Fried, D. G., Killian, T. C., Willmann, L., Landhuis, D., Moss, S. C., Kleppner, D., and Greytak, T. J. 1998. Bose-Einstein condensation of atomic hydrogen. *Phys. Rev. Lett.*, **81**, 3811–3814.
[18] Greytak, T. J. 1995. Prospects for Bose-Einstein condensation in magnetically trapped atomic hydrogen. In Griffin, A., Snoke, D. W., and Stringari, S. (eds), *Bose-Einstein Condensation*. Cambridge: Cambridge University Press, pp. 131–159.
[19] Greytak, T. J., and Kleppner, D. 1984. Lectures on spin-polarized hydrogen. In Grynberg, G., and Stora, R. (eds), *New Trends in Atomic Physics, Les Houches Summer School 1982*. Amsterdam: North-Holland, p. 1125.
[20] Silvera, I. F., and Walraven, J. T. M. 1986. Spin-polarized atomic hydrogen. In Brewer, D. F. (ed), *Progress in Low Temperature Physics*, vol. X. Amsterdam: Elsevier, p. 139.
[21] Walraven, J. T. M. 1996. Atomic hydrogen in magnetostatic traps. Oppo, G. L., Barnett, S. M., Riis, E., and Wilkinson, M. (eds), *Quantum Dynamics of Simple Systems*. London: Institute of Physics Publ., pp. 315–352.
[22] Silvera, I. F. 1995. Spin-polarized hydrogen: prospects for Bose-Einstein condensation and two-dimensional superfluidity. In Griffin, A., Snoke, D. W., and Stringari, S. (eds), *Bose-Einstein Condensation*. Cambridge: Cambridge University Press, pp. 160–172.
[23] Mosk, A. P., Reynolds, M. W., Hijmans, T. W., and Walraven, J. T. M. 1998. Optical observation of atomic hydrogen in the surface of liquid helium. *J. Low Temp. Phys.*, **113**, 217–222.
[24] Matsubara, A., Arai, T., Hotta, S., Korhonen, J. S., Mizusaki, T., and Hiraj, A. 1995. Quest for Kosterlitz-Thouless transition in two-dimensional hydrogen. In Griffin, A., Snoke, D. W., and Stringari, S. (eds), *Bose-Einstein Condensation*. Cambridge: Cambridge University Press, pp. 478–486.

[25] Safonov, A. I., Vasilyev, S. A., Yasnikov, I. S., Lukashevich, I. I., and Jaakola, S. 1998. Observation of quasicondensate in two-dimensional atomic hydrogen. *Phys. Rev. Lett.*, **81**, 4545.

[26] Goldman, V. V., Silvera, I. F., and Leggett, A. J. 1981. Atomic hydrogen in an inhomogeneous magnetic field: density profile and Bose-Einstein condensation. *Phys. Rev. B*, **24**, 2870–2873.

[27] Huse, D. A., and Siggia, E. 1982. The density distribution of a weakly interacting Bose gas in an external potential. *J. Low Temp. Phys.*, **46**, 137.

[28] Oliva, J. 1989. Density profile of the weakly interacting Bose gas confined in a potential well: nonzero temperature. *Phys. Rev. B*, **39**, 4197–4203.

[29] Stoof, H. T. C., Koelman, J. M. V. A., and Verhaar, B. J. 1988. Spin-exchange and dipole relaxation rates in atomic hydrogen: rigorous and simplified calculations. *Phys. Rev. B*, **38**, 4688.

[30] Hess, H. F. 1986. Evaporative cooling of magnetically trapped and compressed spin-polarized hydrogen. *Phys. Rev. B*, **34**, 3476.

[31] Masuhara, N., Doyle, J. M., Sandberg, J. C., Kleppner, D., Greytak, T. J., Hess, H. F., and Kochanski, G. P. 1988. Evaporative cooling of spin-polarized atomic hydrogen. *Phys. Rev. Lett.*, **61**, 935–938.

[32] Griffin, A., Snoke, D. W., and Stringari, S. 1995. *Bose-Einstein Condensation*. Cambridge: Cambridge University Press.

[33] Lin, J. L., and Wolfe, J. P. 1993. Bose-Einstein condensation of paraexcitons in stressed Cu_2O. *Phys. Rev. Lett.*, **71**, 1222.

[34] O'Hara, K. E., Silleabhin, L., and Wolfe, J. P. 1999. Strong nonradiative recombination of excitons in Cu_2O and its impact on Bose-Einstein statistics. *Phys. Rev. B*, **60**, 10565.

[35] Taubes, G. 1994. Hot on the trail of a cold mystery. *Science*, **265**, 184–186.

[36] Ketterle, W., Davis, K. B., Joffe, M. A., Martin, A., and Pritchard, D. E. 1993. High-densities of cold atoms in a dark spontaneous-force optical trap. *Phys. Rev. Lett.*, **70**, 2253–2256.

[37] Cornell, E. A., Monroe, C., and Wieman, C. E. 1991. Multiply loaded, AC magnetic trap for neutral atoms. *Phys. Rev. Lett.*, **67**, 2439.

[38] Kasevich, M. A. 1995. Evaporative cooling and Raman cooling in a crossed dipole trap. *Bull. Am. Phys. Soc.*, **40**, 1270.

[39] Willems, P. A., and Libbrecht, K. G. 1995. Creating long-lived neutral-atom traps in a cryogenic environment. *Phys. Rev. A*, **51**, 1403.

[40] Guery-Odelin, D., Soding, J., Desbiolles, P., and Dalibard, J. 1998. Is Bose-Einstein condensation of atomic cesium possible? *Europhys. Lett.*, **44**, 25–30.

[41] Hijmans, T. W., Kagan, Y., Shlyapnikov, G. V., and Walraven, J. T. M. 1993. Bose condensation and relaxation explosion in magnetically trapped atomic hydrogen. *Phys. Rev. B*, **48**, 12886–12892.

[42] Davis, K. B., Mewes, M.-O., Joffe, M. A., Andrews, M. R., and Ketterle, W. 1995. Evaporative cooling of sodium atoms. *Phys. Rev. Lett.*, **74**, 5202–5205.

[43] High, A. A., Leonard, J. R., Hammack, A. T., Fogler, M. M., Butov, L. V., Kavokin, A. V., Campman, K. L., and Gossard, A. C. 2012. Spontaneous coherence in a cold exciton gas. *Nature*, **483**, 584–588.

[44] Snoke, D., and Kavoulakis, G. M. 2014. Bose-Einstein condensation of excitons in Cu_2O: progress over 30 years. *Rep. Prog. Phys.*, **77**, 116501.

[45] Demokritov, S. O., Demidov, V. E., Dzyapko, O., Melkov, G. A., Serga, A. A., Hillebrands, B., and Slavin, A. N. 2006. Bose-Einstein condensation of quasi-equilibrium magnons at room temperature under pumping. *Nature*, **443**, 430–433.

[46] Klaers, J., Schmitt, J., Vewinger, F., and Weitz, M. 2010. Bose-Einstein condensation of photons in an optical microcavity. *Nature*, **468**, 545–548.

[47] Kasprzak, J., Richard, M., Kundermann, S., Baas, A., Jeambrun, P., Keeling, J. M. J., Marchetti, F. M., Szymanska, M. H., Andre, R., Staehli, J. L., Savona, V., Littlewood, P. B., Deveaud, B., and Dang, L. S. 2006. Bose-Einstein condensation of exciton polaritons. *Nature*, **443**, 409–414.

[48] Lagoudakis, K. G., Wouters, M., Richard, M., Baas, A., Carusotto, I., Andre, R., Dang, L. S., and Deveaud-Pledran, B. 2008. Quantized vortices in an exciton-polariton condensate. *Nat Phys*, **4**, 706–710.

[49] Sanvitto, D., Marchetti, F. M., Szymanska, M. H., Tosi, G., Baudisch, M., Laussy, F. P., Krizhanovskii, D. N., Skolnick, M. S., Marrucci, L., Lemaitre, A., Bloch, J., Tejedor, C., and Vina, L. 2010. Persistent currents and quantized vortices in a polariton superfluid. *Nat Phys*, **6**, 527–533.

[50] Lagoudakis, K. G., Ostatnick, T., Kavokin, A. V., Rubo, Y. G., Andr, R., and Deveaud-Pledran, B. 2009. Observation of half-quantum vortices in an exciton-polariton condensate. *Science*, **326**, 974–976.

[51] Amo, A., Lefrere, J., Pigeon, S., Adrados, C., Ciuti, C., Carusotto, I., Houdre, R., Giacobino, E., and Bramati, A. 2009. Superfluidity of polaritons in semiconductor microcavities. *Nat Phys*, **5**, 805–810.

[52] Abbarchi, M., Amo, A., Sala, V. G., Solnyshkov, D. D., Flayac, H., Ferrier, L., Sagnes, I., Galopin, E., Lemaitre, A., Malpuech, G., and Bloch, J. 2013. Macroscopic quantum self-trapping and Josephson oscillations of exciton polaritons. *Nat Phys*, **9**, 275–279.

[53] Sun, Y., Wen, P., Yoon, Y., Liu, G., Steger, M., Pfeiffer, L. N., West, K., Snoke, D. W., and Nelson, K. A. 2017. Bose-Einstein condensation of long-lifetime polaritons in thermal equilibrium. *Phys. Rev. Lett.*, **118**, 016602.

[54] Borovik-Romanov, A. S., Bunkov, Y. M., Dmitriev, V. V., and Mukharskiy, Y. M. 1984. Long-lived induction signal in superfluid ^{3}He-B. *JETP Lett.*, **40**, 1033.

[55] Bunkov, Y. M., and Volovik, G. E. 2007. Bose-Einstein condensation of magnons in superfluid ^{3}He. *J. Low Temp. Phys.*, **150**, 135–144.

[56] Nikuni, T., Oshikawa, M., Oosawa, A., and Tanaka, H. 2000. Bose-Einstein condensation of dilute magnons in $TlCuCl_3$. *Phys. Rev. Lett.*, **84**, 5868–5871.

[57] Ruegg, C., Cavadini, N., Furrer, A., Gudel, H. U., Kramer, K., Mutka, H., Wildes, A., Habicht, K., and Vorderwisch, P. 2003. Bose-Einstein condensation of the triplet states in the magnetic insulator $TlCuCl_3$. *Nature*, **423**, 62–65.

[58] Giamarchi, T., Ruegg, C., and Tchernyshyov, O. 2008. Bose-Einstein condensation in magnetic insulators. *Nat Phys*, **4**, 198–204.

[59] Volovik, G. E. 2008. Twenty years of magnon Bose condensation and spin current superfluidity in ^{3}He-B. *J. Low Temp. Phys.*, **153**, 266–284.

[60] Kohn, W., and Sherrington, D. 1970. Two kinds of bosons and Bose condensates. *Rev. Mod. Phys.*, **42**, 1–11.

[61] Anderson, M. H., Ensher, J. R., Matthews, M. R., Wieman, C. E., and Cornell, E. A. 1995. Observation of Bose-Einstein condensation in a dilute atomic vapor. *Science*, **269**, 198–201.

[62] Davis, K. B., Mewes, M.-O., Andrews, M. R., van Druten, N. J., Durfee, D. S., Kurn, D. M., and Ketterle, W. 1995. Bose-Einstein condensation in a gas of sodium atoms. *Phys. Rev. Lett.*, **75**, 3969–3973.

[63] Bradley, C. C., Sackett, C. A., and Hulet, R. G. 1997. Bose-Einstein condensation of lithium: observation of limited condensate number. *Phys. Rev. Lett.*, **78**, 985–989.

[64] Bradley, C. C., Sackett, C. A., Tollet, J. J., and Hulet, R. G. 1995. Evidence of Bose-Einstein condensation in an atomic gas with attractive interactions. *Phys. Rev. Lett.*, **75**, 1687–1690.

[65] Andrews, M. R., Townsend, C. G., Miesner, H.-J., Durfee, D. S., Kurn, D. M., and Ketterle, W. 1997. Observation of interference between two Bose condensates. *Science*, **275**, 637–641.

[66] Hall, D. S., Matthews, M. R., Wieman, C. E., and Cornell, E. A. 1998. Measurements of relative phase in two-component Bose-Einstein condensates. *Phys. Rev. Lett.*, **81**, 1543–1546.

[67] Han, D. J., Wynar, R. H., Courteille, P., and Heinzen, D. J. 1998. Bose-Einstein condensation of large numbers of atoms in a magnetic time-averaged orbiting potential trap. *Phys. Rev. A*, **57**, R4114.

[68] Hau, L. V., Busch, B. D., Liu, C., Dutton, Z., Burns, M. M., and Golovchenko, J. A. 1998. Near resonant spatial images of confined Bose-Einstein condensates in the "4D" magnetic bottle. *Phys. Rev. A*, **58**, R54.

[69] Anderson, B. P., and Kasevich, M. A. 1998. Macroscopic quantum interference from atomic tunnel arrays. *Science*, **282**, 1686.

[70] Ernst, U., Marte, A., Schreck, F., Schuster, J., and Rempe, G. 1998. Bose-Einstein condensation in a pure Ioffe-Pritchard field configuration. *Europhys. Lett.*, **41**, 1–6.

[71] Stwalley, W. C. 1976. Stability of spin-aligned hydrogen at low temperatures and high magnetic fields: new field-dependent scattering resonances and predissociations. *Phys. Rev. Lett.*, **37**, 1628.

[72] Tiesinga, E., Verhaar, B. J., and Stoof, H. T. C. 1993. Threshold and resonance phenomena in ultracold ground-state collisions. *Phys. Rev. A*, **47**, 4114–4122.

[73] Chin, C., Grimm, R., Julienne, P., and Tiesinga, E. 2010. Feshbach resonances in ultracold gases. *Rev. Mod. Phys.*, **82**, 1225–1286.

[74] Inouye, S., Andrews, M. R., Stenger, J., Miesner, H.-J., Stamper-Kurn, D. M., and Ketterle, W. 1998. Observation of Feshbach resonances in a Bose-Einstein condensate. *Nature*, **392**, 151–154.

[75] Courteille, P., Freeland, R. S., Heinzen, D. J., van Abeelen, F. A., and Verhaar, B. J. 1998. Observation of a Feshbach resonance in cold atom scattering. *Phys. Rev. Lett.*, **81**, 69–72.

[76] Stenger, J., Inouye, S., Andrews, M. R., Miesner, H.-J., Stamper-Kurn, D. M., and Ketterle, W. 1999. Strongly enhanced inelastic collisions in a Bose-Einstein condensate near Feshbach resonances. *Phys. Rev. Lett.*, **82**, 2422–2425.

[77] Navon, N., Piatecki, S., Günter, K., Rem, B., Nguyen, T. C., Chevy, F., Krauth, W., and Salomon, C. 2011. Dynamics and thermodynamics of the low-temperature strongly interacting Bose gas. *Phys. Rev. Lett.*, **107**, 135301.

[78] Makotyn, P., Klauss, C. E., Goldberger, D. L., Cornell, E. A., and Jin, D. S. 2014. Universal dynamics of a degenerate unitary Bose gas. *Nat Phys*, **10**, 116–119.

[79] Cubizolles, J., Bourdel, T., Kokkelmans, S. J. J. M. F., Shlyapnikov, G. V., and Salomon, C. 2003. Production of long-lived ultracold Li_2 molecules from a Fermi gas. *Phys. Rev. Lett.*, **91**, 240401.

[80] Petrov, D. S., Salomon, C., and Shlyapnikov, G. V. 2004. Weakly bound dimers of fermionic atoms. *Phys. Rev. Lett.*, **93**, 090404.

[81] Pollack, S. E., Dries, D., Junker, M., Chen, Y. P., Corcovilos, T. A., and Hulet, R. G. 2009. Extreme tunability of interactions in a ^{7}Li Bose-Einstein condensate. *Phys. Rev. Lett.*, **102**, 090402.

[82] Newbury, N. R., Myatt, C. J., and Wieman, C. E. 1995. s-wave elastic collisions between cold ground-state ^{87}Rb atoms. *Phys. Rev. A*, **51**, R2680.

[83] Maier, T., Kadau, H., Schmitt, M., Wenzel, M., Ferrier-Barbut, I., Pfau, T., Frisch, A., Baier, S., Aikawa, K., Chomaz, L., Mark, M. J., Ferlaino, F., Makrides, C., Tiesinga, E., Petrov, A., and Kotochigova, S. 2015. Emergence of chaotic scattering in ultracold Er and Dy. *Phys. Rev. X*, **5**, 041029.

[84] Donley, E. A., Claussen, N. R., Cornish, S. L., Roberts, J. L., Cornell, E. A., and Wieman, C. E. 2001. Dynamics of collapsing and exploding Bose-Einstein condensates. *Nature*, **412**, 295–299.

[85] Weber, T., Herbig, J., Mark, M., Nägerl, H.-C., and Grimm, R. 2003. Bose-Einstein condensation of cesium. *Science*, **299**, 232–235.

[86] Jin, D. S., and Regal, C. A. 2008. Fermi gas experiments. In Inguscio, M., Ketterle, W., and Salomon, C. (eds), *Ultracold Fermi Gases, Proceedings of the International School of Physics Enrico Fermi, Course CLXIV*. Amsterdam: IOS Press, pp. 1–51.

[87] Ketterle, W., and Zwierlein, M. W. 2008. Making, probing and understanding ultracold Fermi gases. In Inguscio, M., Ketterle, W., and Salomon, C. (eds), *Ultracold Fermi Gases, Proceedings of the International School of Physics Enrico Fermi, Course CLXIV*. Amsterdam: IOS Press, pp. 95–287.

[88] Nascimbne, S., Navon, N., Jiang, K. J., Chevy, F., and Salomon, C. 2010. Exploring the thermodynamics of a universal Fermi gas. *Nature*, **463**, 1057–1060.

[89] Ku, M. J. H., Sommer, A. T., Cheuk, L. W., and Zwierlein, M. W. 2012. Revealing the superfluid lambda transition in the universal thermodynamics of a unitary Fermi gas. *Science*, **335**, 563–567.

[90] Bloch, I., Dalibard, J., and Zwerger, W. 2008. Many-body physics with ultracold gases. *Rev. Mod. Phys.*, **80**, 885–880.

[91] Takasu, Y., Maki, K., Komori, K., Takano, T., Honda, K., Kumakura, M., Yabuzaki, T., and Takahashi, Y. 2003. Spin-singlet Bose-Einstein condensation of two-electron atoms. *Phys. Rev. Lett.*, **91**, 040404.

[92] Barrett, M. D., Sauer, J. A., and Chapman, M. S. 2001. All-optical formation of an atomic Bose-Einstein condensate. *Phys. Rev. Lett.*, **87**, 010404–4.

[93] Stamper-Kurn, D. M., and Ueda, M. 2013. Spinor Bose gases: symmetries, magnetism, and quantum dynamics. *Rev. Mod. Phys.*, **85**, 1191–1244.

[94] Stellmer, S., Pasquiou, B., Grimm, R., and Schreck, F. 2013. Laser cooling to quantum degeneracy. *Phys. Rev. Lett.*, **110**, 263003.

[95] Aikawa, K., Frisch, A., Mark, M., Baier, S., Grimm, R., and Ferlaino, F. 2014. Reaching Fermi degeneracy via universal dipolar scattering. *Phys. Rev. Lett.*, **112**, 010404.

[96] Doret, S. C., Connolly, C. B., Ketterle, W., and Doyle, J. M. 2009. Buffer-gas cooled Bose-Einstein condensate. *Phys. Rev. Lett.*, **103**, 103005.

[97] Abo-Shaeer, J. R., Raman, C., and Ketterle, W. 2002. Formation and decay of vortex lattices in Bose-Einstein condensates at finite temperatures. *Phys. Rev. Lett.*, **88**, 070409.

[98] van der Stam, K. M. R., van Ooijen, E. D., Meppelink, R., Vogels, J. M., and van der Straten, P. 2007. Large atom number Bose-Einstein condensate of sodium. *Rev. Sci. Inst.*, **78**, 013102.

[99] Bakr, W. S., Gillen, J. I., Peng, A., Fölling, S., and Greiner, M. 2009. A quantum gas microscope for detecting single atoms in a Hubbard regime optical lattice. *Nature*, **462**, 74–77.

[100] Ketterle, W., Durfee, D. S., and Stamper-Kurn, D. M. 1999. Making, probing and understanding Bose-Einstein condensates. In Inguscio, M., Stringari, S., and Wieman, C. E. (eds), *Bose-Einstein Condensation in Atomic Gases, Proceedings of*

the International School of Physics Enrico Fermi, Course CXL. Amsterdam: IOS Press, pp. 67–176.

[101] Bloch, I., Dalibard, J., and Nascimbene, S. 2012. Quantum simulations with ultracold quantum gases. *Nature Physics*, **9**, 267–276.

[102] Goldman, N., Juzelinas, G., Öhberg, P., and Spielman, I. B. 2014. Light-induced gauge fields for ultracold atoms. *Rep. Prog. Phys.*, **77**, 126401.

[103] Kennedy, C. J., Burton, W. C., Chung, W. C., and Ketterle, W. 2015. Observation of Bose-Einstein condensation in a strong synthetic magnetic field. *Nature Physics*, **11**, 859–864.

4

Introduction to Polariton Condensation

PETER B. LITTLEWOOD
Argonne National Laboratory, Lemont, Illinois, USA
James Franck Institute and Department of Physics,
University of Chicago, Illinois, USA

ALEXANDER EDELMAN
James Franck Institute and Department of Physics,
University of Chicago, Illinois, USA

Polaritons are coherent superpositions of light and electronic excitations inside a solid, and they have some appealing characteristics for the study of condensed bosonic gases on account of their light mass (inherited from the photon) and strong interactions (inherited from the excitonic component). This introduction to polaritonic condensates sets out the basic experimental phenomena and discusses theory from the point of view of both microscopic models and collective degrees of freedom.

4.1 Introduction

The lowest energy electronic excitation in a semiconductor is to promote an electron from the top of the valence band to the bottom of the conduction band. If the Coulomb interaction can be ignored, the excitation energy is just equal to the band gap, but in reality the particle and hole will attract forming a bound state, known as an exciton, whose energy is typically somewhat below the gap. When the binding energy is dominated by the long-range Coulomb component, the binding energy is just the renormalized hydrogenic Rydberg scaled by the ratio of the effective mass to the square of the dielectric constant. In compound semiconductors such as GaAs or CdTe, the binding energy is then a few tenths of an eV or smaller, but nonetheless there are (a series of) bound states well separated from the continuum of states above the gap. The excitons have a dispersion $\varepsilon_{\vec{k}} = \varepsilon_0 + k^2/2M$, where M is the total exciton mass.

If the semiconducting gap is direct and there is an electric dipole matrix element (as in the materials above), then excitons are quite short-lived, because there is a direct decay channel for the emission of a photon. More formally, the eigenstates are quantum mechanical superpositions of exciton and photon, an object which has been dubbed a polariton [1, 2]. The short exciton lifetime is then a consequence of mixing with a photon that will naturally escape rapidly from the crystal. In order to make polaritons long-lived, one has to produce confinement of the photon modes, which for the systems we shall discuss in this chapter has been typically to make a two-dimensional (2D) waveguide using a dielectric Bragg grating [3]. Such a 2D cavity confines photon modes in the vertical direction, and the dispersion of these modes is then approximately determined by an intersection of the light cone ($\omega = c|k|$) with the plane determined by fixing k_z – namely a parabola $\omega_{\vec{k}} = (c/n)\sqrt{k^2 + (2\pi N/L_w)^2}$, although more exotic polaritons have been engineered in other nanostructures [4, 5]. (Here n is the refractive index, and N counts the transverse modes in the cavity.) With some care, one can tune the photon modes to be close to the exciton energy. Although in a clean two-dimensional system the excitons will have the quadratic dispersion above, the effective mass of the photon mode is typically some four orders of magnitude smaller than the exciton mass, and thus in practice the excitons may be regarded as infinitely heavy. A last point is that in the dilute limit, when the probability of excitons overlapping is small, the exciton itself can be regarded as a structureless boson. In consequence, there will be a low excitation limit that is described by two coupled Bose fields, and the dispersion shown in Fig. 4.1.

To be explicit, writing a Hamiltonian in terms of photon creation operators $\psi^\dagger_{\vec{k}}$ and $D^\dagger_{\vec{k}}$ for excitons, one has (using $\hbar = 1$ here and throughout):

$$H = \begin{pmatrix} \psi^\dagger_{\vec{k}} & D^\dagger_{\vec{k}} \end{pmatrix} \begin{pmatrix} \omega_{\vec{k}} & \Omega_R/2 \\ \Omega_R/2 & \varepsilon_{\vec{k}} \end{pmatrix} \begin{pmatrix} \psi_{\vec{k}} \\ D_{\vec{k}} \end{pmatrix}. \tag{4.1}$$

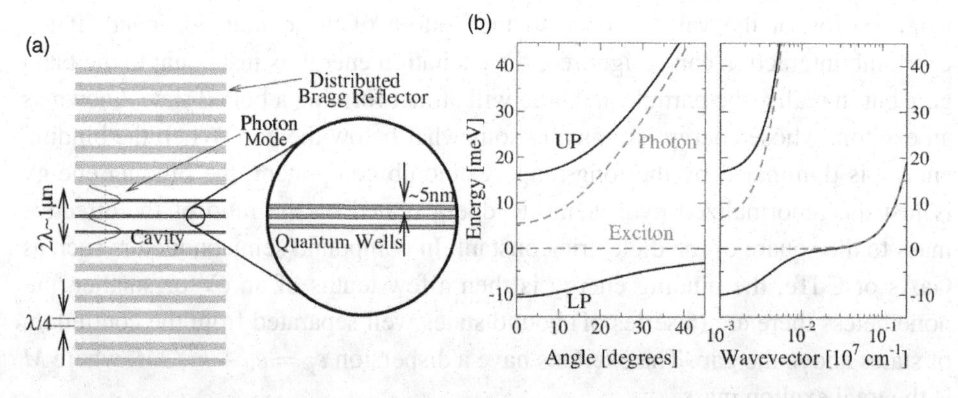

Figure 4.1 (a) Schematic diagram of a microcavity, formed by a pair of distributed Bragg reflector stacks, with quantum wells at the antinodes of the cavity photon mode. (b) Schematic polariton spectrum, plotted as a function of emission angle, $\theta = \sin^{-1}(ck/\omega_0)$ (left), or as a function of momentum on a logarithmic scale, showing the full exciton dispersion (right). (Plotted for $M = 0.08m_e$, $m = 3 \times 10^{-5}m_e$, $\Omega_R = 26$ meV, $\delta = 5.4$ meV, and $\omega_0 = 1.7$ eV.) Both figures reprinted with permission from Keeling, J., et al. (2007), Collective coherence in planar semiconductor microcavities, *Semiconductor Science Technology*, **22**, 1 [6]. Copyright (2007) by the Institute of Physics.

For small k, the photon energy can be written as $\omega_{\vec{k}} = \omega_0 + k^2/2m$, where m is an effective photon mass $m = (n/c)(2\pi N/L_w)$. We define the bottom of the exciton band, ε_0, as the zero of energies; hence the detuning between exciton and photon bands is $\delta = \omega_0 - \varepsilon_0$. The off-diagonal term $\Omega_R/2$ describes the exciton–photon coupling, where Ω_R is the Rabi frequency. Then, diagonalizing the quadratic form in Eq. (4.1) gives the polariton spectrum:

$$E_{\vec{k}}^{\mathrm{LP,UP}} = \frac{1}{2}\left[\left(\delta + \frac{k^2}{2M} + \frac{k^2}{2m}\right) \mp \sqrt{\left(\delta + \frac{k^2}{2M} - \frac{k^2}{2m}\right)^2 + \Omega_R^2}\,\right]. \tag{4.2}$$

(The two branches LP and UP are the lower and upper polariton, respectively.) This spectrum is illustrated in Fig. 4.1.

Pumping of this system can be nonresonant (excitation of electron–hole pairs above the band gap) or resonant with an external laser tuned to the dispersion curve. Note that a polariton tunneling out through a flat mirror will conserve the 2D momentum $\vec{k}$ and therefore be emitted at an angle.

One can make more complex structures than the planar cavity described above, and this has included patterned arrays of pillar microcavities and microcavity wires, as well as superposed gratings. A potential can be induced by applied stress, as well as (inevitably) a cavity wedge. Polaritons in the lower branch at large momentum are, furthermore, nothing more than excitons that interact weakly with light and are therefore are easily trapped by disorder to provide an interaction potential with mobile small-k polaritons.

There is a considerable literature on polariton physics which it will not be possible to review here in detail. Two books [7, 8] and several review articles [6, 9, 10, 11, 12, 13, 14, 15] cover much of it in detail, and we will try instead in this chapter to present a short explication of the subject, especially in order to connect to other chapters in this volume. Our focus will inevitably be on condensation, at the expense of many other interesting aspects of polariton systems. Chapters 10–12 and 20–24 in this volume review many of the current topics of research with polariton condensates.

In such an ensemble of interacting bosons of extremely light mass (in comparison to excitons and certainly to cold atoms), one could expect the formation of a Bose condensate at a temperature comparable to the degeneracy temperature $k_B T_0 \approx \hbar^2 n/m^*$, where m^* is the polariton effective mass, and n the number density of polaritons. However, there are a number of special features:

- The system is open, suffering slow decay through the mirrors and lateral expansion through traps, and by the same token requiring a pump to establish. The decay processes naturally lead to decoherence which may suppress condensate formation. Moreover, the establishment of thermal equilibrium cannot be taken for granted.

- Because the polaritons do decay by tunneling into photon modes in vacuum, there is a window to study population dynamics (measuring the distribution $n(k)$, for example), and also direct measurement of condensate coherence through interference measurements.
- Equally, one can drive the system with an external laser, thereby establishing coherent population(s) in either momentum or real space. Combined with the ability to apply an external potential, this provides considerable flexibility to look at nonequilibrium states and generate polariton circuits ("polaritonics").
- The physics of two-dimensional condensation should be evident.
- The interaction between polaritons is of course produced by the excitonic component (short-ranged), but as light mass particles polaritons are physically "large" (in the sense of the uncertainty principle). In consequence, the interaction between polaritons is long-ranged, and for all but the lowest densities, the system is not well described as a dilute Bose gas.

As we shall see, there are potentially a number of different regimes of behavior, depending on density and other parameters. One natural density parameter is the dimensionless $\rho = na_0^2$, where a_0 is the characteristic size of the exciton. For hydrogenic excitons, the topic of much research, $a_0 = a_B\epsilon/m_r$, with a_B the Bohr radius, ϵ the static dielectric constant, and m_r the reduced mass of the electron–hole pair. a_0 may be of order tens of nanometers, but is generally much smaller than the wavelength of the relevant photons. The exciton binding energy is of order $Ry^* \approx \hbar^2/m_r a_0^2$, which is generally a few tens of meV at most and typically comparable to the strength of the Rabi splitting Ω_R.

The polaritons can be approximated as featureless bosons when their separation, $1/\sqrt{n}$, greatly exceeds their size, which is set by the cavity wavelength (entering through the effective mass) and the electron–hole separation (entering through the Rabi splitting), giving the condition

$$\rho \ll \Omega_R \left(\frac{\hbar^2}{m^* a_0^2} \right)^{-1} = \frac{m^*}{m_r} \frac{\Omega_R}{Ry^*} \approx 10^{-4} \,, \tag{4.3}$$

where the last estimate is a typical number for CdTe cavities. On the other hand, as the system approaches $\rho = 1$, it becomes "Mott insulating" to polariton transport as the number of available exciton states drops to zero; beyond this density, additional polaritons are essentially photonic and remain dilute, with the exciton length scale replaced by a background electron–hole plasma that mediates their interactions [16]. (A similar regime should exist slightly below the Mott density but at temperatures high enough to thermally excite such a background.) Therefore, in both these regimes the BEC transition temperature is $k_B T_0 \approx \hbar^2 n/m^*$ and the dimensionality of the system makes the transition Berezinskii-Kosterlitz-Thouless (BKT)–like.

Depending on the system parameters, there is a possible intermediate regime,

$$\frac{m^*}{m_r}\frac{\Omega_R}{Ry^*} \ll \rho \ll 1, \tag{4.4}$$

in which polaritons overlap strongly without saturating the excitons. The condensation here is into a bosonic coherent state and is mean field–like with $k_B T_0 \approx \Omega_R$.

The conditions above assume perfect translational invariance. Of course, real systems have disorder, or perhaps a deliberate quasiperiodic arrangement of exciton traps. In such a case, it is natural to scale the density not to the exciton radius, but to the trap size (if it is larger), and in dimensionless units ρ may then exceed unity by saturating the excitonic component and generating a mostly photonic condensate.

If one includes the effects of dissipation as an effective polariton linewidth γ, then (crudely) the following can ensue. If $\gamma/\Omega_R > 1$, known as weak coupling, the polariton splitting is destroyed and the system behaves as a conventional semiconductor laser, with a buildup of coherence at sufficiently high densities. In the strong-coupling regime with well-defined polaritons, $\gamma/\Omega_R < 1$, and the condensate should tolerate nonequilibrium fluctuations roughly as well as thermal ones, though this statement is by no means rigorous. In the absence of a condensate, coherence is again determined by population inversion. All this is sketched in Fig. 4.2.

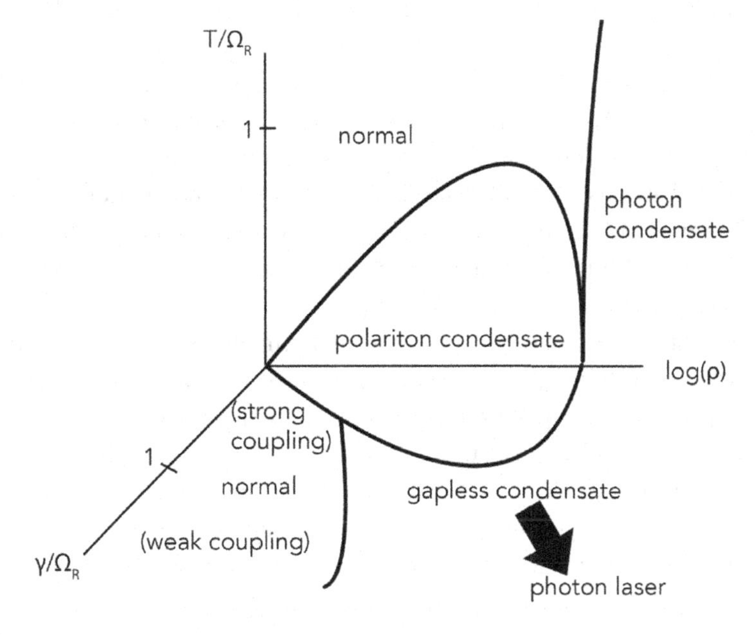

Figure 4.2 Schematic polariton phase diagram. For negative detuning of the exciton level exceeding the Rabi coupling, there is a Mott lobe of re-entrant filled exciton band, as shown here.

4.2 Principal Experimental Features

Strong coupling – defined as the appearance of well-resolved polariton branches so that the Rabi splitting is much larger than the polariton lifetime – is now routinely achievable in a number of systems. Before 2006, a number of experiments had demonstrated nonlinear thresholds for emission (e.g., [17, 18, 19, 20]) and nonlinear emission under four-wave mixing [21]. Polariton condensation (in the sense of coherence developed under near-equilibrium conditions) has been observed in semiconductor microcavities based on II–VI [22], III–V [23], and also organic compounds [24]. Because the systems are open, sometimes one restricts the description of BEC to those systems where the populations are well thermalized, though such a determination is not easy. (For a discussion of early experiments, see [6, 10].)

Note that there are at least three distinct ways to introduce excitations into the system. Incoherent pumping at optical frequencies well above the polariton spectrum generates electron–hole pairs, of which a few excitations find their way to a population of polaritons near the LP minimum. Such pumping inevitably generates a (potentially large) population of trapped excitons that are weakly coupled to light, but interact with the polaritons collisionally, and are usually confined in the vicinity of the excitation spot. It is natural to expect that the polaritons so generated have no initial coherence, so any coherence must result from stimulated scattering. A second method is to tune a laser onto the lower polariton branch; this of course generates an initially coherent wavepacket, but these experiments can be used to study the dynamics of propagating quasicondensates in a potential, or indeed to produce multiple interacting condensates. If the laser is tuned near a point of inflection of the lower polariton dispersion curve, a four-wave mixing process can produce stimulated scattering to the polariton minimum [21]. Such a process can be tuned to conserve energy and momentum, and thus is highly efficient.

One characteristic of a condensate is a macroscopic population of the lowest mode, so at threshold the optical emission narrows substantially below a thermal gas, as in Fig. 4.3. While this is of course the general characteristic of a laser, in the polariton case the energy of the emission corresponds to the lower polariton, not to the bare cavity mode. Interestingly, some experiments show a second threshold at much higher pump powers when the emission shifts close to the cavity mode [25, 26]. It is natural to equate this to a conventional semiconductor laser, where because of inversion of the electronic system, a macroscopic photon occupancy can be maintained even if the electronic excitations are decohered. It is not clear whether in such systems the excitons themselves remain bound. Since the condensate has a polarization direction, as well as a conventional phase, it was expected [27, 28] (and experimentally reported) that the condensate is polarized, with the broken symmetry direction pinned to optical anisotropy. The crossover and connection of lasing and polariton condensation is further discussed in Chapter 20.

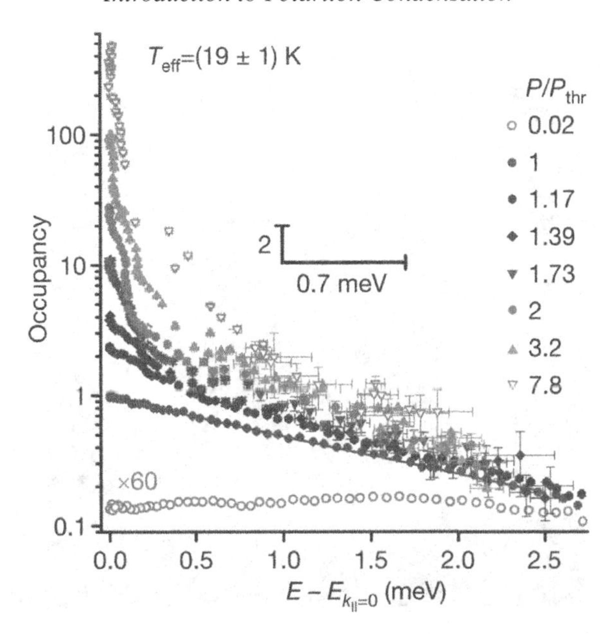

Figure 4.3 Polariton occupation versus energy, showing evolution from below to above threshold power, with the distribution remaining thermalized at a temperature of 19 K. Reprinted with permission from Kasprzak, J., et al. (2006), Bose-Einstein condensation of exciton polaritons, *Nature*, **443** (7110), 409 [22]. Copyright (2006) by the Nature Publishing Group.

The second characteristic of a condensate is of course phase coherence, and first-order coherence can be measured by interfering photons emitted from the condensate as in Fig. 4.4. Interfering an image with its reflection in a Michelson interferometer allows one to measure $g_1(\vec{r}, -\vec{r})$ and thus to track the long-range correlations. The expectation for a thermalized system would be that in two dimensions, this would fall off as a power law in distance, following the predictions of BKT, with an exponent that increases linearly with temperature. In practice, most well-established condensates are too small to track the long-distance behavior easily, as substantial coherence extends across the cloud. Experiments that have studied the power law dependence indicate that an effective temperature, should it exist, is determined by the characteristics of the pump rather than thermal baths [29, 30]. Chapters 10–11 review the experiments and theory of correlation experiments and BKT transitions of polariton condensates.

Related to the above is the appearance of a linear Bogoliubov spectrum, seen in thermal emission of polaritons – though it is challenging to resolve because the coherent emission is so strong [32]. Theory predicts that in the presence of dissipation the spectrum will become diffusive at long wavelengths, though this has not been observed.

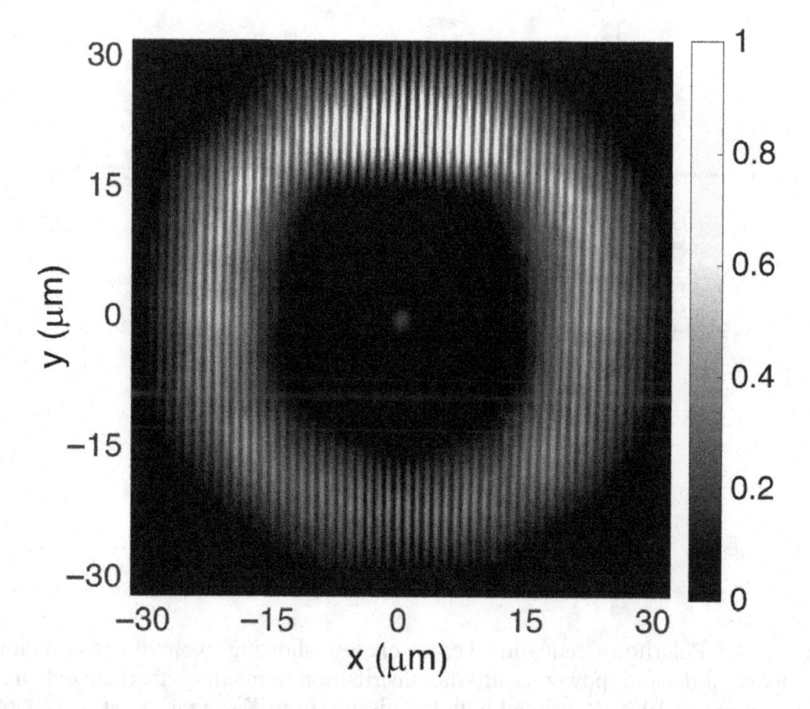

Figure 4.4 Interference between the emission from a polariton condensate and its reflection, exhibiting fringes that are a signature of phase coherence in the condensate. The ring-shaped potential for polaritons is created by a combination of cavity geometry, applied stress, and off-resonant laser that pumps excitons into the center of the ring. The top of the image has one more interference fringe than the bottom, from a π phase shift around the ring due to half-quantized clockwise superflow. Reprinted with permission from Liu, G., et al. (2015), A new type of half-quantum circulation in a macroscopic polariton spinor ring condensate, *Proc. Natl. Acad. Sciences*, **112**, 2676 [31]. Copyright (2015) National Academy of Sciences, USA.

Phase coherence is robust enough to support vortices (and in fact half-vortices since the condensate is a spinor) [33]. Such vortices (or half-vortex pairs) have been seen in experiments where the condensate is pumped incoherently [34] (note that the conditions of a typical incoherent-pumping experiment are such that only pinned immobile vortices are seen). Another class of experiments involves coherent driving, where a laser is tuned to near the bottom of the polariton spectrum, which allows one to generate a nonequilibrium condensate, perhaps with a nonzero momentum. After the laser is turned off, the condensate is seen to propagate ballistically, and can be seen to flow without vorticity around a localized potential disturbance. Changing conditions (i.e., density or velocity) superflow is seen to be disrupted by the shedding of vortices [35]. Chapter 21 reviews experimental observations of vortices in polariton condensates.

In systems where single polaritons have a long lifetime, one should distinguish ballistic single-particle propagation from collective motion. Experiments in III–V quantum wells have demonstrated that polariton wavepackets can propagate for hundreds of microns, in potential landscapes formed by a combination of nearly static excitons, shifts of the band gap by stress, and a wedge in the optical cavity thickness [31, 36]. (Chapter 22 reviews some of the methods of controlling the potential profiles for polaritons.) As well as allowing one to create multiply connected fluid states, as in Fig. 4.4, these studies have shown a sharp transition between ballistic single particle and collective dynamics that is not fully understood.

4.3 Microscopic Theory of Condensation

A simplified model that replaces the excitons by localized two-level systems is a good way to exhibit the physics; this keeps the internal dynamics of excitation creation and annihilation but neglects the center of mass exciton dynamics, with the justification that the exciton mass is too large for exciton motion to be relevant.

The model is the Dicke model of atomic physics [37, 38]:

$$H_{2\text{level}} = \sum_{q} \omega(\vec{q})\psi^{\dagger}_{\vec{q}}\psi_{\vec{q}} + \sum_{j=1}^{N} \frac{\epsilon_j}{2}\left(b^{\dagger}_j b_j - a^{\dagger}_j a_j\right) + \frac{g}{\sqrt{N}}\sum_{j\vec{q}}\left(b^{\dagger}_j a_j \psi_{\vec{q}} + \psi^{\dagger}_{\vec{q}} a^{\dagger}_j b_j\right). \tag{4.5}$$

$H_{2\text{level}}$ describes an ensemble of N two-level oscillators with an energy ϵ_j dipole coupled to cavity modes. b and a are fermionic annihilation operators for a fermion in an upper and lower state, respectively, with a local constraint $b^{\dagger}_j b_j + a^{\dagger}_j a_j = 1$ ensuring that only one state is occupied. These operators are only a device for modeling a two-level system for calculational convenience, and should not be confused with the electron and hole of the physical excitonic bound state. ψ is a bosonic photon annihilation operator. The operator that counts the number of excitations in the system, $N_{\text{ex}} = \sum_q \psi^{\dagger}_{\vec{q}}\psi_{\vec{q}} + \frac{1}{2}\sum_j (b^{\dagger}_j b_j - a^{\dagger}_j a_j) = N_{\psi} + N_X$, commutes with $H_{2\text{level}}$ so is conserved. The photon wavelength is assumed to be large compared with the system size, so phase factors in the last term are ignored. Note that $H_{2\text{level}}$ can also be written as the excitonic two-level systems treated as spin-$\frac{1}{2}$ variables, or equivalently mapped to a Bose-Hubbard model with an infinite hard-core repulsion.

If one has Frenkel excitons, each two-level system is bound to a single molecule (when one may need also to include coupling to local phonon modes [39]). For a system with disorder, trapping the excitons, one would work with a distribution of levels determined by solving for the excitons in a disordered potential [40]; hence $N \approx (L/l)^2$ with l the localization length and L the excitation radius. For a perfect crystal, or in a disordered system when the exciton density exceeds l^{-2}, the levels are really determined by interactions between the excitons, but in practice it is reasonable to assume an exclusion radius of order of the Bohr radius, hence $N \approx (L/a_B)^2$. If $N_X \ll N$, the excitonic occupation can be bosonized, and Eq. (4.5) reduces to Eq. (4.1).

Because the photon wavelength is so long, this model has infinite range interactions, so it is natural to propose a ground state of the coherent state form, viz.

$$|\lambda, u, v\rangle = e^{\lambda \psi_0^\dagger} \prod_j (v_j b_j^\dagger + u_j a_j^\dagger)|0\rangle. \tag{4.6}$$

with the (real) variational parameter λ and variational functions $v_j = v(\epsilon_j)$. (The vacuum state is here defined to be empty of both levels.) The constraint is satified by setting $u_j^2 + v_j^2 = 1$, and the variational functions are obtained by minimizing $H_{2\text{level}} - \mu N_{\text{ex}}$. For detailed results, see [41, 42]. One can extend the mean-field theory to finite temperatures by solving the self-consistent equations assuming a thermal occupancy of quasiparticles, as in Bardeen-Cooper-Schrieffer theory. At low overall density, $\rho < 1/2$, one gets a mean-field estimate of the transition temperature

$$\ln \frac{T_{\text{mf}}}{g} \approx \frac{-1}{g} \nu(\mu), \tag{4.7}$$

where $\nu(\mu)$ is the density of states at the chemical potential. This is a weak-coupling result; the transition temperature otherwise saturates at $T_{\text{mf}} \approx g$ (so long as the photon occupancy is not dominant).

An important feature of this model is that the photon number is unbounded. One consequence is to stabilize a coherent ground state at *any* temperature when $\rho > 1$. (See also Fig. 4.2 and the associated discussion.) The outcome is very different from a treatment of polaritons as a dilute Bose gas. As the population is pumped higher, the effect of saturation makes the added excitations increasingly photon like – and this effect is more and more pronounced with detuning of the exciton level below the photon, as often is the experimental situation.

Nonetheless, there should be a dilute Bose gas limit that is not captured by the mean-field theory above. This can be recovered by including the finite momentum fluctuations implicit in the existence of a Bogoliubov mode, as discussed in Chapter 22. The outcome [43, 6] is that for a low-density range given in Eq. (4.3), the "standard" BEC is restored, as we indicated in the introduction. Note that this density regime is far smaller than the naive condition one would get for zero-range interacting bosons. And unlike a BCS superconductor, where a coherent state is a consequence of weak interactions, for polaritons the coherent state arises together with strong interactions on account of their very long range.

When a system is close to a coherent state, one naturally expects to have a macroscopic description in terms of a Gross-Pitaevskii equation. Under some approximations this can be derived for the photon component of the order parameter in the following form:

$$\left[\omega_0 - \mu - \frac{\nabla^2}{2m} + \mathcal{O}(\nabla^4) - \sum_j \frac{g^2}{N} \frac{\tanh(\beta E_j)}{2E_j}\right] \psi_0 \simeq 0\,, \tag{4.8}$$

where the energy $E_j^2 = (\epsilon_j - \mu)^2 + g^2|\psi_0(\vec{r})|^2$ depends on the local value of the slowly varying $\psi_0(\vec{r})$. The nonlinear last term is very different from a featureless contact interaction as in the dilute Bose gas, and incorporates both energy shifts and saturation terms.

One last technical matter is the inclusion of coupling to the environment, because the system is open. One should allow for external photon decays and balance the population by external pumping (for example, in steady state). From a microscopic theory, this can be accomplished by coupling the Hamiltonian to external baths representing the pumping and decay and then integrating out the couplings using, e.g., the Keldysh formalism [6, 13, 44]. It is clear on physical grounds that any incoherent pump will induce decoherence of the condensate – a phase-coherent polariton will decay into an external photon and then be repumped by the creation of an exciton which enters the system with a random phase. This means that Bogoliubov modes will not propagate to infinity but instead are expected to become damped at the longest wavelengths. Nonetheless, it turns out that at least at the mean-field level, the condensate itself persists.

An equivalent representation of the macroscopic dynamics generalizes the GP equation to a damped, driven version [15, 45]. For a $\psi_0(r,t)$ which varies slowly in space and time, and in the local density approximation, one has after linearization

$$i\partial_t\psi_0 = \left(-\frac{\nabla^2}{2m} + V(r) + U|\psi_0|^2 + i\left[\gamma_{\text{eff}}(\mu_B) - \kappa - \Gamma|\psi_0|^2\right]\right)\psi_0, \tag{4.9}$$

where κ is the photon decay rate, γ is the effective pump rate, and Γ represents the simplest form of nonlinearity of the imaginary part, taking a form that will ensure stability. Chapter 11 reviews many of the theoretical issues of driven, damped condensates, with a related decription also applicable for atomic gases with significant outcoupling.

If the pumping is into an excitonic reservoir, with independent (and possibly slow) dynamics, it may be necessary to model the reservoir separately; see for example Ref. [15].

4.4 Further Issues

This is a rapidly evolving topic, and there are many new lines of attack, of which we pick out just a few.

New Materials and Systems

Polaritons can exist in a wide variety of systems, whereas the focus of this review has just been conventional 2D semiconductors. Evidence for nonlinear cooperative effects with plasmon polariton structures has been seen, for example, in organic semiconductors patterned with metallic superlattices [46]. Metamaterials formed by lattice structures offer particular opportunities for engineering interactions, band-structures [47], and spin-orbit coupling [48]. The newly emerged perovskite photo-voltaics $CH_3NH_3PbX_3$ clearly recycle photons very effectively and are also laser materials with polaritonic transport [49]. Transition metal dichalcogenides have strong exciton–photon coupling, and device structures have been made which have demonstrated polaritonic transport [50] though not yet lasing or condensation. Systems such as BN on graphene couple phonon polaritons with surface plasmons [51]. There has been considerable interest in what has been termed "photon BEC," arising from the coupling of photons in a cavity with excitations in a dye [52, 53, 54, 55] (see Chapter 19). Moving away from conventional solid-state systems, Rydberg atoms in cavities have been tuned into the strong coupling limit [56], and in these systems there is a long-range interaction between the excited atoms, allowing the competition between superfluid states and solids [57]. Indeed, the full suite of "quantum simulation" technologies enabled by precise control over cold atom Hamiltonians will lead to light-matter systems in regimes completely unlike the polaritons discussed in this chapter [58].

Dynamics

Because polariton systems are open, there are of course unavoidable dynamics due to particle flows. But the nonlinearity (and in particular the vorticity) offers some interesting features for driven condensates, including Josephson oscillations [59], dynamically generated vortex-antivortex lattices [60], "petal" states of counter rotating condensates [61], the stabilization of unusual half-vortex states [33, 31], predicted vortex instabilities due to flows [62], "Higgs"-like dynamics under a quench [63], and possibly also Floquet dynamics with resonant pumping [64]. A number of chapters in this volume are devoted to such dynamical issues, namely Chapters 7–8, 10–11, and 20–23.

Phase Transitions and Thermalization

Finally, we note that from a fundamental point of view, polariton systems offer some opportunities to follow theoretical concepts to unexpected places. In what sense can one have a phase transition to a superfluid in an open system with dissipation [65]? (See Chapter 11.) Can one have a phase transition in a system that is explicitly nonthermal, and indeed what changes when one adjusts the spectrum of the bath away from thermal? Can we follow the transition from grand-canonical to canonical thermodynamics [53]? How do two condensates phase lock to each other when they come in contact? (See Chapter 23.)

Polaritons indeed seem tailor-made to probe interesting limits of basic physics. From the small taste in this chapter alone, it is clear that they are a remarkably versatile platform for exploring condensation, at least. This is because they can inherit desirable properties from their excitonic and photonic components while avoiding undesirable competing effects, for instance those that have long complicated the realization of purely excitonic condensates. The virtue of these hybrid systems is their enlarged parameter space, where tuning one component has a benign effect on the other individually, while their collective behavior moves toward the unprecedented.

References

[1] Agranovich, V. M. 1957. On the influence of reabsorption on the decay of fluorescence in molecular crystals. *Optika i Spektroskopiya*, **3**, 84.

[2] Hopfield, J. J. 1958. Theory of the contribution of excitons to the complex dielectric constant of crystals. *Phys. Rev.*, **112**, 1556.

[3] Weisbuch, C., Nishioka, M., Ishikawa, A., and Arakawa, Y. 1992. Observation of the coupled exciton-photon mode splitting in a semiconductor quantum microcavity. *Phys. Rev. Lett.*, **69**, 3314.

[4] Jacob, Zubin. 2014. Nanophotonics: hyperbolic phonon-polaritons. *Nat Mater*, **13**(12), 1081–1083.

[5] High, Alexander A., Devlin, Robert C., Dibos, Alan, Polking, Mark, Wild, Dominik S., Perczel, Janos, de Leon, Nathalie P., Lukin, Mikhail D., and Park, Hongkun. 2015. Visible-frequency hyperbolic metasurface. *Nature*, **522**(7555), 192–196.

[6] Keeling, J., Marchetti, F. M., Szymanska, M. H., and Littlewood, P. B. 2007. Topical review: collective coherence in planar semiconductor microcavities. *Semiconductor Science Technology*, **22**, 1.

[7] Yamamoto, Y., Tassone, F., and Cao, H. 2000. *Semiconductor Cavity Quantum Electrodynamics*. Berlin: Springer-Verlag.

[8] Kavokin, A., and Malpuech, G. 2003. *Cavity Polaritons*, Vol. 32, Thin Films and Nanostructures. New York: Elsevier.

[9] Littlewood, P. B., Eastham, P. R., Keeling, J. M. J., Marchetti, F. M., Simons, B. D., and Szymanska, M. H. 2004. Models of coherent exciton condensation. *Journal of Physics Condensed Matter*, **16**, 3597.

[10] Deng, Hui, Haug, Hartmut, and Yamamoto, Yoshihisa. 2010. Exciton-polariton Bose-Einstein condensation. *Rev. Mod. Phys.*, **82**, 1489–1537.

[11] Keeling, J., Szymańska, M. H., and Littlewood, P. B. 2010. Keldysh Green's function approach to coherence in a non-equilibrium steady state: connecting Bose-Einstein condensation and lasing. Pages 293–329 of: *Nanosscience and Technology,*

vol. 0: Optical Generation and Control of Quantum Coherence in Semiconductor Nanostructures. Edited by Slavcheva, Gabriela, and Roussignol, Philippe. Berlin, Heidelburg: Springer. Page 293.

[12] Richard, Maxime, Kasprzak, Jacek, Baas, Augustin, Kundermann, Stefan, Lagoudakis, Konstantinos, Wouters, Michiel, Carusotto, Iacopo, Andre, Regis, Deveaud-Pledran, Benoit, and Dang, Le. 2010. Exciton-polariton Bose-Einstein condensation: advances and issues. *International Journal of Nanotechnology*, **7**(4–8), 668–685.

[13] Szymańska, M. H., Keeling, J., and Littlewood, P. B. 2013. Non-equilibrium Bose-Einstein condensation in a dissipative environment. Pages 447–459 of: *Quantum Gases: Finite Temperature and Non-Equilibrium Dynamics*. Edited by Proukakis, Nick, et al. World Scientific Publishing.

[14] Byrnes, T., Kim, N. Y., and Yamamoto, Y. 2014. Exciton-polariton condensates. *Nature Physics*, **10**, 803–813.

[15] Carusotto, Iacopo, and Ciuti, Cristiano. 2013. Quantum fluids of light. *Rev. Mod. Phys.*, **85**, 299–366.

[16] Kamide, Kenji, and Ogawa, Tetsuo. 2010. What determines the wave function of electron–hole pairs in polariton condensates? *Phys. Rev. Lett.*, **105**, 056401.

[17] Dang, Le Si, Heger, D., André, R., Bœuf, F., and Romestain, R. 1998. Stimulation of polariton photoluminescence in semiconductor microcavity. *Phys. Rev. Lett.*, **81**, 3920–3923.

[18] Senellart, P., and Bloch, J. 1999. Nonlinear emission of microcavity polaritons in the low density regime. *Phys. Rev. Lett.*, **82**, 1233–1236.

[19] Dasbach, G., Baars, T., Bayer, M., Larionov, A., and Forchel, A. 2000. Coherent and incoherent polaritonic gain in a planar semiconductor microcavity. *Phys. Rev. B*, **62**, 13076–13083.

[20] Deng, Hui, Weihs, Gregor, Santori, Charles, Bloch, Jacqueline, and Yamamoto, Yoshihisa. 2002. Condensation of semiconductor microcavity exciton polaritons. *Science*, **298**(5591), 199–202.

[21] Savvidis, P. G., Baumberg, J. J., Stevenson, R. M., Skolnick, M. S., Whittaker, D. M., and Roberts, J. S. 2000. Angle-resonant stimulated polariton amplifier. *Phys. Rev. Lett.*, **84**, 1547–1550.

[22] Kasprzak, J., Richard, M., Kundermann, S., Baas, A., Jeambrun, P., Keeling, J. M., Marchetti, F. M., Szymanska, M. H., André, R, Staehli, J. L., Savona, V., Littlewood, P. B., Deveaud, B., and Dang, Le Si. 2006. Bose-Einstein condensation of exciton polaritons. *Nature*, **443**(7110), 409–414.

[23] Balili, R., Hartwell, V., Snoke, D., Pfeiffer, L., and West, K. 2007. Bose-Einstein condensation of microcavity polaritons in a trap. *Science*, **316**(5827), 1007–1010.

[24] Kna-Cohen, S., and Forrest, S. R. 2010. Room-temperature polariton lasing in an organic single-crystal microcavity. *Nature Photonics*, **4**, 371–375.

[25] Nelsen, B., Balili, R., Snoke, D. W., Pfeiffer, L., and West, K. 2009. Lasing and polariton condensation: two distinct transitions in GaAs microcavities with stress traps. *Journal of Applied Physics*, **105**, 122414.

[26] Yamaguchi, M., Kamide, K., Nii, R., Ogawa, T., and Yamamoto, Y. 2013. Second thresholds in BEC-BCS-laser crossover of exciton–polariton systems. *Phys. Rev. Lett.*, **111**, 026404.

[27] Shelykh, I. A., Rubo, Yuri G., Malpuech, G., Solnyshkov, D. D., and Kavokin, A. 2006. Polarization and propagation of polariton condensates. *Phys. Rev. Lett.*, **97**, 066402.

[28] Laussy, Fabrice P., Shelykh, Ivan A., Malpuech, Guillaume, and Kavokin, Alexey. 2006. Effects of Bose-Einstein condensation of exciton polaritons in microcavities on the polarization of emitted light. *Phys. Rev. B*, **73**, 035315.

[29] Roumpo, Georgios, Lohs, Michael, Nitsche, Wolfgang H., Jonathan Keeling, Marzena Szymaska, Littlewood, Peter B., Löffler, Andreas, Höfling, Sven, Worschech, Lukas, Forchel, Alfred, and Yamamoto, Yoshihisa. 2012. Power-law decay of the spatial correlation function in exciton–polariton condensates. *PNAS*, **109**, 6467–6472.

[30] Nitsche, W. H., Kim, N. Y., Roumpos, G., Schneider, C., Kamp, M., Höfling, S., Forchel, A., and Yamamoto, Y. 2014. Algebraic order and the Berezinskii-Kosterlitz-Thouless transition in an exciton–polariton gas. *Phys. Rev. B*, **90**, 205430.

[31] Liu, Gangqiang, Snoke, David W., Daley, Andrew, Pfeiffer, Loren N., and West, Ken. 2015. A new type of half-quantum circulation in a macroscopic polariton spinor ring condensate. *Proceedings of the National Academy of Sciences*, **112**(9), 2676–2681.

[32] Utsunomiya, S., Tian, L., Roumpos, G., Lai1, C. W., Kumada, N., Fujisawa, T., Kuwata-Gonokami, M., Löffler, A., Höfling, S., Forchel, A., and Yamamoto, Y. 2008. Observation of Bogoliubov excitations in exciton–polariton condensates. *Nature Physics*, **4**, 700–705.

[33] Rubo, Yuri G. 2007. Half vortices in exciton polariton condensates. *Phys. Rev. Lett.*, **99**, 106401.

[34] Lagoudakis, K. G., Ostatnický, T., Kavokin, A. V., Rubo, Y. G., André, R., and Deveaud-Plédran, B. 2009. Observation of half-quantum vortices in an exciton–polariton condensate. *Science*, **326**(Nov.), 974.

[35] Amo, Alberto, Lefrère, Jérôme, Pigeon, Simon, Adrados, Claire, Ciuti, Cristiano, Carusotto, Iacopo, Houdr, Romuald, Giacobino, Elisabeth, and Bramati, Alberto. 2009. Superfluidity of polaritons in semiconductor microcavities. *Nature Physics*, **5**, 805–810.

[36] Steger, M., Liu, G., Nelsen, B., Gautham, C., Snoke, D. W., Balili, R., Pfeiffer, L., and West, K. 2013. Long-range ballistic motion and coherent flow of long-lifetime polaritons. *Phys. Rev. B*, **88**, 235314.

[37] Dicke, R. H. 1954. Coherence in spontaneous radiation processes. *Phys. Rev.*, **93**, 99–110.

[38] Garraway, Barry M. 2011. The Dicke model in quantum optics: Dicke model revisited. *Philosophical Transactions of the Royal Society of London A: Mathematical, Physical and Engineering Sciences*, **369**(1939), 1137–1155.

[39] Ćwik, J. A., Reja, S., Littlewood, P. B., and Keeling, J. 2014. Polariton condensation with saturable molecules dressed by vibrational modes. *EPL (Europhysics Letters)*, **105**(Feb.), 47009.

[40] Marchetti, F. M., Keeling, J., Szymańska, M. H., and Littlewood, P. B. 2006. Thermodynamics and excitations of condensed polaritons in disordered microcavities. *Phys. Rev. Lett.*, **96**, 066405.

[41] Eastham, P. R., and Littlewood, P. B. 2000. Bose condensation in a model microcavity. *Solid State Communications*, **116**, 357–361.

[42] Eastham, P. R., and Littlewood, P. B. 2001. Bose condensation of cavity polaritons beyond the linear regime: the thermal equilibrium of a model microcavity. *Phys. Rev. B*, **64**, 235101.

[43] Keeling, J., Eastham, P. R., Szymanska, M. H., and Littlewood, P. B. 2005. BCS-BEC crossover in a system of microcavity polaritons. *Phys. Rev. B*, **72**, 115320.

[44] Szymańska, M. H., Keeling, J., and Littlewood, P. B. 2006. Nonequilibrium quantum condensation in an incoherently pumped dissipative system. *Phys. Rev. Lett.*, **96**, 230602.

[45] Wouters, Michiel, and Carusotto, Iacopo. 2007. Excitations in a nonequilibrium Bose-Einstein condensate of exciton polaritons. *Phys. Rev. Lett.*, **99**, 140402.

[46] Rodriguez, S. R. K., Chen, Y. T., Steinbusch, T. P., Verschuuren, M. A., Koenderink, A. F., and Rivas, J. G. 2014. From weak to strong coupling of localized surface plasmons to guided modes in a luminescent slab. *Phys. Rev. B*, **90**, 235406.

[47] Jacqmin, T., Carusotto, I., Sagnes, I., Abbarchi, M., Solnyshkov, D. D., Malpuech, G., Galopin, E., Lemaître, A., Bloch, J., and Amo, A. 2014. Direct observation of Dirac cones and a flatband in a honeycomb lattice for polaritons. *Phys. Rev. Lett.*, **112**, 116402.

[48] Sala, V. G., Solnyshkov, D. D., Carusotto, I., Jacqmin, T., Lemaître, A., Terças, H., Nalitov, A., Abbarchi, M., Galopin, E., Sagnes, I., Bloch, J., Malpuech, G., and Amo, A. 2015. Spin-orbit coupling for photons and polaritons in microstructures. *Phys. Rev. X*, **5**, 011034.

[49] Xing, G. N., Mathews, S. S., Lim, N., Yantara, X., Liu, D., Sabba, M., Gratzel, S., Mhaisalkar, S. and, Sum, T. Z. 2014. Low-temperature solution-processed wavelength-tunable perovskites for lasing. *Nature Materials*, **13**, 476–480.

[50] Fei, Z., Scott, M., Gosztola, D. J., Foley, J. J., Yan, J., Mandrus, D. G., Wen, H., Zhou, P., Zhang, D. W., Sun, Y., Guest, J. R., Gray, S. K., Bao, W., Wiederrecht, G. P., and Xu, X. 2016. Nano-optical imaging of exciton polaritons inside WSe2 waveguides. *Phys. Rev. B* **94**, 081402(R).

[51] Dai, S., Ma, Q., Liu, M. K., Andersen, T., Fei, Z., Goldflam, M. D., Wagner, M., Watanabe, K., Taniguchi, T., Thiemens, M., Keilmann, F., Janssen, G. C. A. M., Zhu, S-E., Jarillo-Herrero, P., Fogler, M. M., and Basov, D. N. 2015. Graphene on hexagonal boron nitride as a tunable hyperbolic metamaterial. *Nat Nano*, **10**(8), 682–686.

[52] Klaers, J., Schmitt, J., Vewinger, F., and Weitz, M. 2010. Bose-Einstein condensation of photons in an optical microcavity. *Nature*, **468**(Nov), 545–548.

[53] Schmitt, Julian, Damm, Tobias, Dung, David, Vewinger, Frank, Klaers, Jan, and Weitz, Martin. 2014. Observation of grand-canonical number statistics in a photon Bose-Einstein condensate. *Phys. Rev. Lett.*, **112**, 030401.

[54] Kirton, P., and Keeling, J. 2013. Nonequilibrium model of photon condensation. *Phys. Rev. Lett.*, **111**, 100404.

[55] Kirton, P., and Keeling, J. 2015. Thermalization and breakdown of thermalization in photon condensates. *Phys. Rev. A*, **91**, 033826.

[56] Ningyuan, J., Georgakopoulos, A., Ryou, A., Schine, N., Sommer, A., and Simon, J. 2016. Observation and characterization of cavity Rydberg polaritons. *Phys. Rev. A* **93**, 041802(R).

[57] Edelman, A., and Littlewood, P. B. 2015. *Physica B – Condensed Matter*, **460**, 260–263.

[58] Sommer, A., Büchler H. P. and Simon, J. 2015. Quantum crystals and Laughlin droplets of cavity Rydberg polaritons. arXiv:1506.00341.

[59] Abbarchi, M., Amo, A., Sala, V. G., Solnyshkov, D. D., Flayac, H., Ferrier, L., Sagnes, I., Galopin, E., Lemaitre, A., Malpuech, G., and Bloch, J. 2013. Macroscopic quantum self-trapping and Josephson oscillations of exciton polaritons. *Nat Phys*, **9**(5), 275–279.

[60] Tosi, G., Christmann, G., Berloff, N. G., Tsotsis, P., Gao, T., Hatzopoulos, Z., Savvidis, P. G., and Baumberg, J. J. 2012. Geometrically locked vortex lattices in semiconductor quantum fluids. *Nat Commun*, **3**(12), 1243.

[61] Dreismann, Alexander, Cristofolini, Peter, Balili, Ryan, Christmann, Gabriel, Pinsker, Florian, Berloff, Natasha G., Hatzopoulos, Zacharias, Savvidis, Pavlos G., and Baumberg, Jeremy J. 2014. Coupled counterrotating polariton condensates in optically defined annular potentials. *Proceedings of the National Academy of Sciences*, **111**(24), 8770–8775.

[62] Berloff, N. G., and Keeling, J. 2013. Universality in modelling non-equilibrium pattern formation in polariton condensates. *ArXiv e-prints*, arXiv:1303.6195.

[63] Eastham, P. R., and Phillips, R. T. 2009. Quantum condensation from a tailored exciton population in a microcavity. *Phys. Rev. B*, **79**, 165303.

[64] Brierley, R. T., Littlewood, P. B., and Eastham, P. R. 2011. Amplitude-mode dynamics of polariton condensates. *Phys. Rev. Lett.*, **107**, 040401.

[65] Keeling, J. 2011. Superfluid density of an open dissipative condensate. *Phys. Rev. Lett.*, **107**, 080402.

Part II

General Topics

Editorial Notes

In this part, we discuss general themes related to universal features of Bose-Einstein condensates, which appear in some form in all relevant physical systems. Although some of the chapters have a particular physical system in mind, the topics covered here are broad, and adaptable (possibly with some reformulation or taking account of system-specific features) to other physical systems exhibiting Bose-Einstein condensation.

Some of the most important questions revolve around the process by which systems come to equilibrium and the physics around the critical region itself. As discussed in Chapter 1, one of the main early controversial topics was how a condensate could form in a system which initially had no condensate and the role of symmetry-breaking and interaction-induced modifications to this process. Although many of the basic questions of this formation process have been answered in the past two decades, numerous questions remain. The important question of spontaneous $U(1)$ symmetry breaking is reviewed in Chapter 5, while Chapter 6 performs a systematic characterization of the critical region for an interacting system. Our current understanding of the condensate formation process is then reviewed at length in Chapter 7, with extensive references to key experiments.

In keeping with the theme of this book, several chapters review attempts to define universal behaviors of Bose-Einstein condensates. It is well known that the model of the ideal Bose gas taught in many elementary textbooks is not only unphysical, but misses entirely some of the essential physics of real condensates, such as the appearance of superfluidity. Chapter 5 reviews the subtle point, raised by Noziérés in the green book,[1] that interactions play a crucial role in stabilizing a condensate in a single macroscopic state, with Chapter 6 reviewing complementary aspects of the general theory of weakly interacting condensates. Chapters 6–7 also address the

[1] Noziérés, P., in *Bose-Einstein Condensation*, Griffin, A., D. Snoke, D. W., and Stringari, S. eds. (Cambridge University Press, 1995).

so-called Kibble-Zurek mechanism characterizing the emergence of defects during a finite-duration quench to below the BEC phase transition. This mechanism, associated with universal behavior across a phase transition, was originally proposed in cosmology[2] and subsequently adapted to the condensed matter context[3] (as also discussed for ^{3}He in Chapter 30), and, more recently, to ultracold atomic gases.

Another long-standing question is how to justify the concept of equilibrium at all, since it seems to imply irreversibility, and an isolated quantum system undergoing unitary evolution should not have time-irreversibility. This topic goes all the way back to Boltzmann in the context of classical particle systems, and has been addressed generically in many theoretical works, including recent work on the question of how quantum dephasing enters in.[4] Chapter 8 specifically looks at how an isolated quantum system can reach equilibrium, following a "quench" in which a system in equilibrium is suddenly given an external change which drives it away from equilibrium – focusing here on the one-dimensional limit as a specific example. Chapter 9 looks at the concept of universality and scale invariance in quantum gases and suggests a new class of quantum gas systems with "intrinsic scale invariance."

Another type of nonequilibrium is a steady-state system with continuous generation and decay, known as a "driven-dissipative" system. This is naturally studied in two-dimensional polariton systems with optical pumping (although proposals have also been made to create an "atom laser" with continuous feeding and emission of coherent atoms). It has long been known that two-dimensional interacting systems cannot have infinite long-range order, with the term "Bose-Einstein condensate" usually restricted to three-dimensional systems, and the comparable transition in two-dimensional systems being identified as the "Berezinskii-Kosterlitz-Thouless" (BKT) transition. The BKT transition involves an abrupt change in the nature of vortices, as reviewed in Chapter 10. Two-dimensional bosonic systems have many properties in common with three-dimensional systems, however, and in finite systems the difference between the two can become nearly moot. Chapter 11 addresses some of these issues in the context of driven-dissipative systems. Finally, Chapter 12 looks at the universal theory of BCS to BEC crossover in fermionic systems.

[2] Kibble, T. W. B. (1976) Topology of cosmic domains and strings, *J. Phys. A: Math. Gen.* **9**, 1387.
[3] Zurek, W. H. (1985) Cosmological experiments in superfluid helium?, *Nature* **505**, 317.
[4] Snoke, D. W., Liu, G., and Girvin, S. (2012) The basis of the second law of thermodynamics in quantum field theory, *Annals of Physics* **327**, 1825.

5

The Question of Spontaneous Symmetry Breaking in Condensates

DAVID W. SNOKE

Department of Physics and Astronomy, University of Pittsburgh, Pennsylvania, USA

ANDREW J. DALEY

Department of Physics and SUPA, University of Strathclyde, Glasgow, Scotland, UK

The question of whether Bose-Einstein condensation involves spontaneous symmetry breaking is surprisingly controversial. We review the theory of spontaneous symmetry breaking in ferromagnets, compare it with the theory of symmetry breaking in condensates, and discuss the different viewpoints on the correspondence to experiments. These viewpoints include alternative perspectives in which we can treat condensates with fixed particle numbers, and where coherence arises from measurements. This question relates to whether condensates of quasiparticles such as polaritons can be viewed as "real" condensates.

5.1 Introduction

Spontaneous symmetry breaking is a deep subject in physics with long historical roots. At the most basic level, it arises in the field of cosmology. Physicists have long had an aesthetic principle that leads us to expect symmetry in the all of the basic equations of physical law. Yet the universe is manifestly full of asymmetries. How does a symmetric system acquire asymmetry merely by evolving in time? Starting in the 1950s, cosmologists began to borrow the ideas of spontaneous symmetry breaking from condensed matter physics, which were originally developed to explain spontaneous magnetization in ferromagnetic systems.

Spontaneous coherence in all its forms (e.g., Bose-Einstein condensation [BEC], superconductivity, and lasing) can be viewed as another type of symmetry breaking. The Hamiltonian of the system is symmetric, yet under some conditions, the energy of the system can be reduced by putting the system into a state with asymmetry, namely, a state with a common phase for a macroscopic number of particles. The symmetry of the system implies that it does not matter what the exact choice of that phase is, as long as it is the same for all the particles.

It is not obvious, however, whether the symmetry breaking which occurs in spontaneous coherence of the type seen in lasers or in Bose-Einstein condensation is the same as that seen in ferromagnetic systems. There are similarities in the systems which encourage the same view of all types of symmetry breaking, but there are also differences. In fact, there is a substantial school of thought that symmetry breaking in Bose-Einstein condensates with ultracold atoms is a "convenient fiction" (a term applied to optical coherence by Mølmer [1]). That is, in contrast to ferromagnets, we do not have direct experimental access to observe symmetry breaking itself, and the experimental consequences of this theory can be equally reproduced in theories using fixed atom number [2, 3, 4, 5, 6, 7, 8, 9, 10, 11, 12], without spontaneous symmetry breaking.

In what follows, we will review the theory of spontaneous symmetry breaking as applied to ferromagnets and then discuss the different viewpoints on spontaneous symmetry breaking in condensates. We will touch on theories involving fixed particle numbers, where coherence arises in the measurement process, and also discuss the relationship to the question of whether polariton systems can be "real" condensates.

Related issues of the BEC phase transition and formation dynamics are further discussed in Chapters 6 and 7, the nature of the phase transition in polaritons is addressed in Chapter 10, while a discussion of lasing versus BEC appears in Chapter 20.

5.2 Review of Elementary Spontaneous Symmetry-Breaking Theory

The canonical example in condensed matter physics for spontaneous symmetry breaking is the ferromagnetic spin system, represented in simple form by the Ising Hamiltonian for a lattice of localized electrons,

$$H = \alpha B \sum_i \sigma_i - J \sum_{\langle i,j \rangle} \sigma_i \sigma_j, \tag{5.1}$$

where $\sigma_i = a^\dagger_{i\uparrow} a_{i\uparrow} - a^\dagger_{i\downarrow} a_{i\downarrow}$ is the spin operator for site i and the sum $\langle i,j \rangle$ is for nearest neighbors. The first term gives the effect of an external magnetic field B, and the second term the effect of spin interactions which favor alignment. The order parameter for the system is defined as

$$m = \frac{1}{N} \sum_i \sigma_i, \tag{5.2}$$

which is the average magnetization. More generally, in a system in which the spin can point in any direction in three dimensions, the order parameter is a vector

$$\vec{m} = \frac{1}{N}\sum_i \vec{\sigma}_i, \tag{5.3}$$

where $\vec{\sigma}_i = (\sigma_{ix}, \sigma_{iy}, \sigma_{iz})$, for the standard Pauli spin matrices.

The mean-field solution for the free energy of (5.1) as a function of m in the absence of external magnetic field can be exactly calculated [13], and is

$$\begin{aligned} F &= -Nk_BT\left(\ln 2 - \frac{T_c}{2T}m^2 + \ln\cosh\left(\frac{T_c}{T}m\right)\right) \\ &\simeq F_0 + Nk_BT\left(\frac{T_c}{2T}\left(\frac{T-T_c}{T}\right)m^2 + \frac{1}{12}\left(\frac{T_c}{T}\right)^4 m^4\right), \end{aligned} \tag{5.4}$$

where T_c is the critical temperature for the ferromagnetic phase transition. Fig. 5.1a shows this free energy for two temperatures, above and below T_c. As seen in this

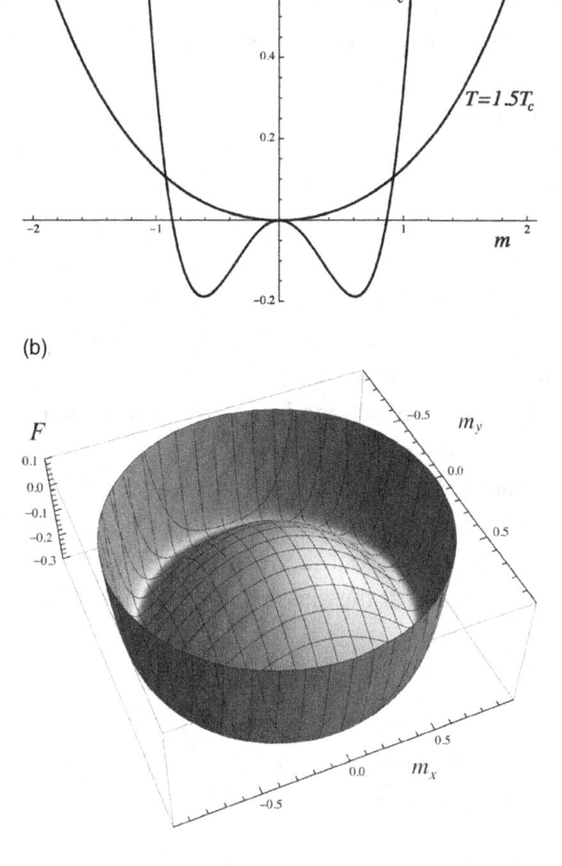

Figure 5.1 (a) Helmholtz free energy (5.4) for the Ising model, for two temperatures, with $Nk_BT = 1$. (b) Free energy profile for the case $T = 0.5T_c$ with two degrees of freedom.

figure, at $T = T_c$ the shape of the curve switches from a single minimum at $m = 0$ to two minima at finite m. The value of m in equilibrium for a homogeneous system is found by solving

$$\frac{\partial F}{\partial m} = Nk_B T \left(\frac{T_c}{T} \left(\frac{T - T_c}{T} \right) m + \frac{1}{3} \left(\frac{T_c}{T} \right)^4 m^3 \right) = 0. \tag{5.5}$$

The solution at $m = 0$ is unstable when $T < T_c$.

The notion of spontaneous symmetry breaking can be seen by thinking of how the system behaves as the temperature passes through T_c from above. The free energy is perfectly symmetric with respect to m, since there is no preferred direction for the spins in the absence of external magnetic field. Below T_c, the free energy curve remains symmetric, but the system can move to lower energy by breaking this symmetry, picking an energy minimum with $m \neq 0$.

How does the system choose a particular value of m and not another? In condensed matter physics, it is quite easy to suppose that there is some stray magnetic field B from outside the system which gives the system a kick in one direction or another. The system then amplifies this small asymmetry until it reaches a macroscopic average value of m.

The equivalent curve for a system which allows spin in two dimensions is illustrated in Fig. 5.1b, sometimes called the "Mexican hat" or "wine bottle" potential. In this case, instead of just two choices for m with minimum free energy, there is a continuous range of values of $\vec{m}$ which satisfy the minimum-energy condition. A stable value is found anywhere on a ring with fixed amplitude, that is, $\vec{m} = m(\hat{x}\cos\theta + \hat{y}\sin\theta)$, where θ is arbitrary. If the system settles into a specific value of $\vec{m}$, this is known as a continuously broken symmetry. Zero-energy variation of θ in this case is known as a Goldstone mode, whereas oscillation of the magnitude m in the radial direction is known as a Higgs mode.

This type of spontaneous symmetry breaking is a model for numerous systems. For example, it can be applied to the onset of lasing. In this case, the control parameter is not the temperature, but the pump power, or optical gain. One writes the Maxwell's wave equation for the classical electric field E

$$-\omega^2 E = \frac{\partial^2 E}{\partial t^2} + \frac{1}{\epsilon_0}\frac{\partial^2 P}{\partial t^2}, \tag{5.6}$$

where P is the average polarization of the medium. For an ensemble of two-level quantum oscillators, one can write the polarization as

$$P(t) = \mathrm{Re}\, \tilde{d}\frac{N}{V}U_1, \tag{5.7}$$

where $\tilde{d}$ is an intrinsic dipole moment for the electric field coupling to the two-level oscillators, and U_1 is a component of the standard average Bloch vector $\vec{U} = (U_1, U_2, U_3)$, where

$$\begin{aligned} U_1 &= \langle a_e^\dagger a_g + a_g^\dagger a_e \rangle \\ U_2 &= i\langle a_e^\dagger a_g - a_g^\dagger a_e \rangle \\ U_3 &= \langle a_e^\dagger a_e - a_g^\dagger a_g \rangle, \end{aligned} \tag{5.8}$$

for the excited (e) and ground (g) states of the two-level oscillator. Assuming the existence of a coherent electric field $E(t) = E_0 e^{-i\omega t}$ and incoherent gain G, the Bloch equations for the evolution of this vector are

$$\begin{aligned} \frac{\partial U_1}{\partial t} &= -\frac{U_1}{T_2} + \omega_0 U_2 - \omega_R U_3 \sin \omega t \\ \frac{\partial U_2}{\partial t} &= -\frac{U_2}{T_2} - \omega_0 U_1 - \omega_R U_3 \cos \omega t \\ \frac{\partial U_3}{\partial t} &= -\frac{U_3 + 1}{T_1} + \omega_R U_1 \sin \omega t + \omega_R U_2 \cos \omega t + G(1 - U_3), \end{aligned} \tag{5.9}$$

where T_1 and T_2 are the relaxation and dephasing time constants, respectively; $\hbar\omega$ is the energy gap between the ground and excited states; and $\omega_R = \tilde{d}E/\hbar$ is the Rabi frequency, proportional to the electric field amplitude.

Solving the Bloch equations in steady state for the amplitude of U_1 and using this in the polarization (5.7), which in turn is used in the Maxwell wave equation (5.6), gives [14]

$$\frac{\partial E_0}{\partial t} = \frac{\omega}{2\epsilon_0} \left(A E_0 - B E_0^3 \right), \tag{5.10}$$

where

$$\begin{aligned} A &= \frac{\tilde{d}^2}{\hbar} \frac{N}{V} \left(\frac{G\tau - 1}{G\tau + 1} \right) \\ B &= A \tilde{d}^2 \frac{\tau^2}{G\tau + 1}, \end{aligned} \tag{5.11}$$

in which we have set $T_1 = T_2 = \tau$.

The steady-state solution of Eq. (5.10) has the same form as (5.5). When A is positive, that is, when the gain exceeds a critical threshold, a small coherent part of the electric field will be amplified, growing in magnitude until it saturates at a nonzero value. If there is no coherent electric field, there will be nothing to amplify, but as in the case of the ferromagnet, we assume that some stray external field may impinge on the system. Since the $E_0 = 0$ point is unstable, like the $m = 0$ point of the ferromagnet, any tiny fluctuation will cause it to evolve toward a stable point.

If the electric field polarization direction is allowed to vary, the laser field will also have a continuously broken symmetry in two dimensions.

This model of spontaneous symmetry breaking has also been applied to the early universe [15, 16, 17]. In this case, it is harder to imagine what might count as an "external" field that gives the tiny kick needed. Typically, in the condensed matter context, the nature of the tiny fluctuation is of little concern, since the instability of the symmetric point is assumed to always make it impossible for the system to remain there; at the unstable point, the system will amplify even the tiniest fluctuation. With Bose condensates, however, the question has arisen whether there must be an external kick to bring about broken symmetry, and if so, where it comes from.

5.3 Coherence in Condensates as Spontaneous Symmetry Breaking

The above analysis can be mapped entirely to the case of Bose-Einstein condensation. The Hamiltonian in this case is

$$H = \sum_{\vec{k}} E_k a_{\vec{k}}^\dagger a_{\vec{k}} + \frac{1}{2V} \sum_{\vec{p},\vec{q},\vec{k}} U(k) a_{\vec{p}}^\dagger a_{\vec{q}}^\dagger a_{\vec{q}+\vec{k}} a_{\vec{p}-\vec{k}}. \tag{5.12}$$

The Einstein [18] argument ignores the interaction energy term and computes the total kinetic energy of a population of bosons. In this limit, one has simply

$$N = \sum_{\vec{k}} \langle a_{\vec{k}}^\dagger a_{\vec{k}} \rangle = \sum_{\vec{k}} \langle N_{\vec{k}} \rangle \tag{5.13}$$

where $N_{\vec{k}}$ is the occupation number of state $\vec{k}$. Converting the sum to an integral for a three-dimensional system, we have

$$N_{\text{ex}} = \frac{V}{2\pi^2} \frac{m^{3/2}}{\hbar^3} \int_0^\infty \bar{N}(E) \sqrt{E} dE \tag{5.14}$$

where $\bar{N} = 1/(e^{(E-\mu)/k_B T} - 1)$ is the Bose-Einstein average occupation number. This gives the well-known result that below a critical temperature T_c, the integral (5.14) cannot account for all the particles; there must be an additional population not accounted for by this sum over excited states, with zero kinetic energy, which we call the condensate.

It has been pointed out by Nozieres [19] and others [20] that the Einstein argument does not address the stability of the condensate. In an infinite system, there is an infinite number of k-states near $k = 0$ with negligible kinetic energy. In a noninteracting system, we could spread the condensate over any number of these states with negligible energy penalty.

To see how the interactions affect the stability of the condensate, imagine that we have a system in which the kinetic interactions have already caused a macroscopic number of particles to accumulate in states with negligible kinetic energy near $\vec{k} = 0$, but they are not in the same quantum state. We now imagine varying the fraction of this population of particles which is in the ground state, that is, the true condensate, and calculate the free energy of the system as a function of this fraction, using the Hamiltonian (5.12).

Computing this is nontrivial for a general interaction potential $U(k)$, but we can see the general behavior if we assume $U(k) = U = \text{constant}$. In this case, the expectation value of the Hamiltonian (5.12) is

$$\langle H \rangle = \frac{U}{2V}N_0^2 - \frac{U}{V}N_0 \sum_{\vec{p}} N_{\vec{p}} + \frac{U}{2V}\sum_{\vec{p},\vec{q}} \left(\langle a^\dagger_{\vec{p}} a^\dagger_{\vec{q}} a_{\vec{q}} a_{\vec{p}} \rangle + \langle a^\dagger_{\vec{p}} a^\dagger_{\vec{q}} a_{\vec{p}} a_{\vec{q}} \rangle \right), \tag{5.15}$$

where N_0 is the number in the ground state. Neglecting terms in the last sum arising from commutation when $\vec{p} = \vec{q}$, since these will be small compared with the whole sum, this becomes

$$\begin{aligned} \langle H \rangle &= \frac{U}{2V}N_0^2 + \frac{U}{V}N_0 N_{\text{ex}} + \frac{U}{V}N_{\text{ex}}^2 \\ &= \frac{U}{V}\left(N^2 - NN_0 + \frac{1}{2}N_0^2 \right), \end{aligned} \tag{5.16}$$

where $N_{\text{ex}} = \sum N_{\vec{k}} = N - N_0$, and N is the total number of particles.

We can see already from (5.16) that $\langle H \rangle$ will be lower when $N_0 > 0$. To account for the entropy of particles leaving the condensate, we write the free energy

$$F = \langle H \rangle - TS, \tag{5.17}$$

where

$$S = -k_B \sum_{\vec{k}} \left(N_{\vec{k}} \ln N_{\vec{k}} - (1 + N_{\vec{k}}) \ln(1 + N_{\vec{k}}) \right) \tag{5.18}$$

is the entropy of a boson gas [21]. We suppose that particles leaving the condensate move to a region in k-space near $k = 0$ with $N_{\vec{k}} \gg 1$. Then we can approximate

$$S \simeq k_B \sum_{\vec{k}} \frac{\partial}{\partial N_{\vec{k}}} N_{\vec{k}} \ln N_{\vec{k}} \simeq k_B \sum_{\vec{k}} \ln N_{\vec{k}}. \tag{5.19}$$

Assuming that the value of $N_{\vec{k}}$ does not deviate too strongly from its average value $\bar{N}_{\vec{k}} = N_{\text{ex}}/N_s$, where N_s is the total number of states in the selected region of k-space, we obtain

$$TS \simeq k_B T N_s \ln(N_{\text{ex}}/N_s) = k_B T N_s [\ln(N - N_0) - \ln N_s], \tag{5.20}$$

which allows us to write

$$\frac{F}{V} = \frac{F_0}{V} - U\frac{N}{V}|\psi_0|^2 + \frac{1}{2}U|\psi_0|^4 - k_BT\frac{N_s}{V}[\ln(N/V - |\psi_0|^2) - \ln N_s/V], \tag{5.21}$$

where we have defined the wave function ψ_0 of the condensate as the order parameter, with $|\psi_0|^2 = N_0/V$. We can estimate N_s, the number of states in the region around $k = 0$ with $N_{\vec{k}} \gg 1$, using the relation

$$N_s = \int_0^{E_{\rm cut}} D(E)dE, \tag{5.22}$$

where $E_{\rm cut}$ is the energy at which $N_k \gg 1$, which we can take as $E_{\rm cut} \sim k_BT$. In three dimensions, this gives us

$$\begin{aligned}\frac{N_s}{V} &= \frac{\sqrt{2}m^{3/2}}{2\pi^2\hbar^3}\int_0^{k_BT} dE\sqrt{E} \\ &= \frac{\sqrt{2}(mk_BT)^{3/2}}{\pi^2\hbar^3} \sim \frac{1}{\lambda_{dB}^3},\end{aligned} \tag{5.23}$$

where λ_{dB} is the deBroglie wavelength determined by setting $(\hbar k)^2/2m = k_BT$, with $k = 2\pi/\lambda_{dB}$.

The free energy (5.21) is plotted in Fig. 5.2, which has the same generic form as Fig. 5.1a. The symmetric center point at $\psi_0 = 0$ is unstable, because increasing the number of particles in the condensate reduces the interaction energy. This occurs because exchange in a bosonic system favors having particles in the same state. This is true for composite bosons as well [20]. Eventually the entropy cost of

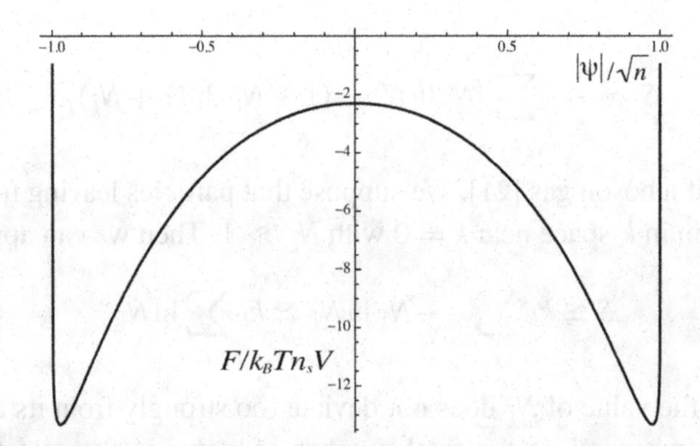

Figure 5.2 Free energy for an interacting Bose-Einstein condensate using the approximation (5.21), with $Un/k_BT = 1$ and $n/n_s = 10$, where $n = N/V$ is the particle density and $n_s = N_s/V$.

adding particles to the zero-entropy condensate state will prevent all the particles from entering it. The free energy for a condensate also has the two-dimensional form of Fig. 5.1b, but here the two components are the real and imaginary parts of $\psi_0 = \sqrt{n_0}e^{i\theta}$, where θ is the phase angle. Goldstone and Higgs modes exist in this case as well.

The depletion of the condensate depends on the ratio of the interaction strength Un to k_BT and the average occupation number of the non-condensed region. The stable point will move closer to $N_0 = N$ for stronger interactions and higher degeneracy of the excited particles. Approximating $\ln(N - N_0)$ at the stable point $N_0 = N_{eq}$ as

$$\ln(N - N_0) \simeq \ln(N - N_{eq}) - \frac{N_0 - N_{eq}}{N - N_{eq}}, \tag{5.24}$$

and taking the first derivative of F with respect to ψ_0^*, we obtain

$$-\left(\frac{U}{V}N + \frac{k_BTN_s}{N - N_{eq}}\right)\psi_0 + \frac{U}{V}|\psi_0|^2\psi_0 = 0. \tag{5.25}$$

This has the same form as (5.10), and is the same as the Landau-Ginzburg equation for a homogeneous boson gas.

When we look at nonequilibrium behavior, the symmetry breaking in a Bose-Einstein condensate also has similarities with the laser system. Starting with the same Hamilitonian (5.12) and deriving the quantum Boltzmann equation using second-order perturbation theory, the equation for the evolution of a coherent state in the ground state of the bosons is found to be [22, 23]

$$\begin{aligned}\frac{d}{dt}\langle a_0\rangle = \langle a_0\rangle\frac{\pi}{\hbar}\left(\frac{2U}{V}\right)^2\sum_{\vec{p},\vec{q}}\Big[&\langle\hat{N}_{\vec{q}}\rangle\langle\hat{N}_{\vec{p}-\vec{q}}\rangle(1 + \langle\hat{N}_{\vec{p}}\rangle) \\ &-\langle\hat{N}_{\vec{p}}\rangle(1 + \langle\hat{N}_{\vec{q}}\rangle)(1 + \langle\hat{N}_{\vec{p}-\vec{q}}\rangle)\Big]\,\delta(E_{\vec{p}} - E_{\vec{q}} - E_{\vec{p}-\vec{q}}).\end{aligned} \tag{5.26}$$

The first term in the square brackets gives the total in-scattering rate, and the second term is the total out-scattering rate.

As with the electric field in the case of a laser controlled by Eq. (5.5), we imagine a tiny coherent part has already been created somehow, and then see that the system can amplify this coherent part until it becomes macroscopic. Eq. (5.26) implies that this amplification will occur whenever there is net influx into the ground state, which occurs when the system approaches the BEC equilibrium kinetic-energy distribution [24, 25]. This implies exponential growth of the amplitude of a phase-coherent part; this growth will end when the system reaches equilibrium, and the influx to the ground state and the outflow balance.

Because the equations for growth of the condensate have the same form as those commonly used for spontaneous symmetry breaking in other systems, it is natural to assume that the same thing occurs in the case of Bose-Einstein condensation. However, many in the field of condensates have paid attention to the crucial role of the fluctuation which seeds the condensate. In the case of a ferromagnet or laser, it is easy to imagine that there is a stray magnetic field or electric field from outside the system. In the case of a condensate, however, the amplitude a_0 (or ψ_0 in the spatial domain) corresponds to the creation or destruction of particles, and in a strictly number-conserved system there is no external field that couples into the system to do this. The system is analogous to the early-universe scenario of spontaneous symmetry breaking, in which there appears to be no "outside" to give the small kick to break the symmetry.

On the other hand, we do not strictly know that there is no coupling term to the order parameter in a matter wave such as a cold atom condensate. If the proton decayed every 10^{30} years, we would have number conservation for all intents and purposes, but there would still be a tiny term that could give a fluctuation which is amplified.

5.4 Coherence in Condensates as a Measurement Phenomenon

As a solution to this problem, the predominant approach in the cold atom community is to view spontaneous symmetry breaking as one of many possible descriptions for the system. The question is most commonly posed in the case of interference experiments, where two Bose-Einstein condensates are released and the resulting interference patterns are measured [26]. An interference pattern between two condensates seems to imply a definite phase relation between the two, which could be most straightforwardly described with each condensate in a coherent state. The problem arises, however, that the standard definition of a coherent state [27] has an indefinite number of particles. At $T = 0$, however, a condensate has a definite number of particles, namely all the particles in the system. Does this then imply that at $T = 0$, in a truly isolated system, there would be no interference pattern? Of course, all real experiments are performed at finite temperature, so that there is fluctuation in the number of condensate particles. But the standard atom–atom interactions which give number fluctuations of number of atoms in the condensate still conserve total particle number and therefore do not give the type of superpositions needed to construct coherent states. For example, a collisional interaction of the form $Ua^{\dagger}_{\vec{k}_3}a^{\dagger}_{\vec{k}_2}a_{\vec{k}_1}a_0$ causes a pure Fock state of the form $|N_0 \ldots N_1 \ldots\rangle$ to evolve into to a superposition of Fock states of the form

$$|\Psi(t)\rangle = \alpha|N_0 \ldots N_1 \ldots\rangle + \beta|N_0 - 1 \ldots N_1 - 1 \ldots N_2 + 1 \ldots N_3 + 1 \ldots\rangle.$$

This has indefinite condensate number N_0, but the expectation value $\langle a_0 \rangle$ for this state is still zero. Do we need a tiny non-number-conserving term such as proton decay to give the needed kick into a coherent state?

In the experiments performed, it was unknown whether the condensates had a well-defined phase before the measurement was made or only afterward. In the spontaneous symmetry-breaking scenario, one can suppose that each condensate spontaneously could acquire a well-defined amplitude prior to interacting. But the experiments which have been performed allow for the possibility that the amplitude was well defined only after the measurement was made.

There are two related setups in which this has been discussed at length: one is where two condensates that are are later measured are coupled with each other, e.g., in the case of Bose-Einstein condensates in a double well with fixed total particle number [9, 10, 28, 29, 30], and the other involves the interference of two independent condensates, each initially with a fixed total particle number [8, 2, 7, 31]. In each case, it can be shown that number-conserving approaches give rise to the same experimental predictions for interference patters as the assumption of spontaneous symmetry breaking. This can be extended beyond interference experiments to a range of processes described by a Gross-Pitaevskii equation or Bogoliubov theory in a number-conserving approach [3, 4, 5].

In the case of the double well, the well-defined phase description could be applied with a justification that the exchange of particles between the coupled condensates leads to an indeterminate number between the individual wells. In both cases, however, it can be equally assumed that the phase of the sample is indeterminate until the condensate is actually measured. If two condensates with fixed particle number are released, and the resulting interference pattern is detected by absorption imaging, then the first particle that is measured could have come from either of the two independent condensates. This measurement process sets up a superposition state between the two condensates, and it is shown in Refs. [8, 2, 7, 31] that a proper analysis of the resulting continuous measurement process gives rise to the same interference pattern that would be expected from symmetry-broken condensates with well-defined (but unknown) initial phases. This same argument, that number states can give rise to interference identical to that in a fixed-phase representation (e.g., a coherent state), was applied to optical coherence by Mølmer [1, 32]. He summarized these ideas in a short poem for the abstract of Ref. [1]: "Coherent states may be of use, so they say, but they wouldn't be missed if they didn't exist." The standard approach in optics, however, is to view coherent states as physically real. This view is supported by the fact that in systems without number conservation, pure number states (a.k.a. Fock states) are unstable to becoming coherent states [23], by the same type of calculation which led to Eq. (5.26) above. Contra Mølmer, the standard approach treats *number* states as mostly dispensable, as for example in this quote from a standard laser textbook [33]: "We have hardly mentioned photons yet in this book ... The problem with the simple photon description ... is that it leaves out and even hides the important wave aspects of the laser interaction process." Quantum mechanics, of course, allows either coherent states or Fock states as bases.

A simple example for comparison between the symmetry-broken description with coherent states and a description that conserves the number of particles is the case of collapse and revival of matter-wave interference in an optical lattice (see also Chapter 13 by Bloch). For atoms confined in three dimensions (3D), this was first realized by Greiner et al. in 2002 [34], and then studied in more detail by Will et al. [35]. The basic context is that a very weakly interacting Bose-Einstein condensate is loaded into a 3D optical lattice. We can then see how this depends on atom number and system size, taking the lattice to have M sites, and N homogeneously distributed atoms. These sites are initially coupled by tunneling, but this is switched off by suddenly making the lattice deep. Onsite energy shifts that are dependent on the particle number then dephase correlations between different sites, leading to a collapse in interference peaks when the atoms are released from the lattice.

In the spontaneous symmetry breaking case, the picture is very clear. We can write the initial state of the noninteracting BEC as a product of coherent states in the local particle number,

$$|\mathrm{BEC_{SB}}\rangle = \prod_l^M e^{-\frac{|\beta_l|^2}{2}} e^{\beta_l b_l^\dagger}|\mathrm{vac}\rangle, \tag{5.27}$$

where β_l is the mean particle number on site l.

Onsite energy shifts due to the number of particles can be described by an interaction Hamiltonian $H_{\mathrm{int}} = (U/2)\sum_l \hat{n}_l(\hat{n}_l - 1)$, where $\hat{n}_l = b_l^\dagger b_l$ is the number operator for bosons on site l, b_l is the annihilation operator for a bosonic atom on site l, and U is the two-particle collisional energy shift. As a function of time, we can then see that the correlations behave as

$$\langle b_i^\dagger b_j\rangle_t = \langle b_i\rangle_t^* \langle b_j\rangle_t = \beta_i^*\beta_j e^{|\beta_i|^2(e^{iUt}-1)} e^{|\beta_j|^2(e^{-iUt}-1)} = |\beta|^2 e^{|\beta|^2(2\cos(Ut)-2)}, \tag{5.28}$$

with $\beta_l = \beta$ in the homogeneous system.

In the alternative case, with fixed particle number, we can write the initial state of N particles in a homogeneous system of M sites as

$$|\mathrm{BEC}_N\rangle = \frac{1}{\sqrt{N!N^N}}\left(\sum_i^M \beta\hat{b}_i^\dagger\right)^N |\mathrm{vac}\rangle.$$

It is helpful to rewrite this state as a number projection of the coherent states. We can see how this is possible by writing

$$|\{\beta\}\rangle \equiv \prod_k^M |\beta\rangle_k \equiv \prod_k^M e^{-\frac{|\beta|^2}{2}} e^{\beta b_k^\dagger}|\text{vac}\rangle, \tag{5.29}$$

where with a projection operator P_N, we obtain

$$P_N|\{\beta\}\rangle = e^{-\frac{1}{2}\sum_k^M |\beta|^2} P_N e^{\sum_k^M \beta b_k^\dagger}|\text{vac}\rangle \tag{5.30}$$

$$= e^{-\frac{1}{2}\sum_k^M |\beta|^2} P_N \sum_{n=0}^{\infty} \frac{1}{n!}\left(\sum_k^M \beta b_k^\dagger\right)^n |\text{vac}\rangle \tag{5.31}$$

$$= \frac{1}{\sqrt{\mathcal{N}}} e^{-\frac{1}{2}\sum_k^M |\beta|^2} \frac{1}{N!}\left(\sum_k^M \beta b_k^\dagger\right)^N |\text{vac}\rangle, \tag{5.32}$$

which is identical to the ground state with fixed particle numbers. We now make use of a trick in which we write the Fock state of fixed particle number as a phase averaged coherent state [7],

$$P_N = \frac{1}{2\pi\sqrt{\mathcal{N}}} \int_{-\pi}^{\pi} d\phi\, e^{i(\hat{N}-N)\phi}, \tag{5.33}$$

where the total particle number operator is given by $\hat{N} = \sum_i b_i^\dagger b_i$.

Writing the state with fixed total particle number in this form means that the time-evolved single-particle density matrix can then be written exactly as

$$F\langle b_i^\dagger b_j\rangle_t = \langle \text{BEC}_N|e^{iHt} b_i^\dagger b_j e^{-iHt}|\text{BEC}_N\rangle \tag{5.34}$$

$$= \frac{1}{2\pi\mathcal{N}} \int_{-\pi}^{+\pi} d\phi\, e^{-iN\phi} \langle\{\beta\}|e^{iHt} b_i^\dagger b_j e^{-iHt} e^{\hat{N}\phi}|\{\beta\}\rangle \tag{5.35}$$

$$= \frac{1}{2\pi\mathcal{N}} \int_{-\pi}^{+\pi} d\phi\, e^{-iN\phi} \langle\{\beta\}|e^{iHt} b_i^\dagger b_j e^{-iHt}|\{\beta e^{i\phi}\}\rangle. \tag{5.36}$$

In this integral, the diagonal elements of the single-particle density matrix remain constant in time, where, as expected from the density, $\langle b_i^\dagger b_i\rangle_t = |\beta|^2$. The key to understanding the time dependence of the interference pattern is to evaulate the off-diagonal elements, which factorize as

$$\langle\{\beta\}\big|e^{iHt} b_i^\dagger b_j e^{-iHt}\big|\{\beta e^{i\phi}\}\rangle \tag{5.37}$$

$$= \langle\beta|e^{iU\hat{n}_i(\hat{n}_i-1)t/2} b_i^\dagger e^{-iU\hat{n}_i(\hat{n}_i-1)t/2}|\beta e^{i\phi}\rangle_i \tag{5.38}$$

$$\times \langle\beta|e^{iU\hat{n}_i(\hat{n}_j-1)t/2} b_j e^{-iU\hat{n}_i(\hat{n}_j-1)t/2}|\beta e^{i\phi}\rangle_j \prod_{k\neq i,j} \langle\beta|\beta e^{i\phi}\rangle_k. \tag{5.39}$$

We can then compute

$$\langle\beta|e^{iU\hat{n}_i(\hat{n}_j-1)t/2} b_j e^{-iU\hat{n}_i(\hat{n}_j-1)t/2}|\beta e^{i\phi}\rangle_j = \beta e^{i\phi} e^{-i\varepsilon_j t} e^{|\beta|^2[e^{i\phi}e^{-iUt}-1]}, \tag{5.40}$$

and

$$\langle\beta|e^{iU\hat{n}_i(\hat{n}_i-1)t/2}b_i^\dagger e^{-iU\hat{n}_i(\hat{n}_i-1)t/2}|\beta e^{i\phi}\rangle_i = \beta^* e^{i\varepsilon_i t} e^{|\beta|^2[e^{i\phi}e^{iUt}-1]}, \tag{5.41}$$

so that for $i \neq j$,

$$\langle b_i^\dagger b_j\rangle_t = \frac{\beta^*\beta}{2\pi N} e^{-N} e^{i(\varepsilon_i-\varepsilon_j)t} \int_{-\pi}^{+\pi} d\phi\, e^{-i(N-1)\phi} \tag{5.42}$$

$$\times e^{e^{i\phi}\left[|\beta|^2 e^{-iUt}+|\beta|^2 e^{iUt}+\sum_{k\neq i,j}|\beta|^2\right]} \tag{5.43}$$

$$= |\beta|^2\left[1+\frac{|\beta|^2}{N}(2\cos(Ut)-2)\right]^{(N-1)}. \tag{5.44}$$

In the limit where the particle number is much larger than the onsite density, i.e., $N \gg |\beta|^2$, we find the reduction

$$|\beta|^2\left[1+\frac{|\beta|^2}{N}(2\cos(Ut)-2)\right]^{(N-1)} \longrightarrow |\beta|^2 e^{|\beta|^2[2\cos(Ut)-2]}, \tag{5.45}$$

which reproduces the result from coherent states in Eq. (5.28).

To see how the interference pattern observed in an experiment depends on the system size and particle number, we compute the height of the zero-quasimomentum peak, $n_{q=0} = (1/M)\sum_{i,j}^{M}\langle b_i^\dagger b_j\rangle_t = (M-1)\langle b_i^\dagger b_j\rangle + |\beta|^2$. This corresponds to the visibility of an interference pattern after a long time of flight. In Fig. 5.3, we show a plot of the time evolution of these visibilities beginning from a ground state with density $|\beta|^2 = 1$. As the particle number increases, we see that the values converge rapidly to the values from coherent states: already for $N \sim 5$, the results are difficult to distinguish from each other. Defining the relative difference, we can show that the difference decreases proportional to $1/M$ [12].

As a final comment on measurement-based treatments with fixed initial numbers, we note that one way to look at the standard spontaneous symmetry-breaking scenario is to say that some form of outside kick has perturbed the system. In the case discussed here, or also in interference of independent condensates, one might suggest that the outside influence is the interaction with the measuring apparatus. This raises the question, addressed by a whole host of philosophers, of what measurement really is. Perhaps the measurement itself involves spontaneous symmetry breaking of the standard kind, when a stray field external to the detector causes it to respond to a matter wave with a collapse in one direction or another. But many studies have shown that if this is the case, some degree of nonlocality must also enter. Detectors seem to coordinate their responses across lightlike separations.

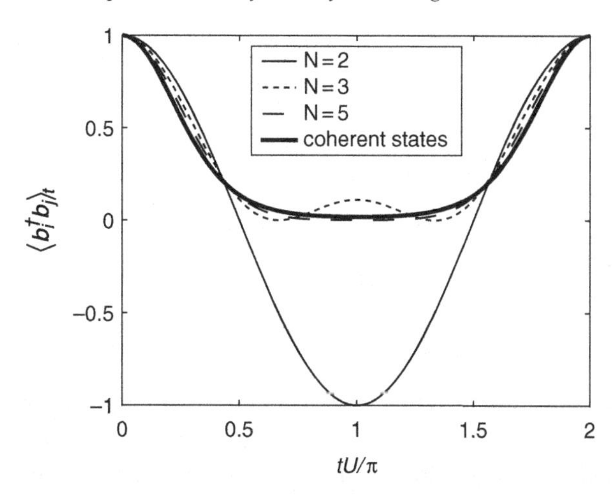

Figure 5.3 Plot of the time evolution of the off-diagonal elements of the single-particle density matrix in a homogeneous system (i.e., periodic boundary conditions) at unit average filling. We clearly see the first collapse and revival of these correlation functions and compare the predictions from the symmetry-broken solution of Eq. (5.28) (solid heavy line) and the number-conserving calculation in Eq. (5.44). Adapted with permission from Schachenmayer, J., et al. (2011), *Phys. Rev. A*, **83**, 043614 [12]. Copyright (2011) by the American Physical Society.

5.5 Spontaneous Symmetry Breaking in Photon and Polariton Condensates

As we have seen in Section 5.2, the standard model of spontaneous symmetry-breaking envisions a small fluctuation which is amplified. In the case of photon [36] and polariton condensates [37, 38, 39, 40, 41] (respectively reviewed in this book in Chapters 19 and 4), this fluctuation can come from an external electromagnetic field.

A polariton condensate is fundamentally no different from a photon condensate; we can see this in the following derivation of the polariton wave equation. We start with Maxwell's wave equation in a nonlinear isotropic medium,

$$\nabla^2 E = \frac{n^2}{c^2}\frac{\partial^2 E}{\partial t^2} + 4\mu_0 \chi^{(3)} \frac{\partial^2}{\partial t^2}|E|^2 E, \tag{5.46}$$

where $\chi^{(3)}$ is the standard nonlinear optical constant, and we ignore frequency-mixing terms in the general E^3 nonlinear response. In the standard polariton scenario, this nonlinear term is produced by a sharp electronic resonance, namely a two-level oscillation. In condensed matter systems, this is usually an excitonic excitation of the valence band electrons.

Writing the solution in the form $E = \psi(x,t)e^{-i\omega t}$, and keeping only leading terms in frequency (known as the slowly varying envelope approximation), we have for the time derivative of E,

$$\frac{\partial^2 E}{\partial t^2} \simeq \left(-\omega^2\psi - 2i\omega\frac{\partial\psi}{\partial t}\right)e^{-i\omega t}, \tag{5.47}$$

and for the time derivative of the nonlinear term

$$\frac{\partial^2}{\partial t^2}|E|^2E \simeq -\omega^2|\psi|^2\psi e^{-i\omega t}. \tag{5.48}$$

The standard polariton structure uses a planar or nearly planar cavity to give one confined direction of the optical mode. We therefore distinguish between the component of momentum k_z in the direction of the cavity confinement, which is fixed by the cavity length, and the momentum $k_\parallel$ for motion in the two-dimensional plane perpendicular to this direction, which is free. We therefore write $\psi = \psi(\vec{x})e^{i(k_\parallel\cdot\vec{x}+k_z z)}$. The full Maxwell wave equation (5.46) then becomes

$$\begin{aligned}&(-(k_z^2+k_\parallel^2)\psi + \nabla_\parallel^2\psi)\\ &\quad = (n/c)^2\left(-\omega^2\psi - 2i\omega\frac{\partial\psi}{\partial t}\right) - 4\mu_0\chi^{(3)}\omega^2|\psi|^2\psi.\end{aligned} \tag{5.49}$$

Since $\omega^2 = (c/n)^2(k_z^2+k_\parallel^2)$, this becomes

$$\nabla_\parallel^2\psi = (n/c)^2\left(-2i\omega\frac{\partial\psi}{\partial t}\right) - 4\mu_0\chi^{(3)}\omega^2|\psi|^2\psi. \tag{5.50}$$

Near $k_\parallel = 0$, we can approximate

$$\hbar\omega = \hbar(c/n)\sqrt{k_z^2+k_\parallel^2} \simeq \hbar(c/n)k_z\left(1+\frac{k_\parallel^2}{2k_z^2}\right) \equiv \hbar\omega_0 + \frac{\hbar^2k_\parallel^2}{2m}, \tag{5.51}$$

which gives an effective mass for the photon motion in the plane. For the first term on the right-hand side of (5.50), we approximate

$$\omega \simeq \omega_0 = \frac{m(c/n)^2}{\hbar}. \tag{5.52}$$

Therefore, we can rewrite (5.50) as

$$i\hbar\frac{\partial\psi}{\partial t} = -\frac{\hbar^2}{2m}\nabla_\parallel^2\psi - \frac{2\mu_0\chi^{(3)}(\hbar\omega)^2}{m}|\psi|^2\psi, \tag{5.53}$$

which we can rewrite as

$$i\hbar\frac{\partial\psi}{\partial t} = -\frac{\hbar^2}{2m}\nabla_\parallel^2\psi + U|\psi|^2\psi. \tag{5.54}$$

This is a Gross-Pitaevskii equation, or nonlinear Schrödinger equation. Note that although the Maxwell wave equation is second order in the time derivative, this equation is first order in the time derivative, as in a typical Schrödinger equation.

The polariton and photon condensates therefore can follow the standard scenario of spontaneous symmetry breaking as occurs in a laser, in which one imagines a small stray electromagnetic field which is amplified. The equations which govern the polariton condensate, however, are identical to those of a standard condensate. In general, in polariton condensates one can add generation and decay terms to the Gross-Pitaevskii equation (5.54), as done by Carusotto and Keeling and coworkers [42, 43] (see also Chapter 11) but this distinction from standard condensates has become less significant in recent years. On one hand, to be strictly accurate, the same type of term should be written for cold atom condensates, because these condensates have particle loss mechanisms due to evaporation from their traps. On the other hand, the lifetime of polaritons in microcavities has been steadily increasing, so that the ratio of the lifetime of the particles in the system to their collision time can be several hundred, comparable to the ratio for cold atoms in traps. Therefore, in both the cold atom and polariton condensate systems, it is reasonable to drop the generation and decay terms as negligible in many cases.

The fact that polaritons decay into photons which leak out of the cavity mirrors means that it is possible to directly observe the phase amplitude of the polaritons in interference measurements. In this case, the interference is not between two condensates, but is between two different regions of the same condensate, more similar to the interference in multiwell systems discussed above. These results are discussed further in Chapters 21–23.

5.6 Conclusions

A question one can ask is whether condensates such as polariton and photon condensates (and magnon condensates [44]) can be viewed as "real" condensates, if they are known to have weak coupling to the outside world which allows spontaneous symmetry breaking of the ferromagnetic type. On this, it would seem strange to treat them as entirely different phenomena when the equations governing their behavior, once the symmetry has been broken, are identical to those governing the behavior of atom condensates.

Another question one can ask is whether *all* spontaneous symmetry breaking must intrinsically be of the ferromagnetic type, that is, whether there must be some external fluctuation, no matter how tiny, which is amplified by the instability of the system. On this, it is clear that a scenario in which there is no outside field to break the symmetry can still result in broken symmetry when quantum measurement is taken into account. However, since we do not fully understand the measurement process, this may simply beg the question, because it cannot be ruled out that measurement itself involves nonlocal broken symmetry somehow in the measuring apparatus.

All of this thinking applies to the cosmology of the early universe. While it is clear that scenarios exist in which a state with broken symmetry can have lower energy than a symmetric state, we do not know how to introduce an "external" fluctuation to break the symmetry in the ferromagnetic analogy, but it is also hard to apply the measurement-broken-symmetry scenario to the early universe without knowing what might count as an observer.

References

[1] Mølmer, K. 1997. Optical coherence: a convenient fiction. *Phys. Rev. A*, **55**, 3195–3203.

[2] Castin, Y., and Dalibard, J. 1997. Relative phase of two Bose-Einstein condensates. *Phys. Rev. A*, **55**, 4330–4337.

[3] Gardiner, C. W. 1997. Particle-number-conserving Bogoliubov method which demonstrates the validity of the time-dependent Gross-Pitaevskii equation for a highly condensed Bose gas. *Phys. Rev. A*, **56**, 1414–1423.

[4] Castin, Y., and Dum, R. 1998. Low-temperature Bose-Einstein condensates in time-dependent traps: beyond the $U(1)$ symmetry-breaking approach. *Phys. Rev. A*, **57**, 3008–3021.

[5] Leggett, A. J. 2001. Bose-Einstein condensation in the alkali gases: some fundamental concepts. *Rev. Mod. Phys.*, **73**, 307–356.

[6] Leggett, A. J., and Sols, F. 1991. On the concept of spontaneously broken gauge symmetry in condensed matter physics. *Foundations of Physics*, **21**(3), 353–364.

[7] Javanainen, J., and Yoo, S. M. 1996. Quantum phase of a Bose-Einstein condensate with an arbitrary number of atoms. *Phys. Rev. Lett.*, **76**, 161–164.

[8] Naraschewski, M., Wallis, H., Schenzle, A., Cirac, J. I., and Zoller, P. 1996. Interference of Bose condensates. *Phys. Rev. A*, **54**, 2185–2196.

[9] Wong, T., Collett, M. J., and Walls, D. F. 1996. Interference of two Bose-Einstein condensates with collisions. *Phys. Rev. A*, **54**, R3718–R3721.

[10] Wright, E. M., Walls, D. F., and Garrison, J. C. 1996. Collapses and revivals of Bose-Einstein condensates formed in small atomic samples. *Phys. Rev. Lett.*, **77**, 2158–2161.

[11] Stenholm, S. 2002. The question of phase in a Bose-Einstein condensate. *Physica Scripta*, **2002**(T102), 89.

[12] Schachenmayer, J., Daley, A. J., and Zoller, P. 2011. Atomic matter-wave revivals with definite atom number in an optical lattice. *Phys. Rev. A*, **83**, 043614.

[13] Snoke, D.W. 2009. *Essential Concepts of Solid State Physics*. Pearson, section 10.2.1.

[14] Ibid., section 11.3.1.

[15] Kazanas, D. 1980. Dynamics of the universe and spontaneous symmetry breaking. *Astrophysical J.*, **241**, L59–L63.

[16] Kibble, T. W. B. 1980. Some implications of a cosmological phase transition. *Physics Reports*, **67**, 183–199.

[17] Zurek, W. H. 1996. Cosmological experiments in condensed matter systems. *Physics Reports*, **276**, 177–221.

[18] Einstein, A. 1924. Quantum theory of the single-atom ideal gas. *Absitz. Pruss. Akad. Wiss. Berlin, Kl. Math.*, **22**, 261.

[19] Noziéres, P. 1995. Some comments on Bose-Einstein condensation. Pages 16–21 of: *Bose-Einstein Condensation*. Cambridge University Press, Griffin, A., Snoke, D. W., and Stringari, S., eds.

[20] Combescot, M., and Snoke, D. W. 2008. Stability of a Bose-Einstein condensate revisited for composite bosons. *Phys. Rev. B*, **78**, 144303.

[21] Band, W. 1955. *An Introduction to Quantum Statistics*. D. Van Nostrand, pp. 154, 162.

[22] Snoke, D. W., Liu, G., and Girvin, S. M. 2012. The basis of the second law of thermodynamics in quantum field theory. *Annals of Physics*, **327**, 1825–1851.

[23] Snoke, D. W., and Girvin, S. M. 2013. Dynamics of phase coherence onset in Bose condensates of photons by incoherent phonon emission. *J. Low Temperature Phys.*, **171**, 1.

[24] Snoke, D., Wolfe, J. P., and Mysyrowicz, A. 1987. Quantum saturation of a Bose gas: excitons in Cu_2O. *Phys. Rev. Lett.*, **59**, 827.

[25] Semikoz, D. V., and Tkachev, I. I. 1995. Kinetics of Bose condensation. *Phys. Rev. Lett.*, **74**, 3093–3097.

[26] Andrews, M. R., Townsend, C. G., Miesner, H.-J., Durfee, D. S., Kurn, D. M., and Ketterle, W. 1997. Observation of interference between two Bose condensates. *Science*, **275**(5300), 637–641.

[27] Mandel, L., and Wolf, E. 1995. *Optical Coherence and Quantum Optics*. Cambridge University Press.

[28] Imamoḡlu, A., Lewenstein, M., and You, L. 1997. Inhibition of coherence in trapped Bose-Einstein condensates. *Phys. Rev. Lett.*, **78**, 2511–2514.

[29] Javanainen, J., and Wilkens, M. 1997. Phase and phase diffusion of a split Bose-Einstein condensate. *Phys. Rev. Lett.*, **78**, 4675–4678.

[30] Milburn, G. J., Corney, J., Wright, E. M., and Walls, D. F. 1997. Quantum dynamics of an atomic Bose-Einstein condensate in a double-well potential. *Phys. Rev. A*, **55**, 4318–4324.

[31] Cirac, J. I., Gardiner, C. W., Naraschewski, M., and Zoller, P. 1996. Continuous observation of interference fringes from Bose condensates. *Phys. Rev. A*, **54**, R3714–R3717.

[32] Mølmer, K. 1997. Quantum entanglement and classical behaviour. *Journal of Modern Optics*, **44**(10), 1937–1956.

[33] Siegman, A. E. 1986. *Lasers*. University Science Books.

[34] Greiner, M., Mandel, O., Hansch, T. W., and Bloch, I. 2002. Collapse and revival of the matter wave field of a Bose-Einstein condensate. *Nature*, **419**(6902), 51–54.

[35] Will, S., Best, T., Schneider, U., Hackermuller, L., Luhmann, D.-S., and Bloch, I. 2010. Time-resolved observation of coherent multi-body interactions in quantum phase revivals. *Nature*, **465**(7295), 197–201.

[36] Klaers, J., Schmitt, J., Vewinger, F., and Weitz, M. 2010. Bose-Einstein condensation of photons in an optical microcavity. *Nature*, **468**, 548.

[37] Kasprzak, J., Richard, M., Kundermann, S., Baas, A., Jeanbrun, P., Keeling, J. M. J., André, R., Staehli, J. L., Savona, V., Littlewood, P. B., Deveaud, B., and Dang, L. S. 2006. Bose-Einstein condensation of exciton polaritons. *Nature*, **443**, 409.

[38] Balili, R., Hartwell, V., Snoke, D. W., Pfeiffer, L., and West, K. 2007. Bose-Einstein condensation of microcavity polaritons in a trap. *Science*, **316**, 1007.

[39] Liu, G., Snoke, D. W., Daley, A. J., Pfeiffer, L. N., and West, K. 2015. A new type of half-quantum circulation in a macroscopic polariton spinor ring condensate. *Proc. National Acad. Sci. (USA)*, **112**, 2676.

[40] Baumberg, J. J., Kavokin, A. V., Christopoulos, S., Grundy, A. J. D., Butte, R., Christmann, G., Solnyshkov, D. D., Malpuech, G., von Hogersthal, G. B. H., Feltin, E., Carlin, J. F., and Grandjean, N. 2008. Spontaneous polarization buildup in a room-temperature polariton laser. *Phys. Rev. Lett.*, **101**, 136409.

[41] Abbarchi, M., Amo, A., Sala, V. G., Solnyshkov, D. D., Flayac, H., Ferrier, L., Sagnes, I., Galopin, E., Lemaître, A., Malpuech, G., and Bloch, J. 2013. Macroscopic quantum self-trapping and Josephson oscillations of exciton polaritons. *Nature Phys.*, **9**, 275.

[42] Wouters, M., and Carusotto, I. 2010. Superfluidity and critical velocities in nonequilibrium Bose-Einstein condensates. *Phys. Rev. Lett.*, **105**, 020602.

[43] Keeling, J. 2011. Superfluid density of an open dissipative condensate. *Phys. Rev. Lett.*, **107**, 080402.

[44] Demidov, V. E., Dzyapko, O., Demokritov, S. O., Melkov, G. A., and Slavin, A. N. 2008. Observation of spontaneous coherence in Bose-Einstein condensate of magnons. *Phys. Rev. Lett.*, **100**, 047205.

6

Effects of Interactions on Bose-Einstein Condensation

ROBERT P. SMITH
Cavendish Laboratory, University of Cambridge, UK

Bose-Einstein condensation is a unique phase transition in that it is not driven by interparticle interactions, but can theoretically occur in an ideal gas, purely as a consequence of quantum statistics. This chapter addresses the question, *'How is this ideal Bose gas condensation modified in the presence of interactions between the particles?'* This seemingly simple question turns out to be surprisingly difficult to answer. Here we outline the theoretical background to this question and discuss some recent measurements on ultracold atomic Bose gases that have sought to provide some answers.

6.1 Introduction

Unlike the vast majority of phase transitions, Bose-Einstein condensation (BEC) is not driven by interparticle interactions but can theoretically occur in an ideal (non-interacting) gas, purely as a consequence of quantum statistics. However, in reality, interactions are needed for a Bose gas to remain close to thermal equilibrium. It is thus interesting to discuss if something close to ideal gas BEC can be observed in a real system and what happens in the vicinity of the BEC transition in the presence of interparticle interactions. These simple questions have not been easy to answer, either theoretically or experimentally.

The theoretical foundations for studying the effect of interactions on Bose condensed systems were laid over half a century ago by Bogoliubov [1], Penrose and Onsager [2], and Belieav [3], among others. These works initially focused on zero-temperature properties and were extended to nonzero temperature in the pioneering papers of Lee, Huang, and Yang [4, 5, 6, 7]. At that time, the main experimental system was liquid He-4 in which the inter-particle interactions are strong, making connections with theory difficult. The realisation in 1995 of BEC in weakly interacting ultracold atomic gases [8, 9] thus opened up the possibility to experimentally revisit some of these long-discussed questions. This was further aided by, among other advances, the use of Feshbach resonances to tune the interaction strength in atomic gases [10, 11]. Thus, in the last twenty years the study of ultracold Bose gases has been very successful (see Chapter 3 for an insightful account of selected topics and surprising developments over this exciting period). However, the fact that, until recently, ultracold atoms were confined using harmonic potentials has hindered the study of the BEC transition itself. This is due to the resulting inhomogeneous density distribution, which often masks the most interesting interaction effects and also makes direct comparison with theory challenging.

In this chapter, we review some recent experimental investigations using atomic Bose gases that have sought to study the role of interactions on BEC. We will mainly focus on measurements in harmonic traps, paying particular attention to how these results relate to the physics in a homogeneous system. We also discuss some of the first measurements on a homogeneous atomic Bose gas and the possibilities that such measurements present in the future. Note that we will focus here on three-dimensional systems close to the BEC transition temperature. Some parts of the present chapter were also discussed in a previous chapter written by the author [12].

The outline of the chapter is as follows. In Section 6.2, we recap some theoretical background related to homogeneous Bose gases in the presence of repulsive contact interactions and outline how these results may be applied to a harmonically trapped system using the local density approximation (LDA). Then in Sections 6.3, 6.4, and 6.5, we focus on the experimental investigations of three aspects of the BEC transition in the presence of interactions: (i) the statistical mechanism of the BEC transition, (ii) the transition temperature, and (iii) the critical behaviour near the transition. (The related topic of condensate formation is also discussed in Chapter 7 by Davis et al.)

6.2 Theoretical Background

In this section, we provide the key theoretical points necessary to understand and interpret the experimental results presented in the rest of the chapter. The treatment we give here is by necessity brief and we refer the interested reader to more comprehensive reviews [13, 14, 15, 16]. We first outline the expected behaviour for an ideal homogeneous gas before going on to consider the effect of weak interactions on such a gas. Finally, we consider how these homogeneous results may be applied to a harmonically trapped gas.

6.2.1 Homogeneous Ideal Bose Gas

In a gas of bosons of mass m in equilibrium at temperature T, the occupation of momentum state $\mathbf{p}$ is given by the Bose distribution function,

$$f_{\mathbf{p}} = \frac{1}{\mathrm{e}^{(p^2/2m-\mu)/k_\mathrm{B}T} - 1} , \tag{6.1}$$

where μ is the chemical potential. The total particle number N can be found by summing over all possible momentum states,

$$N = \sum_p \frac{g_p}{\mathrm{e}^{(p^2/2m-\mu)/k_\mathrm{B}T} - 1} , \tag{6.2}$$

where g_p is the number of states with a given p. The requirement for all terms in the sum to be real positive numbers constrains $\mu \leq p_0^2/2m$, where p_0 is the momentum of the ground state. As μ approaches $p_0^2/2m$ from below (which is achieved by adding particles), the ground-state occupation can become arbitrarily large (see Eq. (6.1)) whereas the sum of the remainder of the states (the excited states) tends to a finite number. This helps us to understand the mechanism for Bose-Einstein condensation, namely the statistical saturation of excited states. Saturation can be graphically represented as shown in Fig. 6.1a by plotting both the ground-state population and excited-state population versus the total atom number for a fixed temperature. As particles are added to the system, they initially populate excited states until a critical atom number is reached above which the excited state population is saturated and any additional particles must enter the ground state.

In the thermodynamic limit, in which both the volume of the system and total atom number become large, we may use the semiclassical approximation. That is, we replace the sum over excited states by an integral in order to calculate the critical atom number (or density). The excited-state density n', which we also call the thermal density, is given by

$$n' = \int \frac{\mathrm{d}\mathbf{p}}{(2\pi\hbar)^3} \frac{1}{\mathrm{e}^{(p^2/2m-\mu)/k_\mathrm{B}T} - 1} = \frac{g_{3/2}(\mathrm{e}^{\mu/k_\mathrm{B}T})}{\lambda^3} , \tag{6.3}$$

where $\lambda = [2\pi\hbar^2/(mk_\mathrm{B}T)]^{1/2}$ is the thermal wavelength and $g_{3/2}(x) = \sum_{k=1}^{\infty} x^k/k^{3/2}$ is a polylogarithm function. Note that in this limit $p_0 \rightarrow 0$, and so now $\mu \leq 0$. We can re-express this result in terms of the phase space density $\mathcal{D} = n\lambda^3$ as

$$\mathcal{D}' \equiv n'\lambda^3 = g_{3/2}(\mathrm{e}^{\beta\mu}) , \tag{6.4}$$

where $\beta = 1/k_\mathrm{B}T$. The maximum value that $\mathcal{D}'$ can take is reached when $\mu = 0$ and occurs when the total density reaches a critical value $\mathcal{D}_c^0 = n_c^0\lambda^3 = g_{3/2}(1) = \zeta(3/2) \approx 2.612$ (where ζ is the Riemann function). Here the superscript ^{0} refers to the fact this is an ideal gas result. We can also invert this result to give the BEC transition temperature for a fixed density:

$$k_\mathrm{B}T_c^0 = \frac{2\pi\hbar^2}{m}\left(\frac{n}{\zeta(3/2)}\right)^{2/3} . \tag{6.5}$$

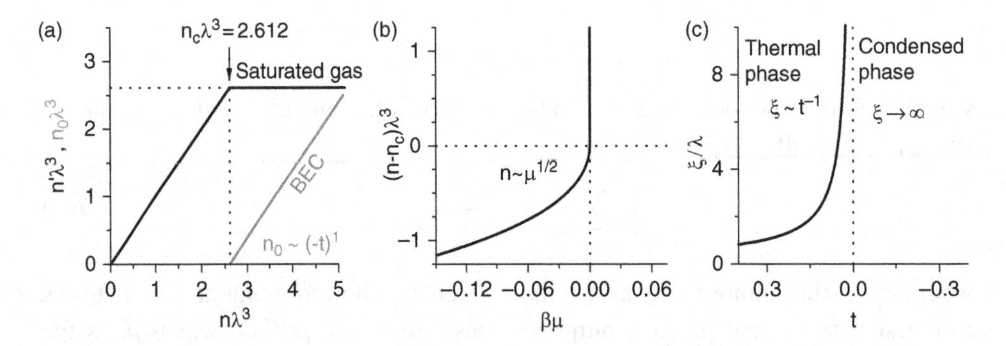

Figure 6.1 Ideal Bose gas condensation. (a) Thermal n' (black line) and condensed n_0 (grey line) density plotted versus the total density n, at a fixed temperature. As atoms are added $n' = n$ and $n_0 = 0$ until the critical atom density n_c is reached. At this point, the excited states of the system saturate, and for $n > n_c$ we have $n' = n_c$ and $n_0 = n - n_c \propto -t$ grows linearly. (b) The density as a function of $\beta\mu$ close to the critical density. (c) The correlation length diverges as $t^{-\nu}$ as we approach the transition from above with $\nu = 1$; below T_c the correlation length remains infinite.

In this large N limit, the transition to a BEC is a well-defined second-order phase transition and is thus characterised by a set of critical exponents which describe how various quantities diverge when approaching the transition. Here we focus on three exponents as summarised in Fig. 6.1.

(i) The growth of the condensate density n_0 below the transition temperature is described by $n_0 = n - n_c \propto (-t)^1$, where t is the *reduced temperature* given by $t = (T - T_c)/T_c$.

(ii) Above T_c the dependence of n on $\beta\mu$ can be found by expanding Eq. (6.4) for small $\beta\mu$; up to first order, this expansion (at constant volume) gives $\mathcal{D} = \mathcal{D}_c - 2\sqrt{\pi}\sqrt{-\beta\mu}$ and thus $n - n_c \propto (\mu_c - \mu)^{1/2}$, where μ_c is the critical chemical potential (for an ideal gas $\mu_c = 0$). The dependence of n on μ is particularly important when discussing inhomogeneous systems.

(iii) The correlation length ξ quantifies the range over which fluctuations in the gas are correlated and its divergence, described by the critical exponent ν, is defined by $\xi \sim |t|^{-\nu}$. The correlation length can defined by the first-order two-point correlation function, $g_1(r) \propto \langle\hat{\Psi}^\dagger(r)\hat{\Psi}(0)\rangle$, where $\hat{\Psi}(\mathbf{r})$ is the Bose field. This correlation function is formally related to the Fourier transform of the momentum distribution. Far from the BEC transition ($\beta\mu \ll -1$), the Bose momentum distribution (Eq. (6.1)) is approximately gaussian and thus $g_1(r)$ is short ranged and given by a gaussian of width $\lambda/\sqrt{2\pi}$. As we approach the transition, long-range correlations begin to develop and for $r > \lambda$ we can approximate the correlation function as [17]

$$g_1(r) \propto \frac{1}{r}\exp(-r/\xi)\,. \tag{6.6}$$

For an ideal gas, $\xi/\lambda = \sqrt{1/(4\pi\beta|\mu|)}$; combining this result with the expansion from (ii) and the fact that for small t we can write $(\mathcal{D}_c - \mathcal{D})/\mathcal{D}_c \approx \frac{3}{2}t$ gives $\xi/\lambda = \frac{2}{3D_c}t^{-1}$. Thus, for an ideal gas BEC transition (at constant volume), $\nu = 1$.

6.2.2 Interacting Homogeneous Bose Gas

For a dilute atomic gas the effective low-energy interaction between two atoms at $\mathbf{r}$ and $\mathbf{r}'$ can be approximated as a contact interaction $g\delta(\mathbf{r} - \mathbf{r}')$ with $g = 4\pi\hbar^2 a/m$, where a is the s-wave scattering length. Note that the dimensionless parameter, which usually defines the relative strength of interactions, a/λ, is typically $\sim 10^{-2}$ for ultracold atomic gases.

A simple theoretical framework in which to understand the effects of contact interactions on a Bose gas is the Hartree-Fock (HF) approximation [13]. In this mean-field (MF) model one treats the thermal atoms as a "non-interacting" gas of density n' that experiences an MF interaction potential

$$V_{\mathrm{int}} = g(2n_0 + 2n'), \tag{6.7}$$

where n_0 is the condensate density. Thus, $p^2/2m$ in Eq. (6.1) is replaced by $p^2/2m + V_{\mathrm{int}}$. Meanwhile, the condensed atoms feel an interaction potential

$$V^0_{\mathrm{int}} = g(n_0 + 2n'), \tag{6.8}$$

where the factor of two difference in the condensate self-interaction comes about due to the lack of the exchange interaction term for particles in the same state. This HF approach does not take into account the modification of the excitation spectrum due to the presence of the condensate, which is included in more elaborate MF theories such as those of Bogoliubov [1] and Popov [18]. However, it is often sufficient to give the correct leading-order MF results. In a homogeneous system, above T_c, the MF interaction potential leads to a uniform energy offset which is simply compensated for by a shift in the chemical potential and thus makes no physical difference to the system. The effects of interactions on a homogeneous system at an MF-level only result from the factor of two difference in the condensate self-interaction and therefore only arise when a condensate is present (or about to appear). All these MF theories are expected to break down as we approach T_c and should only provide a good description outside the critical region ($|t| \gg a/\lambda$). Within the critical region, we must revert to beyond mean-field descriptions.

Fig. 6.2 summarises the effect of interactions on Bose-Einstein condensation. A comparison of Figs. 6.1 and 6.2 allows us to highlight several notable differences:

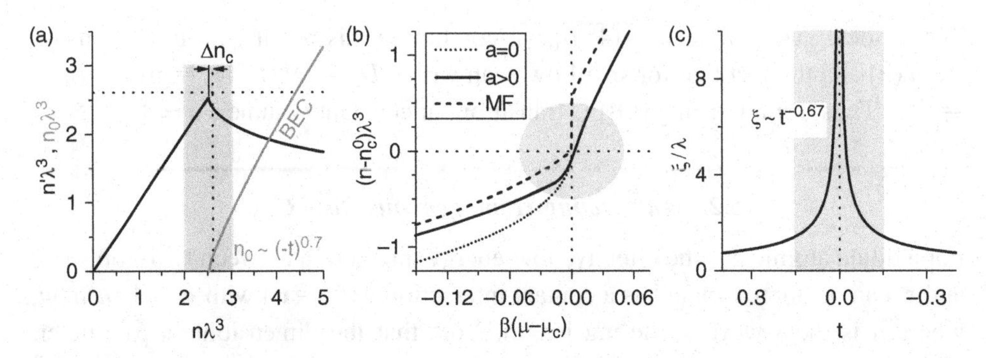

Figure 6.2 Effects of interactions on a Bose gas. Illustrated for $a/\lambda = 0.03$; the critical region (here defined by $|t| < 5a/\lambda$) is shown in grey; the solid lines in (a) and (b) are based on classical field Monte Carlo results [19] within the critical region and an extrapolation onto the Popov approximation outside it. (a) Interactions both shift the critical point and also modify the condensed and thermal densities for $n > n_c$. (b) The dependence of $n - n_c^0$ on $(\mu - \mu_c)$ in the presence of interactions. MF theory (dashed line) predicts $n - n_c^0 \propto \beta(\mu - \mu_c)$ and an erroneous first-order transition whereas beyond-MF theory (solid line) predicts a second-order transition with an exponent between the ideal and MF values. (c) Interactions change the correlation length critical exponent from $\nu = 1$ to $\nu \simeq 0.67$.

1. The gas is no longer saturated after passing through the transition but rather the excited-state density decreases as we continue to increase n above n_c. This can be understood at a mean-field level and is due to the factor of two reduction in the condensate self-interaction, which means that an atom can lower its interaction energy by an amount gn_0 by entering the condensate.
2. Condensation occurs at a phase space density below the ideal gas critical value of 2.612. Qualitatively, this can be understood to be due to the same effect as in point 1 – that interactions favour the occupation of the condensed state. However, MF theory predicts an erroneous first-order transition and cannot predict the shift quantitatively. Theoretically calculating the shift proved notoriously difficult and took several decades for consensus to be reached (for an overview, see e.g. Refs. [16, 20, 21, 22]). It is now generally accepted that the shift is given by [20, 23]:

$$\frac{\Delta n}{n_c} \approx -2.7\frac{a}{\lambda}, \tag{6.9}$$

where $\Delta n = n_c - n_c^0$. Equivalently, the T_c shift at constant n is

$$\frac{\Delta T_c}{T_c^0} \approx -2/3\frac{\Delta n_c}{n_c^0} \approx 1.8\frac{a}{\lambda}. \tag{6.10}$$

3. The critical exponents of the transition are also modified. In fact, the addition of interactions results in a change of universality class to that of the so-called three-dimensional (3D) XY model. The beyond-MF critical exponents expected in this universality class have been calculated to high accuracy [24, 25]. Most notably, the correlation length critical exponent changes from $\nu = 1$ to $\nu \simeq 0.67$.

6.2.3 Bose Gas in a Harmonic Potential

Our approach to tackling inhomogeneous potentials is to apply the local density approximation (LDA). The effect of a potential $V(\mathbf{r})$ is to change the energy (e.g. in Eq. (6.3)) from $p^2/2m$ to $p^2/2m + V(\mathbf{r})$. Within the LDA, we subsume $V(\mathbf{r})$ within the chemical potential such that we have a local chemical potential,

$$\mu(\mathbf{r}) = \mu - V(\mathbf{r}), \tag{6.11}$$

and then assume that the density and momentum distribution of the gas at a point $\mathbf{r}$ is that of a homogeneous system with chemical potential $\mu(\mathbf{r})$.

The LDA is generally valid if $V(\mathbf{r})$ is changing slowly relative to any other relevant lengthscales. For an ideal thermal gas well above T_c the only relevant lengthscales are the thermal wavelength λ and the interparticle spacing $d < \lambda$. This suggests the LDA should be good for a thermal gas as long as $k_B T \gg \hbar\omega$. However, the soundness of LDA becomes less clear as ξ diverges upon approaching the transition. Within LDA the critical point is reached when the maximal local $\mathcal{D}$ reaches the critical phase space density. However, it usually makes sense to define the critical point in terms of the critical total particle number N_c (as the local density is harder to measure than the total atom number in the trap).

For an ideal Bose gas in a harmonic potential, $V(\mathbf{r}) = \sum(1/2)m\omega_i^2 r_i^2$, we can calculate N_c by inserting Eq. (6.11) with this potential into Eq. (6.4), setting $\mu = 0$ and integrating over all space, to give

$$N_c^0 = \zeta(3)\left(\frac{k_B T}{\hbar\bar{\omega}}\right)^3, \tag{6.12}$$

where $\bar{\omega}$ is the geometric mean of the trapping frequencies and $\zeta(3) \approx 1.202$. Equivalently, for a fixed particle number the transition temperature is given by

$$k_B T_c^0 = \hbar\bar{\omega}\left(\frac{N}{\zeta(3)}\right)^{1/3}. \tag{6.13}$$

In a similar fashion to the ideal homogeneous case, if we increase the total atom number N at constant temperature, then for $N < N_c^0$ no condensate is present and the thermal atom number $N' = N$. However, for $N > N_c^0$ the thermal component is saturated at $N' = N_c^0$. Thus for an ideal gas, the basic mechanism of saturation of excited states leading to a BEC transition remains.

However, the difference between homogeneous and harmonically trapped gases is much more fundamental than a simple change in the expression for T_c might suggest. Even for an ideal gas, the inhomogeneous density distribution means that as we approach the transition only the central region of the cloud is close to critical density and so the critical behaviour of the gas is quite different. For example, the divergence of ξ is constrained by a short lengthscale determined by the trap. In the presence of interactions, the differences become even more manifold due to two important factors.

First, unlike in a homogeneous system, when we apply LDA in a trapped system we are no longer under the constraint of constant n but of constant μ. This is because we are only free to vary the global chemical potential to fix the total atom number N, and then locally the chemical potential is set by Eq. (6.11). Therefore, the chemical potential shifts that we could dismiss in the case of a homogeneous system can now have large effects and so a trapped system can display large mean-field effects which were completely absent in the homogeneous case. In fact, as we will see, these effects often go in the opposite direction to those in a homogeneous gas.

Second, in a trapped system, only a small region (at the trap centre) is in the critical regime. This means that the magnitude of any beyond-MF effects are significantly reduced as compared with the homogeneous case. Also, due to our first point above, the beyond-MF effects that we do see are more likely to be related to beyond-MF shifts in μ rather than the homogeneous system density shifts.

6.3 Statistical Mechanism of BEC in an Interacting Bose Gas

In this section, we discuss the concept of the saturation of the excited states as the underlying mechanism driving the BEC transition, and how interactions modify this saturated-gas picture. For superfluid ^{4}He, which is conceptually associated with BEC, strong interactions preclude direct observation of purely statistical effects expected for an ideal Bose gas. This is generally thought to be in contrast to weakly interacting atomic gases in which close-to-textbook ideal BEC is expected. One might therefore expect that the saturation inequality $N' \leq N_c^0$ should be essentially satisfied in these systems. However, careful examination [26] revealed that this is far from being the case for a harmonically trapped gas as shown in Fig. 6.3. The drastic violation of the saturation inequality seen in Fig. 6.3, which is at first sight surprising, results from the combination of repulsive interactions and harmonic trapping. To first order, it can be explained in a simple MF picture in which we just consider the interaction of the thermal atoms with the condensate and not with

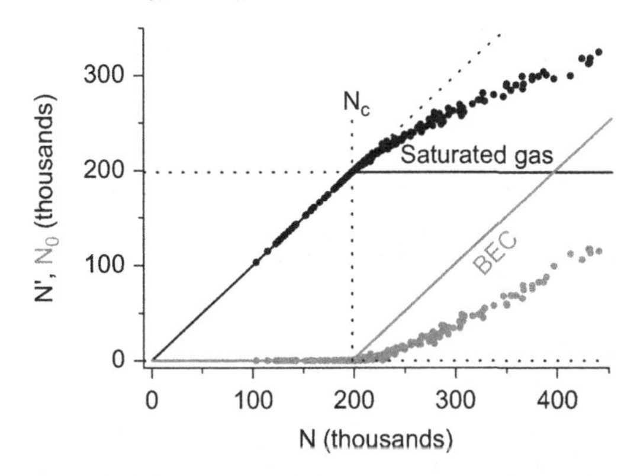

Figure 6.3 Lack of saturation of the thermal component in a quantum degenerate Bose gas with $a/\lambda = 0.01$. Thermal atom number N' (black points) and condensed atom number N_0 (grey points) are plotted versus the total atom number N for a fixed T. The predictions for a saturated gas are shown by black and grey solid lines. Adapted with permission from Tammuz, N., et al. (2011), Can a Bose gas be saturated? *Phys. Rev. Lett.*, **106**, 230401 [26]. Copyright (2011) by the American Physical Society.

other thermal atoms. This approximation works because, due to the harmonic trap, as a condensate is formed and grows, the change in the density of the condensed atoms is much larger than the change of the thermal density. Within the LDA, this leads to a spatially varying interaction potential $V_{\mathrm{int}}(\mathbf{r}) = 2gn_0(\mathbf{r})$ and results in a uniform chemical potential shift everywhere outside the condensate of

$$\mu_0 = gn_0(\mathbf{r}=0) = \frac{\hbar\bar{\omega}}{2}\left(15N_0\frac{a}{a_{\mathrm{ho}}}\right)^{2/5}, \tag{6.14}$$

where N_0 is the condensed atom number and $a_{\mathrm{ho}} = (\hbar/m\bar{\omega})^{1/2}$ is the spatial extension of the ground state of the harmonic oscillator. This shift in chemical potential effects the density everywhere and by integrating over the whole trap one can predict a linear variation of N'/N_c^0 with the small parameter $\beta\mu_0$:

$$\frac{N'}{N_c^0} = 1 + \alpha\,(\beta\mu_0)\,, \tag{6.15}$$

with $\alpha = \zeta(2)/\zeta(3) \approx 1.37$. Using Eq. (6.14), we can equivalently write

$$N' = N_c + S_0 N_0^{2/5}\,, \tag{6.16}$$

where $S_0 = \alpha X$, with X being the dimensionless interaction parameter:

$$X = \frac{\zeta(3)}{2}\left(\frac{k_{\mathrm{B}}T}{\hbar\bar{\omega}}\right)^2\left(\frac{15\,a}{a_{\mathrm{ho}}}\right)^{2/5}. \tag{6.17}$$

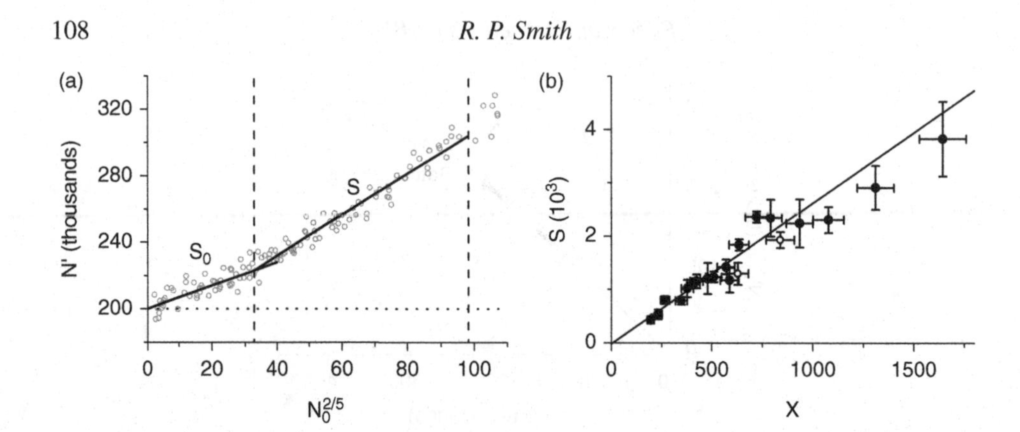

Figure 6.4 Quantifying the lack of saturation. (a) Here N' is plotted as a function of $N_0^{2/5}$ for data in Fig. 6.3. The horizontal dotted line is the saturation prediction $N' = N_c$. The two black lines show the initial slope S_0 and the slope S for $0.1 < \mu_0 / k_B T < 0.3$. (b) The nonsaturation slope S is plotted versus the dimensionless interaction parameter $X \propto T^2 a^{2/5}$ (see text). A linear fit (black line) gives $dS/dX = 2.6 \pm 0.3$ and an intercept $S(0) = -20 \pm 100$, consistent with complete saturation in the ideal-gas limit. Adapted with permission from Tammuz, N., et al. (2011), Can a Bose gas be saturated? *Phys. Rev. Lett.*, **106**, 230401 [26]. Copyright (2011) by the American Physical Society.

This first-order nonsaturation result is identical to that obtained in more elaborate MF approximations, which only modify higher-order terms.

Guided by this theory, Fig. 6.4a shows N' as a function of $N_0^{2/5}$ for the data shown in Fig. 6.3. The initial growth of N' with $N_0^{2/5}$ is seen to agree well with this mean-field prediction. Similar agreement of the initial slope, $dN'/dN_0^{2/5}|_{N_0 \to 0}$, with S_0, was observed for a wide range of interaction strength and temperature [26]. Fig. 6.4a also shows an increase of the slope for higher N_0, which was quantified by a (coarse-grained) slope $S = \Delta[N']/\Delta[N_0^{2/5}]$ for $0.1 < \mu_0 / k_B T < 0.3$ [26] and Fig. 6.4b summarises this nonsaturation slope S depends on the interaction parameter X.

The first and most important thing to notice is that both nonsaturation slopes, S_0 and S, tend to zero for $X \to 0$. These experiments thus confirmed the concept of a saturated Bose gas, and Bose-Einstein condensation as a purely statistical phase transition in the noninteracting limit.

They also highlighted the large effects that an inhomogeneous trapping potential can introduce in the presence of MF interactions (compare Figs. 6.2a and 6.3). While the majority of the observed nonsaturation could be explained by MF theory, a significant discrepancy still remained for larger N_0 [12], the origin of which is still an open question.

We have seen that in a harmonic trap the dominant nonsaturation effect is 'geometric', arising from an interplay of the mean-field repulsion and the

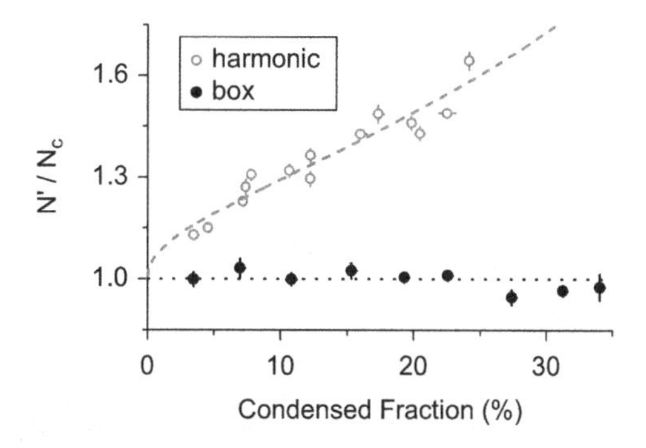

Figure 6.5 Saturation of the thermal component in a partially condensed gas of ^{87}Rb atoms. In the box trap, the gas follows the ideal-gas prediction $N' = N_c$, whereas in the harmonic trap the thermal component is strongly nonsaturated. Adapted with permission from Schmidutz, T., et al. (2014), Quantum Joule-Thomson effect in a saturated homogeneous Bose gas, *Phys. Rev. Lett.* **112**, 040403 [27]. Copyright (2014) by the American Physical Society.

inhomogeneous potential. More recently, the achievement of BEC in a uniform box potential [27] allowed the concept of saturation to be checked for a homogeneous system where this geometric effect is absent. Fig. 6.5 directly compares the harmonically trapped and homogeneous cases and clearly shows that the saturation inequality is much more closely obeyed for a homogeneous gas. The weak interaction strength ($a/\lambda = 0.006$) for these homogeneous measurements means that the expected reduction in N' seen in Fig. 6.2 is not expected to be visible over this range. In the future, it would be very interesting to examine the issue of saturation in more strongly interacting homogeneous gases.

6.4 Transition Temperature of an Interacting Bose Gas

Having considered the effect of interactions on the saturation of the thermal component, we now consider the location of the critical point itself.

The problem of the T_c shift in a harmonically trapped gas is even more complex than for the homogeneous case that we have already discussed. Now, as well as the expected (within LDA) reduction of the critical density which would act to increase the transition temperature[1] we also have an additional mean-field geometric effect that reduces T_c [29]. This negative T_c shift is due to the broadening of the density distribution by repulsive interactions. It arises due to the fact that while the chemical potential is shifted across the whole trap by $V_{\text{int}}(\mathbf{r} = 0) = 2gn_c$, the interaction

[1] For a harmonically trapped gas, the shift of the critical point can be equivalently expressed as $\Delta T_c(N)$ or $\Delta N_c(T)$, with $\Delta N_c(T)/N_c^0 \approx -3\Delta T_c/T_c^0$.

potential itself decreases with the density for $\mathbf{r} > 0$. To second order, it can be calculated analytically using MF theory [29, 30] to give

$$\frac{\Delta T_c^{\mathrm{MF}}}{T_c^0} \approx -3.426 \frac{a}{\lambda_0} + 12.9 \left(\frac{a}{\lambda_0}\right)^2, \tag{6.18}$$

where $\Delta T_c = T_c - T_c^0$ and λ_0 is the thermal wavelength defined at T_c^0. The two opposing effects of repulsive interactions on the critical point of a trapped gas are visually summarised in Fig. 6.6a, where we sketch the density distribution at the condensation point for an ideal and an interacting gas at the same temperature. For weak interactions, the MF effect is dominant, making the overall interaction shift $\Delta N_c(T)$ positive, or equivalently $\Delta T_c(N)$ negative.

The dominance of the negative MF shift of T_c over the positive beyond-MF one goes beyond the difference in the numerical factors in Eqs. (6.10) and (6.18). In a harmonic trap, at the condensation point only the central region of the cloud is close to criticality; this reduces the net effect of critical correlations so that they affect T_c only at a higher order in a/λ_0. The MF result of Eq. (6.18) should therefore be exact at first order in a/λ_0. The higher-order beyond-MF shift is still expected to be positive, but the theoretical consensus on its value has not been reached [31, 32, 33, 34, 35].

Since the early days of atomic BECs, there have been several measurements of the interaction T_c shift in a harmonically trapped gas [37, 38, 39]. These experiments nicely confirmed the theoretical prediction for the linear MF shift of Eq. (6.18), but could not discern the beyond-MF effects of critical correlations.

More recent measurements [36], shown in Fig. 6.6, which employed a Feshbach resonance, were able to discern the beyond-MF T_c-shift in a trapped atomic gas. The MF prediction agrees very well with the data for $a/\lambda_0 \lesssim 0.01$, but for larger a/λ_0 there is a clear deviation from this prediction. All the data are fitted well by a second-order polynomial

$$\frac{\Delta T_c}{T_c^0} \approx b_1 \frac{a}{\lambda_0} + b_2 \left(\frac{a}{\lambda_0}\right)^2, \tag{6.19}$$

with $b_1 = -3.5 \pm 0.3$ and $b_2 = 46 \pm 5$. The value of b_1 is in agreement with the MF prediction of -3.426 whereas b_2 strongly excludes the MF result and its larger magnitude is consistent with the expected effect of beyond-MF critical correlations.

In order to make a connection between the experiments on trapped atomic clouds and the theory of a uniform Bose gas, we also need to consider the effect of interactions on the critical chemical potential μ_c. In a uniform gas, the interactions differently affect T_c (or equivalently n_c) and μ_c at both the MF and beyond-MF

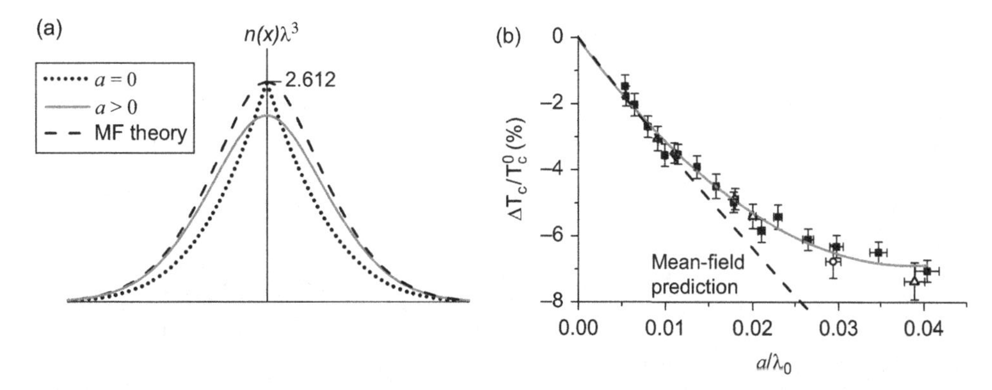

Figure 6.6 Interaction shift of the critical temperature. (a) Opposing effects of interactions on the critical point of a Bose gas in a harmonic potential. Compared with an ideal gas (dotted line) at the same temperature, repulsive interactions reduce the critical density, but also broaden the density distribution (solid line). Mean-field theory (dashed line) captures only the latter effect and predicts an increase of the critical atom number N_c at fixed T, equivalent to a decrease of T_c at fixed N. (b) Measured interaction shift of the critical temperature. The solid line shows a second-order polynomial fit to the data (see text). Figures adapted with permission from Smith, R., et al. (2011), Effects of interactions on the critical temperature of a trapped Bose gas, *Phys. Rev. Lett.* **106**, 250403 [36]. Copyright (2011) by the American Physical Society.

level. The simple MF shift $\beta\mu_c^{\text{MF}} = 4\zeta(3/2)\,a/\lambda_0$ has no effect on condensation. To lowest beyond-MF order, we have[2]

$$\beta\mu_c \approx \beta\mu_c^{\text{MF}} + B_2\left(\frac{a}{\lambda_0}\right)^2. \tag{6.20}$$

We see that there is a qualitative difference between Eqs. (6.10) and (6.20). Specifically, we have $n_c^{\text{MF}} - n_c \propto a/\lambda_0$, but $\mu_c^{\text{MF}} - \mu_c \propto (a/\lambda_0)^2$. This difference highlights the nonperturbative nature of the problem – near criticality, the equation of state does not have a regular expansion in μ, otherwise one would get $\Delta n_c \propto \mu_c - \mu_c^{\text{MF}}$.

For a harmonic trap, within LDA the uniform-system results for n_c and μ_c apply in the centre of the trap, and elsewhere the local $\mu(\mathbf{r})$ is given by Eq. (6.11). The result for the T_c shift, however, does not carry over easily to the nonuniform case. Examination of Eqs. (6.19) and (6.20) reveals that the experimentally observed T_c shift qualitatively mirrors the expected shift in μ_c. This similarity can be explained as follows: (i) The interaction shift of μ_c affects the density everywhere in the trap. (ii) Outside the small critical region, the local density shift is simply proportional to the local μ shift. (iii) The N_c (or T_c) shift from the noncritical region is thus proportional to the μ_c shift and greatly outweighs the contribution from the n_c shift

[2] Note that B_2 is not just a constant but contains logarithmic corrections in a/λ_0 [20]. We neglect these in our discussion since they are not discernible at the current level of experimental precision.

within the critical region. So the beyond-MF T_c shift observed in a trapped gas is directly related to the quadratic beyond-MF μ_c shift rather than the linear n_c shift. We are thus still lacking a direct measurement of the historically most debated n_c shift. The achievement of homogeneously trapped gases has now brought such a measurement within reach.

6.5 Critical Exponents of an Interacting Bose Gas

Having discussed the location of the critical point, we now briefly discuss the critical behaviour around that point.

The smallness of the critical region for a harmonically trapped gas places limitations on the critical behaviour that can be measured in these systems. This issue can be partly overcome by performing local measurements; such an approach was put to beautiful effect by Donner et al. [40], who used an RF out-coupling technique to measure the divergence of the correlation length close to T_c and obtained the critical exponent $\nu = 0.67 \pm 0.13$, in agreement with the expected beyond-MF exponent for the 3D XY model.

The advent of homogeneous atom traps has opened up many more possibilities for the measurement of critical phenomena. The first of these measurements for a 3D atomic Bose gas focused on the dynamics of spontaneous symmetry breaking at the BEC transition [41]. As we approach a second-order transition, the relaxation time (τ), required to establish the diverging correlation length, also diverges. This divergence is described by $\tau \sim \xi^z \sim |t|^{-\nu z}$, where z is the dynamical critical exponent. The consequence of this diverging τ is that as the transition is approached at any finite rate the system cannot adiabatically follow the diverging equilibrium ξ. As a result, the transition is crossed at a finite value of ξ, leading to the formation of domains with independent choices of the symmetry-breaking order parameter, as shown in Fig. 6.7a. In a Bose gas, this results in domains each of which is characterised by a wavefunction with a different phase. Kibble-Zurek theory describes how the lengthscale l associated with these domains scales with the speed of the quench, and predicts that $l \sim \tau_Q^b$, where τ_Q defines the quench rate across the transition via $\dot{t} = 1/\tau_Q$ and $b = \nu/(1 + \nu z)$. Beyond-MF dynamical critical theory [42] predicts $z = 3/2$; combining this with the established $\nu = 0.67$ gives $b \approx 1/3$. Measurements on a homogeneous Bose gas of ^{87}Rb atoms with $a/\lambda = 0.008$ shown Fig. 6.7 give $b = 0.35 \pm 0.04$ in agreement with this expected scaling. This work not only confirmed the expected critical exponents for the BEC transition but also provided one of the first quantitative tests for Kibble-Zurek theory.

Further discussion of our current theoretical understanding of critical behaviour and condensate formation dynamics is given in Chapter 7, whereas Kibble-Zurek in superfluid ^{3}He is discussed in Chapter 30.

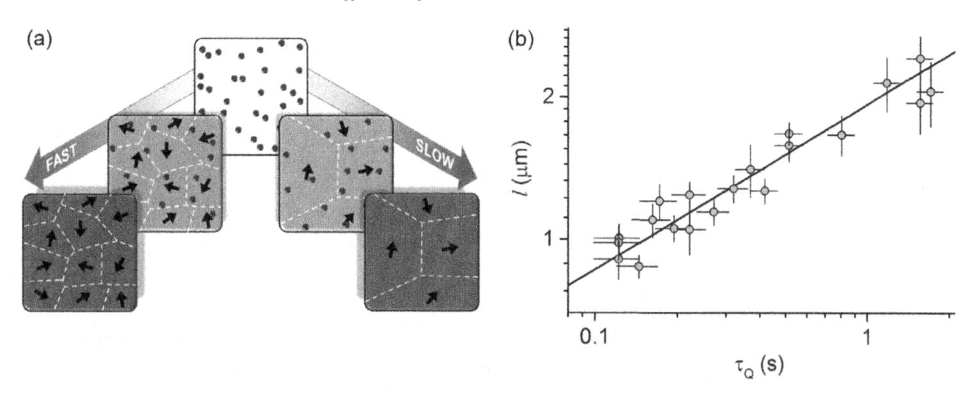

Figure 6.7 Critical exponents of the interacting BEC transition. (a) The average size of the domains formed on crossing the critical point depends on the cooling rate. (b) Scaling of domain size ($\propto l$) with quench time τ_Q. The data show the expected Kibble-Zurek scaling $l \sim t_Q^b$ with $b = 0.35(4)$ (solid line). This is in agreement with the beyond-MF prediction of $b \approx 1/3$, corresponding to $\nu \approx 0.67$ and $z = 3/2$. Figures adapted with permission from the American Association for the Advancement of Science (AAAS) from Navon, N., et al. (2015), Critical dynamics of spontaneous symmetry breaking in a homogeneous Bose gas, *Science*, **347**, 167 [41].

6.6 Conclusion and Outlook

In this chapter, we have explored the effects of weak repulsive interactions on the condensation of atomic Bose gases. We have seen that the consequences of interactions depend strongly on whether we have a homogeneous system or one that is harmonically trapped. In general, the presence of a harmonic trapping potential tends to magnify the effect of mean-field interactions while making the more interesting beyond-MF critical behaviour harder to observe. This makes the recent advances in studying Bose gases in homogeneous potentials particularly exciting. These advances promise the ability to study, in greater depth than has so far been possible, many interesting effects of interactions on both the thermodynamics and dynamics of Bose-Einstein condensation.

We have focused in this chapter on weak interactions, but a very interesting open question is what happens in the other extreme when interactions are as strong as possible. This regime, known as the unitary regime, happens when the scattering length a tends to infinity and thus ceases to be a relevant scale in the problem. At this point, the behaviour of the gas should be universal – only depending on the density. Experimental studies in this unitary regime are more difficult due to the rapid three-body losses that occur for large a, and only recently are any results beginning to emerge [43, 44, 45].

Acknowledgements: The author thanks the Royal Society for financial support and Alex Gaunt and Zoran Hadzibabic for a critical reading of the chapter.

References

[1] Bogoliubov, N. N. 1947. On the theory of superfluidity. *J. Phys. (USSR)*, **11**, 23.

[2] Penrose, O., and Onsager, L. 1956. Bose–Einstein condensation and liquid helium. *Phys. Rev.*, **104**, 576.

[3] Beliaev, S. T. 1958. *Sov. Phys. JETP*, **34**, 323.

[4] Lee, T. D., Huang, K., and Yang, C. N. 1957. Eigenvalues and eigenfunctions of a Bose system of hard spheres and its low-temperature properties. *Phys. Rev.*, **106**, 1135–1145.

[5] Lee, T. D., and Yang, C. N. 1957. Many-body problem in quantum mechanics and quantum statistical mechanics. *Phys. Rev.*, **105**, 1119–1120.

[6] Lee, T. D., and Yang, C. N. 1958. Low-temperature behavior of a dilute Bose system of hard spheres. I. Equilibrium properties. *Phys. Rev.*, **112**, 1419–1429.

[7] Huang, K., Yang, C. N., and Luttinger, J. M. 1957. Imperfect Bose gas with hard-sphere interaction. *Phys. Rev.*, **105**, 776–784.

[8] Anderson, M. H., Ensher, J. R., Matthews, M. R., Wieman, C. E., and Cornell, E. A. 1995. Observation of Bose-Einstein condensation in a dilute atomic vapor. *Science*, **269**, 198.

[9] Davis, K. B., Mewes, M. O., Andrews, M. R., van Druten, N. J., Durfee, D. S., Kurn, D. M., and Ketterle, W. 1995. Bose-Einstein condensation in a gas of sodium atoms. *Phys. Rev. Lett.*, **75**, 3969–3973.

[10] Inouye, S., Andrews, M., Stenger, J., Miesner, H. J., Stamper-Kurn, D. M., and Ketterle, W. 1998. Observation of Feshbach resonances in a Bose-Einstein condensate. *Nature*, **392**, 151.

[11] Courteille, Ph., Freeland, R. S., Heinzen, D. J., van Abeelen, F. A., and Verhaar, B. J. 1998. Observation of a Feshbach resonance in cold atom scattering. *Phys. Rev. Lett.*, **81**, 69–72.

[12] Smith, R. P., and Hadzibabic, Z. 2013. *Physics of Quantum Fluids*. New York: Springer. Chap. Effects of interactions on Bose-Einstein condensation of an atomic gas, pages 341–359.

[13] Dalfovo, F. S., Pitaevkii, L. P., Stringari, S., and Giorgini, S. 1999. Theory of Bose-Einstein condensation in trapped gases. *Rev. Mod. Phys.*, **71**, 463.

[14] Pethick, C. J., and Smith, H. 2002. *Bose-Einstein Condensation in Dilute Gases*. Cambridge: Cambridge University Press.

[15] Pitaevskii, L., and Stringari, S. 2003. *Bose-Einstein Condensation*. Oxford: Oxford University Press.

[16] Andersen, J. O. 2004. Theory of the weakly interacting Bose gas. *Rev. Mod. Phys.*, **76**(2), 599–639.

[17] Huang, K. 1987. *Statistical Mechanics*. New York: Wiley.

[18] Popov, V. N. 1987. *Functional Integrals and Collective Modes*. Cambridge: Cambridge University Press.

[19] Prokof'ev, N., Ruebenacker, O., and Svistunov, B. 2004. Weakly interacting Bose gas in the vicinity of the normal-fluid–superfluid transition. *Phys. Rev. A*, **69**, 053625.

[20] Arnold, P., and Moore, G. 2001. BEC transition temperature of a dilute homogeneous imperfect Bose gas. *Phys. Rev. Lett.*, **87**, 120401.

[21] Baym, G., Blaizot, J.-P., Holzmann, M., Laloë, F., and Vautherin, D. 2001. Bose-Einstein transition in a dilute interacting gas. *Eur. Phys. J. B*, **24**, 107–124.

[22] Holzmann, M., Fuchs, J. N., Baym, G., Blaizot, J. P., and Laloë, F. 2004. Bose-Einstein transition temperature in a dilute repulsive gas. *Comptes Rendus Physique*, **5**, 21.

[23] Kashurnikov, V. A., Prokof'ev, N. V., and Svistunov, B. V. 2001. Critical temperature shift in weakly interacting Bose gas. *Phys. Rev. Lett.*, **87**, 120402.

[24] Campostrini, M., Hasenbusch, M., Pelissetto, A., and Vicari, E. 2006. Theoretical estimates of the critical exponents of the superfluid transition in ^{4}He by lattice methods. *Phys. Rev. B*, **74**, 144506.

[25] Burovski, E., Machta, J., Prokof'ev, N., and Svistunov, B. 2006. High-precision measurement of the thermal exponent for the three-dimensional *XY* universality class. *Phys. Rev. B*, **74**, 132502.

[26] Tammuz, N., Smith, R. P., Campbell, R. L. D., Beattie, S., Moulder, S., Dalibard, J., and Hadzibabic, Z. 2011. Can a Bose gas be saturated? *Phys. Rev. Lett.*, **106**, 230401.

[27] Gaunt, A. L., Schmidutz, T. F., Gotlibovych, I., Smith, R. P., and Hadzibabic, Z. 2013. Bose-Einstein condensation of atoms in a uniform potential. *Phys. Rev. Lett.*, **110**, 200406.

[28] Schmidutz, T. F., Gotlibovych, I., Gaunt, A. L., Smith, R. P., Navon, N., and Hadzibabic, Z. 2014. Quantum Joule-Thomson effect in a saturated homogeneous Bose gas. *Phys. Rev. Lett.*, **112**, 040403.

[29] Giorgini, S., Pitaevskii, L. P., and Stringari, S. 1996. Condensate fraction and critical temperature of a trapped interacting Bose gas. *Phys. Rev. A*, **54**, R4633.

[30] Gaunt, A., and Smith, R. P. Private communication.

[31] Houbiers, M., Stoof, H. T. C., and Cornell, E. A. 1997. Critical temperature of a trapped Bose gas: mean-field theory and fluctuations. *Phys. Rev. A*, **56**, 2041.

[32] Holzmann, M., Krauth, W., and Naraschewski, M. 1999. Precision Monte Carlo test of the Hartree-Fock approximation for a trapped Bose gas. *Phys. Rev. A*, **59**, 2956–2961.

[33] Arnold, P., and Tomášik, B. 2001. T_c for trapped dilute Bose gases: a second-order result. *Phys. Rev. A*, **64**, 053609.

[34] Davis, M. J., and Blakie, P. B. 2006. Critical temperature of a trapped Bose gas: comparison of theory and experiment. *Phys. Rev. Lett.*, **96**, 060404.

[35] Zobay, O. 2009. Phase transition of trapped interacting Bose gases. *Laser Physics*, **19**, 700–724.

[36] Smith, R. P., Campbell, R. L. D., Tammuz, N., and Hadzibabic, Z. 2011. Effects of interactions on the critical temperature of a trapped Bose gas. *Phys. Rev. Lett.*, **106**, 250403.

[37] Ensher, J. R., Jin, D. S., Matthews, M. R., Wieman, C. E., and Cornell, E. A. 1996. Bose-Einstein condensation in a dilute gas: measurement of energy and ground-state occupation. *Phys. Rev. Lett.*, **77**, 4984.

[38] Gerbier, F., Thywissen, J. H., Richard, S., Hugbart, M., Bouyer, P., and Aspect, A. 2004. Critical temperature of a trapped, weakly interacting Bose gas. *Phys. Rev. Lett.*, **92**, 030405.

[39] Meppelink, R., Rozendaal, R. A., Koller, S. B., Vogels, J. M., and van der Straten, P. 2010. Thermodynamics of Bose-Einstein–condensed clouds using phase-contrast imaging. *Phys. Rev. A*, **81**, 053632.

[40] Donner, T., Ritter, S., Bourdel, T., Ottl, A., Köhl, M., and Esslinger, T. 2007. Critical behavior of a trapped interacting Bose gas. *Science*, **315**(5818), 1556–1558.

[41] Navon, N., Gaunt, A. L., Smith, R. P., and Hadzibabic, Z. 2015. Critical dynamics of spontaneous symmetry breaking in a homogeneous Bose gas. *Science*, **347**(6218), 167–170.

[42] Hohenberg, P. C., and Halperin, B. I. 1977. Theory of dynamic critical phenomena. *Rev. Mod. Phys.*, **49**(Jul), 435–479.

[43] Rem, B. S., Grier, A. T., Ferrier-Barbut, I., Eismann, U., Langen, T., Navon, N., Khaykovich, L., Werner, F., Petrov, D. S., Chevy, F., and Salomon, C. 2013. Lifetime of the Bose gas with resonant interactions. *Phys. Rev. Lett.*, **110**, 163202.

[44] Fletcher, R. J., Gaunt, A. L., Navon, N., Smith, R. P., and Hadzibabic, Z. 2013. Stability of a unitary Bose gas. *Phys. Rev. Lett.*, **111**, 125303.

[45] Makotyn, P., Klauss, C. E., Goldberger, D. L., Cornell, E. A., and Jin, D. S. 2014. Universal dynamics of a degenerate unitary Bose gas. *Nature Physics*, **10**, 116–119.

7

Formation of Bose-Einstein Condensates

MATTHEW J. DAVIS
School of Mathematics and Physics, The University of Queensland, Australia
JILA, University of Colorado, Boulder, USA

TOD M. WRIGHT
School of Mathematics and Physics, The University of Queensland, Australia

THOMAS GASENZER
Kirchhoff-Institut für Physik, Ruprecht-Karls-Universität Heidelberg, Germany
ExtreMe Matter Institute, GSI, Darmstadt, Germany

SIMON A. GARDINER
Joint Quantum Centre (JQC) Durham-Newcastle, Durham University, UK

NICK P. PROUKAKIS
Joint Quantum Centre (JQC) Durham-Newcastle, Newcastle University, UK

The problem of understanding how a coherent, macroscopic Bose-Einstein condensate (BEC) emerges from the cooling of a thermal Bose gas has attracted significant theoretical and experimental interest over several decades. The pioneering achievement of BEC in weakly interacting dilute atomic gases in 1995 was followed by a number of experimental studies examining the growth of the BEC number, as well as the development of its coherence. More recently, there has been interest in connecting such experiments to universal aspects of nonequilibrium phase transitions, in terms of both static and dynamical critical exponents. Here, the spontaneous formation of topological structures such as vortices and solitons in quenched cold-atom experiments has enabled the verification of the Kibble-Zurek mechanism predicting the density of topological defects in continuous phase transitions, first proposed in the context of the evolution of the early universe. This chapter reviews progress in the understanding of BEC formation and discusses open questions and future research directions in the dynamics of phase transitions in quantum gases.

7.1 Introduction

The equilibrium phase diagram of the dilute Bose gas exhibits a continuous phase transition between condensed and noncondensed phases. The order parameter characteristic of the condensed phase vanishes above some critical temperature T_c and grows continuously with decreasing temperature below this critical point. However, the dynamical process of condensate formation has proved to be a challenging phenomenon to address both theoretically and experimentally. This formation process is a crucial aspect of Bose systems and of direct relevance to all condensates discussed in this book, despite their evident system-specific properties. Important questions leading to intense discussions in the early literature include the time scale for condensate formation and the role of inhomogeneities and finite-size effects in "closed" systems. These issues are related to the concept of spontaneous symmetry breaking, its causes, and implications for physical systems (see, for example, Chapter 5 by Snoke and Daley).

In this chapter, we give an overview of the dynamics of condensate formation and describe the present understanding provided by increasingly well-controlled cold-atom experiments and corresponding theoretical advances over the past twenty years. We focus on the growth of BECs in cooled Bose gases, which, from a theoretical standpoint, requires a suitable nonequilibrium formalism. A recent book provides a more complete introduction to a number of different theoretical approaches to the description of nonequilibrium and nonzero-temperature quantum gases [1].

The past decade has seen the observation of BEC in a number of diverse experimental systems beyond ultracold atoms, including exciton-polaritons, photons, and magnons, which are covered in other chapters of this volume (see, e.g., Chapters 4, 19, and 25–26, respectively). Many of the universal aspects of condensate formation also apply to these systems.

7.2 The Physics of BEC Formation

The essential character of the excitations and collective response of a condensed Bose gas is well described by perturbative approaches that take as their starting point the breaking of the $U(1)$ gauge symmetry of the Bose quantum field. This approach can be extended further to provide a kinetic description of excitations in a condensed gas weakly perturbed away from equilibrium [2]. The description of the process of formation of a Bose-Einstein condensate in a closed system begins, however, in the opposite regime of kinetics of a noncondensed gas. Over the past decades, there have been many studies using methods of kinetic theory to investigate the initiation of Bose-Einstein condensation. It is now well established that these descriptions break down near the critical point, and in particular in any situation in which the formation process is far from adiabatic. A number of different theoretical methodologies have been applied to the issue of condensate formation, but most have converged to a similar description of the essential physics. The prevailing view is that a classical nonlinear wave description – a form of Gross-Pitaevskii equation – can describe the nonequilibrium dynamics of the condensation process, which involves in general aspects of weak-wave turbulence and, in more aggressive cooling scenarios, strong turbulence. The classical field describes the highly occupied modes of the gas at finite temperature and out of equilibrium.

A summary of the consensus picture of condensate formation in a Bose gas cooled from above the critical temperature is as follows. Well above the critical point, the coherences between particles in distinct eigenstates of the appropriate single-particle Hamiltonian are negligible and the system is well described by a quantum Boltzmann kinetic equation for the occupation numbers of these single-particle modes. As cooling of the gas proceeds due to interparticle collisions and interactions with an external bath, if one is present, the occupation numbers of lower-energy modes increase. Once phase correlations between these modes become significant, the system is best described in terms of an emergent quasiclassical field, which may in general exhibit large phase fluctuations, topological structures, and turbulent dynamics, the nature of which may vary over time and depend on the specific details of the system – including its dimensionality, density, and strength of interactions. This regime is sometimes referred to as a nonequilibrium *quasicondensate*, in analogy to the phase-fluctuating equilibrium regimes of low-dimensional Bose systems [3, 4]. The eventual relaxation of this quasicondensate establishes phase coherence across the sample, producing the state that we routinely call a Bose-Einstein condensate.

As a final aside, we note that while one might be led to believe that use of the Gross-Pitaevskii equation is equivalent to assuming that $U(1)$ gauge symmetry (and therefore conservation of the total particle number) is broken, it is in fact more subtle than this. It is quite possible to derive a Gross-Pitaevskii equation, in both its time-dependent and time-independent forms, while formally having perfect global number conservation [5, 6, 7]. The Gross-Pitaevskii wavefunction itself is then defined as something that is in principle different from that deduced from symmetry breaking (clearly it cannot be the expectation value of a field operator). At the level of describing the whole Bose-condensed system with a single Gross-Pitaevskii wavefunction, the equations of motion are identical; however, differences become apparent when, for example, accounting for quantum fluctuations beyond this lowest-order description.

7.2.1 The Precondensation Kinetic Regime

Early investigations of the kinetics of condensation of a gas of massive bosons began with studies of such a system coupled to a thermal bath with infinite heat capacity, consisting of phonons [8] or fermions [9, 10, 11]. These works inherited ideas from earlier studies of condensation of photons in cosmological scenarios [12]. In a homogeneous system, condensation is signified by a delta-function singularity of the momentum distribution at zero momentum (see, e.g., Ref. [13]). Levich and Yakhot found [9] that an initially nondegenerate equilibrium ideal Bose gas brought in contact with a bath with a temperature below T_c would develop such a singularity at zero momentum only in the limit of an infinite evolution time (see also Ref. [14]). These same authors subsequently found that the introduction of collisions between the bosons led to the "explosive" development of a singular peak at small momenta after a finite evolution time [10, 11]. They pointed out, in particular, that already in the precondensation regime the appearance of a growing and narrowing low-wavenumber peak leads to the eventual invalidation of the Boltzmann-equation description, after which the further ordering and condensation process is more appropriately described by the nonlinear Gross-Pitaevskii equation for a classical field [11]. They were also careful to point out that, since the development of coherence invalidates the assumptions underlying the quantum Boltzmann description, "the system in the course of phase transition passes through a stage which may be identified as a period of strong turbulence" [11].

Experimental attempts in the 1980s to achieve Bose condensation of spin-polarized hydrogen (see Chapter 2 by Greytak and Kleppner for an overview and recent developments) and excitons in semiconductors such as Cu_2O inspired renewed theoretical interest in Bose-gas kinetics. Eckern developed a kinetic theory [15] for Hartree-Fock-Bogoliubov quasiparticles appropriate to the relaxation of the system on the condensed side of the transition. Snoke and Wolfe revisited the question of the kinetics of approach to the condensation transition by undertaking numerical calculations of the quantum Boltzmann equation [16]. They found in particular that the bosonic enhancement of scattering rates in the degenerate regime offset the increased number of scattering events required for rethermalization in this regime, such that re-equilibration of a shock-cooled thermal distribution takes place on the order of three to four kinetic collision times, $\tau_{\text{kin}} = (\rho\sigma v_T)^{-1}$, where ρ is the particle density, σ is the collisional cross section, and the mean thermal velocity $v_T = (3k_B T/m)^{1/2}$. These results imply that a Boltzmann-equation description of this early kinetic regime is valid even for short-lived particles such as excitons, as the particle lifetime is long compared with this equilibration time scale.

Over time a comprehensive picture of the process of condensation of a quench-cooled gas has emerged, and comprises three distinct stages of nonequilibrium dynamics: a kinetic redistribution of population towards lower energy modes in the noncondensed phase, development of an instability that leads to nucleation of the condensate and a subsequent buildup of coherence, and finally condensate growth and phase ordering. In the midst of increasingly intensive efforts to achieve Bose condensation in dilute atomic gases, by then including the new system of alkali-metal vapors, these stages were analyzed in more detail in the early 1990s, beginning with a series of papers by Stoof [17, 18, 19, 20, 21] and by Svistunov, Kagan, and Shlyapnikov [22, 23, 24, 25].

In Ref. [22], Svistunov discussed condensate formation in a weakly interacting, dilute Bose gas, with so-called gas parameter $\zeta = \rho^{1/3}a \ll 1$, where a is the scattering length. In a closed system, a cooling quench generically leads to a particle distribution which, below some energy scale ε_0, exceeds the equilibrium occupation number corresponding to the total energy and particle content. Energy and momentum conservation then imply that a few particles scattered to high-momentum modes carry away a large fraction of the excess energy associated with this over-occupation, allowing the momentum of a majority of the particles to decrease. Should the characteristic energy scale of the overpopulated regime be sufficiently small, $\varepsilon_0 \ll \hbar^2\rho^{2/3}/m \sim k_BT_c$, mode-occupation numbers in this regime will be much larger than unity, and the subsequent particle transport in momentum space towards lower energies is described by the quantum Boltzmann equation in the classical-wave limit [22, 23, 24]. This is valid for modes with energies above the scale set by the chemical potential $\mu = g\rho \sim \zeta k_BT_c$ of the ultimate equilibrium state, where $g = 4\pi\hbar^2a/m$ is the interaction constant for particles of mass m. At lower energies, the phase correlations between momentum modes become significant, and a description beyond the quantum Boltzmann equation is required.

We note that for open systems such as exciton-polariton condensates (reviewed by Littlewood and Edelman in Chapter 4), the quasicoherent dynamics of such low-energy modes will in general be sensitive to the driving and dissipation corresponding to the continual decay and replenishment of the bosons. Such external coupling can dramatically alter the behavior of the system, and its effects on condensate formation dynamics are a subject of current research – see, e.g., Refs. [26, 27, 28] and Chapter 11 by Keeling et al. Hereafter, unless otherwise specified, the theoretical developments we discuss pertain to closed systems in which the bosons undergoing condensation are conserved in number during the formation process.

By assuming the scattering matrix elements in the wave Boltzmann equation to be independent of the mode energies, Svistunov discussed several different transport scenarios within the framework of weak-wave turbulence, in analogy to similar processes underlying Langmuir-wave turbulence in plasmas [29]. He concluded that the initial kinetic transport stage of the condensation process evolves as a weakly nonlocal particle wave in momentum space. Specifically, he proposed that the particle-flux wave followed the self-similar form $n(\varepsilon, t) \sim \varepsilon_1(t)^{-7/6}f(\varepsilon/\varepsilon_1(t))$, with $\varepsilon_1(t) \sim (t - t_*)^3$, and scaling function f falling off as $f(x) \propto x^{-\alpha}$ for $x \gg 1$, with $\alpha = 7/6$. Following the arrival of this wave at time $t_* \simeq t_0 + \hbar\varepsilon_0/\mu^2$, a quasistationary wave-turbulent cascade forms in which particles are transported locally, from momentum shell to momentum shell, from the scale ε_0 of the energy concentration in the initial state to the low-energy regime $\varepsilon \lesssim \mu$, where coherence formation sets in.

The wave-kinetic (or weak-wave turbulence) stage of condensate formation following a cooling quench was investigated in more detail by Semikoz and Tkachev [30, 31], who solved the wave Boltzmann equation numerically and found results consistent with the above scenario, albeit with a slightly shifted power-law exponent $\alpha \simeq 1.24$ for the wave-turbulence spectrum. Later dynamical classical-field simulations of the condensation formation process by Berloff and Svistunov [32] further corroborated the above picture.

7.2.2 The Formation of Coherence and Condensation

It has been known for some time that a kinetic Boltzmann equation model is unable to describe the development of a macroscopic zero-momentum occupation in the absence of seeding or other modifications [9, 10, 11, 16, 22]. In any event, the quantum Boltzmann equation ceases to be valid in the high-density, low-energy regime in which condensation occurs. The two-body scattering receives significant many-body corrections once the interaction energy $g\int_{k\lesssim p} \mathrm{d}\mathbf{k}\, n_k$ of particles with momenta below a given scale p exceeds the kinetic energy at that scale, and these are indeed the prevailing conditions when phase coherence emerges and the condensate begins to grow [22].

In a series of papers [17, 18, 20, 21], Stoof took account of these many-body corrections and developed a theory of condensate formation resting on kinetic equations incorporating a ladder-resummed many-body T-matrix determined from a one-particle-irreducible (1PI) effective-action or free-energy functional. In the 1PI formalism, the propagators appearing in the effective action are taken as fixed, determined in this case by the initial thermal Bose number distribution and the spectral properties of a free gas.

Constructed within the Schwinger-Keldysh closed-time-path framework, the method allows the determination of the time evolution of the self-energy and thus of an effective chemical potential for the zero-momentum mode through the phase transition. During the kinetic stage, once the system has reached temperatures below the interaction-renormalized critical temperature, the self-energy renders the vacuum state of the zero-momentum mode metastable. Stoof found that this modification of the self-energy occurs on a time scale $\sim \hbar/k_\mathrm{B}T_\mathrm{c}$ and gives rise to a small seed population in the zero mode, $n_0 \sim \zeta^2\rho$, within the coherence time scale $\tau_\mathrm{coh} \sim \hbar/(\zeta^2 k_\mathrm{B}T_\mathrm{c})$. He argued that, following this seeding, the system undergoes an unstable semiclassical evolution of the low-energy modes. Taking interactions between quasiparticles into account, he found that the squared dispersion $\omega(\mathbf{p})^2$ becomes negative for $p \lesssim \hbar\sqrt{an_0(t)}$, i.e., below a momentum scale of the order of the inverse healing length associated with the density $n_0(t)$ of the existing condensed fraction. As a result, the condensate grows linearly in time over the kinetic time scale τ_{kin}. The growth process eventually ceases due to the conservation of total particle number, whereafter the final kinetic equilibration of quasiparticles takes place over a time scale $\sim \hbar/(\zeta^3 k_B T_c)$ as discussed previously by Eckern [15] and by Semikoz and Tkachev [31].

Turbulent Condensation

The semiclassical scenario of Stoof is built on the assumptions that the cooling quench has driven the system to the critical point in a quasiadiabatic fashion, and that the neglect of thermal fluctuations and nonequilibrium over-occupations in the self-energy is justified [21]. However, as previously pointed out in Refs. [10, 11], in general a peak in the low-momentum regime will grow first, reflecting the formation of coherence and rendering the further condensation an essentially classical process. Moreover, a more vigorous quench may drive the system into an intermediate stage of strong turbulence [11], where the coherences between wave frequencies lead to the formation of coherent structures, such as vortices, that have a significant influence on the subsequent dynamics. The main processes and scales governing this stage were discussed in detail by Kagan and Svistunov [24, 25]. As a result of excess particles being transported kinetically into the coherent regime (wavenumbers below the final inverse healing length, $k \lesssim \xi^{-1} \sim \sqrt{a\rho}$), the density and phase of the Bose field fluctuate strongly on length scales shorter than ξ. The growing population at even smaller wavenumbers then implies, according to Refs. [22, 23, 24], the formation of a *quasicondensate* over the respective length scales, as the coherent evolution of the field according to the Gross-Pitaevskii equation causes the density fluctuations to strongly decrease at the expense of phase fluctuations. This short-range phase-ordering occurs on a time scale $\tau_c \sim \hbar/\mu \sim \hbar/(\zeta k_B T_c)$. Depending on the flux of excess particles entering the coherent regime, this leads to quasicondensate formation over a minimum length scale $l_v > \xi$ (see Sections 7.4.1 and 7.4.2) [33]. The phase, however, remains strongly fluctuating on larger length scales due to the formation of topological defects – vortex lines and rings. These vortices appear in the form of clumps of strongly tangled filaments [34] with an average distance between filaments of order l_v. If the cooling quench is sufficiently strong to drive the system near a nonthermal fixed point, cf. Section 7.4.2, this quasicondensate is characterized by new universal scaling laws in space and time.

The work of Kagan and Svistunov laid the foundations for studying the role of superfluid turbulence in the process of Bose-Einstein condensation. Kozik and Svistunov have subsequently elucidated the decay of the vortex tangle via the transport of Kelvin waves created on the vortex filaments through their reconnections, which can itself assume a wave-turbulent structure [35, 36, 37, 38]. (The related topic of controlled studies of superfluid, or quantum, turbulence and its decay is reviewed in Chapter 17.)

7.3 Condensate Formation Experiments

7.3.1 Growth of Condensate Number

We now provide a historical overview of both experiments and theory related to condensate formation in ultracold atomic gases. The first experiments to achieve Bose-Einstein condensation in 1995 [39, 40] reached the phase-space density necessary for quantum degeneracy using the technique of evaporative cooling [41] – the steady removal of the most energetic atoms, followed by rethermalization to a lower temperature via atomic collisions. These experiments, which concentrated on the BEC atom number as the conceptionally simplest observable, provided an indication of the time scale for condensation in trapped atomic gases, in the range of milliseconds to seconds. This gave the impetus for the development of a quantum kinetic theory by Gardiner and Zoller using the techniques of open quantum systems. They first considered the homogeneous Bose gas [42] before extending the formalism to trapped gases [43, 44]. Their methodology split the system into a "condensate band," containing modes significantly affected by the presence of a BEC, and a "noncondensate band," containing all other levels. A master equation was derived for the condensate band using standard techniques [45], yielding equations of motion for the occupations of the condensate mode and the low-lying excited states contained in the condensate band. A simple BEC growth equation derived from this approach provided a reasonable first estimate of the time of formation for the ^{87}Rb and ^{23}Na BECs of the Joint Institute for Laboratory Astrophysics (JILA) [39] and Massachusetts Institute of Technology (MIT) [40] groups, respectively.

The first experiment to explicitly study the formation dynamics of a BEC in a dilute weakly interacting gas was performed by the Ketterle group at MIT, using its newly developed technique of nondestructive imaging [46]. Beginning with an equilibrium gas just above the critical temperature, the group performed a sudden evaporative cooling "quench" by removing all atoms above a certain energy. The subsequent evolution led to the formation of a condensate, with a characteristic S-shaped curve for the growth in condensate number. This was interpreted as evidence of bosonic stimulation in the growth process, and they fitted the simple

BEC growth equation of Ref. [47] to their experimental observations. However, the measured growth rates did not fit the theory all that well.

Gardiner and coworkers subsequently developed an expanded rate-equation approach incorporating the dynamics of a number of quasiparticle levels [48, 49]. This formalism predicted faster growth rates, mostly due to the enhancement of collision rates by bosonic stimulation, but still failed to agree with the experimental data. One limitation of this approach was that it neglected the evaporative cooling dynamics of the thermal cloud, instead treating it as being in a supersaturated thermal equilibrium.

The details of the evaporative cooling were simulated in two closely related works by Davis et al. [50] and Bijlsma et al. [51]. The former was based on the quantum kinetic theory of Gardiner and Zoller, while the latter emerged as a limit of the field-theoretical approach of Stoof [20, 21] and the "ZNG" formalism previously developed for nonequilibrium trapped Bose gases [52] by Zaremba, Nikuni, and Griffin. The latter authors used a broken-symmetry approach to derive a quantum Boltzmann equation for noncondensed atoms coupled to a Gross-Pitaevskii equation for the condensate [52, 53], thereby extending their two-fluid model for trapped BECs [53], which was based on the pioneering work of Kirkpatrick and Dorfman [54, 55, 56, 57]. The ZNG methodology has since been used successfully and extensively to study a variety of nonequilibrium phenomena in partially condensed Bose gases, such as the temperature dependence of collective excitations, as reviewed in Ref. [2]. As this methodology is explicitly based on symmetry breaking, it cannot address the initial seeding of a BEC or any critical physics arising from fluctuations. However, it can model continued growth once a BEC is present.

The works of Davis et al. [50] and Bijlsma et al. [51] both introduced approximations to the formalisms they were built on, assuming that the condensate grew adiabatically in its ground state, and treating all noncondensed atoms in a Boltzmann-like approach. Both papers boiled down to simulating the quantum Boltzmann equation in the ergodic approximation, in which the phase-space distribution depends on the phase-space variables only through the energy [58]. Despite the different approaches, the calculations were in excellent agreement with one another – yet still quantitatively disagreed with the MIT experimental data [46]. This disagreement has remained unexplained.

A second study of evaporative cooling to BEC in a dilute gas was performed by the group of Esslinger and Hänsch in Munich [59]. In this experiment, the Bose cloud, which was again initially prepared in an equilibrium state slightly above T_c, was subjected to a continuous radio-frequency (rf) field, inducing the ejection of high-energy atoms from the sample. By adjusting the frequency of the applied field and thus the energies of the removed atoms, these authors were

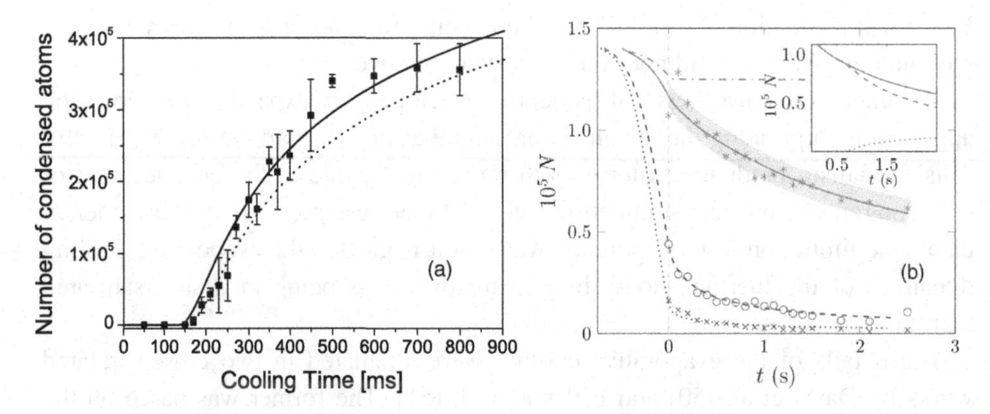

Figure 7.1 (a) Growth of an atomic Bose-Einstein condensate modeled with the quantum Boltzmann equation. The experiment began with a ^{87}Rb Bose gas in an elongated harmonic trap with $N_\mathrm{i} = (4.2 \pm 0.2) \times 10^6$ atoms at a temperature of $T_\mathrm{i} = (640 \pm 30)$ nK, before turning on rf evaporative cooling with a truncation energy of $1.4k_\mathrm{B}T$. The solid and dotted lines show the theoretical calculations with a starting number of $N_\mathrm{i} = 4.2 \times 10^6$ and $N_\mathrm{i} = 4.4 \times 10^6$ atoms, respectively. Figure reprinted with permission from Köhl, M., et al. (2002), Growth of Bose-Einstein condensates from thermal vapor, *Phys. Rev. Lett.*, **88**, 080402 [59]. Copyright (2002) by the American Physical Society. (b) Surface evaporation leading to the formation of a BEC, showing the total atom number for three different cloud-surface distances. The lines are the results of ZNG simulations, the points are from experiment. The dot-dashed, line is for a ZNG simulation neglecting collisions in the thermal cloud, demonstrating that modeling the full dynamics of the thermal cloud is necessary for a quantitative understanding of the experiment. The inset shows the total number, thermal cloud number, and condensate number, from top to bottom, respectively, as a function of time. Figure reprinted with permission from Märkle, J., et al. (2014), Evaporative cooling of cold atoms at surfaces, *Phys. Rev. A*, **90**, 023614 [68]. Copyright (2014) by the American Physical Society.

able to investigate the growth of the condensate for varying rates of evaporative cooling. Davis and Gardiner extended their earlier approach [50] to include the effects of three-body loss and gravitational sag on the cooling of the ^{87}Rb cloud in this experiment [60]. Their calculations yielded excellent agreement with the experimental data of Ref. [59] within its statistical uncertainty for all but the slowest cooling scenarios considered. An example is shown in Fig. 7.1a.

In 1997, Pinkse et al. [61] experimentally demonstrated that adiabatically changing the trap shape could increase the phase-space density of an atomic gas by up to a factor of two and conjectured that this effect could be exploited to cross the BEC transition in a thermodynamically reversible fashion. This scenario was subsequently realized in the MIT group by Stamper-Kurn et al. [62] by slowly ramping on a tight "dimple" trap formed from an optical dipole potential on top of a weaker harmonic magnetic trap. This experiment was the setting for the first application of a stochastic Gross-Pitaevskii methodology [63], previously developed from a nonequilibrium formalism for Bose gases by Stoof [21]. This is based on the many-body T-matrix approximation, and uses the Schwinger-Keldysh path integral formulation of nonequilibrium quantum field theory to derive a Fokker-Planck equation for both the coherent and incoherent dynamics of a Bose gas. The classical modes of the gas were represented by a Gross-Pitaevskii equation, with additional dissipative and noise terms resulting from a collisional coupling to a thermal bath with a temperature T and chemical potential μ.

Proukakis et al. [64] subsequently used this methodology to study the formation of quasicondensates in a one-dimensional dimple trap. A much later experiment [65] investigated the dynamics of condensate formation following the sudden introduction of a dimple trap and included quantum-kinetic simulations that were in good agreement with the data.

A novel method of cooling a bosonic cloud to condensation was introduced in 2003 by the Cornell group at JILA [66], who demonstrated the evaporative cooling of an atomic Bose cloud brought in close proximity to a dielectric surface due to the selective adsorption of high-energy atoms. More recently, similar experiments have been undertaken by the Durham [67] and Tübingen groups [68], with the observed rates of loss in the latter case explained accurately by nonergodic ZNG-method calculations of the evaporative cooling dynamics. Example results are shown in Fig. 7.1b.

7.3.2 Other Theories for Condensate Formation

For completeness, here we briefly outline other theoretical methods that can be applied to condensate formation. A generalized kinetic equation for thermally excited Bogoliubov quasiparticles was obtained by Imamovic-Tomasovic and Griffin [69] based on the application of the Kadanoff-Baym nonequilibrium Green's function approach [70] to a trapped Bose gas. This kinetic equation reduces to that of Eckern [15] in the homogeneous limit and to that of ZNG [52] when the quasiparticle character of the excitation spectrum is neglected. Walser et al. [71, 72] derived a kinetic theory for a weakly interacting condensed Bose gas in terms of a coarse graining of the N-particle density operator over configurational variables. Neglecting short-lived correlations between colliding atoms in a Markov approximation, they obtained kinetic equations for the condensate and noncondensate mean fields, which were subsequently shown to be microscopically equivalent [73] to the nonequilibrium Green's function approach of Ref. [69]. Exactly the same kinetic equations were derived by Proukakis [74], within the formalism of his earlier quantum-kinetic formulation [75, 76], based on the adiabatic elimination of rapidly evolving averages of noncondensate operators, ideas which fed into the development of the ZNG kinetic model [77]. Although elegant, these formalisms have not provided a tractable computational methodology for modeling condensate formation away from the quasistatic limit. Barci et al. [78] have applied nonequilibrium field theory to study the instability leading to condensate formation in a homogeneous gas.

A nonperturbative method for the many-body dynamics of the Bose gas far from equilibrium has been developed by Berges, Gasenzer, and co-workers [79, 80, 81, 82]. This two-particle irreducible (2PI) effective-action approach provides a systematic way to derive approximate Kadanoff-Baym equations consistent with conservation laws such as those for energy and particle number. In contrast to 1PI methods, single-particle correlators are determined self-consistently by these equations. This approach allows the description of strongly correlated systems and has been exploited in the context of turbulent condensation [83, 84, 85], where it provides a self-consistently determined many-body T matrix. The 2PI effective-action approach is useful for studying strongly interacting systems such as one-dimensional (1D) gases with large coupling constants [86], or relaxation and (pre-)thermalization of strongly correlated spinor gases [87].

7.3.3 Other Pioneering Condensate-Formation Experiments

There are a number of experimental methods other than evaporative cooling to increase the phase-space density of a quantum gas and form a condensate. We briefly mention them here for completeness.

An experimental technique that has proved to be extremely useful for multicomponent quantum gases is the method of sympathetic cooling, in which an atomic gas is cooled by virtue of its collisional interaction with a second gas of atoms, distinguished from the first either isotopically or by internal quantum numbers, which is itself subject to, e.g., evaporative cooling. This technique was first demonstrated by Myatt et al. [88] in a gas comprising two distinct spin states of ^{87}Rb, and was subsequently employed to cool a single-component Fermi gas to degeneracy by Schreck et al. [89].

In a similar spirit, in 2009 the Inguscio group in Florence used entropy exchange between components of a two-species ^{87}Rb-^{41}K Bose gas mixture to induce BEC in one of the components [90]. The two gases were brought close to degeneracy by cooling, after which the strength of the ^{41}K trapping potential was adiabatically increased, by introducing an optical dipole potential to which the ^{87}Rb component was largely insensitive. In a single-component system, this would lead to an increase in the temperature and leave the phase-space density unaffected. However, in the dual-species setup the ^{87}Rb cloud acted as a thermal reservoir, suppressing the temperature increase of the ^{41}K component and causing it to cross the BEC threshold.

In 2004, the Sengstock group observed the formation of a BEC at constant temperature [91]. Working with a spin-1 system, they prepared a partially condensed gas consisting of $m_F = \pm 1$ states. Spin collisions within the BEC components populated the $m_F = 0$ state, which then quickly thermalized. When the population of the $m_F = 0$ component reached the critical number, a new BEC emerged. The experiment was modeled with a simple rate equation.

In the same year, Ketterle's MIT group performed an experiment in which they distilled a BEC from one trap minimum to another [92]. A nonzero-temperature BEC was formed in an optical dipole trap, before a second trap with a greater potential depth was brought nearby. Atoms of sufficient thermal energy were able to cross the barrier between the two potential minima, populating the second trap. Eventually the first condensate evaporated, and a second condensate formed in the new global trap minimum.

Finally, we mention a recent experiment by the group of Schreck at Innsbruck, who demonstrated the first experimental production of a BEC solely by laser cooling [93]. This feat was made possible by laser cooling on a narrow-linewidth transition of ^{84}Sr, resulting in a low Doppler-limit temperature of just 350 nK. A "light-shift" laser beam was introduced at the center of the trap so that the atoms in that region no longer responded to the laser cooling, after which an additional dimple trap was introduced to confine the atoms. Repeatedly cycling the dimple trap on and off resulted in the formation of several condensates [93].

7.3.4 Low-Dimensional Bose Systems and Phase Fluctuations

The experiments described above were in the three-dimensional (3D) realm, in which long-wavelength phase fluctuations are strongly suppressed away from the vicinity of the phase transition. In lower-dimensional systems, such fluctuations are enhanced, leading to dramatic modifications to the physics of the degenerate regime. In a two-dimensional (2D) system, thermal fluctuations of the phase erode the long-range order associated with true condensation, leaving only so-called quasi–long-range order characterized by correlation functions that decay algebraically with spatial separation [3]. A more complete analysis reveals the importance of vortex-antivortex pairs in this phase-fluctuating "quasicondensed" regime [94]. Such pairs undergo a so-called Berezinskii-Kosterlitz-Thouless (BKT) deconfinement transition at some finite temperature, above which even quasi–long-range order is lost and superfluidity is extinguished. Two-dimensional Bose systems are of particular interest due to their natural realization in systems such as liquid-helium films and the fact that the degenerate Bose quasiparticles such as excitons and polaritons in semiconductor systems are typically confined in a planar geometry. An insightful overview of BKT physics can be found in Chapter 10 by Kim, Nitsche, and Yamamoto.

There have been numerous experimental realizations of (quasi-)2D Bose gases in cold-atom experiments [95, 96, 97, 98, 99, 100], with most notable the observations of thermally activated vortices via interferometric measurements [95] and the direct probing of the equation of state and scale invariance of the 2D system [100] (see Chapter 9 by Chin and Refs. [101, 102, 103, 104, 105] for related theoretical considerations). Further details and a lengthy discussion of the interplay between BKT and BEC in homogeneous and trapped systems can be found in Ref. [106]. Although theoretical works on the dynamics of such systems have existed for some time, little experimental work on the formation dynamics of condensates in these systems has been undertaken (aside from the quasi-2D Kibble-Zurek works discussed in the following section). Considerable discussion is currently taking place regarding the emergence and nature of the BKT transition in driven-dissipative polariton condensates: experimentalists have observed evidence for quasi–long-range order [107, 108] (see Chapter 10), but the nature of the transition and its nonequilibrium features are topics of current debate [27, 28] (see also Chapter 11 by Keeling et al.).

In one dimension, the effects of phase fluctuations are even more pronounced, leading to the complete destruction of long-range order and superfluidity at any finite temperature. Many experiments with cold atoms in elongated "cigar-shaped" traps have investigated the physics of such (quasi-)one-dimensional systems, though again, little work has been done on the formation dynamics of these degenerate samples. We note, however, that quasicondensate regimes somewhat analogous to those of (quasi-)one-dimensional systems can be realized in elongated 3D traps [109]. In such a regime, the Bose gas behaves much as a conventional three-dimensional Bose condensate, except that the coherence length of the gas is shorter than the system extent along the long axis of the trap. A study of condensate formation in this regime was performed by the Amsterdam group of Walraven [110] in 2002 in an elongated ^{23}Na cloud. Similar to the MIT experiment [46], they performed rapid quench cooling of their sample from just above the critical temperature. However, the system was in the hydrodynamic regime in the weakly trapped dimension; i.e., the mean distance between collisions was much shorter than the system length.

It was argued that the system rapidly came to a local thermal equilibrium in the radial direction, resulting in cooling of the cloud below the *local* degeneracy temperature over a large spatial region and generating an elongated quasicondensate. However, the extent of this quasicondensate along the long axis of the trap was larger than that expected at equilibrium, leading to large amplitude oscillations. The momentum distribution of the cloud was imaged via "condensate focusing," with the breadth of the focal point giving an indication of the magnitude of the phase fluctuations present in the sample. This interesting experiment was somewhat ahead of its time, with theoretical techniques unable to address many of the nonequilibrium aspects of the problem.

In 2007, the group of Aspect from Institut d'Optique also studied the formation of a quasicondensate in an elongated three-dimensional trap via continuous evaporative cooling [111] in a similar fashion to the earlier work by Köhl et al. [59]. As well as measuring the condensate number, they also performed Bragg spectroscopy during the growth to determine the momentum width and hence the coherence length of the system. They found that the momentum width they measured rapidly decreased with time to the width expected in equilibrium for the instantaneous value of the condensate number. Modeling of the growth of the condensate population using the methodology of Ref. [60] produced results in good agreement with the experimental data, apart from an unexplained delay of 10–50 ms, depending on the rate of evaporation.

7.4 Criticality and Nonequilibrium Dynamics

As Bose-Einstein condensation is a continuous phase transition, the theory of critical phenomena [112] predicts that in the vicinity of the critical point the correlations of the Bose field obey universal scaling relations. In particular, the scaling of correlations at and near equilibrium is governed by a set of universal critical exponents and scaling functions, independent of the microscopic parameters of the gas. For a homogeneous system close to criticality, standard theory predicts that the correlation length ξ, relaxation time τ, and first-order correlation function $G(x) = \langle \psi^\dagger(x)\psi(0)\rangle$ obey scaling laws

$$\xi = \frac{\xi_0}{|\epsilon|^\nu}, \qquad \tau = \frac{\tau_0}{|\epsilon|^{\nu z}}, \qquad G(x) = \epsilon^{\nu(d-2+\eta)}\mathcal{F}(\epsilon^\nu x), \tag{7.1}$$

with $\epsilon = T/T_c - 1$ the reduced temperature, ν and z the correlation length and dynamical critical exponents, η the scaling dimension of the Bose field, and $\mathcal{F}$ a universal scaling function. The static Bose gas belongs to the XY (or $O(2)$) universality class, and is thus expected to have the same critical exponents as superfluid helium, i.e., in 3D, $\nu \simeq 0.67$ and $\eta = 0.038(4)$ [113]. The critical dynamics of the system are expected to conform to those of the diffusive model denoted by F in the classification of Ref. [114], implying a value $z = 3/2$ for the dynamical critical exponent.

The influence of critical physics is significantly reduced in the conditions of harmonic confinement typical of experimental Bose-gas systems, as compared with homogeneous systems. Within a local-density approximation, the inhomogeneous thermodynamic parameters of the system imply that only a small fraction of atoms in the gas enter the critical regime, and so global observables are relatively insensitive to the effects of criticality. Nevertheless, a few experiments have attempted to observe aspects of the critical physics of trapped Bose gases.

In a homogeneous gas the introduction of interparticle interactions has no effect on the critical temperature at the mean-field level, but the magnitude and even the sign of the shift due to critical fluctuations were debated for several decades (see Ref. [115] and references therein) before being settled by classical-field Monte Carlo calculations [116, 117]. An experiment by the Aspect group carefully measured a shift in critical temperature of the trapped gas, but was unable to unambiguously infer any beyond-mean-field contribution to this shift [118, 119]. A later experiment by the group of Hadzibabic made use of a Feshbach resonance to control the interaction strength in ^{41}K, and found clear evidence of a positive beyond-mean-field shift [120] (see also Chapter 6 by Smith).

In 2007, the ETH Zürich group of Esslinger revisited their experiments on condensate formation and the coherence of a three-dimensional BEC with a new tool: the ability to count single atoms passing through an optical cavity below their ultracold gas [121]. They outcoupled atoms from two different vertical locations from their sample as it was cooled, realizing interference in the falling matter waves. By monitoring the visibility of the fringes, they were able to measure the growth of the coherence length as a function of time. Using the same optical-cavity setup, the Esslinger group subsequently measured the coherence length of their Bose gas as it was driven through the critical temperature by a small background heating rate and determined the correlation-length critical exponent to be $\nu = 0.67 \pm 0.13$ [122]. Their results are shown in Fig. 7.2a. Classical-field simulations of their experiment were in reasonable agreement, determining $\nu = 0.80 \pm 0.12$ [123].

Although an important topic in its own right, the greatest significance of the equilibrium theory of critical fluctuations to studies of condensate formation is that it provides a basis for generalizations of concepts such as critical scaling laws and universality classes to the domain of nonequilibrium physics. In the remainder of this section, we discuss two such extensions: the Kibble-Zurek mechanism (KZM) and the theory of nonthermal fixed points.

7.4.1 The Kibble-Zurek Mechanism

The theory of the Kibble-Zurek mechanism leverages the well-established results of the equilibrium theory of criticality to make immediate predictions for universal scaling behavior in the nonequilibrium dynamics of passage through a second-order phase transition. The underlying idea – that causally disconnected regions of space break symmetry independently, leading to the formation of topological defects – was first discussed by Kibble [125], who predicted that the distribution

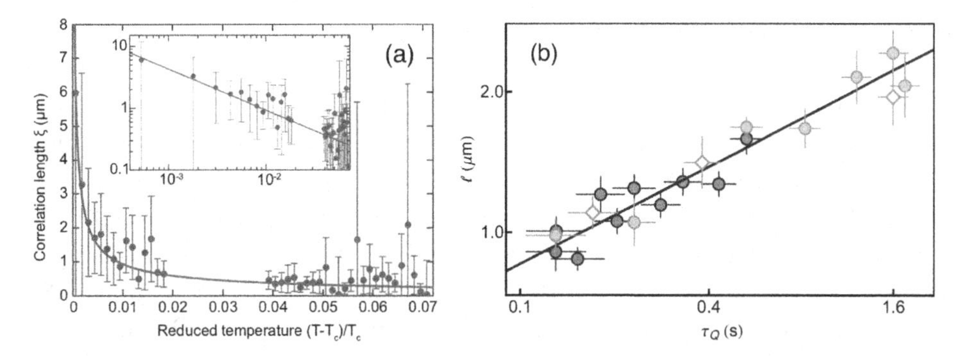

Figure 7.2 Critical phenomena in BECs. (a) Divergence of the equilibrium correlation length ξ as a function of the reduced temperature, and the fitting of the critical exponent, giving the result $\nu = 0.67 \pm 0.13$. Inset: Double logarithmic plot of the same data. Reprinted with permission from the American Association for the Advancement of Science (AAAS) from Donner, T., et al. (2007), Critical behavior of a trapped interacting Bose gas, *Science*, **315**, 1556 [122]. (b) Log-log plot of the dependence of the correlation length, here labeled ℓ, as a function of the characteristic time τ_{Q} of the quench through the BEC phase transition. The solid line corresponds to a Kibble-Zurek power-law scaling $\ell \propto \tau_{\mathrm{Q}}^{b}$ with $b = 0.35 \pm 0.04$, in agreement with the beyond-mean-field prediction $b = 1/3$ of the so-called F model [114] and inconsistent with the mean-field value $b = 1/4$. This in turn implies a value $z = 1.4 \pm 0.2$ for the dynamical critical exponent. Reprinted with permission from AAAS from Navon, N., et al. (2015), Critical dynamics of spontaneous symmetry breaking in a homogeneous Bose gas, *Science*, **347**, 167 [124].

of defects following the transition would be determined by the instantaneous correlation length of the system as it passes through the Ginzburg temperature [126]. Zurek later emphasized [127] the importance of dynamic critical phenomena [114] in such a scenario. In particular, the scaling relations (7.1) imply that both the correlation length and the characteristic relaxation time of the system diverge as the critical point is approached ($\epsilon \to 0$), imposing a limit to the size of spatial regions over which order can be established during the transition. Topological defects will thus be seeded, with a density determined by the correlation length at the time the system "freezes" during the transition, and will subsequently decay in the symmetry-broken phase. The more rapidly the system passes through the critical point, the shorter the correlation length that is frozen in, and therefore more topological defects will form. A dimensional analysis predicts that a linear ramp $\epsilon(t) = -t/\tau_{\mathrm{Q}}$ of the reduced temperature through the critical point on a characteristic time scale τ_{Q} results in a distribution of spontaneously formed defects with a density n_{d} that scales as [128]

$$n_{\mathrm{d}} \propto \tau_{\mathrm{Q}}^{(p-d)\nu/(1+\nu z)}, \tag{7.2}$$

where d is the dimensionality of the sample and p is the intrinsic dimensionality of the defects.

Zurek initially described the KZM in the context of vortices in the λ transition of superfluid ^{4}He [127]. Although vortices are observed in the wake of this transition, it is difficult to identify them as having formed due to the KZM rather than being induced by, e.g., inadvertent stirring [128] (see also Chapter 30 by Pickett). The prospect of generating vorticity in atomic BECs by means of the KZM was first discussed by Anglin and Zurek in 1999 [129]. However, it was not until the 2008 experiment of the Anderson group at the University of Arizona [130] that spontaneously formed vortices were first observed in such a system (see also Ref. [131]).

The observations of spontaneous vortices in Ref. [130] were supported by numerical simulations using the stochastic projected Gross-Pitaevskii equation description of Gardiner and Davis [132]. Their results are shown in Fig. 7.3. This formalism is essentially a variant of the Gardiner-Zoller quantum-kinetic theory, obtained by making a high-temperature approximation to the condensate-band master equation and then exploiting the quantum-classical correspondence of the

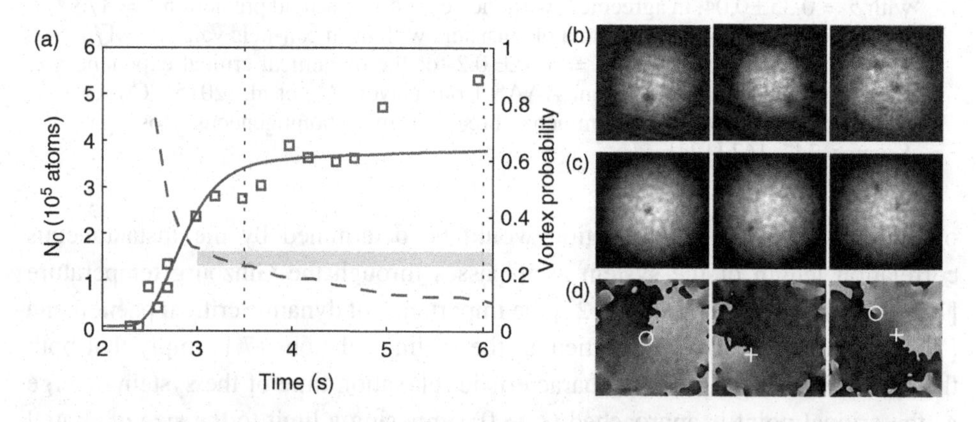

Figure 7.3 Spontaneous vortices in the formation of a Bose-Einstein condensate. (a) Squares: experimentally measured condensate population as a function of time. Solid line: condensate number from stochastic Gross-Pitaevskii simulations. Dashed line: probability of finding one or more vortices in the simulations as a function of time, averaged over 298 trajectories. The shaded area indicates the statistical uncertainty in the experimentally measured vortex probability at $t = 6.0$ s. It was observed in experiment that there was no discernible vortex decay between 3.5 s and 6.0 s. (b) Experimental absorption images taken after 59 ms time-of-flight showing the presence of vortices. (c) Simulated in-trap column densities at $t = 3.5$ s (indicated by the left vertical dotted line in (a)). (d) Phase images through the $z = 0$ plane, with plusses (open circles) representing vortices with positive (negative) circulation. Adapted with permission from Weiler, C. N., et al. (2008), Spontaneous vortices in the formation of Bose-Einstein condensates, *Nature*, **455**, 948 [130]. Copyright (2008) by the Nature Publishing Group.

Wigner representation to obtain a stochastic classical-field description of the condensate band [132, 133]. Although derived using different theoretical techniques, the resulting description is similar to the stochastic Gross-Pitaevskii equation of Stoof [21, 63], both in terms of its physical content and its computational implementation – see, e.g., discussion in Refs. [134, 135]. A related phase-space method originating in quantum optics known as the positive-P representation has also been applied to ultracold gases [136]. This has been used to investigate cooling of a small system towards BEC by Drummond and Corney [137], who observed features consistent with spontaneously formed vortices. Despite formally being a statistically exact method, for interacting systems it tends to suffer from numerical divergences after a relatively short evolution time.

It seems likely that spontaneously formed vortices and other defects were present in earlier BEC-formation experiments, but not observed due to the practical difficulties inherent in resolving these defects in experimental imaging – and indeed the fact that these experiments were not attempting to investigate whether such structures were present. Another difficulty in identifying quantitative signatures of the KZM in experimental BECs is the inhomogeneity of the system in the experimental trapping potential, which is typically harmonic. From the point of view of a local-density approximation, this inhomogeneity implies that the instantaneous coherence length and relaxation time scale are spatially varying quantities, and that the transition occurs at different times in different regions of space as the system is cooled. Following preliminary reports of the experimental observation of dark solitons following the formation of a quasi–one-dimensional BEC by the Engels group at the University of Washington [138], Zurek applied the framework of the KZM to a quasi-1D BEC in a cigar-shaped trap to estimate the scaling of the number of spontaneously generated solitons as a function of the quench time [139, 140]. Witkowska et al. [141] numerically studied cooling leading to solitons in a comparable one-dimensional geometry. Zurek's methodology for inhomogeneous systems was applied by del Campo et al. [142] to strongly oblate geometries in which vortex filaments behave approximately as point vortices in the plane, an idealization of the geometry of the experiment of Weiler et al. [130].

Lamporesi et al. [143] recently reported the spontaneous creation of Kibble-Zurek dark solitons in the formation of a BEC in an elongated trap, and found the scaling of the number of observed defects with cooling rate in good agreement with the predictions of Zurek [139]. It was later realized that the apparent solitons were actually solitonic vortices [144]. The effects of inhomogeneity in such experiments can be mitigated by the realization of "boxlike" flat-bottomed trapping geometries. The Dalibard group in Paris has observed the formation of spontaneous vortices in a quasi-2D boxlike geometry, and found scaling of the vortex number with quench rate in good agreement with the predictions of the KZM [145]. We also note further work by the Dalibard group [146] verifying the production of quench-induced supercurrents in a toroidal or "ring-trap" geometry [147] analogous to the annular sample of superfluid helium considered in Zurek's original proposal [127].

Experimental investigations of the KZM in dilute atomic gases have largely focused on the imaging of defects in the wake of the phase transition – either following time-of-flight expansion [130, 143, 145] or *in situ* [144]. However, the accurate extraction of critical scaling behavior from such observations is hampered by the large background excitation of the field near the transition, and the relaxation (or "coarsening") dynamics of defects in the symmetry-broken phase. (See Ref. [148] for a study of this issue in a holographic superfluid. The coarsening process is a special case of dynamics near a nonthermal fixed point discussed in Section 7.4.2.) An alternative approach is to make quantitative measurements of global properties of the system following the quench. Performing quench experiments in a three-dimensional boxlike geometry, the Hadzibabic group in Cambridge [124] made careful measurements of the scaling of the correlation length with quench time. From the measured scaling law, these authors were able to infer a beyond-mean-field value $z = 1.4 \pm 0.2$ for the dynamical critical exponent for this universality class. Some of the results of Ref. [124] are displayed in Fig. 7.2b (see also Chapter 6).

The possibilities for the trapping and cooling of multicomponent systems in atomic physics experiments have naturally led to investigations of the spontaneous formation of more complicated topological defects during a phase transition. Although such experiments have so far largely focused on the formation of defects following a quench of Hamiltonian parameters [149, 150], the formation of nontrivial domain structures following gradual sympathetic cooling in immiscible ^{85}Rb-^{87}Rb [151] and ^{87}Rb-^{133}Cs [152] Bose-Bose mixtures has also been observed. The competing growth dynamics of the two immiscible components in the formation of such a binary condensate have recently been investigated theoretically in the limit of a sudden temperature quench [153] (see also Refs. [154, 155, 156] for related critical scaling in other Hamiltonian quenches). These investigations indicate the rich nonequilibrium dynamics possible in these systems, including strong memory effects on the coarsening of spontaneously formed defects and the potential "microtrapping" of one component in spontaneous defects formed in the other.

7.4.2 Nonthermal Fixed Points

A general characterisation of the relaxation dynamics of quantum many-body systems quenched far out of equilibrium remains a largely open problem. In particular, it is interesting to ask to what extent analogues of the universal descriptions arising from the equilibrium theory of critical fluctuations may exist for nonequilibrium systems. A recent advance toward answering such questions has been made in the development of the theory of nonthermal fixed points: universal nonequilibrium configurations showing scaling in space and (evolution) time, characterised by a small number of fundamental properties. The theory of such fixed points transposes the concepts of equilibrium and diffusive near-equilibrium renormalisation-group theory to the real-time evolution of nonequilibrium systems. These developments provide, for example, a framework within which to understand the turbulent, coarsening, and relaxation dynamics following the creation of various kinds of defects and nonlinear patterns in a Kibble-Zurek quench.

The existence and significance of nonthermal scaling solutions in space and time were discussed by Berges and collaborators in the context of reheating after early-universe inflation [83, 84] and then generalized by Berges, Gasenzer, and coworkers to scenarios of strong matter-wave turbulence [85, 157]. For the condensation dynamics of the dilute Bose gas discussed here, the presence of a nonthermal fixed point can exert a significant influence in the case of a strong cooling quench [33, 158, 159].

As an illustration, we consider a particle distribution that drops abruptly above the healing-length scale $k_\xi = \sqrt{8\pi a\rho}$, as depicted on a double-logarithmic scale in Fig. 7.4 (dashed line). In order for the influence of the nonthermal fixed point to be

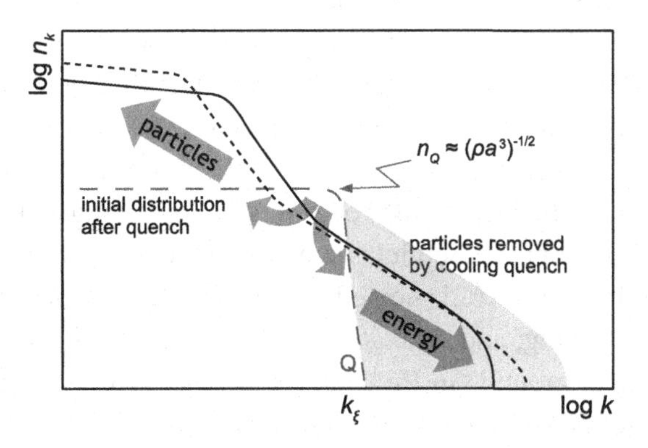

Figure 7.4 Sketch of the evolution of the single-particle momentum distribution $n_k(t)$ of a Bose gas close to a nonthermal fixed point. Starting from the extreme initial distribution (dashed line; see main text for details) produced, e.g., by a strong cooling quench, a bidirectional redistribution of particles and energy in momentum space (arrows, solid- to dashed-line distributions) builds up a quasicondensate in the infrared while refilling the thermal tail at large momenta. Both the particle transport toward zero momentum and the energy transport toward large momenta are characterised by a self-similar scaling evolution in space and time, $n(k,t) = (t/t_0)^\alpha n([t/t_0]^\beta k, t_0)$, with, in general, distinct values for both of the characteristic scaling exponents α, β. Note the double-logarithmic scale.

observed, the decay of $n(k)$ above $Q \simeq k_\xi$ is assumed to be much steeper than the quasithermal scaling that develops in the kinetic stage of condensation following a weak quench [30, 31], as discussed in Section 7.2. Such a distribution would, e.g., result from a severe cooling quench of a thermal Bose gas initially just above the critical temperature where $T > |\mu|/k_B$ such that the Bose-Einstein distribution has developed a Rayleigh-Jeans scaling regime where $n(k) \sim 2mk_BT/(\hbar k)^2$. The modulus of the chemical potential of this state determines the momentum scale Q, where the flat infrared scaling of the distribution goes over to the Rayleigh-Jeans scaling at larger k. If this chemical potential is of the order of the ground-state energy of the postquench fully condensed gas, $(\hbar Q)^2/2m \simeq |\mu| \simeq g\rho$, with $g = 4\pi\hbar^2 a/m$, then the energy of the entire gas is concentrated at the scale $Q \simeq k_\xi$ after the quench. This is a key feature of the extreme nonequilibrium initial state from which a nonthermal fixed point can be approached. We note that, if in this state there is no significant zero-mode occupation n_0, the respective occupation number at Q is on the order of the inverse of the diluteness parameter, $n_Q \sim \zeta^{-3/2}$.

In analogy to the weak-wave-turbulence scenario [22, 23, 30, 31] discussed in Section 7.2.1, the initial overpopulation of modes with energies $\sim (\hbar Q)^2/2m$ leads to inverse particle transport while energy is transported to higher wavenumbers, as indicated by the arrows and the consecutive, solid, and dashed-line distributions in Fig. 7.4 [33, 158, 159]. Note that the inverse transport involves nonlocal scattering and thus does not represent a cascade. Furthermore, in contrast to the case of a weak quench [17, 18, 21, 22, 23, 31, 32], in which weak-wave turbulence produces a quasithermal momentum distribution that relaxes quickly to a thermal equilibrium distribution, here the inverse transport is characterized by a strongly nonthermal power-law scaling in the infrared. Specifically, the momentum distribution $n(k) \sim k^{-d-2} \sim k^{-5}$ in $d = 3$ dimensions [85] provides the "smoking-gun" of the influence of the nonthermal fixed point. Semiclassical simulations by Nowak, Gasenzer, and collaborators [33, 160, 161, 162] showed that this scaling is associated with the creation, dilution, coarsening, and relaxation of a complex vortex tangle, as predicted on phenomenological grounds in Ref. [24], and other types of (quasi-)topological excitations in low-dimensional, spinor, and gauge systems [148, 163, 164, 165, 166, 167, 168].

The dynamics in the vicinity of the fixed point are characterized by an anomalously slow relaxation of the total vortex line length, which exhibits an algebraic decay $\sim t^{-0.88}$ (see Ref. [163] for results in the 2D case where a significantly slower $\sim t^{-0.4}$ decay is seen). At the same time, the condensate population grows as $n_0(t) \sim t^2$ [33, 158], a significant slowing compared with the $\sim t^3$ behavior observed for weakly nonequilibrium condensate formation [33, 169].

In the vicinity of the fixed point, the momentum distribution is expected to follow a self-similar scaling behavior in space and time in the infrared,

$n(k,t) = (t/t_0)^{\alpha} n([t/t_0]^{\beta} k, t_0)$. For a 3D Bose gas, these scaling exponents have recently been numerically determined to be $\alpha = 1.66(12)$, $\beta = 0.55(3)$, in agreement with the analytically predicted values $\alpha = \beta d$, $\beta = 1/(2-\eta)$ [158] if one assumes $\eta \simeq 0$, and consistent with a coarsening process [170]. This behavior, here corresponding to the dilution and relaxation of vortices leading to a buildup of the condensate population, represents the generalization of critical slowing down to real-time evolution far away from thermal equilibrium. At very late times, the system leaves the vicinity of the nonthermal fixed point, typically when the last topological patterns decay, and finally approaches thermal equilibrium [35, 36, 37, 38, 163, 171]. This equilibrium state corresponds to a fully established condensate superimposed with weak sound excitations.

In summary, nonthermal fixed points are nonequilibrium field configurations, exhibiting universal scaling in time and space, to which the system is attracted if suitably forced – e.g., in the case of Bose condensation, following a sufficiently strong cooling quench. In the vicinity of such fixed points, the relaxation of the field is critically slowed down and the dynamics exhibit self-similar time evolution, governed by new critical exponents and scaling functions. The possibility of categorizing systems into generalized "universality classes" associated with the new critical exponents is a fascinating prospect and the subject of current research [26, 27].

Finally in this section, we note that prethermalization [172, 173, 174], i.e., the approach towards a state characterized by a generalized Gibbs ensemble [175, 176, 177, 178, 179], typically expected in near-integrable systems (see Chapter 8 by Langen and Schmiedmayer), represents a special case of a Gaussian nonthermal fixed point, meaning that the effective coupling of the prethermalized modes vanishes. It is expected that the exponents α and β in such a situation can become very small compared with unity. The remaining effects of interactions may only manifest on very long time scales, eventually driving the system away from the fixed point towards a thermal state.

7.5 Conclusions and Outlook

In this chapter, we have provided a brief introduction to the scenario of the formation of a Bose-Einstein condensate and physics related to the dynamics of the BEC phase transition. We have given a fairly comprehensive review of the experiments studying the formation of simple, single-component BECs in three-dimensional atomic gases, with brief mentions of how such features are affected by reduced effective dimensionality or in cases where more than one condensate may coexist. However, the underlying physics described here is relevant to several other systems, most notably exciton-polaritons confined in strictly two-dimensional geometries featuring pumping and decay, where experiments on condensate formation have also been performed [180, 181, 182].

An interesting question is what are the similarities and differences between these systems, and others such as BECs of photons [183] and magnons [184]? Furthermore, what can phase transitions in quantum gases teach us about phase transitions that cannot be accessed experimentally, such as inflationary scenarios of early-universe evolution? This was one of the motivating questions in the formulation of the Kibble-Zurek mechanism, as well as in the development of the theory of nonthermal fixed points. It remains to be seen what we can learn about such matters as quark-gluon plasma formation following a heavy-ion collision [185, 186, 187], the reheating of the postinflationary universe [83], the formation of cosmological topological defects [128], or baryon asymmetry by studying nanokelvin gases here on Earth.

Acknowledgements: We thank B. Anderson, O. Bahat-Treidel, R. Ballagh, J. Berges, A. Bradley, K. Burnett, I. Chantesana, J. Dalibard, R. Duine, S. Erne, C. Ewerz, G. Ferrari, C. Gardiner, M. Garrett, A. Griffin, Z. Hadzibabic, T. Judd, M. Karl, I.-K. Liu, G. Lamporesi, T. Langen, J. Märkle, L. McLerran, N. Navon, B. Nowak, M. Oberthaler, J. Pawlowski, A. Piñeiro Orioli, J. Sabbatini, H. Salman, A. Samberg, J. Schmiedmayer, J. Schole, D. Sexty, H. Stoof, E. Zaremba, and W. Zurek for discussions and collaboration on the topics described here, and B. Svistunov for helpful comments concerning the manuscript. We also thank N. Berloff, P. Comaron, I. Carusotto, J. Keeling, M. Szymanska, and A. Zamora for discussions of related issues in polaritons. We acknowledge funding from the Australian Research Council (DP1094025 and DP110101047; M.J.D. and T.M.W.), the European Union (FET-Proactive grant AQuS, Project No. 640800; T.G.), and the United Kingdom EPSRC (EP/I019413/1, EP/K03250X/1; N.P.P.).

References

[1] Proukakis, N. P., Gardiner, S. A., Davis, M. J., and Szymańska, M. (eds). 2013. *Quantum Gases: Finite Temperature and Non-Equilibrium Dynamics*. London, UK: Imperial College Press.

[2] Griffin, A., Nikuni, T., and Zaremba, E. 2009. *Bose-Condensed Gases at Finite Temperatures*. Cambridge, UK: Cambridge University Press.

[3] Popov, V. N. 1972. On the theory of the superfluidity of two- and one-dimensional Bose systems. *Theor. Math. Phys.*, **11**, 565.

[4] Popov, V. N. 1983. *Functional Integrals in Quantum Field Theory and Statistical Physics*. Dordrecht, Netherlands: Reidel.

[5] Gardiner, C. W. 1997. Particle-number-conserving Bogoliubov method which demonstrates the validity of the time-dependent Gross-Pitaevskii equation for a highly condensed Bose gas. *Phys. Rev. A*, **56**, 1414.

[6] Castin, Y., and Dum, R. 1998. Low-temperature Bose-Einstein condensates in time-dependent traps: beyond the $U(1)$ symmetry-breaking approach. *Phys. Rev. A*, **57**, 3008.
[7] Gardiner, S. A., and Morgan, S. A. 2007. Number-conserving approach to a minimal self-consistent treatment of condensate and noncondensate dynamics in a degenerate Bose gas. *Phys. Rev. A*, **75**, 043621.
[8] Inoue, A., and Hanamura, E. 1976. Emission spectrum from the Bose-condensed excitonic molecules. *J. Phys. Soc. Jpn.*, **41**, 771.
[9] Levich, E., and Yakhot, V. 1977a. Time evolution of a Bose system passing through the critical point. *Phys. Rev. B*, **15**, 243.
[10] Levich, E., and Yakhot, V. 1977b. Kinetics of phase transition in ideal and weakly interacting Bose gas. *J. Low Temp. Phys.*, **27**, 107.
[11] Levich, E., and Yakhot, V. 1978. Time development of coherent and superfluid properties in the course of a λ-transition. *J. Phys. A*, **11**, 2237.
[12] Zeldovich, Ya. B., and Levich, E. V. 1968. Bose condensation and shock waves in photon spectra. *[Zh. Eksp. Teor. Fiz.* **55**, *2423 (1968)] Sov. Phys. JETP*, **28**, 1287.
[13] Pitaevskii, L. P., and Stringari, S. 2003. *Bose-Einstein Condensation*. Oxford, UK: Clarendon Press.
[14] Tikhodeev, S. G. 1990. Bose condensation of finite-lifetime particles with excitons as an example. *[Zh. Eksp. Teor. Fiz.* **97**, *681 (1990)] Sov. Phys. JETP*, **70**, 380.
[15] Eckern, U. 1984. Relaxation processes in a condensed Bose gas. *J. Low Temp. Phys.*, **54**, 333.
[16] Snoke, D. W., and Wolfe, J. P. 1989. Population dynamics of a Bose gas near saturation. *Phys. Rev. B*, **39**, 4030.
[17] Stoof, H. T. C. 1991. Formation of the condensate in a dilute Bose gas. *Phys. Rev. Lett.*, **66**, 3148.
[18] Stoof, H. T. C. 1992. Nucleation of Bose-Einstein condensation. *Phys. Rev. A*, **45**, 8398.
[19] Stoof, H. T. C. 1995. *Bose-Einstein Condensation*. Cambridge, UK: Cambridge University Press. Chap. Condensate formation in a Bose gas, page 226.
[20] Stoof, H. T. C. 1997. Initial stages of Bose-Einstein condensation. *Phys. Rev. Lett.*, **78**, 768.
[21] Stoof, H. T. C. 1999. Coherent versus incoherent dynamics during Bose-Einstein condensation in atomic gases. *J. Low Temp. Phys.*, **114**, 11.
[22] Svistunov, B. V. 1991. Highly nonequilibrium Bose condensation in a weakly interacting gas. *J. Mosc. Phys. Soc.*, **1**, 373.
[23] Kagan, Yu., Svistunov, B. V., and Shlyapnikov, G. V. 1992. Kinetics of Bose condensation in an interacting Bose gas. *Zh. Éksp. Teor. Fiz.*, **101**, 528. [Sov. Phys. JETP **75**, 387 (1992)].
[24] Kagan, Yu., and Svistunov, B. V. 1994. Kinetics of the onset of long-range order during Bose condensation in an interacting gas. *Zh. Éksp. Teor. Fiz.*, **105**, 353. [Sov. Phys. JETP **78**, 187 (1994)].
[25] Kagan, Yu. 1995. *Bose-Einstein Condensation*. Cambridge, UK: Cambridge University Press. Chap. Kinetics of Bose-Einstein condensate formation in an interacting Bose gas, page 202.
[26] Sieberer, L. M., Huber, S. D., Altman, E., and Diehl, S. 2013. Dynamical critical phenomena in driven-dissipative systems. *Phys. Rev. Lett.*, **110**, 195301.
[27] Altman, E., Sieberer, L. M., Chen, L., Diehl, S., and Toner, J. 2015. Two-dimensional superfluidity of exciton polaritons requires strong anisotropy. *Phys. Rev. X*, **5**, 011017.

[28] Dagvadorj, G., Fellows, J. M., Matyjaskiewicz, S., Marchetti, F. M., Carusotto, I., and Szymanska, M. H. 2015. Non-equilibrium Berezinskii-Kosterlitz-Thouless transition in a driven open quantum system. *Phys. Rev. X*, **5**, 041028.

[29] Zakharov, V. E., L'vov, V. S., and Falkovich, G. 1992. *Kolmogorov Spectra of Turbulence I: Wave Turbulence*. Berlin, Germany: Springer-Verlag.

[30] Semikoz, D. V., and Tkachev, I. I. 1995. Kinetics of Bose condensation. *Phys. Rev. Lett.*, **74**, 3093.

[31] Semikoz, D. V., and Tkachev, I. I. 1997. Condensation of bosons in the kinetic regime. *Phys. Rev. D*, **55**, 489.

[32] Berloff, N. G., and Svistunov, B. V. 2002. Scenario of strongly nonequilibrated Bose-Einstein condensation. *Phys. Rev. A*, **66**, 013603.

[33] Nowak, B., Schole, J., and Gasenzer, T. 2014. Universal dynamics on the way to thermalisation. *New J. Phys.*, **16**, 093052.

[34] Schwarz, K. W. 1988. Three-dimensional vortex dynamics in superfluid ^{4}He: homogeneous superfluid turbulence. *Phys. Rev. B*, **38**, 2398.

[35] Kozik, E., and Svistunov, B. 2004. Kelvin-wave cascade and decay of superfluid turbulence. *Phys. Rev. Lett.*, **92**, 035301.

[36] Kozik, E., and Svistunov, B. 2005. Scale-separation scheme for simulating superfluid turbulence: Kelvin-wave cascade. *Phys. Rev. Lett.*, **94**, 025301.

[37] Kozik, E., and Svistunov, B. 2005. Vortex-phonon interaction. *Phys. Rev. B*, **72**, 172505.

[38] Kozik, E. V., and Svistunov, B. V. 2009. Theory of decay of superfluid turbulence in the low-temperature limit. *J. Low Temp. Phys.*, **156**, 215.

[39] Anderson, M. H., Ensher, J. R., Matthews, M. R., Wieman, C. E., and Cornell, E. A. 1995. Observation of Bose-Einstein condensation in a dilute atomic vapor. *Science*, **269**, 198.

[40] Davis, K. B., Mewes, M. O., Andrews, M. R., van Druten, N. J., Durfee, D. S., Kurn, D. M., and Ketterle, W. 1995. Bose-Einstein condensation in a gas of sodium atoms. *Phys. Rev. Lett.*, **75**, 3969.

[41] Ketterle, W., and Van Druten, N. J. 1996. Evaporative cooling of trapped atoms. *Adv. At. Mol. Opt. Phys.*, **37**, 181.

[42] Gardiner, C. W., and Zoller, P. 1997. Quantum kinetic theory: a quantum kinetic master equation for condensation of a weakly interacting Bose gas without a trapping potential. *Phys. Rev. A*, **55**, 2902.

[43] Gardiner, C. W., and Zoller, P. 1998. Quantum kinetic theory III: quantum kinetic master equation for strongly condensed trapped systems. *Phys. Rev. A*, **58**, 536.

[44] Gardiner, C. W., and Zoller, P. 2000. Quantum kinetic theory V: quantum kinetic master equation for mutual interaction of condensate and noncondensate. *Phys. Rev. A*, **61**, 033601.

[45] Gardiner, C. W., and Zoller, P. 2004. *Quantum Noise*, 3rd edn. Berlin and Heidelberg, Germany: Springer-Verlag.

[46] Miesner, H.-J., Stamper-Kurn, D. M., Andrews, M. R., Durfee, D. S., Inouye, S., and Ketterle, W. 1998. Bosonic stimulation in the formation of a Bose-Einstein condensate. *Science*, **279**, 1005.

[47] Gardiner, C. W., Zoller, P., Ballagh, R. J., and Davis, M. J. 1997. Kinetics of Bose-Einstein condensation in a trap. *Phys. Rev. Lett.*, **79**, 1793.

[48] Gardiner, C. W., Lee, M. D., Ballagh, R. J., Davis, M. J., and Zoller, P. 1998. Quantum kinetic theory of condensate growth: comparison of experiment and theory. *Phys. Rev. Lett.*, **81**, 5266.

[49] Lee, M. D., and Gardiner, C. W. 2000. Quantum kinetic theory. VI. The growth of a Bose-Einstein condensate. *Phys. Rev. A*, **62**, 033606.

[50] Davis, M. J., Gardiner, C. W., and Ballagh, R. J. 2000. Quantum kinetic theory. VII. The influence of vapor dynamics on condensate growth. *Phys. Rev. A*, **62**, 063608.

[51] Bijlsma, M. J., Zaremba, E., and Stoof, H. T. C. 2000. Condensate growth in trapped Bose gases. *Phys. Rev. A*, **62**, 063609.

[52] Zaremba, E., Nikuni, T., and Griffin, A. 1999. Dynamics of trapped Bose gases at finite temperatures. *J. Low Temp. Phys.*, **116**, 277.

[53] Zaremba, E., Griffin, A., and Nikuni, T. 1998. Two-fluid hydrodynamics for a trapped weakly interacting Bose gas. *Phys. Rev. A*, **57**, 4695.

[54] Kirkpatrick, T. R., and Dorfman, J. R. 1983. Transport theory for a weakly interacting condensed Bose gas. *Phys. Rev. A*, **28**, 2576.

[55] Kirkpatrick, T. R., and Dorfman, J. R. 1985. Transport coefficients in a dilute but condensed Bose gas. *J. Low Temp. Phys.*, **58**, 399.

[56] Kirkpatrick, T. R., and Dorfman, J. R. 1985. Transport in a dilute but condensed nonideal Bose gas: kinetic equations. *J. Low Temp. Phys.*, **58**, 301.

[57] Kirkpatrick, T. R., and Dorfman, J. R. 1985c. Time correlation functions and transport coefficients in a dilute superfluid. *J. Low Temp. Phys.*, **59**, 1.

[58] Luiten, O. J., Reynolds, M. W., and Walraven, J. T. M. 1996. Kinetic theory of the evaporative cooling of a trapped gas. *Phys. Rev. A*, **53**, 381.

[59] Köhl, M., Davis, M. J., Gardiner, C. W., Hänsch, T. W., and Esslinger, T. W. 2002. Growth of Bose-Einstein condensates from thermal vapor. *Phys. Rev. Lett.*, **88**, 080402.

[60] Davis, M. J., and Gardiner, C. W. 2002. Growth of a Bose-Einstein condensate: a detailed comparison of theory and experiment. *J. Phys. B: At. Mol. Opt. Phys.*, **35**, 733.

[61] Pinske, P. W. H., Mosk, A., Weidemüller, M., Reynolds, M. W., Hijmans, T. W., and Walraven, J. T. M. 1997. Adiabatically changing the phase-space density of a trapped Bose gas. *Phys. Rev. Lett.*, **78**, 990.

[62] Stamper-Kurn, D. M., Miesner, H.-J., Chikkatur, A. P., Inouye, S., Stenger, J., and Ketterle, W. 1998. Reversible formation of a Bose-Einstein condensate. *Phys. Rev. Lett.*, **81**, 2194.

[63] Stoof, H. T. C., and Bijlsma, M. J. 2001. Dynamics of fluctuating Bose-Einstein condensates. *J. Low. Temp. Phys.*, **124**, 431.

[64] Proukakis, N. P., Schmiedmayer, J., and Stoof, H. T. C. 2006. Quasicondensate growth on an atom chip. *Phys. Rev. A*, **73**, 053603.

[65] Garrett, M. C., Ratnapala, A., van Ooijen, E. D., Vale, C. J., Weegink, K., Schnelle, S. K., Vainio, O., Heckenberg, N. R., Rubinsztein-Dunlop, H., and Davis, M. J. 2011. Growth dynamics of a Bose-Einstein condensate in a dimple trap without cooling. *Phys. Rev. A*, **83**, 013630.

[66] Harber, D. M., McGuirk, J. M., Obrecht, J. M., and Cornell, E. A. 2003. Thermally induced losses in ultra-cold atoms magnetically trapped near room-temperature surfaces. *J. Low Temp. Phys.*, **133**, 229.

[67] Marchant, A. L., Händel, S., Wiles, T. P., Hopkins, S. A., and Cornish, S. L. 2011. Guided transport of ultracold gases of rubidium up to a room-temperature dielectric surface. *New J. Phys.*, **13**, 125003.

[68] Märkle, J., Allen, A. J., Federsel, P., Jetter, B., Günther, A., Fortágh, J., Proukakis, N. P., and Judd, T. E. 2014. Evaporative cooling of cold atoms at surfaces. *Phys. Rev. A*, **90**, 023614.

[69] Imamovic-Tomasovic, M., and Griffin, A. 2001. Quasiparticle kinetic equation in a trapped Bose gas at low temperatures. *J. Low Temp. Phys.*, **122**, 616.

[70] Kadanoff, L. P., and Baym, G. 1962. *Quantum Statistical Mechanics*. Menlo Park, CA: W. A. Benjamin.

[71] Walser, R., Williams, J., Cooper, J., and Holland, M. 1999. Quantum kinetic theory for a condensed bosonic gas. *Phys. Rev. A*, **59**, 3878.

[72] Walser, R., Cooper, J., and Holland, M. 2001. Reversible and irreversible evolution of a condensed bosonic gas. *Phys. Rev. A*, **63**, 013607.

[73] Wachter, J., Walser, R., Cooper, J., and Holland, M. 2001. Equivalence of kinetic theories of Bose-Einstein condensation. *Phys. Rev. A*, **64**, 053612.

[74] Proukakis, N. P. 2001. Self-consistent quantum kinetics of condensate and non-condensate via a coupled equation of motion formalism. *J. Phys. B: At. Mol. Opt. Phys.*, **34**, 4737.

[75] Proukakis, N. P., and Burnett, K. 1996. *J. Res. Natl. Inst. Stand. Technol.*, **101**, 457.

[76] Proukakis, N. P., Burnett, K., and Stoof, H. T. C. 1998. Microscopic treatment of binary interactions in the nonequilibrium dynamics of partially Bose-condensed trapped gases. *Phys. Rev. A*, **57**, 1230.

[77] Shi, H., and Griffin, A. 1998. Finite-temperature excitations in a dilute Bose-condensed gas. *Phys. Rep.*, **304**, 187.

[78] Barci, D. G., Fraga, E. S., and Ramos, R. O. 2000. A nonequilibrium quantum field theory description of the Bose-Einstein condensate. *Phys. Rev. Lett.*, **85**, 479.

[79] Gasenzer, T., Berges, J., Schmidt, M. G., and Seco, M. 2005. Nonperturbative dynamical many-body theory of a Bose-Einstein condensate. *Phys. Rev. A*, **72**, 063604.

[80] Berges, J., and Gasenzer, T. 2007. Quantum versus classical statistical dynamics of an ultracold Bose gas. *Phys. Rev. A*, **76**, 033604.

[81] Branschädel, A., and G., Thomas. 2008. 2PI nonequilibrium versus transport equations for an ultracold Bose gas. *J. Phys. B: At. Mol. Opt. Phys.*, **41**, 135302.

[82] Bodet, C., Kronenwett, M., Nowak, B., Sexty, D., and Gasenzer, T. 2012. *Quantum Gases: Finite Temperature and Non-Equilibrium Dynamics*. College Press, London. Chap. Non-equilibrium quantum many-body dynamics: functional integral approaches.

[83] Berges, J., Rothkopf, A., and Schmidt, J. 2008. Non-thermal fixed points: effective weak-coupling for strongly correlated systems far from equilibrium. *Phys. Rev. Lett.*, **101**, 041603.

[84] Berges, J., and Hoffmeister, G. 2009. Nonthermal fixed points and the functional renormalization group. *Nucl. Phys.*, **B813**, 383.

[85] Scheppach, C., Berges, J., and Gasenzer, T. 2010. Matter-wave turbulence: beyond kinetic scaling. *Phys. Rev. A*, **81**, 033611.

[86] Kronenwett, M., and Gasenzer, T. 2011. Far-from-equilibrium dynamics of an ultracold Fermi gas. *Appl. Phys. B*, **102**, 469.

[87] Babadi, M., Demler, E., and Knap, M. 2015. Far-from-equilibrium field theory of many-body quantum spin systems: prethermalization and relaxation of spin spiral states in three dimensions. *Phys. Rev. X*, **5**, 041005.

[88] Myatt, C. J., Burt, E. A., Ghrist, R. W., Cornell, E. A., and Wieman, C. E. 1997. Production of two overlapping Bose-Einstein condensates by sympathetic cooling. *Phys. Rev. Lett.*, **78**, 586.

[89] Schreck, F., Ferrari, G., Corwin, K. L., Cubizolles, J., Khaykovich, L., Mewes, M.-O., and Salomon, C. 2001. Sympathetic cooling of bosonic and fermionic lithium gases towards quantum degeneracy. *Phys. Rev. A*, **64**, 011402.

[90] Catani, J., Barontini, G., Lamporesi, G., Rabatti, F., Thalhammer, G., Minardi, F., Stringari, S., and Inguscio, M. 2009. Entropy exchange in a mixture of ultracold atoms. *Phys. Rev. Lett.*, **103**, 140401.

[91] Erhard, M., Schmaljohann, H., Kronjäger, J., Bongs, K., and Sengstock, K. 2004. Bose-Einstein condensation at constant temperature. *Phys. Rev. A*, **70**, 031602(R).

[92] Shin, Y., Saba, M., Schirotzek, A., Pasquini, T. A., Leanhardt, A. E., Pritchard, D. E., and Ketterle, W. 2004. Distillation of Bose-Einstein condensates in a double-well potential. *Phys. Rev. Lett.*, **92**, 150401.

[93] Stellmer, S., Pasquiou, B., Grimm, R., and Schreck, F. 2013. Laser cooling to quantum degeneracy. *Phys. Rev. Lett.*, **110**, 263003.

[94] Kosterlitz, J. M., and Thouless, D. J. 1973. Ordering, metastability and phase transitions in two-dimensional systems. *J. Phys. C.*, **6**, 1181.

[95] Hadzibabic, Z., Kruger, P., Cheneau, M., Battelier, B., and Dalibard, J. 2006. Berezinskii-Kosterlitz-Thouless crossover in a trapped atomic gas. *Nature*, **441**, 1118.

[96] Schweikhard, V., Tung, S., and Cornell, E. A. 2007. Vortex proliferation in the Berezinskii-Kosterlitz-Thouless regime on a two-dimensional lattice of Bose-Einstein condensates. *Phys. Rev. Lett.*, **99**, 030401.

[97] Krüger, P., Hadzibabic, Z., and Dalibard, J. 2007. Critical point of an interacting two-dimensional atomic Bose gas. *Phys. Rev. Lett.*, **99**, 040402.

[98] Cladé, P., Ryu, C., Ramanathan, A., Helmerson, K., and Phillips, W. D. 2009. Observation of a 2D Bose gas: from thermal to quasicondensate to superfluid. *Phys. Rev. Lett.*, **102**, 170401.

[99] Tung, S., Lamporesi, G., Lobser, D., Xia, L., and Cornell, E. A. 2010. Observation of the presuperfluid regime in a two-dimensional Bose gas. *Phys. Rev. Lett.*, **105**, 230408.

[100] Hung, C.-L., Zhang, X., Gemekle, N., and Chin, C. 2011. Observation of scale invariance and universality in two-dimensional Bose gases. *Nature*, **470**, 236.

[101] Prokof'ev, N., Ruebenacker, O., and Svistunov, B. 2001. Critical point of a weakly interacting two-dimensional Bose gas. *Phys. Rev. Lett.*, **87**, 270402.

[102] Simula, T. P., and Blakie, P. B. 2006. Thermal activation of vortex-antivortex pairs in quasi-two-dimensional Bose-Einstein condensates. *Phys. Rev. Lett.*, **96**, 020404.

[103] Holzmann, M., and Krauth, W. 2008. Kosterlitz-Thouless transition of the quasi-two-dimensional trapped Bose gas. *Phys. Rev. Lett.*, **100**, 190402.

[104] Bisset, R. N., Davis, M. J., Simula, T. P., and Blakie, P. B. 2009. Quasicondensation and coherence in the quasi-two-dimensional trapped Bose gas. *Phys. Rev. A*, **79**, 033626.

[105] Cockburn, S. P., and Proukakis, N. P. 2012. *Ab initio* methods for finite-temperature two-dimensional Bose gases. *Phys. Rev. A*, **86**, 033610.

[106] Hadzibabic, Z., and Dalibard, J. 2011. Nano optics and atomics: transport of light and matter waves. *Proceedings of the International School of Physics "Enrico Fermi,"* vol. CLXXIII, vol. 34. Rivista del Nuovo Cimento. Chap. Two dimensional Bose fluids: an atomic physics perspective, page 389.

[107] Roumpos, G., Lohse, M., Nitsche, W. H., Keeling, J., Szymańska, M. H., Littlewood, P. B., Löffler, A., Höfling, S., Worschech, L., Forchel, A., and Yamamoto, Y.. 2012.

Power-law decay of the spatial correlation function in exciton-polariton condensates. *Proc. Natl. Acad. Sci. U.S.A.*, **109**(17), 6467–6472.

[108] Nitsche, W. H., Kim, N. Y., Roumpos, G., Schneider, C., Kamp, M., Höfling, S., Forchel, A., and Yamamoto, Y. 2014. Algebraic order and the Berezinskii-Kosterlitz-Thouless transition in an exciton-polariton gas. *Phys. Rev. B*, **90**, 205430.

[109] Petrov, D. S., Shlyapnikov, G. V., and Walraven, J. T. M. 2001. Phase-fluctuating 3D Bose-Einstein condensates in elongated traps. *Phys. Rev. Lett.*, **87**, 050404.

[110] Shvarchuck, I., Buggle, Ch., Petrov, D. S., Dieckmann, K., Zielonkowski, M., Kemmann, M., Tiecke, T. G., von Klitzing, W., Shlyapnikov, G. V., and Walraven, J. T. M. 2002. Bose-Einstein condensation into nonequilibrium states studied by condensate focusing. *Phys. Rev. Lett.*, **89**, 270404.

[111] Hugbart, M., Retter, J. A., Varón, A. F., Bouyer, P., Aspect, A., and Davis, M. J. 2007. Population and phase coherence during the growth of an elongated Bose-Einstein condensate. *Phys. Rev. A*, **75**, 011602.

[112] Binney, J. J., Dowrick, N. J., Fisher, A. J., and Newman, M. E. J. 1992. *The Theory of Critical Phenomena: An Introduction to the Renormalization Group*. New York, NY: Oxford University Press.

[113] Zinn-Justin, J. 2002. *Quantum Field Theory and Critical Phenomena*, 4th edn. Oxford, UK: Clarendon Press.

[114] Hohenberg, P. C., and Halperin, B. I. 1977. Theory of dynamic critical phenomena. *Rev. Mod. Phys.*, **49**, 435.

[115] Baym, G., Blaizot, J.-P., Holzmann, M., Laloë, F., and Vautherin, D. 1999. The transition temperature of the dilute interacting Bose gas. *Phys. Rev. Lett.*, **83**, 1703.

[116] Kashurnikov, V. A., Prokof'ev, N. V., and Svistunov, B. V. 2001. Critical temperature shift in weakly interacting Bose gas. *Phys. Rev. Lett.*, **87**, 120402.

[117] Arnold, P., and Moore, G. 2001. BEC transition temperature of a dilute homogeneous imperfect Bose gas. *Phys. Rev. Lett.*, **87**, 120401.

[118] Gerbier, F., Thywissen, J. H., Richard, S., Hugbart, M., Bouyer, P., and Aspect, A. 2004. Critical temperature of a trapped, weakly interacting Bose gas. *Phys. Rev. Lett.*, **92**, 030405.

[119] Davis, M. J., and Blakie, P. B. 2006. Critical temperature of a trapped Bose gas: comparison of theory and experiment. *Phys. Rev. Lett.*, **96**, 060404.

[120] Smith, R. P., Campbell, R. L. D., Tammuz, N., and Hadzibabic, Z. 2011. Effects of interactions on the critical temperature of a trapped Bose gas. *Phys. Rev. Lett.*, **106**, 250403.

[121] Ritter, S., Öttl, A., Donner, T., Bourdel, T., Köhl, M., and Esslinger, T. 2007. Observing the formation of long-range order during Bose-Einstein condensation. *Phys. Rev. Lett.*, **98**, 090402.

[122] Donner, T., Ritter, S., Bourdel, T., Öttl, A., Köhl, M., and Esslinger, T. 2007. Critical behavior of a trapped interacting Bose gas. *Science*, **315**, 1556.

[123] Bezett, A., and Blakie, P. B. 2009. Critical properties of a trapped interacting Bose gas. *Phys. Rev. A*, **79**, 033611.

[124] Navon, N., Gaunt, A. L., Smith, R. P., and Hadzibabic, Z. 2015. Critical dynamics of spontaneous symmetry breaking in a homogeneous Bose gas. *Science*, **347**, 167.

[125] Kibble, T. W. B. 1976. Topology of cosmic domains and strings. *J. Phys. A: Math. Gen.*, **9**, 1387.

[126] Landau, L. D., and Lifshitz, E. M. 1980. *Statistical Physics, Part 1*, 3rd edn. Oxford, UK: Butterworth-Heinemann.

[127] Zurek, W. H. 1985. Cosmological experiments in superfluid helium? *Nature*, **317**, 505.

[128] Zurek, W. H. 1996. Cosmological experiments in condensed matter systems. *Phys. Rep.*, **276**, 177.

[129] Anglin, J. R., and Zurek, W. H. 1999. Vortices in the wake of rapid Bose-Einstein condensation. *Phys. Rev. Lett*, **83**, 1707.

[130] Weiler, C. N., Neely, T. W., Scherer, D. R., Bradley, A. S., Davis, M. J., and Anderson, B. P. 2008. Spontaneous vortices in the formation of Bose-Einstein condensates. *Nature*, **455**, 948.

[131] Freilich, D. V., Bianchi, D. M., Kaufman, A. M., Langin, T. K., and Hall, D. S. 2010. Real-time dynamics of single vortex lines and vortex dipoles in a Bose-Einstein condensate. *Science*, **329**, 1182.

[132] Gardiner, C. W., and Davis, M. J. 2003. The stochastic Gross-Pitaevskii equation: II. *J. Phys. B: At. Mol. Opt. Phys.*, **36**, 4731.

[133] Blakie, P. B., Bradley, A. S., Davis, M. J., Ballagh, R. J., and Gardiner, C. W. 2008. Dynamics and statistical mechanics of ultra-cold Bose gases using c-field techniques. *Adv. Phys.*, **57**, 363.

[134] Proukakis, N. P., and Jackson, B. 2008. Finite temperature models of Bose-Einstein condensation. *J. Phys. B: At. Mol. Opt.*, **41**, 203002.

[135] Cockburn, S. P., and Proukakis, N. P. 2009. The stochastic Gross-Pitaevskii equation and some applications. *Laser Phys.*, **19**, 558.

[136] Steel, M. J., Olsen, M. K., Plimak, L. I., Drummond, P. D., Tan, S. M., Collett, M. J., Walls, D. F., and Graham, R. 1998. Dynamical quantum noise in trapped Bose-Einstein condensates. *Phys. Rev. A*, **58**, 4824.

[137] Drummond, P. D., and Corney, J. F. 1999. Quantum dynamics of evaporatively cooled Bose-Einstein condensates. *Phys. Rev. A.*, **60**, R2661.

[138] Chang, J. J., Hamner, C., and Engels, P. 2009. *Formation of Solitons During the BEC Phase Transition*. 40th Annual Meeting of the APS Division of Atomic, Molecular and Optical Physics.

[139] Zurek, W. H. 2009. Causality in condensates: gray solitons as relics of BEC formation. *Phys. Rev. Lett.*, **102**, 105702.

[140] Damski, B., and Zurek, W. H. 2010. Soliton creation during a Bose-Einstein condensation. *Phys. Rev. Lett.*, **104**, 160404.

[141] Witkowska, E., Deuar, P., Gajda, M., and Rzążewski, K. 2011. Solitons as the early stage of quasicondensate formation during evaporative cooling. *Phys. Rev. Lett.*, **106**, 135301.

[142] del Campo, A, Retzker, A, and Plenio, M B. 2011. The inhomogeneous Kibble-Zurek mechanism: vortex nucleation during Bose-Einstein condensation. *New J. Phys.*, **13**, 083022.

[143] Lamporesi, G., Donadello, S., Serafini, S., Dalfovo, F., and Ferrari, G. 2013. Spontaneous creation of Kibble-Zurek solitons in a Bose-Einstein condensate. *Nat. Phys.*, **9**, 656.

[144] Donadello, S., Serafini, S., Tylutki, M., Pitaevskii, L. P., Dalfovo, F., Lamporesi, G., and Ferrari, G. 2014. Observation of solitonic vortices in Bose-Einstein condensates. *Phys. Rev. Lett.*, **113**, 065302.

[145] Chomaz, L., Corman, L., Bienaimé, T., Desbuquois, R., Weitenberg, C., Beugnon, J., Nascimbène, S., and Dalibard, J. 2015. Emergence of coherence via transverse condensation in a uniform quasi-two-dimensional Bose gas. *Nat. Comm.*, **6**, 6162.

[146] Corman, L., Chomaz, L., Bienaimé, T., Desbuquois, R., Weitenberg, C., Nascimbène, S., Dalibard, J., and Beugnon, J. 2014. Quench-induced supercurrents in an annular Bose gas. *Phys. Rev. Lett.*, **113**, 135302.

[147] Das, A., Sabbatini, J., and Zurek, W. H. 2012. Winding up superfluid in a torus via Bose Einstein condensation. *Sci. Rep.*, **2**, 352.

[148] Chesler, P. M., García-García, A. M., and Liu, H. 2015. Defect formation beyond Kibble-Zurek mechanism and holography. *Phys. Rev. X*, **5**, 021015.

[149] Sadler, L. E., Higbie, J. M., Leslie, S. R., Vengalattore, M., and Stamper-Kurn, D. M. 2006. Spontaneous symmetry breaking in a quenched ferromagnetic spinor Bose-Einstein condensate. *Nature*, **443**, 312.

[150] De, S., Campbell, D. L., Price, R. M., Putra, A., Anderson, B. M., and Spielman, I. B. 2014. Quenched binary Bose-Einstein condensates: spin-domain formation and coarsening. *Phys. Rev. A*, **89**, 033631.

[151] Papp, S. B., Pino, J. M., and Wieman, C. E. 2008. Tunable miscibility in a dual-species Bose-Einstein condensate. *Phys. Rev. Lett.*, **101**, 040402.

[152] McCarron, D. J., Cho, H. W., Jenkin, D. L., Köppinger, M. P., and Cornish, S. L. 2011. Dual-species Bose-Einstein condensate of ^{87}Rb and ^{133}Cs. *Phys. Rev. A*, **84**, 011603.

[153] Liu, I.-K., Pattinson, R. W., Billam, T. P., Gardiner, S. A., Cornish, S. L., Huang, T.-M., Lin, W.-W., Gou, S.-C., Parker, N. G., and Proukakis, N. P. 2015. Stochastic growth dynamics and composite defects in quenched immiscible binary condensates. *Phys. Rev. A* **93**, 023628.

[154] Sabbatini, J., Zurek, W. H., and Davis, M. J. 2011. Phase separation and pattern formation in a binary Bose-Einstein condensate. *Phys. Rev. Lett.*, **107**, 230402.

[155] Swisłocki, T., Witkowska, E., Dziarmaga, J., and Matuszewski, M. 2013. Double universality of a quantum phase transition in spinor condensates: modification of the Kibble-Zurek mechanism by a conservation law. *Phys. Rev. Lett.*, **110**, 045303.

[156] Hofmann, J., Natu, S. S., and Das Sarma, S. 2014. Coarsening dynamics of binary Bose condensates. *Phys. Rev. Lett.*, **113**, 095702.

[157] Mathey, S., Gasenzer, T., and Pawlowski, J. M. 2015. Anomalous scaling at nonthermal fixed points of Burgers' and Gross-Pitaevskii turbulence. *Phys. Rev. A*, **92**, 023635.

[158] Piñeiro Orioli A., Boguslavski, K., and Berges, J. 2015. Universal self-similar dynamics of relativistic and nonrelativistic field theories near nonthermal fixed points. *Phys. Rev. D*, **92**, 025041.

[159] Berges, J., and Sexty, D. 2012. Bose condensation far from equilibrium. *Phys. Rev. Lett.*, **108**, 161601.

[160] Nowak, B., Sexty, D., and Gasenzer, T. 2011. Superfluid turbulence: nonthermal fixed point in an ultracold Bose gas. *Phys. Rev. B*, **84**, 020506(R).

[161] Nowak, B., Schole, J., Sexty, D., and Gasenzer, T. 2012. Nonthermal fixed points, vortex statistics, and superfluid turbulence in an ultracold Bose gas. *Phys. Rev. A*, **85**, 043627.

[162] Nowak, B., Erne, S., Karl, M., Schole, J., Sexty, D., and Gasenzer, T. 2013. Non-thermal fixed points: universality, topology, and turbulence in Bose gases. In: *Proc. Int. School on Strongly Interacting Quantum Systems Out of Equilibrium, Les Houches, 2012* (to appear). arXiv:1302.1448.

[163] Schole, J., Nowak, B., and Gasenzer, T. 2012. Critical dynamics of a two-dimensional superfluid near a non-thermal fixed point. *Phys. Rev. A*, **86**, 013624.

[164] Schmidt, M., Erne, S., Nowak, B., Sexty, D., and Gasenzer, T. 2012. Nonthermal fixed points and solitons in a one-dimensional Bose gas. *New J. Phys.*, **14**, 075005.

[165] Karl, M., Nowak, B., and Gasenzer, T. 2013. Tuning universality far from equilibrium. *Sci. Rep.*, **3**, 2394.

[166] Karl, M., Nowak, B., and Gasenzer, T. 2013. Universal scaling at non-thermal fixed points of a two-component Bose gas. *Phys. Rev. A*, **88**, 063615.

[167] Gasenzer, T., McLerran, L., Pawlowski, J. M., and Sexty, D. 2014. Gauge turbulence, topological defect dynamics, and condensation in Higgs models. *Nucl. Phys.*, **A930**, 163.

[168] Ewerz, C., Gasenzer, T., Karl, M., and Samberg, A. 2015. Non-thermal fixed point in a holographic superfluid. *J. High Energy Phys.*, **05**, 070.

[169] Damle, K., Majumdar, S. N., and Sachdev, S. 1996. Phase ordering kinetics of the Bose gas. *Phys. Rev. A*, **54**, 5037.

[170] Bray, A. J. 1994. Theory of phase-ordering kinetics. *Adv. Phys.*, **43**, 357.

[171] Connaughton, C., Josserand, C., Picozzi, A., Pomeau, Y., and Rica, S. 2005. Condensation of classical nonlinear waves. *Phys. Rev. Lett.*, **95**, 263901.

[172] Aarts, G., Bonini, G. F., and Wetterich, C. 2000. Exact and truncated dynamics in nonequilibrium field theory. *Phys. Rev. D*, **63**, 025012.

[173] Berges, J., Borsanyi, S., and Wetterich, C. 2004. Prethermalization. *Phys. Rev. Lett.*, **93**, 142002.

[174] Gring, M., Kuhnert, M., Langen, T., Kitagawa, T., Rauer, B., Schreitl, M., Mazets, I., Adu Smith, D., Demler, E., and Schmiedmayer, J. 2012. Relaxation and prethermalization in an isolated quantum system. *Science*, **337**, 1318.

[175] Jaynes, E. T. 1957a. Information theory and statistical mechanics. *Phys. Rev.*, **106**, 620.

[176] Jaynes, E. T. 1957b. Information theory and statistical mechanics. II. *Phys. Rev.*, **108**, 171–190.

[177] Rigol, M., Dunjko, V., Yurovsky, V., and Olshanii, M. 2007. Relaxation in a completely integrable many-body quantum system: an *ab inito* study of the dynamics of the highly excited states of 1D lattice hard-core bosons. *Phys. Rev. Lett.*, **98**, 050405.

[178] Rigol, M., Dunjko, V., and Olshanii, M. 2008. Thermalization and its mechanism for generic isolated quantum systems. *Nature*, **452**, 854.

[179] Langen, T., Erne, S., Geiger, R., Rauer, B., Schweigler, T., Kuhnert, M., Rohringer, W., Mazets, I. E., Gasenzer, T., and Schmiedmayer, J. 2015. Experimental observation of a generalized Gibbs ensemble. *Science*, **348**, 207.

[180] Nardin, G., Lagoudakis, K. G., Wouters, M., Richard, M., Baas, A., André, R., Dang, L. S., Pietka, B., and Deveaud-Plédran, B. 2009. Dynamics of long-range ordering in an exciton-polariton condensate. *Phys. Rev. Lett.*, **103**, 256402.

[181] Belykh, V. V., Sibeldin, N. N., Kulakovskii, V. D., Glazov, M. M., Semina, M. A., Schneider, C., Höfling, S., Kamp, M., and Forchel, A. 2013. Coherence expansion and polariton condensate formation in a semiconductor microcavity. *Phys. Rev. Lett.*, **110**, 137402.

[182] Lagoudakis, K. G., Manni, F., Pietka, B., Wouters, M., Liew, T. C. H., Savona, V., Kavokin, A. V., André, R., and Deveaud-Plédran, B. 2011. Probing the dynamics of spontaneous quantum vortices in polariton superfluids. *Phys. Rev. Lett.*, **106**, 115301.

[183] Klaers, J., Schmitt, J., Vewinger, F., and Weitz, M. 2010. Bose-Einstein condensation of photons in an optical microcavity. *Nature*, **468**, 545.

[184] Demokritov, S. O., Demidov, V. E., Dzyapko, O., Melkov, G. A., Serga, A. A., Hillebrands, B., and Slavin, A. N. 2006. Bose-Einstein condensation of quasi-equilibrium magnons at room temperature under pumping. *Nature*, **443**, 430.

[185] Berges, J., Boguslavski, K., Schlichting, S., and Venugopalan, R. 2015. Universality far from equilibrium: from superfluid Bose gases to heavy-ion collisions. *Phys. Rev. Lett.*, **114**, 061601.

[186] Berges, J., Schenke, B., Schlichting, S., and Venugopalan, R. 2014. Turbulent thermalization process in high-energy heavy-ion collisions. *Nucl. Phys.*, **A931**, 348.

[187] Mace, M., Schlichting, S., and Venugopalan, R. 2016. Off-equilibrium sphaleron transitions in the Glasma. *Phys. Rev. D* **93**, 074036.

8

Quenches, Relaxation, and Prethermalization in an Isolated Quantum System

TIM LANGEN

Vienna Center for Quantum Science and Technology, Atominstitut, TU Wien, Austria
Present address: JILA, NIST, and University of Colorado, Boulder, USA

JÖRG SCHMIEDMAYER

Vienna Center for Quantum Science and Technology, Atominstitut, TU Wien, Austria

Does an isolated many-body system that is prepared in a nonthermal initial state relax to thermal equilibrium? In quantum mechanics, unitarity appears to render the very concept of thermalization counterintuitive. Nonetheless, many quantum systems can very successfully be described by a thermal state. With the progress in the control and probing of ultracold quantum gases, this fundamental problem has become within reach of experimental investigations. In this chapter, we present experiments with ultracold one-dimensional Bose gases which provide novel insights into this problem.

8.1 Introduction

The study of relaxation to a thermal state in an isolated system addresses the fundamental relation between the macroscopic description of statistical mechanics and the microscopic quantum world. It plays an important role in such diverse fields as cosmology, high-energy physics, and condensed matter and has been intensely debated since the 1920s [1], leading to important theoretical advances [2, 3, 4, 5, 6, 7, 8]. In contrast to that, actual physical implementations of quantum systems which are both well isolated and accessible to detailed experimental study have remained scarce.

This situation has changed profoundly with the advent of ultracold quantum gases (see Chapter 3 by Ketterle), which provide a unique platform for the study of such problems [9]. With typical system sizes from a few to $\sim 10^7$ atoms and temperatures reaching down to the pico-Kelvin range, the mere existence of these gases demonstrates their near-perfect isolation from the environment. Despite this isolation, the quantum evolution can be observed in great detail and on experimentally accessible time scales, while the dynamical tunability of many parameters

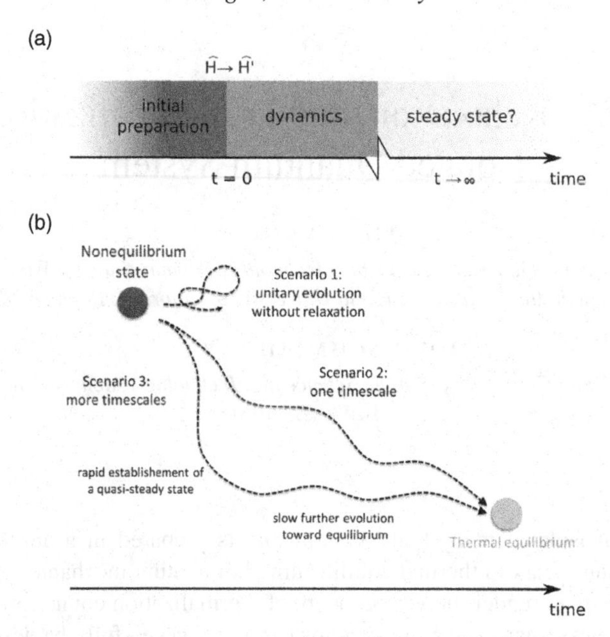

Figure 8.1 (a) Generic protocol of a quench: a rapid change of the Hamiltonian creates a nonequilibrium state. (b) Scenarios for the subsequent dynamics. Unitarity leads to time-reversal symmetry and prevents thermal equilibrium from being reached. Alternative scenarios conjecture that the expectation values of observables become arbitrarily close to their thermal values. The corresponding relaxation could proceed on a single time scale or through more complex paths with intermediate states that already share certain properties with the thermal equilibrium state. Figure adapted with permission from Langen, T. (2013), Non-equilibrium dynamics of one-dimensional Bose gases, Ph.D. thesis, Vienna University of Technology [10].

allows the realization of a multitude of different physical situations (see Chapters 13–18 for a discussion of selected topical issues with ultracold atoms).

A particularly useful protocol in studying nonequilibrium physics is the quantum quench: an instantaneous change of the system's Hamiltonian $\hat{H} \to \hat{H}'$ at time t_0. After the quench, the many-body wave-function $|\psi_0\rangle$ is in general not an eigenstate of the new Hamiltonian $\hat{H}'$ and will evolve in time as $|\psi(t)\rangle = \exp(-i\hat{H}'t/\hbar)|\psi_0\rangle$. For all practically relevant cases, this leads to a complex time evolution, which might or might not lead to a thermal state. Different scenarios are depicted in Fig. 8.1.

The situation is further complicated if the system is integrable, i.e., exhibits nontrivial conserved quantities. As in classical mechanics, the conserved quantities constrain the dynamics of the system after the quench [11, 12]. Consequently, even if thermalization is possible in general, it will be strongly suppressed in this case. Historically, the study of integrable systems was important for the understanding of equilibration and thermalization in the classical world. In the following, we will show that, analogously, important insights into equilibration and thermalization in the quantum world can be obtained from the study of quantum integrable systems [13]. Related concepts of the role of nonthermal fixed points in the relaxation of a quantum many-body system are discussed in Chapter 7 by Davis et al.

8.2 Studying Nonequilibrium Dynamics in 1D Systems

One-dimensional (1D) Bose gases are a versatile testbed to study quantum many-body physics out of equilibrium. They offer a rich variety of interesting many-body effects, while still being tractable with reasonable theoretical effort. This allows for a precise comparison between theoretical predictions and experimental results [14].

In the homogeneous limit, a 1D Bose gas with short-range interactions is described by the Lieb-Liniger Hamiltonian [15]

$$\hat{H} = \frac{\hbar^2}{2m} \int dz \, \frac{\partial \hat{\Psi}^\dagger(z)}{\partial z} \frac{\partial \hat{\Psi}(z)}{\partial z} + \frac{g}{2} \int dz\, dz' \, \hat{\Psi}^\dagger(z) \hat{\Psi}^\dagger(z') \delta(z-z') \hat{\Psi}(z') \hat{\Psi}(z), \quad (8.1)$$

where the interaction strength is given by the Lieb-Liniger parameter $\gamma = mg/\hbar^2 n_{1d}$. Here, $\hat{\Psi}(z)$ are bosonic field operators, m is the mass of the atoms, g is the interaction strength, n_{1d} is the 1D line density, and $\hbar$ denotes the reduced Planck constant. For $\gamma \gg 1$, the gas is strongly interacting (Tonks-Girardeau regime), for $\gamma \ll 1$ the gas is a weakly interacting quasicondensate where density fluctuations are suppressed while the phase fluctuates strongly along the system [16]. One can express the field operators in terms of density and phase operators $\hat{\Psi}(z) = e^{i\hat{\theta}(z)} \sqrt{\hat{n}(z)}$, which satisfy the commutation relation $[\hat{n}(z), \hat{\theta}(z')] = i\delta(z - z')$ [17]. Neglecting higher-order terms then leads to a quadratic model describing the low-energy limit of the Lieb-Liniger model, the Luttinger liquid Hamiltonian:

$$\hat{H} = \frac{\hbar c}{2} \int dz \left[\frac{K}{\pi} \left(\frac{\partial \hat{\theta}(z)}{\partial z} \right)^2 + \frac{\pi}{K} \hat{n}(z)^2 \right] = \sum_k \hbar \omega_k \hat{a}_k^\dagger \hat{a}_k, \quad (8.2)$$

with $K = \sqrt{n_{1d}(\hbar\pi)^2/4gm}$ the Luttinger parameter and $c = \sqrt{g n_{1D}/m}$ the speed of sound. The eigenmodes are noninteracting phonons with momenta k and linear dispersion relation $\omega_k = ck$. Their creation and annihilation operators $\hat{a}_k$ and $\hat{a}_k^\dagger$ are directly related to the Fourier components of density and phase (phase and density quadrature of the phonon) via

$$\hat{n}_k \sim \left(\hat{a}_k(t) + \hat{a}_{-k}^\dagger(t) \right), \qquad \hat{\theta}_k \sim \left(\hat{a}_k(t) - \hat{a}_{-k}^\dagger(t) \right). \quad (8.3)$$

The Lieb-Liniger model and its low-energy limit, the Luttinger liquid, are prime examples of integrable models [15, 11]. As we have outlined above, this has a profound influence on the nonequilibrium dynamics.

Creating an effectively 1D system in a 3D world requires a situation where the dynamics along the radial directions can be frozen out. For cold atoms, this can be realized in traps with a tight confinement in the two *transverse* spatial directions, characterized by the harmonic trapping frequency $\omega_\perp$. The gas will behave effectively one-dimensionally once the temperature T and the chemical potential $\mu = gn_{1D}$ fulfill the condition $k_B T, \mu \ll \hbar\omega_\perp$. The physics in the axial direction can then be described by the Lieb-Liniger and Luttinger liquid Hamiltonians. If the s-wave scattering length a_s is small compared with the ground-state width of the radial confinement, the effective 1D interaction strength that enters these Hamiltonians is given by $g = 2\hbar a_s \omega_\perp$ [18].

The highly anisotropic traps required to implement 1D systems can be created in strongly focused optical dipole traps [19, 20], optical lattices [21, 22], or, as in our experiments, a magnetic microtrap [23, 24]. Typical trap frequencies in our setup are $\omega_\perp = 2\pi \cdot 2\,\text{kHz}$ in the tightly confining radial directions and $\omega_\parallel = 2\pi \cdot 10\,\text{Hz}$ in the weakly confining axial direction. The 1D Bose gas is prepared using standard laser and evaporative cooling. Temperatures are on the order of $T \sim 20\text{--}100\,\text{nK}$ and the chemical potential amounts to $\mu \sim 2\pi \cdot 0.5\text{--}1\,\text{kHz}$. This results in $\gamma \sim 0.03$, such that the gas is a weakly interacting quasicondensate.

8.3 Creating a Nonequilibrium State by Splitting

We create a nonequilibrium state by coherently splitting a single 1D gas into two halves [25, 26] (see Fig. 8.2). The splitting is realized by dressing the Zeeman sublevels of the atoms using near-field radio-frequency radiation (for details, see [27]). If the splitting is performed fast compared with the dynamics in the system ($t_{\text{split}} < \hbar/\mu$), a quench is realized. The splitting can then be described as a *local* beam splitter, where each atom is coherently and independently split into the left and right half of the new system, leading to local, binomial number fluctuations.

Once the two 1D systems are spatially separated, it is convenient to perform a variable transformation to *new* symmetric (common) and antisymmetric (relative) degrees of freedom. Starting from *old* degrees of freedom given by the density and phase fluctuations in the left and right halves (denoted by $\hat{n}_{l,r}(z)$ and $\hat{\theta}_{l,r}(z)$, respectively), we find

$$\hat{\phi}_{\text{rel}}(z) = \hat{\theta}_l(z) - \hat{\theta}_r(z), \qquad \hat{\phi}_{\text{com}}(z) = [\hat{\theta}_r(z) + \hat{\theta}_l(z)]/2$$
$$\hat{\nu}_{\text{rel}}(z) = \hat{n}_r(z) - \hat{n}_l(z), \qquad \hat{\nu}_{\text{com}}(z) = [\hat{n}_r(z) + \hat{n}_l(z)]/2.$$

The common degrees of freedom inherit all initial (thermal) excitations that were present in the gas before the splitting. The relative degrees of freedom only contain the quantum shot-noise that, for the fast splitting discussed here, is introduced by

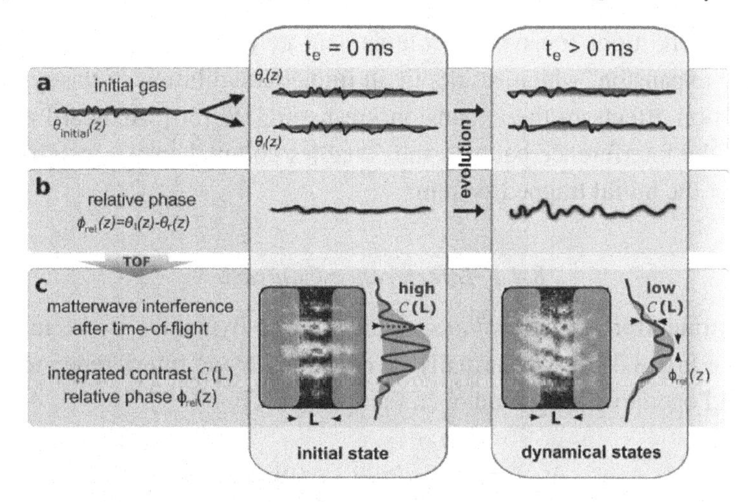

Figure 8.2 Experimental scheme. (a) An initial phase fluctuating 1D Bose gas is split into two uncoupled gases with almost identical phase distributions $\phi_l(z)$ and $\phi_r(z)$ (represented by the black solid lines) and allowed to evolve for a time t. (b) At $t = 0\,\text{ms}$, fluctuations in the local phase difference $\phi_{\text{rel}}(z)$ between the two gases are very small, but start to randomize during the evolution. It is an open question if and how this randomization leads to the thermal equilibrium situation of completely uncorrelated gases. (c) Typical experimental matter-wave interference patterns obtained by overlapping the two gases in time-of-flight. Differences in the local relative phase lead to a locally displaced interference pattern. Extracting the contrast $\mathcal{C}(L)$ over a length L or the relative phase $\phi_{\text{rel}}(z)$ from these interference patterns allows us to probe the dynamics. Figure adapted with permission from Langen, T. (2013), Non-equilibrium dynamics of one-dimensional Bose gases, Ph.D. thesis, Vienna University of Technology [10].

the binomial distribution of atoms into the two halves. The state created by splitting is thus out of equilibrium.

In thermal equilibrium, the relative and the common degrees of freedom contain an equal amount of energy. In the experiment, this thermal equilibrium situation can intentionally be prepared by splitting the trap *before* a degenerate gas is created. Subsequent cooling in both wells creates two independent degenerate gases with no mutual knowledge of each other [27].

This setup thus provides the unique possibility to deliberately prepare and contrast nonequilibrium and thermal states in identical settings.

8.4 Probing the Quantum State

We probe the system and its dynamics after splitting by switching off the double-well trap and allowing for a time-of-flight expansion (for details, see [28]). The pair of condensates expands, overlaps, and forms a matter-wave interference pattern [29]. The tight transversal confinement of the 1D gases leads to a very rapid radial expansion, which results in an immediate dilution of the system such that interaction effects in the expansion are negligible [30]. Standard absorption imaging of the interference patterns thus enables comprehensive insights into the properties of the initial trapped system.

8.4.1 Interference Contrast

The most straightforward technique to analyze the dynamics is the interference contrast (see Fig. 8.2). It can formally be described by an interference term of the bosonic field operators integrated over a length L [31]

$$\hat{A}(L) = \int_{L/2}^{L/2} dz\, \hat{\Psi}_l^{\dagger}(z,t)\hat{\Psi}_r(z,t). \tag{8.4}$$

The magnitude of $\hat{A}(L)$ is related to the integrated contrast of the interference patterns $\langle \mathcal{C}^2(L)\rangle = \langle |\hat{A}(L)|^2\rangle / n_{1d}^2 L^2$. Experimentally, the normalized quantity $\alpha = \mathcal{C}^2/\langle |\mathcal{C}|^2\rangle$ is less prone to systematic deviations.

8.4.2 Full Distribution Functions

Recording the shot-to-shot fluctuations of α gives the full distribution function (FDF) $P(\alpha)d\alpha$, i.e., the probability of observing a contrast in the interval $\alpha + d\alpha$. The FDFs contain the information about all even moments of $\hat{A}(L)$

$$\frac{\langle |\hat{A}|^{2m}\rangle}{\langle |\hat{A}|^2\rangle^m} = \langle \alpha^m \rangle = \int_0^{\infty} P(\alpha)\alpha^m d\alpha, \tag{8.5}$$

and can thus provide deeper insights into the state of the system than the mean contrast alone [32, 33, 31].

8.4.3 Phase Correlation Functions

Similar to the situation in Young's double-slit experiment, the local *in situ* relative phase $\phi_{\text{rel}}(z)$ between the two halves of the system determines the local position of the interference pattern in time-of-flight. The observed interference pattern thus allows the determination of $\phi_{\text{rel}}(z)$. From the measured values of $\phi_{\text{rel}}(z)$, we can extract equal-time N-point correlation functions

$$C(z_1, z_2, \ldots, z_N) \sim \langle \exp[i\phi_{\text{rel}}(z_1) - i\phi_{\text{rel}}(z_2) + \cdots - i\phi_{\text{rel}}(z_N)]\rangle. \tag{8.6}$$

Here, $z_1, z_2, \ldots, z_N$ are coordinates along the 1D direction of the system. In the experiments, the expectation value is realized by averaging over many identical realizations. Assuming that density fluctuations can be neglected, which is a very good approximation in the quasicondensate regime, the correlation functions are related to the field operators:

$$C(z_1, z_2, \ldots, z_N) \sim \langle \hat{\Psi}_l(z_1)\hat{\Psi}_r^\dagger(z_1)\hat{\Psi}_l^\dagger(z_2)\hat{\Psi}_r(z_2)\cdots\hat{\Psi}_l^\dagger(z_N)\hat{\Psi}_r(z_N)\rangle. \quad (8.7)$$

8.5 Probing the Relaxation

To get an overview of the different stages of the relaxation after the splitting quench, it is instructive to first look at the mean squared contrast $\langle C^2(L)\rangle$. As can be seen in Fig. 8.3, the relaxation starts with a fast initial decay followed by the establishment of a quasisteady state.

8.5.1 Prethermalization

We start our detailed discussion with this quasisteady state. Fig. 8.4a,b shows an analysis of the mean squared contrast and the FDFs in this steady state. Surprisingly, the data are very well described by a thermal-like state. However, while showing characteristic thermallike correlations, the quasisteady state is markedly different from thermal equilibrium [35, 36]. This difference is most striking when comparing the data with FDFs that were measured for an identical system in thermal equilibrium (Fig. 8.4c).

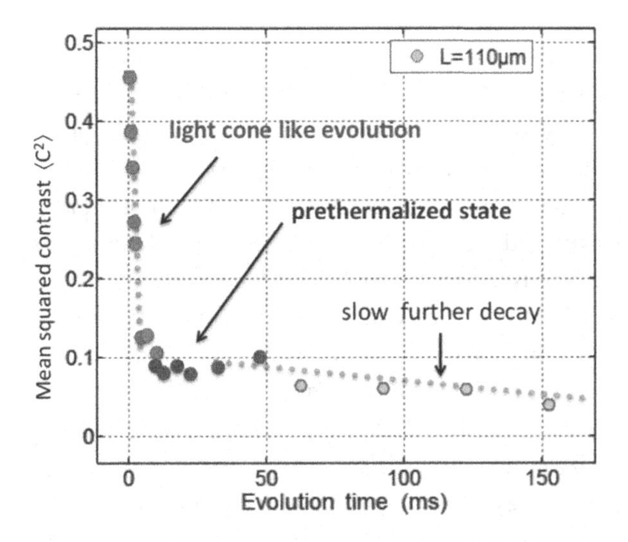

Figure 8.3 Overview of the relaxation process after the splitting quench illustrated through the mean squared contrast $\langle C^2(L)\rangle$ integrated over the central $L = 110\,\mu\text{m}$ of our sample. A fast initial decay, proceeding through a light-cone-like relaxation [34], leads to the establishment of a thermal-like quasisteady state. This *prethermalized* state [35, 36] can be described by a generalized Gibbs ensemble [37].

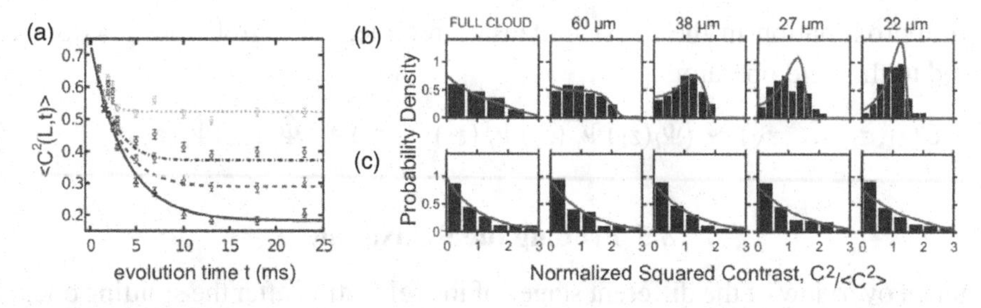

Figure 8.4 Contrast dynamics and full distribution functions of a coherently split 1D Bose gas. (a) Measured values of the mean squared contrast for various integration lengths L (points). From top to bottom: $L = 18, 40, 60, 100\,\mu$m. The lines show the results of a Luttinger liquid calculation with an effective temperature of $T_{\text{eff}} = 14$ nK, which is significantly lower than the true initial temperature of the gas ($T = 120$ nK). Reprinted with permission from Kuhnert, M., et al. (2013), Multimode dynamics and emergence of a characteristic length scale in a one-dimensional quantum system, *Phys. Rev. Lett.*, **110**, 090405 [36]. Copyright (2013) by the American Physical Society. (b) Full distribution functions after relaxation to the prethermalized state. The solid red lines show theoretical equilibrium distributions, also with $T_{\text{eff}} = 14$. The prethermalized nature of the state is clearly revealed by comparing it with the expected thermal equilibrium (c), prepared by creating two independent 1D Bose gases. Plots (b)–(c) reprinted with permission from Adu Smith, D., et al. (2013), Prethermalization revealed by the relaxation dynamics of full distribution functions, *New Journal of Physics*, **15**(7), 075011 [28].

We find that the temperature of the quasisteady state T_{eff} can be identified with the energy introduced by the shot-noise of the splitting process ($gn_{1\text{D}}/2 = k_B T_{\text{eff}}$), which is significantly smaller than the initial temperature T of the system. At the same time, the common degrees of freedom show a temperature comparable to the initial temperature $T > T_{\text{eff}}$. The system has thus not thermalized, but rather reached a prethermalized state [38, 39, 35]. The physics behind this is that common and relative degrees of freedom decouple for a balanced splitting quench. No energy can be exchanged and the system can never fully forget its initial state.

Microscopically, the establishment of the effective temperature $k_B T_{\text{eff}} = gn_{1\text{D}}/2$ can be well understood within the Luttinger liquid description as the dephasing of the phonon modes [25, 26]. The energy introduced by the quench is initially stored in the density quadrature and phonons, initialized at zero phase. During the time evolution, the energy of each mode oscillates between density and phase with the momentum-dependent frequency ω_k, which eventually leads to a dephasing. The thermal nature arises from the occupations of the modes. Because of the linear dispersion relation, the splitting prepares the relative degrees of freedom with occupation numbers that scale as $1/k$. All modes thus obtain the same amount of energy $k_B T_{\text{eff}}$ from shot-noise, which, after dephasing, makes the state indistinguishable from a thermal state with the corresponding temperature. Similar dynamics have recently also been observed in the cooling process of 1D Bose gases. In this case, outcoupling of atoms directly reduces the energy in the density quadrature and dephasing leads to an apparent cooling, far below the temperatures that are achievable by evaporative cooling alone [40, 41].

8.5.2 Light-Cone-like Emergence of the Prethermalized State

More insights about the local properties of the relaxation can be obtained by studying the two-point correlation function $C(z, z') \sim \langle \exp[i\phi(z) - i\phi(z')] \rangle$ (Fig. 8.5) [34]. Immediately after a fast coherent splitting quench, $C(z, z')$ is close to one over all relative distances $\bar{z} = z - z'$. After approximately 15 ms, the system settles into the prethermalized state, where correlations decay exponentially with $\bar{z}$. For a 1D Bose gas, this exponential decay corresponds to thermal correlations, with the characteristic length scale of the decay λ being directly related to the temperature T_{eff} via $\lambda = \hbar^2 n_{1D}/mk_B T_{\mathrm{eff}}$.

For a given point in time, the correlations decay exponentially up to a certain crossover distance z_c beyond which the long-range order of the initial state prevails. The evolution of this crossover point plotted in Fig. 8.5b is linear, revealing that the exponentially decaying correlations spread through the system in a light-cone-like dynamic with a characteristic velocity [34]. The velocity can be identified with

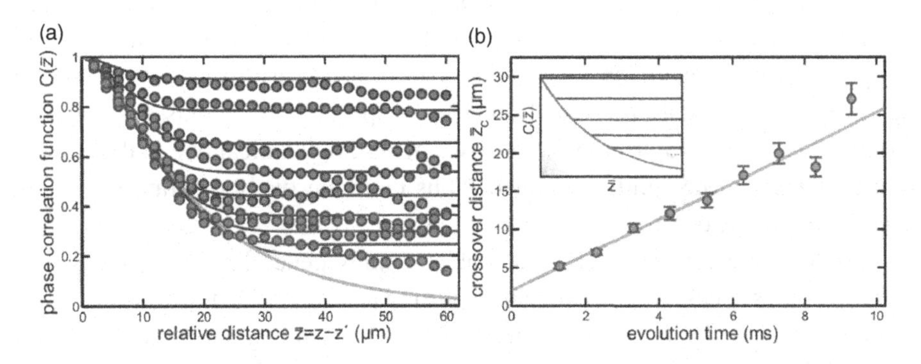

Figure 8.5 (a) Measured phase correlation functions (circles) of the evolution following the splitting together with the Luttinger liquid predictions (solid lines). Time increases from 1 ms after the splitting (top) to 9 ms (bottom). The lowermost exponential line is the prediction for the final prethermalized steady state. (b) Evolution of the crossover distance z_c between the exponentially decaying correlations and the plateau with long-range order (see text for details). The linear behavior shows that the thermal correlations appear locally and spread through the system in a light-cone-like fashion. Adapted with permission from Langen, T., et al. (2013), Local emergence of thermal correlations in an isolated quantum many-body system, *Nat. Physics*, **9**, 1 [34]. Copyright (2013) by the Nature Publishing Group.

the speed of sound of the phonons, which thus act as carriers of information in the system [42, 43]. This observation provides a direct connection between the establishment of thermal properties and the propagation of correlations in a continuous quantum many-body system. The underlying principles are even more general and also govern the distribution of entanglement, with profound implications, e.g., for quantum information science and computer simulations of complex materials [44].

8.5.3 Generalized Gibbs Ensemble

The fact that the phonon occupations of the system are preserved during the dynamics is deeply rooted in the integrability of the Luttinger liquid model that describes the dynamics. Each relative mode acts like a harmonic oscillator that does not interact but dephases with respect to the rest of the system.

Nevertheless, it has been conjectured that such systems still relax to a maximum entropy state which is given by the density matrix of a so-called generalized Gibbs ensemble (GGE) [12]

$$\hat{\rho} = \frac{1}{Z} e^{-\sum \lambda_j \hat{I}_j}. \tag{8.8}$$

Here, Z is the partition function, $\hat{I}_j$ are the operators of the conserved quantities, and λ_j the corresponding Lagrange multipliers. If only energy is conserved, this density matrix reduces to the well-known canonical or Gibbs ensemble, with temperature being the only Lagrange multiplier. If many more conserved quantities exist (e.g., the phonon occupations $\hat{I}_j = \hat{a}_j^\dagger \hat{a}_j$ in the Luttinger liquid model), many Lagrange multipliers, one for each conserved quantity, are necessary to maximize entropy. The prethermalized state presented above is a special case of this ensemble: all Lagrange multipliers describing the relative modes are identical due to the equipartition of energy in the splitting quench. Thus there are two temperatures, one for the common mode, one for the relative mode.

To demonstrate the presence of a more complex generalized Gibbs ensemble (GGE), it is thus necessary to change the splitting process, such that different modes exhibit occupations of the phonon modes different from equipartition. The results are shown in Fig. 8.6. Again, the relative phase correlation function can be used to characterize the dynamical states of the system. While we were showing only one coordinate of this function in Fig. 8.5, plotting the full function $C(z, z')$ provides more insight. The correlation functions show a trivial maximum on the diagonal ($z = z'$), which arises due to the fact that every point is perfectly correlated with itself. However, a second maximum arises on the antidiagonal ($z = -z'$), indicating that points that are located symmetrically around the center of the system are more strongly correlated. This implies that modes which are symmetric around the center are more strongly occupied than modes that are antisymmetric around the center.

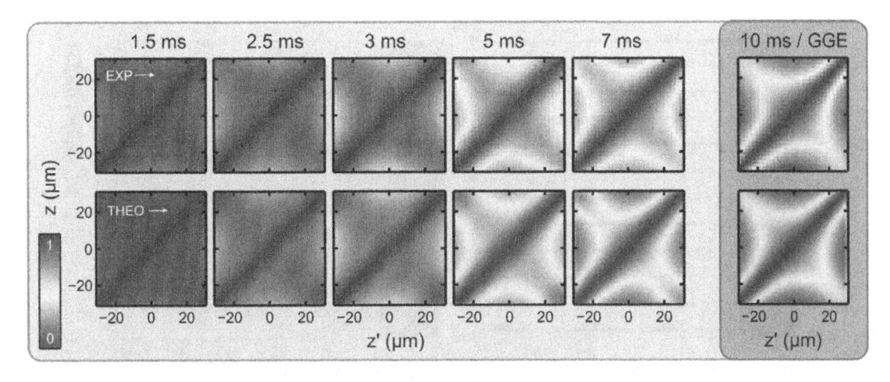

Figure 8.6 Relaxation dynamics of a coherently split 1D Bose gas with different populations for different modes. Two-point correlation functions $C(z,z')$ for increasing evolution time show maxima on the diagonal and the anti-diagonal. The experimental observations (top row) are in very good agreement with the theoretical model (bottom row), demonstrating the presence of many different temperatures in the system [45]. Adapted with permission from Langen, T. (2013), Non-equilibrium dynamics of one-dimensional Bose gases, Ph.D. thesis, Vienna University of Technology [10].

A more detailed analysis of the relaxed state allows the extraction of all mode occupations that are necessary to describe the state [45]. Given these extracted occupation numbers, the dephasing model also provides a detailed description of the dynamics, which proves that the conserved quantities were indeed set during the splitting process.

The measured interference patterns not only contain the two-point correlations described above, but also higher-order correlations up to ten-point functions can be extracted [10, 45, 46]. This shows that the GGE is a good description of the system at least up to ten-body observables.

Moreover, we observe that higher-order correlation functions factorize into two-point functions. Generally speaking, an analysis of whether and under which conditions this factorization is possible allows the characterization of the essential features of a quantum field theory from experimental measurements, detecting the relevant quasiparticles and their interactions [46].

Most importantly, these observations visualize, both experimentally and theoretically, how the unitary evolution of our quantum many-body system connects to a steady state that can be described by a thermodynamic ensemble.

8.5.4 Absence of Recurrences

While the dephasing of the system establishes a steady state that can be described by thermodynamics, one would nevertheless expect that the finite system size would still lead to a (periodic) rephasing to the initial state.

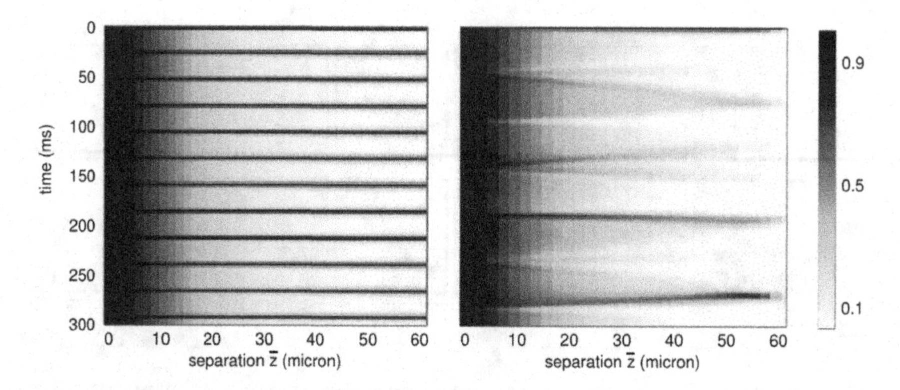

Figure 8.7 Time evolution of the relative-phase correlation function $C(\bar{z})$ for homogeneous (left) and trapped systems (right). The color-scale indicates the degree of correlation (black: high correlation, white: low correlation). In the homogeneous case, the initial state is periodically re-established at times which are multiples of the system length divided by the characteristic velocity. In the trapped case, the recurrences are only partial, and the more complex structure is due to the incommensurate mode frequencies. For the given example, the strongest recurrence is observed at 202 ms. Adapted with permission from Geiger, R., et al. (2014), Local relaxation and light-cone-like propagation of correlations in a trapped one-dimensional Bose gas, *New Journal of Physics*, **16**, 053034 [47]. Copyright (2014) by the Institute of Physics.

In a homogeneous system, the time between recurrences can be estimated as $t_{rec} = L/2c$, which corresponds to twice the time to reach the perfectly dephased prethermalized state. For typical parameters, $t_{rec} \sim 30$ ms. However, such rephasing is not observed in experiments, even on much longer time scales. The reason for this is the longitudinal harmonic confinement, which significantly alters the situation (Fig. 8.7) [47]. While in the ideal homogeneous case the mode energies are equally spaced $\omega_k = ck$, the modes in the experimentally realized harmonically trapped condensate are described by Legendre polynomials [16]. This leads to the modified dispersion relation $\omega_j = \omega_\parallel \sqrt{j(j+1)/2}$, where j is an integer. For typical parameters, this shifts the first significant revival in the trapped case to more than 200 ms [47]. The prethermalized state is thus indeed quasisteady and persists for very long times despite the underlying unitary dynamics of the system.

8.6 Outlook

Our experiments demonstrate the power of many-body dephasing to establish thermal properties in a quantum system, revealing important information about the relation between statistical and quantum mechanics. These results pose a number of new questions, which we discuss in the following.

8.6.1 Dynamics Beyond Prethermalization

In Sections 8.4 and 8.5.3 of this chapter, we demonstrated that the 1D Bose gas after a splitting quench does not relax to thermal equilibrium but to a prethermalized state that can be described by a generalized Gibbs ensemble. This behavior is rooted in the integrability of the Lieb-Liniger model and its low-energy approximation, the Luttinger liquid. However, the 1D Bose gas realized in our experiments is only nearly integrable. One reason for this is, for example, the existence of radially excited states that can affect the 1D dynamics [48].

It has been conjectured that in this case the observed prethermalized state is only an intermediate steady state on the way to thermal equilibrium, its lifetime being directly related to the degree of integrability breaking [49]. This may be related to a quantum version of the Kolmogorov-Arnold-Moser (KAM) theorem [50], which describes the consequences of integrability breaking in classical systems. First indications can be seen in Fig. 8.3, where the prethermalized state is followed by a further slow decay. Experimental and theoretical investigations into this effect are ongoing in our and other groups.

8.6.2 Coupled 1D Superfluids: The Sine-Gordon Model

More complex field theories can be investigated by studying nonequilibrium dynamics and relaxation in pairs of 1D Bose gases that are connected by a tunnel coupling. The evolution of the relative phase $\phi_{\text{rel}}(z)$ and density fluctuations ν_{rel} of the two superfluids can then be described by the sine-Gordon Hamiltonian,

$$\hat{H}_{\text{SG}} = \int \mathrm{d}z \left[g\hat{\nu}_{\text{rel}}^2 + \frac{\hbar^2 n_{1\text{D}}}{4m} (\partial_z \hat{\phi}_{\text{rel}})^2 \right] - \int \mathrm{d}z \, 2\hbar J n_{1\text{D}} \cos \hat{\phi}_{\text{rel}} \,, \tag{8.9}$$

where J denotes the tunnel coupling strength. The first term represents again the quadratic Luttinger liquid Hamiltonian. The second term is non-quadratic and includes all powers of the field $\hat{\phi}_{\text{rel}}$, which leads to many intriguing properties such as a tunable gap, non-Gaussian fluctuations, non-trivial quasiparticles, and topological excitations. A detailed study of the equilibrium properties of tunnel-coupled superfluids using high-order correlation functions can be found in Ref. [46]. The dynamics after a quench and the related nonequilibrium properties are currently under investigation.

8.6.3 Quantum State Tomography

In general, characterizing the quantum states of a many-body system is a formidable task even for systems containing only a few particles. The complexity of the states

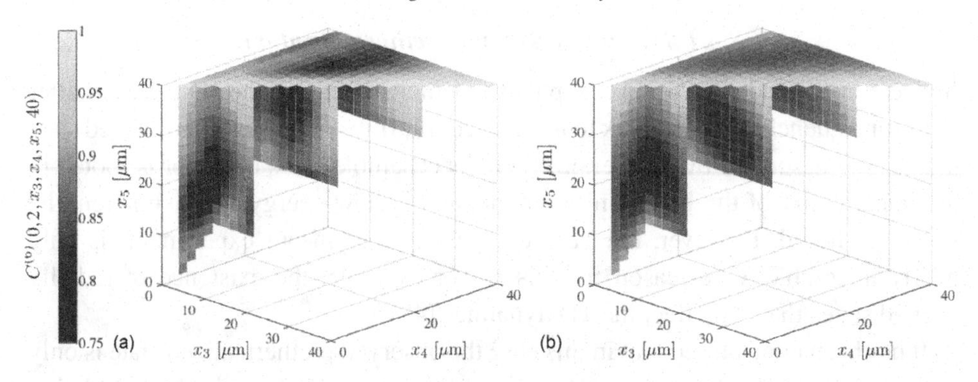

Figure 8.8 Projections of the (a) measured and (b) predicted correlation functions for $t = 3\,\text{ms}$ after the quench. We employ continuous matrix product states with bond dimension $d = 2$ to fit experimentally measured two-point and four-point functions and use the results to predict the six-point functions. Reprinted with permission from Steffens, A., et al. (2015), Towards experimental quantum field tomography with ultracold atoms, *Nature Comm.*, **6**, 7663 [52]. Copyright (2015) by the Nature Publishing Group.

studied here therefore prevents standard quantum state tomography. A different approach is quantum field tomography [51, 52]. It is based on continuous matrix product states [53] which naturally incorporate the locality present in realistic physical settings and are thus prime candidates for describing the physics of locally interacting quantum fields. The idea is to partially reconstruct the quantum fields of a many-body system from measured two-point and four-point correlation functions. An example is shown in Fig. 8.8. We expect this versatile technique to play an important role in future studies of continuous quantum many-body systems.

8.7 Conclusion

The relaxation of isolated quantum many-body systems is an interesting but yet unsolved problem. Experiments with ultracold quantum gases, and 1D Bose gases in particular, allow the realization and manipulation of well-controlled and truly isolated quantum systems. This provides novel opportunities to study and understand nonequilibrium phenomena. For example, the results discussed in this chapter demonstrate several characteristic aspects of these dynamics, including the existence of stable, thermal-like prethermalized states and their dynamical, light-cone-like emergence. Furthermore, the connection of the prethermalized state with generalized statistical ensembles, and thus of the unitary quantum evolution and statistical mechanics, was highlighted. The progress in this field is rapid, and we expect it to continue to have profound implications for our understanding of quantum many-body systems.

Acknowledgments: This work was supported the Austrian Science Fund (FWF) through the SFB Foundations and Application of Quantum Science (FoQuS) and by the European Union (EU) (Simulators and Interfaces with Quantum Systems [SIQS] and European Collaborative Research [ERC] advanced grant Quantum-Relax). T. Langen acknowledges support by the Humboldt Foundation through a Feodor Lynen Research Fellowship.

References

[1] Neumann, J. V. 1929. Beweis des Ergodensatzes und des H-Theorems in der neuen Mechanik. *Z. Phys.*, **57**(Jan), 30–70.

[2] Jaynes, E. T. 1957a. Information theory and statistical mechanics. *Phys. Rev.*, **106**, 620–630.

[3] Jaynes, E. T. 1957b. Information theory and statistical mechanics. II. *Phys. Rev.*, **108**, 171–190.

[4] Srednicki, M. 1994. Chaos and quantum thermalization. *Phys. Rev. E*, **50**, 888–901.

[5] Deutsch, J. M. 1991. Quantum statistical mechanics in a closed system. *Phys. Rev. A*, **43**, 2046–2049.

[6] Rigol, Marcos, Dunjko, Vanja, and Olshanii, Maxim. 2008. Thermalization and its mechanism for generic isolated quantum systems. *Nature*, **452**, 854–858.

[7] Polkovnikov, A., Sengupta, K., Silva, A., and Vengalattore, M. 2011. Colloquium: nonequilibrium dynamics of closed interacting quantum systems. *Rev. Mod. Phys.*, **83**, 863–883.

[8] Eisert, J., Friesdorf, M., and Gogolin, C. 2015. Quantum many-body systems out of equilibrium. *Nature Phys.*, **11**, 124–130.

[9] Langen, T., Geiger, R., and Schmiedmayer, J. 2015. Ultracold atoms out of equilibrium. *Annu. Rev. Cond. Mat. Phys.*, **6**, 201.

[10] Langen, T. 2013. Non-equilibrium dynamics of one-dimensional Bose gases. Ph.D. thesis, Vienna University of Technology.

[11] Sutherland, B. 2004. *Beautiful Models: 70 Years of Exactly Solved Quantum Many-Body Problems*. World Scientific.

[12] Rigol, M., Dunjko, V., Yurovsky, V., and Olshanii, M. 2007. Relaxation in a completely integrable many-body quantum system: An an ab-initio study of the dynamics of the highly excited states of 1D lattice hard-core bosons. *Phys. Rev. Lett.*, **98**, 050405.

[13] Langen, T., Gasenzer, T., and Schmiedmayer, J. Forthcoming (2016). Prethermalization and universal dynamics in near-integrable quantum systems. *J. Stat.* Mech. 064009.

[14] Cazalilla, M., Citro, R., Giamarchi, T., Orignac, E., and Rigol, M. 2011. One dimensional bosons: from condensed matter systems to ultracold gases. *Rev. Mod. Phys.*, **83**, 1405–1466.

[15] Lieb, E. H., and Liniger, W. 1963. Exact analysis of an interacting Bose gas. I. The general solution and the ground state. *Phys. Rev.*, **130**, 1605–1616.

[16] Petrov, D. M., Gangardt, G. V., and Shlyapnikov, D. S. 2004. Low-dimensional trapped gases. *J. Phys. IV France*, **116**, 3–44.

[17] Mora, C., and Castin, Y. 2003. Extension of Bogoliubov theory to quasicondensates. *Phys. Rev. A*, **67**, 053615.

[18] Olshanii, M. 1998. Atomic scattering in presence of an external confinement and a gas of impenetrable bosons. *Phys. Rev. Lett.*, **81**, 938–941.

[19] Dettmer, S., Hellweg, D., Ryytty, P., Arlt, J., Ertmer, W., Sengstock, K., Petrov, D., Shlyapnikov, G., Kreutzmann, H., Santos, L., and Lewenstein, M. 2001. Observation of phase fluctuations in elongated Bose-Einstein condensates. *Phys. Rev. Lett.*, **87**, 160406.

[20] Serwane, F., Zürn, G., Lompe, T., Ottenstein, T. B., Wenz, A. N., and Jochim, S. 2011. Deterministic preparation of a tunable few-fermion system. *Science*, **332**, 336–338.

[21] Paredes, B., Widera, A., Murg, V., Mandel, O., Fölling, S., Cirac, I., Shlyapnikov, G. V., Hänsch, T. W., and Bloch, I. 2004. Tonks-Girardeau gas of ultracold atoms in an optical lattice. *Nature*, **429**, 277–281.

[22] Kinoshita, T., Wenger, T., and Weiss, D. 2006. A quantum Newton's cradle. *Nature*, **440**, 900–903.

[23] Folman, R., Kruger, P., Schmiedmayer, J., Denschlag, J., and Henkel, C. 2002. Microscopic atom optics: from wires to an atom chip. *Adv. At. Mol. Opt. Phys.*, **48**, 263–356.

[24] Reichel, J., and Vuletic, V. (eds). 2011. *Atom Chips*. Wiley-VCH.

[25] Kitagawa, T., Pielawa, S., Imambekov, A., Schmiedmayer, J., Gritsev, V., and Demler, E. 2010. Ramsey interference in one-dimensional systems: the full distribution function of fringe contrast as a probe of many-body dynamics. *Phys. Rev. Lett.*, **104**, 255302.

[26] Kitagawa, T., Imambekov, A., Schmiedmayer, J., and Demler, E. 2011. The dynamics and prethermalization of one-dimensional quantum systems probed through the full distributions of quantum noise. *New. J. Phys.*, **13**, 073018.

[27] Hofferberth, S., Lesanovsky, I., Fischer, B., Verdu, J., and Schmiedmayer, J. 2006. Radio-frequency dressed state potentials for neutral atoms. *Nature Phys.*, **2**, 710–716.

[28] Adu Smith, D., Gring, M., Langen, T., Kuhnert, M., Rauer, B., Geiger, R., Kitagawa, T., Mazets, I., Demler, E., and Schmiedmayer, J. 2013. Prethermalization revealed by the relaxation dynamics of full distribution functions. *New J. Phys.*, **15**, 075011.

[29] Schumm, T., Hofferberth, S., Andersson, L. M., Wildermuth, S., Groth, S., Bar-Joseph, I., Schmiedmayer, J., and Krüger, P. 2005. Matter-wave interferometry in a double well on an atom chip. *Nature Phys.*, **1**, 57–62.

[30] Imambekov, A., Mazets, I. E., Petrov, D. S., Gritsev, V., Manz, S., Hofferberth, S., Schumm, T., Demler, E., and Schmiedmayer, J. 2009. Density ripples in expanding low-dimensional gases as a probe of correlations. *Phys. Rev. A*, **80**, 033604.

[31] Imambekov, A., Gritsev, V., and Demler, E. 2007. *Fundamental noise in matter interferometers in Ultracold Fermi gases, Proc. Internat. School Phys. Enrico Fermi*. IOS Press. Vol. 164, 535-606.

[32] Gritsev, V., Altman, E., Demler, E., and Polkovnikov, A. 2006. Full quantum distribution of contrast in interference experiments between interacting one dimensional Bose liquids. *Nature Phys.*, **2**, 705–709.

[33] Polkovnikov, A., Altman, E., and Demler, E. 2006. Interference between independent fluctuating condensates. *Proc. Natl. Acad. Sci.*, **103**, 6125–6129.

[34] Langen, T., Geiger, R., Kuhnert, M., Rauer, B., and Schmiedmayer, J. 2013. Local emergence of thermal correlations in an isolated quantum many-body system. *Nature Phys.*, **9**, 640–643.

[35] Gring, M., Kuhnert, M., Langen, T., Kitagawa, T., Rauer, B., Schreitl, M., Mazets, I., Adu Smith, D., Demler, E., and Schmiedmayer, J. 2012. Relaxation and prethermalization in an isolated quantum system. *Science*, **337**, 1318–1322.

[36] Kuhnert, M., Geiger, R., Langen, T., Gring, M., Rauer, B., Kitagawa, T., Demler, E., Adu Smith, D., and Schmiedmayer, J. 2013. Multimode dynamics and emergence of a characteristic length scale in a one-dimensional quantum system. *Phys. Rev. Lett.*, **110**, 090405.

[37] Langen, T., Geiger, Remi, and Schmiedmayer, Jörg. 2015. Ultracold atoms out of equilibrium. *Ann. Rev. Cond. Matt. Phys.*, **6**, 201–217.

[38] Berges, J., Borsányi, Sz., and Wetterich, C. 2004. Prethermalization. *Phys. Rev. Lett.*, **93**, 14–17.

[39] Eckstein, M., Kollar, M., and Werner, P. 2009. Thermalization after an interaction quench in the Hubbard model. *Phys. Rev. Lett.*, **103**, 23–26.

[40] Rauer, B., Grišins, P., Mazets, I. E., Schweigler, T., Rohringer, W., Geiger, R., Langen, T., and Schmiedmayer, J. 2016. Cooling of a one-dimensional Bose gas. *Phys Rev Lett.*, **116**, 030402.

[41] Grisins, P., Rauer, B., Langen, T., Schmiedmayer, J., and Mazets, I. E. 2016. Degenerate Bose gases with uniform loss. *Phys. Rev. A*, **93**, 033634.

[42] Lieb, E. H., and Robinson, D. W. 1972. The finite group velocity of quantum spin systems. *Commun. Math. Phys.*, **28**, 251–257.

[43] Cheneau, M., Barmettler, P., Poletti, D., Endres, M., Schauß, P., Fukuhara, T., Gross, C., Bloch, I., Kollath, C., and Kuhr, S. 2012. Light-cone-like spreading of correlations in a quantum many-body system. *Nature*, **481**, 484–487.

[44] Eisert, J., Cramer, M., and Plenio, M. B. 2010. Colloquium: area laws for the entanglement entropy. *Rev. Mod. Phys.*, **82**, 277–306.

[45] Langen, T., Erne, S., Geiger, R., Rauer, B., Schweigler, T., Kuhnert, M., Rohringer, W., Mazets, I. E., Gasenzer, T., and Schmiedmayer, J. 2015. Experimental observation of a generalized Gibbs ensemble. *Science*, **348**, 207.

[46] Schweigler, T., Kasper, V., Erne, S., Rauer, B., Langen, T., Gasenzer, T., Berges, J., and Schmiedmayer, J. 2015. On solving the quantum many-body problem. *arXiv:1505.03126.*

[47] Geiger, R., Langen, T., Mazets, I. E., and Schmiedmayer, J. 2014. Local relaxation and light-cone-like propagation of correlations in a trapped one-dimensional Bose gas. *New J. Phys.*, **16**, 053034.

[48] Mazets, I. E., Schumm, T., and Schmiedmayer, J. 2008. Breakdown of integrability in a quasi-1D ultracold Bosonic gas. *Phys. Rev. Lett.*, **100**, 210403.

[49] Kollar, M., Wolf, F. A., and Eckstein, M. 2011. Generalized Gibbs ensemble prediction of prethermalization plateaus and their relation to nonthermal steady states in integrable systems. *Phys. Rev. B*, **84**, 054304.

[50] Kolmogorov, A. N. 1954. *Dokl. Akad. Nauk SSSR*, **98**, 527–553.

[51] Steffens, A., Riofrio, C. A., Hübener, R., and Eisert, J. 2014. Quantum field tomography. *New J. Phys.*, **16**, 123010.

[52] Steffens, A., Friesdorf, M., Langen, T., Rauer, B., Schweigler, T., Hübener, R., Schmiedmayer, J., Riofrio, C. A., and Eisert, J. 2015. Towards experimental quantum field tomography with ultracold atoms. *Nature Comm.*, **6**, 7663.

[53] Verstraete, F., and Cirac, J. I. 2010. Continuous matrix product states for quantum fields. *Phys. Rev. Lett.*, **104**, 190405.

9

Ultracold Gases with Intrinsic Scale Invariance

CHENG CHIN
James Franck Institute, Enrico Fermi Institute,
University of Chicago, Illinois, USA

Scale invariance is conventionally discussed in the context of critical phenomena near a continuous phase transition. In atomic gases, a new class of scale invariant systems emerges, which we call "intrinsic scale invariance." In these systems, the scaling symmetry does not rely on a phase transition, but comes from the balance of kinetic and interaction energy at different length scales. Three examples are discussed here: three-dimensional unitary Fermi gas, two-dimensional dilute Bose gas, and unitary Bose gas. The first two are invariant under continuous scaling transformation, and thus they share the same universal thermodynamics. Unitary Bose gas is a special case with expected discrete scaling symmetry because of Efimov three-body physics. We show that the discrete scale invariance can be captured by a complex scaling dimension and suggest a universal form for the observables.

9.1 Introduction

The achievement of atomic Bose-Einstein condensation (BEC) in 1995 initiated a new adventure to explore quantum many-body phenomena in the gas phase, a journey that has yielded much excitement and is still progressing fast. Because of their diluteness, ultracold atomic gases can be precisely modeled and characterized with high accuracy. In addition, new opportunities have emerged in the past decade to prepare cold atoms in the strong coupling regime. By loading the samples into optical lattices [1] or tuning them near a Feshbach resonance [2], for instance, cold atoms offer a unique platform to investigate strongly correlated quantum phenomena. Examples include the superfluid-Mott insulator transition [3] (see also Chapter 13), BEC-BCS (Bardeen-Cooper-Schrieffer) crossover [4, 5, 6, 7] (see also related overview in Chapter 12), Efimov trimer states [8], and Berezinskii-Kosterlitz-Thouless transition in two dimensions [9] (see also Chapter 10). One key motivation in these explorations is to uncover new, universal laws that underpin the behavior of strong-coupled few- and many-body quantum systems.

In this chapter, we discuss "scale invariance" as a generic attribute in various cold atom systems, including all of the aforementioned examples. While scale invariance is conventionally discussed in the context of critical phenomena, the cold atom systems that interest us belong to a new class, which we call "intrinsic scale invariance."

In the following, we first overview the main features of different types of scale invariance and discuss related examples in cold atom systems.

9.1.1 Critical Scale Invariance

Scale invariance in this category results from the divergent correlation length of a many-body system when it is sufficiently close to a continuous phase transition. Here the long wavelength behaviors of the systems are insensitive to the details of microscopic physics and the system is hypothesized to develop a universal behavior captured by a few critical exponents. In cold atom experiments, divergence of correlation length near the superfluid phase transition [10] and scaling symmetry across a superfluid-Mott insulator quantum phase transition in optical lattices [3] are two examples. In the latter case, critical exponents have been extracted from the equation of state measurement [11].

9.1.2 Intrinsic Scale Invariance (Continuous Scaling Symmetry)

A number of cold atom systems are scale invariant at all temperatures and chemical potentials. Here the scaling symmetry does not require the proximity of a phase transition or long wavelength approximation, but comes from the balance of kinetic and interaction energies at all length scales. The Hamiltonians of the systems are invariant under scaling transformation, and so are the physical observables. Experimental examples include three-dimensional (3D) unitary Fermi gases with resonant two-body interactions [12] and two-dimensional (2D) Bose gases with a constant coupling constant [9].

9.1.3 Intrinsic Scale Invariance (Discrete Scaling Symmetry)

Unitary Bose gases, the bosonic counterparts of unitary Fermi gas, belong to this special category. While their Hamiltonians are invariant under continuous scaling transformation, the Bose systems acquire a discrete scaling symmetry due to the Efimov three-body physics [13]. The discrete scaling symmetry can be alternatively understood as a result of a complex scaling dimension.

In addition to the above examples, there are increasingly more interests in new types of cold atom systems which can possess scale invariant behavior, e.g., one-dimensional Tonks gas [14], two-dimensional Fermi gas [15, 16], and nonequilibrium systems [17, 18, 19, 20]. In this chapter, we will focus our discussion on the equilibrium systems with intrinsic scale invariance (both with continous and discrete scaling symmetry mentioned above), which are unique to cold atoms.

9.2 What Is Scale Invariance?

Scale invariance refers to a special property of a system whose behavior can be described in the same manner at different length or energy scales. To understand the scaling behavior of a system, we first introduce the scaling transformation, which stretches all spatial coordinates x_i of the system by a factor of Λ

$$x_i \to \Lambda x_i, \tag{9.1}$$

and the system is scale invariant if its observables transform as

$$f(x_1, x_2...) \to f(\Lambda x_1, \Lambda x_2...) = \Lambda^{-\nu} f(x_1, x_2...), \tag{9.2}$$

where ν is called the scaling dimension of the observable f.

We begin with a simple example. In mechanics, a point particle moving frictionlessly in a uniform gravitational field g follows a projectile trajectory $z(t) = z(0) + v_0 t - gt^2/2$. The parabolic trajectory is universal regardless of the scale of the motion. We can show this by performing the following scaling transformation: $z \to \Lambda z$, $t \to \Lambda^{1/2} t$ and $v_0 \to \Lambda^{1/2} v_0$, which leaves the form of the trajectory $z(t)$ intact. This transformation follows directly from the scaling behavior of the three lengths in the system: z, $v_0 t$, and gt^2. Alternatively, it can be shown that the underlying equation of motion $d^2z/dt^2 = -g$ is invariant under the above transformation.

A similar but less trivial example is the scaling symmetry of a many-body system near a phase transition. Because of the divergence of correlation length, the long-wavelength behavior of the system is dominated by only a few parameters. A successive scale transformation of the system results in an evolution of the system in the parameter space, which is hypothesized to have fixed points near continuous phase transitions. At a fixed point, the system is invariant under scaling transformation. Powerful concepts such as universality class and renormalization group flow describe the scaling symmetry of the system near the critical point.

9.3 Generic Scale Invariance in Quantum Systems

We can systematically identify cold atom systems that are scale invariant based on the scaling analysis described in the previous section. We start with the Schrödinger equation of a few- or many-body system:

$$i\hbar\frac{\partial\Psi(x)}{\partial t} = \left(\frac{\hat{p}^2}{2m} + \hat{V}\right)\Psi(x), \tag{9.3}$$

where x refers to the positions of the particles, $\hat{p} = -i\hbar\nabla$ is the total momentum operator, $2\pi\hbar$ is the Planck constant, m is the atomic mass, and $\hat{V}$ is the interaction.

An obvious choice of the scale transformation that preserves the above equation is given by $x \to \Lambda x$, $p \to \Lambda^{-1}p$, $t \to \Lambda^2 t$, and $\hat{V}(x) \to \hat{V}(\Lambda x) = \Lambda^{-2}\hat{V}(x)$. This choice reduces all terms in Eq. (9.3) by Λ^2. The transformation on $\hat{V}$, simplified as $V(x) = \Lambda^2 V(\Lambda x)$, imposes a condition on the form of interactions that permits scaling symmetry.

9.3.1 Contact Interaction in Two Dimensions

Contact interactions apply to many dilute atomic gas systems. Under mean-field approximation, the interaction is effectively given by $\hat{V} = g|\psi|^2$, where the wavefunction ψ in a D-dimensional system transforms as $\psi \to \Lambda^{-D/2}\psi$, and g is the coupling constant. Thus the scale invariance condition reduces to $g = \Lambda^{2-D}g'$, where g' is the coupling constant in the transformed system.

One obvious way to satisfy the above condition is to consider a two-dimensional system ($D = 2$) with a constant coupling constant $g = g' = \text{const}$. This case is first discussed in Ref. [21]. Apart from a logarithmic correction which is small in typical experiments [9, 22], the coupling constant g of a 2D Bose gas is a constant. (An additional Lorentz symmetry for 2D gas in a harmonic trap is further discussed in Ref. [21].)

9.3.2 Unitary Gas in Three Dimensions

The second example is a three-dimensional atomic gas with diverging scattering length $a = \pm\infty$. First discussed in Ref. [23], the unitarity limit of s-wave scattering leads to a scale transformation of $g \to g' = \Lambda g$, which satisfies the scaling condition $g = \Lambda^{2-D}g'$ for $D = 3$.

Remarkably, both the 2D Bose gas and 3D unitary Fermi gas are scale invariant regardless of the state of the system being superfluid or normal gas. Despite their very different types of interaction and underlying quantum statistics, both systems are scale invariant.

9.3.3 Long-Range Interactions

Long-range $V(r) \propto r^{-2}$ interaction potential clearly satisfies the scaling condition $V(r) = \Lambda^2 V(\Lambda r)$. A prominent example is the three-body Efimov potential for bosons with pairwise resonant interactions. The Efimov potential can be expressed in the hyperspherical coordinate as $V_E(R) = -(\hbar^2/m)(s_0 + 1/4)R^{-2}$ [13], where R is the hyperspherical radius and s_0 is a dimensionless constant. Interestingly, the Efimov physics leads to a unique discrete scaling symmetry in strongly interacting Bose gases, distinctly different from the continuous symmetry for 2D Bose gas and 3D unitary Fermi gas.

In the following, we will outline recent works to test and explore scaling symmetry and universality in cold atom systems. One interesting consequence of the scaling symmetry is the universal thermodynamics shared by both 2D Bose gas and 3D unitary Fermi gas; see Sections 9.4 and 9.5. Finally, the special case of the unitary Bose gas with discrete scaling symmetry is discussed in Section 9.6.

9.4 Continuous Scale Invariance: 3D Unitary Fermi Gas and 2D Bose Gas

Unitary Fermi gas refers to fermionic atomic vapor with interspin interactions tuned to the unitary limit. In cold atom experiments, unitary Fermi gas has been extensively studied with two-spin component fermionic atoms of ^{6}Li or ^{40}K, where the atomic scattering length between atoms in different spin states is tuned to infinity $a = \pm\infty$ via a Feshbach resonance [2]. Interactions between fermions in the same spin state, on the other hand, are strongly suppressed at low temperatures, and can be assumed zero in an ultracold Fermi gas.

Ho [23] first pointed out the scaling symmetry of a unitary Fermi gas and that the divergence of scattering length eliminates the interaction length scale in the system. Together with the fact that molecular potential length scales are negligibly small, the only remaining length scale in the ground state is the interpaticle separation, and the associated energy scale is Fermi energy $E_F = \hbar^2 k_F^2/2m = k_B T_F$, where k_B is the Boltzmann constant, $k_F = (3\pi^2 n)^{1/3}$ is the Fermi wavenumber, T_F is the Fermi temperature, and n is the atomic density per spin state. Given only one length scale in the ground state, all other observables measured in the unit of this scale are hypothesized to be universal constants [23]. For example, ground-state chemical potential is ξE_F, where ξ is called the Bertsch parameter [24, 25] and is measured to be $\xi \approx 0.37$ [26, 27]; the superfluid critical temperature is $T_c = \beta T_F$, where $\beta \approx 0.17$ is a constant [27]. (It should be noted that the universality does not hold for unitary Bose gas; see Section 9.6.)

The two-dimensional atomic gas with a constant coupling constant is our second example. Experimentally two-dimensional atomic gases refer to atoms confined in a strongly anisotropic trap with a very tight axial confinement and weak confinement in the two transverse directions. This is realized by loading atoms into a highly oblate dipole trap [9] or a one-dimensional lattice [28]. The sample is in the "quasi-2D" regime when the axial motions of the atoms are quantum pressure limited to the lowest vibrational ground state, while microscopic atomic collisions are still three-dimensional [29]. The mean-field interactions in a dilute 2D Bose gas is given by $(\hbar^2/m)gn$, where n is the 2D atomic density and g is a dimensionless coupling constant.

The two-dimensional atomic gas with a constant coupling g is scale invariant, as pointed out in Ref. [30]. An intuitive picture to realize why this occurs in 2D is that the kinetic energy and the 2D atomic density scale identically as 1/(length scale)^2. The coupling constant g, defined as the ratio of the interaction and the kinetic energy, thus remains constant under scale transformation. Up to a logarithmic correction, Petrov et al. showed that $g = \sqrt{8\pi}a/l_z$ [29] for a 2D Bose gas with 3D scattering length $a \ll l_z$, where l_z is the harmonic length in the tightly confined direction. The logarithmic correction is small, below 1% in typical 2D Bose gas experiments [22].

The scale invariance shared by 3D unitary Fermi gas and 2D Bose gas can be understood in a simple thought experiment. When one adiabatically expands the sample size by a factor of Λ, both the kinetic energy and the interaction energy of the system reduce by a factor of Λ^2. The balance of the two energy scales leaves the sample in the same quantum phase. Thus all other observables of the system are expected to scale accordingly.

At finite temperatures T, a second length scale is present in the system, thermal de Broglie wavelength $\lambda_{dB} = h/(2\pi m k_B T)^{1/2}$, and the new energy scale is $k_B T$. Given two length scales in a system, one can hypothesize that their ratio is the only independent parameter that determines the properties of the system. For example, in a canonical ensemble, the ratio T/T_F is frequently chosen as the single parameter that characterizes a unitary Fermi gas.

In a grand canonical ensemble, one has thermal energy $k_B T$ and chemical potential μ as two independent energy scales. We can show that the scale invariance greatly simplifies the functional form of thermodynamic variables. Choosing the scaling factor $\Lambda = 1/(k_B T)$ in Eq. (9.2), we show that any observable $f(\mu, T)$ satisfies

$$f(\mu, T) = (k_B T)^{-\nu} F(\tilde{\mu}), \tag{9.4}$$

where ν is the scaling dimension and the universal function $F(\tilde{\mu})$ only depends on the dimensionless ratio of $\tilde{\mu} = \mu/k_B T$, or equivalently the fugacity $z = \exp(\mu/k_B T)$.

The scale invariance can be experimentally tested. An example on 2D Bose gases is shown in Fig. 9.1. Based on *in situ* measurement of optically trapped atoms, both the density and fluctuations of the sample can be measured as a function of chemical potential μ and temperature T: the chemical potential dependence is derived from

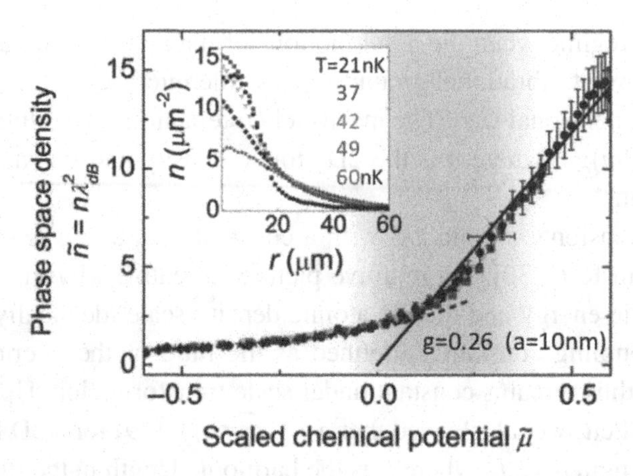

Figure 9.1 Scale invariance of 2D Bose gases. Scaled densities $n\lambda_{dB}^2$ (phase space density) versus reduced chemical potential $\tilde{\mu} = \mu/k_BT$ at five different temperatures ($T \approx$ 20–60 nK) collapse to a single curve, which supports the scale invariance of the system. Mean-field expectations for normal gas (dashed line) and Thomas-Fermi approximation (solid line) are shown for comparison. Inset shows the raw density profiles in the trap, with the temperature increasing from narrower to broader profiles. Figure adapted with permission from Hung, C.-L., et al. (2011), Observation of scale invariance and universality in two-dimensional Bose gases, *Nature*, **470**, 236. Copyright (2011) by the Nature Publishing Group [22].

the density profile in the trap assuming local density approximation, while the temperature dependence is given by a controlled heating of the sample [22]. To test the scale invariance, a range of temperatures and chemical potentials is surveyed and the measurement shows that density n and fluctuations δn in dimensionless units collapse to a single curve, conforming to the scale-invariant form of Eq. (9.4); see Fig. 9.1 and Ref. [22]. The measured equation of state is consistent with various theoretical calculations [31, 32, 33].

For a unitary Fermi gas, thermodynamic measurements also show universal behavior [34, 35]. The equation of state of 3D unitary Fermi gases has been measured experimentally [36, 37, 27] and all observables are expected to depend on the ratio $\tilde{\mu} = \mu/k_BT$. An example is shown in Fig. 9.4. Here density n and pressure p are measured or derived over a large range of $\tilde{\mu}$ and are compared with the theoretical calculation. The thermodynamics of Fermi gas provides a wealth of information to identify the superfluid transition and to test various theoretical models.

In addition to simplifying the thermodynamic description of a many-body system, scale invariance also offers a set of universal thermodynamic relations, which can be experimentally tested. We will summarize these relations below.

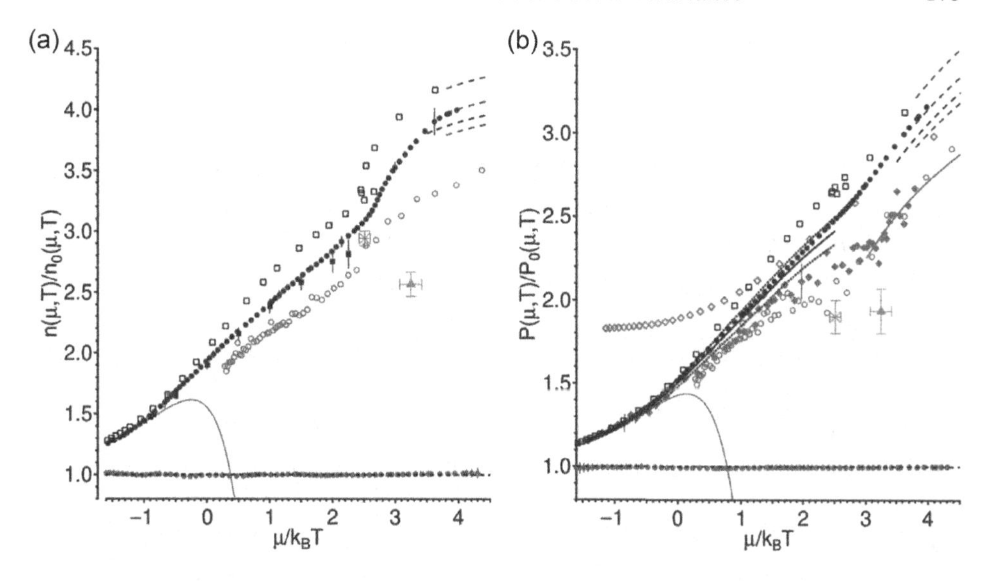

Figure 9.2 Equation of state of a unitary Fermi gas: normalized density (a) and pressure (b) of a two-component unitary Fermi gas of lithium-6 atoms from various experiments (filled circles, open/filled diamonds); ideal Fermi gas (filled circles at 1.0) is compared with various theoretical calculations (other symbols). Due to scale invariance, the only independent variable is $\tilde{\mu} = \mu/k_B T$, the ratio of the chemical potential μ and thermal energy $k_B T$. Figure reprinted with permission from the American Association for the Advancement of Science (AAAS) from Ku, M. J. H., et al. (2012), Revealing the superfluid lambda transition in the universal thermodynamics of a unitary Fermi gas, *Science*, **335**, 563 [27], where further details are described.

9.5 Universal Thermodynamics of Scale-Invariant Systems

In this section, we show that based merely on the scaling symmetry a number of universal thermodynamic relationships can be derived that are insensitive to the details of microscopic physics. Two examples we will focus on here are 3D unitary Fermi gas and 2D Bose gas.

We start our discussion with the thermodynamic pressure $p(\mu, T)$, which is equivalent to the grand potential $\Omega = -pV$, and follows the fundamental thermodynamic equation: $dp(\mu, T) = n(\mu, T)d\mu + s(\mu, T)dT$. Since all quantities take the form of Eq. (9.4) in a scale-invariant system, the above relation reduces to an ordinary differential equation with $x = \mu/k_B T$ being the only independent variable. Thus the knowledge of density n or entropy s alone is sufficient to determine pressure p. From pressure, all other thermodynamic variables can be derived. See Table 9.1.

From Table 9.1, one can immediately identify the following universal thermodynamic relations that hold for all temperatures T and chemical potentials μ:

Table 9.1 *Universal thermodynamic quantities in a D-dimensional scale invariant system. Here density is defined as* $n = \partial p/\partial\mu|_T$*, entropy density* $s = \partial p/\partial T|_\mu$*, speed of sound* $v_s = \sqrt{m^{-1}\partial p/\partial n|_{s/n}}$*, compressibility* $\kappa_T = \partial n/\partial\mu|_T$*, and specific heat capacity* $c_V = n^{-1}\partial\epsilon/\partial T|_{N,V}$*. By taking* λ_{dB} *and* $k_B T$ *as the length and energy scales, respectively, all thermodynamic quantities in the dimensionless form are uniquely linked through the universal function* $F(x)$*, where we define* $x = \mu/k_B T$*.*

Thermodynamic quantity	Dimensionless form	
chemical potential μ	$\tilde{\mu} = \mu/k_B T$	$\equiv x$
pressure p	$\tilde{p} = p\lambda_{dB}^D/k_B T$	$\equiv F(x)$
energy density ϵ	$\tilde{\epsilon} = \epsilon\lambda_{dB}^D/k_B T$	$= \frac{D}{2}F(x)$
density n	$\tilde{n} = n\lambda_{dB}^D$	$= F'(x)$
entropy density s	$\tilde{s} = s\lambda_{dB}^D/k_B$	$= \frac{D+2}{2}F(x) - xF'(x)$
sound speed v_s	$\tilde{v}_s = v_s\sqrt{m/k_B T}$	$= \sqrt{\frac{(D+2)F(x)}{DF'(x)}}$
compressibility κ_T	$\tilde{\kappa}_T = \kappa_T\lambda_{dB}^D k_B T$	$= F''(x)$
specific heat capacity c_V	$\tilde{c}_V = c_V/k_B$	$= \frac{D(D+2)F(x)}{4F'(x)} - \frac{D^2F'(x)}{4F''(x)}$

$$\epsilon = \frac{D}{2}p \tag{9.5}$$

$$sT = \frac{D+2}{2}p - \mu n \tag{9.6}$$

$$mv_s^2 = \frac{D+2}{D}\frac{p}{n} \tag{9.7}$$

$$c_V T = \frac{D(D+2)}{4}\frac{p}{n} - \frac{D^2}{4}\frac{n}{\kappa_T}. \tag{9.8}$$

These results are universal in the sense that our only assumption here is the scale invariance. Thus the results apply to both the 3D unitary Fermi gas [35] and the 2D dilute Bose gas even through their interactions and the underlying quantum statistics are very different.

In the zero temperature limit $T = 0$, fundamental laws of thermodynamics demand that entropy per particle $s/n \to 0$ vanishes. Combining Eqs. (9.6)–(9.8), we obtain $s = c_V = 0$ and

$$p = \frac{2}{D+2}\mu n = \frac{D}{D+2}\frac{n^2}{\kappa_T} = \frac{D}{D+2}nmv_s^2. \tag{9.9}$$

In this limit, the universal function $F(x)$ also takes a particularly simple form since the atomic separation is the only length scale. Starting from the known results that in the ground state chemical potential is $\mu = \xi E_F$ for a 3D unitary Fermi gas, where $\xi \approx 0.37$ [27], and $\mu = \sqrt{8\pi}(a/l_z)(\hbar^2/m)n$ for a 2D Bose gas, we can express the chemical potential in the scale-invariant form $\tilde{\mu} = (g_0\tilde{n})^{2/D}$, where $g_0 = 3\sqrt{\pi}\xi^{3/2}/8 \approx 0.15$ for the Fermi gas, and $g_0 = \sqrt{2/\pi}(a/l_z)$ for the 2D Bose gas. The scale-invariant form allows us to determine the asymptotic behavior of the universal function $F(x)$ as

$$\lim_{x\to\infty} F(x) = \frac{2}{D+2} g_0^{-1} x^{\frac{D+2}{2}}. \tag{9.10}$$

Away from $T = 0$, we consider finite-temperature contributions to leading order based on the following consideration. For both 3D unitary Fermi gas and 2D Bose gas, thermal excitations sufficiently below the superfluid critical temperature T_c are dominated by the long-wavelength (Bogoliubov) phonons. Given Eq. (9.10) and that the entropy density of a D-dimensional phonon gas is [38]

$$s = k_B\beta\left(\frac{k_BT}{\hbar v_s}\right)^D, \tag{9.11}$$

where $\beta = 2\pi^2/45$ for $D = 3$, $\beta = 3\zeta(3)/2\pi$ for $D = 2$, and $\zeta(x)$ is the Riemann zeta function, we can express the entropy in the scale-invariant form as $\tilde{s} = s_0\tilde{\mu}^{-D/2}$ with $s_0 = \beta(D\pi)^{D/2}$ being a constant. This form preserves the temperature dependence of entropy $s \propto T^D$. Using Eqs. (9.6) and (9.10), one derives the universal function $F(x)$ including the finite temperature correction to leading order as

$$F(x) = \frac{2}{D+2} g_0^{-1} x^{\frac{D+2}{2}} + \frac{s_0}{D+1} x^{-\frac{D}{2}} + \cdots. \tag{9.12}$$

Other thermodynamic quantities at low temperatures can be derived using Table 9.1. As an example, the entropy of a particle s/n and the specific heat c_V are given by

$$\frac{s}{n} = \frac{1}{D} c_V = k_B g_0 s_0 \left(\frac{k_BT}{\mu}\right)^D. \tag{9.13}$$

The universal form shown in Eq. (9.12) has several interesting consequences that are relevant to experiments. For a 2D Bose gas, we may derive the phase space density $n\lambda_{dB}^2 = F'(x)$ as

$$n\lambda_{dB}^2 = \frac{\mu}{g_0 k_B T}\left[1 - \frac{s_0 g_0}{3}\left(\frac{k_BT}{\mu}\right)^3 + \cdots\right], \tag{9.14}$$

where the second term in the expansion shows a suppression of the density due to finite temperature as compared with the Thomas-Fermi result. This correction is also revealed in recent equation of state measurements [22, 39]; see Fig. 9.1, where the density in the superfluid regime approaches the Thomas-Fermi limit (solid line) from below. An interesting consequence is that in a thermal junction with uniform chemical potential, the colder side has a higher atomic density. Thus a 2D Bose superfluid is expected to diffuse toward the hotter side (superfluid counterflow).

For a 3D unitary Fermi gas, we can use Eq. (9.12) to expand the chemical potential at low temperatures $T \ll T_F$ and obtain

$$\mu = \xi E_F \left[1 + \gamma \left(\frac{T}{T_F} \right)^4 + \cdots \right], \tag{9.15}$$

where $\gamma = g_0 s_0 \xi^{-4}/4 \approx 25.3$ is a positive constant. The upshift in chemical potential at finite temperatures is interesting since it contrasts the downshift in an ideal Fermi gas: $\mu_{\text{ideal}} = E_F[1-(\pi^2/12)(T/T_F)^2]$. As temperature increases from zero, we thus expect that chemical potential μ of a unitary Fermi gas first increases in the superfluid phase and declines at high temperatures when the system becomes normal. The temperature at which μ reaches the maximum can be associated with the superfluid phase transition. This behavior is indeed observed in the recent experiment on the unitary Fermi gas of ^{6}Li atoms [27].

Our last case of investigation is the scale-invariant system at zero chemical potential $\mu = 0$. This regime is of particular interest of research. At $\mu = 0$, the unitary Fermi gas is normal, but remains strongly correlated. The Fermi gas in this regime has been speculated to acquire a very low entropy-to-viscosity ratio [40]. At $\mu = 0$, 2D Bose gas is also normal and is in the so-called quasicondensate regime with strong fluctuations.

Remarkably, a scale-invariant system with $\mu = 0$ has only one independent length scale, the thermal de Broglie wavelength λ_{dB}. From Table 9.1, one can easily express other length scales in terms of λ_{dB}. For example, the mean interparticle separation is $d = n^{-1/D} = A\lambda_{dB}$, where $A = F'(0)^{-1/D}$ is a constant. Similarly, internal energy per particle is proportional to the thermal energy $\epsilon/n = Bk_BT$, where $B = (D/2)F(0)/F'(0)$. The sole dependence on the temperature at $\mu = 0$ resembles that in a quantum critical system for which all observables scale with the temperature. As examples, pressure and entropy at $\mu = 0$ are given by Eq. (9.6) and

$$p = \frac{2}{D+2} sT = \zeta n k_B T, \tag{9.16}$$

where the last equality shows that the system resembles an ideal gas with $p = nk_BT$. Interactions and correlations can be effectively encapsulated into a dimensionless constant $\zeta = F(0)/F'(0) = [2/(D+2)](s/nk_B)$, which is given by the entropy per particle s/n. Even 3D unitary Fermi gas and 2D Bose gas remain strongly correlated at $\mu = 0$; their equations of state are expected to be indistinguishable from an ideal gas up to a universal constant.

In addition to offering universal thermodynamic relationships, the scale invariance also greatly facilitates experimental efforts to measure thermodynamic quantities. This is because the essential information of an observable $F(\mu, T)$ depends only on the ratio $\mu/k_B T$; the complete information of F is fully contained on a half-circle in the parameter space $(\mu, T \geq 0)$; see Fig. 9.3a. If one measures $F(\mu, T)$ along a section of chemical potential at a constant temperature, the measurement can be immediately converted into the knowledge in the entire area within the angle covered by the measurement. Shown in Fig. 9.3a as an example, the measurements A and B that cover the same angle contain the same information.

We may generalize the idea of scale invariance to systems with more degrees of freedom. One interesting application would be the 3D Fermi gas in the entire BEC-BCS crossover regime. In this system, there are three dominant length scales: scattering length a, thermal de Broglie wavelength λ_{dB}, and the interparticle separation $n^{-1/3}$, and all other microscopic length scales are comparably small and assumed negligible. Here we hypothesize that a simultaneous scaling of all three length scales by a factor of Λ will leave the system in the same state. Thus all observables transform as

$$f\left(\Lambda\lambda_{dB}, \Lambda a, \Lambda n^{-1/3}\right) = \Lambda^{-\nu} f\left(\lambda_{dB}, a, n^{-1/3}\right), \tag{9.17}$$

where ν is the scaling dimension of the observable f. In the parameter space of $(\lambda_{dB}, a, n^{-1/3})$, the information of f is contained on the surface of a unit sphere. Thus any measurement of f over an area on the surface can reveal the complete information of f in the parameter space within the solid angle subtended by the area; see Fig. 9.3b.

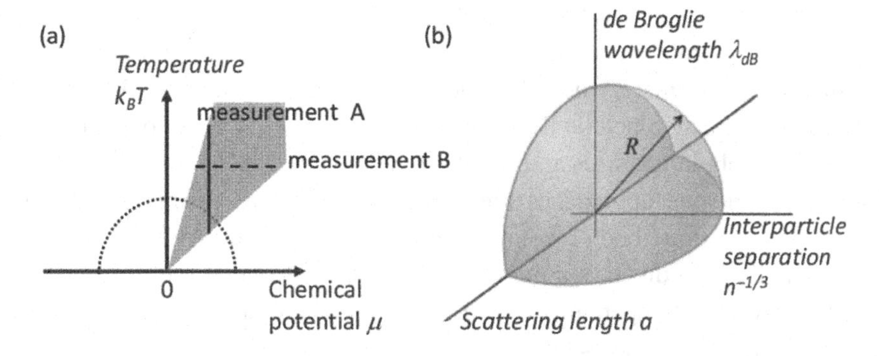

Figure 9.3 Intrinsic scaling symmetry. (a) In a scale-invariant system with two degrees of freedom μ and k_BT, its information is fully contained on a unit half-circle (dotted line) in the parameter space. A temperature scan (measurement A, solid line) and a chemical potential scan (measurement B, dashed line) provide the same information if they subtend the same angle relative to the origin (shaded area). (b) Given a system (example here is a Fermi gas in the BEC-BCS crossover regime) that is described by three dominant length scales – scattering length a, interparticle spacing $n^{-1/3}$, and de Broglie wavelength λ_{dB} – its information is fully constrained on the surface of the sphere in the parameter space.

9.6 Discrete Scale Invariance: Unitary Bose Gas

Unitary Bose gas refers to cold bosonic atoms with resonant or near resonant pairwise interactions characterized by a diverging scattering length $a \to \pm\infty$. While the fermionic counterpart, unitary Fermi gas, is scale invariant, the situation for unitary Bose gas is more subtle since the system is expected to be only invariant under discrete scaling transformation. The scaling property of a resonantly interacting Bose gas has been an active topic in recent theoretical research [41, 42], as well as in experiments [43, 44, 45].

The difference between the scaling symmetries of unitary Bose and Fermi gases lays in the three-body interaction sector. For two-component Fermi gases, three-body processes are strongly suppressed at low temperatures due to the Pauli exclusion principle. For bosons, on the other hand, three-body interactions are allowed and dominate the behavior of the system in two significant ways that determines the underlying scaling symmetry.

First of all, for bosons with resonant pairwise interaction $a = \pm\infty$, there exists an infinite series of three-body bound states, predicted by V. Efimov [13]. The Efimov states have a peculiar property that the size L_n and the binding energy E_n of the nth lowest Efimov state follow a geometric progression ($i = 1, 2, 3...$):

$$L_i = \Lambda L_{i-1} \tag{9.18}$$

$$E_i = \Lambda^{-2} E_{i-1}, \tag{9.19}$$

where the scaling factor is $\Lambda = e^{\pi/s_0}$ and, for identical bosons, $s_0 \approx 1.00624$ is a constant [13]. The emergence of Efimov states can be understood as a result of an attractive Efimov potential $V_E(R) = -\hbar^2(s_0^2 + 1/4)/mR^2$ in the three-body hyperspherical coordinate, where R is the hyperspherical radius [46]. Interestingly, it is the $1/R^2$ attractive potential that supports an infinite number of bound states with progressively larger sizes and smaller binding energies.

Second, when one approaches the unitary limit by increasing the scattering length $a \to \pm\infty$, three-body recombination leads to an instability of the Bose gas and the reaction rate displays a unique log-periodic dependence on the scattering length [47]. At zero temperature $T = 0$, the recombination coefficient K_3 is given to a good approximation as [41]:

$$\lim_{a\to\infty} K_3 = Aa^4 \left[\sin^2(s_0 \ln a + \phi_+) + \sinh^2 \eta\right] \tag{9.20}$$

$$\lim_{a\to-\infty} K_3 = Ba^4 \left[\sin^2(s_0 \ln |a| + \phi_-) + \sinh^2 \eta\right]^{-1} \tag{9.21}$$

where $A = 67.1e^{-2\eta}\hbar/m$ and $B = 4590\sinh(2\eta)\hbar/m$ are constants. Here the width η parameter and the phase factor $\phi_\pm$ depend on the short-range molecular potential. The resonant structure of the recombination loss for $a < 0$ is linked to

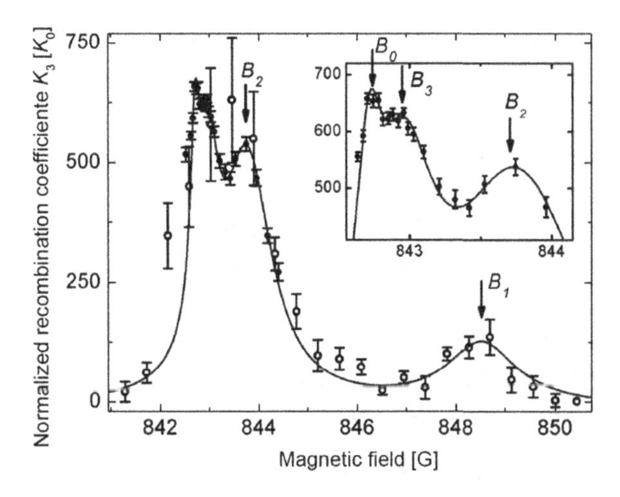

Figure 9.4 Discrete scale invariance in a mixture of ^{6}Li and ^{133}Cs. Recombination rate shows resonances, features when an Efimov state is tuned near the continuum on the negative scattering length side $a < 0$ (magnetic field $B > 842.75$ G). Three consecutive resonances at B_1, B_2, and B_3 are observed [48, 49] and the corresponding scattering lengths follow a geometric progression and support the discrete scaling symmetry [48]. In the expanded view, B_0 is the location of the Li-Cs Feshbach resonance. Reprinted with permission from Tung, S.-K. (2014), Geometric scaling of Efimov states in a ^{6}Li–^{133}Cs mixture, *Phys. Rev. Lett.*, **113**, 240402. Copyright (2014) by the American Physical Society.

the emergence of an Efimov state and is the key observable in the first experimental confirmation of an Efimov state in a Bose gas [8].

Both the sizes and binding energies of Efimov states, as well as the log-periodic recombination rate, suggest a discrete scaling symmetry of the system. In particular, the scaling of the recombination rate has been observed recently in experiments with ultracold mixture of Li and Cs atoms near an interspecies Feshbach resonance [48, 49], where three consecutive Efimov resonances follow a geometric scaling; see Fig. 9.4.

It is intriguing to understand the origin of the discrete scaling symmetry. Given that the Efimov potential preserves the continuous scaling symmetry $V_E(R) = \Lambda^2 V_E(\Lambda R)$ for arbitrary scaling factor of Λ, the reduced discrete symmetry in its observables presents a case of spontaneous symmetry breaking. One theoretical interpretation is based on effective field theory renormalization [50], which relates the discrete scaling symmetry to the limit cycle of Wilson's renormalization group flow [41, 42, 50]. An alternative picture is that the three-body phase factor $\phi_\pm$ is given by the short-range molecular potential. Since the phase factor determines the locations of the Efimov states and Efimov resonances, the continuous scaling symmetry breaks down. A recent survey on Efimov resonances in various cold atom experiments further suggests that the three-body phase factor is indeed given by the molecular van der Waals length [51], a finding that was later supported by a model calculation [52] and a molecular potential calculation [53].

The discrete scaling symmetry of unitary Bose gas suggests a unique scaling behavior. By expressing the log-periodic function in terms of the power-law function with a complex exponent using $x^{u+iv} = x^u \cos(v \ln x) + ix^u \sin(v \ln x)$ for $x > 0$, we suggest that the discrete scaling property of a unitary Bose gas can be captured by a complex scaling dimension $\nu = u + iv$, where $u =$ and v are its real and imaginary parts, respectively. As an example, the recombination rate coefficients in Eqs. (9.20)–(9.21) can be written in terms of power-law functions of scattering length with a complex scaling exponent $\nu = 4 + i2s_0$:

$$K_3(a > 0) = A\left|\mathrm{Im}(a^{\nu/2}e^{i\phi_+})\right|^2 + C\left|a^{\nu}\right| \tag{9.22}$$

$$K_3(a < 0)^{-1} = B^{-1}\left|\mathrm{Im}(|a|^{-\nu/2}e^{i\phi_-})\right|^2 + D^{-1}\left|a^{-\nu}\right|, \tag{9.23}$$

where $C = A\sinh^2\eta$ and $D = B/\sinh^2\eta$ are constants.

The discrete scaling symmetry can also apply to finite temperature Bose gas near a Feshbach resonance. In such system, there are two dominant length scales: scattering length a and scattering energy length scale. The latter can be chosen as de Broglie wavelength λ_{dB} for energy above the dissociation threshold and Feshbach molecule size $h/\sqrt{mE_B}$ for energy below the threshold; see Fig. 9.5. Under successive scaling transformations $x \to \Lambda x$ of all length scales with $\Lambda > 1$, a Bose system will asymptotically approach the unitary Bose gas at zero temperature with $a = \lambda_{dB} = h/\sqrt{mE_B} = \infty$, which is the fixed point of the transformation. The discrete scaling symmetry suggests that a system moving along any ray approaching the fixed point (as an example, see the shaded area in Fig. 9.5) would also acquire a log-periodic behavior. In general, an observable f in a system with scaling dimension $\nu = u + iv$ should follow the universal form:

$$f(a, \lambda_{dB}) = \lambda_{dB}^{u} F\left(\rho^{iv}, a/\lambda_{dB}\right), \tag{9.24}$$

where the function F is a universal function, and is log-periodic in the radial distance $\rho = (a^2 + \lambda_{dB}^2)^{1/2}$ from the fixed point when a/λ_{dB} is held constant.

In comparing systems with continuous scaling symmetry where their information is contained on the surface of a sphere in the parameter space, one expects that the information of a system with discrete scaling symmetry is on the shell of a sphere with a finite thickness. For strongly interacting Bose gas, one needs to vary the length scale of the shell by a factor of $\Lambda = e^{\pi/s_0}$ to access the complete information of the system; see Fig. 9.5. The discrete scale invariance provides constraints as well as predictions that any observables should vary log-periodically when the system approaches the fixed point.

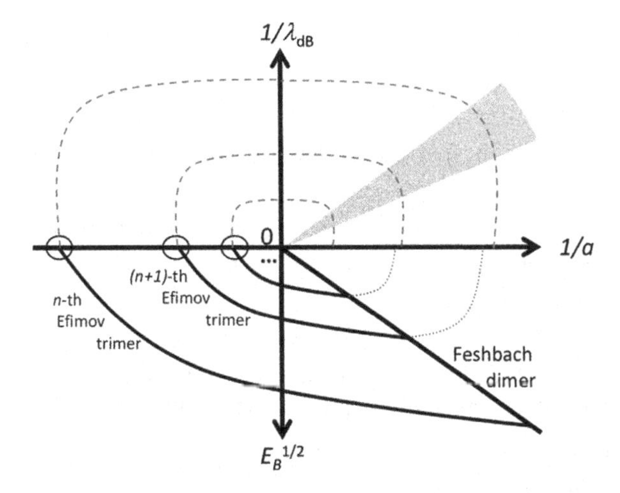

Figure 9.5 Discrete scale invariance of a unitary Bose system. In the parameter space of inverse scattering length $1/a$ and inverse thermal de Broglie wavelength $1/\lambda_{dB}$ for positive scattering energy and $\sqrt{mE_B}/h$ for positive binding energy E_B, a discrete scaling symmetry is anticipated with respect to the origin. The symmetry has been theoretically suggested and experimentally tested for the locations of the Efimov states (solid curved lines), Efimov resonances (circles) and the recombination loss in the unitarity limit $1/a = 0$. Atom-Feshbach dimer resonances are also expected to scale geometrically along the dimer dissociation threshold (diagonal solid line in lower-right quadrant). Our conjecture is that all observables acquire a discrete scaling symmetry along any ray that passes through the origin (the shaded area is an example). One potential observable on the entire upper half plane is the recombination coefficient (dashed lines mark the local maxima of the coefficients). Between the atom and dimer thresholds, the observable can be the atom-dimer collision cross section (the gray dotted lines mark the local maxima).

9.7 Conclusion

Three unique cold atom systems are discussed here that possess the new *intrinsic scaling symmetry*: 3D unitary Fermi gas, 2D Bose gas, and unitary Bose gas. The scaling symmetry of these systems comes from the balance between the kinetic and interaction energies under scaling transformation, and thus does not require the proximity of a phase transition. The scaling symmetry provides great simplification to both theoretical description and experimental characterization of the system. In particular, 3D unitary Fermi gas and 2D Bose gas are invariant under continuous scale transformation, which has been experimentally tested based on the equation of state measurement. Because of the underlying symmetry, the thermodynamics of these systems are greatly simplified and both systems share a large number of universal relationships in their thermodynamics. We show that all thermodynamic quantities are connected through a single universal function.

Strongly interacting Bose gas is a special interesting case which is invariant only under discrete scaling transformation. The discrete scaling symmetry is evident in the emergence of three-body Efimov states when the system is near a Feshbach resonance, as well as the log-periodic behavior of the recombination rate observed in recent experiments. The discrete scaling symmetry can be captured by a complex scaling dimension $\nu = 4 + i2s_0$. By extending the symmetry to near-resonant Bose gases at finite temperature, we conjecture that the log-periodic dependence on all observables is both generic and universal.

Acknowledgments: We thank Yujun Wang for useful discussion. This work is supported by the National Science Foundation under Grant No. PHY-1511696, Materials Research Science and Engineering Centers (MRSEC) No. DMR-1420709, and ARO-MURI W911NF-14-1-0003.

References

[1] Bloch, Immanuel. 2005. Ultracold quantum gases in optical lattices. *Nat. Phys.*, **1**, 23–30.

[2] Chin, C., Grimm, R., Julienne, P., and Tiesinga, E. 2010. Feshbach resonances in ultracold gases. *Rev. Mod. Phys.*, **82**, 1225–1286.

[3] Greiner, M., Mandel, O., Esslinger, T., Hansch, T. W., and Bloch, I. 2002. Feshbach resonances in ultracold gases. *Nature*, **415**, 39–44.

[4] Regal, C. A., Greiner, M., and Jin, D. S. 2004. Observation of resonance condensation of fermionic atom pairs. *Phys. Rev. Lett.*, **92**, 040403.

[5] Zwierlein, M. W., Stan, C. A., Schunck, C. H., Raupach, S. M. F., Kerman, A. J., and Ketterle, W. 2004. Condensation of pairs of fermionic atoms near a Feshbach resonance. *Phys. Rev. Lett.*, **92**, 120403.

[6] Bartenstein, M., Altmeyer, A., Riedl, S., Jochim, S., Chin, C., Denschlag, J. Hecker, and Grimm, R. 2004. Crossover from a molecular Bose-Einstein condensate to a degenerate Fermi gas. *Phys. Rev. Lett.*, **92**, 120401.

[7] Bourdel, T., Khaykovich, L., Cubizolles, J., Zhang, J., Chevy, F., Teichmann, M., Tarruell, L., Kokkelmans, S. J. J. M. F., and Salomon, C. 2004. Experimental study of the BEC-BCS crossover region in lithium 6. *Phys. Rev. Lett.*, **93**, 050401.

[8] Kraemer, T., Mark, M., Waldburger, P., Danzl, J. G., Chin, C., Engeser, B., Lange, A. D., Pilch, K., Jaakkola, A., Ngerl, H.-C., and Grimm, R. 2006. Evidence for Efimov quantum states in an ultracold gas of caesium atoms. *Nature*, **440**, 315–318.

[9] Hadzibabic, Z., Kruger, P., Cheneau, M., Battelier, B., and Dalibard, J. 2006. Berezinskii-Kosterlitz-Thouless crossover in a trapped atomic gas. *Nature*, **441**, 1118–1121.

[10] Donner, T., Ritter, S., Bourdel, T., Öttl, A., Köhl, M., and Esslinger, T. 2007. Critical behavior of a trapped interacting Bose gas. *Science*, **315**, 1556–1558.

[11] Zhang, X., Hung, C.-L., Tung, S.-K., and Chin, C. 2012. Observation of quantum criticality with ultracold atoms in optical lattices. *Science*, **335**, 1070.

[12] O'Hara, K. M., Hemmer, S. L., Gehm, M. E., Granade, S. R., and Thomas, J. E. 2002. Observation of a strongly interacting degenerate Fermi gas of atoms. *Science*, **298**, 2179–2182.

[13] Efimov, V. 1970. Energy levels arising from resonant two-body forces in a three-body system. *Phys. Lett. B*, **33**, 563–564.
[14] Kinoshita, T., Wenger, T., and Weiss, D. S. 2004. Observation of a one-dimensional Tonks-Girardeau gas. *Science*, **305**, 1125–1128.
[15] Martiyanov, K., Makhalov, V., and Turlapov, A. 2010. Observation of a two-dimensional Fermi gas of atoms. *Phys. Rev. Lett.*, **105**, 030404.
[16] Vogt, E., Feld, M., Fröhlich, B., Pertot, D., Koschorreck, M., and Köhl, M. 2012. Scale invariance and viscosity of a two-dimensional Fermi gas. *Phys. Rev. Lett.*, **108**, 070404.
[17] Cao, C., Elliott, E., Joseph, J., Wu, H., Petricka, J., Schäfer, T., and Thomas, J. E. 2011. Universal quantum viscosity in a unitary Fermi gas. *Science*, **331**, 58–61.
[18] Gritsev, V., Barmettler, P., and Demler, E. 2010. Scaling approach to quantum non-equilibrium dynamics of many-body systems. *New J. Phys.*, **12**, 113005.
[19] Hung, C.-L., Gurarie, V., and Chin, C. 2013. From cosmology to cold atoms: observation of Sakharov oscillations in a quenched atomic superfluid. *Science*, **341**, 1213.
[20] Rohringer, W., Fischer, D., Steiner, F., Mazets, I. E., Schmiedmayer, J., and Trupke, M. 2015. Non-equilibrium scale invariance and shortcuts to adiabaticity in a one-dimensional Bose gas. *Sci. Rep.*, **5**, 1–7.
[21] Pitaevskii, L. P., and Rosch, A. 1997. Breathing modes and hidden symmetry of trapped atoms in two dimensions. *Phys. Rev. A*, **55**, R853–R856.
[22] Hung, C.-L., Zhang, X., Gemelke, N., and Chin, C. 2011. Observation of scale invariance and universality in two-dimensional Bose gases. *Nature*, **470**, 236.
[23] Ho, T.-L. 2004. Universal thermodynamics of degenerate quantum gases in the unitarity limit. *Phys. Rev. Lett.*, **92**, 090402.
[24] Bishop, R. A. 2001. Preface *Int. J. Mod. Phys. B*, **15**, iii.
[25] Baker, G. A. Jr. 1999. Neutron matter model *Phys. Rev. C*, **60**, 054311.
[26] Gehm, M. E., Hemmer, S. L., Granade, S. R., O'Hara, K. M., and Thomas, J. E. 2003. Mechanical stability of a strongly interacting Fermi gas of atoms. *Phys. Rev. A*, **68**, 011401.
[27] Ku, M. J. H., Sommer, A. T., Cheuk, L. W., and Zwierlein, M. W. 2012. Revealing the superfluid lambda transition in the universal thermodynamics of a unitary Fermi gas. *Science*, **335**, 563–567.
[28] Gemelke, N., Zhang, X., Hung, C.-L., and Chin, C. 2009. In situ observation of incompressible Mott-insulating domains in ultracold atomic gases. *Nature*, **460**, 995.
[29] Petrov, D. S., Holzmann, M., and Shlyapnikov, G. V. 2000. Bose-Einstein condensation in quasi-2D trapped gases. *Phys. Rev. Lett.*, **84**, 2551–2555.
[30] Pitaevskii, L. P., and Rosch, A. 1997. Breathing modes and hidden symmetry of trapped atoms in two dimensions. *Phys. Rev. A*, **55**, R853–R856.
[31] Prokof'ev, N., and Svistunov, B. 2002. Two-dimensional weakly interacting Bose gas in the fluctuation region. *Phys. Rev. A*, **66**, 43608.
[32] Holzmann, Markus, Chevallier, Maguelonne, and Krauth, Werner. 2010. Universal correlations and coherence in quasi-two-dimensional trapped Bose gases. *Phys. Rev. A*, **81**, 043622.
[33] Cockburn, S. P., and Proukakis, N. P. 2012. *Ab initio* methods for finite-temperature two-dimensional Bose gases. *Phys. Rev. A*, **86**, 033610.
[34] Kinast, J., Turlapov, A., Thomas, J. E., Chen, Q., Stajic, J., and Levin, K. 2005. Heat capacity of a strongly interacting Fermi gas. *Science*, **307**, 1296–1299.
[35] Thomas, J. E., Kinast, J., and Turlapov, A. 2005. Virial theorem and universality in a unitary Fermi gas. *Phys. Rev. Lett.*, **95**, 120402.

[36] Horikoshi, M., Nakajima, S., Ueda, M., and Mukaiyama, T. 2010. Measurement of universal thermodynamic functions for a unitary Fermi gas. *Science*, **327**, 442–445.

[37] Nascimbène, S., Navon, N., Jiang, K. J., Chevy, F., and Salomon, C. 2010. Exploring the thermodynamics of a universal Fermi gas. *Nature*, **463**, 1057–1060.

[38] Pitaevskii, L.P., and Lifshitz, E. M. 1980. *Statistical Physics, Part 2: Theory of the Condensed State*. Vol. 9. Butterworth-Heinemann.

[39] Desbuquois, R., Yefsah, T., Chomaz, L., Weitenberg, C., Corman, L., Nascimbène, S., and Dalibard, J. 2014. Determination of scale-invariant equations of state without fitting parameters: application to the two-dimensional Bose gas across the Berezinskii-Kosterlitz-Thouless transition. *Phys. Rev. Lett.*, **113**, 020404.

[40] Son, D. T., and Wingate, M. 2006. General coordinate invariance and conformal invariance in nonrelativistic physics: unitary Fermi gas. *Ann. Phys.*, **321**(1, special issue), 197–224.

[41] Braaten, Eric, and Hammer, H.-W. 2006. Universality in few-body systems with large scattering length. *Phys. Rep.*, **428**, 259–390.

[42] Horinouchi, Y., and Ueda, M. 2015. Onset of a limit cycle and universal three-body parameter in Efimov physics. *Phys. Rev. Lett.*, **114**, 025301.

[43] Rem, B. S., Grier, A. T., Ferrier-Barbut, I., Eismann, U., Langen, T., Navon, N., Khaykovich, L., Werner, F., Petrov, D. S., Chevy, F., and Salomon, C. 2013. Lifetime of the Bose gas with resonant interactions. *Phys. Rev. Lett.*, **110**, 163202.

[44] Fletcher, R. J., Gaunt, A. L., Navon, N., Smith, R. P., and Hadzibabic, Z. 2013. Stability of a unitary Bose gas. *Phys. Rev. Lett.*, **111**, 125303.

[45] Makotyn, P., Klauss, C. E., Goldberger, D. L., Cornell, E. A., and Jin, D. S. 2014. Universal dynamics of a degenerate unitary Bose gas. *Nat. Phys.*, **10**, 116.

[46] Efimov, V. 1979. Low-energy properties of three resonantly-interacting particles. *Sov. J. Nucl. Phys.*, **29**, 546.

[47] Esry, B. D., Greene, Chris H., and Burke, James P. 1999. Recombination of three atoms in the ultracold limit. *Phys. Rev. Lett.*, **83**, 1751–1754.

[48] Tung, S.-K., Jiménez-García, K., Johansen, J., Parker, C. V., and Chin, C. 2014. Geometric scaling of Efimov states in a $^6\text{Li}-^{133}\text{Cs}$ mixture. *Phys. Rev. Lett.*, **113**, 240402.

[49] Pires, R., Ulmanis, J., Häfner, S., Repp, M., Arias, A., Kuhnle, E. D., and Weidemüller, M. 2014. Observation of Efimov resonances in a mixture with extreme mass imbalance. *Phys. Rev. Lett.*, **112**, 250404.

[50] Bedaque, P. F., Hammer, H.-W., and van Kolck, U. 1999. Renormalization of the three-body system with short-range interactions. *Phys. Rev. Lett.*, **82**, 463–467.

[51] Berninger, M., Zenesini, A., Huang, B., Harm, W., Nägerl, H.-C., Ferlaino, F., Grimm, R., Julienne, P. S., and Hutson, J. M. 2011. Universality of the three-body parameter for Efimov states in ultracold cesium. *Phys. Rev. Lett.*, **107**, 120401.

[52] Chin, C. Universal scaling of Efimov resonance positions in cold atom systems. *arXiv:1111.1484*.

[53] Wang, J., D'Incao, J. P., Esry, B. D., and Greene, C. H. 2012. Origin of the three-body parameter universality in Efimov physics. *Phys. Rev. Lett.*, **108**, 263001.

10

Berezinskii-Kosterlitz-Thouless Phase of a Driven-Dissipative Condensate

NA YOUNG KIM

Edward L. Ginzton Laboratory, Stanford University, California, USA
Present address: Institute for Quantum Computing, Department of Electrical and Computer Engineering, University of Waterloo, Canada

WOLFGANG H. NITSCHE

Edward L. Ginzton Laboratory, Stanford University, California, USA
Present address: Halliburton, Houston, Texas, USA

YOSHIHISA YAMAMOTO

Edward L. Ginzton Laboratory, Stanford University, California, USA
Japan Science and Technology Agency, Chiyoda-ku, Tokyo, Japan

Microcavity exciton-polaritons are interacting Bose particles which are confined in a two-dimensional (2D) system suitable for studying coherence properties in an inherently nonequilibrium condition. A primary question of interest here is whether a true long-range order exists among the 2D exciton-polaritons in a driven open system. We give an overview of theoretical and experimental works concerning this question, and we summarize the current understanding of coherence properties in the context of Berezinskii-Kosterlitz-Thouless transition.

10.1 Introduction

Strange but striking phenomena, which are accessed by advanced experimental techniques, become a fuel to stimulate both experimental and theoretical research. Experimentalists concoct new tools for sophisticated measurements, and theorists establish models in order to explain the surprising observation, ultimately expanding our knowledge boundary. A classic example of the seed to the knowledge expansion is the feature of abnormally high heat conductivity in liquid helium reported by Kapitza and Allen's group, who used cryogenic liquefaction techniques in 1938 [1, 2]. It is a precursor to a "resistance-less flow" a new phase of matter, coined as superfluidity in the He-II phase. Immediately after this discovery, London conceived a brilliant insight between superfluidity and Bose-Einstein condensation (BEC) of noninteracting ideal Bose gases [3], which has led to establish the concept of coherence as off-diagonal long-range order emerging in the exotic states of matter. Since then, it is one of the core themes in equilibrium Bose systems to elucidate the intimate link of superfluidity and BEC in natural and artificial materials, where dimensionality and interaction play a crucial role in determining the system phase.

Let us consider the noninteracting ideal Bose gases whose particle number N is fixed in a three-dimensional box with a volume V. According to the Bose-Einstein statistics, the average occupation number N_i in the state i with energy ϵ_i is given by $N_i = 1/(e^{\beta(\epsilon_i - \mu)} - 1)$ with the chemical potential μ and a temperature parameter $1/\beta = k_B T$ (Boltzmann constant k_B and temperature T). For the positive real number of N_i, μ is restricted to be smaller than ϵ_i, and the ground-state particle number N_0 diverges as μ approaches the lowest energy ϵ_0. Its thermodynamic phase transition refers to BEC, in which the macroscopic occupation in the ground state is represented by the classical field operator $\Psi(\mathbf{r}, t) = \sqrt{n(\mathbf{r}, t)} e^{i\phi(\mathbf{r}, t)}$, where $n(\mathbf{r}, t)$ is the particle density and $\phi(\mathbf{r}, t)$ is the phase. The critical temperature T_c is defined as a temperature at which particles are accumulated in the ground state, while the number of particles in excited states would be finite [4, 5]. The definite phase in the BEC manifests spontaneous spatial coherence by virtue of spontaneous symmetry breaking in the system. The spatial coherence is quantified by the first-order coherence function $g^{(1)}(\mathbf{r}_1, t_1, \mathbf{r}_2, t_2)$ at different spatial $(\mathbf{r}_1, \mathbf{r}_2)$ and time (t_1, t_2) coordinates. By definition, we express the $g^{(1)}(\mathbf{r}_1, t_1, \mathbf{r}_2, t_2)$ function in the normalized form in terms of $\Psi(\mathbf{r}, t)$:

$$g^{(1)}(\mathbf{r}_1, t_1, \mathbf{r}_2, t_2) = \frac{\langle \Psi^\dagger(\mathbf{r}_1, t_1)\, \Psi(\mathbf{r}_2, t_2) \rangle}{\sqrt{\langle \Psi^\dagger(\mathbf{r}_1, t_1)\, \Psi(\mathbf{r}_1, t_1) \rangle \langle \Psi^\dagger(\mathbf{r}_2, t_2)\, \Psi(\mathbf{r}_2, t_2) \rangle}}. \tag{10.1}$$

In equilibrium, the time dependence is suppressed in Eq. (10.1), and we apply the particle distribution in momentum space $n(\mathbf{k})$ via Fourier transform using $\Psi(\mathbf{k}) = (2\pi\hbar)^{-\frac{3}{2}} \int d\mathbf{r} \Psi(\mathbf{r}) \exp(i\mathbf{k} \cdot \mathbf{r}))$. Then, the numerator of Eq. (10.1) is reduced to a simple formula, $\langle \Psi^\dagger(\mathbf{r}_1)\, \Psi(\mathbf{r}_2) \rangle = V^{-1} \int d\mathbf{k} \langle n(\mathbf{k}) \rangle \exp(-i\mathbf{k} \cdot (\mathbf{r}_1 - \mathbf{r}_2))$. Below the critical temperature, the momentum distribution $n(\mathbf{k}) = N_0 \delta(\mathbf{k}) + \sum_{\mathbf{k} \neq 0} n(\mathbf{k})$ has the dominant ground-state term, which reduces the integral in the numerator to be N_0. Consequently, the $g^{(1)}$-function converges to a constant value of N_0/N even at a large distance limit, $r = |\mathbf{r}_1 - \mathbf{r}_2| \to \infty$. This attribute is known to be off-diagonal long-range order (ODLRO) arising from the constant condensate fraction.

In real systems, we should consider various sources of fluctuations, which populate particles at excited states apart from the condensates. Furthermore, the particle interaction and the system dimension are crucial to determine ODLRO. Theoretically, Hohenberg, Mermin, and Wagner reached a conclusion that the true long-range order is impossible in low-dimensional noninteracting Bose gases at finite temperatures, which is known as the Hohenberg-Mermin-Wagner theorem [6, 7].

Namely, the spatial coherence in the large distance limit is zero in infinite 1D and 2D systems, where thermal fluctuations completely destroy the ODLRO. However, Berezinskii, Kosterlitz, and Thouless in the early 1970s conjectured a topological phase order which exhibits coherence and superfluidity in 2D systems [8, 9]. The Berezinskii-Kosterlitz-Thouless (BKT) theory is considered to be central for identifying the relation of superfluidity in diverse materials systems. This chapter is organized as follows: in Section 10.2, we review the fundamental physics of the BKT phase and introduce major material systems to unveil the BKT phase. Sections 10.3 and 10.4 discuss both theoretical and experimental activities on the BKT phase in microcavity exciton-polaritons under the nonequilibrium condition. We finish by examining the current status of the BKT research efforts followed by prospective remarks on proposed experimental schemes in order to reach the conclusive results. A closely related theoretical discussion of the nature of the phase transition in 2D polariton condensates is given in Chapter 11 by Keeling et al.

10.2 Berezinskii-Kosterlitz-Thouless Physics

At zero temperature, the perfect phase coherence exists in the free bosons as the true ODLRO associated with BEC; however, at finite temperature, we pay attention to several factors for identifying coherence properties: the system dimensionality, the system size, and particle interaction. Given the focus of the chapter, we limit our discussion to 2D, whose energy density of states is constant even at the zero energy. Hence, in principle, BEC is impossible in a 2D uniform system with continuous symmetry due to a large number of particles occupied in all the excited states. On the other hand, the density of states in spatially confined systems can vanish for a specific potential profile; consequently, BEC can be restored in a finite-sized 2D system.

When we take the repulsive interaction among particles into account in 2D Bose fluids, the story becomes profound. Berezinskii, Kosterlitz, and Thouless studied the 2D XY model for low-temperature ordered states in the infinite system, which may apply to spin crystals and superfluid helium, superconducting materials, and 2D atoms [8, 9]. The so-called quasicoherence in the BKT phase forms in the infinite system for weakly interacting Bose fluids and the coherence nature of weakly interacting Bose fluids in the finite-sized system are currently studied for trapped atoms and microcavity exciton-polaritons. Recently, Hadzibabic's group managed to control the interaction strengths in a gas trapped in a harmonic potential. They confirmed the observation of the BKT phase in this 2D finite system consisting of interacting Bose atoms [10]. Table 10.1 summarizes the system order with respect to the system size and the particle interactions at nonzero temperature in 2D.

Table 10.1 *Summary of 2D phases in terms of system size and interaction at finite temperatures.*

$T \neq 0$	2D finite system	2D infinite system
zero interaction	ODLRO/BEC	no ODLRO/no BEC
weak repulsive interaction	BKT	BKT

10.2.1 Quasi-Long-Range Order

Thermal fluctuations indeed disturb a well-defined phase of coherent particles, and the degree of phase fluctuations becomes appreciable if the density fluctuations are suppressed by the particle–particle interaction. Two major mechanisms to control the spatial coherence are thermally excited phonons and topological defects, "vortices" [11, 12]. Vortices have zero particle density at their core and quantized angular momentum with a continuous phase rotation of $2\pi n$ around the core. Singly charged vortices with $n = 1$ (for a vortex) and $n = -1$ (for an antivortex) are primarily relevant in coherent condensates since multicharged vortices with $|n| > 1$ will rapidly dissociate into several single-charged vortices. Two forms of vortices exist in the condensates: free vortices or bound vortex–antivortex pairs. Their influence over the system phase is dissimilar: whereas the circulating phase of free vortices kills the spatial coherence of the system where free vortices flow around, the bound pairs of vortex–antivortex have no effect on the phase as a result of the mutual cancellation between oppositely rotating phases. Hence, such pairs do not alter the coherence over large distances.

From a thermodynamic point of view, a vortex–antivortex pair is energetically favorable over a free vortex. Both a single free vortex and a vortex–antivortex pair increase the entropy S by an amount approximately proportional to the logarithm of the system size. However, while a free vortex also increases the energy E by an amount proportional to the logarithm of the system size, a vortex–antivortex pair gives a finite energy contribution which is independent of the system size. With respect to the free energy $F = E - TS$ in a large 2D condensate, the presence of vortex–antivortex pairs is always advantageous, and free vortices exist only above a critical temperature, T_{BKT} [11]. The low-temperature state of a condensate, where all vortices are paired up, is called the BKT phase [8, 9]. The only remaining mechanism to contribute to a decay of the spatial coherence in the BKT phase is the thermal excitation of phononic phase fluctuations, which leads to the decay of the spatial coherence with a slow power law of the form:

$$g^{(1)}(\mathbf{r}_1, t, \mathbf{r}_2, t) \propto |\mathbf{r}_1 - \mathbf{r}_2|^{-a_{\mathrm{p}}}, \tag{10.2}$$

where the exponent $a_p = (n_s\lambda_T^2)^{-1}$ depends only on the superfluid density n_s and the thermal wavelength $\lambda_T = h/\sqrt{2\pi m_{\text{eff}} k_B T}$ [11].

We can derive the upper limit of a_p in equilibrium by assuming a vortex with a core radius ξ in the center of a superfluid with radius R and effective particle mass m_{eff}. The energy E of a free vortex is calculated by integrating the local kinetic energy of the superfluid flow as

$$E = \int \frac{m_{\text{eff}} n_s(\mathbf{r})}{2} (v(\mathbf{r}))^2 d^2\mathbf{r} = n_s \lambda_T \frac{k_B T}{2} \ln\left(\frac{R}{\xi}\right), \tag{10.3}$$

and the entropy S of a vortex is proportional to the logarithm of the number of positions as

$$S = k_B \ln\left(\frac{\pi R^2}{\pi \xi^2}\right) = 2k_B \ln\left(\frac{R}{\xi}\right). \tag{10.4}$$

The overall free energy F of the vortex has a simple expression,

$$F = E - TS = \left(n_s\lambda_T^2 - 4\right) \frac{k_B T}{2} \ln\left(\frac{R}{\xi}\right). \tag{10.5}$$

If $n_s\lambda_T^2 < 4$, F is negative since the rest terms are all positive, thus the BKT threshold is determined by $n_s\lambda_T^2 = 4$, which separates two phases drawn in Fig. 10.1. The excitation of the first free vortices reduces n_s [13], making the creation of more free vortices even easier, and eventually superfluidity disappears [11]. The BKT state with the bound vortex–antivortex pairs possesses superfluidity, and a flowing condensate experiences friction through the excitation

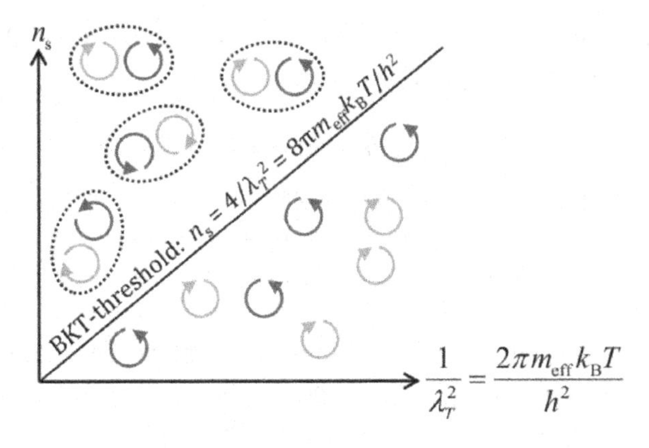

Figure 10.1 Diagram of equilibrium BKT transition. At high superfluid density n_s and low-temperature T, spatial coherence appears over long distances where only vortex–antivortex pairs exist. Decreasing n_s or increasing T causes the system to cross the BKT threshold where the vortices unbind, which destroys the spatial coherence.

Table 10.2 *Decay mechanisms of the coherence in 2D superfluids.*

$0 < T < T_{\mathrm{BKT}}$		$T > T_{\mathrm{BKT}}$
phonons	vortex–antivortex pairs	free vortices
slow decay of coherence	no decay of coherence	fast decay of coherence
quasi-long-range order		no ODLRO

of quasiparticles (i.e., phonons) whose dispersion relation is a form of $\epsilon_{\mathrm{excitation}} \approx v\,|\mathbf{p}_{\mathrm{excitation}}|$. The Landau criterion [14, 15, 16, 17] tells us that such excitations can only be created if the condensate flows with a velocity above v. Decay mechanisms of the coherence in 2D superfluids are summarized in Table 10.2.

Four representative physical systems are under the active investigation to search the BKT phase both in experiments and theories, and Fig. 10.2 collects the captured features of the BKT phases in their measurable parameters: liquid He, superconductors, photons, and cold atoms.

Liquid Helium

The first direct observation of the BKT state succeeded in liquid helium films by Bishop and Reppy [18, 21]. In these experiments, thin ^{4}He films have been absorbed on a substrate with a high-quality torsional oscillator. If the system is cooled below the transition temperature, the oscillation period of the mechanical system increases since the superfluid He decouples from the rest of the mechanical system. From the period shift, superfluid density and dissipation are determined unambiguously. Fig. 10.2a collects the experimental data in the superfluid jump along transition temperatures, which agrees well with BKT theory.

Superconductors

Different from the previously discussed superfluids consisting of electric neutral particles, the related concept of superfluidity in superconductors has been described as the resistance-free flow of charged particles, in which the onset of resistance corresponds to the movement of free phase vortices through the system [22]. Measurements on aluminum films [19] confirmed the applicability of the BKT theory, and the DC resistance is proportional to the number of free vortices shown in Fig. 10.2b. Superconductivity occurs below the transition temperature, and it has also been suggested [23] that at low temperatures, the vortices and antivortices in a superconductor form a 2D crystal structure, similar to ions in a standard crystal. In this case, the system becomes resistive once the vortex–antivortex crystal starts to melt.

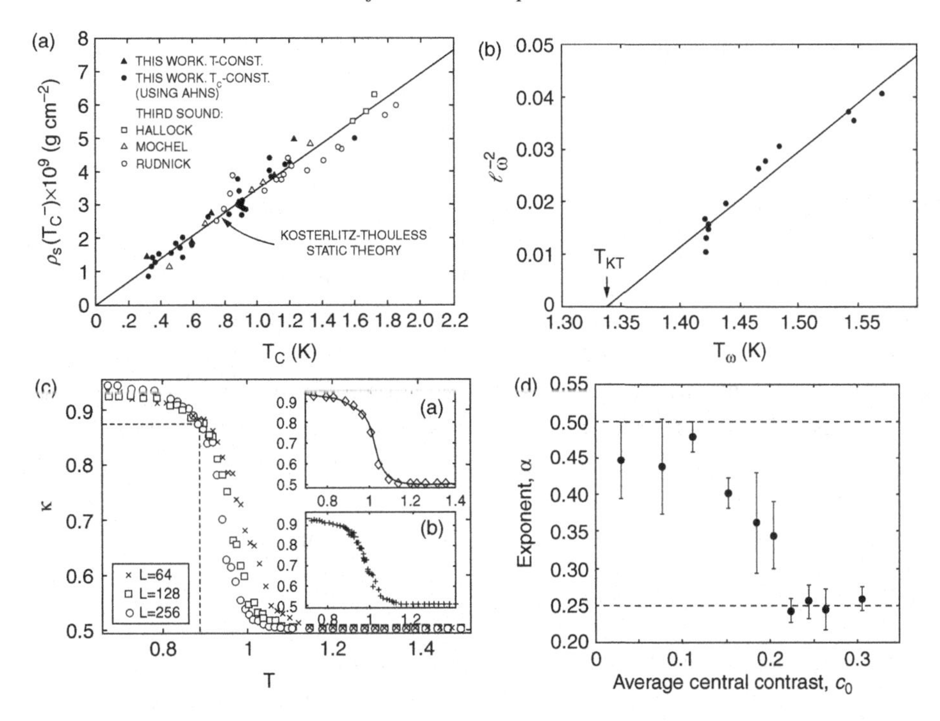

Figure 10.2 Experimental signatures of the BKT crossover captured in different physical systems: (a) He-II phase; reprinted with permission from Bishop, D. J., and Reppy, J. D. (1978), Study of the superfluid transition in two-dimensional ^{4}He films, *Phys. Rev. Lett.*, **40**, 1727 [18]. Copyright (1978) by the American Physical Society. (b) Superconductors; reprinted with permission from Hebard, A. F., and Fiory, A. T. (1980), Evidence for the Kosterlitz-Thouless transition in thin superconducting aluminum films, *Phys. Rev. Lett.*, **44**, 291 [19]. Copyright (1980) by the American Physical Society. (c) 2D photon lattices with Kerr nonlinearity; reprinted with permission from Small, et al. (2011), Kosterlitz-Thouless crossover in a photonic lattice. *Phys. Rev. A*, **83**, 013806 [20]. Copyright (2011) by the American Physical Society. (d) Trapped ultracold atoms; reprinted with permission from Hadzibabic, Z., et al. (2006), Berezinskii-Kosterlitz-Thouless crossover in a trapped atomic gas, Nature **441**, 1118 [12]. Copyright (2006) by the Nature Publishing Group.

2D XY Magnetic Crystals

Despite the fact that it has been a great challenge to search the ideal material to study the 2D XY model in nature, there have been tremendous efforts to grow layered magnetic crystals, which would be suitable to explore the BKT phase transition in 2D [24]. The same 2D BKT physics applies to these materials such as $BaNi_2(PO_4)_2$; namely, a long-range order induced by the vortex–antivortex pair would exhibit signatures of diverging magnetic susceptibility and a broad peak in the magnetic specific heat. Recently, there is a report by Putsch et al. that similar signatures in magnetic properties are observed in a Cu-based spin-dimer crystal, $C_{36}H_{48}Cu_2F_6N_8O_{12}S_2$ induced by a magnetic field [25].

Photonic Systems

The BKT crossover has been also addressed in a 2D nonlinear photonic medium. Since the nonlinear Schrödinger equation to describe the nonlinear photonics system is mathematically equivalent to the 2D spin XY model, we can engineer 2D photonic lattices to create the optical analog BKT phase transition and to observe the associated unbinding of vortices. Such lattices can be realized as a nonlinear 2D waveguide array, where the effective temperature is controlled by the initial randomness. Simulations confirmed the proliferation of free vortices above the transition temperature [20]. Experimental work was reported in a nonlinear optical system comprising light traveling through a nonlinear crystal, with the observation of the creation of free vortices above a critical temperature [26].

Atomic Bose Gases

Interference measurements between 2D gases of ultracold atoms demonstrated quasicoherence at low temperatures. The temperature rise induces a loss of the coherence, which is accompanied by the appearance of vortices identified in the interference patterns, confirming the predictions of the BKT theory [12]. These measurements used a strongly interacting Bose gas. Weakly interacting Bose gases start with the normal state and first go through a transition into an intermediate state which exhibits quasi-long-range order without superfluidity before finally undergoing a second transition into the BKT state with superfluidity and a quasi-long-range order [27].

The superfluidity of a 2D atomic BKT condensate has clearly been demonstrated by stirring such a condensate with a laser beam which behaves like a solid obstacle due to repulsive interaction [28, 29]. Finally, the bound vortex–antivortex pairs in an atomic 2D BKT gas have been observed by measuring the shadow which the condensate casts when illuminated with light at the resonance frequency of the atoms [30]. The exponent values of the power-law decay are plotted against the average interference contrast, reaching $1/4$, the theoretical value in equilibrium BKT physics (Fig. 10.2d). Related concepts of scale invariance in 2D atomic gases are discussed in Chapter 9 by Chin.

10.2.2 BKT Physics in Exciton-Polariton Systems

Besides the aforementioned platforms, microcavity exciton-polaritons (reviewed in Chapter 4 by Littlewood and Edelman) are primary quasiparticles in the strongly coupled photon and quantum-well (QW) exciton systems, which are regarded as quantum fluids. In low-density limit, they are composite bosons governed by Bose-Einstein statistics. Since the discovery in the monolithically grown semiconductor microcavity-QW structure [31], condensation and its coherence character have been studied using inorganic and organic semiconductors for the last two decades [32, 33]. Inherited from underlying entities, photons and excitons, the resulting exciton-polaritons are short-lived due to the extremely short photon lifetime, and they are scattered by repulsive Coulomb exchange interaction among fermions. Their effective mass is on the order of 10^{-4} electron mass, and they can dwell on the order of 1–100 ps inside the cavity. Hence, it is inevitable to refill exciton-polaritons, which naturally leak through the cavity as the open-dissipative system. This very nature of the 2D confinement in a planar QW-microcavity and the open-dissipativeness adds a rich context to explore the BKT physics in microcavity exciton-polaritons. Both theorists and experimentalists have been actively scrutinizing the search of steady-state coherent states and superfluidity.

10.3 Theoretical Studies of Exciton-Polaritons

Appreciating the nonequilibrium character of exciton-polaritons, theorists have addressed a series of fundamental questions: the existence of steady-state spontaneous condensation by repeated pump-decay processes, distinct signatures of nonequilibrium condensation compared with the equilibrium case, and the effect of fluctuations, which lead to the computation of spatial correlations in the context of BKT physics. Szymanska et al. set a Hamiltonian to describe a system, a bath and the system-bath interaction including pump and decay terms. The self-consistent solution for steady-state spontaneous condensation emerges even in the nonequilibrium condition. The excitation spectra induced by the diffusive phase mode are calculated as shown in Fig. 10.3a; the different correlations, decay would be expected from the diffusive phase mode [38]. The authors presented the extensive follow-up work in the mean-field theory to construct the phase map [34] (see also Chapter 4).

Discussed in the introduction, the spatial coherence is enumerated by the first-order correlation function. The principal mechanisms to affect the coherence property yield the different asymptotic behavior of the first-order correlation function in the long-distance limit. Thus, the long-distance decay will be a clue to identifying the prevailing mechanism. Tsyplyatyev and Whittaker computed the shape of the first-order coherence function in 1D depending on the known mechanisms [39]. Following the standard procedure, the authors found that $g^{(1)}(x)$ has a Gaussian form for nonzero momentum distributions in a trap, whereas

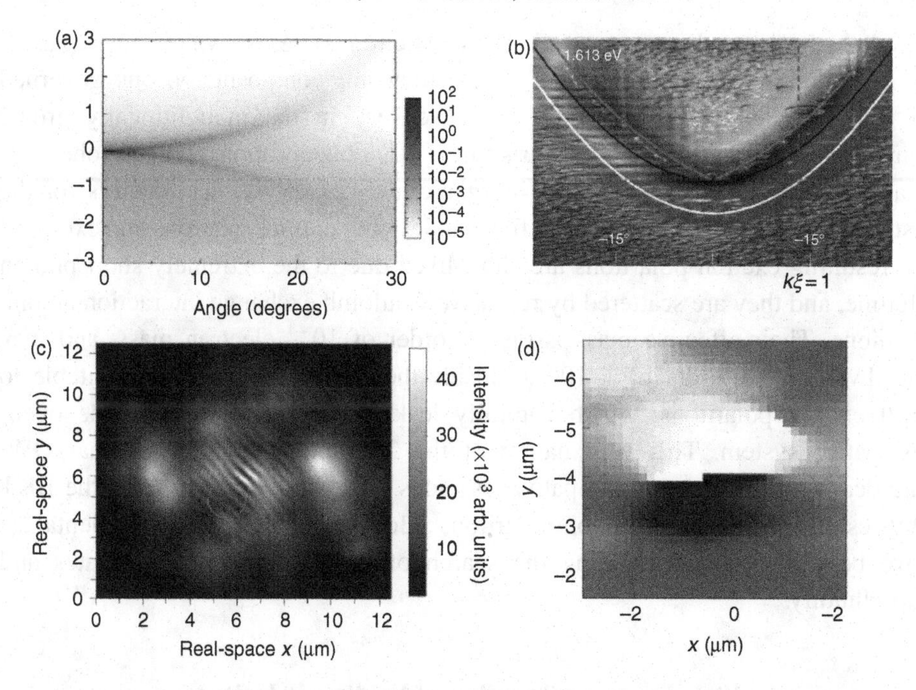

Figure 10.3 (a) Theoretically computed excitation spectra of exciton-polaritons. Reprinted with permission from Szymańska, M. H., et al. (2007), Mean-field theory and fluctuation spectrum of a pumped decaying Bose-Fermi system across the quantum condensation transition, *Phys. Rev. B*, **75**, 195331 [34]. Copyright (2007) by the American Physical Society. (b) Experimentally measured Bogoliubov linear spectra from a trapped polariton. Reprinted with permission from Utsunomiya, S., et al. (2008), Observation of Bogoliubov excitations in exciton-polariton condensates, *Nat. Phys.*, **4**, 700 [35]. Copyright (2008) by the Nature Publishing Group. (c) Interferograms of a pinned vortex in a CdTe-polariton system. Reprinted with permission from Lagoudakis, K. G., et al. (2008), Quantized vortices in an exciton-polariton condensate, *Nat. Phys.*, **4**, 706 [36]. Copyright (2008) by the Nature Publishing Group. (d) The extracted phase map of the vortex–antivortex pair in a GaAs-microcavity sample. Reprinted with permission from Roumpos, G., et al. (2011), Single vortex-antivortex pair in an exciton-polariton condensate, *Nat. Phys.*, **7**, 129 [37]. Copyright (2011) by the Nature Publishing Group.

$g^{(1)}(x)$ decays exponentially both for localization and polariton-interaction with the coherence length.

There are also several recent theoretical papers which target 2D superfluidity with the emphasis of nonequilibrium nature in exciton-polariton condensates [40, 41, 42]. Altman et al. inquired if effective equilibrium ODLRO emerges in driven 2D isotropic Bose systems or not. By mapping a nonequilibrium condensate system to an effective equilibrium system, the authors concluded that algebraic order appears at an intermediate length scale and the true ODLRO completely disappears by nonlinearity to cause nonequilibrium fluctuations [40] (see also discussion in Chapter 11). On the other hand, Chiocchetta and Carusotto developed a phenomenological model in terms of the linearized stochastic Gross-Pitaevskii equation to compute a spatial correlation function given density and phase fluctuations. They claimed that in 2D, the power-law form of the correlation function appears at the long-distance limit as a BKT phase transition [41]. In the optical parametric oscillator regime, Dagvadorj and colleagues were able to simulate the BKT phase transition to trace vortex–antivortex pairs and the unbound vortices across the pump thresholds [42]. They also calculated first-order spatial correlations, whose long-range behavior shows the exponential decay below threshold but the algebraic decay above threshold. The power-law exponent values are pump-power dependent and approximately 1.4, which is much higher than 0.25, the maximum value in the equilibrium counterpart near the threshold, to 0 at a higher threshold, indicating the true ODLRO in the exciton-polariton systems.

10.4 Experimental Work with Exciton-Polaritons

Exciton-polaritons are captured from reflectivity measurements on a semiconductor QW-microcavity based on GaAs, where the anticrossing of energy states appears at the coupled exciton-photon regions [31]. Soon after the first report in 1992, there have been numerous theoretical works to predict the quantum Bose nature and experimental ones to observe its experimental signatures for last two decades. We encourage readers to refer to Chapter 1 for a brief overview of the early history and associated challenges by Snoke et al. Exciton-polaritons naturally decay through the leakage of photons out of the cavity, and these photons preserve the microscopic information of decaying exciton-polaritons in terms of energy, in-plane momentum, and polarization. Due to the extremely light effective mass of $10^{-4} \sim 10^{-5}$ times the elementary electron mass, the condensation temperatures of GaAs and CdTe semiconductors lie around $4 \sim 10$ K, and they can be at room temperatures for GaN and inorganic semiconductors. Exciton-polaritons indeed exhibit common features similar to equilibrium quantum fluids in addition to exclusive signatures of the nonequilibrium setting.

10.4.1 Superfluidity

Quantum mechanical effects become pronounced when the primary length scale of particle is on the order of thermal de Broglie wavelength given by λ_T. At a low density, exciton-polaritons undergo a phase transition to Bose condensates with spontaneous phase symmetry breaking. Experimental signatures are accumulated to understand dynamical nature of condensation due to a finite lifetime: nonlinear threshold behavior in the occupation of the zero momentum state and spectral narrowing above threshold, and spontaneous coherence [46, 43, 47]. Subsequently, the search for superfluidity in exciton-polaritons began. Fig. 10.3b displays the Bogoliubov-like phonon excitation spectra reported in trapped exciton-polaritons by an incoherent pumping [35]. In a coherent excitation pumping to set a parametric oscillator regime, the flow velocity of exciton-polariton fluids is measured against the defect [15], which would be well explained by the famous Landau criterion. Hydrodynamic solitons propagate at the defects in exciton-polariton superfluids [16]. Quantized vortices (Fig. 10.3c) [36], half-quantum vortices [48], and a bound vortex–antivortex pair (Fig. 10.3d) [37] are clearly seen in coherent exciton-polariton condensates through interferometric measurement. Unfortunately, time-integrated interferometric measurements are a limitation to capturing spontaneous formation of free vortices and bound vortex–antivortex pairs, which would be the direct evidence of the BKT phase transition.

10.4.2 Spatial Coherence

Across the BKT phase transition, in addition to the spontaneous formation of bound vortex–antivortex pairs, another manifestation occurs in the long-distance asymptotic behavior of the first-order spatial correlation function introduced in Section 10.1. In order to construct the phase map and to measure the spatial coherence of the condensate, two standard types of interferometers were commonly used to measure spatial coherence: the Michelson interferometer (Fig. 10.4a) [43, 44] and Young's double-slit interferometer (Fig 10.4b) [50, 45]. The visibility of the interferograms is directly mapping to the $g^{(1)}(x, -x)$ functions presented in Fig. 10.4c, d. Although both methods provide spatial dependence, the Michelson one is convenient to yield the 2D map with improved spatial resolution fixed by a detector pixel size. Hence, the spatial dependence in the $g^{(1)}(x, -x)$ function is extensively measured using the Michelson interferometer in general whose simple schematic is drawn in Fig. 10.4a.

The dense $g^{(1)}(x, -x)$-function 2D maps with GaAs QW-embedded microcavity samples are reported in Refs. [44, 49]. Note that the two works deploy different pumping profiles: a top-hat pump spot and a Gaussian pump spot. In both excitation schemes, at the first-order phase transition, the sharp emergence of superfluid density occurs at a specific pump power to excite exciton-polaritons, and a power-law decay of the spatial coherence is observed in both measurements. However, the exponent values of the power law from the fitting are very different: the power-law exponent is above 1 for the top-hat profile, which exceeds the maximum

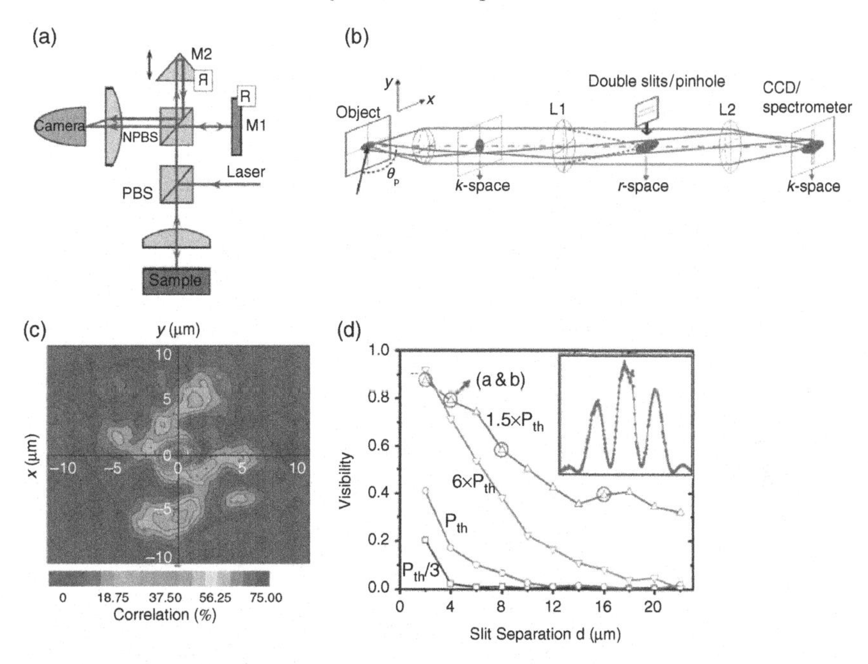

Figure 10.4 Schematics of (a) a Michelson interferometer and (b) Young's double-slit interferometer. (c) Spontaneous spatial coherence of the CdTe condensates taken by the Michelson interferometer in Ref. [43]. (d) The visibility of Young's interference peaks from the GaAs condensates at different pump-power values. (a) reprinted with permission from Roumpos, G., et al. (2012), Power-law decay of the spatial correlation function in exciton-polariton condensates, *Proc. Natl. Acad. Sci.*, **109**, 6467 [44]. Copyright (2012) National Academy of Sciences, USA. (b) and (d) reprinted with permission from Lai, C. W., et al. (2007), Coherent zero-state and π-state in an exciton-polariton condensate array, *Nature*, **450**, 529 [45]. Copyright (2007) by the Nature Publishing Group. (c) reprinted with permission from Kasprzak, J., et al. (2006), Bose-Einstein condensation of exciton polaritons. *Nature*, **443**, 409. [43]. Copyright (2006) by the Nature Publishing Group.

value, $1/4$ for a BKT condensate in thermal equilibrium [44]. On the other hand, the observed exponent of the power-law signals using the Gaussian pump profile is $\approx 1/4$ near the condensation threshold and becomes $<1/4$ at higher particle densities [49].

In addition, the exponent in the earlier work [44] by Roumpos et al. seemed to increase slightly at higher exciton-polariton densities, whereas the theory predicts it to decrease. This article argued that the unexpected large exponent might be a result of having a nonequilibrium condensate due to the continuous creation and decay of exciton-polaritons. Later, Nitsche and the team found that a top-hat excitation profile invariably results in simultaneous condensation in multiple modes, which leads to a faster decay of the measured spatial coherence since any interference

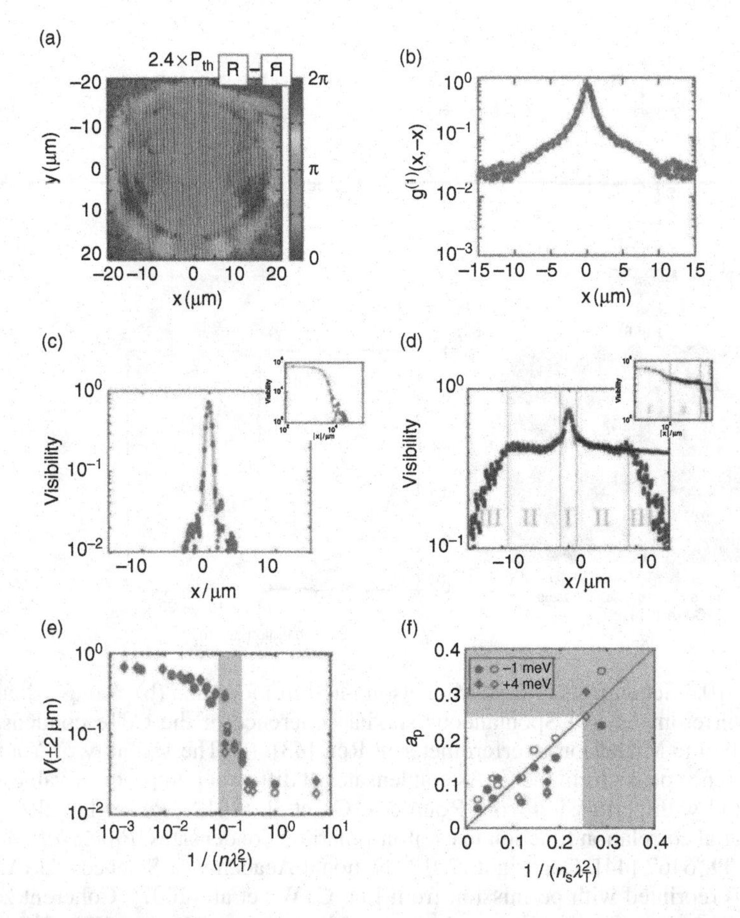

Figure 10.5 (a) A representative visibility above threshold from the interference between the reference image and the retro-reflective image. Reprinted with permission from Roumpos, G., et al. (2012), Power-law decay of the spatial correlation function in exciton-polariton condensates, *Proc. Natl. Acad. Sci.*, **109**, 6467 [44]. Copyright (2012) National Academy of Sciences, USA. (b) A semi-log scale plot of the first-order spatial correlation function under the top-hat pump profile with the Gaussian and power-law fit (dotted line) in Ref. [44]. Using the Gaussian pump profile, the interference visibility shows different shapes in space (c) below and (d) above threshold. The lines in (c) and (d) are the Gaussian fit, and the line in (d) is the power-law fit. (e) Visibility and (f) the power-law exponent a_p are plotted in terms of the total density n and the superfluid density n_s. Panels (b)–(f) are reprinted with permission from Nitsche, W. H., et al. (2014), Algebraic order and the Berezinskii-Kosterlitz-Thouless transition in an exciton-polariton gas, Phys. Rev. B **90**, 205430 [49]. Copyright (2014) by the American Physical Society.

between light coming from different modes of the condensate averages out to zero during the integration time of the camera [51]. However, the Gaussian laser profile seems to create a condensate only in the ground-state mode with which it maximizes the overlap, thereby avoiding artifacts related to multimode condensation. In the latter [49], a sample with only four quantum wells is used and excited by a laser with fewer power fluctuations than in the work of Ref. [44].

10.5 Conclusions

Interacting microcavity exciton-polaritons, as quantum fluids, experience a thermodynamic phase transition with coherence even in 2D despite their driven-dissipative nature. The finite system size defined by either a laser spot size or a trap potential combined with their short lifetime makes exciton-polaritons an interesting system to discuss the BKT physics, where a quasi-long-range order occurs below the transition temperatures. We learn that the excitation pump profile, pumping scheme, and the spatial detuning values affect the spatial coherence properties.

Even under the nonequilibrium condition, exciton-polaritons exhibit algebraic order with the power-decay form. Recently, theoretical models in a particular pumping scheme [40, 41] offer support for the experimental findings regarding the power-law decay of correlation functions; yet a microscopic theory needs to be developed in order to apply for nonresonant excitation cases, which still remains as a calculation-heavy challenge. Experimentally, we have had no luck with observing the spontaneous formation of vortices and bound vortex pairs across phase transition for the unequivocal claim of the BKT transition. This is the next target in decoding the clear relation between the true long-range-ordered state and the quasi-long-range order in dynamical exciton-polariton condensates.

References

[1] Allen, J. F., and Misener, A. D. 1938. Flow of liquid helium II. *Nature*, **141**, 75.

[2] Kapitza, P. 1938. Viscosity of liquid helium below the λ-point. *Nature*, **141**, 74.

[3] London, F. 1938. The λ-phenomenon of liquid helium and the Bose-Einstein degeneracy. *Nature*, **141**, 643–644.

[4] Anderson, M. H., Ensher, J. R., Matthews, M. R., Wieman, C. E., and Cornell, E. A. 1995. Observation of Bose-Einstein condensation in a dilute atomic vapor. *Science*, **269**, 198–201.

[5] Davis, K. B., Mewes, M.-O., Andrews, M. R., van Druten, N. J., Durfee, D. S., Kurn, D. M., and Ketterle, W. 1995. Bose-Einstein condensation in a gas of sodium atoms. *Phys. Rev. Lett.*, **75**, 3969–3973.

[6] Mermin, N. D., and Wagner, H. 1966. Absence of ferromagnetism or antiferromagnetism in one- or two-dimensional isotropic Heisenberg models. *Phys. Rev. Lett.*, **17**, 1133–1136.

[7] Hohenberg, P. C. 1967. Existence of long-range order in one and two dimensions. *Phys. Rev.*, **158**, 383–386.

[8] Berezinskii, V. L. 1972. Destruction of long-range order in one-dimensional and two-dimensional systems possessing a continuous symmetry group. II. Quantum systems. *Soviet Journal of Experimental and Theoretical Physics*, **34**, 610–616.

[9] Kosterlitz, J. M., and Thouless, D. J. 1973. Ordering, metastability and phase transitions in two-dimensional systems. *Journal of Physics C Solid State Physics*, **6**, 1181–1203.

[10] Fletcher, R. J., Robert-de Saint-Vincent, M., Man, J., Navon, N., Smith, R. P., Viebahn, K. G. H., and Hadzibabic, Z. 2015. Connecting Berezinskii-Kosterlitz-Thouless and BEC phase transitions by tuning interactions in a trapped gas. *Phys. Rev. Lett.*, **114**, 255302.

[11] Hadzibabic, Z., and Dalibard, J. 2011. Two-dimensional Bose fluids: an atomic physics perspective. *Riviesta del Nuovo Cimento*, **34**, 389–433.

[12] Hadzibabic, Z., Krüger, P., Cheneau, M., Battelier, B., and Dalibard, J. 2006. Berezinskii-Kosterlitz-Thouless crossover in a trapped atomic gas. *Nature*, **441**, 1118–1121.

[13] Leggett, A. J. 1999. Superfluidity. *Rev. Mod. Phys.*, **71**, S318–S323.

[14] Landau, L. D., and Lifshitz, E. M. 1959. *Fluid Mechanics*. Pergamon Press.

[15] Amo, A., Lefrère, J., Pigeon, S., Adrados, C., Ciuti, C., Carusotto, I., Houdré, R., Giacobino, E., and Bramati, A. 2009. Superfluidity of polaritons in semiconductor microcavities. *Nature Physics*, **5**, 805–810.

[16] Amo, A., Pigeon, S., Sanvitto, D., Sala, V. G., Hivet, R., Carusotto, I., Pisanello, F., Leménager, G., Houdré, R., Giacobino, E., Ciuti, C., and Bramati, A. 2011. Polariton superfluids reveal quantum hydrodynamic solitons. *Science*, **332**, 1167–1170.

[17] Kavokin, A., Baumberg, J. J., Malpuech, G., and Laussy, F. P. 2011. *Microcavities*. Oxford Science Publications. Oxford University Press.

[18] Bishop, D. J., and Reppy, J. D. 1978. Study of the superfluid transition in two-dimensional ^{4}He films. *Phys. Rev. Lett.*, **40**, 1727–1730.

[19] Hebard, A. F., and Fiory, A. T. 1980. Evidence for the Kosterlitz-Thouless transition in thin superconducting aluminum films. *Phys. Rev. Lett.*, **44**, 291–294.

[20] Small, Eran, Pugatch, Rami, and Silberberg, Yaron. 2011. Berezinskii-Kosterlitz-Thouless crossover in a photonic lattice. *Phys. Rev. A*, **83**, 013806.

[21] Bishop, D. J., and Reppy, J. D. 1980. Study of the superfluid transition in two-dimensional ^{4}He films. *Phys. Rev.*, **22**, 5171–5185.

[22] Beasley, M. R., Mooij, J. E., and Orlando, T. P. 1979. Possibility of vortex-antivortex pair dissociation in two-dimensional superconductors. *Phys. Rev. Lett.*, **42**, 1165–1168.

[23] Gabay, Marc, and Kapitulnik, Aharon. 1993. Vortex–antivortex crystallization in thin superconducting and superfluid films. *Phys. Rev. Lett.*, **71**, 2138–2141.

[24] Regnault, L. P., and Rossat-Mignod, J. 1990. *Magnetic Properties of Layered Transition Metal Compounds*. Physics and Chemistry of Materials with Low-Dimensional Structures, vol. 9. Kluwer Academic Publishers. Chap. Phase transitions in quasi-two-dimensional planar magnets, pages 271–321.

[25] Tutsch, U., Wolf, B., Wessle, S., Postulka, L., Tsui, Y., Jeschke, H.O., Opahle, I., Saha-Dasgupta, T., Valenti, R., Bruhl, A., Removic-Langer, K., Kretz, T., Lerner, H.-W., Wagner, M., and Lang, M. 2014. Evidence of a field-induced Berezinskii-Kosterlitz-Thouless scenario in a two-dimensional spin-dimer system. *Nature Comm.*, **5**, 5169.

[26] Situ, G., Muenzel, S., and Fleischer, J. W. 2013. Berezinskii-Kosterlitz-Thouless transition in a photonic lattice. *arXiv:1304.6980*, Apr.

[27] Cladé, P., Ryu, C., Ramanathan, A., Helmerson, K., and Phillips, W. D. 2009. Observation of a 2D Bose gas: from thermal to quasicondensate to superfluid. *Phys. Rev. Lett.*, **102**, 170401.

[28] Campbell, G. K. 2012. Quantum gases: superfluidity goes 2D. *Nature Physics*, **8**, 643–644.

[29] Desbuquois, R., Chomaz, L., Yefsah, T., Léonard, J., Beugnon, J., Weitenberg, C., and Dalibard, J. 2012. Superfluid behaviour of a two-dimensional Bose gas. *Nature Physics*, **8**, 645–648.

[30] Choi, J.-y., Seo, S. W., and Shin, Y.-i. 2013. Observation of thermally activated vortex pairs in a quasi-2D Bose gas. *Phys. Rev. Lett.*, **110**, 175302.

[31] Weisbuch, C., Nishioka, M., Ishikawa, A., and Arakawa, Y. 1992. Observation of the coupled exciton–photon mode splitting in a semiconductor quantum microcavity. *Phys. Rev. Lett.*, **69**, 3314–3317.

[32] Deng, H., Haug, H., and Yamamoto, Y. 2010. Exciton-polariton Bose-Einstein condensation. *Rev. Mod. Phys.*, **82**, 1489–1537.

[33] Byrnes, T. Kim, N. Y., and Yamamoto, Y. 2014. Exciton-polariton condensates. *Nature Physics*, **10**, 803–813.

[34] Szymańska, M. H., Keeling, J., and Littlewood, P. B. 2007. Mean-field theory and fluctuation spectrum of a pumped decaying Bose-Fermi system across the quantum condensation transition. *Phys. Rev. B*, **75**, 195331.

[35] Utsunomiya, S., Tian, L., Roumpos, G., Lai, C. W., Kumada, N., Fujisawa, T., Kuwata-Gonokami, M., Löffler, A., Höfling, S., Forchel, A., and Yamamoto, Y. 2008. Observation of Bogoliubov excitations in exciton-polariton condensates. *Nature Physics*, **4**, 700–705.

[36] Lagoudakis, K. G., Wouters, M., Richard, M., Baas, A., Carusotto, I., André, R., Dang, L. S., and Deveaud-Plédran, B. 2008. Quantized vortices in an exciton-polariton condensate. *Nature Physics*, **4**, 706.

[37] Roumpos, G., Fraser, M. D., Löffler, A., Höfling, S., Forchel, A., and Yamamoto, Y. 2011. Single vortex–antivortex pair in an exciton-polariton condensate. *Nature Physics*, **7**, 129–133.

[38] Szymańska, M. H., Keeling, J., and Littlewood, P. B. 2006. Nonequilibrium quantum condensation in an incoherently pumped dissipative system. *Phys. Rev. Lett.*, **96**, 230602.

[39] Tsyplyatyev, O., and Whittaker, D. M. 2012. Spatial coherence of a polariton condensate in 1D acoustic lattice. *Physica Status Solidi B Basic Research*, **249**, 1692–1697.

[40] Altman, Ehud, Sieberer, Lukas M., Chen, Leiming, Diehl, Sebastian, and Toner, John. 2015. Two-dimensional superfluidity of exciton polaritons requires strong anisotropy. *Phys. Rev. X*, **5**, 011017.

[41] Chiocchetta, A., and Carusotto, I. 2013. Non-equilibrium quasi-condensates in reduced dimensions. *Europhysics Letters*, **102**, 67007.

[42] Dagvadorj, G., Fellows, J. M., Matyjaśkiewicz, S., Marchetti, F. M., Carusotto, I., and Szymańska, M. H. 2015. Nonequilibrium phase transition in a two-dimensional driven open quantum system. *Phys. Rev. X*, **5**, 041028.

[43] Kasprzak, J., Richard, M., Kundermann, S., Baas, A., Jeambrun, P., Keeling, J. M. J., Marchetti, F. M., Szymańska, M. H., André, R., Staehli, J. L., Savona, V., Littlewood, P. B., Deveaud, B., and Dang, L. S. 2006. Bose-Einstein condensation of exciton polaritons. *Nature*, **443**, 409–414.

[44] Roumpos, G., Lohse, M., Nitsche, W. H., Keeling, J., Szymańska, M. H., Littlewood, P. B., Löffler, A., Höfling, S., Worschech, L., Forchel, A., and Yamamoto, Y. 2012. Power-law decay of the spatial correlation function in exciton-polariton condensates. *Proceedings of the National Academy of Science*, **109**, 6467–6472.

[45] Lai, C. W., Kim, N. Y., Utsunomiya, S., Roumpos, G., Deng, H., Fraser, M. D., Byrnes, T., Recher, P., Kumada, N., Fujisawa, T., and Yamamoto, Y. 2007. Coherent zero-state and π-state in an exciton-polariton condensate array. *Nature*, **450**, 529–532.

[46] Deng, H., Weihs, G., Santori, C., Bloch, J., and Yamamoto, Y. 2002. Condensation of semiconductor microcavity exciton polaritons. *Science*, **298**, 199–202.

[47] Balili, R., Hartwell, V., Snoke, D., Pfeiffer, L., and West, K. 2007. Bose-Einstein condensation of microcavity polaritons in a trap. *Science*, **316**, 1007–1010.

[48] Lagoudakis, K. G., Ostatnický, T., Kavokin, A. V., Rubo, Y. G., André, R., and Deveaud-Plédran, B. 2009. Observation of half-quantum vortices in an exciton-polariton condensate. *Science*, **326**, 974–976.

[49] Nitsche, Wolfgang H., Kim, Na Young, Roumpos, Georgios, Schneider, Christian, Kamp, Martin, Höfling, Sven, Forchel, Alfred, and Yamamoto, Yoshihisa. 2014. Algebraic order and the Berezinskii-Kosterlitz-Thouless transition in an exciton-polariton gas. *Phys. Rev. B*, **90**, 205430.

[50] Deng, H., Solomon, G. S., Hey, R., Ploog, K. H., and Yamamoto, Y. 2007. Spatial coherence of a polariton condensate. *Phys. Rev. Lett.*, **99**, 126403.

[51] Nitsche, Wolfgang H., Kim, Na Young, Roumpos, Georgios, Schneider, Christian, Höfling, Sven, Forchel, Alfred, and Yamamoto, Yoshihisa. 2016. Spatial correlation of two-dimensional bosonic multimode condensates. *Phys. Rev. A*, **93**, 053622.

11

Superfluidity and Phase Correlations of Driven Dissipative Condensates

JONATHAN KEELING
SUPA, School of Physics and Astronomy, University of St Andrews, UK

LUKAS M. SIEBERER
Department of Physics, University of California, Berkeley, USA;
Department of Condensed Matter Physics, Weizmann Institute of Science, Rehovet, Israel;
Institute for Theoretical Physics, University of Innsbruck, Austria

EHUD ALTMAN
Department of Physics, University of California, Berkeley, USA;
Department of Condensed Matter Physics, Weizmann Institute of Science, Rehovet, Israel

LEIMING CHEN
College of Science, China University of Mining and Technology, Xuzhou, Jiangsu, People's Republic of China

SEBASTIAN DIEHL
Institut für Theoretische Physik, Universität zu Köln, Cologne, Germany

JOHN TONER
Department of Physics and Institute of Theoretical Science, University of Oregon, Eugene, USA

We review recent results on the coherence and superfluidity of driven dissipative condensates, i.e., systems of weakly interacting nonconserved bosons, such as polariton condensates. The presence of driving and dissipation has dramatically different effects depending on dimensionality and anisotropy. In three dimensions, equilibrium behaviour is recovered at large scales for static correlations, while the dynamical behaviour is altered by the microscopic driving. In two dimensions, for an isotropic system, drive and dissipation destroy the algebraic order that would otherwise exist; however, a sufficiently anisotropic system can still show algebraic phase correlations. We discuss the consequences of this behaviour for recent experiments measuring phase coherence and outline potential measurements that might directly probe superfluidity.

11.1 Introduction

This chapter is dedicated to superfluidity and its relation to Bose-Einstein condensation (BEC), a topic with a long history. Many reviews of the concepts of condensation and superfluidity in thermal equilibrium can be found; see for example

Refs. [1, 2, 3, 4]. The focus of this chapter is on how these concepts apply (or fail to apply) to driven dissipative condensates – systems of bosons with a finite lifetime, in which loss is balanced by continuous pumping. We focus entirely on the steady state of such systems, neglecting transient, time-dependent behaviour.

Experimentally, the most studied example of a driven dissipative condensate has been microcavity polaritons, an overview of which is given in Ref. [5] and Chapter 4. However, similar issues can arise in many other systems, most obviously photon condensates [6] (see also Chapter 19), magnon condensates [7] (see also Chapters 25–26), and potentially exciton condensates (although typical exciton lifetimes are much longer than for polaritons). Even experiments on cold atoms (see, e.g., Chapter 3) could be driven into a regime in which such physics occurs, when considering continuous loading of atoms balancing three-body losses [8] or atom laser setups [9, 10, 11].

Experiments on polaritons are two-dimensional, and in two dimensions it is particularly important to clearly distinguish three concepts often erroneously treated as equivalent: superfluidity, condensation, and phase coherence. This is because no true Bose-Einstein condensate exists in a homogeneous two-dimensional system. Before addressing superfluidity and phase coherence in the steady state of a driven dissipative system, we review in Section 11.2 the essential ideas of condensation, superfluidity, and phase coherence for systems in thermal equilibrium. In Section 11.3, we set up a generic microscopic model for weakly interacting driven dissipative Bose gases and make precise the sense in which these systems are nonequilibrium. Section 11.4 reviews the connection of these driven dissipative systems to the Kardar-Parisi-Zhang equation, and explains the absence, for isotropic systems, of algebraic order at large distances based on this mapping. We also show that algebraic order is possible in the strongly anisotropic case. We frame this discussion in the context of current experiments, which have all been done in the weakly anisotropic regime. In Section 11.5, we discuss the meaning of superfluidity in a driven, number-nonconserving setup and discuss experimental probes. Section 11.6 gives a brief account of vortices in such open systems. (See also Chapter 21 for an overview of vortex studies in such systems.) Conclusions and challenges for future research are given in Section 11.7.

11.2 Bose-Einstein Condensation and Superfluidity

Bose-Einstein condensation for a gas of weakly interacting bosons is a phase transition associated with the appearance of off-diagonal long-range order (ODLRO) [12]. (See also Chapter 7.) This means that correlation functions such as $\langle \psi^\dagger(\mathbf{r})\psi(\mathbf{r}')\rangle$ remain nonzero even between distant points, $|\mathbf{r} - \mathbf{r}'| \to \infty$. These correlations indicate the spontaneous symmetry breaking of the $U(1)$ phase of the condensate wavefunction; i.e., writing $\psi = \sqrt{\rho}e^{i\theta}$, ODLRO corresponds to the phase θ being correlated at distant points. In such a symmetry-broken phase there is a 'phase stiffness', i.e., there is an energy cost, $E[\theta(\mathbf{r})] = (K_s/2)\int d^2\mathbf{r}(\nabla\theta)^2$ for phase twists of the condensate.

A gedanken experiment to determine this phase stiffness is to measure the change of energy as one imposes a phase twist between the boundaries of the system. Alternatively, since phase gradients correspond to currents, a more practical way to measure the phase stiffness is to apply a force that tries to drive a current, and observe the condensate's response. The behaviour of a condensate in a rotating container illustrates the role of the phase stiffness very clearly [2, 13]. The condensate cannot be made to rotate except by creating quantised vortices, and these cost energy due to the phase stiffness. Thus, for slow rotation, the condensate fails to rotate. Similar behaviour can be seen in a ring trap, where the core of a vortex can be located outside the condensate. This is the Hess-Fairbank [14] effect and is a defining property of a superfluid; i.e., when a condensate has nonzero phase stiffness, it becomes superfluid, as seen by its reduced response to rotation.

However, superfluidity is not equivalent to Bose-Einstein condensation; as discussed below, in two dimensions, Bose-Einstein condensation and ODLRO do not exist, yet superfluidity persists. It is therefore useful to be able to define superfluidity directly without reference to condensation. This can be done by defining a superfluid density as the part of the system that fails to respond to slow rotations. This definition also clarifies another important point: except at zero temperature, a system will have both superfluid and normal components, as thermally excited quasiparticles can respond normally to rotations. To identify the superfluid density, we must consider the response function $\chi_{ij}(\mathbf{q},\omega)$, which relates the current $\langle\hat{j}_i(\mathbf{q},\omega)\rangle$ to the force that induces it, $f_j(\mathbf{q},\omega)$:

$$\langle\hat{j}_i(\mathbf{q},\omega)\rangle = \chi_{ij}(\mathbf{q},\omega)f_j(\mathbf{q},\omega), \tag{11.1}$$

where i,j refer to Cartesian components. The operator $\hat{j}_i$ appearing here is the standard particle current written in momentum space:

$$\hat{j}_i(\mathbf{q}) = \sum_{\mathbf{k}} \hat{\psi}^\dagger_{\mathbf{k}+\mathbf{q}}\gamma_i(2\mathbf{k}+\mathbf{q})\hat{\psi}_{\mathbf{k}}, \qquad \gamma_i(\mathbf{K}) = \frac{K_i}{2m}.$$

We consider systems in which this current is conserved, i.e., $\partial_t\rho + \nabla\cdot\mathbf{j} = 0$, where ρ is the particle density. According to Noether's theorem, current conservation corresponds to the existence of a $U(1)$ phase symmetry in the Hamiltonian – this is the same symmetry which is spontaneously broken on Bose-Einstein condensation. Considering static (i.e., $\omega = 0$) long-wavelength (i.e., $\mathbf{q} \to 0$) currents, the most general response function possible for an isotropic system is

$$\chi_{ij}(\mathbf{q} \to 0, \omega = 0) = \chi_L \frac{q_i q_j}{q^2} + \chi_T \left(\delta_{ij} - \frac{q_i q_j}{q^2} \right). \tag{11.2}$$

The terms χ_L, χ_T describe the response to longitudinal and transverse forces, i.e., $\mathbf{f} \parallel \mathbf{q}$, which occurs for a potential force, and $\mathbf{f} \perp \mathbf{q}$, which occurs for rotational or magnetic forces. Current conservation can be shown to mean that $(q_i q_j / q^2) m \chi_{ij}(\mathbf{q}, \omega) = \rho(\mathbf{q}, \omega)$, where $\rho(\mathbf{q}, \omega)$ is the single particle Green's function, so that indeed χ_L is related to the total particle number. The transverse part of χ_{ij} describes the response to rotations; therefore, superfluidity corresponds to a reduction of χ_T. In a nonsuperfluid, current is parallel to force, which means $\chi_T = \chi_L$. The superfluid fraction of a system is therefore given by $(\chi_L - \chi_T)/\chi_L$, and the normal fraction by χ_T/χ_L.

The above definition of superfluidity depends on the *equilibrium* effect that a system in true thermal equilibrium has a reduced response to rotational forces. This effect is conceptually distinct from the *metastable* persistent flow that can also be seen in a non-simply connected (e.g., annular) geometry [2]. Metastable persistent flow occurs if one first sets the fluid in motion by rotating the container while above the critical temperature and then cools the fluid until it becomes superfluid. If the container then stops rotating, the fluid remains in motion, as the lifetime for the current to decay is exponentially large.

It is also important to note that the superfluid density defined above is a static property of the system; if excited dynamically, it is always possible to create excitations out of the condensate. Because the elementary spectrum of an interacting Bose gas has a linearly dispersing Bogoliubov sound mode, this sound velocity c_s acts as a critical velocity [2, 4, 13]: for a defect moving at a lower velocity, no excitations can be created beyond the condensed component. However, at nonzero temperatures, thermally excited quasiparticles can respond to flow at any velocity, and a normal component will occur, as discussed above. In equilibrium, there is a fundamental connection between the existence of a nonzero sound velocity and the presence of a nonzero superfluid fraction: as discussed in Ref. [15], the finite frequency response function $\chi_{ij}(\mathbf{q}, \omega)$ has the same poles as the single particle Green's function, and so the finite frequency generalisation of the superfluid part of Eq. (11.2) is $\chi_{ij}^{SF}(\mathbf{q}, \omega) = c_s^2 q_i q_j / (c_s^2 q^2 - \omega^2)$. The fact that this is finite at $\omega = 0, \mathbf{q} \to 0$ is thus connected to the form of the dispersion and the existence of a nonzero sound velocity.

11.2.1 Two Dimensions

In two dimensions, the distinction between superfluidity and BEC becomes even more important, since a homogeneous two-dimensional system is unable to show true long-range order due to the Mermin-Wagner theorem [16]. This can be seen by considering the correlation function $\langle\psi^\dagger(\mathbf{r})\psi(\mathbf{r}')\rangle \simeq \rho_0\langle\exp[i(\theta(\mathbf{r}')-\theta(\mathbf{r}))]\rangle$. Even for a system with a nonzero phase stiffness, $E[\theta(\mathbf{r})] = (K_s/2)\int d^2\mathbf{r}(\nabla\theta)^2$, the vanishing energy cost of long wavelength phase twists leads to a thermal expectation which, at long distances, takes the form

$$\left\langle e^{i(\theta(\mathbf{r}')-\theta(\mathbf{r}))}\right\rangle \propto \exp\left(-\alpha_s \ln|\mathbf{r}-\mathbf{r}'|\right) \propto |\mathbf{r}-\mathbf{r}'|^{-\alpha_s}, \qquad \alpha_s = \frac{k_B T}{2\pi K_s}, \tag{11.3}$$

which vanishes algebraically as $|\mathbf{r}-\mathbf{r}'| \to \infty$. Therefore, there is no ODLRO, and so no single mode is macroscopically occupied; i.e., there is no BEC. Nonetheless, superfluidity can survive, because the phase stiffness K_s implies a resistance to rotation of the low-energy modes of the condensate. In fact, either by directly comparing the calculation of phase stiffness and superfluid density [15] or by calculating the current–current response function from the parametrisation and energy functional above [1, 13], one finds that the phase stiffness (and thus the power-law decay) is directly related to the superfluid stiffness, specifically, $K_s = \rho_s/m$, where m is the quasiparticle mass.

In addition to the distinct nature of the low-temperature phase in two dimensions, the transition to this phase is also unusual. As noted above, a superfluid can be made to rotate by creating quantised vortices, in which the phase winds by 2π around a point, and the density of the condensate vanishes at that point. As discussed by Kosterlitz and Thouless [17, 18] and Berezinskii [19], the transition out of the superfluid phase occurs through the proliferation of these vortices. The phase boundary can be found by a renormalisation group approach [18, 20, 21], which predicts that if $K_s < (2/\pi)k_BT$, vortices proliferate and correlations decay exponentially, whereas for $K_s > (2/\pi)k_BT$, vortices are irrelevant at large scales and correlations decay algebraically. As a result, the exponent for the algebraic decay, Eq. (11.3), takes the universal value $\alpha_s = 1/4$ at the phase boundary, and the superfluid density undergoes a universal jump [20].

11.3 Modelling Driven Dissipative Condensates

A driven dissipative system is by definition one that is coupled to more than one environment or is driven by a time-dependent pumping field. One can therefore no longer apply equilibrium concepts, such as the minimisation of free energy or the fluctuation-dissipation theorem. For a given system, such as a weakly interacting dilute Bose gas, there are many different types of driving and dissipation. Some of these conserve particle number; others do not. In this chapter, we consider the latter case, motivated by systems such as microcavity polaritons (for a review, see Ref. [5] and Chapter 4).

Microcavity polaritons are superpositions of excitons and photons, and the photon part can leak out through the mirrors. To reach a steady state, this loss must be replenished by a compensating pump. Several models can be written to describe this, with varying descriptions of the reservoir that replenishes the condensate [22, 23, 24]. All such models lead to the same essential differences between the equilibrium and driven dissipative cases. We consider the model starting from the weakly interacting dilute Bose gas,

$$\hat{H} = \int d^d r \hat{\psi}^\dagger(r) \left(-\frac{\nabla^2}{2m}\right) \hat{\psi}(r) + \frac{U}{2} \hat{\psi}^\dagger(r) \hat{\psi}^\dagger(r) \hat{\psi}(r) \hat{\psi}(r),$$

and consider loss terms described by the quantum master equation

$$\frac{\partial \rho}{\partial t} = -i[\hat{H}, \rho] + \int d^d r \left(\frac{\kappa}{2} \mathcal{L}[\hat{\psi}(r), \rho] + \frac{\gamma}{2} \mathcal{L}[\hat{\psi}^\dagger(r), \rho] + \frac{\Gamma}{4} \mathcal{L}[\hat{\psi}^2(r), \rho] \right), \tag{11.4}$$

where $\mathcal{L}[\hat{X}, \rho] = 2\hat{X}\rho\hat{X}^\dagger - [\hat{X}^\dagger \hat{X}, \rho]_+$ is the usual Lindblad operator. The terms in Eq. (11.4) describe single-particle loss, single-particle incoherent pump, and two-particle losses at rates κ, γ, and Γ, respectively.

11.3.1 Mean-Field Description and Collective Excitations

The role of the dissipative terms in Eq. (11.4) can be seen by considering the corresponding mean-field equation of motion, found by replacing $\langle \psi(\mathbf{r}) \rangle = \varphi(\mathbf{r})$, and decoupling all correlators. This gives:

$$i\frac{\partial \varphi}{\partial t} = \left[-\frac{\nabla^2}{2m} + U|\varphi|^2 + \frac{i}{2}\left(\gamma - \kappa - \Gamma|\varphi|^2\right) \right] \varphi, \tag{11.5}$$

which is a modified Gross-Pitaevskii equation (GPE) [25], including dissipative terms describing particle gain and loss. The nonlinear term with coefficient Γ describes gain saturation, or feedback between the condensate and the reservoir, which prevents the particle density from diverging. For small fluctuations around the steady state, $\Gamma|\varphi_0|^2 = \gamma_{\text{net}} \equiv \gamma - \kappa$, one finds the fluctuations have a complex spectrum $\omega_{\mathbf{k}}$ of the form

$$\omega_{\mathbf{k}} = -i\left(\frac{\gamma_{\text{net}}}{2}\right) \pm \sqrt{\xi_{\mathbf{k}}^2 - \left(\frac{\gamma_{\text{net}}}{2}\right)^2}, \quad \xi_{\mathbf{k}}^2 = \frac{k^2}{2m}\left(\frac{k^2}{2m} + 2U|\varphi_0|^2\right). \tag{11.6}$$

The quantity $\xi_{\mathbf{k}}$ reduces to the equilibrium Bogoliubov excitation spectrum in the limit $\kappa, \gamma, \Gamma \to 0$. As discussed in Section 11.2, $\xi_{\mathbf{k}}$ has a linear dispersion,

$\xi_{\mathbf{k}} \simeq c_s|\mathbf{k}|$ at low momentum, where $c_s^2 = U|\varphi_0|^2/m$. In contrast, $\omega_{\mathbf{k}}$ is diffusive at low momentum, $\omega_{\mathbf{k}} = -iDk^2$ with $D = c_s^2/\gamma_{\text{net}}$.

This modified dispersion would appear to mean that the critical velocity vanishes. Nonetheless, signatures of a nonvanishing critical velocity can survive in static correlation functions – albeit washed out by dissipation. To see this, one may calculate theoretically the drag force on a defect immersed in a steady flow pattern around the defect potential $V_{\text{defect}}(\mathbf{r})$. The static drag force on the defect is given by [26] $\mathbf{F}_{\text{drag}} \propto \int d^d\mathbf{r}\rho(\mathbf{r})\nabla V_{\text{defect}}(\mathbf{r})$. For a perfect superfluid below the critical velocity, the flow pattern is symmetric, so that the effects of pressure ahead of and behind the defect cancel: i.e., 'd'Alembert's paradox' occurs, that an irrotational flow produces no drag. The GPE, with $\rho(r) = |\varphi(r)|^2$, describes such a perfect superfluid and would thus normally predict zero drag; however, if one calculates the drag using the complex GPE given above, a finite drag force exists at all velocities [27, 28]. There does, however, remain a marked threshold at $v = c_s$ above which the drag increases more rapidly with velocity. This is shown in Fig. 11.1, taken from Ref. [27]. This behaviour is similar to an equilibrium superfluid at finite temperature; however, calculating the finite temperature drag for an equilibrium superfluid requires including fluctuations beyond the mean-field theory.

The modified dispersion also has effects on the linear-response calculation of phase correlations [29, 30], but as discussed below, a more dramatic change arises

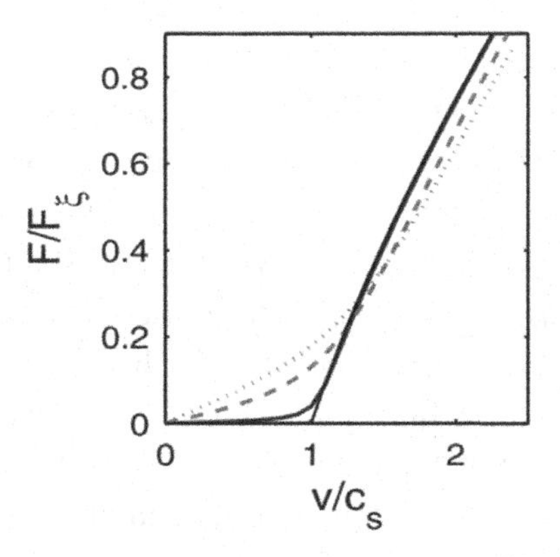

Figure 11.1 Drag force as a function of steady flow velocity past a defect. Drag force in units of $F_\xi = 1/m\xi^3$, where $\xi = 1/\sqrt{mU|\varphi_0|^2}$ is the healing length. The thin solid line is without dissipation; the other solid, dashed, and dotted lines correspond to $\kappa/U|\varphi_0|^2 = 0.1$, 1, and 2, respectively, with constant γ/κ, and $\Gamma|\varphi_0|^2/\kappa$. Reprinted with permission from Wouters, M., and Carusotto, I. (2010), Superfluidity and critical velocities in nonequilibrium Bose-Einstein condensates, *Phys. Rev. Lett.*, **105**, 020602 [27]. Copyright (2010) by the American Physical Society. (Note that the model used to produce this figure in Ref. [27] differs slightly from Eq. (11.5).)

because a linearised theory becomes inadequate in calculating correlations of the driven dissipative Bose gas in two dimensions.

11.3.2 Beyond Mean-Field Description

In order to calculate correlations and response functions, the mean-field description is not sufficient. Various methods for dealing with driven dissipative systems such as Eq. (11.4) exist. We consider an approach starting from the Schwinger-Keldysh path integral (see Ref. [31] for an introduction), defined by the 'partition sum'

$$\mathcal{Z} = \int \mathcal{D}(\psi_C, \psi_Q) e^{iS[\psi_C, \psi_Q]}, \tag{11.7}$$

where S is the Schwinger-Keldysh action written in terms of 'classical' and 'quantum' fields ψ_C, ψ_Q. For the model in Eq. (11.4), this takes the form

$$S[\psi_C, \psi_Q] = \int dt d^d r \left\{ \begin{pmatrix} \bar{\psi}_C & \bar{\psi}_Q \end{pmatrix} \begin{pmatrix} 0 & [D_0^A]^{-1} \\ [D_0^R]^{-1} & [D_0^{-1}]_K \end{pmatrix} \begin{pmatrix} \psi_C \\ \psi_Q \end{pmatrix} \right.$$
$$\left. - \left[\left(\tfrac{U}{2} + i\tfrac{\Gamma}{4}\right) \left((\bar{\psi}_C^2 + \bar{\psi}_Q^2) \psi_C \psi_Q \right) + \text{c.c.} \right] + i\Gamma \bar{\psi}_C \psi_C \bar{\psi}_Q \psi_Q \right\}, \tag{11.8}$$

where the bare inverse retarded Green's function is given by $[D_0^R]^{-1} = i\partial_t - \left[-\nabla^2/(2m) + i(\gamma - \kappa)/2 \right]$ in time and real space domain, or $\omega - \left[\epsilon_k + i(\gamma - \kappa)/2 \right]$ in the frequency and momentum domain. The inverse advanced Green's function $[D_0^A]^{-1}$ is the complex conjugate of this, and the Keldysh component of the inverse bare Green's function $[D_0^{-1}]_K = i(\kappa + \gamma)$ describes the noise associated with both pumping and decay. Despite the existence of terms which create and destroy particles, Eq. (11.8) still possesses a $U(1)$ symmetry under simultaneous phase rotations of the fields ψ_C, ψ_Q; as such, there still exists the possibility of spontaneous symmetry breaking, phase stiffness, and superfluidity.

Eqs. (11.7) and (11.8) are fully equivalent to the quantum master equation, Eq. (11.4), but written in a functional integral formulation. This formulation allows one to apply a wide range of techniques from quantum field theory. A first, rather generic simplification consists in taking the *semiclassical limit*, which can be justified by a power counting argument. This is strictly justified close to the condensation threshold, where $\gamma - \kappa \to 0$, but provides a useful approximation also away from this limit. At threshold, the retarded and advanced inverse Green's functions scale as $\sim k^2$ with $\omega \sim k^2$, while there is no scaling of the Keldysh component, $[D_0^{-1}]_K = i(\kappa + \gamma) \sim k^0$. Using these, along with the natural scaling $dr \sim k^{-1}, dt \sim \omega^{-1}$, we can then determine the scaling dimensions of the fields $\psi_C \sim k^{(d-2)/2}$, $\psi_Q \sim k^{(d+2)/2}$ required in order that the quadratic contributions to the action are dimensionless. This then allows determination of the scaling of the various quartic terms. Due to this scaling, any quartic term involving more than a single quantum field is irrelevant – i.e., such terms scale to zero at long wavelength $k \to 0$, and can thus be omitted in the semiclassical limit. This provides direct contact to the dissipative GPE (11.5). The field equation obtained from the saddle point, $\delta S/\delta\bar{\psi}_Q = 0$, of Eq. (11.8) in the semiclassical limit reads

$$i\frac{\partial \psi_C}{\partial t} = \left[-\frac{\nabla^2}{2m} + \frac{U}{2}|\psi_C|^2 + \frac{i}{2}\left(\gamma - \kappa - \frac{\Gamma}{2}|\psi_C|^2\right)\right]\psi_C + i(\kappa+\gamma)\psi_Q. \quad (11.9)$$

This almost matches Eq. (11.5) if one identifies $\varphi = \psi_C/\sqrt{2}$, but with an extra term involving the quantum field ψ_Q.

While the 'classical' field can acquire a finite expectation in the condensed state, the 'quantum' field has to vanish on average and describes the noise. As a saddle point equation, Eq. (11.9) neglects the fluctuations of the fields ψ_C, ψ_Q, appearing in the full functional integral. Remarkably, this equation can be upgraded to a full description of the problem by means of the Martin-Siggia-Rose construction [31]. This shows that the functional integral Eq. (11.7) in the semiclassical limit is *equivalent* to a stochastic partial differential equation. In our case, this is the driven dissipative stochastic Gross-Pitaevskii equation (DSGPE) which is equivalent to Eq. (11.9) with the replacement $i(\kappa + \gamma)\psi_Q \to \xi(\mathbf{r}, t)$, where $\xi(\mathbf{r}, t)$ describes a Gaussian white noise process characterised by $\langle\xi(\mathbf{r}, t)\rangle = 0$ and $\langle\xi(\mathbf{r}, t)\bar{\xi}(\mathbf{r}', t')\rangle = ((\gamma + \kappa)/2)\delta(t - t')\delta(\mathbf{r} - \mathbf{r}')$, and vanishing off-diagonal correlators. In this sense, the DSGPE corresponds to a fully fluctuating (semiclassical) many-body problem – in stark contrast to the deterministic GPE (11.5).

11.3.3 Equilibrium versus Nonequilibrium Dynamics

The action (11.8) in the semiclassical limit or the DSGPE allows us to state precisely the sense in which the driven dissipative system is a genuinely nonequilibrium situation. To this end, we rewrite Eq. (11.9) by splitting the deterministic parts on the right-hand side into coherent (reversible) and dissipative (irreversible) contributions labelled c, d. To avoid confusion with these labels, we omit the suffix C on the classical field ψ_C. The DSGPE takes the form

$$i\frac{\partial \psi}{\partial t} = \frac{\delta H_c}{\delta \bar{\psi}} - i\frac{\delta H_d}{\delta \bar{\psi}} + \xi \quad (11.10)$$

with effective coherent and dissipative Hamiltonians

$$H_{\alpha=c,d} = \int d^d r \left(K_\alpha|\nabla\psi|^2 + r_\alpha|\psi|^2 + \frac{u_\alpha}{2}|\psi|^4\right). \quad (11.11)$$

The coefficients are $K_c = 1/2m, K_d = 0, r_c = 0, r_d = (\gamma - \kappa)/2, u_c = U/2$ in our problem, and $u_d = \Gamma/4$. Note, however, that the value of r_c is adjustable by a gauge transformation $\psi \mapsto \psi e^{-i\omega t}$ such that $r_c \mapsto r_c - \omega$. It can be shown [32, 33] that if the system is to relax to thermodynamic equilibrium (if the steady state of Eq. (11.10) is to be described by a Gibbs distribution), then the coherent and dissipative Hamiltonians must be proportional to each other; that is,

$$H_c = \nu H_d \quad \Leftrightarrow \quad \nu = \frac{K_c}{K_d} = \frac{u_c}{u_d}, \tag{11.12}$$

where ν is a constant.

This requirement is in general not satisfied for a driven dissipative system because the microscopic origins of reversible and irreversible dynamics are independent. For example, in the microscopic description of Eq. (11.4), the rates can be tuned independently from the Hamiltonian parameters, as they have completely different physical origins.

We have thus far made two important observations pertaining to a driven condensate. One is that, in spite of particle number nonconservation, the equations of motion describing an incoherently driven condensate enjoy a $U(1)$ phase symmetry. The second observation is that the steady state cannot be in general described by an equilibrium ensemble. So, while a condensation transition involving spontaneous symmetry breaking is possible, the nature of the transition and the conditions under which such a condensate would be indeed a stable fixed point of the dynamics may be different from the equilibrium case.

The question concerning the nature of the condensation transition in three-dimensional driven condensates was addressed in Refs. [34, 33, 35]. One main result of this analysis was that the equilibrium symmetry, $H_c = \nu H_d$, is emergent in the low-frequency limit even though it is not present microscopically. Hence the *correlation* functions correspond to an effective equilibrium description with a well-defined emergent temperature. Nevertheless, the drive conditions affect the long wavelength dynamical *response* functions, guaranteed by the existence of a new and independent critical exponent.

In two dimensions, as we will show next, the long wavelength behaviour is changed much more dramatically: in isotropic systems, the slow algebraic decay of correlation functions that occurs in equilibrium is replaced by far faster exponential or stretched exponential decay. Only for sufficiently anisotropic systems can the quasi-long-ranged algebraic decay found in equilibrium be recovered.

11.4 Long-Wavelength Fluctuations and Phase Coherence

As discussed in Section 11.2, the Mermin-Wagner theorem states that a homogeneous equilibrium system with a continuous symmetry cannot show long-range order that breaks that continuous symmetry in two or fewer dimensions. Rather, in two dimensions, long-wavelength phase fluctuations lead to algebraic decay of order parameter correlations. Naturally, the question arises whether this statement remains true out of equilibrium. The answer to this question proves to depend on both dimensionality and anisotropy. In three dimensions, the deviation from effective equilibrium, which is encoded in the difference between the ratios K_c/K_d and u_c/u_d (cf. Eq. (11.12)), vanishes in the long-wavelength limit, both close to criticality [34] and in the ordered (Bose-condensed) phase [36]. However, in isotropic two-dimensional systems, these nonequilibrium effects are relevant [36] and ultimately lead to the destruction of algebraic order. There is, however, a loophole: a spatially *anisotropic* system can support an algebraically ordered phase. These conclusions follow from a hydrodynamic description of the order parameter dynamics, which we review in the following.

In order to allow for spatial anisotropy, we consider a generalisation of the model described by Eq. (11.10), in which the gradient terms in the effective Hamiltonians (11.11) are replaced by $\sum_{i=x,y} K_\alpha^i |\partial_i \psi|^2$. As described in Section 11.2, a hydrodynamic description can be obtained by employing a density-phase representation, $\psi = \sqrt{\rho} e^{i\theta}$, leading to coupled equations of motion for the density ρ and phase θ. Eliminating the gapped fluctuations of the density around its mean value ρ_0, and keeping only the leading terms in a low-frequency and low-momentum expansion in the remaining equation for the phase, we obtain the anisotropic Kardar-Parisi-Zhang (KPZ) equation [36, 37],

$$\frac{\partial \theta}{\partial t} = \sum_{i=x,y} \left[D_i \partial_i^2 \theta + \frac{\lambda_i}{2} (\partial_i \theta)^2 \right] + \eta, \tag{11.13}$$

where $\eta(\mathbf{r}, t)$ is a Gaussian stochastic noise with zero mean, $\langle \eta(\mathbf{r}, t) \rangle = 0$, and second moment $\langle \eta(\mathbf{r}, t) \eta(\mathbf{r}', t') \rangle = 2\Delta \delta(\mathbf{r} - \mathbf{r}') \delta(t - t')$, with

$$\Delta = \frac{(\kappa + \gamma)\left(u_c^2 + u_d^2\right)}{2u_d(\kappa - \gamma)}. \tag{11.14}$$

The effective diffusion constants in Eq. (11.13) and the nonlinear couplings are given by

$$D_i = K_c^i \left(\frac{K_d^i}{K_c^i} + \frac{u_c}{u_d} \right), \qquad \lambda_i = -2K_d^i \left(\frac{K_c^i}{K_d^i} - \frac{u_c}{u_d} \right). \tag{11.15}$$

Evidently the nonlinear terms in the KPZ equation vanish when the equilibrium condition $K_c^x/K_d^x = K_c^y/K_d^y = u_c/u_d$, which generalises Eq. (11.12) to the spatially anisotropic case, is met. The degree of anisotropy is measured by the anisotropy parameter $\Phi = \lambda_y D_x / \lambda_x D_y$: when $\Phi \neq 1$, the system is anisotropic.

An important difference exists between our KPZ model and the original context of this equation, as an equation for the interface height in a model of randomly growing interfaces [37]. The analogue of the interface height in our model is actually a phase, θ, and the phase is compact, i.e., $\theta \equiv \theta + 2\pi$. This means that topological defects in this field – vortices – are possible. This difference with the conventional KPZ equation also arises in active smectics [38]. Analysis of Eq. (11.13) in the absence of vortices is the analogue of the low-temperature spin-wave (linear phase fluctuation) theory of the equilibrium *XY* model. Indeed, without the nonlinear terms, the KPZ equation reduces to linear diffusion, which would bring the field to an effective thermal equilibrium with power-law off-diagonal correlations (in $d = 2$). A transition to the disordered phase in this equilibrium situation can occur only as a Berezinskii-Kosterlitz-Thouless (BKT) transition through proliferation of topological defects in the phase field.

Our aim is to obtain the behaviour of correlations of the condensate field ψ at large distances. We have taken the first step of reducing this task to finding the correlations of the phase field θ, whose dynamics are given by the KPZ equation (11.13). But this equation contains more information than is actually required: in particular, it involves fluctuations with all wavelengths, ranging from the microscopic condensate healing length ξ_0 up to the largest scales of order of the linear system size L (we consider a square 2D system of area L^2). Because of the nonlinear term, fluctuations at different wavelengths couple to each other. Our goal, therefore, is to eliminate the short-scale fluctuations, and in this way obtain an effective description of the system at large scales.

This idea is implemented by the renormalisation group procedure [21]. To this end, we decompose the phase field into short- and long-wavelength components, above and below some length scale ℓ. We then integrate out the short-wavelength components, treating the nonlinear terms $\lambda_{x,y}$ which couple short- and long-wavelength components perturbatively. As such, this approach is limited to close-to-equilibrium conditions in which the couplings $\lambda_{x,y}$ are small.[1] This procedure is then iterated for increasing length scales ℓ, successively integrating out the short-wavelength components. In real space, this corresponds to repeated coarse-graining, eliminating fine details on scales shorter than ℓ. Performing this program for the anisotropic KPZ equation [39, 38], the resulting equation for the long-scale components of the phase field again takes the form (11.13), but with coefficients that are modified by the short-scale fluctuations.

The modification of the coefficients as one goes to long-length scales can be characterised entirely by the 'flow' of two dimensionless parameters: these are a

[1] More precisely, the perturbation theory is valid when a suitable dimensionless measure of the ratio of the nonlinear $\lambda_{x,y}$ terms in the KPZ equation to the linear ones is small; this measure proves to be the parameter g defined below.

dimensionless form of the nonlinearity, $g \equiv \lambda_x^2 \Delta / D_x^2 \sqrt{D_x D_y}$, and the anisotropy parameter $\Phi = \lambda_y D_x / \lambda_x D_y$ introduced below Eq. (11.15). The parameter g describes the importance of the nonlinear terms λ_x, for 'typical' (i.e., root mean squared) fluctuations of the field θ. The mean-squared fluctuations of θ, according to the linear theory, are proportional to the noise strength Δ, which drives the fluctuations, and inversely proportional to the geometric mean of the diffusion coefficients $D_{x,y}$, which smooth out those fluctuations. Knowing the anisotropy parameter Φ, together with g, then allows us to estimate the importance of the other nonlinearity λ_y. The microscopic parameters of the system determine the 'bare' values g_0 and Φ_0 at the starting length scale $\ell = \xi_0$. The values of g, Φ at some other scale $\ell > \xi_0$ are obtained by integrating the renormalisation group (RG) flow equations

$$\begin{aligned} \frac{dg}{dl} &= \frac{g^2}{32\pi}\left(\Phi^2 + 4\Phi - 1\right), \\ \frac{d\Phi}{dl} &= \frac{\Phi g}{32\pi}\left(1 - \Phi^2\right), \end{aligned} \tag{11.16}$$

with the logarithmic scale $l = \ln(\ell/L)$. The resulting RG flow is illustrated in Fig. 11.2. There are two distinct flow patterns to the left and right of the line $\Phi = 0$, and we discuss these in turn.

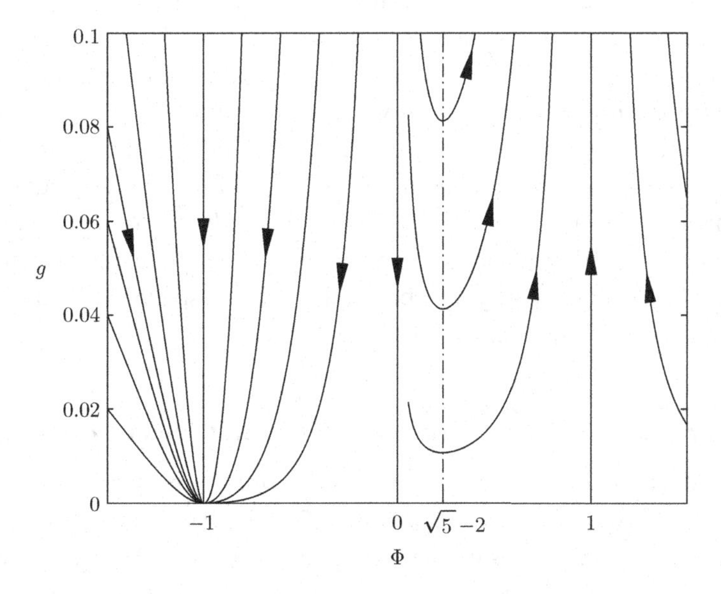

Figure 11.2 RG flow for the anisotropic KPZ equation in 2D. In the weakly anisotropic regime $\Phi > 0$, the flow lines approach the isotropic line $\Phi = 1$ and flow towards $g \to \infty$; for $\Phi < 0$, in the strongly anisotropic regime, they converge to a stable effective equilibrium fixed point with $\Phi = -1$ and $g = 0$.

For $\Phi > 0$, the flow with increasing ℓ is towards strong coupling, $g \to \infty$, and isotropy, $\Phi \to 1$. Thus, in this regime, which we denote *weak anisotropy*, the approximation of treating g perturbatively eventually fails. Simulations of the isotropic KPZ equation [40, 41, 42, 43, 44, 45] find that correlations of the phase field $\langle(\theta(\mathbf{r}) - \theta(\mathbf{r}'))^2\rangle$ scale algebraically with separation $|\mathbf{r} - \mathbf{r}'|$. This suggests that ultimately the flow of g must terminate at a strong coupling fixed point g_*, which is, however, beyond the scope of the perturbatively derived flow equations (11.16). This scaling of $\langle(\theta(\mathbf{r}) - \theta(\mathbf{r}'))^2\rangle$ would lead, according to Eq. (11.3), to stretched exponential decay of condensate field correlations.

Note, however, that in our analysis of the KPZ equation (11.13) we have so far neglected topological defects, which can exist in the compact field θ. Proliferation of such defects at the strong coupling fixed point would lead to exponential (i.e., even faster) decay of correlations of ψ.

The regime of *strong anisotropy*, on the other hand, corresponds to the region $\Phi < 0$. There the flow lines terminate for $\ell \to \infty$ in an effective *equilibrium* fixed point with $g = 0$ and $\Phi = -1$. Thus, in this regime, the effective description of the system at large scales approaches the equilibrium description (note that $g \propto \lambda_x^2$ and hence $g = 0$ in equilibrium), and so algebraic correlations of the condensate field can survive (as long as one is below the BKT transition temperature).

The possibility of a flow to strong coupling is in stark contrast to the 3D case in which even in isotropic systems with $\Phi = 1$ small deviations $g \ll 1$ from equilibrium are irrelevant and flow to zero as $\ell \to \infty$, leading to the effective equilibrium physics discussed at the end of Section 11.3.3. Note, however, that even in 3D, for sufficiently strong drive and dissipation, i.e., values of g larger than some critical value, there may be a nonequilibrium transition to the disordered phase, described by the strong coupling fixed point of the 3D KPZ equation [46]. In one dimension, even equilibrium systems show only short-range (exponential) correlations at nonzero temperatures. Nonetheless, it is still possible to see the effect of the KPZ nonlinearity on the scaling of the spatial [47] and temporal coherence [48, 49] of a one-dimensional condensate.

11.4.1 Current Experiments, Weak Anisotropy, BKT Physics, and Crossovers

As discussed in Chapter 10, by Kim, Nitsche, and Yamamoto, experiments on incoherently pumped polariton condensates [29, 50] have measured an apparent algebraic decay of correlations by measuring the fringe visibility in an interference experiment. Similar results have also been seen in numerical experiments on a parametrically pumped (OPO) system [51]. Since, as discussed above, algebraic order is destroyed at large scales, a question arises about the interpretation of these experiments. In this section, we discuss how, although the asymptotic behaviour at

$\ell \to \infty$ does lead to the strong coupling fixed point, the length scales ℓ required to see this may be very large, particularly when the condensate is well developed.

Current experiments with exciton-polaritons fall into the regime of weak anisotropy. This anisotropy results from the interplay between polarisation pinning to the crystal structure, and the splitting of transverse electric and transverse magnetic cavity modes [5, 52] – taken together these mean that there is anisotropy between directions parallel and perpendicular to this pinned lattice direction. As discussed above, for weak anisotropy, the flow is to strong coupling, and algebraic order is absent on the largest scales. However, at intermediate scales, $g(\ell) \ll 1$, and so correlations can still decay algebraically. It is therefore natural to ask how large the system must be to see the breakdown of algebraic correlations – i.e., what system size L is required such that $g(L) \simeq 1$. Setting $\Phi = 1$ in the flow equation for g in Eq. (11.16), we find that if the microscopic parameters in Eq. (11.11) correspond to close-to-equilibrium conditions, i.e., if the bare value $g_0 \ll 1$, then the renormalised value $g = 1$ is reached only at the exponentially large characteristic KPZ scale $L = L_* = \xi_0 \exp(8\pi/g_0)$. Starting from a microscopic model for polariton condensation [5] that models the excitonic reservoir explicitly, and which provides a more faithful description of the condensation dynamics than the model of Eq. (11.4), we obtain the expression for the bare nonlinearity [36]

$$g_0 = \frac{2u_c\bar{\gamma}^2}{K_c}\frac{\bar{\gamma}^2 + (1+x)^2}{x(1+x)^3}, \tag{11.17}$$

which depends on the dimensionless net pumping rate $x = \gamma/\kappa - 1$, and the dimensionless combination $\bar{\gamma} = \kappa R/\gamma_R u_c$, where R and γ_R are the rate of scattering between the excitonic reservoir and condensate, and the reservoir relaxation rate, respectively. The dimensionless parameter x gives a measure of how far above 'threshold' the system is pumped. For high pump rates, the KPZ scale L_* grows rapidly beyond any reasonable system size, so that a sufficiently strongly pumped system will always appear algebraically ordered up to system size L. Such a system resides effectively in equilibrium, with a temperature set by the noise strength. The form of Eq. (11.17) implies that for *any* finite system size L, it is always possible to choose a value x large enough that algebraic correlations are seen over the entire system. As such, the experimental observation of algebraic correlations is perfectly consistent with the results here.

At weak pump rates, near the threshold $x \to 0$, the bare coupling g_0 grows and so the critical size L_* decreases, and may become comparable to the system size L. This then prompts a second question: as the pump strength is reduced, does the algebraic order break down before the BKT transition occurs? As long as $L_* \gg L$, the system is effectively thermal and consequently it can undergo a BKT transition to a disordered phase as described in Section 11.2 if x is decreased below a critical value x_{BKT}. One can then evaluate the length scale L_* at this pump rate; i.e., we denote $L' \equiv L_*(x = x_{BKT}) \approx \xi_0 \exp(2/\bar{\gamma}^2)$. If the system is much smaller than this critical size (i.e., $L \ll L'$), then even at the BKT point, $L \ll L_*$, and so the KPZ physics does not become apparent. This is consistent with recent experiments on exciton polaritons [50] which showed the behaviour expected of a BKT transition as discussed in Section 11.2.1 – i.e., a transition between algebraic order with exponent $\alpha_s = 1/4$ right at the transition and short-ranged, exponentially decaying order. Such behaviour is consistent with a system of size $L \ll L'$. However, if the system is sufficiently large [36], i.e., $L \gg L'$, then algebraic order at length scale L will break down before the BKT transition. The dependence of L' on $\bar{\gamma}$ can be understood intuitively: as $\bar{\gamma} \to 0$, then the polariton lifetime diverges, and thermalisation is perfect for any finite-size system. Therefore, $L' \to \infty$ and strong coupling with $g \approx 1$ is never reached. Thus, in order to clearly see the breakdown of algebraic order, one should increase the polariton loss rate κ.

11.4.2 Strong Anisotropy, Re-entrant Phase Transition

While the above analysis shows that the algebraic order observed in recent experiments with (nearly) isotropic polariton systems (see Chapter 10 and Refs. [29, 50]) must be an intermediate scale crossover phenomenon, true algebraic order in the thermodynamic limit is nevertheless possible in the *strong anisotropy* regime, $\Phi_0 < 0$, as discussed above. We discuss here the experimental consequences this would have for a sufficiently anisotropic polariton system.

The effective temperature of the system at the strong anisotropy fixed point is given by the *renormalised* value of the dimensionless noise $\tau \equiv \Delta/\sqrt{D_x D_y}$. Because the phase is a compact variable, algebraic order only exists if this effective noise temperature is low enough. Specifically, the BKT transition occurs if the renormalised value of this dimensionless noise reaches π. One may then derive a phase diagram by identifying the location of this condition $\tau(\ell \to \infty) = \pi$ is in the manifold of *bare* couplings, i.e., in terms of the microscopic experimental parameters. To do this, we must complement the RG flow equations (11.16) for g and Φ by additional equations for τ and $D_{x,y}$ (see Ref. [38] for details), and follow the flow to the effective equilibrium regime $g \approx 0$. We must also assume that, up to the length scale at which the latter regime is reached, vortices are sufficiently dilute that their influence on the RG flows can be neglected. It turns out that the renormalised value τ of the dimensionless noise strength crosses the critical value for the BKT transition for bare values τ_0 and Φ_0 that are located on the curve determined by

$$\tau_0 = -\frac{4\pi\Phi_0}{(1-\Phi_0)^2}. \tag{11.18}$$

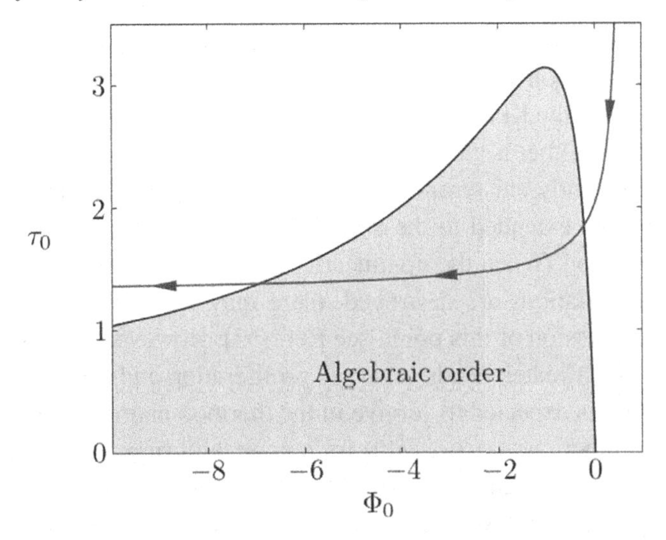

Figure 11.3 Phase diagram of a 2D anisotropic driven dissipative system for low noise. The thin line corresponds to values of Φ_0 and τ_0 derived from a microscopic model for polariton condensation [36], with the arrows indicating the direction of increasing pump rate γ. Note that the ordered phase is first entered and then left again as γ is increased. Values for the other dimensionless parameters are $\bar{\gamma} = 2, \bar{u} = 2, K_d^x/K_c^x = 1/8$, and $K_d^y/K_c^y = 1/4$.

The resulting phase diagram of strongly anisotropic driven dissipative systems in the $\Phi_0 - \tau_0$ plane is depicted in Fig. 11.3.

It is interesting to work out the bare parameters Φ_0 and τ_0 starting from the same microscopic model for polariton condensation that led to the estimate of g_0 in Eq. (11.17). This leads to a particular trajectory through the $\Phi_0 - \tau_0$ phase diagram as one increases the microscopic pump rate γ; this trajectory is shown on Fig. 11.3. An initially moderate anisotropy, i.e., $\Phi_0 > 0$ but different from 1, becomes more substantial as the pump rate is increased, so that Φ_0 first becomes negative and then, if the value of the dimensionless interaction strength $\bar{u} = u_c/\sqrt{K_c^x K_c^y}$ is sufficiently small (for details, see Ref. [36]), crosses the boundary to the algebraically ordered phase. Remarkably, upon pumping the system at an even higher rate the ordered phase is then left again; that is, the transition is *re-entrant.*

11.5 How to Define and Measure Superfluidity in a Dissipative System

In the previous section, we discussed how the presence of drive and dissipation affect the low-energy effective theory of the two-dimensional system, and how these changes are manifested in the correlation functions of the system. In this section, we focus instead on the current–current response function, $\chi_{ij}(\mathbf{q}, \omega)$, for a dissipative system and how to identify whether a superfluid fraction survives. The discussion in this section is focused on the case where phase fluctuations remain small, i.e., when the nonlinear term in the KPZ equation does not dominate the physics. As discussed above, this requires either a small finite system (as in the current experiments) or a system with sufficiently anisotropic interactions. While the calculations presented below can easily be extended to the anisotropic case, we present results only for the isotropic situation. When the nonlinearity in the KPZ equation becomes large, and algebraic correlations are destroyed, there may still exist a finite superfluid fraction; for a discussion of this point, see Ref. [53]. However, if the growth of the KPZ nonlinearity ultimately leads to vortex proliferation and short-range order, no superfluid fraction is expected to survive in the thermodynamic limit.

The response function $\chi_{ij}(\mathbf{q}, \omega = 0)$ for a nonequilibrium system can be calculated by defining a generating functional for correlations of currents. As noted in Section 11.3.2, the driven dissipative system no longer has a conserved current, but does still show a $U(1)$ phase symmetry.

How then does the system respond to a phase twist between the boundaries of the system [54]? Such a physical phase twist couples to the unphysical 'quantum' current $j_{Q,i}(\mathbf{q}) = \sum_{\mathbf{k}} \left[\bar{\psi}_{C,\mathbf{k}+\mathbf{q}}\psi_{Q,\mathbf{k}} + \bar{\psi}_{Q,\mathbf{k}+\mathbf{q}}\psi_{C,\mathbf{k}}\right]\gamma_i(2\mathbf{k}+\mathbf{q})$. To measure a response function, we must see how some physical quantity responds to such a phase twist. For this, we measure the standard particle current. This leads us to the generating functional

$$\mathcal{Z}[\mathbf{f}, \mathbf{g}] = \int \mathcal{D}(\bar{\psi}, \psi) \exp\left(iS[\bar{\psi}, \psi] + iS_j[\bar{\psi}, \psi]\right), \tag{11.19}$$

$$S_j[\bar{\psi}, \psi] = \sum_{\mathbf{k},\mathbf{q}} \bar{\Psi}^T_{\mathbf{k}+\mathbf{q}} \begin{pmatrix} g_i(\mathbf{q}) & f_i(\mathbf{q}) + g_i(\mathbf{q}) \\ f_i(\mathbf{q}) - g_i(\mathbf{q}) & -g_i(\mathbf{q}) \end{pmatrix} \Psi_{\mathbf{k}}\, \gamma_i(2\mathbf{k}+\mathbf{q}),$$

where $\Psi^T = (\psi_C \;\; \psi_Q)$ and $S[\bar{\psi}, \psi]$ is the Schwinger-Keldysh action, e.g., Eq. (11.8).[2] The source field $\mathbf{f}$ corresponds to the phase twist and couples to the quantum current. The field $\mathbf{g}$ couples to the observable particle current; the strange form in Keldysh space occurs in order to calculate normal-ordered expectations from the Schwinger-Keldysh path integral.[3] Taking derivatives with respect to these fields, we find the current–current correlation function:

$$\chi_{ij}(\mathbf{q}, \omega = 0) = -\frac{i}{2} \left.\frac{d^2\mathcal{Z}[\mathbf{f}, \mathbf{g}]}{df_i(\mathbf{q})dg_j(-\mathbf{q})}\right|_{\mathbf{f},\mathbf{g}\to 0}. \tag{11.20}$$

[2] In Ref. [55], a different model action was taken, involving frequency-dependent gain. This difference does not affect the general points discussed below, but does affect several details of the calculation.

[3] One can alternatively use a simpler form at the expense of introducing causality factors to ensure normal ordering.

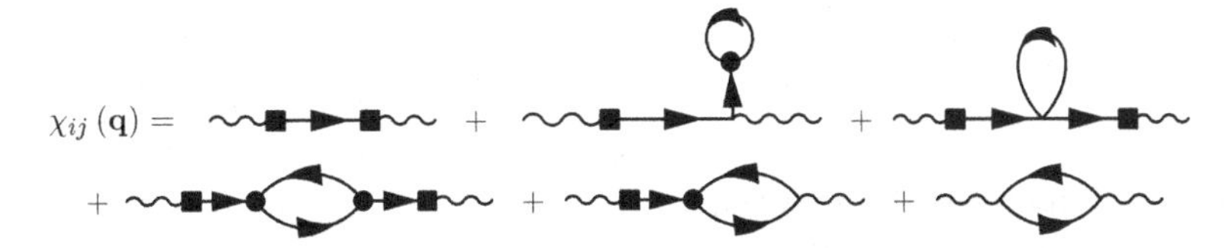

Figure 11.4 Feynman diagrams representing contributions to the response function at one-loop order. Solid circles and squares involve factors of the quasicondensate density, and wavy lines represent coupling to currents. The first five terms all contribute to the superfluid component, and the last gives the normal density. Reprinted with permission from Keeling, J. (2011), Superfluid density of an open dissipative condensate, *Phys. Rev. Lett.*, **107**, 080402 [55]. Copyright (2011) by the American Physical Society.

At this stage, the calculation is exact; however, evaluating this requires the ability to calculate the partition function exactly, taking for the bare action an expression such as Eq. (11.8). If a linearised approach is valid, one can proceed by evaluating the path integral via a saddle point approach, first minimising over $\bar{\psi}, \psi$, and then including quadratic fluctuations about this saddle point. For an equilibrium system to respect sum rules, it is necessary that the saddle point should be calculated *in the presence of the fields* $\mathbf{f}, \mathbf{g}$. Repeating this in the nonequilibrium case leads to a generating function of the form $\mathcal{Z} \propto \exp(iS_0[\mathbf{f}, \mathbf{g}] - \mathrm{Tr}\ln(1 + DA[\mathbf{f}, \mathbf{g}]))$, with contributions of the source terms to both the saddle point action and the action for fluctuations. Calculating the response function gives a sequence of terms that can represented by the Feynman diagrams shown in Fig. 11.4. The first of these diagrams corresponds to the contribution from the saddle point action S_0; the others arise from terms such as $\mathrm{Tr}[D(d^2A/df_i dg_j)]$ and $\mathrm{Tr}[D(dA/df_i)D(dA/dg_j)]$.

While the full expression coming from these diagrams [55] is rather involved, two simple conclusions can be drawn without reference to these details. The most important conclusion stemming from the above formalism is that (as long as the linearised theory is valid) a superfluid fraction will exist, due to the form of the first five diagrams shown in Fig. 11.4. Crucially, each of these diagrams involves a term in which a single line carries the entire incoming and outgoing momentum, and so they give expressions of the form

$$\chi_{ij}^{SF}(\mathbf{q}, \omega) \propto -|\psi_0|^2 q_i q_j D^R(\mathbf{q}, \omega),$$

where $D^R(\mathbf{q}, \omega)$ is the full retarded Green's function (including the normal and anomalous self energies arising due to the interaction terms in Eq. (11.8)). This structure reflects the fact that each diagram contains current vertices with one line having zero momentum, while the other carries the full momentum $\mathbf{q}$. Hence each diagram includes a factor $\gamma_i(\mathbf{q}) \propto q_i$. The retarded Green's function has poles corresponding to the normal mode frequencies given in Eq. (11.6). One may in fact show from the Keldysh action that the Green's function has the form $D^R(\mathbf{q},\omega) \propto C/[\omega(\omega+i\gamma_{\text{net}}) - \xi_{\mathbf{q}}^2]$, which has the crucial feature that $D^R(\mathbf{q}\to 0, \omega = 0) \propto 1/q^2$. This feature (along with the structure of the diagrams) ensures a superfluid contribution to the response function.

The second important conclusion comes from the contribution of the last diagram, giving the normal fraction. This gives an expression

$$m\chi_T = -\frac{i}{4}\int\frac{d\omega}{2\pi}\iint\frac{d^dk}{(2\pi)^d}\epsilon_k \mathrm{Tr}\left[\sigma_z D_{\mathbf{k}}^K \sigma_z \left(D_{\mathbf{k}}^R + D_{\mathbf{k}}^A\right)\right],$$

which can be evaluated and shown not to vanish unless the loss terms vanish. This has a simple physical interpretation: the noise associated with pumping and dissipation leads to the excitation of quasiparticles, and thus the creation of a normal fraction in all cases.

11.5.1 Experimental Probes of Superfluidity

As emphasised in the preceding sections, the superfluid density calculated above is a measure of the difference in how a system responds to transverse (rotational) versus longitudinal forces. As such, measurement of this superfluid response requires measuring the response to a rotational perturbation of some kind. For superfluid helium, the classic experiment is that of Andronikashvilli, using a torsional oscillator formed of parallel discs, and studying the changing inertia of the fluid as the temperature varies.

For polaritons, several issues arise: the particles are quasiparticles, which do not strongly couple to a rotating inertial frame; they live inside a semiconductor, and so applying a stirring force is challenging; and their rotation does not provide any measurable contribution to the mechanical angular momentum. What is therefore required is a way to engineer an effective rotating frame *as seen by the polaritons* and to measure the response of the polaritons to this.

Engineering a rotating frame for polaritons can be achieved in the same spirit as proposed for cold atoms [56, 57], namely manufacturing a real space Berry curvature arising from spatially varying spin structure of polariton eigenstates. The essence of such a synthetic rotation is to use the two-component spinor structure of the polariton to construct a spatially varying ground state $|\chi(\mathbf{r})\rangle$ such that the Berry connection $\mathbf{A}(\mathbf{r}) = i\langle\chi|\nabla\chi\rangle$ takes the form $\mathbf{A}(\mathbf{r}) = m\omega(r)\hat{\mathbf{z}}\times\mathbf{r}$, corresponding to a synthetic rotating frame. Such a configuration arises from the ground state of the spin Hamiltonian $H = \lambda[\ell_0^2\sigma^z + r^2(e^{2i\theta}\sigma^- + \text{H.c.})]$, where r,θ are in-plane polar coordinates, and σ^i are Pauli matrices for the spinor basis. The term $\lambda\ell_0^2\sigma^z$ corresponds to a Zeeman splitting induced by an external field. The other term requires either an induced strain or a strong radial magnetic field (i.e., diverging in the plane of the polariton system). This leads to an angular velocity peaking at $\omega \simeq 0.3\hbar/m\ell_0^2$ around $r \sim \ell_0$.

Measuring the response to such a field is in principle possible in a variety of ways, since an advantage of the polariton system is the ability to directly probe polariton correlation functions in both real and momentum space. In particular, it is in principle possible to measure correlations such as $\langle a^\dagger_{\mathbf{k}+\mathbf{q}} a_{\mathbf{k}} \rangle$ as a function of $\mathbf{k}, \mathbf{q}$ by taking the interference between two momentum-space images of the condensate, displaced in real space, with a variable phase delay, i.e., calculating $I(\phi_d) = \langle (a^\dagger_{\mathbf{k}} + e^{-i\phi_d} a^\dagger_{\mathbf{k}+\mathbf{q}})(a_{\mathbf{k}} + e^{i\phi_d} a_{\mathbf{k}+\mathbf{q}}) \rangle$ and mapping the fringe visibility as one varies ϕ_d. Real space displaced fringe visibility maps are routinely measured (see., e.g., [58]); the equivalent momentum space tomography would allow access to the current induced by a given force and thus reconstruction of the response function $\chi_{ij}(\mathbf{q})$.

Existing experiments probing aspects of superfluidity in polaritons are still far from this limit. Indeed, what has been observed thus far is suppression of drag rather than the difference between transverse and longitudinal response. In addition, such experiments to date have included measurements of the suppression of scattering from defects, as a function of velocity and intensity of a coherently driven condensate [59, 60].

11.6 Vortices and Metastable Flow in a Driven Dissipative System

As discussed in Section 11.2, the appearance of quantised vortices demonstrates the existence of short-range coherence in a system and is closely related to metastable persistent flow in a non-simply-connected geometry. We review here how these features are modified in a dissipative condensate. It is notable that the structure of vortices changes in the presence of drive and dissipation. Because the dissipation depends on density, the 'continuity' equation derived from Eq. (11.4) takes the form

$$\frac{\partial \rho}{\partial t} + \nabla \cdot \mathbf{j} = (\gamma - \kappa - \Gamma\rho)\rho, \qquad \mathbf{j} \equiv -\frac{i}{2m}\left(\varphi^* \nabla \varphi - \varphi \nabla \varphi^*\right). \tag{11.21}$$

This has the consequence that near the core of a vortex, where density is depleted, there is net gain, and so there must be a current with nonvanishing divergence. Given that current can be rewritten as $\mathbf{j} = \frac{\rho}{m}\nabla[\arg(\varphi)]$, this diverging current implies that vortices in a driven dissipative system must have a spiral structure, with radial as well as angular variation of phase. Such spiral vortices have been discussed in the context of nonlinear optics [61]. The existence of a spiral structure can modify the force between a pair of vortices, and so may modify the nature of the vortex binding/unbinding at the BKT transition. This provides a further complication – in addition to that provided by the strong nonlinearity of the KPZ equation – in understanding the BKT transition in a driven dissipative system.

There can also be cases in which combinations of spatial variation of drive, dissipation, and potential trapping destabilise the vortex free configuration and instead stabilise configurations such as vortex lattices [62, 63]. However, understanding whether such configurations occur in practice requires analysis of the normal state, going beyond the scope of the complex GPE [5].

Because of the photon component of a polariton condensate, it is also possible to directly imprint vortices on the condensate, by using a coherent Gauss-Laguerre beam. Calculations using stochastic classical field methods have shown [64] how such pulses can create metastable vortex states in both simply and non-simply-connected geometries. The additional noise associated with pumping and decay means that the time scale for a vortex to move out of such a condensate can be relatively short: rather than quantum tunnelling, it can diffuse across a small annulus, driven by noise from the pump and decay terms.

Vortices have been clearly observed in polariton condensates [58]; however, since most images of polariton condensates require long integration times, vortices can only be seen if they are either stationary (pinned on disorder) or if they move along a repeatable path [65] (so that averages over many realisations recover the same trajectory). As such, indirect methods of imaging vortices may be necessary, such as interference measurements with angular offsets [64] or, for vortex lattices, energy resolved and interference images [63].

11.7 Future Directions, Open Questions

The most profound open question is how the KPZ roughening interacts with the fact that phase is a compact variable and can thus support topological defects. As discussed previously, two scenarios seem possible. It may be that as the KPZ equation flows to strong nonlinearity, the destruction of algebraic order also destroys any resistance to vortex proliferation. In this case, correlations are always exponential, and there is no ordered phase, no superfluidity, and no phase transition. In this scenario, all experiments on polariton condensates would be finite-size effects, although potentially exceptionally large system sizes needed before the 'true' behaviour becomes visible. The other possible scenario is that the strong nonlinearity is compatible with vortex binding. In this case, there would be a phase transition between a high-temperature phase showing exponential decay of correlations and a low-temperature phase with stretched exponential decay. If this scenario holds, then despite the absence of algebraic order, there can be a nonzero superfluid stiffness [53]. If such a phase transition exists, it could potentially require a new universality class, distinct from the equilibrium BKT universality class.

While the above questions concern the fundamental physics of the thermodynamic limit, a second set of questions concern the signatures of this behaviour visible in finite experimental systems. Physical [29, 50] and numerical [51] experiments have observed power-law correlations, but with surprising values of the power-law exponent; understanding the physical origin of this behaviour may help us to understand the nature of the driven dissipative system. Finally, experiments directly probing the superfluid response function – i.e., measuring the superfluid fraction – of a driven dissipative condensate have yet to be realised. Even in the context of cold atom systems, direct measurements of the superfluid fraction remain challenging [56]. Realising such measurements for driven dissipative systems can provide confirmation and guidance to the theoretical question of whether superfluidity exists in these systems and whether it is a finite-size or thermodynamic phenomenon.

Acknowledgements: J. Keeling acknowledges financial support from the Engineering and Physical Science Research Council (EPSRC) programme Topological Protection and Non-Equilibrium States in Strongly Correlated Electron Systems (TOPNES) (EP/I031014/1) and from the Leverhulme Trust (IAF-2014-025). E. Altman and L. Sieberer acknowledge support by the European Research Council (ERC) under the European Unions Horizon 2020 synergy grant Ultracold Quantum Matter (UQUAM). S. Diehl also acknowledges support by the ERC under the European Unions Horizon 2020 research and innovation programme (grant agreement No 647434).

References

[1] Huang, Kerson. 1995. Bose-Einstein condensation and superfluidity. Page 31 of: Griffin, A., Snoke, D., and Stringari, S. (eds), *Bose-Einstein Condensation*, Cambridge: Cambridge University Press.

[2] Leggett, A. J. 1999. Superfluidity. *Rev. Mod. Phys.*, **6861**, 318–323.

[3] Hohenberg, P. C., and Martin, P. C. 2000. Microscopic theory of superfluid helium. *Ann. Phys. (N.Y).*, **281**, 636–705.

[4] Leggett, A. J. 2006. *Quantum Liquids: Bose Condensation and Cooper Pairing in Condensed-Matter Systems*. Oxford: Oxford University Press.

[5] Carusotto, Iacopo, and Ciuti, Cristiano. 2013. Quantum fluids of light. *Rev. Mod. Phys.*, **85**, 299.

[6] Klaers, J., Schmitt, J., Vewinger, F., and Weitz, M. 2010. Bose-Einstein condensation of photons in an optical microcavity. *Nature*, **468**, 545–548.

[7] Demokritov, S. O., Demidov, V. E., Dzyapko, O., Melkov, G. A., Serga, A. A., Hillebrands, B., and Slavin, A. N. 2006. Bose-Einstein condensation of quasi-equilibrium magnons at room temperature under pumping. *Nature*, **443**, 430–433.

[8] Falkenau, M., Volchkov, V. V., Rührig, J., Griesmaier, A., and Pfau, T. 2011. Continuous loading of a conservative potential trap from an atomic beam. *Phys. Rev. Lett.*, **106**, 163002.

[9] Mewes, M.-O., Andrews, M. R., Kurn, D. M., Durfee, D. S., Townsend, C. G., and Ketterle, W. 1997. Output coupler for Bose-Einstein condensed atoms. *Phys. Rev. Lett.*, **78**, 582–585.

[10] Robins, N. P., Figl, C., Jeppesen, M., Dennis, G. R., and Close, J. D. 2008. A pumped atom laser. *Nat. Phys.*, **4**, 731.

[11] Robins, N. P., Altin, P. A., Debs, J. E., and Close, J. D. 2013. Atom lasers: production, properties and prospects for precision inertial measurement. *Phys. Rep.*, **529**, 265–296.

[12] Yang, C. N. 1962. Concept of off-diagonal long-range order and the quantum phases of liquid He and of superconductors. *Rev. Mod. Phys.*, **34**, 694–704.

[13] Pitaevskii, L. P., and Stringari, S. 2003. *Bose-Einstein Condensation*. Oxford: Clarendon Press.

[14] Hess, G. B., and Fairbank, W. M. 1967. Measurements of angular momentum in superfluid helium. *Phys. Rev. Lett.*, **19**, 216–218.

[15] Griffin, Allan. 1994. *Excitations in a Bose-Condensed Liquid*. Cambridge: Cambridge University Press.

[16] Mermin, N. D., and Wagner, H. 1966. Absence of ferromagnetism or antiferromagnetism in one- or two- dimensional isotropic Heisenberg models. *Phys. Rev. Lett.*, **17**, 1133.

[17] Kosterlitz, J. M., and Thouless, D. J. 1973. Ordering, metastability and phase transitions in two-dimensional systems. *J. Phys. C Solid State Phys.*, **6**, 1181.

[18] Kosterlitz, J. M. 1974. The critical properties of the two-dimensional XY model. *J. Phys. C Solid State Phys.*, **7**, 1046–1060.

[19] Berezinskii, V. L. 1972. Destruction of long-range order in one-dimensional and two-dimensional systems possessing a continuous symmetry group. II. Quantum systems. *Sov. Phys. JETP*, **34**, 610 [*Zh. Eksp. Teor. Fiz.* **61**, 1144–1156 (1971)].

[20] Nelson, David R. 1977. Universal jump in the superfluid density of two-dimensional superfluids. *Phys. Rev. Lett.*, **39**, 1201–1205.

[21] Chaikin, P. M., and Lubensky, T. C. 1995. *Principles of Condensed Matter Physics*. Cambridge: Cambridge University Press.

[22] Wouters, M., and Carusotto, I. 2006. Absence of long-range coherence in the parametric emission of photonic wires. *Phys. Rev. B*, **74**, 245316.

[23] Szymańska, M. H., Keeling, J., and Littlewood, P. B. 2006. Nonequilibrium quantum condensation in an incoherently pumped dissipative system. *Phys. Rev. Lett.*, **96**, 230602.

[24] Szymańska, M. H., Keeling, J., and Littlewood, P. 2007. Mean-field theory and fluctuation spectrum of a pumped decaying Bose-Fermi system across the quantum condensation transition. *Phys. Rev. B*, **75**, 195331.

[25] Aranson, Igor, and Kramer, Lorenz. 2002. The world of the complex Ginzburg-Landau equation. *Rev. Mod. Phys.*, **74**, 99–143.

[26] Astrakharchik, G. E., and Pitaevskii, L. P. 2004. Motion of a heavy impurity through a Bose-Einstein condensate. *Phys. Rev. A*, **70**, 013608.

[27] Wouters, Michiel, and Carusotto, Iacopo. 2010. Superfluidity and critical velocities in nonequilibrium Bose-Einstein condensates. *Phys. Rev. Lett.*, **105**, 020602.

[28] Cancellieri, E., Marchetti, F. M., Szymańska, M. H., and Tejedor, C. 2010. Superflow of resonantly driven polaritons against a defect. *Phys. Rev. B*, **82**, 224512.

[29] Roumpos, G., Lohse, M., Nitsche, W. H., Keeling, J., Szymańska, M. H., Littlewood, P. B., Löffler, A., Höfling, S., Worschech, L., Forchel, A., and Yamamoto, Y. 2012. Power-law decay of the spatial correlation function in exciton-polariton condensates. *Proc. Nat. Acad. Sci*, **109**, 6467.

[30] Chiocchetta, A., and Carusotto, I. 2013. Non-equilibrium quasi-condensates in reduced dimensions. *Eur. Phys. Lett.*, **102**, 67007.

[31] Kamenev, A. 2011. *Field Theory of Non-Equilibrium Systems*. Cambridge: Cambridge University Press.

[32] Graham, R., and Tel, T. 1990. Steady-state ensemble for the complex Ginzburg-Landau equation with weak noise. *Phys. Rev. A*, **42**, 4661.

[33] Sieberer, L. M., Huber, S. D., Altman, E, and Diehl, S. 2014. Nonequilibrium functional renormalization for driven-dissipative Bose-Einstein condensation. *Phys. Rev. B*, **89**, 134310.

[34] Sieberer, L. M., Huber, S. D., Altman, E., and Diehl, S. 2013. Dynamical critical phenomena in driven-dissipative systems. *Phys. Rev. Lett.*, **110**, 195301.

[35] Täuber, U. C., and Diehl, S. 2014. Perturbative field-theoretical renormalization group approach to driven-dissipative Bose-Einstein criticality. *Phys. Rev. X*, **4**, 021010.

[36] Altman, E., Sieberer, L. M., Chen, L., Diehl, S., and Toner, J. 2015. Two-dimensional superfluidity of exciton polaritons requires strong anisotropy. *Phys. Rev. X*, **5**, 011017.

[37] Kardar, M., Parisi, G., and Zhang, Y.-C. 1986. Dynamic scaling of growing interfaces. *Phys. Rev. Lett.*, **56**, 889–892.

[38] Chen, L., and Toner, J. 2013. Universality for moving stripes: a hydrodynamic theory of polar active smectics. *Phys. Rev. Lett.*, **111**, 088701.

[39] Wolf, D. E. 1991. Kinetic roughening of vicinal surfaces. *Phys. Rev. Lett.*, **67**, 1783–1786.

[40] Kim, J., and Kosterlitz, J. 1989. Growth in a restricted solid-on-solid model. *Phys. Rev. Lett.*, **62**, 2289–2292.

[41] Miranda, V. G., and Aarão Reis, F. D. A. 2008. Numerical study of the Kardar-Parisi-Zhang equation. *Phys. Rev. E*, **77**, 031134.

[42] Marinari, E., Pagnani, A., and Parisi, G. 2000. Critical exponents of the KPZ equation via multi-surface coding numerical simulations. *J. Phys. A: Math. Gen.*, **33**, 8181–8192.

[43] Ghaisas, S. 2006. Stochastic model in the Kardar-Parisi-Zhang universality class with minimal finite size effects. *Phys. Rev. E*, **73**, 022601.

[44] Chin, C.-S., and den Nijs, M. 1999. Stationary-state skewness in two-dimensional Kardar-Parisi-Zhang type growth. *Phys. Rev. E*, **59**, 2633–2641.

[45] Tang, L.-H., Forrest, B., and Wolf, D. 1992. Kinetic surface roughening. II. Hypercube-stacking models. *Phys. Rev. A*, **45**, 7162–7179.

[46] Fisher, M. P. A., and Grinstein, G. 1992. Nonlinear transport and $1/f^{\alpha}$ noise in insulators. *Phys. Rev. Lett.*, **69**, 2322–2325.

[47] Gladilin, V. N., Ji, K., and Wouters, M. 2014. Spatial coherence of weakly interacting one-dimensional nonequilibrium bosonic quantum fluids. *Phys. Rev. A*, **90**, 023615.

[48] Ji, K., Gladilin, V. N., and Wouters, M. 2015. Temporal coherence of one-dimensional nonequilibrium quantum fluids. *Phys. Rev. B*, **91**, 045301.

[49] He, L., Sieberer, L. M., Altman, E., and Diehl, S. 2015. Scaling properties of one-dimensional driven-dissipative condensates. *Phys. Rev. B* **92**, 155307.

[50] Nitsche, W. H., Kim, N. Y., Roumpos, G., Schneider, C., Kamp, M., Höfling, S., Forchel, A., and Yamamoto, Y. 2014. Algebraic order and the Berezinskii-Kosterlitz-Thouless transition in an exciton-polariton gas. *Phys. Rev. B*, **90**, 205430.

[51] Dagvadorj, G., Fellows, J. M., Matyjaskiewicz, S., Marchetti, F. M., Carusotto, I., and Szymańska, M. H. 2015. Non-equilibrium Berezinskii-Kosterlitz-Thouless transition in a driven open quantum system, *Phys. Rev. X* **5**, 041028 .

[52] Shelykh, I. A., Kavokin, A. V., Rubo, Yuri G., Liew, T. C. H., and Malpuech, G. 2010. Polariton polarization-sensitive phenomena in planar semiconductor microcavities. *Semiconductor Science and Technology*, **25**, 013001.

[53] Wachtel, G., Sieberer, L. M., Diehl, S. and Altman E. 2016. Electrodynamic duality and vortex unbinding in driven-dissipative condensates. *Phys. Rev. B* **94**, 104520.

[54] Janot, A., Hyart, T., Eastham, P., and Rosenow, B. 2013. Superfluid stiffness of a driven dissipative condensate with disorder. *Phys. Rev. Lett.*, **111**, 230403.

[55] Keeling, J. 2011. Superfluid density of an open dissipative condensate. *Phys. Rev. Lett.*, **107**, 080402.

[56] Cooper, N. R., and Hadzibabic, Z. 2010. Measuring the superfluid fraction of an ultracold atomic gas. *Phys. Rev. Lett.*, **104**, 030401.

[57] John, S. T., Hadzibabic, Z., and Cooper, N. R. 2011. Spectroscopic method to measure the superfluid fraction of an ultracold atomic gas. *Phys. Rev. A*, **83**, 023610.

[58] Lagoudakis, K. G., Wouters, M., Richard, M., Baas, A., Carusotto, I., André, R., Dang, Le Si, and Deveaud-Plédran, B. 2008. Quantized vortices in an exciton-polariton condensate. *Nat. Phys.*, **4**, 706–710.

[59] Amo, A., Sanvitto, D., Laussy, F. P., Ballarini, D., del Valle, E., Martin, M. D., Lemaître, A., Bloch, J., Krizhanovskii, D. N., Skolnick, M. S., Tejedor, C., and Viña, L. 2009a. Collective fluid dynamics of a polariton condensate in a semiconductor microcavity. *Nature*, **457**, 291–295.

[60] Amo, A., Lefrère, J., Pigeon, S., Adrados, C., Ciuti, C., Carusotto, I., Houdré, R., Giacobino, E., and Bramati, A. 2009b. Superfluidity of polaritons in semiconductor microcavities. *Nat. Phys.*, **5**, 805–810.

[61] Staliunas, K., and Sanchez-Morcillo, V. J. 2003. *Transverse Patterns in Nonlinear Optical Resonators*. Berlin: Springer-Verlag.

[62] Keeling, J., and Berloff, N. G. 2008. Spontaneous rotating vortex lattices in a pumped decaying condensate. *Phys. Rev. Lett.*, **100**, 250401.

[63] Borgh, M. O., Keeling, J., and Berloff, N. G. 2010. Spatial pattern formation and polarization dynamics of a nonequilibrium spinor polariton condensate. *Phys. Rev. B*, **81**, 235302.

[64] Wouters, M., and Savona, V. 2010. Superfluidity of a nonequilibrium Bose-Einstein condensate of polaritons. *Phys. Rev. B*, **81**, 054508.

[65] Lagoudakis, K. G., Manni, F., Pietka, B., Wouters, M., Liew, T. C. H., Savona, V., Kavokin, A. V., André, R., and Deveaud-Plédran, B. 2011. Probing the dynamics of spontaneous quantum vortices in polariton superfluids. *Phys. Rev. Lett.*, **106**, 115301.

12

BEC to BCS Crossover from Superconductors to Polaritons

ALEXANDER EDELMAN
James Franck Institute and Department of Physics, University of Chicago, Illinois, USA

PETER B. LITTLEWOOD
Argonne National Laboratory, Lemont, Illinois, USA
James Franck Institute and Department of Physics, University of Chicago, Illinois, USA

We briefly review aspects of the crossover from Bardeen-Cooper-Schrieffer (BCS) theory to Bose-Einstein condensation (BEC), two limits of a theory of attracting fermions. We sketch the theory in its simplest terms and some crucial elaborations with the goal of developing the minimal phenomenology to practically distinguish the two limits. Throughout, we identify systems that display crossover physics and emphasize the competing effects that inhibit the usefulness of the framework and its predictions. We focus in particular on the dynamical signatures which remain robust in realistic systems.

12.1 Introduction

A system of two kinds of fermion with an attraction between them will spontaneously order, pairing off its constituents and further conspiring to keep the pairs in phase; it is phase coherence that is characteristic of the superconductive broken symmetry. In a dilute system, pairing occurs independently of the onset of phase coherence, whereas in the conventional weakly coupled superconductors pairing and phase coherence appear coincidently. The two limits of weak coupling BCS and strong coupling BEC have long been discussed as limits of a single theory [1], but it is only in the last decade that there are physical systems which can reliably span the space: conventional superconductors, cold atoms, and exciton-polaritons.

12.2 Generic Crossover Physics

12.2.1 The Vanilla Theory

Why should a crossover occur between two states of matter with substantially different phenomenology, and what are the universal parameters which could describe it without recourse to detailed materials properties? We begin heuristically with a trial wave function in the Bose or BEC limit of strong interactions or low densities, and motivate additional formalism as we develop an understanding for a generic fermionic system.

The ground state of a weakly interacting Bose gas is well described by a condensate wave function $|\Psi\rangle = \mathcal{N} e^{\lambda b_0^\dagger}|0\rangle$, where $b_0^\dagger$ creates a boson in its lowest-energy state, λ is a variational parameter that describes the condensate amplitude, $|0\rangle$ is the vacuum for the bosons, and $\mathcal{N}$ is a normalization constant. To generalize the condensate to a limit of paired fermion states, we write the "bosonic" creation operator in terms of fermion operators $c_k^{(\dagger)}$:

$$b_0^\dagger = \begin{cases} \sum_{\mathbf{k}} \phi(\mathbf{k}) c_{\mathbf{k}\uparrow}^\dagger c_{-\mathbf{k}\downarrow}^\dagger, & \text{Cooper Pairs} \\ \sum_{\mathbf{k}} \phi(\mathbf{k}) c_{\alpha\mathbf{k}}^\dagger c_{\beta\mathbf{k}}, & \text{Excitons} \\ \sum_{\mathbf{k}} \phi(\mathbf{k}) c_{\alpha\mathbf{k}+\mathbf{Q}/2}^\dagger c_{\beta\mathbf{k}-\mathbf{Q}/2}, & \text{Charge Density Waves (CDW)} \\ \vdots & \end{cases} \tag{12.1}$$

where $\phi(k)$ is a so far unspecified internal wave function for the fermion pairs, $|0\rangle$ should now be understood as the Fermi sea, and of course the bosonic character is corrupted. For superconductivity, we have particle–particle pairing, and for excitons and CDWs one has particle and hole paired in different electronic bands α, β. In what follows, we will use the traditional example of Cooper pairs in a superconductor, but the same analysis qualitatively applies to any system of fermions with a pairing interaction. Expanding the condensate wave function using (12.1) and applying the anticommutation relations, we obtain

$$|\Psi\rangle = \prod_{\mathbf{k}} (u_{\mathbf{k}} + v_{\mathbf{k}} c_{\mathbf{k}\uparrow}^\dagger c_{-\mathbf{k}\downarrow}^\dagger) \tag{12.2}$$

where one recovers the Bardeen-Cooper-Schrieffer (BCS) variational wavefunction proposed to explain superconductivity at weak interactions and high densities, with $\phi(\mathbf{k}) = v_{\mathbf{k}}/\lambda u_{\mathbf{k}}$, and we normalize so that $|u|^2 + |v|^2 = 1$. There is no distinction between the BCS and BEC limits except in the form of the wavefunction; for BEC, one expects $\phi(\mathbf{k})$ to be the wavefunction of a single bound pair, but in the context of this theory, it is a *variational function*.

To be concrete, we follow the traditional route to minimize the free energy $F = \langle \hat{H} - \mu\hat{N}\rangle$ of a model Hamiltonian, subject to the constraint $N = \langle \hat{N}\rangle$:

$$\hat{H} - \mu\hat{N} = \sum_{\mathbf{k}\sigma} c_{\mathbf{k}\sigma}^\dagger \varepsilon_{\mathbf{k}} c_{\mathbf{k}\sigma} + \sum_{\mathbf{k}\mathbf{k}'} c_{\mathbf{k}\uparrow}^\dagger c_{-\mathbf{k}\downarrow}^\dagger V_{\mathbf{k}\mathbf{k}'} c_{\mathbf{k}'\downarrow} c_{-\mathbf{k}'\uparrow} \tag{12.3}$$

where σ, σ' are spin indices, $\varepsilon_{\mathbf{k}} = \hbar^2\mathbf{k}^2/2m$ is the single-particle energy, and $V_{\mathbf{kk}'}$ is an (attractive) interaction strength. This yields

$$\Delta_{\mathbf{k}} = -\sum_{\mathbf{k}'} V_{\mathbf{kk}'} \frac{\Delta_{\mathbf{k}'}}{2E_{\mathbf{k}'}}, \qquad N = \sum_{\mathbf{k}'} \left(1 - \frac{\xi_{\mathbf{k}'}}{E_{\mathbf{k}'}}\right), \tag{12.4}$$

known respectively as the gap and number equations, where $E_{\mathbf{k}} = \sqrt{\xi_{\mathbf{k}}^2 + \Delta_{\mathbf{k}}^2}$, $\xi_{\mathbf{k}} = \varepsilon_{\mathbf{k}} - \mu$ and the variational parameters in the wavefunction are given by $v_{\mathbf{k}} = (1 - \xi_{\mathbf{k}}/E_{\mathbf{k}})/2$, and the normalization condition $|v_{\mathbf{k}}|^2 + |u_{\mathbf{k}}|^2 = 1$. $\Delta_{\mathbf{k}} \equiv \Delta$, assumed to be uniform, is the condensate order parameter [2].

It is clear from intensively rescaling (12.4) that the crossover is controlled by the particle number density n and the interaction strength. We shall parameterize the former by the Fermi momentum $k_F = (3\pi^2 n)^{1/3}$ (in three dimensions). Parameterizing the interaction is more subtle, physically because an arbitrary $V_{\mathbf{kk}'}$ can encode a great deal of structure absent some simplifying assumptions, and technically because (12.3) must be understood as a low-energy approximation which may contain unregularized divergences. We will remedy the technical issue first by working within the so-called T-matrix approximation, replacing the bare interaction with an effective two-particle vertex given by the solution of the diagrams in Fig. 12.1 [3]. Carrying out this procedure also naturally clarifies the physical issue, distilling the interaction to a small set of relevant parameters. (As we shall see, there is often little hope for a detailed microscopic understanding of the pairing interaction in a particular experimental realization, while it is possible to control a_S.) For instance, an attractive point-contact interaction of strength g is reduced to a scattering length a_S[1] as defined by the relation

$$\frac{m}{4\pi a_S} = \frac{1}{g} + \sum_{\mathbf{k}}^{\Lambda} \frac{1}{2\varepsilon_{\mathbf{k}}} \tag{12.5}$$

where the cutoff Λ can be taken to ∞ in a controlled fashion.

A clarification is in order: in the so-called zero-range approximation, where the interaction potential does not vary over any length scale, it is always possible to

Figure 12.1 Ladder digrams for the T-matrix approximation.

[1] In two dimensions and below, scattering theory is not this simple, and in particular there exists a bound state for all purely attractive interactions. The basic principle of defining a new scale and eliminating the bare interaction through the T-matrix nevertheless still holds [4]. In fact, the crossover picture as a whole holds remarkably well, although additional transitions associated with the low-dimensional order appear, and although the control parameters of the theory are quantitatively modified compared with the 3D case [5, 6, 7].

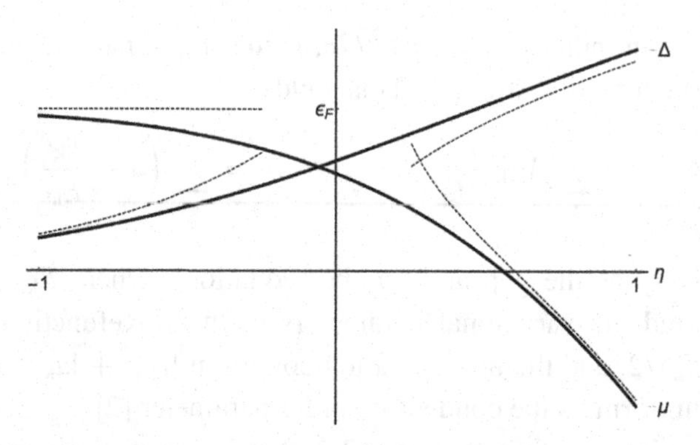

Figure 12.2 μ and Δ for the vanilla theory, as a function of the dimensionless coupling constant η. Dotted lines show approximations in the $\eta \to \pm\infty$ limits.

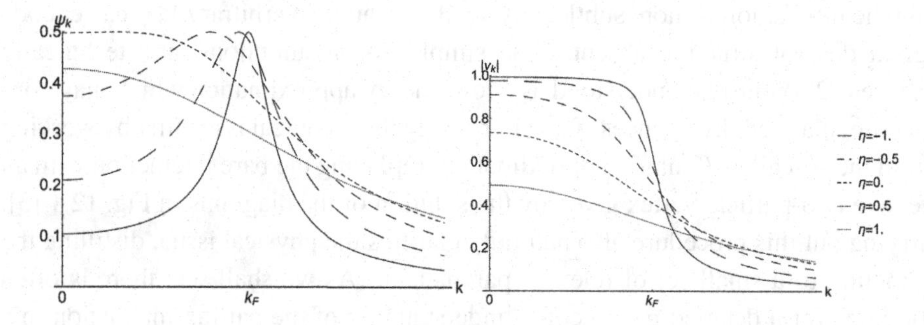

Figure 12.3 Occupancy and pair wavefunctions for the vanilla theory.

parameterize it only by a_S, as can be seen from solving the Schrödinger equation directly, and no diagrammatic techniques are necessary. The utility of the T-matrix technique and its extensions is in the ability to systematically incorporate more interactions with more physical content.

The entire crossover for this vanilla theory – three dimensional, isotropic, and s-wave – will depend only on a single dimensionless parameter $\eta = 1/k_F a_S$ [8]. In Fig. 12.2, we plot as a function of η the order parameter (identifiable with the gap in the excitation spectrum) and chemical potential obtained from solving (12.4). The chemical potential's zero crossing between its weak-coupling asymptote at the Fermi energy and its plunge downward as half the pair binding energy in the Bose limit is usually taken to demarcate the two regimes. The order parameter becomes exponentially small but remains finite deep into the BCS limit.

In Fig. 12.3, we illustrate the behavior of the pair wavefunction across the crossover. It is important to distinguish the occupancy of states v_k, which is simply a Fermi step function at weak coupling and subsequently broadens, from the interal wavefunction $\psi_k \sim v_k u_k$, which characterizes the pairing and is restricted to a thin shell around the Fermi surface in the BCS limit, while incorporating more momenta in the BEC limit as the pairs become more tightly bound in real space.

12.2.2 Deviations from Vanilla

As desired, the theory so far interpolates sensibly between a weakly paired BCS state and a Bose-Einstein condensate of tightly bound pairs. Before we develop it further, it is worth questioning the approximations we have made to render it tractable.

One assumption relevant to experimental realizations is already present in the contact interaction (12.5). Encoded in the cutoff Λ is a diluteness assumption: in order for the crossover to be controlled only by the ratio of average interparticle distance to scattering length (or equivalently pair size), neither scale can probe the short-distance behavior of the interaction. Thus if $1/a_S \ll \Lambda$ or $k_F \ll \Lambda$ is violated, we must distinguish a "density-driven" crossover controlled by k_F from "interaction-driven" crossover driven by a_S, which correspond to two experimental paths toward its realization.

Cold Atoms: An Interaction-Driven Crossover

In the past two decades, ultracold atomic gases have opened new experimental frontiers as physicists have gained unprecedented microscopic control over their attributes. Particularly interesting for observing crossover phenomena is a scattering property known as the Feshbach resonance. In these systems, the pairing is between two species of atoms polarized into particular angular momentum and spin states – for instance two hyperfine states of lithium – and thus scattering between them is substantially complicated by the presence of additional channels. As shown in Fig. 12.4a, this property is exploited by magnetically tuning the Zeeman energy of the "open channel" scattering atoms with energy E close to a bound state in a "closed channel" (with internal energy exceeding E), producing a field-dependent scattering length $a_S = a_{\text{bg}}(1-\Delta/(B-B_0))$, where B is the magnetic field, B_0 denotes the field at which the bound state is resonant, a_{bg} is a "background" scattering length, and Δ is a resonance width that depends on microscopic parameters [9]. Thus one can change η by directly varying a_S at a fixed density of particles. Starting with a BEC and moving to weak coupling, a superfluid has been observed to persist across the crossover by the existence of a vortex lattice induced by rotating the system [10].

Fig. 12.4b shows the chemical potential as a function of η for fixed density and varying scattering length, for a model Gaussian interaction potential with fixed range $\langle r \rangle$ [11]. As the diluteness assumption is violated, the crossover point is pushed to stronger interactions. A finite range is not relevant for the current

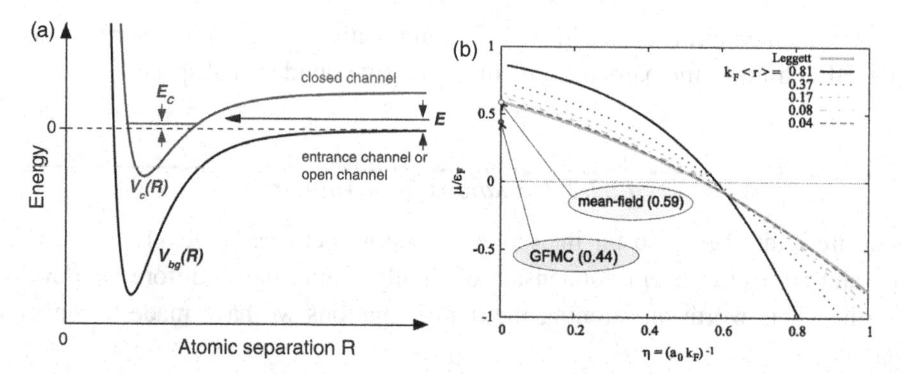

Figure 12.4 (a) Schematic illustration of the Feshbach resonance mechanism. The bound state with energy E_c is magnetically tuned close to zero energy to resonantly couple to atoms scattering with energy E. Reprinted with permission from Chin, C., et al. (2010), Feshbach resonances in ultracold gases, *Rev. Mod. Phys.*, **82**, 1225 [9]. Copyright (2010) by the American Physical Society. (b) Chemical potential in an interaction-driven crossover at fixed density, with finite-range potential. Notice that the zero crossing is shifted compared with the universal ("Leggett") curve as the zero-range approximation is violated. Reprinted with permission from Parish, M. M., et al. (2005), BCS-BEC crossover with a finite-range interaction, *Phys. Rev. B*, **71**, 064513 [11]. Copyright (2005) by the American Physical Society.

generation of cold atom experiments, but the general lesson holds that corrections to the "universal" mean-field theory can shift the crossover point substantially. In two dimensions, mean-field theory places the crossover at $\ln(k_F a_S) = 0$ (where this parameter plays the role of η owing to the perpetual presence of a bound state), while a high-temperature strong-coupling expansion sees signatures of Bose-like behavior well beyond this point, and in particular for the parameters accessible to experiment [12, 13, 14]. Roughly speaking, capturing these effects in the formalism developed in the first section is akin to adding to the variational wavefunction "dimer" terms that capture energetically favorable boson–boson correlations.

Excitons: A Density-Driven Crossover

Electronic systems have of course classically been the province of the BCS state, and a thus natural place to look for the condensation of bosonic pairs as well. Rather than Cooper pairing, we consider here the condensation of excitons, which depends on the screened Coulomb interaction between charge carriers. A key material property is therefore the effective Bohr radius, in analogy to the hydrogen Rydberg: $a_B = 4\pi\epsilon\epsilon_0\hbar^2/(m^* e^2)$, where m^* is the effective (reduced) mass of the charge carriers in question, and ϵ is the dielectric constant. The latter quantity can be orders of magnitude larger than unity, and in particular cause a_B to exceed the lattice

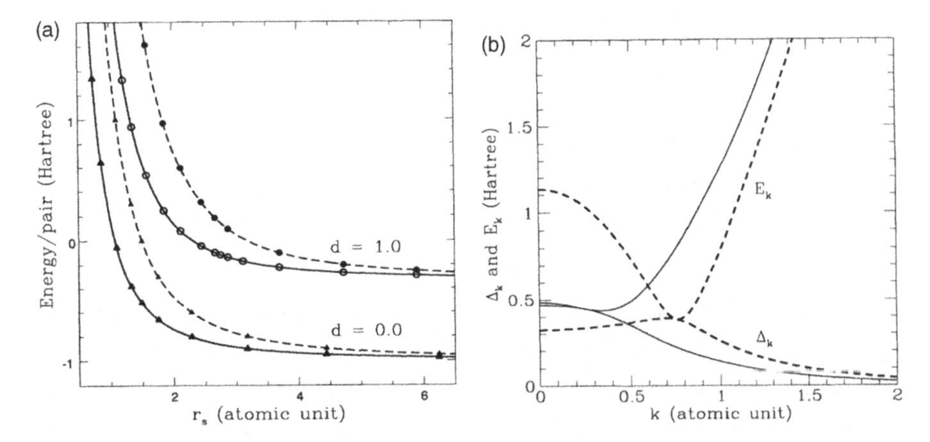

Figure 12.5 (a) Energy per pair (solid line) and chemical potential (dashed) as a function of the electron gas parameter r_S in a one-dimensional bilayer structure (see text), for electron-hole separations d. (b) Excitation spectrum (upper curve, and see Section 12.2.3) and gap function (lower) at $r_S = 2.66, 5.90$ (dashed, solid, respectively). Reprinted with permission from Zhu, X., et al. (1995), Exciton condensate in semiconductor quantum well structures, *Phys. Rev. Lett.*, **74**, 1633 [29]. Copyright (1995) by the American Physical Society.

spacing substantially so that an interacting gas becomes a reasonable model [15]. The other parameter relevant to crossover physics is $r_s \propto 1/k_F$, a dimensionless measure of interparticle distance in units of a_B. It is the latter quantity that is experimentally tunable by changing the carrier density.

The Coulomb interaction binds the excitons constituents together while the Pauli repulsion of the latter ensures the stability of the system [16]. We may incorporate all of these effects within the same formalism as in the first section and ask whether the system can condense [17]. Fig. 12.5 shows the chemical potential and order parameter at various densities parameterized by r_s for a one-dimensional model that constrains the electrons and holes to opposite sides of an insulating layer. The hallmarks of a crossover are visible in the growing gap and the chemical potential meeting the binding energy with growing r_S. Lest one worry that these results are an artifact of the geometry, variational Monte Carlo simulations in three dimensions with unrestricted particle positions confirm the crossover scenario and indeed verify that a pair wavefunction is energetically favored over a simple plasma of electrons and holes [18]. (As before, the improvement over the vanilla calculation comes from the explicit inclusion of additional pair correlation terms in the variational wavefunction.)

Excitonic systems were early experimental candidates for observation of BEC. Excitons are easily formed as "preformed pairs" on the BEC side of the crossover, and the experimental challenge to observing condensation lies in achieving a sufficiently *high* density that condensation may occur at experimentally accessible temperatures (see the next section for a discussion of the thermodynamics of the crossover) [19]. There has been for several decades experimental work aimed at achieving excitonic BEC (recent results include [2, 20, 21, 22, 23]), although the achievement is not yet unambiguous [24, 25, 26, 27, 28].

Anisotropic Pairing

At face value, some interactions do not admit a crossover scenario: a sufficiently strong short-range repulsive component, for instance, might favor a BCS state with p- or d-wave symmetry, in principle necessitating a phase transition to the isotropic BEC. These cases have been studied in detail and it appears, remarkably, that crossover physics continues to provide a good description [30]. The key is in recognizing that the order parameter Δ can only be identified with the gap in excitation spectrum in the vanilla theory (see the next section for a discussion of spectral signatures of the crossover) [4]. Thus the order parameter can retain the symmetry favored by the interaction while an isotropic gap opens in the spectrum as the chemical potential is lowered below $\mu = 0$. To the extent that this is a phase transition, it is an unusual one, involving weak singularities that leave first and second derivatives of the free energy continuous. It has been argued these are associated with a change in ground-state topology [31].

The Unitary Gas

It is necessary to distinguish the "universality" of the vanilla theory, which we have seen abundantly violated above, from the area of the phase diagram near which $\eta \to 0^{\pm}$, where the scattering length becomes infinite (assuming that the interaction range remains small), known as the unitarity limit. This is a strongly interacting limit (in three dimensions), where from dimensional analysis the physics may depend only on η (and a temperature). This truly universal regime, in principle accessible in cold atom experiments, is fertile and active ground for testing new theory [32]. Some results can be interpolated from ϵ expansions in four and two dimensions, where remarkably the unitary gas maps onto, respectively, a noninteracting Bose gas of infinitely tightly bound dimers or a noninteracting Fermi gas at precisely the threshold for the appearance of a bound state [33]. In three dimensions, the strong-coupling physics can be described by a number of universal relations in terms of a quantity known as the contact, which roughly speaking is a measure of pair density that encodes the anomalous scaling behavior induced by the strong correlations between particles [34]. This fact has prompted the development of much new theory for this regime where neither a Cooper pair nor tightly bound dimer description is appropriate.

12.2.3 Finite Temperature and Excitations

Thermal Physics

The natural generalization of the variational theory is to a functional integral formalism, which is amply reviewed elsewhere [35]. Our approach is to write the thermodynamic partition function as a path integral over configurations of the order parameter Δ,

$$Z = \mathcal{N} \int \mathcal{D}(\Delta, \bar{\Delta})\, e^{-S_{\text{eff}}[\Delta,\bar{\Delta}]}, \tag{12.6}$$

where $S_{\text{eff}}[\Delta, \bar{\Delta}]$ is an effective action of the complex order parameter field Δ (and may be obtained by a Hubbard-Stratonovich decoupling and integration over the fermion fields of the full action (12.8)). In principle, no approximations have been made at this stage, but in practice S_{eff} must be computed by expanding order by order in some small parameter. Our choice of decoupling field Δ implies that the theory will be valid only with a weak coupling $V_{\mathbf{kk}'}$ favoring Cooper pairing.

The systematic approach to solution is to first obtain the mean-field solutions (12.4) by a saddle point minimization of S_{eff} to lowest order, and to accommodate pairing fluctuations beyond mean field by expanding S_{eff} to second order and applying the prescription once more [36]. Physically, the additional terms incorporate the effects of pair-breaking thermal fluctuations about the mean field; at strong coupling, these dephase the condensate into a gas of bosons at the BEC transition temperature $T_c \sim \epsilon_F$ for a gas of bosons of mass $2m$, well before it becomes energetically favorable to dissociate into a Fermi liquid at $T^* \sim \epsilon_F \exp(-\pi|\eta|/2)$ [37]. At weak coupling, this distinction is lost. The resulting phase diagram is sketched in Fig. 12.6.

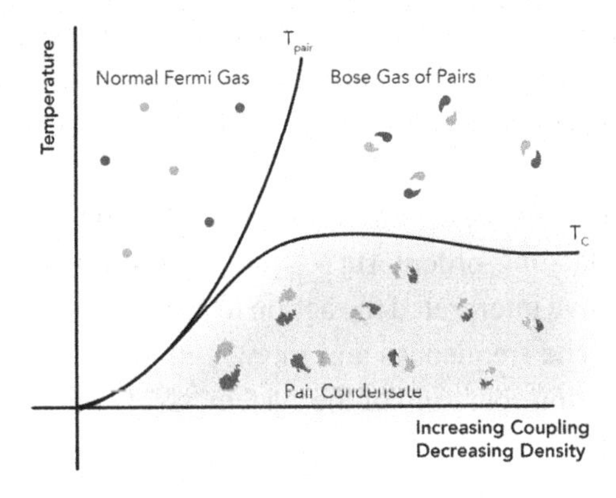

Figure 12.6 Schematic finite-temperature phase diagram of the crossover. T_{pair} is the mean-field weak-coupling result.

This theory, equivalent to the T-matrix approximation (Fig. 12.1), becomes that of a free Bose gas in the BEC limit, and an expansion to higher orders produces an effective pair–pair repulsion of scattering length $2a_S$ between bosons of mass $2m$, which stabilizes the condensate [38]. Direct solution of the Schrödinger equation in fact finds a dimer–dimer scattering length $a_{dd} \sim 0.6a_S$, which repulsion is responsible for the maximum in T_c near the crossover as the transition temperature is modified by $\Delta T_c/T_c \sim n_b^{1/3} a_{dd}$, where n_b is the boson density [39, 40, 41, 42, 43]. Another important class of corrections arises from including exchange of spin fluctuations between fermions, which suppresses the transition temperature in the low-density limit [44, 45]. For mass-imbalanced fermions (relevant to cold atom experiments where different atomic species take the place of spins), the theory predicts an unphysical double-valued transition temperature and vanishing superfluidity at weak coupling, which has only recently been remedied with a self-consistent approach [46, 47]. This calculation, then, should be taken as its original authors advised: an interpolation scheme, albeit one which remains remarkably qualitatively correct for its simplicity.

Excitations

In this section, we offer two complementary methods for deriving the excitation spectrum of the vanilla theory, for the separate fermionic (single quasiparticle) and bosonic (collective mode) sectors. Excitations in the BCS mean-field theory are described by fermionic quasiparticles defined by the Bogoliubov transformation

$$(\hat{H} - \mu\hat{N})^{\text{M.F.}} = \sum_{\mathbf{k}} \begin{pmatrix} c^\dagger_{\mathbf{k}\uparrow} \\ c_{-\mathbf{k}\downarrow} \end{pmatrix}^{\mathrm{T}} \begin{pmatrix} \xi_{\mathbf{k}} & -\Delta \\ -\bar{\Delta} & -\xi_{\mathbf{k}} \end{pmatrix} \begin{pmatrix} c_{\mathbf{k}\uparrow} \\ c^\dagger_{-\mathbf{k}\downarrow} \end{pmatrix} + \text{const.} \tag{12.7}$$

with spectrum $E_{\mathbf{k}} = \sqrt{\xi_{\mathbf{k}}^2 + \Delta^2}$. In the limit $\Delta \to 0$, the quasiparticles become ordinary electrons and holes, with gap 2Δ at the Fermi surface. As Δ grows and the chemical potential crosses zero, the Fermi surface is destroyed and the minimum quasiparticle energy is half the binding energy at $k = 0$ and grows with the chemical potential.

Collective excitations are best derived from the functional integral formalism. We have seen that the leading-order term $S^{(0)}_{\text{eff}}[\Delta, \bar{\Delta}]$ sets the mean-field value of the order parameter, and have interpreted the action to second-order $S^{(2)}_{\text{eff}}[\Delta, \bar{\Delta}]$ as incorporating fluctuations. The frequency- and momentum-dependent zero eigenvalues of $S^{(2)}_{\text{eff}}$ define the excitation spectrum of the system.[2] Note that these are fluctuations

[2] This is nothing more than the statement that the excitations are the poles of the Green function.

in the *bosonic* order parameter field and correspond to a collective response of the condensate rather than individual quasiparticles.

This analysis finds two modes: as Goldstone's theorem predicts, the existence of a macroscopic order parameter Δ that breaks the symmetry of (12.3) to global $U(1)$ phase rotations of the fermion operators implies the existence of a sound mode with a dispersion linear in k at long wavelengths. This mode, with stiffness $v_F/\sqrt{3}$ where v_F is the Fermi velocity, corresponds to modulations of the order parameter phase, while the second, massive amplitude mode, with spectrum $\omega(k) = 2\Delta + v_F^2k^2/4\alpha\Delta$ where α is of order unity, reflects modulations in condensate population [48]. These spectra are derived without assuming a value of η (except insofar as they come from expanding a weak-coupling theory), yet their quantitative forms yield qualitatively different behavior on different sides of the crossover. (Note, however, that inclusion of other effects, particularly the Coulomb interaction for charged fermions, modifies these results [49, 50].)

In the BCS limit, the sound mode is quite stiff in the sense that already at low momenta it intersects the quasiparticle continuum (i.e., its energy exceeds 2Δ) where scattering processes are allowed that cause it to quickly decay. (This provides the connection to the BCS/Pippard coherence length $\xi = \hbar v_F/2\Delta$. Hence a restatement of the BCS limit is the condition $k_F\xi \gg 1$.) The amplitude mode likewise quickly decays for the same reason, though note that the precise coincidence $\omega(k=0) = 2\Delta$ is an artifact of the level of approximation. We conclude therefore that gapped quasiparticle excitations will dominate the observed spectra in the BCS limit.

In the BEC limit, in contrast, where the quasiparticle gap is the binding energy and Δ is large compared with ϵ_F, the superfluid sound mode is soft in comparison, and is the dominant thermally occupied low-energy excitation. Quantitatively, this limit is perhaps better described by the Bogoliubov theory of the weakly interacting Bose gas, which predicts a sound velocity that depends sensitively on the boson–boson interaction strength, which can be incorporated directly as a parameter of the theory. As we have seen in Section 12.2.3, this differs substantially from the values obtained in our low-order expansion. Note, though, that at finite temperatures pair-breaking excitations that would not exist in a purely bosonic system continue to be important [51].

The spectra in both limits are sketched in Fig. 12.7. It appears that in these dynamical signatures we have finally obtained a quantity that differs substantially between the two limits. The experimental implications of these results taken at face value are nevertheless misleading – the remarkable collective response that enables superconductivity occurs in the BCS regime, for instance, where we have claimed that the dynamics are dominated by quasiparticles. We turn to the issue of collective mode dynamics that distinguish the two regimes in the next section.

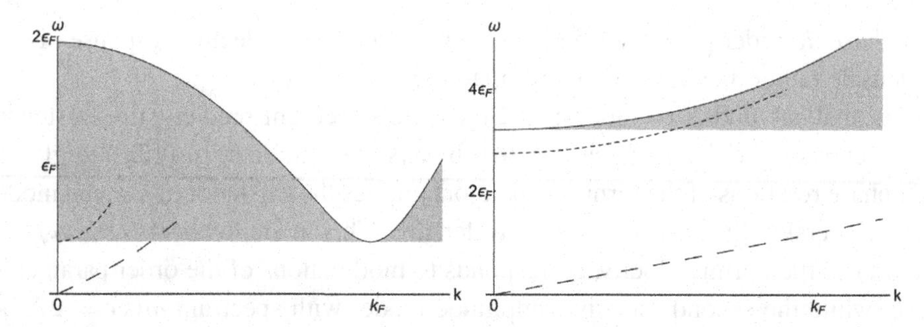

Figure 12.7 Excitation spectra on the BCS ($\eta = -1$, left) and BEC ($\eta = 1$, right) sides of the crossover. Shading below the quasiparticle spectrum $E_{\mathbf{k}}$ indicates the continuum where the collective Higgs (short-dashed) and Bogoliubov (long-dashed) modes decay.

12.3 Collective Mode Dynamics

12.3.1 Higgs Modes and the Higgs Mechanism

We have seen above that the collective response of the broken-symmetry state consists of a sound-like phase mode, often called the Goldstone mode, and a massive amplitude mode, often called the Higgs mode. First, let us clarify some (largely semantic) ambiguity surrounding these terms. In our discussion so far, the symmetry in question has been a global $U(1)$ phase rotation corresponding to charge conservation, $c^{\dagger}_{\mathbf{k}\sigma} \to e^{i\theta} c^{\dagger}_{\mathbf{k}\sigma}$, which is broken by the choice of a particular phase for $\Delta \sim \sum_{\mathbf{k}} \langle c_{\mathbf{k}\uparrow} c_{-\mathbf{k}\downarrow} \rangle$, which introduces non-number-conserving $\Delta c^{\dagger} c^{\dagger}$ terms. (One can explicitly compare how (12.7) and (12.3) transform.)

There is a second discussion, particularly relevant to superconductors, concerning the Higgs *mechanism* and gauge symmetries, which is best illustrated by explicitly including the electromagnetic potential $A_{\mu} = (\phi, \mathbf{A})$ in the model:

$$\begin{aligned} S[A_{\mu}, c, \bar{c}] = & \int_0^{\beta} d\tau \int d^d\mathbf{r}\, F^{\mu\nu} F_{\mu\nu} \\ & + \int_0^{\beta} d\tau \int d^d\mathbf{r}\, \bar{c}_{\sigma}(\tau, \mathbf{r}) \left(\partial_{\tau} + ie\phi + \frac{1}{2m}(-i\nabla - e\mathbf{A})^2 - \mu \right) c_{\sigma}(\tau, \mathbf{r}) \\ & + \int_0^{\beta} d\tau \sum_{\mathbf{k}\mathbf{k}'} \bar{c}_{\mathbf{k}\uparrow}(\tau) \bar{c}_{-\mathbf{k}\downarrow}(\tau) V_{\mathbf{k}\mathbf{k}'} c_{\mathbf{k}'\downarrow}(\tau) c_{-\mathbf{k}'\uparrow}(\tau) \end{aligned} \tag{12.8}$$

where the field-strength tensor $F_{\mu\nu} = \partial_{\mu} A_{\nu} - \partial_{\nu} A_{\mu}$. Under gauge transformations that take the charged fields $c^{\dagger}_{\mathbf{k}\sigma} \to e^{i\theta(\tau,\mathbf{r})} c^{\dagger}_{\mathbf{k}\sigma}$ and the gauge field $A_{\mu} \to A_{\mu} - \partial_{\mu}\theta(\tau, \mathbf{r})$ for an arbitrary $\theta(\mathbf{r}, \tau)$, the theory is invariant, whether in the normal or broken-symmetry state, as the order parameter Δ transforms as a matter field with charge $2e$. This invariance, required for any physical theory, simply accounts for the excess degrees of freedom in the field theory compared with physical electromagnetism, and is guaranteed by the electromagnetic coupling in the fermion kinetic term.

Taking (12.8) as a starting point and repeating the procedure of Section 12.2.3 to obtain an effective action $S_{\text{eff}}[A_\mu, \Delta, \bar{\Delta}]$, we may further decompose $\Delta = |\Delta|e^{i\theta}$ to isolate the low-energy physics of the phase fluctuations $\delta\theta$. (The massive fluctuations $\delta|\Delta|$ are the Higgs mode.) Two limits are useful: considering only the electric potential ϕ, one finds that the sound mode dispersion acquires a gap $\Omega = \sqrt{4\pi ne^2/m}$, known as the plasma frequency. Physically, our previous analysis had only considered the attractive interaction that promotes pairing, whereas now inclusion of the electric field that mediates the Coulomb interaction has revealed an additional collective response of the charged fluid to density fluctuations. We may also consider a purely magnetic field, $\mathbf{B} = \nabla \times \mathbf{A}^\perp$, whereupon integration over the phase fluctuations produces

$$S[\mathbf{A}^\perp] = \frac{\beta}{2}\sum_{\mathbf{k}}\left(\frac{n_s}{m} + \mathbf{k}^2\right)\mathbf{A}^\perp_{\mathbf{k}} \cdot \mathbf{A}^\perp_{-\mathbf{k}}, \tag{12.9}$$

where n_s is a measure of the superfluid density. Remarkably, the electromagnetic field carried in vacuum by the massless photon with dispersion $\omega = ck$ has acquired a mass inside the superconductor, which manifests physically as magnetic flux expulsion. This exemplifies the Higgs mechanism: a gauge field coupled to a matter Hamiltonian acquires a mass when the latter enters its broken-symmetry state, at the expense of the massless sound mode that would otherwise be present. We stress the limits of this scenario: the appropriate coupling to a gauge field requires charged particles, and the condensate order parameter must be in the Cooper channel.

12.3.2 Higgs Modes in Charge Density Wave Systems

The physics that enables the Higgs mechanism in superconductors at once vitiates direct observation of the Higgs mode by coupling to an electromagnetic probe. In charge density wave (CDW) systems, in contrast, the order parameter $u \sim \sum_{\mathbf{k}}\langle c^\dagger_{\mathbf{k}+\mathbf{Q}}c_{\mathbf{k}}\rangle$ is simply a spatially modulated electronic density, and may be thought of as a frozen phonon mode. As before, there is a collective phase mode, associated physically with translations of the CDW, and a Higgs mode corresponding to modulations of $|u|$. Since u controls the gap, it is possible to observe the evolution of its amplitude by time-resolved measurements of the electronic structure. Such experiments have been performed in the rare earth tritellurides, using time- and angle-resolved photoemission spectroscopy (trARPES) to track the evolution of the electronic structure of the CDW material as a function of time after a powerful pump pulse is used to transiently destroy the condensate [52, 53]. Not only do the experimental spectra show oscillations of the Higgs mode, the data for two successive pump pulses in or out of phase demonstrate that they are coherent, with the collective motion of the electron fluid Rabi flopping like a single spin.

This "quantum quench" response with oscillations in the Higgs mode is typical of the BCS side of the crossover. A quench in the strong-coupling limit is in contrast dominated by interference between the sound modes that dominate the low-energy spectrum leading to the observation of so-called Sakharov oscillations [54]. We also note that while all the observed dynamics are damped, the damping cannot be readily interpreted in terms of interaction between quasiparticles and collective modes as in Section 12.2.3, since the experimental systems involved in quench experiments are generally highly out of equilibrium and coupled to external sources of dissipation [55].

The ease of optically coupling to the CDW order parameter also provides an opportunity to observe the Higgs mode of the *superconducting* order parameter in materials that exhibit both types of order, such as $NbSe_2$. As we have seen, the order parameter Δ depends on the density of particles, or more carefully in the BCS limit, the density of states at the Fermi surface. But the formation of a CDW depletes precisely this quantity, providing a coupling between the two order parameters u and Δ. In particular, there is a linear coupling between the phonon and superconducting Higgs mode (which, owing to the phonon interaction, is pushed below the quasiparticle continuum), so that the phonon self-energy depends on Δ and becomes singular in the vicinity of the superconducting gap [56, 57, 58]. Raman spectroscopy, which measures this quantity by perturbing the CDW order, shows the development of a second peak in the susceptibility as the material (already CDW-ordered) is cooled below the superconducting transition temperature, at the expense of the peak associated with the CDW [59, 60, 61]. In this way, the superconducting Higgs mode, which we predicted above should not couple to electromagnetic probes (to linear order), becomes visible in the Raman spectrum: shaking one condensate at once shakes the other. Of course, at sufficiently high orders of expansion, a nonlinear electromagnetic coupling directly to the superconductor Higgs mode appears and seems to be recently accessible to high-intensity THz experiments [62].

12.3.3 Polaritons

An alternate strategy for observing the dynamics of the order parameter in the Cooper channel with an electromagnetic coupling is to pick a model system around this feature. Polaritons, quasiparticles hybridizing an ensemble of two-level systems with a photonic degree of freedom, were first realized between quantum-well excitons and a cavity photon confined by a distributed Bragg reflector, although implementations in cold atoms are currently being attempted as well. Polaritons are introduced elsewhere in this volume (see Chapter 4 by Littlewood and Edelman), and we will just make some brief remarks in the context of BCS-to-BEC crossover. A polariton consists of a superposition of a bound state of an exciton with a photon: thus in the conventional sense they are a dilute gas of bosonic degrees of freedom – the excitons are dilute. However, the polariton has a very light mass due to the photonic component, so the Bogoliubov sound mode is about four orders of magnitude stiffer than an excitonic condensate of the same density. Consequently, the dynamical spectrum is dominated by the "Higgs" mode, to be identified here as the collective oscillation of exciton-polaritons close to the Rabi frequency.

12.4 Concluding Remarks

In this brief survey of the BEC-BCS crossover, we began with the observation that a coherent superposition of large and overlapping weakly interacting pairs may be described by the same wave function as a condensate of tightly bound dimers. The simplest version of this theory is controlled by a single parameter – either the range of the interaction or the scaled density. Real physical systems can be more complex.

The reliable signatures of crossover are in the dynamics. At strong coupling when the cost of breaking pairs is high, the only low-energy physics remaining is in the collective sound mode. The BCS limit is in contrast the province of "more scales," which are liable to scramble low energies sufficiently that only amplitude mode dynamics emerge unscathed – the only strategy is to shake the condensate and watch.

References

[1] Randeria, M. 1995. Crossover from BCS theory to Bose-Einstein condensation. In: A. Griffin, D. W. Snoke, S. Stringari (eds), *Bose-Einstein Condensation*. Cambridge: Cambridge University Press.

[2] Alloing, M., Beian, M., Lewenstein, M., Fuster, D., González, Y., González, L., Combescot, R., Combescot, M., and Dubin, F. 2014. Evidence for a Bose-Einstein condensate of excitons. *Europhys. Lett.*, **107**, 10012.

[3] Randeria, M., Duan, J.-M., and Shieh, L.-Y. 1989. Bound states, Cooper pairing, and Bose condensation in two dimensions. *Phys. Rev. Lett.*, **62**, 981–984.

[4] Randeria, M., Duan, J.-M., and Shieh, L.-Y. 1990. Superconductivity in a two-dimensional Fermi gas: evolution from Cooper pairing to Bose condensation. *Phys. Rev. B*, **41**, 327–343.

[5] Fisher, Daniel, and Hohenberg, P. 1988. Dilute Bose gas in two dimensions. *Phys. Rev. B*, **37**, 4936–4943.

[6] Fuchs, J. N., Recati, A., and Zwerger, W. 2004. Exactly solvable model of the BCS-BEC crossover. *Phys. Rev. Lett.*, **93**, 090408.

[7] Tokatly, I. V. 2004. Dilute Fermi gas in quasi-one-dimensional traps: from weakly interacting fermions via hard core bosons to a weakly interacting Bose gas. *Phys. Rev. Lett.*, **93**, 090405.

[8] Leggett, A. J. 1980. Diatomic molecules and Cooper pairs. In: A. Pkalski, J. A. Przystawa (eds), *Modern Trends in the Theory of Condensed Matter, Proceedings of the XVI Karpacz Winter School*. Berlin, Heidelberg, New York: Springer.

[9] Chin, C., Grimm, R., Julienne, P., and Tiesinga, E. 2010. Feshbach resonances in ultracold gases. *Rev. Mod. Phys.*, **82**, 1225–1286.

[10] Zwierlein, M. W., Abo-Shaeer, J. R., Schirotzek, A., Schunck, C. H., and Ketterle, W. 2005. Vortices and superfluidity in a strongly interacting Fermi gas. *Nature*, **435**, 1047–1051.

[11] Parish, M. M., Mihaila, B., Timmermans, E. M., Blagoev, K. B., and Littlewood, P. B. 2005. BCS-BEC crossover with a finite-range interaction. *Phys. Rev. B*, **71**, 064513.

[12] Ngampruetikorn, V., Levinsen, J., and Parish, M. M. 2013. Pair correlations in the two-dimensional Fermi gas. *Phys. Rev. Lett.*, **111**, 265301.

[13] Sommer, A. T., Cheuk, L. W., Ku, M. J. H., Bakr, W. S., and Zwierlein, M. W. 2012. Evolution of fermion pairing from three to two dimensions. *Phys. Rev. Lett.*, **108**, 045302.

[14] Levinsen, J., and Parish, M. M. 2015. Strongly interacting two-dimensional Fermi gases. In: Madison, K. W., Bongs, K., Carr, L., Rey, A. M. and Zhai, H. (Ed.), *Annual Review of Cold Atoms and Molecules*, Volume 3. World Scientific.

[15] Keldysh, L. V. 1995. Macroscopic coherent states of excitons in superconductors. In: A. Griffin, D. W. Snoke, S. Stringari (eds), *Bose-Einstein Condensation*. Cambridge: Cambridge University Press.

[16] Keldysh, L. V., and Kozlov, A. N. 1968. Collective properties of excitons in semiconductors. *Soviet Physics JETP*, **27**, 521.

[17] Littlewood, P. B., and Zhu, X. 1996. Possibilities for exciton condensation in semiconductor quantum-well structures. *Phys. Scr.*, **1996**, 56–67.

[18] Zhu, X., Hybertsen, M. S., and Littlewood, P. B. 1996. Electron-hole system revisited: a variational quantum Monte Carlo study. *Phys. Rev. B*, **54**, 13575–13580.

[19] Hanamura, E., and Haug, H. 1977. Condensation effects of excitons. *Physics Reports*, **33**, 209–284.

[20] Yoshioka, K., Morita, Y., Fukuoka, K., and Kuwata-Gonokami, M. 2013. Generation of ultracold paraexcitons in cuprous oxide: a path toward a stable Bose-Einstein condensate. *Phys. Rev. B*, **88**, 041201.

[21] Stolz, H., Schwartz, R., Kieseling, F., Som, S., Kaupsch, M., Sobkowiak, S., Semkat, D., Naka, N., Koch, T., and Fehske, H. 2012. Condensation of excitons in Cu_2O at ultracold temperatures: experiment and theory. *New J. Phys.*, **14**, 105007.

[22] Yoshioka, Kosuke, Chae, Eunmi, and Kuwata-Gonokami, Makoto. 2011. Transition to a Bose-Einstein condensate and relaxation explosion of excitons at sub-Kelvin temperatures. *Nat. Comm.*, **2**, 328.

[23] High, A. A., Leonard, J. R., Hammack, A. T., Fogler, M. M., Butov, L. V., Kavokin, A. V., Campman, K. L., and Gossard, A. C. 2012. Spontaneous coherence in a cold exciton gas. *Nature*, **483**, 584–588.

[24] O'Hara, K. E., and Wolfe, J. P. 2000. Relaxation kinetics of excitons in cuprous oxide. *Phys. Rev. B*, **62**, 12909–12922.

[25] Jang, J. I., O'Hara, K. E., and Wolfe, J. P. 2004. Spin-exchange kinetics of excitons in Cu_2O: transverse acoustic phonon mechanism. *Phys. Rev. B*, **70**, 195205.

[26] Jang, J. I., and Wolfe, J. P. 2006. Auger recombination and biexcitons in Cu_2O: a case for dark excitonic matter. *Phys. Rev. B*, **74**, 045211.

[27] Jang, J. I., and Wolfe, J. P. 2005. Biexcitons in the semiconductor Cu_2O: an explanation of the rapid decay of excitons. *Phys. Rev. B*, **72**, 241201.

[28] Wolfe, J. P., and Jang, J. I. 2014. The search for Bose-Einstein condensation of excitons in Cu_2O: exciton-Auger recombination versus biexciton formation. *New J. Phys.*, **16**, 123048.

[29] Zhu, Xuejun, Littlewood, P. B., Hybertsen, Mark S., and Rice, T. M. 1995. Exciton condensate in semiconductor quantum well structures. *Phys. Rev. Lett.*, **74**, 1633–1636.

[30] Ohashi, Y. 2005. BCS-BEC crossover in a gas of Fermi atoms with a p-wave Feshbach resonance. *Phys. Rev. Lett.*, **94**, 050403.

[31] Duncan, R., and Sá de Melo, C. 2000. Thermodynamic properties in the evolution from BCS to Bose-Einstein condensation for a d-wave superconductor at low temperatures. *Phys. Rev. B*, **62**, 9675–9687.

[32] Bloch, I., Dalibard, J., and Nascimbène, S. 2012. Quantum simulations with ultracold quantum gases. *Nat Phys*, **8**, 267–276.

[33] Nishida, Y., and Son, D. T. 2012. Unitary Fermi gas, ϵ expansion, and nonrelativistic conformal field theories. In: Zwerger, W. (ed), *The BCS-BEC Crossover and the Unitary Fermi Gas*. Heidelberg: Springer-Verlag.

[34] Braaten, E. 2012. Universal relations for fermions with large scattering length. In: Zwerger, Wilhelm (ed), *The BCS-BEC Crossover and the Unitary Fermi Gas*. Heidelberg: Springer-Verlag.

[35] Altland, A., and Simons, B. 2010. *Condensed Matter Field Theory*, 2nd edn. New York: Cambridge University Press.

[36] Nozières, P., and Schmitt-Rink, S. 1985. Bose condensation in an attractive fermion gas: from weak to strong coupling superconductivity. *J. Low Temp. Phys.*, **59**, 195–211.

[37] Engelbrecht, J. R., Randeria, M., and Sá de Melo, C. A. R. 1997. BCS to Bose crossover: broken-symmetry state. *Phys. Rev. B*, **55**, 15153–15156.

[38] Sá de Melo, C., Randeria, M., and Engelbrecht, J. 1993. Crossover from BCS to Bose superconductivity: transition temperature and time-dependent Ginzburg-Landau theory. *Phys. Rev. Lett.*, **71**, 3202–3205.

[39] Petrov, D. S., Salomon, C., and Shlyapnikov, G. V. 2005. Scattering properties of weakly bound dimers of fermionic atoms. *Phys. Rev. A*, **71**, 012708.

[40] Petrov, D. S., Salomon, C., and Shlyapnikov, G V. 2004. Weakly bound dimers of fermionic atoms. *Phys. Rev. Lett.*, **93**, 090404.

[41] Burovski, E., Kozik, E., Prokof'ev, N., Svistunov, B., and Troyer, M. 2008. Critical temperature curve in BEC-BCS crossover. *Phys. Rev. Lett.*, **101**, 090402.

[42] Baym, G., Blaizot, J.-P., Holzmann, M., Laloë, F., and Vautherin, D. 1999. The transition temperature of the dilute interacting Bose gas. *Phys. Rev. Lett.*, **83**, 1703–1706.

[43] Haussmann, R., Rantner, W., Cerrito, S., and Zwerger, W. 2007. Thermodynamics of the BCS-BEC crossover. *Phys. Rev. A*, **75**, 023610.

[44] Gorkov, L. P., and Melik-Barkhudarov, T. K. 1961. Contribution to the theory of superfluidity in an imperfect Fermi gas. *Soviet Physics JETP*, **13**, 1018–1022 [J. Exptl. Theoret. Phys. (U.S.S.R.) 40, 1452–1458 (1961)].

[45] Heiselberg, H., Pethick, C. J., Smith, H., and Viverit, L. 2000. Influence of induced interactions on the superfluid transition in dilute Fermi gases. *Phys. Rev. Lett.*, **85**, 2418–2421.

[46] Tajima, H., Kashimura, T., Hanai, R., Watanabe, R., and Ohashi, Y. 2014. Uniform spin susceptibility and spin-gap phenomenon in the BCS-BEC-crossover regime of an ultracold Fermi gas. *Phys. Rev. A*, **89**, 033617.

[47] Hanai, R., and Ohashi, Y. 2014. Heteropairing and component-dependent pseudogap phenomena in an ultracold Fermi gas with different species with different masses. *Phys. Rev. A*, **90**, 043622.

[48] Popov, V. N. 1987. *Functional Integrals and Collective Excitations*. Cambridge: Cambridge University Press.

[49] Côté, R., and Griffin, A. 1993. Cooper-pair-condensate fluctuations and plasmons in layered superconductors. *Phys. Rev. B*, **48**, 10404–10425.

[50] Belkhir, L., and Randeria, M. 1992. Collective excitations and the crossover from Cooper pairs to composite bosons in the attractive Hubbard model. *Phys. Rev. B*, **45**, 5087–5090.

[51] Kosztin, I., Chen, Q., Kao, Y.-J., and Levin, K. 2000. Pair excitations, collective modes, and gauge invariance in the BCS–Bose-Einstein crossover scenario. *Phys. Rev. B*, **61**, 11662–11675.

[52] Schmitt, F., Kirchmann, P. S., Bovensiepen, U., Moore, R. G., Rettig, L., Krenz, M., Chu, J. H., Ru, N., Perfetti, L., Lu, D. H., Wolf, M., Fisher, I. R., and Shen, Z. X. 2008. Transient electronic structure and melting of a charge density wave in $TbTe_3$. *Science*, **321**, 1649–1652.

[53] Rettig, L., Chu, J. H., Fisher, I. R., Bovensiepen, U., and Wolf, M. 2014. Coherent dynamics of the charge density wave gap in tritellurides. *Faraday Discuss.*, **171**, 299–310.

[54] Hung, C.-L., Gurarie, V., and Chin, C. 2013. From cosmology to cold atoms: observation of Sakharov oscillations in a quenched atomic superfluid. *Science*, **341**, 1213–1215.

[55] Rançon, A., Hung, Chen-Lung, Chin, Cheng, and Levin, K. 2013. Quench dynamics in Bose-Einstein condensates in the presence of a bath: theory and experiment. *Phys. Rev. A*, **88**, 031601.

[56] Littlewood, P., and Varma, C. 1981. Gauge-invariant theory of the dynamical interaction of charge density waves and superconductivity. *Phys. Rev. Lett.*, **47**, 811–814.

[57] Browne, D. A., and Levin, K. 1983. Collective modes in charge-density-wave superconductors. *Phys. Rev. B*, **28**, 4029–4032.

[58] Littlewood, P., and Varma, C. 1982. Amplitude collective modes in superconductors and their coupling to charge-density waves. *Phys. Rev. B*, **26**, 4883–4893.

[59] Sooryakumar, R., and Klein, M. V. 1980. Raman scattering by superconducting-gap excitations and their coupling to charge-density waves. *Phys. Rev. Lett.*, **45**, 660–662.

[60] Pekker, D., and Varma, C. M. 2015. Amplitude/Higgs modes in condensed matter physics. *Annu. Rev. Condens. Matter Phys.*, **6**, 269–297.

[61] Méasson, M. A., Gallais, Y., Cazayous, M., Clair, B., Rodière, P., Cario, L., and Sacuto, A. 2014. Amplitude Higgs mode in the 2H–$NbSe_2$ superconductor. *Phys. Rev. B*, **89**, 060503.

[62] Matsunaga, R., Hamada, Y. I., Makise, K., Uzawa, Y., Terai, H., Wang, Z., and Shimano, R. 2013. Higgs amplitude mode in the BCS superconductors $Nb_{1-x}Ti_xN$ induced by terahertz pulse excitation. *Phys. Rev. Lett.*, **111**, 057002.

Part III

Condensates in Atomic Physics

Editorial Notes

The field of ultracold atoms is simply too large and diverse to survey with completeness in one volume. Much of the general theory on universal themes of condensates has been studied in the context of cold atoms over the past two decades, and has already been discussed in Part I of this book. In this part, we discuss a subset of active research topics within this ultracold environment, focusing on aspects facilitating further analogies to conventional condensed matter systems.

For the first few years after the successful demonstration of Bose-Einstein condensation (BEC) of cold atoms, the field remained focused on a detailed characterization of the properties of weakly-interacting Bose gases. Two experimental developments which then revolutionized the field were the utilization of the Feshbach resonance to control the strength and sign of the interatomic interactions with magnetic fields and the development and control of optical lattices to the level of being able to confine a single atom to each site.

Examples of the usefulness of such developments include fruitful studies of the BEC-BCS crossover and the unitary gas (already reviewed in Chapter 12), phase mixing studies for multicomponent condensates, and a whole host of experiments in which cold atoms in lattices could be used to simulate classic many-body problems in condensed matter physics. In particular, they could be used to make in reality the idealized models which had been developed for condensed matter systems, but which were never actually true descriptions of real condensed matter systems, because real condensed matter systems have contributions from effects such as disorder, impurities, and lattice vibrations (which can actually be controllably introduced in ultracold systems, thus expanding the spectrum of their capabilities in matching other physical systems). Chapter 13 surveys past experiments and the state of the art in optical lattice experiments.

The controlled experimental preparation and study of topological matter has been a long-standing goal. Numerous schemes based on atom-light coupling, controlled rotation, or periodic potential modulations have been proposed and implemented to realize synthetic magnetic fields on the neutrally charged atoms, thereby modifying the conventional condensates and opening the possibility of exploring novel forms of strongly-correlated phases. Condensed-matter analogue examples discussed here include, among others, the generation of topological Bloch bands (Chapter 14) or magnetic-like phase transitions arising in systems with spin-orbit coupling (Chapter 15). Controlled experiments of this type enable the probing of non-Abelian gauge fields, which could even lead to indirect simulations of hot quark matter in particle physics.[1]

Among the numerous other cold atom developments, we highlight here two topics identified due to their strong connections to other condensed matter systems: first, Chapter 16 gives a joint experimental/theoretical account of cold atom experiments with second sound, the study of which has shaped our current understanding of superfluid helium.[2] Another important topic in the liquid helium community focuses on the emergence and decay of turbulence in two- and three-dimensional superfluids.[3] Chapter 17 gives an overview of recent cold atom activity on this topic, which ties long-range, macroscopic properties to the microscopic quantum nature of condensates.

Finally, the ability to generate multicomponent atomic condensates with different spin degrees of freedom, and the presence of long-range interactions due to dipolar aspects, leads to a rich phase diagram, enabling the study of diverse phenomena such as spin dynamics and quantum magnetism, supersolidity, and ferrofluidity. Such features are reviewed in Chapter 18, which also connects to the spinor nature of several condensed matter condensates, e.g., Chapters 21 and 24.

[1] Banerjee, D., et al. (2013), Atomic quantum simulation of U(N) and SU(N) non-Abelian lattice gauge theories, *Phys. Rev. Lett.*, **110**, 125303.
[2] See Griffin, A., *Excitations in a Bose-Condensed Liquid* (Cambridge University Press, 1993).
[3] See Barenghi, C.F., and Sergeev, Y. (eds.) *Vortices and Turbulence at Very Low Temperatures* (Springer, 2008).

13

Probing and Controlling Strongly Correlated Quantum Many-Body Systems Using Ultracold Quantum Gases

IMMANUEL BLOCH

Max-Planck Institute of Quantum Optics, Garching, Germany, and
Ludwig-Maximilians University, Munich, Germany

Ultracold atoms in optical lattices provide an extremely clean and controllable setting to explore quantum many-body phases of matter. Here we give a brief review of the strong-correlation physics that has been realized using such ultracold atoms in optical lattices ranging from the realization of Hubbard models to studies of quantum magnetism and the detection of single atoms with lattice site resolution. All this has opened up fundamentally new opportunities for the investigation of quantum many-body systems.

13.1 Introduction

Over the past years, ultracold atoms in optical lattices have emerged as versatile new systems to explore the physics of quantum many-body systems. On the one hand, they can be helpful in gaining a better understanding of known phases of matter and their dynamical behavior; on the other hand, they allow one to realize completely novel quantum systems that have not been studied before in nature [1, 2, 3]. Commonly, the approach of exploring quantum many-body systems in such a way is referred to as "quantum simulations." Examples of some of the first strongly interacting many-body phases that have been realized both in lattices and in the continuum include the quantum phase transition from a superfluid to a Mott insulator [4, 5, 6], fermionic Mott insulators [7, 8], the achievement of a Tonks-Girardeau gas [9, 10], and the realization of the Bose-Einsten condensate (BEC)–Bardeen-Cooper-Schrieffer (BCS) crossover (see also Chapter 12) in Fermi gas mixture [11] using Feshbach resonances [12].

In almost all of these experiments, detection was limited to time-of-flight imaging or more refined derived techniques that mainly characterized the momentum distribution of the quantum gas [2]. For several years, researchers in the field have therefore aspired to employ in situ single-particle detection methods for the analysis of ultracold quantum gases. Only recently has it become possible to implement such imaging techniques, marking a milestone for the characterization and manipulation of ultracold quantum gases [13, 14, 15, 16, 17]. In our discussion, we will focus on one of the most successful of these techniques based on high-resolution fluorescence imaging. Despite being a rather new technique, such quantum gas microscopy has already proven to be an enabling technology for probing and manipulating quantum many-body systems. For the first time, controllable and strongly interacting many-body systems, as realized with ultracold atoms, could be observed on a local scale [17, 16]. The power of the technique became even more apparent with the advent of local hyperfine state-specific addressing in optical lattices [18]. Together with the local detection this provides a complete toolbox for the manipulation of one- and two-dimensional lattice gases on the scale of a few hundred nanometers. The short review will concentrate on the experiments employing ultracold bosonic and fermionic atoms in optical lattices, showing applications in Mott insulator physics [6, 7, 8] and quantum magnetism [19, 20, 21, 22, 23].

13.2 Bose and Fermi Hubbard Models

The single-band Hubbard models play a paramount role in the context of condensed matter physics. Although they are among the simplest models used to describe interacting particles on a lattice, in several cases not even the phase diagram of the system is known and analytical solutions of the Hubbard model have not yet been found. In the case of the fermionic Hubbard model, it is also widely believed that it contains the essential physics for the explanation of high-temperature superconductivity [24, 25]. The fact that both models can be realized efficiently with ultracold atoms [5, 26] and the fact that all the underlying parameters of the Hubbard model may be tuned and controlled in cold atom and molecule experiments have led to widespread interest of ultracold gases as efficient quantum simulators of these foundational Hamiltonians. Current research is driven by the quest to explore the low-temperature (entropy) phases of these models, but has also opened a new path to studying nonequilibrium phenomena in strongly correlated quantum systems beyond linear response [27, 28, 29, 30, 31].

As one of the most striking phenomena in both the case of bosons and fermions, for strong repulsive interactions between the particles compared with their kinetic energy $U \gg J$ and integer fillings, the many-body system forms a Mott insulating state, with strongly suppressed density fluctuations. Mott insulators form the basis for states with magnetic order, when the temperature of the system becomes sufficiently lower than the superexchange coupling between two spin states on neighboring lattice sites. The quest to realize such magnetically ordered states and the novel possibilities offered by ultracold atoms and molecules is discussed in the subsequent chapters.

Before we proceed, we will need to discuss a fundamental difference between typical condensed matter and cold quantum gas experiments. Typical condensed matter experiments are carried out under conditions where a probe sample is held at a constant temperature through a connection to a reservoir. Lowering the temperature then allows one to access novel phases of the many-body system. In the context of ultracold quantum gases, we are dealing with quantum systems that are completely isolated from their environment. A change of the underlying trapping or lattice parameters typically also leads to a change in the temperature of the isolated gas. A temperature reduction caused by a trap deformation, for example, might not bring one any closer to the transition point of a new phase, as the associated transition temperature typically is lowered as well. A much more useful system variable that is invariant to such adiabatic changes of the Hamiltonian parameters is the total entropy per particle of the quantum gas $S/(Nk_B)$. Whether or not a specific many-body phase can thus be reached via adiabatic change of system parameters is only a question of whether the initial entropy of the initial system is low enough. For example, experiments typically begin with a fermionic or bosonic quantum gas cooled via evaporative cooling to a certain temperature in the harmonic trapping potential of a magnetic or an optical dipole trap. This temperature determines the entropy of the quantum gas. Then optical lattices are turned on by increasing the intensity of the corresponding light fields. Under the assumption of adiabatic changes of the lattice potential, the entropy of the systems remains constant. Typically experiments are carried out under this assumption of conserved entropy; however, it is not always fulfilled, as time scales for reaching equilibrium in interacting many-body systems have been shown to increase with interaction strength [32, 33, 29, 30].

13.2.1 Bose-Hubbard Model

The theory proposal [4, 5] and the subsequent realization of the Bose-Hubbard model with ultracold atoms [6] marks the starting point for strong correlation physics with ultracold quantum gases. The model considers bosonic particles with onsite interactions U, hopping in the lowest energy band from site to site with a tunnel amplitude J:

$$\hat{H} = -J \sum_{\langle \mathbf{R},\mathbf{R}' \rangle,\sigma} \hat{a}^{\dagger}_{\mathbf{R}} \hat{a}_{\mathbf{R}'} + \frac{1}{2} U \sum_{\mathbf{R}} \hat{n}_{\mathbf{R}} \left(\hat{n}_{\mathbf{R}} - 1 \right) + V_t \sum_{\mathbf{R}} \mathbf{R}^2 \hat{n}_{\mathbf{R}}. \tag{13.1}$$

Here $\hat{a}_{\mathbf{R}}(\hat{a}^{\dagger}_{\mathbf{R}})$ denote the particle destruction (creation) operators on lattice site $\mathbf{R}$, $\hat{n}_{\mathbf{R}} = \hat{a}^{\dagger}_{\mathbf{R}}\hat{a}_{\mathbf{R}}$, counts the number of particles on site $\mathbf{R}$ and the last term characterizes the underlying harmonic trapping potential, typically present in cold atom experiments.

Within this model, a gas of interacting bosons occupying the lowest Bloch band of a periodic potential forms a superfluid below a critical ratio of interaction to kinetic energy, i.e., when $U/J < (U/J)_c$. For integer filling of the lattice and for $U/J > (U/J)_c$, the system turns into a strongly correlated Mott insulator. By increasing the optical lattice depth, the ratio of interaction to kinetic energy of the system U/J can be tuned to increasingly large values. Even without employing scattering resonances to tune U [12], it is thus possible to bring the bosonic quantum gas into a strongly correlated regime of a Mott insulator simply by quenching the kinetic energy of the system. The transition from a superfluid to a Mott insulator has by now been the focus of numerous theoretical and experimental investigations, and it is beyond the scope of this chapter to give a complete survey of these. The interested reader may find more detailed reviews on the topic in Refs. [1, 2, 3].

In the following, we will instead try to highlight a few characteristic properties that exemplify the dramatic changes occurring when the quantum gas is converted from a superfluid into a Mott insulator (see, e.g., Fig. 13.1). One of the most prominent features is the measurement of the change of coherence properties when the transition is crossed. For a superfluid state, the underlying condensate exhibits long-range phase coherence and thus shows sharp matter-wave interference peaks when

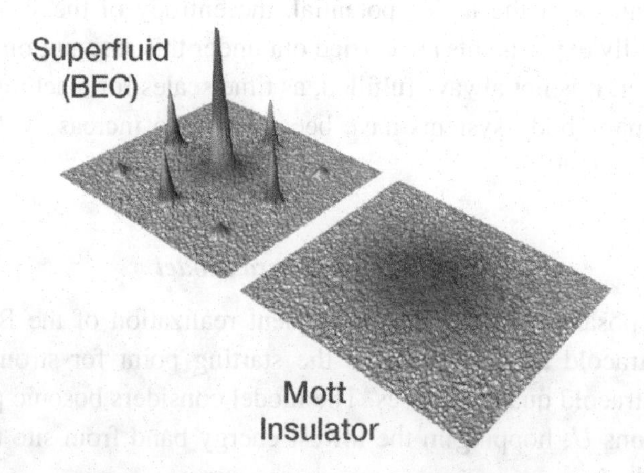

Figure 13.1 Several characteristic properties in the transition from a superfluid to a Mott insulator have been investigated experimentally. Among these are the different coherence properties, number statistics, density distributions, and transport properties. The distinctly different phase coherence properties of the superfluid (BEC) state and an atomic limit Mott insulator become apparent through time-of-flight imaging of the corresponding states, as shown (for more details, see Ref. [37]).

the quantum gas is released from the optical lattice. Deep in the Mott insulating phase $U/J \gg (U/J)_c$, the system is comprised of pure Fock states of integer onsite density and thus no interference pattern is observed [6]. For finite tunneling, particle–hole fluctuations induced by the kinetic energy of the quantum gas can allow even a Mott insulator to exhibit finite coherence properties [34]. However, as the particle–hole pairs are created only on nearest neighbor lattice sites, this coherence is rather short ranged in nature and distinctly different from the long-range phase coherence of a BEC. As the transition point to the superfluid is approached, the particle–hole pairs begin to extend over larger and larger distances, eventually becoming deconfined and inducing the formation of a superfluid at the transition point [35, 36]. It is thus natural to view the Mott insulator as a gas of bound particle–hole pairs, on top of a fixed density background.

13.2.2 Fermi-Hubbard Model

Restricting our discussion to the lowest energy band of a simple cubic three-dimensional (3D) optical lattice, a fermionic quantum gas mixture can be modeled via the Hubbard-Hamiltonian [38] with an additional term describing the underlying harmonic potential and σ indicating the spin state of the particles:

$$\begin{aligned}\hat{H} = &-J \sum_{\langle \mathbf{R},\mathbf{R}'\rangle,\sigma} \hat{a}^{\dagger}_{\mathbf{R},\sigma}\hat{a}_{\mathbf{R}',\sigma} + U\sum_{\mathbf{R}} \hat{n}_{\mathbf{R},\downarrow}\hat{n}_{\mathbf{R},\uparrow} \\ &+V_t \sum_{\mathbf{R},\sigma} \mathbf{R}^2 \hat{n}_{\mathbf{R},\sigma}.\end{aligned} \tag{13.2}$$

The quantum phases of the fermionic Hubbard model with harmonic confinement are governed by the interplay between three energy scales: kinetic energy, whose scale is given by the lattice bandwidth $12J$; interaction energy U; and the strength of the harmonic confinement, which can conveniently be expressed by the characteristic trap energy $E_t = V_t(N_\sigma/(4\pi/3))^{2/3}$, denoting the Fermi energy of a noninteracting cloud in the zero-tunneling limit, with N_σ being the number of atoms per spin state ($N_\downarrow = N_\uparrow$). The characteristic trap energy depends on both atom number and trap frequency via $E_t \propto \omega_\perp^2 N_\sigma^{2/3}$ and describes the effective compression of the quantum gas, controlled by the trapping potential in the experiment.

Depending on which term in the Hamiltonian dominates, different kinds of many-body ground states can occur in the trap center. For weak interactions in a shallow trap $U \ll E_t \ll 12J$, the Fermi energy is smaller than the lattice bandwidth ($E_F < 12J$) and the atoms are delocalized in order to minimize their kinetic energy. This leads to compressible metallic states with central filling $n_{0,\sigma} < 1$, where the local filling factor $n_{\mathbf{R},\sigma} = \langle \hat{n}_{\mathbf{R},\sigma} \rangle$ denotes the average occupation per spin state of a given lattice site. A dominating repulsive interaction $U \gg 12J$ and $U \gg E_t$

suppresses the double occupation of lattice sites and can lead to Fermi-liquid ($n_{0,\sigma} < 1/2$) or Mott-insulating ($n_{0,\sigma} = 1/2$) states at the trap center, depending on the ratio of kinetic to characteristic trap energy. Stronger compressions lead to higher filling factors, ultimately ($E_t \gg 12J, E_t \gg U$) resulting in an incompressible band insulator with unity central filling at $T = 0$.

Finite temperature reduces all filling factors and enlarges the cloud size, as the system needs to accommodate the corresponding entropy. Furthermore, in the trap the filling always varies smoothly from a maximum at the center to zero at the edges of the cloud. For a dominating trap and strong repulsive interactions at low temperature ($E_t > U > 12J$), the interplay between the different terms in the Hamiltonian gives rise to a wedding-cake-like structure consisting of a band-insulating core ($n_{0,\sigma} \approx 1$) surrounded by a metallic shell ($1/2 < n_{\mathbf{R},\sigma} < 1$), a Mott-insulating shell ($n_{\mathbf{R},\sigma} = 1/2$), and a further metallic shell ($n_{\mathbf{R},\sigma} < 1/2$) [39]. The outermost shell remains always metallic, independent of interaction and confinement; only its thickness varies.

Recent experiments on ultracold fermionic spin mixtures of ^{40}K atoms have been able to reach a paramagnetic Mott insulating phase for increasing interactions in the quantum gases in the range of $U/(12J) \simeq 1.5-4$ [7, 8]. In the experiments, the suppression of double occupancy (doublons) or the incompressible nature of an insulating phase have been used in order to identify the Mott insulating phase. For example, for $k_BT < U$ and strong interactions, one expects doubly occupied sites $D = \langle \hat{n}_{\mathbf{R},\uparrow}\hat{n}_{\mathbf{R},\downarrow}\rangle / (\langle n_{\mathbf{R},\uparrow}\rangle + \langle n_{\mathbf{R},\downarrow}\rangle)$ to be strongly suppressed compared with the noninteracting case. Furthermore, when the system is in an insulating phase, the compressibility of the system will drop to a minimum. Both these quantities can be compared with ab initio dynamical mean field theory (DMFT) calculations [8, 40]. As the lowest achieved temperatures of the quantum gases are still above the single-particle hopping $k_BT \gtrsim J$, a high-temperature series expansion of the partition function has also been shown to be useful for comparison with the experimental results. The detailed comparison with theory has allowed one to extract the entropies per particle of current experiments being in the range of $S/N = (1 - 1.8)k_B$, for which Mott insulating behavior could be observed. Interestingly, for the upper-limit values, these entropies are larger than the maximum entropy that can be stored in a homogeneous single-band Hubbard model system of $S_{\max}/N = k_B 2 \log 2$. For a homogeneous system, one would thus not expect the system to show Mott insulating behavior; however, for the trapped quantum gas the entropy per particle is distributed inhomogeneously throughout the system, such that in the metallic wings the excess entropy above $S/N = k_B \log 2$ can be efficiently stored as configurational entropy of the particles. This inhomogeneous entropy distribution is in fact key to novel cooling (or better entropy reduction) methods that have been proposed for ultracold atoms in optical lattices [41, 42].

13.3 Quantum Magnetism with Ultracold Atoms in Optical Lattices

Magnetically ordered quantum phases play an important role in the low-temperature regime of the Hubbard model. The underlying spin–spin interactions responsible for the magnetically ordered phases arise due to superexchange mediated coupling of neighboring spins, and our primary goal will be to understand how such superexchange couplings emerge in two-component quantum gases on a lattice in the regime of strong interactions. For this, it will be useful to first introduce a "toy model" of two spins in a double well – a system that can be in effect realized in the lab using superlattice potentials and that plays an important role in the context of realizing solid-state qubits in electronic double-well quantum dots. Extensions of these results to larger plaquette-sized systems and the formation of resonating valence bond states will also be discussed.

13.3.1 Superexchange Spin Interactions

Superexchange Interactions in a Double Well

Spin–spin interactions between neighboring atoms are mediated via so-called superexchange processes. They directly arise from within the Hubbard model in the regime of strong interactions, leading to an effective Hamiltonian that couples the spin of neighboring atoms in a lattice. Let us first discuss how such superexchange interactions can be derived theoretically (see also [43]). As a starting point, we discuss the case of an atom with spin-up $|\uparrow\rangle$ and another atom with $|\downarrow\rangle$ loaded into a double-well potential. In the regime of strong repulsive interactions $U \gg J$, doubly occupied sites are energetically suppressed and our system can be described by the following basis states of the left and right well $\mathcal{S} = \{|\downarrow,\uparrow\rangle, |\uparrow,\downarrow\rangle, |\uparrow,\uparrow\rangle, |\downarrow,\downarrow\rangle\}$.

The action of the tunneling operator of the Hubbard Hamiltonian can be evaluated in the strongly interacting regime via perturbation theory. First-order tunneling processes lead out of the energetically allowed subspace and are therefore forbidden. However, second-order tunneling processes that leave the system within $\mathcal{S}$ lead to an effective coupling between the different spin states. We can describe such processes via an effective Hamiltonian of the system, whose matrix elements within $\mathcal{S}$ can be evaluated via second-order perturbation theory:

$$\hat{H}^{\text{eff}}_{a,b} = -\left\langle a \left| \hat{H}_J \frac{1-\hat{P}_{\mathcal{S}}}{U} \hat{H}_J \right| b \right\rangle = -\sum_{n \notin \mathcal{S}} \left\langle a|\hat{H}_J|n\right\rangle \frac{1}{\left\langle n|\hat{H}^{\text{int}}|n\right\rangle} \left\langle n|\hat{H}_J|b\right\rangle. \quad (13.3)$$

Here $\hat{H}_J$ denotes the tunneling, $\hat{H}^{\text{int}}$ the interaction part of the Hubbard Hamiltonian, and $\hat{P}_{\mathcal{S}}$ represents the projector into the subspace $\mathcal{S}$.

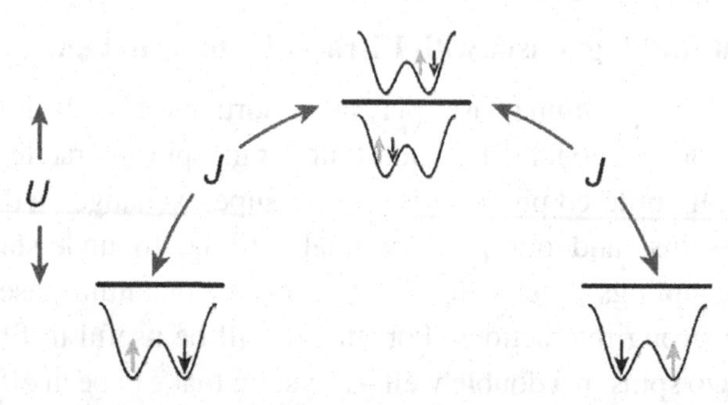

Figure 13.2 Schematic of superexchange interactions in a double well. For the case of strong repulsive interactions between the particles, atoms can mediate interactions between neighboring sites via second-order tunneling processes, depicted in the image. The second-order hopping of the particles via an intermediate state can lead to an exchange of the spins in the system.

Let us consider, for example, the process that can lead to an exchange of two spins (see also Fig. 13.2):

$$|\downarrow,\uparrow\rangle \xrightarrow{J} \overbrace{|0,\uparrow\downarrow\rangle}^{-1/U} \xrightarrow{J} |\uparrow,\downarrow\rangle$$
$$\text{or } \underbrace{|\uparrow\downarrow,0\rangle}_{-1/U} \xrightarrow{J} |\uparrow,\downarrow\rangle \tag{13.4}$$

The two possible pathways thus sum up to an effective coupling strength $-J_{ex} = -2J^2/U$. Evaluating the other matrix elements in the same way, we obtain for the case of bosonic atoms the following matrix form of H^{eff} in the basis $\mathcal{S}$:

$$\hat{H}^{\text{eff}}_{\text{bosons}} = J_{ex} \begin{pmatrix} -1 & -1 & 0 & 0 \\ -1 & -1 & 0 & 0 \\ 0 & 0 & -1 & 0 \\ 0 & 0 & 0 & -1 \end{pmatrix}. \tag{13.5}$$

Diagonalizing the effective Hamiltonian yields the new eigenstates and eigenenergies:

$$\{|t_+\rangle, |t_0\rangle, |t_-\rangle\} \quad \text{with} \quad E = -2J_{ex} \tag{13.6}$$
$$|s_0\rangle \quad \text{with} \quad E = 0. \tag{13.7}$$

Here $|t_+\rangle = |\uparrow,\uparrow\rangle, |t_-\rangle = |\downarrow,\downarrow\rangle$, and $|t_0\rangle = 1/\sqrt{2}(|\downarrow,\uparrow\rangle + |\uparrow,\downarrow\rangle)$ are the spin-triplet eigenstates, whereas $|s_0\rangle = 1/\sqrt{2}(|\downarrow,\uparrow\rangle - |\uparrow,\downarrow\rangle)$ denotes the spin-singlet eigenstate of the two atoms. We may thus write $\hat{H}^{\text{eff}}$ via a projector into the spin-triplet subspace $\hat{P}_T$, as

$$\hat{H}^{\text{eff}} = -2J_{ex}\hat{P}_T = -J_{ex}\left(\hat{1} + \hat{X}_{LR}\right), \tag{13.8}$$

where $\hat{X}_{LR}$ denotes the exchange operator between the left and right well. The projection operator into the singlet and triplet subspace can be expressed via Dirac notation as

$$\hat{P}_T = \frac{3}{4} + \hat{S}_L \cdot \hat{S}_R. \tag{13.9}$$

We can thus write the effective Hamiltonian as an interaction term between spins on the neighboring wells:

$$\hat{H}^{\text{eff}} = -2J_{ex}\hat{S}_L \cdot \hat{S}_R. \tag{13.10}$$

The minus sign for the case of bosons indicates ferromagnetic interactions, as the energy of the two spins is lowered if they align along the same direction.

For the case of fermionic spin mixtures, we can essentially follow the same derivation; however, for equal spin on neighboring sites, second-order hopping processes are forbidden, due to Pauli blocking in the intermediate state, where both particles with identical spin would occupy the same spin state. Also, when two particles are exchanged, we obtain an additional minus sign in the coupling owing to the odd exchange symmetry of fermionic particles, leading to an overall antiferromagentic superexchange spin Hamiltonian.

Superexchange Interactions on a Lattice

The above derivation can be extended to the case of a lattice system in a straightforward manner. For a bosonic or fermionic quantum system consisting of an equal mixture of two spin components in a Mott insulating regime with $\langle \hat{n}_{i,\uparrow} \rangle + \langle \hat{n}_{i,\downarrow} \rangle = 1$ and low enough temperatures, one expects magnetically ordered quantum phases due to such superexchange spin–spin interactions. In the simplest case, such spin interactions take the form of an isotropic Heisenberg model:

$$\hat{H} = \pm J_{ex} \sum_{\langle \mathbf{R},\mathbf{R}' \rangle} \hat{S}_{\mathbf{R}} \cdot \hat{S}_{\mathbf{R}'}, \tag{13.11}$$

with effective spin-1/2 operators $\hat{S}^x_{\mathbf{R}} = (\hat{a}^\dagger_{\mathbf{R},\uparrow}\hat{a}_{\mathbf{R},\downarrow} + \hat{a}^\dagger_{\mathbf{R},\downarrow}\hat{a}_{\mathbf{R},\uparrow})/2$, $\hat{S}^y_{\mathbf{R}} = (\hat{a}^\dagger_{\mathbf{R},\uparrow}\hat{a}_{\mathbf{R},\downarrow} - \hat{a}^\dagger_{\mathbf{R},\downarrow}\hat{a}_{\mathbf{R},\uparrow})/2i$, $\hat{S}^z_{\mathbf{R}} = (\hat{n}_{\mathbf{R},\uparrow} - \hat{n}_{\mathbf{R},\downarrow})/2$, and exchange coupling constant $J_{ex} = 2J^2/U$.

It is instructive to rewrite the Heisenberg Hamiltonian using the spin raising and lowering operators $\hat{S}^+_{\mathbf{R}} = \left(\hat{S}^x_{\mathbf{R}} + i\hat{S}^y_{\mathbf{R}}\right) = \hat{a}^\dagger_{\mathbf{R},\uparrow}\hat{a}_{\mathbf{R},\downarrow}$ and $\hat{S}^-_{\mathbf{R}} = \left(\hat{S}^x_{\mathbf{R}} - i\hat{S}^y_{\mathbf{R}}\right) = \hat{a}^\dagger_{\mathbf{R},\downarrow}\hat{a}_{\mathbf{R},\uparrow}$. We find

$$\hat{H} = \pm\frac{J_{ex}}{2} \sum_{\langle \mathbf{R},\mathbf{R}' \rangle} \left(\hat{S}^+_{\mathbf{R}}\hat{S}^-_{\mathbf{R}'} + \hat{S}^-_{\mathbf{R}}\hat{S}^+_{\mathbf{R}'}\right) \pm J_{ex} \sum_{\langle \mathbf{R},\mathbf{R}' \rangle} \hat{S}^z_{\mathbf{R}}\hat{S}^z_{\mathbf{R}'} \tag{13.12}$$

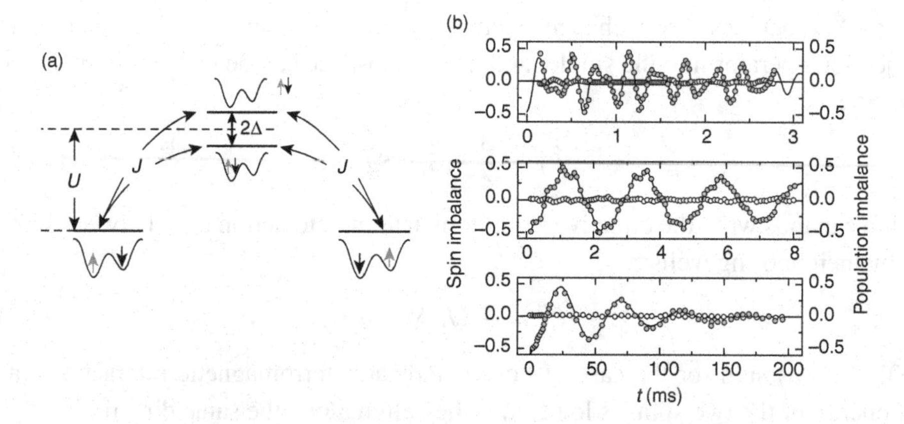

Figure 13.3 Detecting and controlling superexchange interactions. (a) Superexchange interactions are mediated via second-order hopping processes. By introducing an energy offset Δ between neighboring lattice sites, one may tune the exchange coupling J_{ex}. (b) Dynamical observation of superexchange interaction in double wells, initially prepared in a z-Néel order (see (a)). For increasing interactions (top row to bottom row), one observes how Heisenberg-type superexchange spin–spin interactions emerge and single-particle hopping becomes increasingly suppressed due to the increased repulsive interactions between the particles. Adapted with permission from the American Association for the Advancement of Science (AAAS) from Trotzky, S., et al. (2008), Time-resolved observation and control of superexchange interactions with ultra-cold atoms in optical lattices, *Science*, **319**, 295 [19].

Note that in this form it becomes especially apparent that the first part of the spin Hamiltonian has exactly the same structure as the tunneling operator in the Hubbard Hamiltonian, while the second term acts like a nearest neighbor interaction term.

Using optical superlattice techniques, it has been possible both to prepare magnetic quantum correlations and to probe them for both bosonic [20] and fermionic [22] Hubbard systems. Another possibility to reveal magnetic ordering is to make use of state selective Bragg scattering. This was most recently used to reveal short-range magnetic correlations in the fermionic Hubbard model [23]. The experiment indicates that temperatures close to the transition temperature to long-range antiferromagentic ordering have been reached. Next to probing static properties induced by superexchange-mediated quantum magnetism, one can also probe dynamical properties of quantum magnetism. By flipping a single or few spins in the quantum gas (see Section 13.5), it was possible to reveal the coherent propagation of single or bound magnon pairs [21, 44].

13.3.2 Resonating Valence Bond States in a Plaquette

The concept of valence bond resonance plays a fundamental role in the theory of the chemical bond [45, 46] and is believed to lie at the heart of many-body

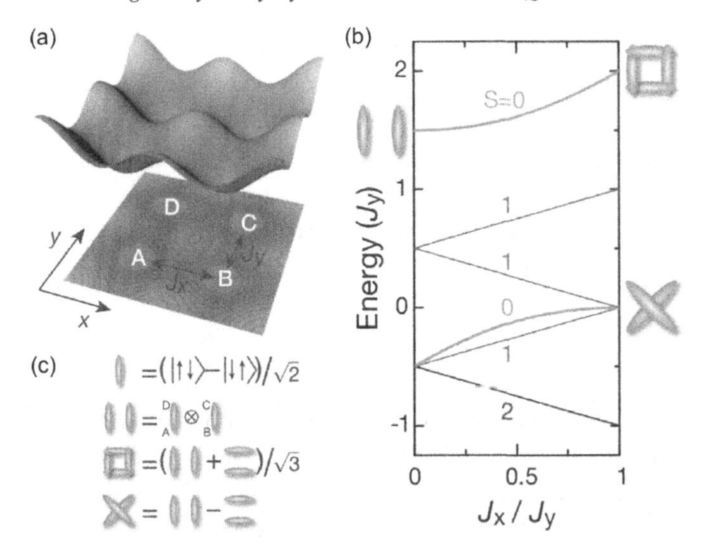

Figure 13.4 Schematics of a single plaquette and energy levels at half filling. (a) Scheme of the lattice potential in the x, y plane, created by a pair of bichromatic optical lattices. The elementary cell is made of four wells arranged in a square configuration. (b) Energy levels of four atoms on a plaquette in a Mott insulating state at half filling, with superexchange spin couplings along x (y) denoted by $J_x(J_y)$. For any ratio J_x/J_y, the highest energy state is a total spin-$\frac{1}{2}$ singlet. In the case of $J_x/J_y = 0$, it corresponds to the valence bond state with singlets aligned along the vertical direction, whereas for $J_x/J_y = 1$ it is the s-wave RVB state. The other total singlet for $J_x = J_y$, lower in energy, is the d-wave RVB state, with singlets along the diagonals. (c) Symbols used for a singlet bond and for the s-wave and d-wave plaquette RVB states. Reprinted with permission from Nascimbène, S., et al. (2012), Experimental realization of plaquette resonating valence-bond states with ultracold atoms in optical superlattices, *Phys. Rev. Lett.*, **108**, 205301 [50]. Copyright (2012) by the American Physical Society.

quantum physical phenomena [47, 48]. By making use of optical superlattices in two orthogonal directions, it has recently become possible to create such resonating valence bond (RVB) states of different symmetry types in arrays of plaquettes (see Fig. 13.4) [49, 50]. In the experiment, one could for example begin with spin singlets along the vertical direction (A–D and B–C in Fig. 13.4a) with suppressed exchange coupling in the horizontal direction, i.e., $J_x = 0$. By adiabatically turning on the exchange coupling in the horizontal direction to a point where $J_x = J_y$, the state could be transformed into an RVB state with s-wave symmetry. Such a state can be viewed as a coherent superposition of spin singlets in the horizontal and vertical direction, very much in analogy to the electronic binding configuration in a benzene molecule. If the exchange coupling along the horizontal direction was turned on abruptly to $J_x = J_y$, however, the systems started to exhibit valence bond oscillations between the two configurations where the singlets are oriented along the vertical and horizontal direction. Using more elaborate preparation techniques, it has also been possible to realize RVB states with d-wave symmetry [49, 50].

Having such control possibilities at hand for the local creation of plaquette RVB states opens the path for coupling these plaquettes to larger system sizes and thereby extending the RVB state over a larger area of the two-dimensional spin system. Different protocols for achieving this have been discussed in the literature [51, 52].

13.4 Site-Resolved Imaging

One of the standard imaging techniques in ultracold quantum gases – absorption imaging – cannot be easily extended to the regime of single atom sensitivity. This is mainly due to the limited absorption a laser beam experiences when interacting with a single atom. For typical experimental conditions, the absorption signal is always smaller than the accompanying photon shot noise. While high-resolution images of down to 1 μm resolution have been successfully used to record *in situ* absorption images of trapped quantum gases [53], they have not reached the single-atom sensitive detection regime. Fluorescence imaging can, however, overcome this limited signal-to-noise ratio and therefore provides a viable route for combining high-resolution imaging with single-atom sensitivity. By using laser-induced fluorescence in an optical molasses configuration and by trapping the atoms in a very deep potential, several hundred thousand photons can be scattered from a single atom, of which a few thousand are ultimately detected. An excellent signal-to-noise ratio in the detection of a single atom can therefore be achieved.

This idea was first pioneered for the case of optical lattices by the group of D. Weiss, who loaded atoms from a magneto-optical trap into a three-dimensional lattice with a lattice constant of 6 μm [13]. However, for typical condensed matter oriented experiments, such large spaced lattices are of limited use due to their almost vanishing tunnel coupling between neighboring potential wells. Extending fluorescence imaging to a regime where the resolution can be comparable to typical submicron lattice spacings thus requires large numerical aperture (NA) microscope objectives, as the smallest resolvable distances in classical optics are determined by $\sigma = \lambda/(2\text{NA})$.

In recent works, Bakr et al. [15, 16] and Sherson et al. [17] have demonstrated such high-resolution imaging and applied it to image the transition of a superfluid to a Mott insulator in 2D. In the experiments, 2D Bose-Einstein condensates were first created in tightly confining potential planes. Subsequently, the depth of a two-dimensional simple-cubic type lattice was increased, leaving the system in either a superfluid or Mott insulating regime. The lattice depths were then suddenly increased to very deep values of approximately 300 μK, essentially freezing out

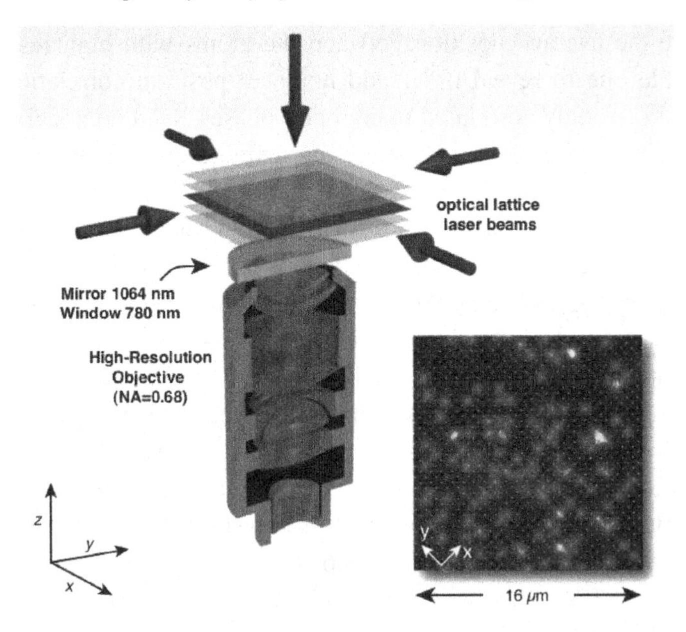

Figure 13.5 Schematic setup for high-resolution fluorescence imaging of a two-dimensional (2D) quantum gas. Two-dimensional bosonic quantum gases are prepared in a single 2D plane of an optical standing wave along the *z*-direction, which is created by retro-reflecting a laser beam ($\lambda = 1064$ nm) on the coated vacuum window. Additional lattice beams along the *x*- and *y*-directions are used to bring the system into the strongly correlated regime of a Mott insulator. The atoms are detected using fluorescence imaging via a high-resolution microscope objective. Fluorescence of the atoms was induced by illuminating the quantum gas with an optical molasses that simultaneously laser cools the atoms. The inset shows a section from a fluorescence picture of a dilute thermal cloud (points mark the lattice sites). Adapted with permission from Sherson, J. F., et al. (2010) Single-atom-resolved fluorescence imaging of an atomic Mott insulator, *Nature*, **467**, 68 [17]. Copyright (2010) by the Nature Publishing Group.

the density distribution of the atoms in the lattice. A near-resonant optical molasses was then used to induce fluorescence of the atoms in the deep lattice and also provide laser cooling, such that atoms remained on lattice sites while fluorescing. High-resolution microscope objectives with numerical apertures of NA ≈ 0.7–0.8 were used to record the fluorescence and image the atomic density distribution on charged coupled device (CCD) cameras (see Fig. 13.5). In the detection process, a (possibly) highly complex and correlated many-body wavefunction $|\Psi\rangle = \sum_{\{n_i\}} \alpha_{\{n_1,\ldots,n_N\}} |n_1, \ldots, n_N\rangle$, consisting of different superposition states $|n_1, \ldots, n_N\rangle$ of occupational configurations of the N lattice sites, will collapse onto one of those spatial configurations, which in the end is detected in the experiment. Such single "snapshots" of a quantum many-body system can then not only be used, e.g., to image the average position between the atoms with high resolution, but also can enable one to reveal order and nonlocal position correlations between the particles. In strongly correlated many-body phases, such nonlocal correlations very often are very characteristic and are the defining quantity that characterizes them [54, 55]. Using single-atom imaging, they have now also become accessible in experiments [35, 36]. A limitation of the detection method currently employed is that so-called inelastic light-induced collisions occurring during the illumination period only allow one to record the parity of the onsite atom number. Whenever pairs of atoms are present on a single lattice site, both atoms are rapidly lost within the first few hundred microseconds of illumination due to a large energy release caused by radiative escape and fine-structure changing collisions [56].

In both experiments, high-resolution imaging has allowed one to reconstruct the atom distribution (modulo 2) on the lattice down to a single-site level. Results for the case of a Bose-Einstein condensate and Mott insulators of such a particle number reconstruction are displayed in Fig. 13.6. The fidelity of the imaging process is

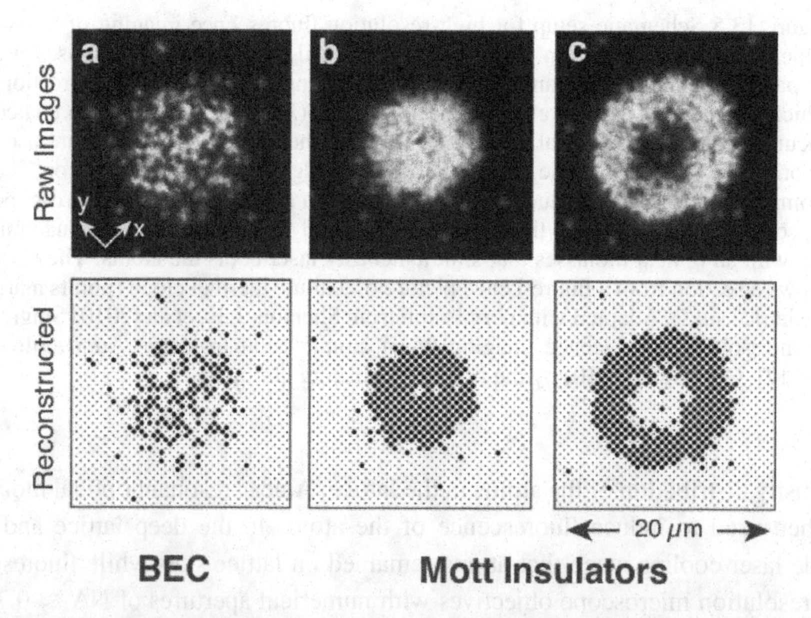

Figure 13.6 High-resolution fluorescence images of a weakly interacting Bose-Einstein condensate and Mott insulators. (a) Bose-Einstein condensate exhibiting large particle number fluctuations and (b, c) wedding-cake structure of $n = 1$ and $n = 2$ Mott insulators. Using a numerical algorithm, the corresponding atom distribution on the lattice can be reconstructed. The reconstructed images can be seen in the row below (small points mark lattice sites, large points mark position of a single atom). Adapted with permission from Sherson, J. F., et al. (2010) Single-atom-resolved fluorescence imaging of an atomic Mott insulator, *Nature*, **467**, 68 [17]. Copyright (2010) by the Nature Publishing Group.

currently limited to approximately 99% by atom loss during the illumination due to background gas collisions.

13.5 Single-Site-Resolved Addressing of Individual Atoms

Being able to spatially resolve single lattice sites also allows researchers to manipulate atoms with single-site resolution. A laser beam can be sent in reverse through the high-resolution objective and, hence, is focused onto the atoms. Thereby the high-resolution objective is used twice – for imaging and for local addressing. In typical cases, the resulting spot size of the laser beam will still be on the order of a lattice spacing and for most applications too large in order to reliably address atoms on single lattice sites. One possibility to increase the spatial resolution is to make use of a resonance imaging technique: the focused laser is tuned to such a wavelength that it creates a differential energy shift between two internal hyperfine ground states of an atom. Then global microwave radiation will be resonant only at the position of the focused beam and thus can be used to control the spin state of the atom [57, 18]. The spatial resolution for the addressing of single atoms can thereby be increased up to a limit given by (often magnetic field–driven) fluctuations of the energy splitting between the two hyperfine states. For typical parameters, this corresponds to an increase by almost an order of magnitude down to $\simeq$ 50 nm, well below the optical diffraction limit.

In the experiment, such addressing was demonstrated in a 2D Mott insulator with unity occupation per lattice site [18]. In order to prepare an arbitrary pattern of spins in the array, the laser beam was moved to a specific site and a Landau-Zener microwave sweep was applied in order to flip the spin of the atom located at the lattice site. The laser beam was then moved to the next lattice site and the procedure was repeated. In order to detect the resulting spin pattern, unflipped atoms were removed by applying a resonant laser beam that rapidly expelled these atoms from the trap [18]. The remaining spin-flipped atoms were then detected using standard high-resolution fluorescence imaging, as described above. The resulting atomic patterns can be seen in Fig. 13.7, showing that almost arbitrary atomic orderings can be produced in this way. The described scheme can be enhanced to allow for simultaneous addressing of multiple lattice sites using an intensity shaped laser beam instead of a focused Gaussian beam. Such a beam can be prepared in the lab using spatial light modulators [44].

13.6 Outlook

The preparation of ultracold atoms in optical lattices marked the first realization of strong correlations physics with cold quantum gases. Today, this research direction

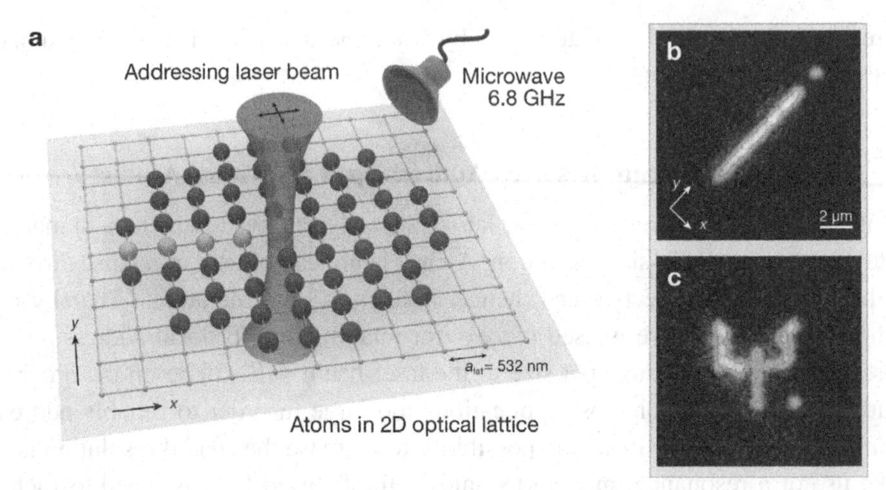

Figure 13.7 High-resolution addressing of single atoms. (a) Atoms in a Mott insulator with unity filling arranged on a square lattice with period $a_{\text{lat}} = 532\,\text{nm}$ were addressed using an off-resonant laser beam. The beam was focused onto individual lattice sites by a high-aperture microscope objective (not shown) and could be moved in the *xy* plane with an accuracy of better than $0.1\,a_{\text{lat}}$. (b,c) Fluorescence images of spin-flipped atoms following the addressing procedure. Reprinted with permission from Weitenberg, C., et al. (2011), Single-spin addressing in an atomic Mott insulator, *Nature*, **471**, 319 [18]. Copyright (2011) by the Nature Publishing Group.

offers many novel applications in various fields of physics ranging from condensed matter physics [3] to statistical physics [58] to quantum field theories of high-energy physics [59, 60]. The new probing and control techniques introduced by single atoms have allowed for completely new ways of characterizing nonlocal quantum correlations in these systems that had seemed useful only as theoretical concepts previously [35, 36]. The possibility of revealing hidden order parameters of topological phases of matter in higher dimensions [61, 62], the ability to measure the full counting statistics in a many-body setting, or the possibility to directly measure entanglement entropies [63, 64, 65] will open new avenues for our understanding of correlated quantum phases of matter. In addition to new observation and control techniques, the formation of quantum gases with long-range interactions [66, 67, 68] promises to significantly extend the regime of quantum simulations with ultracold quantum gases. Engineering of topological Bloch bands with ultracold atoms offers another exciting route [69, 70, 71, 72, 73, 74] for probing new strong correlation phenomena under the action of extremely strong effective magnetic fields (see also Chapter 14). Finally, understanding nonequilibrium quantum dynamics and novel interacting localization phenomena in high-energy states such as many-body localization [75, 76, 77] is ideally suited for cold quantum gas experiments, owing to their almost perfect isolation from the environment.

References

[1] Jaksch, D., and Zoller, P. 2005. The cold atoms Hubbard toolbox. *Ann. Phys.*, **315**, 52.

[2] Bloch, I., Dalibard, J., and Zwerger, W. 2008. Many-body physics with ultracold gases. *Rev. Mod. Phys.*, **80**, 885.

[3] Lewenstein, M., Sanpera, A., Ahufinger, V., Damski, B., De, A. Sen, and Sen, U. 2007. Ultracold atomic gases in optical lattices: mimicking condensed matter physics and beyond. *Adv. Phys.*, **56**, 243–379.

[4] Fisher, M. P. A., Weichman, P. B., Grinstein, G, and Fisher, D. S. 1989. Boson localization and the superfluid-insulator transition. *Phys. Rev. B*, **40**, 546–570.

[5] Jaksch, D., Bruder, C., Cirac, J. I., Gardiner, C. W., and Zoller, P. 1998. Cold bosonic atoms in optical lattices. *Phys. Rev. Lett.*, **81**, 3108–3111.

[6] Greiner, M., Mandel, O., Esslinger, T., Hänsch, T. W., and Bloch, I. 2002. Quantum phase transition from a superfluid to a Mott insulator in a gas of ultracold atoms. *Nature*, **415**, 39–44.

[7] Jördens, R., Strohmaier, N., Günter, K., Moritz, H., and Esslinger, T. 2008. A Mott insulator of fermionic atoms in an optical lattice. *Nature*, **455**, 204–207.

[8] Schneider, U., Hackermüller, L., Will, S., Best, Th., Bloch, I., Costi, T. A. A, Helmes, R. W., Rasch, D., and Rosch, A. 2008. Metallic and insulating phases of repulsively interacting fermions in a 3D optical lattice. *Science*, **322**, 1520–1525.

[9] Paredes, B., Widera, A., Murg, V., Mandel, O., Fölling, S., Cirac, J. I., Shlyapnikov, G. V., Hänsch, T. W., and Bloch, I. 2004. Tonks-Girardeau gas of ultracold atoms in an optical lattice. *Nature*, **429**, 277–281.

[10] Kinoshita, T., Wenger, T., and Weiss, D. S. 2004. Observation of a one-dimensional Tonks-Girardeau gas. *Science*, **305**, 1125–1128.

[11] Randeria, M., Zwerger, W., and Zwierlein, M. (eds). 2012. *The BCS-BEC Crossover and the Unitary Fermi Gas*. Lecture Notes in Physics, vol. 836. Springer.

[12] Chin, C., Grimm, R., Julienne, P., and Tiesinga, E. 2010. Feshbach resonances in ultracold gases. *Rev. Mod. Phys.*, **82**, 1225–1286.

[13] Nelson, K. D., Li, X., and Weiss, D. S. 2007. Imaging single atoms in a three-dimensional array. *Nat. Phys.*, **3**, 556–560.

[14] Gericke, T., Würtz, P., Reitz, D., Langen, T., and Ott, H. 2008. High-resolution scanning electron microscopy of an ultracold quantum gas. *Nature Phys.*, **4**, 949–953.

[15] Bakr, W. S., Gillen, J. I., Peng, A., Fölling, Si., and Greiner, M. 2009. A quantum gas microscope for detecting single atoms in a Hubbard-regime optical lattice. *Nature*, **462**, 74–77.

[16] Bakr, W. S., Peng, A., Tai, M. E., Ma, R., Simon, J., Gillen, J. I., Fölling, S., Pollet, L., and Greiner, M. 2010. Probing the superfluid-to-Mott insulator transition at the single-atom level. *Science*, **329**, 547–550.

[17] Sherson, Jacob F., Weitenberg, Christof, Endres, Manuel, Cheneau, Marc, Bloch, Immanuel, and Kuhr, Stefan. 2010. Single-atom-resolved fluorescence imaging of an atomic Mott insulator. *Nature*, **467**, 68–72.

[18] Weitenberg, C., Endres, M., Sherson, J. F., Cheneau, M., Schauß, P., Fukuhara, T., Bloch, I., Kuhr, S., and Schauss, P. 2011. Single-spin addressing in an atomic Mott insulator. *Nature*, **471**, 319–324.

[19] Trotzky, S., Cheinet, P., Fölling, S., Feld, M., Schnorrberger, U., Rey, A. M. M, Polkovnikov, A., Demler, E. A. A, Lukin, M. D. D, and Bloch, I. 2008. Time-resolved observation and control of superexchange interactions with ultracold atoms in optical lattices. *Science*, **319**, 295–299.

[20] Trotzky, S., Chen, Y.-A., Schnorrberger, U., Cheinet, P., and Bloch, I. 2010. Controlling and detecting spin correlations of ultracold atoms in optical lattices. *Phys. Rev. Lett.*, **105**, 265303.

[21] Fukuhara, T., Schauß, P., Endres, M., Hild, S., Cheneau, M., Bloch, I., and Gross, C. 2013. Microscopic observation of magnon bound states and their dynamics. *Nature*, **502**, 76–79.

[22] Greif, D., Uehlinger, T., Jotzu, G., Tarruell, L., and Esslinger, T. 2013. Short-range quantum magnetism of ultracold fermions in an optical lattice. *Science*, **340**, 1307–10.

[23] Hart, R., Duarte, P., Yang, T., Liu, X., Paiva, T., Khatami, E., Scalettar, R. T., Trivedi, N., Huse, D. A., and Hulet, R. 2014. Observation of antiferromagnetic correlations in the Hubbard model with ultracold atoms. *Nature*, **519**, 211–214.

[24] Lee, P. A., Nagaosa, N., and Wen, X.-G. 2006. Doping a Mott insulator: physics of high-temperature superconductivity. *Rev. Mod. Phys.*, **78**, 17–85.

[25] Le Hur, K., and Maurice Rice, T. 2009. Superconductivity close to the Mott state: from condensed-matter systems to superfluidity in optical lattices. *Ann. Phys.*, **324**, 1452–1515.

[26] Hofstetter, W., Cirac, J. I., Zoller, P., Demler, E., and Lukin, M. D. 2002. High-temperature superfluidity of fermionic atoms in optical lattices. *Phys. Rev. Lett.*, **89**, 220407.

[27] Polkovnikov, A., Sengupta, K., Silva, A., and Vengalattore, M. 2011. Colloquium: nonequilibrium dynamics of closed interacting quantum systems. *Rev. Mod. Phys.*, **83**, 863–883.

[28] Trotzky, S., Chen, Y-a., Flesch, A., McCulloch, I. P., Schollwöck, U., Eisert, J., and Bloch, I. 2012. Probing the relaxation towards equilibrium in an isolated strongly correlated one-dimensional Bose gas. *Nature Phys.*, **8**, 325–330.

[29] Schneider, U., Hackermüller, L., Ronzheimer, J.-P., Will, S., Braun, S., Best, T., Bloch, I., Demler, E., Mandt, S., Rasch, D., and Rosch, A. 2012. Fermionic transport and out-of-equilibrium dynamics in a homogeneous Hubbard model with ultracold atoms. *Nature Phys.*, **8**, 213–218.

[30] Ronzheimer, J. P., Schreiber, M., Braun, S., Hodgman, S. S., Langer, S., McCulloch, I. P., Heidrich-Meisner, F., Bloch, I., and Schneider, U. 2013. Expansion dynamics of interacting bosons in homogeneous lattices in one and two dimensions. *Phys. Rev. Lett.*, **110**, 205301.

[31] Hild, S., Fukuhara, T., Schauß, P., Zeiher, J., Knap, M., Demler, E., Bloch, I., and Gross, C. 2014. Far-from-equilibrium spin transport in Heisenberg quantum magnets. *Phys. Rev. Lett.*, **113**, 147205.

[32] Strohmaier, N., Greif, D., Jördens, R., Tarruell, L., Moritz, H., and Esslinger, T. 2010. Observation of elastic doublon decay in the Fermi-Hubbard model. *Phys. Rev. Lett.*, **104**, 080401.

[33] Hung, C.-L., Zhang, X., Gemelke, N., and Chin, C. 2010. Slow mass transport and statistical evolution of an atomic gas across the superfluid-Mott-insulator transition. *Phys. Rev. Lett.*, **104**, 160403.

[34] Gerbier, F., Widera, A., Fölling, S., Mandel, O., Gericke, T., and Bloch, I. 2005. Phase coherence of an atomic Mott insulator. *Phys. Rev. Lett.*, **95**, 050404.

[35] Endres, M., Cheneau, M., Fukuhara, T., Weitenberg, C., Schauss, P., Gross, C., Mazza, L., Banuls, M. C., Pollet, L., Bloch, I., and Kuhr, S. 2011. Observation of correlated particle–hole pairs and string order in low-dimensional Mott insulators. *Science*, **334**, 200–203.

[36] Endres, M., Cheneau, M., Fukuhara, T., Weitenberg, C., Schauß, P., Gross, C., Mazza, L., Bañuls, M. C., Pollet, L., Bloch, I., and Kuhr, S. 2013. Single-site- and single-atom-resolved measurement of correlation functions. *Appl. Phys. B*, **113**, 27–39.

[37] Greiner, M., Mandel, O., Esslinger, T., Hänsch, T. W., and Bloch, I. 2002. Quantum phase transition from a superfluid to a Mott insulator in a gas of ultracold atoms. *Nature*, **415**, 39–44.

[38] Hubbard, J. 1963. Electron correlations in narrow energy bands. *Proc. Roy. Soc. A*, **276**, 238–257.

[39] Helmes, R., Costi, T., and Rosch, A. 2008. Mott transition of fermionic atoms in a three-dimensional optical trap. *Phys. Rev. Lett.*, **100**, 056403.

[40] Jördens, R., Tarruell, L., Greif, D., Uehlinger, T., Strohmaier, N., Moritz, H., Esslinger, T., De Leo, L., Kollath, C., Georges, A., Scarola, V., Pollet, L., Burovski, E., Kozik, E., and Troyer, M. 2010. Quantitative determination of temperature in the approach to magnetic order of ultracold fermions in an optical lattice. *Phys. Rev. Lett.*, **104**, 180401.

[41] Bernier, J.-S., Kollath, C., Georges, A., De Leo, L., Gerbier, F., Salomon, C., and Köhl, M. 2009. Cooling fermionic atoms in optical lattices by shaping the confinement. *Phys. Rev. A*, **79**, 061601(R).

[42] Ho, T.-L., and Zhou, Q. 2009. Universal cooling scheme for quantum simulation. *arXiv:0911.5506*, Nov.

[43] Auerbach, A. 1994. *Interacting Electrons and Quantum Magnetism*. New York: Springer-Verlag NY.

[44] Fukuhara, T., Kantian, A., Endres, M., Cheneau, M., Schauß, P., Hild, S., Bellem, D., Schollwöck, U., Giamarchi, T., Gross, C., Bloch, I., and Kuhr, S. 2013. Quantum dynamics of a mobile spin impurity. *Nature Phys.*, **9**, 235–241.

[45] Pauling, L. 1931. *J. Am. Chem. Soc.*, **53**, 1367.

[46] Hückel, E. 1931. Quantentheoretische Beiträge zum Benzolproblem. *Z. Phys. A*, **70**, 204.

[47] Anderson, P. 1973. Resonating valence bonds: a new kind of insulator? *Mat. Res. Bull.*, **8**, 153.

[48] Anderson, P. W. 1987. The resonating valence bond state in La_2CuO_4 and superconductivity. *Science*, **235**, 1196–1198.

[49] Paredes, Belén, and Bloch, I. 2008. Minimum instances of topological matter in an optical plaquette. *Phys. Rev. A*, **77**, 23603.

[50] Nascimbène, S., Chen, Y.-A., Atala, M., Aidelsburger, M., Trotzky, S., Paredes, B., and Bloch, I. 2012. Experimental realization of plaquette resonating valence-bond states with ultracold atoms in optical superlattices. *Phys. Rev. Lett.*, **108**, 205301.

[51] Trebst, S., Schollwöck, U., Troyer, M., and Zoller, P. 2006. d-wave resonating valence bond states of fermionic atoms in optical lattices. *Phys. Rev. Lett.*, **96**, 250402.

[52] Rey, A. M., Sensarma, R., Fölling, S., Greiner, M., Demler, E., and Lukin, M. D. 2009. Controlled preparation and detection of d-wave superfluidity in two-dimensional optical superlattices. *Europhys. Lett.*, **87**, 60001.

[53] Gemelke, N., Zhang, X., Hung, Ch.-L., and Chin, Ch. 2009. In situ observation of incompressible Mott-insulating domains in ultracold atomic gases. *Nature*, **460**, 995–998.

[54] Anfuso, F., and Rosch, A. 2007. String order and adiabatic continuity of Haldane chains and band insulators. *Phys. Rev. B*, **75**, 144420.

[55] Berg, E., Dalla Torre, E., Giamarchi, T., and Altman, E. 2008. Rise and fall of hidden string order of lattice bosons. *Phys. Rev. B*, **77**, 245119.

[56] DePue, M. T., McCormick, C., Winoto, S. L., Oliver, S., and Weiss, D. S. 1999. Unity occupation of sites in a 3D optical lattice. *Phys. Rev. Lett.*, **82**, 2262–2265.

[57] Weiss, D. S., Vala, J., Thapliyal, A. V., Myrgren, S., Vazirani, U., and Whaley, K. B. 2004. Another way to approach zero entropy for a finite system of atoms. *Phys. Rev. A*, **70**, 40302.

[58] Braun, S., Ronzheimer, J. P., Schreiber, M., Hodgman, S. S., Rom, T., Bloch, I., and Schneider, U. 2013. Negative absolute temperature for motional degrees of freedom. *Science*, **339**, 52–55.

[59] Podolsky, D., Auerbach, A., and Arovas, D. 2011. Visibility of the amplitude (Higgs) mode in condensed matter. *Phys. Rev. B*, **84**, 174522.

[60] Endres, M., Fukuhara, T., Pekker, D., Cheneau, M., Schauss, P., Gross, C., Demler, E., Kuhr, S., and Bloch, I. 2012. The "Higgs" amplitude mode at the two-dimensional superfluid/Mott insulator transition. *Nature*, **487**, 454–8.

[61] Wen, X. G. 2004. *Quantum Field Theory of Many-Body Systems*. Oxford Graduate Texts. Oxford: Oxford University Press.

[62] Rath, S. P., Simeth, W., Endres, M., and Zwerger, W. 2013. Non-local order in Mott insulators, duality and Wilson loops. *Annals of Physics*, **334**, 256–271.

[63] Alves, C., and Jaksch, D. 2004. Multipartite entanglement detection in bosons. *Phys. Rev. Lett.*, **93**, 1–4.

[64] Daley, A. J., Pichler, H., Schachenmayer, J., and Zoller, P. 2012. Measuring entanglement growth in quench dynamics of bosons in an optical lattice. *Phys. Rev. Lett.*, **109**, 020505.

[65] Pichler, H., Bonnes, L., Daley, A. J., Läuchli, A. M., and Zoller, P. 2013. Thermal versus entanglement entropy: a measurement protocol for fermionic atoms with a quantum gas microscope. *New J. Phys.*, **15**, 063003.

[66] Ni, K.-K., Ospelkaus, S., de Miranda, M. H. G., Pe'er, A., Neyenhuis, B., Zirbel, J. J., Kotochigova, S., Julienne, P. S., Jin, D. S., and Ye, J. 2008. A high phase-space-density gas of polar molecules. *Science*, **322**, 231–235.

[67] Schauß, P., Cheneau, M., Endres, M., Fukuhara, T., Hild, S., Omran, A., Pohl, T., Gross, C., Kuhr, S., and Bloch, I. 2012. Observation of spatially ordered structures in a two-dimensional Rydberg gas. *Nature*, **490**, 87–91.

[68] Schauß, P., Zeiher, J., Fukuhara, F., Hild, S., Cheneau, M., Macri, T., Pohl, T., Bloch, I., and Gross, C. 2015. Crystallization in Ising quantum magnets. *Science*, **347**, 1455–1458.

[69] Atala, M., Aidelsburger, M., Barreiro, J. T., Abanin, D. A., Kitagawa, T., Demler, E., and Bloch, I. 2013. Direct measurement of the Zak phase in topological Bloch bands. *Nature Phys.*, **9**, 795–800.

[70] Aidelsburger, M., Atala, M., Lohse, M., Barreiro, J. T., Paredes, B., and Bloch, I. 2013. Realization of the Hofstadter Hamiltonian with ultracold atoms in optical lattices. *Phys. Rev. Lett.*, **111**, 185301.

[71] Miyake, H., Siviloglou, G., Kennedy, C. J., Burton, W. C., and Ketterle, W. 2013. Realizing the Harper Hamiltonian with laser-assisted tunneling in optical lattices. *Phys. Rev. Lett.*, **111**, 185302.

[72] Atala, M., Aidelsburger, M., Lohse, M., Barreiro, J. T., Paredes, B., and Bloch, I. 2014. Observation of chiral currents with ultracold atoms in bosonic ladders. *Nature Phys.*, **10**, 13–15.

[73] Jotzu, G., Messer, M., Desbuquois, R., Lebrat, M., Uehlinger, T., Greif, D., and Esslinger, T. 2014. Experimental realisation of the topological Haldane model. *Nature*, **515**, 237–240.

[74] Aidelsburger, M., Lohse, M., Schweizer, C., Atala, M., Barreiro, J. T., Nascimbène, S., Cooper, N. R., Bloch, I., and Goldman, N. 2015. Measuring the Chern number of Hofstadter bands with ultracold bosonic atoms. *Nature Phys.*, **11**, 162–166.

[75] Basko, D. M., Aleiner, I. L., and Altshuler, B. L. 2006. Metal-insulator transition in a weakly interacting many-electron system with localized single-particle states. *Annals of Physics*, **321**, 1126–1205.

[76] Gornyi, I. V., Mirlin, A. D., and Polyakov, D. G. 2005. Dephasing and weak localization in disordered Luttinger liquid. *Phys. Rev. Lett.*, **95**(4).

[77] Nandkishore, R., and Huse, D. A. 2015. Many-body localization and thermalization in quantum statistical mechanics. *Ann. Rev. Cond. Mat. Phys.*, **6**, 15–38.

14

Preparing and Probing Chern Bands with Cold Atoms

NATHAN GOLDMAN

Center for Nonlinear Phenomena and Complex Systems,
Université Libre de Bruxelles, Belgium

NIGEL R. COOPER

T.C.M. Group, Cavendish Laboratory,
University of Cambridge, UK

JEAN DALIBARD

Laboratoire Kastler Brossel, Collège de France, CNRS,
ENS-PSL Research University, UPMC-Sorbonne Universités, Paris, France

The present chapter discusses methods by which topological Bloch bands can be prepared in cold-atom setups. Focusing on the case of Chern bands for two-dimensional systems, we describe how topological properties can be triggered by driving atomic gases, either by dressing internal levels with light or through time-periodic modulations. We illustrate these methods with concrete examples, and we discuss recent experiments where geometrical and topological band properties have been identified.

14.1 Introduction

Ultracold atoms constitute a promising physical platform for the preparation and exploration of novel states of matter [1, 2, 3]. In particular, the engineering of topological band structures with cold-atom systems, together with the capability of tuning interactions between the particles, opens an interesting route toward the realization of intriguing strongly correlated states with topological features, such as fractional topological insulators and quantum Hall liquids [4].

This chapter is dedicated to the preparation and the detection of topological band structures characterized by nonzero Chern numbers [5]. Such *Chern bands*, which constitute the building blocks for realizing (fractional) Chern insulators [6, 7], arise in two-dimensional (2D) systems presenting time-reversal-symmetry (TRS) breaking effects. For instance, nontrivial Chern bands naturally appear in the Harper-Hofstadter model [8], a lattice penetrated by a uniform flux, where they generalize the (nondispersive) Landau levels to the lattice framework. Additionally, Chern bands also appear in staggered flux configurations, such as in Haldane's honeycomb-lattice model [9] or in lattice systems combining Rashba spin-orbit coupling and Zeeman (exchange) fields.

The atoms being charge neutral, "magnetic" fluxes cannot be simply produced by subjecting optical lattices to "real" magnetic fields. It is the aim of this chapter to review several schemes that have been recently implemented in laboratories with the goal of realizing synthetic magnetic fields leading to Chern bands for cold atoms. Our presentation is complementary to that of Chapter 15, as we focus on synthetic magnetic fields in lattice-based systems for which the flux density can be made very large (with magnetic length comparable to interparticle spacing). The chapter is structured as follows: Section 14.2 describes how the Chern number is related to physical observables defined in a lattice framework. In particular, it clarifies the link between recent Chern-number measurements performed in cold bosonic gases and the more conventional (electronic) quantum Hall effect. Section 14.3 reviews a few lattice models, and it relates their properties to the well-known Landau levels of the continuum. The Section 14.4 describes diverse schemes by which Chern bands can be prepared and probed for cold atoms; a special emphasis is placed upon experimentally realized schemes. The last Section 14.5 is devoted to final remarks and discussions.

14.2 Topological Characteristics of Two-Dimensional Bloch Bands

The Chern number – a topological invariant $\nu_{\text{ch}} \in \mathbb{Z}$ classifying fiber bundles – naturally enters the description of particles moving in two-dimensional lattices, where it offers an elegant interpretation for the (anomalous) quantum Hall effect [10, 11, 9]. This section relates this topological invariant to physical observables and discusses methods to measure it in experiments.

14.2.1 Bloch Bands, the Anomalous Velocity, and the Chern Number

The general problem of a particle subjected to a 2D space-periodic potential $U(\boldsymbol{r}+\boldsymbol{a}) = U(\boldsymbol{r})$ starts by invoking Bloch's theorem, which stipulates that the eigenstates of the system can be decomposed as $\psi_{\lambda\boldsymbol{k}}(\boldsymbol{r}) = \exp(i\boldsymbol{r}\cdot\boldsymbol{k})\, u_{\lambda\boldsymbol{k}}(\boldsymbol{r})$, where $u_{\lambda\boldsymbol{k}}(\boldsymbol{r})$ has the periodicity of the lattice $u_{\lambda\boldsymbol{k}}(\boldsymbol{r}+\boldsymbol{a}) = u_{\lambda\boldsymbol{k}}(\boldsymbol{r})$ and where $\boldsymbol{k} = (k_x, k_y)$ is the quasimomentum. The associated eigenenergies $E_\lambda(\boldsymbol{k})$ display bands, labeled by the index λ, over the first Brillouin zone (FBZ) of the quasimomentum $\boldsymbol{k}$. In the absence of additional potentials, a state $u_{\lambda\boldsymbol{k}}$ in a given Bloch band λ with quasimomentum $\boldsymbol{k}$ is characterized by the averaged velocity $\boldsymbol{v}_\lambda(\boldsymbol{k}) \equiv \langle u_{\lambda\boldsymbol{k}}|\hat{\boldsymbol{v}}|u_{\lambda\boldsymbol{k}}\rangle = (1/\hbar)\partial E_\lambda(\boldsymbol{k})/\partial\boldsymbol{k}$, where $\hat{\boldsymbol{v}}$ denotes the velocity operator.

Subjecting the lattice system to a constant force F generates Bloch oscillations. Indeed, restricting the dynamics to one dimension (1D) for now, and considering for simplicity a semiclassical approach, the system is described by the equations of motion

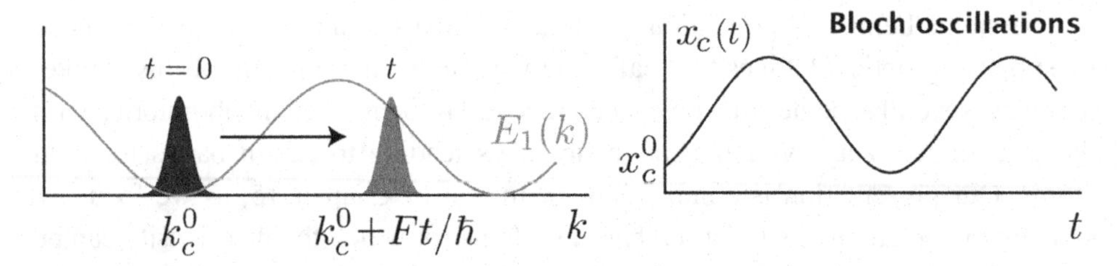

Figure 14.1 Bloch oscillations for a wave packet initially prepared in the lowest band E_1, with center-of-mass position x_c^0 and momentum k_c^0. The trajectory $x_c(t)$ results from the semiclassical equations of motion (14.1).

$$\hbar \dot{x}_c(t) = \hbar v_\lambda(k_c) = \frac{\partial E_1(k_c)}{\partial k_c}, \quad \hbar \dot{k}_c(t) = F, \tag{14.1}$$

where $x_c(t)$ and $\hbar k_c(t)$ denote the center-of-mass position and momentum of a wave packet prepared in the lowest band E_1 (i.e., $\lambda = 1$); see Fig. 14.1. The equations of motion (14.1) are valid when the force F is weak enough to preclude any interband transitions.

When considering 2D lattices, an interesting effect adds to the standard Bloch oscillations, which involves the Berry curvature of the band [12]. Indeed, applying a force along a given direction, e.g., $\boldsymbol{F} = F_y \mathbf{1}_y$, modifies the averaged velocity $v_\lambda(\boldsymbol{k})$ along the transverse (x) direction according to

$$v_\lambda^x(\boldsymbol{k}) = \frac{\partial E_\lambda(\boldsymbol{k})}{\hbar \partial k_x} - \frac{F_y}{\hbar}\Omega_\lambda(\boldsymbol{k}), \tag{14.2}$$

$$\Omega_\lambda(\boldsymbol{k}) = i\left(\langle \partial_{k_x} u_{\lambda \boldsymbol{k}} | \partial_{k_y} u_{\lambda \boldsymbol{k}} \rangle - \langle \partial_{k_y} u_{\lambda \boldsymbol{k}} | \partial_{k_x} u_{\lambda \boldsymbol{k}} \rangle\right), \tag{14.3}$$

where $\Omega_\lambda(\boldsymbol{k})$ denotes the Berry curvature of the band λ. Thus, the velocity in a state $u_\lambda(\boldsymbol{k})$ has two contributions: the usual *band* velocity, responsible for Bloch oscillations (Fig. 14.1); and the so-called *anomalous* velocity, which can produce a net drift transverse to the applied force. As for Eq. (14.1), the general result (14.2) assumes the absence of interband transitions (i.e., weak-force regime).

The anomalous (transverse) velocity in Eq. (14.2) can be isolated and observed experimentally by canceling any contribution from the band velocity, which can otherwise dominate. This can be achieved by comparing trajectories for opposite forces, $\pm F_y$, noting that the anomalous velocity changes sign under reversal of the force [13]. Alternatively, the average anomalous velocity can be isolated by uniformly populating the bands, namely by averaging the velocity in Eq. (14.2) over the entire FBZ, since

$$\int_{\text{FBZ}} \left(\partial E_\lambda(\boldsymbol{k})/\partial k_x\right) d^2k = 0, \tag{14.4}$$

due to the periodicity of the energies in k-space.

We now compute the total transverse velocity in the case of uniformly populated bands using Eq. (14.2); explicit physical implementations will be discussed in the next Sections 14.2.2–14.2.3. In the following, we consider a general square lattice system of size $A_{\text{syst}} = L_x \times L_y$, characterized by a unit cell of size $A_{\text{cell}} = d_x a \times d_y a$, where a denotes the primitive lattice spacing.[1] The number of states within each band is $N_{\text{states}} = A_{\text{syst}}/A_{\text{cell}}$. We write the total number of particles as $N_{\text{tot}} = \sum_\lambda N^{(\lambda)}$, where $N^{(\lambda)}$ is the number of particles occupying a given band λ. We now make the assumption that each band is populated homogeneously, so that the average number of particles in a state $u_{\lambda \boldsymbol{k}}$ is uniform over the FBZ, and it is given by

$$\rho^{(\lambda)}(\boldsymbol{k}) = \rho^{(\lambda)} = \frac{N^{(\lambda)}}{N_{\text{states}}}. \tag{14.5}$$

Using Eqs. (14.2) and (14.5), we obtain the total averaged velocity along the direction transverse to the force

$$v_{\text{tot}}^x = \sum_\lambda \rho^{(\lambda)} \sum_{\boldsymbol{k}} v_\lambda^x(\boldsymbol{k}) = -\frac{F_y A_{\text{cell}}}{h} \sum_\lambda N^{(\lambda)}\, \nu_{\text{ch}}^{(\lambda)}, \tag{14.6}$$

$$\nu_{\text{ch}}^{(\lambda)} = \frac{1}{2\pi} \sum_{\boldsymbol{k}} \Omega_\lambda(\boldsymbol{k})\, \Delta k_x \Delta k_y \longrightarrow \frac{1}{2\pi} \int_{\text{FBZ}} \Omega_\lambda(\boldsymbol{k})\, d^2k \in \mathbb{Z}, \tag{14.7}$$

where $\Delta k_{x,y} = 2\pi/L_{x,y}$. The latter equations reveal that the total velocity is related to the quantities $\nu_{\text{ch}}^{(\lambda)}$, which converge toward the Chern number of the bands λ when taking the thermodynamic limit $L_{x,y} \to \infty$. The Chern number $\nu_{\text{ch}}^{(\lambda)}$ is an integer, obtained by averaging the Berry curvature $\Omega_\lambda(\boldsymbol{k})$ over the FBZ (Eq. (14.7)); it is a topological invariant, meaning that $\nu_{\text{ch}}^{(\lambda)}$ remains a constant as long as the spectral gaps to other bands do not vanish; see Refs. [10, 11]. As announced above, any contribution from the band velocity cancels under the homogeneous-population condition (14.5), as a direct consequence of Eq. (14.4).

The following Sections 14.2.2–14.2.3 discuss two different physical realizations that revealed the Chern numbers $\nu_{\text{ch}}^{(\lambda)}$ in experiments, through the homogeneous population of energy bands (Fig. 14.2). These sections aim to clarify the link between the quantum Hall effect [10, 11, 5], as observed in electronic systems since the 1980s, and the Chern-number measurement recently performed with ultracold bosonic atoms [15].

14.2.2 Fermions, the Quantum Hall Effect, and the TKNN Formula

The first situation that we consider is a 2D noninteracting polarized Fermi gas at zero temperature (Fig. 14.2). Setting the Fermi energy E_{F} within a spectral gap

[1] For instance, $d_x = d_y/2 = 1$ for the brick-wall lattice [14], and $d_x = d_y = 2$ for the magnetic unit cell of a square lattice penetrated by a uniform flux $\Phi = \pi/2$ per plaquette [15].

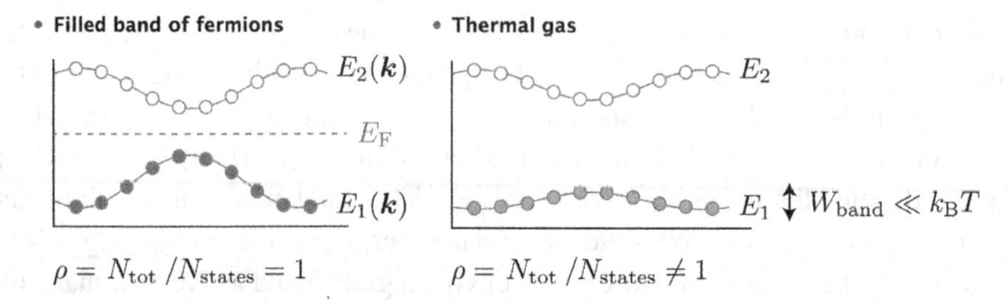

Figure 14.2 Two realizations of uniformly populated energy bands, suitable to reveal the Chern number in experiments. Left: Considering fermions at zero temperature, the band is perfectly filled by setting the Fermi energy within the spectral gap. Right: A system of bosons uniformly populate the band when the temperature is large compared with the bandwidth W_{band}, but small compared with the gap.

naturally leads to a perfect filling of the bands $E_\lambda < E_F$ located below the gap: the average number of particles in a state $u_\lambda(\boldsymbol{k})$, Eq. (14.5), is exactly $\rho^{(\lambda)} = N^{(\lambda)}/N_{states} = 1$ for $E_\lambda < E_F$. Setting the latter condition into Eq. (14.6) yields

$$v^x_{tot} = -\frac{F_y A_{syst}}{h} \sum_{E_\lambda < E_F} \nu^{(\lambda)}_{ch}, \tag{14.8}$$

which indicates that the total velocity of the Fermi gas is directly related to the sum of Chern numbers associated with populated bands.

In systems presenting time-reversal symmetry, the Berry curvature satisfies $\Omega_\lambda(\boldsymbol{k}) = -\Omega_\lambda(-\boldsymbol{k})$, in which case all the Chern numbers $\nu^{(\lambda)}_{ch} = 0$; see Eq. (14.7). Hence, observing the transverse drift associated with v^x_{tot} requires a system without time-reversal symmetry, which is the case, e.g., in electronic systems subjected to magnetic fields [5]. In these electronic setups, the transverse transport predicted by Eq. (14.8) is measured through the Hall conductivity σ_{xy}, which relates the electric field E_y to the current density $j_x = e v^x_{tot}/A_{syst}$, where e is the electron charge. Using these definitions together with Eq. (14.8), we recover the well-known Thouless-Kohmoto-Nightingale-den Nijs (TKNN) formula [10] for the electric Hall conductivity

$$j_x = \sigma_{xy} E_y, \quad \sigma_{yx} = \frac{e^2}{h} \sum_{E_\lambda < E_F} \nu^{(\lambda)}_{ch} = -\sigma_{xy}, \tag{14.9}$$

where we introduced the electric field[2] $E_y = F_y/e$ acting on the electrons. The TKNN formula (14.9) expresses the fact that transport measurements directly reveal

[2] In real quantum Hall samples the electric field is not uniform, being affected by charging of the edge states and by screening (which can be inhomogeneous due to disorder), leading to a nonuniform current. Because of topological robustness these effects, which are absent for atomic gases, do not affect the quantization of the Hall conductance.

the Chern numbers $\nu_{\text{ch}}^{(\lambda)}$ in electronic systems, through the quantization of the Hall conductivity. In particular, the quantized value $\sigma_{yx} = (e^2/h) \times$ (integer) remains constant as long as the Fermi energy stays in an open gap [10, 11]. Note that populating a Chern band with a Fermi gas is not limited to electronic systems, as it could also be performed by trapping fermionic atoms in an optical lattice subjected to artificial magnetic fields [16].

14.2.3 Thermal Bose Gas and the Center-of-Mass Displacement

Let us now consider a radically different configuration: a thermal gas of noninteracting polarized bosons, whose temperature is large compared with the bandwidth W_{band} of the lowest band E_1, but small compared with the spectral gap Δ above it (Fig. 14.2). In this case, the average number of bosons in a state of the lowest band is homogeneous but density dependent, $\rho^{(1)} = N_{\text{tot}}/N_{\text{states}} \neq 1$, and $\rho^{(\lambda)} = 0$ for $\lambda > 1$. The homogeneity of the band filling, which is a reasonable assumption when the lowest band presents a large flatness ratio $\Delta/W_{\text{band}} \gg 1$, can be tested in cold-atom experiments using band-mapping techniques [1, 15]. Setting the particle filling condition into Eq. (14.6) yields

$$v_{\text{tot}}^x = -\frac{F_y A_{\text{syst}}}{h} \rho\, \nu_{\text{ch}}^{(1)}, \quad \rho = \frac{N_{\text{tot}}}{N_{\text{states}}} \neq 1, \tag{14.10}$$

which differs from the Fermi-gas result in Eq. (14.8) in that only the lowest band contributes, but also through the additional density dependence.

The Bose gases considered in cold-atom experiments are charge neutral. However, in analogy with the quantum Hall effect discussed in Section 14.2.2, one could consider a transport measurement relating the particle current density $j_x = v_{\text{tot}}^x/A_{\text{syst}}$ to the applied force F_y. In this case, using Eq. (14.10), we find that the analogue of the Hall conductivity would read

$$j_x = \sigma_{xy} F_y, \quad \sigma_{xy} = -\frac{1}{h} \rho\, \nu_{\text{ch}}^{(1)}. \tag{14.11}$$

In contrast with the TKNN formula (14.9), the relation between the measured transport coefficient σ_{xy} and the topological Chern number involves a density-dependent factor ρ/h. Thus, in such an experiment, identifying the Chern number of the lowest band would require combining simultaneous transport and density measurements, which constitutes a severe drawback of this approach. To overcome this difficulty, one can consider another physical observable: the center-of-mass (CM) displacement of the gas. Indeed, using Eq. (14.10), we find that the transverse velocity of the CM is given by

$$v_{\text{CM}}^x = \frac{v_{\text{tot}}^x}{N_{\text{tot}}} = -\frac{F_y A_{\text{cell}}}{h} \nu_{\text{ch}}^{(1)}. \tag{14.12}$$

Since both the unit cell area $A_{\rm cell}$ and the strength of the applied force F_y can be determined with precision, the transverse displacement of the CM $\Delta x_{\rm CM}(t) = v^x_{\rm CM} t$ offers a direct measure of the Chern number of the lowest band. This Chern-number measurement was successfully implemented in Munich [15] in 2014; see Section 14.4.

14.3 Models: Harper-Hofstadter Bands versus Haldane-like Bands

We now turn to the presentation of two lattice models that have been recently implemented with cold-atom setups and that lead to a topologically nontrivial lowest band. For the sake of completeness, we start our discussion with a short reminder of the well-known Landau levels, which characterize the quantum motion of a free particle in a uniform magnetic field. This Landau-level structure, with its nontrivial bulk topological features and the associated edge currents, is crucial to explain quantum-Hall-type phenomena, in both the integer and the fractional cases.

14.3.1 Landau Levels

The Landau-level spectrum [17] emerges when one looks for the eigenstates of the Hamiltonian $\hat{H} = (\hat{\boldsymbol{p}} - q\boldsymbol{A}(\hat{\boldsymbol{r}}))^2/2m$, where $\hat{\boldsymbol{p}}$ (and $\hat{\boldsymbol{r}}$), respectively, denote the canonical momentum and position operators for a particle of charge q and mass m. The vector potential $\boldsymbol{A}(\boldsymbol{r})$, defined up to a gauge transformation, satisfies $\nabla \times \boldsymbol{A} = \boldsymbol{B}$, where $\boldsymbol{B}$ is uniform. In order to derive the spectrum of $\hat{H}$, we note that it can be written as $\hat{H} = (\hat{\Pi}_x^2 + \hat{\Pi}_y^2)/2m$, where we introduced the kinetic momentum $\hat{\boldsymbol{\Pi}} = \hat{\boldsymbol{p}} - q\boldsymbol{A}(\hat{\boldsymbol{r}})$. The kinetic momentum operators satisfy the simple commutation relation $[\hat{\Pi}_x, \hat{\Pi}_y] = i\hbar qB$. The corresponding operator algebra is thus formally equivalent to that of a harmonic oscillator, where $\hat{H} = (\hat{X}^2 + \hat{P}^2)/2$ with $[\hat{X}, \hat{P}] = i$. We infer from this equivalence that the energy levels form an equidistant set $\hbar\omega_c(n + 1/2)$, where $\omega_c = qB/m$ denotes the cyclotron frequency and n is a nonnegative integer. Each Landau level has a macroscopic degeneracy $A_{syst}/(2\pi\ell^2)$, where A_{syst} is the area of the sample and $\ell = (\hbar/qB)^{1/2}$ is the magnetic length. One thus obtains a bandlike spectrum as for a particle in a periodic lattice, each band being infinitely narrow. Each Landau level has a Chern number equal to 1, so that the notion of anomalous velocity is also relevant here. Actually, it has in this case a simple classical interpretation in terms of Hall current. It is indeed well known that when a particle in a magnetic field along z is acted upon by an additional uniform force F_y, its motion consists in the combination of the circular cyclotron motion and a uniform translation motion at the constant velocity $v_x = F_y/(qB)$; the latter is nothing but the anomalous velocity described above.

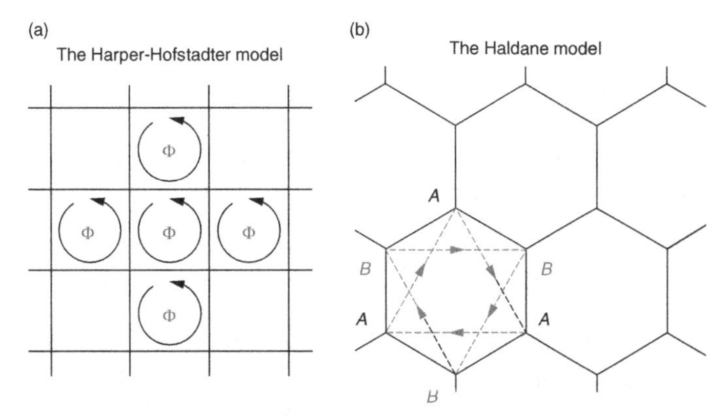

Figure 14.3 Two lattice models leading to topologically nontrivial bands. (a) The Harper-Hofstadter model: a square lattice operating in the tight-binding regime is placed in a uniform field, with the flux Φ across each unit cell of the lattice [18, 8]. (b) The Haldane model: a periodic honeycomb lattice with nearest-neighbor couplings $A \leftrightarrow B$ (solid lines) corresponding to a real tunneling matrix element, and next-to-nearest neighbor couplings $A \leftrightarrow A$ and $B \leftrightarrow B$ (dashed lines) corresponding to a nonreal tunneling matrix element. The arrow directions encode the sign of the argument of this matrix element [9].

14.3.2 Lattices and Tight-Binding Models

We now turn to the Harper-Hofstadter problem [18, 8], which is the transposition of the Landau problem to a discretized space. It models the motion of a charged particle in a square two-dimensional lattice of period a, normal to the uniform magnetic field (Fig. 14.3a). This motion is described using the single band approximation, assuming that only nearest neighbor couplings are relevant (Hubbard model). The quantum state corresponding to the particle localized on the lattice site $\boldsymbol{r} = a(m, n)$ is denoted $|\boldsymbol{r}\rangle$, with m, n integers. The Hamiltonian reads in zero magnetic field

$$\hat{H}_0 = -J \sum_{\langle \boldsymbol{r},\boldsymbol{r}'\rangle} \hat{a}^\dagger_{\boldsymbol{r}'} \hat{a}_{\boldsymbol{r}}, \tag{14.13}$$

where $\hat{a}_{\boldsymbol{r}}$ annihilates a particle in the state $|\boldsymbol{r}\rangle$ and $J > 0$ is the tunneling matrix element. The eigenstates of this Hamiltonian are the Bloch functions $|\psi_{\boldsymbol{k}}\rangle = \sum_{\boldsymbol{r}} e^{i\boldsymbol{k}\cdot\boldsymbol{r}} |\boldsymbol{r}\rangle$ with the energy $E(\boldsymbol{k}) = -2J\left[\cos(k_x a) + \cos(k_y a)\right]$, corresponding to an allowed band of width $8J$.

The presence of the magnetic field is taken into account in the single-band approximation by assigning a complex value to the tunneling matrix elements between adjacent sites. The individual phases of these matrix elements have to be chosen such that the total Aharonov-Bohm phase [19] accumulated along a closed contour is equal to $2\pi\Phi/\Phi_0$, where Φ is the flux of the magnetic field through the contour and $\Phi_0 = h/q$ is the flux quantum [8, 3]. Because of gauge invariance, there are an infinite number of choices for individual phases matching this prescription. Here we choose the Landau gauge, which amounts to taking zero phase (a real matrix element) for tunneling along the y direction, and the y-dependent phase $2\pi\alpha n$ for the tunneling along the x direction, where we introduced the dimensionless flux parameter $\alpha = \Phi/\Phi_0$. The Hamiltonian is now

$$\hat{H}_\alpha = -J \sum_{m,n} \left(\hat{a}^\dagger_{m,n+1} \hat{a}_{m,n} + e^{i2\pi\alpha n} \hat{a}^\dagger_{m+1,n} \hat{a}_{m,n} \right) + \text{h.c.} \tag{14.14}$$

which is periodic in α with period 1. The term $\exp(i2\pi\alpha n)$, which captures the effect of the magnetic field, is generally referred to as the Peierls phase-factor [18, 8].

The single-particle spectrum plotted as a function of α acquires a fractal structure known as the *Hofstadter butterfly*. The calculation of this spectrum for an arbitrary value of α is a difficult mathematical task. If one restricts to rational values $\alpha = p/q$ (p and q relatively prime integers), the problem is simpler since one recovers a two-dimensional periodic problem, with a period a along x and qa along y. The corresponding unit cell (the so-called *magnetic cell*) now contains q sites, and the energy band of width $8J$ for $\Phi = 0$ splits in q nonoverlapping subbands (two adjacent subbands may touch each other via Dirac points). In general, each subband has a nonzero Chern index, which can be obtained via the solution of a Diophantine equation [10], whereas the sum of all Chern indices over the q subbands is null. In the particular case $\alpha = 1/q$, the Chern index of the lowest band is 1, so that this band is topologically equivalent to the lowest Landau level found in the absence of a lattice.

In the Harper-Hofstadter problem, the presence of the uniform magnetic field breaks the translational symmetry of the initial lattice. Haldane proposed in 1988 another model [9], in which a nontrivial band topology could appear without any modification of the lattice unit cell nor breaking of the translational symmetry. The starting point is a graphene-like honeycomb lattice with nearest-neighbor couplings (Fig. 14.3b). The unit cell thus consists of two equivalent sites, denoted A and B, with tunneling from sites A (resp. B) to the three neighboring sites B (resp. A) with equal amplitude. At this stage, the single-particle spectrum consists of two subbands, touching each other in Dirac points. Haldane's crucial insight was to add nonreal next-to-nearest-neighbor (NNN) couplings, i.e., $A \to A$ and $B \to B$, which break time-reversal symmetry. By contrast to the Harper-Hofstadter model, these couplings are constant over the whole lattice, and could in principle be created by a staggered magnetic field with a zero flux through the honeycomb unit cell. These additional couplings lift the initial degeneracy at the Dirac points, and the two subbands are now separated by a gap with nonzero opposite Chern indices, $+1$ and -1. With the lowest band filled with spinless noninteracting fermions,

Haldane's model constitutes a prototype of a Chern insulator that goes beyond quantum Hall setups. It played a key role for the subsequent discovery of time-reversal invariant topological insulators, where spin-orbit coupling replaces the complex next-to-nearest-neighbour tunnel coupling. The optical flux lattice concept that will be presented later in this chapter is also directly related to Haldane's model, since it provides topologically nontrivial energy bands for configurations where the atom–light interaction is periodic.

14.4 Implementation: Driving Atoms into Topological Matter

This section describes several schemes realizing Chern bands in cold-atom systems. For the sake of clarity, this section makes the distinction between methods based on a tight-binding approach (Section 14.4.1) and those applicable in the continuum or weak-lattice regimes (Section 14.4.2). A special emphasis is set upon recent theoretical works and experimental implementations, which led to unambiguous signatures of topological properties associated with nonzero Chern numbers in 2D optical lattices.

14.4.1 Schemes Based on a Tight-Binding Approach

Noninteracting cold atoms moving in a deep optical lattice [1] are well described by the single-band tight-binding Hamiltonian in Eq. (14.13). Starting with this topologically trivial tight-binding band, it appears that a natural way to generate Chern bands consists in implementing the Harper-Hofstadter or the Haldane model (Section 14.3.2), which can be achieved by controlling the tunneling matrix elements within the optical-lattice setup; see Eq. (14.14). Indeed, as was discussed in Section 14.3.2, Peierls phase factors can be associated with (local) magnetic fluxes, which break TRS and potentially lead to Chern bands. We now review several methods by which Peierls phase factors and magnetic fluxes can be engineered using different aspects of cold-atom technology; see Fig. 14.4.

Using Internal States of the Atoms

A natural way to induce complex tunneling matrix elements in optical lattices is to exploit the spatial dependence of the optical phase in photon-assisted tunneling [20, 21, 22, 23]. Consider two internal states of an atom, denoted $|g\rangle$ and $|e\rangle$, respectively trapped in two independent (state-dependent) optical lattices $V_{g,e}$ along the x direction; see Fig. 14.4b. For large enough lattice potentials, and in the absence of coupling between the two states, the hopping is completely inhibited. Adding a coherent coupling, with wave vector $\boldsymbol{q}$ and frequency ω_{ge} matching the energy difference between the internal states, effectively activates tunneling

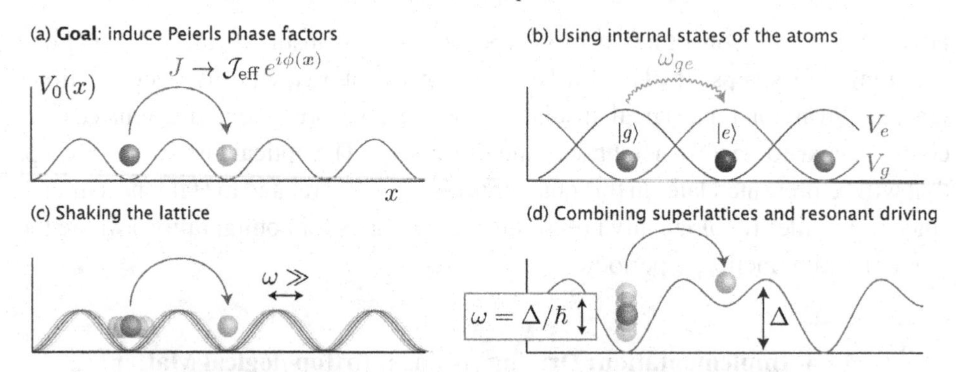

Figure 14.4 (a) Effective magnetic fluxes can be created in deep optical lattices by inducing complex tunneling matrix elements. (b) Method using state-dependent optical potentials $V_{g,e}$, trapping atoms in two internal states $|g\rangle$ and $|e\rangle$, combined with a resonant atom–light coupling with frequency $\omega_{ge} = (E_e - E_g)/\hbar$. (c) Shaking a lattice, with an arbitrarily large driving frequency ω. (d) Method using a superlattice with energy offset Δ, large compared with the bare hopping energy, combined with a resonant onsite energy modulation with frequency $\omega = \Delta/\hbar$.

processes between the two sublattices. The tunneling matrix elements between two nearest-neighboring lattice sites $\boldsymbol{r}_g$ and $\boldsymbol{r}_e$ are of the form $J(\boldsymbol{r}_g) = \mathcal{J}_{\text{eff}} \exp(i\boldsymbol{q} \cdot \boldsymbol{r}_g)$, where the amplitude $\mathcal{J}_{\text{eff}}$ is proportional to the coupling's Rabi frequency and to the overlap between Wannier functions defined at the two sites involved in the process [20, 21, 3, 2]. Hence, space-dependent Peierls phase factors, and the corresponding magnetic fluxes penetrating lattice plaquettes, can be controlled by tuning the coupling's wave vector $\boldsymbol{q}$. This method can be applied to square optical lattices, leading to an effective Harper-Hofstadter Hamiltonian (14.14), but also to triangular/honeycomb lattices in view of realizing the Haldane model [24, 25, 26]. The scheme is general and can be applied to diverse atom-light configurations: for instance, it may involve two hyperfine states in the ground-state manifold coupled through a (two-photon) Raman coupling [20], or a one-photon coupling between a ground state and a long-lived excited state [21]. This method has not yet been implemented experimentally.

The scheme described above concerns the coupling between atoms living on a 2D optical lattice. However, it is intimately related to the concept of "synthetic magnetic fields in synthetic dimensions" [27], which has been implemented at the European Laboratory for Nonlinear Spectroscopy (LENS) [28] and the National Institute of Standards and Technology (NIST) [29] in 2015. Here, atoms in M internal states live on a standard 1D optical lattice, characterized by real-valued tunneling matrix elements. Atom–light coupling then drives onsite transitions between the M internal states, such that each transition $m \to m+1$ is accompanied with a space-dependent phase $\boldsymbol{q} \cdot \boldsymbol{r}$ (i.e., a momentum transfer). Interpreting these internal-state transitions as hopping processes along a synthetic (internal-state) dimension, which is spanned by $m = 1, \dots, M$, this system is found to be equivalent to the effective Harper-Hofstadter system described above. The propagation of chiral edge states, an unambiguous signature of synthetic magnetic fluxes in 2D lattices, has been measured in these experiments [28, 29].

Superlattices and Resonant Time Modulations

Complex tunneling matrix elements can also be engineered without considering the internal structures of the atoms. In direct analogy with the scheme described above, this can be achieved by imposing an energy offset between neighboring sites of an optical lattice and modulating the system with a resonant time-periodic modulation of the onsite energies; see Fig. 14.4d. In this case, the Peierls phase factors are directly related to the phase of the modulation. To be explicit, let us consider a minimal lattice system: two lattice sites (a and b), separated in energy by a large offset Δ, and subjected to a resonant modulation acting on the lower site (a) only. We write the tight-binding Hamiltonian in the form

$$\hat{H}(t) = -J\left(\hat{a}^\dagger \hat{b} + \hat{b}^\dagger \hat{a}\right) + \Delta\, \hat{b}^\dagger \hat{b} + \kappa \cos(\omega t + \phi)\hat{a}^\dagger \hat{a}, \tag{14.15}$$

where J is the bare hopping amplitude between the sites. In the large-frequency regime $\Delta = \hbar\omega \gg J$, the long-time dynamics resulting from the time-dependent Hamiltonian $\hat{H}(t)$ is found to be captured by a time-independent (effective) Hamiltonian, which in this case, takes the simple form [30, 31, 32, 33, 34]

$$\hat{H}_{\text{eff}} = -J\mathcal{J}_1\left(\frac{\kappa}{\hbar\omega}\right) \exp(i\phi)\, \hat{a}^\dagger \hat{b} + \text{h.c.}, \tag{14.16}$$

where $\mathcal{J}_1$ denotes the Bessel function of the first kind. According to Eq. (14.16), the tunneling between the sites is effectively restored and the corresponding tunneling matrix elements include a Peierls phase factor related to the phase of the modulation ϕ, which can be made space dependent. The result in Eq. (14.16) constitutes the building blocks for the generation of magnetic fluxes in time-modulated optical lattices, as implemented in Munich [15, 35, 36] and at the Massachusetts Institute of Technology (MIT) [37]. The following of the section describes these schemes in more detail.

In order to treat time-modulated optical lattices, we first introduce a set of useful equations which offer a powerful theoretical framework to analyze the physics of time-dependent problems. Let us consider a general Hamiltonian of the form $\hat{H}(t) = \hat{H}_0 + \hat{V}(t)$, where $\hat{V}(t+T) = \hat{V}(t)$ and where $T = 2\pi/\omega$ is the period of the driving. In the following, the period T is considered to be small compared with all characteristic time scales [38]. We are interested in describing the dynamics of a initial state $\psi(t_0)$ under the driving, which is assumed to start at time t_0. Formally, the state at a given time $\psi(t)$ is obtained by acting on the initial state $\psi(t_0)$ with the time-evolution operator $\hat{U}(t;t_0)$, which can be partitioned as [39, 38]

$$\hat{U}(t;t_0) = \exp\left[-i\hat{K}(t)\right]\exp\left[-i(t-t_0)\hat{H}_{\text{eff}}/\hbar\right]\exp\left[i\hat{K}(t_0)\right], \tag{14.17}$$

where $\hat{H}_{\text{eff}}$ is a time-independent (effective) Hamiltonian describing the long-time dynamics, and where the "kick" operator $\hat{K}(t)$ captures the micro motion and the effects related to the initial phase of the modulation. The dynamics is thus completely captured by the operators $\hat{H}_{\text{eff}}$ and $\hat{K}(t)$, which, according to Refs. [39, 38, 34], can be calculated systematically through a perturbative expansion in powers of $1/\omega$. Considering a single-harmonic modulation, $\hat{V}(t) = \hat{V}\exp(i\omega t) + \text{h.c.}$, these operators are approximatively given by [38]

$$\begin{aligned}\hat{H}_{\text{eff}} &= \hat{H}_0 + \left(\frac{1}{\hbar\omega}\right)\left[\hat{V},\hat{V}^\dagger\right] + \frac{1}{(\sqrt{2}\hbar\omega)^2}\left(\left[\left[\hat{V},\hat{H}_0\right],\hat{V}^\dagger\right] + \text{h.c.}\right) + \mathcal{O}\left(\frac{1}{\omega^3}\right),\\ \hat{K}(t) &= \left(\frac{1}{i\hbar\omega}\right)\left(\hat{V}\exp(i\omega t) - \text{h.c.}\right) + \mathcal{O}\left(\frac{1}{\omega^2}\right).\end{aligned} \tag{14.18}$$

In some cases, the infinite series in Eq. (14.18) can be (partially or totally) resummed; this allows convergence of the expansion even in the strong-driving regime [34, 38, 40]. The equations (14.17)–(14.18) offer a systematic way to build the time-evolution operator, which can be directly applied to arbitrarily complicated systems.

Having defined a theoretical framework to treat time-dependent problems, we now apply it to time-modulated optical superlattices. The Hamiltonian is taken in the form $\hat{H}(t) = \hat{H}_0 + \hat{V}(t)$, where the static part consists of a 2D tight-binding Hamiltonian with an additional superlattice potential directed along the x direction

$$\hat{H}_0 = -\sum_{m,n}\left(J_x\hat{a}^\dagger_{m+1,n}\hat{a}_{m,n} + J_y\hat{a}^\dagger_{m,n+1}\hat{a}_{m,n} + \text{h.c.}\right) + \Delta\sum_{m,n} s(m)\hat{a}^\dagger_{m,n}\hat{a}_{m,n}, \tag{14.19}$$

where $\hat{a}_{m,n}$ creates an atom at lattice site $\boldsymbol{r} = a(m,n)$, a is the spacing, and (m,n) are integers. The superlattice function $s(m)$ is assumed to create energy offsets $\Delta \gg J_x$ between all neighboring sites, i.e., $s(m+1) - s(m) = \pm 1$ for all m. The tunneling is then restored by applying an onsite time-periodic modulation of the form

$$\hat{V}(t) = \hat{V}\exp(i\omega t) + \text{h.c.}, \quad \hat{V} = \kappa\sum_{m,n} v(m,n)\hat{a}^\dagger_{m,n}\hat{a}_{m,n}. \tag{14.20}$$

Here, the complex numbers $v(m,n)$ capture the space dependence of the modulation, which turns out to be crucial, and we impose the resonance condition $\omega = \Delta/\hbar$. The long-time dynamics of the modulated 2D lattice (Eqs. (14.19)–(14.20)) is captured by the effective Hamiltonian $\hat{H}_{\text{eff}}$ in Eq. (14.18), which after partial resummation of the series yields a Harper-Hofstadter–like Hamiltonian [34]

$$\hat{H}_{\text{eff}} = \sum_{m,n} \mathcal{J}_x(m,n) e^{\pm i\theta(m,n)} \hat{a}^{\dagger}_{m+1,n} \hat{a}_{m,n} + \mathcal{J}_y(m,n) \hat{a}^{\dagger}_{m,n+1} \hat{a}_{m,n} + \text{h.c.}, \qquad (14.21)$$

where $\theta(m,n) = \arg[v(m+1,n) - v(m,n)]$, and where the sign $\pm$ depends on whether the hopping $m \to m+1$ starts from a low- or a high-energy site of the superlattice potential $s(m)$. Hence, the Peierls phase factors in the effective Hamiltonian (and the corresponding magnetic fluxes per plaquette) depend on the time-modulation $\hat{V}(t)$, through $v(m,n)$ (Eq. (14.20)), but also on the static superlattice potential $s(m)$ (Eq. (14.19)). The effective hopping amplitudes $\mathcal{J}_{x,y}$ are *a priori* space dependent (see [41] and the discussion in [42]); they are explicitly given by

$$\mathcal{J}_x(m,n) = J_x \mathcal{J}_1 \left[K_0 |\delta_x v(m,n)|\right], \quad \mathcal{J}_y(m,n) = J_y \mathcal{J}_0 \left[K_0 |\delta_y v(m,n)|\right],$$

where $K_0 = 2\kappa/\hbar\omega$ and $\delta_{x,y} v$ denote finite-difference operations along the x and y directions, e.g., $\delta_x v(m,n) = v(m+1,n) - v(m,n)$. The kick operator can be obtained along the same line; its impact on physical observables (e.g., momentum distributions) was discussed in Ref. [34].

Based on the result (14.21), two different features of the system should be designed so as to eventually realize the Harper-Hofstadter Hamiltonian (14.14): the superlattice function $s(m)$ and the time-modulation function $v(m,n)$. The main difficulty in achieving the uniform-magnetic flux configuration is handling the sign of the phase $\pm\theta(m,n)$, which can typically lead to staggered flux patterns in arbitrary superlattices [35]. To solve this issue, the Munich [36] and MIT [37] teams first combined a Wannier-Stark ladder potential generated by a magnetic-field gradient, i.e., $s(m) = m$, with a simple moving potential, $v(m,n) = \exp(i\boldsymbol{q} \cdot \boldsymbol{r})$, created by a single pair of laser beams. In this configuration, a uniform flux $2\pi\alpha = q_y a$, together with homogeneous hopping amplitudes $\mathcal{J}_{x,y}$, was achieved. By setting $q_y = \pi/2a$, the Munich team obtained a topological band structure, whose lowest band had a Chern number $\nu_1 = 1$. However, this setup was found to be unsuitable for the Chern-number measurement described in Section 14.2.3, due to instabilities arising from the linear gradient $s(m) = m$. In order to achieve uniform flux with an all-optical superlattice [43, 15], the Munich team then developed a novel setup, based on a two-site superlattice, $s(m) = (-1)^m$, combined with a more sophisticated time modulation of the lattice.[3] The latter was induced by two pairs of laser beams, so as to generate a uniform flux over the entire lattice [15, 34], hence providing a stable platform to measure the Chern number, as we now describe.

The Chern-number measurement [15] was achieved by loading bosonic atoms into the lowest band of the Harper-Hofstadter spectrum for a synthetic magnetic

[3] Combining a two-site superlattice to a simple moving potential with $v(m,n) = \exp(i\boldsymbol{q} \cdot \boldsymbol{r})$ leads to a staggered flux configuration, which is associated with zero Chern numbers [35]. The time modulation used in [15] allowed researchers to rectify the flux, by individually addressing successive links with two independent pairs of lasers.

flux $\alpha = 1/4$. The transverse drift of the cloud (Eq. (14.12)) was then detected *in situ*, as a response to a weak optical gradient applied along the y direction. For short times, the transverse motion of the cloud was found to be linear, in agreement with the prediction in Eq. (14.12). For longer times, heating processes, which were found to be independent of the applied force, promoted atoms to higher bands and led to a saturation of the cloud's transverse drift. A careful analysis taking into account the dynamical repopulations of the bands revealed the expected Chern number with a precision at the 1% level. This Chern-number measurement was a direct probe for the topological order associated with the effective bulk energy bands, which resulted from the modulated optical superlattice.

Off-Resonant Shaken Optical Lattices

Before ending this section on tight-binding realizations of Chern bands, let us briefly describe another promising strategy for creating and observing topological properties in optical-lattice systems. Similarly to the schemes discussed above, this method is based on time-periodic driving of the lattice. However, the driving frequency is now considered to be *off-resonant* with respect to any energy separations in the problem; see Fig. 14.4c. Let us compare this idea with the resonant-modulation approach, by rewriting the two-site minimal model in Eq. (14.15) as

$$\hat{H}(t) = -J\left(\hat{a}^\dagger \hat{b} + \hat{b}^\dagger \hat{a}\right) + \kappa \cos(\omega t + \phi)\hat{a}^\dagger \hat{a}, \tag{14.22}$$

and by computing the corresponding effective Hamiltonian (14.18) [38, 44]

$$\hat{H}_{\text{eff}} = -J\mathcal{J}_0\left(\frac{\kappa}{\hbar\omega}\right)\left(\hat{a}^\dagger \hat{b} + \hat{b}^\dagger \hat{a}\right). \tag{14.23}$$

Contrary to the resonant-modulation case (Eq. (14.16)), the driving phase ϕ does not contribute to any Peierls phase factor. In fact, a TRS-breaking driving is required to generate effective complex tunneling matrix elements. This was demonstrated in Hamburg in 2012, where a Peierls phase factor was shown to appear in a shaken 1D optical lattice [45]. The shaking strategy was also applied in 2D triangular lattices by the same team, in view of studying frustrated magnetism in optical lattices [46, 47]. While these setups indeed produce local magnetic fluxes, their corresponding band structure is associated with zero Chern numbers.

In 2014, Jotzu et al. achieved the shaking of a 2D honeycomb lattice in a circular manner [48]. Similarly to the case of rotating traps (Section 14.4.2), this fast circular motion induced a chirality within the system, which formally breaks time-reversal symmetry. In fact, the effective Hamiltonian was shown to be equivalent to the Haldane model presented in Section 14.3.2: the circular shaking effectively induces complex NNN tunneling matrix elements. This opens a bulk gap in the honeycomb-lattice (Dirac) spectrum, and it generates Chern bands. The Berry curvature of the bands was probed in the system [48] through the observation of an anomalous velocity.

14.4.2 Schemes in the Continuum

A natural route to the formation of topological bands with nonzero Chern number is to simulate the orbital effect of a uniform magnetic field. The atom should move continuously through space, subject only to this uniform field. This leads to the Landau-level spectrum: highly degenerate bands, each with a unit Chern number. There exist several ways to achieve this goal for cold atoms, at least to a good approximation. (Deviations can arise from nonuniformity of the field and/or the presence of additional potentials.)

Rotation

One very direct approach is to cause the atomic cloud to rotate, for example, by applying a rotating deformation and bringing the system to equilibrium in a frame of reference rotating at angular frequency $\boldsymbol{\Omega}$ [4, 49]. In this rotating frame, an atom of mass m moving with velocity $\boldsymbol{v}$ experiences a Coriolis force $\boldsymbol{F}_{\mathrm{C}} = 2m\boldsymbol{v} \times \boldsymbol{\Omega}$. This plays the same role as the Lorentz force on a charged particle in a uniform magnetic field, $\boldsymbol{F}_{\mathrm{L}} = q\boldsymbol{v} \times \boldsymbol{B}$. Equating coefficients, $q\boldsymbol{B} = 2m\boldsymbol{\Omega}$, one deduces an effective flux density $n_\phi \equiv q|\boldsymbol{B}|/h = 2m|\boldsymbol{\Omega}|/h$. For a Bose-Einstein condensate, this manifests itself as a lattice of quantized vortex lines of areal density n_ϕ. Such vortex lattices have been observed, and their properties studied in detail, in experiments dating back over fifteen years [50, 51, 52, 53]. For a circularly symmetric harmonic confinement of frequency ω_0, mechanical stability under the centrifugal force sets an upper limit on rotation rate $|\boldsymbol{\Omega}| \leq \omega_0$. At the point of balance, $|\boldsymbol{\Omega}| = \omega_0$, the spectrum for the 2D motion is exactly that of Landau levels. This condition $|\boldsymbol{\Omega}| \simeq \omega_0$ has been reached experimentally to within 1% [54]. This rotation technique can also be implemented with optical lattices, see, e.g., Refs. [55, 56, 57].

Raman Dressing

An important feature of the physics of cold gases is that atoms (or molecules) can be prepared in different internal states (spin states or electronic excited states) which can have long lifetimes. It is then possible to use optical fields to drive atoms into coherent superpositions of these internal states. Optical dressing of internal degrees of freedom provides very powerful ways to generate artificial gauge fields and to form optical lattices with topological bands.

The physics underlying the use of internal states to generate artificial gauge fields can be understood in terms of a Berry connection in real space [58, 59]. Consider an atom with two internal states, $\alpha = 1, 2$. We denote the eigenstates at position $\boldsymbol{r}$ for these internal states by $|1\rangle_{\boldsymbol{r}}$ and $|2\rangle_{\boldsymbol{r}}$. In the presence of (optical) fields that couple the internal states, and in the rotating wave approximation, the local energy eigenstates are dressed states $|n\rangle_{\boldsymbol{r}} = a_{n,1}(\boldsymbol{r})|1\rangle_{\boldsymbol{r}} + a_{n,2}(\boldsymbol{r})|2\rangle_{\boldsymbol{r}}$. The coefficients $a_{n,\alpha}(\boldsymbol{r})$ are determined by the fields at position $\boldsymbol{r}$, so they vary in space. Hence, the adiabatic motion of an atom in the dressed state $|n\rangle_{\boldsymbol{r}}$ is associated with a Berry connection $\boldsymbol{A}_n(\boldsymbol{r}) = i\hbar\langle n|\nabla|n\rangle_{\boldsymbol{r}} = i\hbar\left(a^*_{n,1}\nabla a_{n,1} + a^*_{n,2}\nabla a_{n,2}\right)$. This plays the role of a vector potential coupling to the motion of the atom. By careful choice of the spatial dependence of the fields, one can arrange that not only $\boldsymbol{A}_n$ but also $\nabla \times \boldsymbol{A}_n$ is nonzero (for example for the lowest energy dressed state), such that the particle experiences a nonzero artificial magnetic field [2]. This method of generating artificial gauge fields has been demonstrated in experiments at NIST [60] using Raman coupling of spin states of rubidium; see the contribution by L. J. LeBlanc and I. B. Spielman (Chapter 15).

For the method of Ref. [60], the vector potential is limited to $|\boldsymbol{A}| \lesssim h/\lambda_{\rm r}$, where $\lambda_{\rm r}$ is the optical wavelength. By Stokes's theorem, the total number of flux quanta through a circular atomic cloud of radius R is then $N_\phi \equiv \frac{1}{h}\int \nabla \times \boldsymbol{A}\, d^2r = \frac{1}{h}\oint \boldsymbol{A} \cdot d\boldsymbol{l} \lesssim R/\lambda_{\rm r}$. Hence, the flux density $n_\phi \equiv N_\phi/(\pi R^2)$ is limited to $n_\phi \lesssim 1/(R\lambda_{\rm r})$. In Ref. [61], it was shown that periodic optical dressing of internal states can allow the formation of "optical flux lattices," with very much increased flux densities. These lattices lead to flux densities $n_\phi \sim 1/\lambda_{\rm r}^2$, larger by a factor of $R/\lambda_{\rm r}$ over previous proposals [2], which is $R/\lambda_{\rm r} \sim 50-100$ for typical experimental settings.

Optical Flux Lattices

The original motivation to consider periodic optical lattices coupling internal states [61] was to find a way in which to increase the strength of the artificial magnetic fields to $n_\phi \sim 1/\lambda_{\rm r}^2$. Here we shall not focus on the real-space magnetic field. Instead, we concentrate on the band structure, defined in reciprocal space. We shall show that two-dimensional optical lattices involving the coupling of internal states provide a very powerful way to generate topological (Chern) bands [62]. We shall return at the end to comment on the connection to real-space magnetic fields.

Consider an atom with $N_{\rm s}$ long-lived internal states. We denote the (plane-wave) state of an atom of internal state $\alpha = 1 \ldots N_{\rm s}$ and momentum $\hbar\boldsymbol{q}$ by $|\alpha, \boldsymbol{q}\rangle$. The coupling of the atom to optical fields will lead to processes which change the internal state and/or the momentum of the atom. We denote the optical coupling for $|\alpha, \boldsymbol{q}\rangle \to |\alpha', \boldsymbol{q}'\rangle$ by $V^{\alpha'\alpha}_{\boldsymbol{q}'-\boldsymbol{q}}$. We consider the case of periodic lattices, for which the momentum transfers of the set of all such couplings, $V^{\alpha'\alpha}_{\boldsymbol{\kappa}}$, are commensurate.

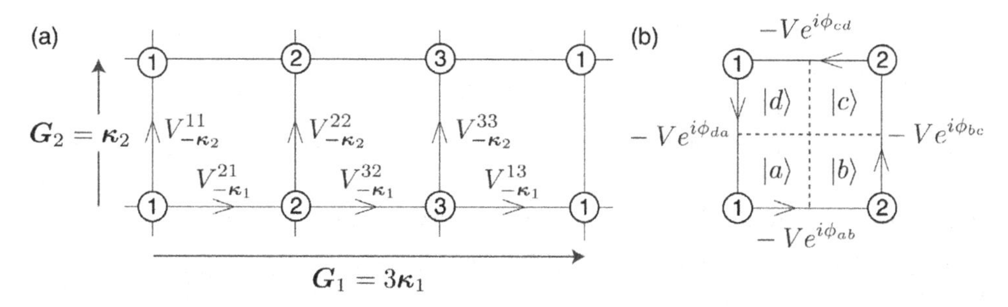

Figure 14.5 (a) Example of the reciprocal space representation of an optical lattice involving couplings of $N_s = 3$ internal states. The sites denote the different values of $(\alpha, \boldsymbol{G})$, and the bonds the couplings $V^{\alpha\alpha'}_{\boldsymbol{g}_{\alpha'}+\boldsymbol{G}'-\boldsymbol{g}_\alpha-\boldsymbol{G}}$. The offset momenta are $\boldsymbol{g}_1 = 0, \boldsymbol{g}_2 = \boldsymbol{\kappa}_1, \boldsymbol{g}_3 = 2\boldsymbol{\kappa}_1$, and the reciprocal lattice has basis vectors $\boldsymbol{G}_1 = 3\boldsymbol{\kappa}_1$, $\boldsymbol{G}_2 = \boldsymbol{\kappa}_2$. (b) A representation of one plaquette of the reciprocal space lattice. For vanishing coupling $V = 0$, the lowest energy state is one of the plane wave states $(\{|a\rangle, |b\rangle, |c\rangle, |d\rangle\})$ defined in the text and is degenerate for quasimomenta along the dashed lines. For nonzero V, gaps open along these lines. To first order in V, the integral of the Berry curvature of the lowest band over this plaquette is by Eq. (14.26).

The wave vector of any component α is then only conserved up to the addition of reciprocal lattice vectors $\boldsymbol{G}$. Similarly, by Bloch's theorem, the energy eigenstates can be assigned a band index n and a quasimomentum $\hbar\boldsymbol{k}$, and decomposed as $|\psi^{n\boldsymbol{k}}\rangle = \sum_{\alpha,\boldsymbol{G}} c^{n\boldsymbol{k}}_{\alpha,\boldsymbol{G}} |\alpha, \boldsymbol{k} - \boldsymbol{g}_\alpha - \boldsymbol{G}\rangle$, where $\boldsymbol{g}_\alpha$ accounts for possible momentum offsets of the different internal states.[4] The band energies $E_n(\boldsymbol{k})$ follow from

$$E_n(\boldsymbol{k}) c^{n\boldsymbol{k}}_{\alpha\boldsymbol{G}} = \epsilon_{\boldsymbol{k}-\boldsymbol{g}_\alpha-\boldsymbol{G}}\, c^{n\boldsymbol{k}}_{\alpha\boldsymbol{G}} + \sum_{\alpha',\boldsymbol{G}'} V^{\alpha\alpha'}_{\boldsymbol{g}_{\alpha'}+\boldsymbol{G}'-\boldsymbol{g}_\alpha-\boldsymbol{G}}\, c^{n\boldsymbol{k}}_{\alpha'\boldsymbol{G}'}, \tag{14.24}$$

where $\epsilon_{\boldsymbol{q}} \equiv \hbar^2 |\boldsymbol{q}|^2 / 2m$ is the kinetic energy for an atom of momentum $\hbar\boldsymbol{q}$.

The structure of the couplings in Eq. (14.24) can be conveniently represented by a lattice in reciprocal space. An example for $N_s = 3$ is shown in Fig. 14.5: the sites denote the different values of $(\alpha, \boldsymbol{G})$; the bonds represent the couplings $V^{\alpha'\alpha}_{\boldsymbol{\kappa}}$. (We provide here no experimental implementation of this lattice, but note that it can be generated by a variant of the triangular lattice in Ref. [63].)

It is instructive to construct the lowest energy band using this picture in the "nearly free particle" limit of small $|V^{\alpha'\alpha}_{\boldsymbol{\kappa}}|$. Consider first vanishing coupling $|V^{\alpha'\alpha}_{\boldsymbol{\kappa}}| = 0$. From Eq. (14.24), at a given quasimomentum $\hbar\boldsymbol{k}$, the lowest energy

[4] This labeling of Bloch states takes advantage of an enhanced spatial symmetry of the system under combined translations and internal-state gauge changes. For the model illustrated in Fig. 14.5a, the reciprocal lattice vector $\boldsymbol{G}_1$ is linked to a symmetry under the spatial translation by $\frac{1}{3}\frac{2\pi}{|\boldsymbol{\kappa}_1|}\hat{\kappa}_1$ combined with the unitary transformation $\left[|1\rangle\langle 1| + e^{i2\pi/3}|2\rangle\langle 2| + e^{i4\pi/3}|3\rangle\langle 3|\right]$.

state is that state with smallest $|\boldsymbol{k} - \boldsymbol{g}_\alpha - \boldsymbol{G}|^2$: that is, it is the plane wave state $|\alpha, \boldsymbol{k} - \boldsymbol{g}_\alpha - \boldsymbol{G}\rangle$ associated with the lattice point $(\alpha, \boldsymbol{G})$ closest to $\boldsymbol{k}$. This divides each square plaquette into four regions, shown in Fig. 14.5b for the leftmost plaquette of Fig. 14.5a, with the lowest energy state in these regions being $|a\rangle = |1, \boldsymbol{k} - \boldsymbol{g}_1\rangle$, $|b\rangle = |2, \boldsymbol{k} - \boldsymbol{g}_2\rangle$, $|c\rangle = |2, \boldsymbol{k} - \boldsymbol{g}_2 - \boldsymbol{G}_2\rangle$, $|d\rangle = |1, \boldsymbol{k} - \boldsymbol{g}_1 - \boldsymbol{G}_2\rangle$. The dashed lines in Fig. 14.5b show the locations where the lowest energy state is degenerate for $V = 0$: twofold degenerate on each dashed line and fourfold degenerate at the point $\boldsymbol{k}^*$ where the dashed lines cross.

For $|V_{\boldsymbol{\kappa}}^{\alpha'\alpha}| \neq 0$, the degeneracies along the dashed lines are lifted. We describe the behavior in the case where the couplings around this plaquette are $-Ve^{i\phi}$ with $\phi = \phi_{ab,bc,cd,da}$ as illustrated in Fig. 14.5b. Working to lowest order in V involves first-order (degenerate) perturbation. At the point of fourfold degeneracy $\boldsymbol{k}^*$, this leads to the Hamiltonian

$$\begin{pmatrix} \epsilon_{\boldsymbol{k}^*} & -Ve^{-i\phi_{ab}} & 0 & -Ve^{i\phi_{da}} \\ -Ve^{i\phi_{ab}} & \epsilon_{\boldsymbol{k}^*} & -Ve^{-i\phi_{bc}} & 0 \\ 0 & -Ve^{i\phi_{bc}} & \epsilon_{\boldsymbol{k}^*} & -Ve^{-i\phi_{cd}} \\ -Ve^{-i\phi_{da}} & 0 & -Ve^{i\phi_{cd}} & \epsilon_{\boldsymbol{k}^*} \end{pmatrix}. \tag{14.25}$$

The spectrum of this Hamiltonian depends on the gauge invariant phase

$$\Phi_{\mathrm{p}} \equiv \phi_{ab} + \phi_{bc} + \phi_{cd} + \phi_{da}. \tag{14.26}$$

For $\Phi_{\mathrm{p}} \neq \pi$, the lowest energy state is nondegenerate at $\boldsymbol{k}^*$, and indeed for all $\boldsymbol{k}$ within this plaquette. The lowest band therefore has a well-defined Berry curvature, with no singularities, for $\Phi_{\mathrm{p}} \neq \pi$. One can then compute the integral of the Berry curvature over the plaquette in terms of the line integral of the Berry connection around the four sides of the plaquette. Still working to first order in V, the adiabatic transfer along each of these sides just involves the coupling of a pair of states (e.g., $|a\rangle$ and $|b\rangle$ coupled by $-Ve^{i\phi_{ab}}$). It is straightforward to show that the total integral of the Berry connection around the plaquette is just Φ_{p}. Thus, to first order in V, the integral of the Berry curvature over the plaquette is precisely Φ_{p} for $\Phi_{\mathrm{p}} \neq \pi$. For $\Phi_{\mathrm{p}} = \pi$, the lowest two bands touch at a Dirac point at $\boldsymbol{k} = \boldsymbol{k}^*$, and the Berry curvature of the lowest band is ill defined.

This result provides a very powerful prescription by which to design optical couplings that generate Chern bands. By choosing the phases of the couplings $V_{\boldsymbol{\kappa}}^{\alpha'\alpha}$ around each plaquette in the reciprocal space lattice, one can specify (the integral of) the Berry curvature of the lowest energy band over each of these plaquettes. If the sum over all plaquettes in the first Brillouin zone $\Phi_{\mathrm{tot}} = \sum_{\mathrm{p}\in\mathrm{FBZ}} \Phi_{\mathrm{p}}$ is nonzero, and all $\Phi_{\mathrm{p}} \neq \pi$, then the lowest energy band will have a Chern number of $\mathcal{C} = \Phi_{\mathrm{tot}}/(2\pi)$. Thus the net number of "flux quanta" through this reciprocal space lattice sets the Chern number of the lowest energy band. For example, if the couplings of Fig. 14.5a are chosen

$$V^{21}_{-\kappa_1} = V^{32}_{-\kappa_1} = V^{13}_{-\kappa_1} = -V, \quad V^{\alpha\alpha}_{-\kappa_2} = -Ve^{i\,2\pi\alpha/3}, \tag{14.27}$$

then the lowest energy band will have integrated Berry curvature of $2\pi/3$ in each of the plaquettes in reciprocal space, so Chern number $\mathcal{C} = 1$. Although this result applies in the weak lattice limit $|V^{\alpha'\alpha}_{\kappa}| \ll E_{\mathrm{r}}$, the fact that the Chern number is a topological invariant guarantees robustness of the result up to moderate $|V^{\alpha'\alpha}_{\kappa}|$.

Our considerations of the energy bands, and their topological properties, have been presented in reciprocal space. How do these considerations relate to the real-space magnetic field discussed in Section 14.4.2? The magnetic field is defined in the limit in which the particle moves adiabatically in real space, valid when $|V^{\alpha'\alpha}_{\kappa}| \gg E_{\mathrm{r}}$. In this limit, the kinetic energy term in (14.24) is negligible and the energy eigenstates follow from a tight-binding model on the reciprocal space lattice. An elegant duality emerges [62]: the magnetic flux through the unit cell in real space due to adiabatic motion in the lowest energy dressed state is equal to the Chern number of the lowest energy band of the tight-binding model defined by the couplings $V^{\alpha'\alpha}_{\kappa}$ on the reciprocal space lattice. To form an "optical flux lattice," with nonzero magnetic flux for the lowest energy dressed state through the real-space unit cell, one should just choose the couplings $V^{\alpha'\alpha}_{\kappa}$ on the reciprocal space lattice to give a lowest energy band with a nonzero Chern number. For the example of Fig. 14.5a, with the above couplings (14.27), the reciprocal space model is precisely the Harper-Hofstadter model for flux $1/3$, per plaquette. As discussed in Section 14.3.2, this model readily leads to Chern bands: for flux $1/3$, the lowest band has unit Chern number, so the couplings (14.27) form an optical flux lattice with one magnetic flux quantum through each real-space unit cell.

The nonzero flux densities of optical flux lattices lead to low-energy bands that are very similar to those of the lowest Landau level: with a unit Chern number and with narrow width in energy. The bands depart from the exact degeneracy of Landau levels owing to the fact that the magnetic field and the scalar potential are nonuniform in space. However, for well-chosen parameters, these effects can be made small in practical implementations [63, 64].

14.5 Discussion

In this chapter, we discussed how Chern bands can be created in cold-atom systems, through a wide variety of schemes based on atom–light coupling, rotation, or time-periodic modulations. Several of these techniques have been experimentally implemented in different laboratories, demonstrating physical manifestations of synthetic magnetic fields [15, 35, 36, 37, 46, 47, 48, 54, 60, 65]. Beyond the signatures discussed here, topological energy bands could also be probed in cold-atom systems through the detection of Skyrmion patterns in time-of-flight images [24, 25], interferometry [66, 67], measurements of the spin polarization at highly symmetric points of the Brillouin zone [68], or the presence of chiral edge states [69, 70, 71, 72, 73].

This chapter focused on single-particle phenomena in Chern bands. Theoretical studies have extended the ideas presented here in a variety of ways. The methods for generating Chern bands for cold atoms have been extended to higher dimensions and other symmetry classes, including the $\mathbb{Z}_2$ topological insulator in three dimensions [74, 75] both in lattice-based [76, 77] and in continuum [78] formulations, and to systems with sublattice (chiral) symmetry [79]. It is very interesting to consider the effects of interactions on degenerate bosons or fermions in the flat, or nearly flat, topological bands that we have described above. For the Harper-Hofstadter model, numerical studies have established the existence of fractional quantum Hall states, including the Laughlin state [80, 81, 82, 83] as well as states that exist only on the lattice [82, 84]. The use of internal states to form optical flux lattices leads to nearly flat topological bands, which have been shown to give rise to exotic non-Abelian phases of bosons even for weak two-body repulsion [63, 85], as well as to interesting forms of ferromagnetic-nematic ordering for one-component fermions [86]. Studies of fractional Chern insulators have identified a wealth of tight-binding lattice geometries with Chern bands where strongly correlated phases can appear [6, 7] and which may find realizations in cold-atom setups [87]. These theoretical works show that cold atomic gases have the potential to provide an ideal setting in which to explore novel forms of strongly correlated quantum phases, including fractional Chern insulators and other lattice-based strongly correlated topological phases.

Acknowledgments: N. Goldman is financed by the Fonds de la Recherche Scientifique (FRS-FNRS) Belgium and by Bibliothéque des sciences économiques, sociales, politiques et de communication (BSPO) under PAI Project No. P7/18 Dynamics, Geometry and Statistical Physics (DYGEST). N. R. Cooper acknowledges support from the Engineering and Physical Sciences Research Council (EPSRC) Grant No. EP/K030094/1, and J. Dalibard from the European Research Council (ERC) (Synergy Ultracold Quantum Matter [UQUAM]) and L'Agence nationale de la recherche (ANR) (Artificial Gauge Fields on Neutral Atoms [AGAFON]).

References

[1] Bloch, I., Dalibard, J., and Zwerger, W. 2008. Many-body physics with ultracold gases. *Rev. Mod. Phys.*, **80**, 885.

[2] Dalibard, J., Gerbier, F., Juzeliūnas, G., and Öhberg, P. 2011. Colloquium: artificial gauge potentials for neutral atoms. *Rev. Mod. Phys.*, **83**, 1523.

[3] Goldman, N., Juzeliūnas, G., Öhberg, P., and Spielman, I. B. 2014. Light-induced gauge fields for ultracold atoms. *Rep. Prog. Phys.*, **77**, 126401.

[4] Cooper, N. R. 2008. Rapidly rotating atomic gases. *Advances in Physics*, **57**, 539.

[5] Bernevig, B. A., and Hughes, T. L. 2013. *Topological Insulators and Topological Superconductors*. Princeton, NJ: Princeton University Press.

[6] Parameswaran, S. A., Roy, R., and Sondhi, S. L. 2013. Fractional quantum Hall physics in topological flat bands. *Comptes Rendus Physique*, **14**, 816.

[7] Bergholz, E. J., and Liu, Z. 2013. Topological flat band models and fractional Chern insulators. *Int. J. Mod. Phys. B*, **27**, 1330017.

[8] Hofstadter, D. R. 1976. Energy levels and wave functions of Bloch electrons in rational and irrational magnetic fields. *Phys. Rev. B*, **14**, 2239.

[9] Haldane, F. D. M. 1988. Model for a quantum Hall effect without Landau levels: condensed-matter realization of the "parity anomaly." *Phys. Rev. Lett.*, **61**, 2015.

[10] Thouless, D. J., Kohmoto, M., Nightingale, M. P., and den Nijs, M. 1982. Quantized Hall conductance in a two-dimensional periodic potential. *Phys. Rev. Lett.*, **49**, 405.

[11] Kohmoto, M. 1985. Topological invariant and the quantization of the Hall conductance. *Annals of Physics*, **160**, 343.

[12] Xiao, D., Chang, M.-C., and Niu, Q. 2010. Berry phase effects on electronic properties. *Rev. Mod. Phys.*, **82**, 1959.

[13] Price, H. M., and Cooper, N. R. 2012. Mapping the Berry curvature from semiclassical dynamics in optical lattices. *Phys. Rev. A*, **85**, 033620.

[14] Tarruell, L., Greif, D., Uehlinger, T., Jotzu, G., and Esslinger, T. 2012. Creating, moving and merging Dirac points with a Fermi gas in a tunable honeycomb lattice. *Nature*, **483**, 302.

[15] Aidelsburger, M., Lohse, M., Schweizer, C., Atala, M., Barreiro, J. T., Nascimbene, S., Cooper, N. R., Bloch, I., and Goldman, N. 2015. Measuring the Chern number of Hofstadter bands with ultracold bosonic atoms. *Nature Physics*, **11**, 162.

[16] Dauphin, A., and Goldman, N. 2013. Extracting the Chern number from the dynamics of a Fermi gas: implementing a quantum Hall bar for cold atoms. *Phys. Rev. Lett.*, **111**, 135302.

[17] Cohen-Tannoudji, C., Diu, B., and Laloë, F. 1991. *Quantum Mechanics*. New York: Wiley.

[18] Harper, P. G. 1955. Single band motion of conduction electrons in a uniform magnetic field. *Proc. Phys. Soc. A*, **68**, 879.

[19] Aharonov, Y., and Bohm, D. 1959. Significance of electromagnetic potentials in quantum theory. *Phys. Rev.*, **115**, 485.

[20] Jaksch, D., and Zoller, P. 2003. Creation of effective magnetic fields in optical lattices: the Hofstadter butterfly for cold neutral atoms. *New Journal of Physics*, **5**, 56.

[21] Gerbier, F., and Dalibard, J. 2010. Gauge fields for ultracold atoms in optical superlattices. *New Journal of Physics*, **12**, 3007.

[22] Ruostekoski, J., Dunne, G. V, and Javanainen, J. 2002. Particle number fractionalization of an atomic Fermi-Dirac gas in an optical lattice. *Phys. Rev. Lett.*, **88**, 180401.

[23] Mueller, E. 2004. Artificial electromagnetism for neutral atoms: Escher staircase and Laughlin liquids. *Phys. Rev. A*, **70**, 041603.

[24] Alba, E., Fernandez-Gonzalvo, X., Mur-Petit, J., Pachos, J., and Garcia-Ripoll, J. 2011. Seeing topological order in time-of-flight measurements. *Phys. Rev. Lett.*, **107**, 235301.

[25] Goldman, N., Anisimovas, E., Gerbier, F., Öhberg, P., Spielman, I. B., and Juzeliūnas, G. 2013. Measuring topology in a laser-coupled honeycomb lattice: from Chern insulators to topological semi-metals. *New Journal of Physics*, **15**, 3025.

[26] Anisimovas, E., Gerbier, F., Andrijauskas, T., and Goldman, N. 2014. Design of laser-coupled honeycomb optical lattices supporting Chern insulators. *Phys. Rev. A*, **89**, 013632.

[27] Celi, A., Massignan, P., Ruseckas, J., Goldman, N., Spielman, I. B., Juzeliūnas, G., and Lewenstein, M. 2014. Synthetic gauge fields in synthetic dimensions. *Phys. Rev. Lett.*, **112**, 043001.

[28] Mancini, M., Pagano, G., Cappellini, G., Livi, L., Rider, M., Catani, J., Sias, C., Zoller, P., Inguscio, M., Dalmonte, M., and Fallani, L. 2015. Observation of chiral edge states with neutral fermions in synthetic Hall ribbons. *Science* **349**, 1510.

[29] Stuhl, B. K., Lu, H. I., Aycock, L. M., Genkina, D., and Spielman, I. B. 2015. Visualizing edge states with an atomic Bose gas in the quantum Hall regime. *Science* **349**, 1514.

[30] Eckardt, A., and Holthaus, M. 2007. AC-induced superfluidity. *Europhysics Letters (EPL)*, **80**, 50004.

[31] Lim, L.-K., Hemmerich, A., and Morais Smith, C. 2010. Artificial staggered magnetic field for ultracold atoms in optical lattices. *Phys. Rev. A*, **81**, 023404.

[32] Bermudez, A., Schaetz, T., and Porras, D. 2011. Synthetic gauge fields for vibrational excitations of trapped ions. *Phys. Rev. Lett.*, **107**, 150501.

[33] Hauke, P., Tieleman, O., Celi, A., Ölschläger, C., Simonet, J., Struck, J., Weinberg, M., Windpassinger, P., Sengstock, K., Lewenstein, M., and Eckardt, A. 2012. Non-Abelian gauge fields and topological insulators in shaken optical lattices. *Phys. Rev. Lett.*, **109**, 145301.

[34] Goldman, N., Dalibard, J., Aidelsburger, M., and Cooper, N. R. 2015. Periodically-driven quantum matter: the case of resonant modulations. *Phys. Rev. A* **91**, 033632.

[35] Aidelsburger, M., Atala, M., Nascimbene, S., Trotzky, S., Chen, Y.-A., and Bloch, I. 2011. Experimental realization of strong effective magnetic fields in an optical lattice. *Phys. Rev. Lett.*, **107**, 255301.

[36] Aidelsburger, M., Atala, M., Lohse, M., Barreiro, J. T., Paredes, B., and Bloch, I. 2013. Realization of the Hofstadter Hamiltonian with ultracold atoms in optical lattices. *Phys. Rev. Lett.*, **111**, 85301.

[37] Miyake, H., Siviloglou, G. A., Kennedy, C. J., Burton, W. C., and Ketterle, W. 2013. Realizing the Harper Hamiltonian with laser-assisted tunneling in optical lattices. *Phys. Phys. Lett.*, **111**, 185302.

[38] Goldman, N., and Dalibard, J. 2014. Periodically-driven quantum systems: effective Hamiltonians and engineered gauge fields. *Phys. Rev. X.*, **4**, 031027.

[39] Rahav, S., Gilary, I., and Fishman, S. 2003. Effective Hamiltonians for periodically driven systems. *Phys. Rev. A*, **68**, 013820.

[40] Bukov, M., D'Alessio, L., and Polkovnikov, A. 2015. Universal high-frequency behavior of periodically driven systems: from dynamical stabilization to Floquet engineering. *Adv. Phys.* **64**, 139.

[41] Kolovsky, A. R. 2011. Creating artificial magnetic fields for cold atoms by photon-assisted tunneling. *Europhysics Letters (EPL)*, **93**, 20003.

[42] Creffield, C. E., and Sols, F. 2013. Comment on "Creating artificial magnetic fields for cold atoms by photon-assisted tunneling" by Kolovsky, A. R. *Europhysics Letters (EPL)*, **101**, 40001.

[43] Baur, S. K., Schleier-Smith, M. H., and Cooper, N. R. 2014. Dynamic optical superlattices with topological bands. *Phys. Rev. A* **89**, 051605(R).

[44] Arimondo, E., Ciampinia, D., Eckardt, A., Holthause, M., and Morsch, O. 2012. Kilohertz-driven Bose–Einstein condensates in optical lattices. *Adv. At. Molec. Opt. Phys.*, **61**, 515.

[45] Struck, J., Ölschläger, C., Weinberg, M., Hauke, P., Simonet, J., Eckardt, A., Lewenstein, M., Sengstock, K., and Windpassinger, P. 2012. Tunable gauge potential for neutral and spinless particles in driven optical lattices. *Phys. Rev. Lett.*, **108**, 225304.

[46] Struck, J., Ölschläger, C., Le Targat, R., Soltan-Panahi, P., Eckardt, A., Lewenstein, M., Windpassinger, P., and Sengstock, K. 2011. Quantum simulation of frustrated classical magnetism in triangular optical lattices. *Science*, **333**, 996.

[47] Struck, J., Weinberg, M., Ölschläger, C., Windpassinger, P., Simonet, J., Sengstock, K., Höppner, R., Hauke, P., Eckardt, A., Lewenstein, M., and Mathey, L. 2013. Engineering ising-XY spin-models in a triangular lattice using tunable artificial gauge fields. *Nature Physics*, **9**, 738.

[48] Jotzu, G., Messer, M., Desbuquois, R., Lebrat, M., Uehlinger, T., Greif, D., and Esslinger, T. 2014. Experimental realisation of the topological Haldane model. *Nature*, **515**, 237.

[49] Fetter, A. L. 2009. Rotating trapped Bose-Einstein condensates. *Rev. Mod. Phys.*, **81**, 647.

[50] Madison, K. W., Chevy, F., Wohlleben, W., and Dalibard, J. 2000. Vortex formation in a stirred Bose-Einstein condensate. *Phys. Rev. Lett.*, **84**, 806.

[51] Abo-Shaeer, J. R., Raman, C., Vogels, J. M. and Ketterle, W. 2001. Observation of vortex lattices in Bose-Einstein condensates. *Science*, **292**, 476.

[52] Hodby, E., Hechenblaikner, G., Hopkins, S. A., Marago, O. M., and Foot, C. J. 2002. Vortex nucleation in Bose-Einstein condensates in an oblate, purely magnetic potential. *Phys. Rev. Lett.*, **88**, 010405.

[53] Coddington, I., Haljan, P. C., Engels, P., Schweikhard, V., Tung, S., and Cornell, E. A. 2004. Experimental studies of equilibrium vortex properties in a Bose-condensed gas. *Phys. Rev. A*, **70**, 063607.

[54] Schweikhard, V., Coddington, I., Engels, P., Mogendorff, V. P., and Cornell, E. A. 2004. Rapidly rotating Bose-Einstein condensates in and near the lowest Landau level. *Phys. Rev. Lett.*, **92**, 040404.

[55] Tung, S., Schweikhard, V., and Cornell, E. A. 2006. Observation of vortex pinning in Bose-Einstein condensates. *Phys. Rev. Lett.*, **97**, 240402.

[56] Hemmerich, A., and Morais Smith, C. 2007. Excitation of a d-density wave in an optical lattice with driven tunneling. *Phys. Rev. Lett.*, **99**, 113002.

[57] Williams, R. A., Al-Assam, S., and Foot, C. J. 2010. Observation of vortex nucleation in a rotating two-dimensional lattice of Bose-Einstein condensates. *Phys. Rev. Lett.*, **104**, 050404.

[58] Berry, M. V. 1984. Quantal phase factors accompanying adiabatic changes. *Proc. Roy. Soc. London A*, **392**, 45.

[59] Dum, R., and Olshanii, M. 1996. Gauge structures in atom-laser interaction: Bloch oscillations in a dark lattice. *Phys. Rev. Lett.*, **76**, 1788.

[60] Lin, Y.-J., Compton, R. L., Jiménez-García, K., Porto, J. V., and Spielman, I. B. 2009. Synthetic magnetic fields for ultracold neutral atoms. *Nature*, **462**, 628.

[61] Cooper, N. R. 2011. Optical flux lattices for ultracold atomic gases. *Phys. Rev. Lett.*, **106**.

[62] Cooper, N. R., and Moessner, R. 2012. Designing topological bands in reciprocal space. *Phys. Rev. Lett.*, **109**, 215302.

[63] Cooper, N. R., and Dalibard, J. 2013. Reaching fractional quantum Hall states with optical flux lattices. *Phys. Rev. Lett.*, **110**, 185301.

[64] Cooper, N. R., and Dalibard, J. 2011. Optical flux lattices for two-photon dressed states. *Europhysics Letters*, **95**, 66004.

[65] LeBlanc, L. J., Jiménez-García, K., Williams, R. A., Beeler, M. C., Perry, A. R., Phillips, W. D., and Spielman, I. B. 2012. Observation of a superfluid Hall effect. *Proc. Natl. Acad. Sci. (PNAS)*, **109**, 10811.

[66] Abanin, D. A., Kitagawa, T., Bloch, I., and Demler, E. 2013. Interferometric approach to measuring band topology in 2D optical lattices. *Phys. Rev. Lett.*, **110**, 165304.

[67] Duca, L., Li, T., Reitter, M., Bloch, I., Schleier-Smith, M., and Schneider, U. 2015. An Aharonov-Bohm interferometer for determining Bloch band topology. *Science*, **347**, 288.

[68] Liu, Xiong-Jun, Law, K. T., Ng, T. K., and Lee, Patrick A. 2013. Detecting topological phases in cold atoms. *Phys. Rev. Lett.*, **111**, 120402.

[69] Liu, Xiong-Jun, Liu, Xin, Wu, Congjun, and Sinova, Jairo. 2010. Quantum anomalous Hall effect with cold atoms trapped in a square lattice. *Phys. Rev. A*, **81**, 033622.

[70] Stanescu, T. D., Galitski, V., and Das Sarma, S. 2010. Topological states in two-dimensional optical lattices. *Phys. Rev. A*, **82**, 013608.

[71] Goldman, N., Beugnon, J., and Gerbier, F. 2012. Detecting chiral edge states in the Hofstadter optical lattice. *Phys. Rev. Lett.*, **108**, 255303.

[72] Buchhold, M., Cocks, D., and Hofstetter, W. 2012. Effects of smooth boundaries on topological edge modes in optical lattices. *Phys. Rev. A*, **85**, 63614.

[73] Goldman, N., Dalibard, J., Dauphin, A., Gerbier, F., Lewenstein, M., Zoller, P., and Spielman, I. B. 2013. Direct imaging of topological edge states in cold-atom systems. *Proc. Natl. Acad. Sci. (PNAS)*, **110**, 1.

[74] Hasan, M. Z., and Kane, C. L. 2010. Colloquium: topological insulators. *Rev. Mod. Phys.*, **82**, 3045.

[75] Qi, X.-L., and Zhang, S.-C. 2011. Topological insulators and superconductors. *Rev. Mod. Phys.*, **83**, 1057.

[76] Goldman, N., Satija, I., Nikolic, P., Bermudez, A., Martin-Delgado, M. A., Lewenstein, M., and Spielman, I. B. 2010. Realistic time-reversal invariant topological insulators with neutral atoms. *Phys. Rev. Lett.*, **105**, 255302.

[77] Bermudez, A., Mazza, L., Rizzi, M., Goldman, N., Lewenstein, M., and Martin-Delgado, M. A. 2010. Wilson fermions and axion electrodynamics in optical lattices. *Phys. Rev. Lett.*, **105**, 190404.

[78] Béri, B., and Cooper, N. R. 2011. $\mathbb{Z}_2$ Topological insulators in ultracold atomic gases. *Phys. Rev. Lett.*, **107**, 145301.

[79] Essin, A. M., and Gurarie, V. 2012. Antiferromagnetic topological insulators in cold atomic gases. *Phys. Rev. B*, **85**, 195116.

[80] Sorensen, A. S., Demler, E., and Lukin, M. D. 2005. Fractional quantum Hall states of atoms in optical lattices. *Phys. Rev. Lett.*, **94**, 086803.

[81] Palmer, R. N., and Jaksch, D. 2006. High-field fractional quantum Hall effect in optical lattices. *Phys. Rev. Lett.*, **96**, 180407.

[82] Möller, G., and Cooper, N. R. 2009. Composite fermion theory for bosonic quantum Hall states on lattices. *Phys. Rev. Lett.*, **103**, 105303.

[83] Sterdyniak, A., Regnault, N., and Möller, G. 2012. Particle entanglement spectra for quantum Hall states on lattices. *Phys. Rev. B*, **86**, 165314.

[84] Hormozi, L., Möller, G., and Simon, S. H. 2012. Fractional quantum Hall effect of lattice bosons near commensurate flux. *Phys. Rev. Lett.*, **108**, 256809.

[85] Sterdyniak, A., Bernevig, B. Andrei, Cooper, Nigel R., and Regnault, N. 2015. Interacting bosons in topological optical flux lattices. *Phys. Rev. B*, **91**, 035115.

[86] Baur, S. K., and Cooper, N. R. 2012. Coupled ferromagnetic and nematic ordering of fermions in an optical flux lattice. *Phys. Rev. Lett.*, **109**, 265301.

[87] Yao, N. Y., Gorshkov, A. V., Laumann, C. R., Luchli, A. M., Ye, J., and Lukin, M. D. 2013. Realizing fractional Chern insulators in dipolar spin systems. *Phys. Rev. Lett.*, **110**, 185302.

15

Bose-Einstein Condensates in Artificial Gauge Fields

LINDSAY J. LEBLANC

Department of Physics, University of Alberta, Edmonton, Canada

IAN B. SPIELMAN

Joint Quantum Institute, National Institute of Standards and Technology, and University of Maryland, Gaithersburg, USA

Quantum mechanically, static or applied gauge fields manifest themselves by directly coupling to a system's phase. Artificial gauge field techniques manipulate the phase of a Bose-Einstein condensate's order parameter, yielding effects present in condensed matter systems subject to magnetic fields, electric fields, or spin-orbit coupling. In this chapter, we explore the impact of these artificial gauge fields on Bose-Einstein condensates and explain the consequences. Several analogues to traditional condensed matter systems are enabled by these techniques, including the Hall effect, superfluid vortex nucleation in systems with artificial magnetic fields, and magnetic-like phase transitions in systems with spin-orbit coupling.

15.1 Introduction

Since the first experiments with Bose-Einstein condensates (BECs), researchers have exploited the universality of physical laws to draw analogies between the behavior of ultracold quantum gases and systems ranging from material to astrophysical. These analogous situations are deliberately engineered, giving cold-atom "quantum simulators" [1] that exhibit behavior either observed or predicted in analogous quantum systems.

Superfluidity was one of the first many-body quantum phenomena demonstrated in BECs. For example, several experiments showed superfluid behavior by rapidly rotating the condensates and observing the formation of vortices [2, 3, 4, 5] – quantized point-defects in the BEC's order parameter. In the rotating frame of these experiments, this nucleation of vortices could be understood as the consequence of an artificial magnetic field [6] that arises via the formal equivalence of the rotating-frame Coriolis force and the Lorentz force, which we describe in Section 15.4.1. Together with other measurements, these results proved to be in excellent agreement with the predictions of the Gross-Pitaevskii equation (GPE) [7], a decades-old idealization of superfluid helium.

Buoyed by the prospect for studying magnetic field–driven strongly correlated systems using these nearly ideal quantum gases, proposals for alternate techniques to simulate gauge fields were developed. These methods primarily employ an internal state–selective atom–light coupling, where the photon momentum provides the forces necessary to mimic a Lorentz force [8, 9, 10, 11, 12, 13, 14, 15, 16]. The first demonstration of this type of technique used a spatially dependent Raman-transition scheme to effect a magnetic field [17], and experiments demonstrating artificial electric fields [18] and spin-orbit coupling [19, 20, 21, 22, 23, 24] soon followed.

In this chapter, the effects of artificial gauge fields upon BECs are discussed. Using a GPE description modified to include these fields, the predicted behavior for both artificial magnetic fields and spin-orbit coupling is explained, and three experiments that probe the BEC by exploiting these properties are considered: a Hall effect, a vortex-nucleation transition, and an easy-axes to easy-plane ferromagnetic transition.

15.2 Artificial Gauge Fields Applied to BECs: Formalism

The details of artificial gauge fields created both through rotation [6] and light-induced methods [16] have been reviewed thoroughly. This chapter will focus on the artificial fields' influence on the dynamics and transport properties of BECs, using the Gross-Pitaevskii formalism to describe trapped, interacting, bosonic gases. The discussion here will concentrate on a two-laser Raman-coupling method and explore, in particular, the bulk responses of BECs to changes in their environments while considering how these macroscopic responses reveal the microscopic details of the systems.

15.2.1 Laser-Assisted Raman-Transition Artificial Gauge Fields

Light-assisted artificial gauge techniques rely on the selective transfer of momentum between laser photons and atoms. This modifies the single-particle dispersion relationship [8, 10, 15], as shown in Fig. 15.1a for ^{87}Rb atoms in the $f = 1$ ground state in an applied magnetic field $\mathbf{B} = B_0\mathbf{e}_z$. A two-photon Raman transition couples adjacent magnetic sublevels (Fig. 15.1b), with linear Zeeman splitting $\hbar\omega_Z = g_F\mu_B B_0$. Here μ_B is the Bohr magneton and g_F is the the hyperfine Landé g-factor (in ^{87}Rb's lowest energy $f = 1$ manifold $g_F \approx -1/2$). The laser propagation vectors $\mathbf{k}_1$ and $\mathbf{k}_2$ associated with the two frequencies ω_1 and ω_2 point in different directions so as to create a net momentum transfer $\mathbf{\Delta k} = \mathbf{k}_1 - \mathbf{k}_2$ to the atom (absorption from one beam and re-emission into the other) equal to the

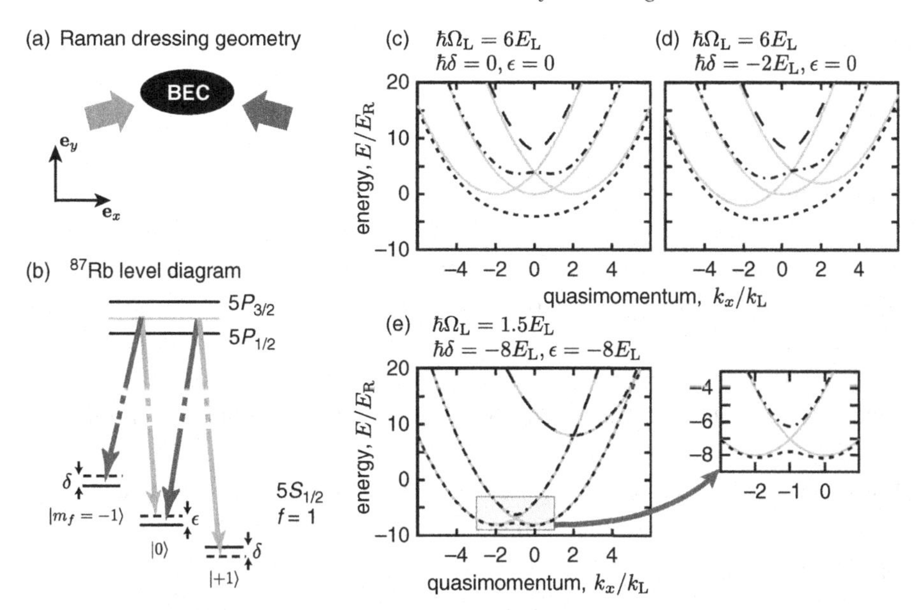

Figure 15.1 Raman dressing and dispersion relations. (a) General geometry used for two-laser Raman dressing in BEC experiments. (b) Level diagram of the $f = 1$ ground state (as for ^{87}Rb) showing two-photon Raman transitions. (c–e) Dispersion relationships for the Raman-dressed states in three regimes, where bare (undressed) states are shown with solid gray curves, and the dressed eigenstates of the coupled system are shown with dashed curves: (c) Dressed states with zero-detuning yield a ground-state dispersion with minimum at $k_x = 0$.(d) Dressed states with nonzero detuning shift the minimum of the ground-state dispersion such that $k_x \neq 0$. This feature is critical for creating artificial gauge fields. (e) With a large quadratic Zeeman shift ϵ and a detuning with a similar energy, two of the three bare states have similar energies, while the third is very different. These two states, when coupled, form a "double-well"–type dispersion that allows for spin-orbit coupling, which is more evident in the callout to the right.

difference. For counterpropagating beams, the net momentum transfer is twice the recoil momentum $|\mathbf{\Delta k}| = 2\hbar k_L$, where $k_L = 2\pi/\lambda_0$ and λ_0 is the Raman laser wavelength.

The coupling between these states is experimentally controlled with three parameters: the detuning $\delta = \omega_Z - (\omega_1 - \omega_2)$ from two-photon Raman resonance, the two-photon Raman Rabi frequency Ω_L (determined by the intensities of the laser beams), and the quadratic Zeeman shift ϵ (determined by the external Zeeman magnetic field). Expressed in terms of these parameters, the single-particle Hamiltonian describing a coupled three-level system is

$$\hat{\mathcal{H}}_3 = \frac{\hbar^2(\hat{k}_x^2 + \hat{k}_y^2 + \hat{k}_z^2)}{2m}\check{1} + \frac{\hbar\Omega_L}{2}\check{F}_1 + \frac{\hbar^2 k_L \hat{k}_x}{m}\check{F}_3 - \frac{\hbar\delta}{2}\check{F}_3 + 4E_L\check{F}_3^2 + \epsilon(\check{1} - \check{F}_3^2), \quad (15.1)$$

where we have assumed without loss of generality that $\mathbf{\Delta k}$ is parallel to $\mathbf{e}_x$. The operators $\{\check{F}_1, \check{F}_2, \check{F}_3, \check{1}\}$ are the $f = 1$ angular momentum and identity operators, and $E_\mathrm{L} = \hbar k_\mathrm{L}^2/2m$ is the recoil energy associated with the Raman transition. Fig. 15.1c–e shows the coupled energies for the three-state system for three experimental configurations.

With appropriate preparation [25], a BEC will be found in the lowest-energy eigenstate of this coupled system. The Raman coupling modifies the kinetic energy to give rise to a dispersion relationship that depends on experimental parameters. Along $\mathbf{\Delta k}$, the dispersion is in general different from its free-space equivalent in terms of curvature (leading to an effective mass), its momentum-dependent eigenstate decomposition, and the location of the dispersion minimum in terms of k_x.

For large coupling strengths ($\hbar\Omega_\mathrm{L} > 4k_\mathrm{L}$), the lowest-energy eigenstate exhibits a single minimum, and the small range of occupied momenta by the BEC permits an approximation of the kinetic energy as

$$\hat{\mathcal{H}}_{\mathcal{B}} \approx \frac{(\hat{\mathbf{p}} - \mathcal{A}\mathbf{e}_x)^2}{2m^*}, \tag{15.2}$$

where $\hat{\mathbf{p}} = \hbar\hat{\mathbf{k}} = -i\hbar\nabla$ is the canonical momentum operator and $\mathcal{A}$ indicates the location of the dispersion's minimum with respect to $\mathbf{p} = \hbar\mathbf{k} = 0$. By identifying $\boldsymbol{\mathcal{A}} = \mathcal{A}\mathbf{e}_x$ as an artificial vector potential,[1] it follows that magnetic and electric fields can be simulated by this configuration if $\boldsymbol{\mathcal{A}}$ is endowed with appropriate spatial or temporal dependence: the artificial magnetic field is $\boldsymbol{\mathcal{B}} = \nabla \times \boldsymbol{\mathcal{A}}$ and the artificial electric field is $\boldsymbol{\mathcal{E}} = \partial_t\boldsymbol{\mathcal{A}}$.

For smaller coupling strengths, the multiple-well character of the original bare-state dispersions persists. For sufficiently large B_0 and with $|\epsilon| \approx |\delta|$ (Fig. 15.1e), an individual two-level system can be isolated, giving an analogue of spin-orbit coupling (SOC). In this case, only two of the three bare states contribute and the approximate Hamiltonian is

$$\hat{\mathcal{H}}_2 = \frac{\hbar^2(\hat{k}_x^2 + \hat{k}_y^2 + \hat{k}_z^2)}{2m}\check{1} + \frac{\hbar\Omega_\mathrm{R}}{2}\check{\sigma}_1 + \frac{\hbar^2 k_\mathrm{L}\hat{k}_x}{m}\check{\sigma}_3 + (E_L)\,\check{1}, \tag{15.3}$$

where we have replaced the $f = 1$ angular momentum operators in Eq. (15.1) by the corresponding Pauli matrices $\check{\sigma}_{1,2,3}$ and defined $\Omega_\mathrm{R} = \sqrt{2}\Omega_\mathrm{L}$. Ignoring the constant-energy offsets, this Hamiltonian may be approximated as

$$\hat{\mathcal{H}}_\mathrm{SOC} \approx \frac{(\hat{\mathbf{p}}\check{1} - \mathcal{A}\check{\sigma}_3\mathbf{e}_x)^2}{2m^*}, \tag{15.4}$$

[1] Since there are no real electric charges in these neutral systems, we adopt the convention where we work in terms of charge-free quantities, e.g., for the vector potential $q\mathbf{A} \to \boldsymbol{\mathcal{A}}$ and $q\mathbf{B} \to \boldsymbol{\mathcal{B}} = \nabla \times \boldsymbol{\mathcal{A}}$.

where $\mathcal{A} = \hbar k_L[1 - (\hbar\Omega_L/2k_L)^2]^{1/2}$ is the magnitude of the vector potential $\boldsymbol{\mathcal{A}} = \mathcal{A}\check{\sigma}_3\mathbf{e}_x$, which in this case is matrix-valued, leading to a system that experiences state-dependent artificial magentic or electric fields when $\nabla \times \boldsymbol{\mathcal{A}} \neq 0$ or $\partial_t\boldsymbol{\mathcal{A}} \neq 0$.

15.2.2 Artificial Fields and the Gross-Pitaevskii Equation

Weakly interacting dilute gas BECs are near-perfect realizations of an idealized model originally conceived for studying the properties of helium superfluids [26, 27]. For harmonically trapped ultracold BECs, this description [28, 7], characterized by the GPE, accounts for the density distribution, collective excitations, and time-of-flight evolution.

Artificial gauge fields are readily incorporated into GPE by modifying the kinetic energy terms in the Hamiltonian. In rotating-system experiments, the GPE was adapted to account for the effective magnetic field experienced by the ultracold gas in the rotating frame [29, 30]. Similarly, Raman-transition artificial fields modify the system Hamiltonian, this time in the laboratory frame. The modified single-particle Hamiltonian from the previous section can be incorporated into the standard time-dependent GPE by including the vector potential $\boldsymbol{\mathcal{A}}$ in the kinetic energy such that

$$\begin{aligned} i\hbar\frac{\partial}{\partial t}\psi(\mathbf{r}) &= \hat{\mathcal{H}}_{\mathrm{GPE}}\psi(\mathbf{r}) \\ &= \left[\frac{1}{2}\left(\hat{p}_i - \mathcal{A}_i\right)\left(m^{-1}\right)_{i,j}\left(\hat{p}_j - \mathcal{A}_j\right) + V(\mathbf{r}) + gn(\mathbf{r})\right]\psi(\mathbf{r}), \end{aligned} \tag{15.5}$$

where $V(\mathbf{r})$ is the external trapping potential, $g = 4\pi\hbar^2 a/m$ is the coupling constant for interatomic scattering length a and atomic mass m, $n = N|\psi(\mathbf{r})|^2$ is the number density for order parameter $\psi(\mathbf{r})$ and total atom number N, and $(m^{-1})_{ij}$ are the tensor elements of the possibly anisotropic (and even nondiagonal) effective mass that accounts for the modification of the dispersion relationship's curvature (the sum over repeated indices is assumed).

From this equation, the set of hydrodynamic equations

$$\frac{\partial n(\mathbf{r},t)}{\partial t} = -\left(m^{-1}\right)_{ij}\frac{\partial}{\partial x_i}\left\{\left[\hat{p}_j(\mathbf{r},t) - \mathcal{A}_j(\mathbf{r})\right]n(\mathbf{r})\right\} \tag{15.6}$$

$$\frac{\partial p_i(\mathbf{r},t)}{\partial t} = -\frac{\partial}{\partial x_i}\left\{\frac{1}{2}\left(\hat{p}_j - \mathcal{A}_j\right)\left(m^{-1}\right)_{j,k}\left(\hat{p}_k - \mathcal{A}_k\right) + V(\mathbf{r},t) + gn(\mathbf{r})\right\} \tag{15.7}$$

describe the time-dependence of the density and canonical momentum. The vector potential is assumed to be time independent, and the "quantum pressure" term was omitted (valid for the case where density varies smoothly). The Heisenberg equation of motion for velocity gives $\hat{v}_i = (m^{-1})_{ij}(-i\hbar\partial_j - \mathcal{A}_j)$ for particles of mass

m, emphasizing that the canonical momentum $\mathbf{p}(\mathbf{r},t)$ is gauge dependent and differs from the gauge-independent mechanical momentum $\mathbf{p}_\mathrm{m}(\mathbf{r},t) = \mathbf{p}(\mathbf{r},t) - \boldsymbol{\mathcal{A}}(\mathbf{r})$.

One final constraint on this system is its irrotationality: the canonical momentum necessarily obeys $\nabla \times \mathbf{p}(\mathbf{r},t) = 0$. The mechanical momentum, however, need not be irrotational, so the application of an appropriate gauge field can lead to rotational motion.

The steady-state solutions of these hydrodynamic equations, found when $\partial n(\mathbf{r},t)/\partial t = 0$ and $\partial \mathbf{p}(\mathbf{r},t)/\partial t = 0$, yield expressions for the *in situ* density distribution. As an example, consider a harmonically trapped BEC with confining potential $V(\mathbf{r}) = m(\omega_x^2 x^2 + \omega_y^2 y^2 + \omega_z^2 z^2)/2$, subject to a uniform artificial magnetic field $\boldsymbol{\mathcal{B}} = \mathcal{B}\mathbf{e}_z$ expressed via a Landau-gauge vector potential $\boldsymbol{\mathcal{A}} = -\mathcal{B}y\mathbf{e}_x$. Irrotationality permits solutions for the canonical momentum with the form $\mathbf{p} = -\alpha\mathcal{B}\nabla(xy)$, where $\alpha = (1+\tilde{\epsilon})/2$ and $\tilde{\epsilon}$ is an anisotropy parameter. This gives

$$n(\mathbf{r},t) = \frac{1}{g}\left\{\mu - \frac{1}{2}m_x^*\tilde{\omega}_x^2 x^2 - \frac{1}{2}m_y^*\tilde{\omega}_y^2 y^2 - \frac{1}{2}m\omega_z^2 z^2\right\}, \tag{15.8}$$

with chemical potential μ, effective frequencies $\tilde{\omega}_x^2 = \omega_x^2 + \alpha^2\Omega_C^2$ and $\tilde{\omega}_y^2 = \omega_y^2 + (1-\alpha)^2\Omega_C^2$, effective masses m_x^* and m_y^* (representing the nonzero elements of the effective mass tensor; here the nondiagonal components are zero), cyclotron frequency $\Omega_C = \mathcal{B}/\bar{m}$ (where $\bar{m} = \sqrt{m_x^* m_y^*}$), and anisotropy $\tilde{\epsilon} = (\tilde{\omega}_x^2 - \tilde{\omega}_y^2)/(\tilde{\omega}_x^2 + \tilde{\omega}_y^2)$. Like the usual trapped BEC, the *in situ* density distribution is of the form of an inverted parabola. Thus, one effect of the uniform artificial magnetic field is to modify the cloud's widths.

Time-Dependent Superfluid Hydrodynamics

Similar to solid-state systems, the transport, or dynamic, properties of the system can be analyzed to reveal the internal properties of the BEC. The presence of artificial fields, such as real magnetic or electric fields, modifies the transport properties in ways that depend on the microscopic details of the system. The transport that emerges in response to an external driving force manifests itself as time-dependent variations in density and velocity; for sufficiently small perturbations, the total density and velocity are equal to their steady-state values plus a time-dependent perturbation: $n(\mathbf{r},t) = n_0(\mathbf{r}) + \delta n(\mathbf{r},t)$ and $\mathbf{p}(\mathbf{r},t) = \mathbf{p}(\mathbf{r}) + \delta\mathbf{p}(\mathbf{r},t)$. Absent vortices, these variations are described through the linearized superfluid hydrodynamic equations,

$$\begin{aligned}
0 &= \frac{\partial}{\partial t}\delta n(\mathbf{r},t) + \left(m^{-1}\right)_{ij}\frac{\partial}{\partial x_i}\left\{n_0(\mathbf{r})\delta\hat{p}_j(\mathbf{r},t) + \left[\hat{p}_j(\mathbf{r},t) - \mathcal{A}_j(\mathbf{r})\right]\delta n(\mathbf{r},t)\right\} \\
0 &= \frac{\partial}{\partial t}\delta p_i(\mathbf{r},t) + \frac{\partial}{\partial x_i}\left\{\frac{1}{2}[\hat{p}_j(\mathbf{r},t) - \mathcal{A}_j(\mathbf{r})]\left(m^{-1}\right)_{j,k}[\delta\hat{p}_k(\mathbf{r},t) - \delta\mathcal{A}_k(\mathbf{r})]\right. \\
&\qquad \left. + \delta U(\mathbf{r}) + g\delta n(\mathbf{r},t)\right\} + \tau^{-1}\delta p_i(\mathbf{r},t),
\end{aligned} \tag{15.9}$$

where damping (whether due to trap imperfections or finite temperature) is phenomenologically included through τ.

In harmonically trapped systems, the spectrum of collective modes can be determined from these equations (or their spin-orbit coupling counterparts). Like the emergence of the scissors mode in rotating systems [3], artificial fields can lead to coupling between the conventional collective modes of the BEC (e.g., dipole, monopole, or quadrupole [28]). For modes that depend on microscopic parameters such as interatomic interactions, changes in amplitude or frequency can reveal phase transitions between many-body states, such as an SOC-driven quantum phase transition between magnetic/nonmagnetic [21] states and a $\mathcal{B}$-driven structural phase transition between continuous-phase/vortex states [31].

Vortex Nucleation

As in "rotating bucket" experiments for superfluid helium, BECs subject to artificial magnetic fields may be able to reduce their total energy by concentrating variations in phase to a singular point – a vortex. Here, the density is zero at a point, around which the phase winds by 2π. The threshold for vortex nucleation depends on the details of the system, including the interaction parameter and the density (including its overall shape). The energetic cost to the system for each vortex is associated with the kinetic energy required to change the density from its nominal value to zero over a short distance; the system undergoes a structural phase transition from a continuous-phase system to one with vortices when the artificial gauge field exceeds a critical value $\mathcal{B}^{\text{crit}} = (5\hbar/R_\perp^2)\ln(R_\perp/\xi)$ in a disk-shaped condensate, where $R_\perp = \left[4\mu/m(\omega_x^2 + \omega_y^2)\right]^{1/2}$ is the mean transverse size in a harmonic potential with frequencies $\omega_{x,y,z}$ and the healing length $\xi = (\hbar^2/2m\mu)^{1/2}$ sets the characteristic vortex core size, where μ is the chemical potential [6]. As in rotating BEC experiments that demonstrated superfluidity via observations of quantized vortices [2, 3, 4, 5], artificial gauge fields have also been used to generate vortices in BECs [17].

Time-of-Flight

As in many other BEC experiments, time-of-flight (TOF) measurements are used to study the momentum distributions of systems subject to artificial fields. In general, TOF is performed by abruptly removing the trapping potential and allowing the system to evolve freely. This free evolution begins with rapid expansion driven by interactions until the density is reduced, after which time it continues ballistically [32]. In experiments with artificial gauge fields, the finite volume over which the artificial field acts is smaller than the final TOF volume of the cloud. To avoid obfuscating the TOF dynamics, the artificial fields are generally removed as TOF begins.

Though artificial field–driven changes in the momentum distribution can modify the shape of the TOF distribution in ways that initially seem to obscure the usual interpretations of BEC momentum distributions, these variations can be used to extract information about the *in situ* distribution or dynamics of the system. Generally, the artificial vector potential $\mathcal{A}(\mathbf{r})$ at the beginning of TOF is spatially dependent, and it is changed to a final spatially uniform value $\mathcal{A}_\mathrm{f} = \mathcal{A}_\mathrm{f}\mathbf{e}_x$. The temporal variation of $\mathcal{A}$ modifies the mechanical momentum due to an artificial electric field $\mathcal{E} = \Delta\mathcal{A}(\mathbf{r})/\Delta t$ and its associated impulse $\Delta\mathbf{p}_\mathrm{m}(\mathbf{r}) = \mathcal{A}(\mathbf{r}) - \mathcal{A}_f$. Up to the spatially dependent final vector potential $\mathcal{A}_f$, the TOF momentum is modified by a spatially dependent contribution that is proportional to the *in situ* vector potential; this contribution can be measured and used to determine $\mathcal{A}(\mathbf{r})$.

The specifics of the TOF expansion, described by the GPE, are well approximated by the linearized superfluid hydrodynamic equations (15.6, 15.7) if the ansatz for the density distribution,

$$n(x, y, z, t) = \frac{\mu(t)}{g}\left\{1 - \left[\frac{x}{R_x(t)}\right]^2 - \left[\frac{y}{R_y(t)}\right]^2 - \left[\frac{z}{R_z(t)}\right]^2 - \frac{s_{xy}(t)}{R_x(t)R_y}xy\right\}, \tag{15.10}$$

includes a "skew" parameter s_{xy} that accounts for $\mathbf{e}_y$ dependent force along $\mathbf{e}_x$ that accompanies the $\mathcal{A}$ turnoff, and $R_i(t)$ are the time-dependent Thomas-Fermi radii that are increasing as the cloud undergoes expansion. Measuring the skew parameter reveals the effect of the well-defined change in $\mathcal{A}$ upon the system's momentum distribution, which lends insight into the *in situ* dynamics of the system.

15.3 BECs with Uniform Synthetic Magnetic Fields

In Raman-dressed systems, artificial magnetic fields are created when the dispersion relationship for the lowest-energy eigenstate of the coupled system exhibits a single minimum, and that minimum in spatially dependent. In the Landau gauge $\mathcal{A} = \mathcal{A}\mathbf{e}_x = -\mathcal{B}y\mathbf{e}_x$, $\mathcal{B}$ is the magnitude of this artificial magnetic field. In low-strength artificial magnetic fields, the order parameter phase is continuous across the system, while for high fields, phase singularities (vortices) are distributed throughout.

In the low-field regime, the experimental signatures of the artificial field are more subtle: *in situ*, the shape of the cloud is slightly modified, as described by Eq. (15.8). While such changes would be difficult to measure experimentally, two other signatures of the artificial magnetic field are evident when dynamics are considered: *in situ*, the superfluid-Hall effect is evident when transport is introduced; in TOF, the bulk shape of the cloud is modified when the removal of the artificial field induces an artificial electric field impulse.

In the high-field regime, the nucleation of vortices significantly modifies the momentum distribution of the system, and the effect of this is evident in both superfluid-Hall and TOF measurements. Significant qualitative differences between these point to the location of the structural phase transition between the two regimes. These experiments demonstrate a new technique for probing such phase transitions in quantum gas environments.

15.3.1 Superfluid Hall Effect

Using a transport measurement akin to conventional Hall methods, the superfluid Hall effect is evident in dynamics measurements of a BEC in an artificial magnetic field. Similar to the initial perpendicular deflection of an electron current moving through a solid-state device in the response to an external magnetic field in the conventional Hall effect, the analogue of the Lorentz force due to an artificial magnetic field causes a deflection of moving atoms in a BEC in a direction perpendicular to both the artificial magnetic field and the momentum. Formally, the dynamics of the electron gas and the atoms in a BEC obey the same equations of motion; written in terms of typical BEC quantities, Eq. (15.7) can be recast in terms of the convective derivative and a "resistivity" matrix:

$$0 = \frac{D\mathbf{p}_{\mathrm{m}}(\mathbf{r})}{Dt} + \begin{pmatrix} \tau^{-1} & -\Omega_{\mathrm{C}} \\ \Omega_{\mathrm{C}} & \tau^{-1} \end{pmatrix} \mathbf{p}_{\mathrm{m}}(\mathbf{r}) + \nabla\left[U(\mathbf{r}) + gn(\mathbf{r})\right]. \qquad (15.11)$$

This same equation represents a Drude model of an electron gas [33], where, for the two-dimensional (2D) electron gas [34], the interaction term $\nabla\left[gn(\mathbf{r})\right]$ is replaced by a Fermi pressure term $(1/n)\nabla P(\mathbf{r})$. The interplay between terms in this equation means the $\mathcal{B}$-dependent dynamics also depend on interaction properties and allow these internal characteristics to be extracted from transport measurements.

To study transport in the BEC system, an elongated BEC was subjected to a periodic modulation of its trapping potential along its long direction (Fig. 15.2a). The compression and expansion of the cloud in this direction constituted an oscillating atomic current, and in the presence of the artificial magnetic field, this current was deflected perpendicularly by an amount proportional to the field strength (Fig. 15.2b). To measure this superfluid Hall effect, the *in situ* density of the atoms was recorded throughout several periods of the modulation cycle (Fig. 15.2c), and the skew periodically induced in the normally elliptical density profile indicated the degree of deflection (Fig 15.2d). By probing this behavior at a frequency near one of the collective resonances of the BEC dynamics (which depend on the interactions in the system), a strong response was observed and could be used to extract the internal properties of the BEC. In particular, by modeling the superfluid hydrodynamics of

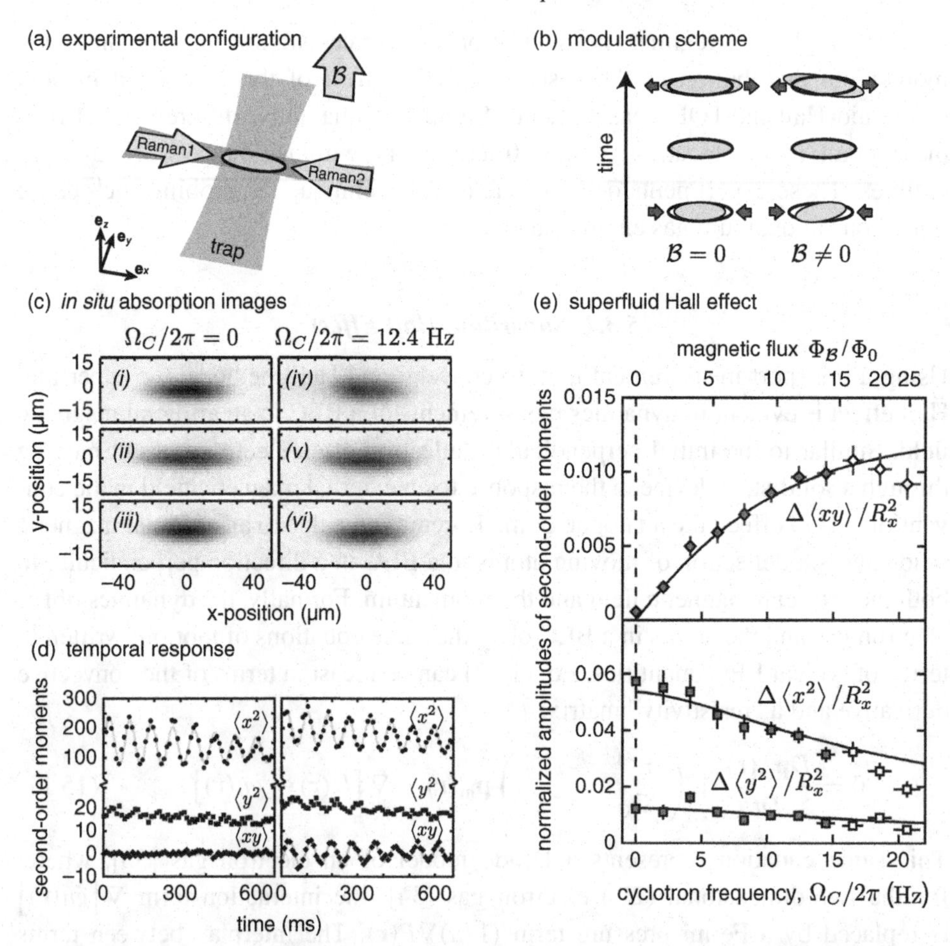

Figure 15.2 The superfluid Hall effect in a BEC. (a) The BEC was confined at the intersection of two optical-dipole trapping beams, and subject to an artificial magnetic field as a consequence of the counterpropagating Raman beams, along $\mathbf{e}_x$ and a magnetic field gradient along $\mathbf{e}_y$ (not shown). The trapping beam along $\mathbf{e}_y$ was intensity modulated, to drive transport along $\mathbf{e}_x$. (b) The modulation stretches the cloud primarily along $\mathbf{e}_x$. When an artificial field is present, the Hall effect causes shearing of the density profile. (c) Absorption images at three instances ((*i,iv*) 157.3 ms, (*ii,v*) 178.1 ms, (*iii,vii*) 198.9 ms), in the 83.3 ms modulation period with (*i–iii*) and without (*iv–vi*) an artificial field present. (d) Measurements of the second-order moments of the density profiles throughout eight cycles of modulation without (left) and with (right) an artificial magnetic field present. Light gray curves are guides to the eye. The amplitudes of these modulations give the signal shown in (e). (e) Measurements of the Hall signal $\Delta\langle xy\rangle/R_xR_y$ as a function of artificial field strength, parameterized as $\Omega_C = \mathcal{B}/\bar{m}$, as well as the signals for the complementary second-order moments $\Delta\langle x^2\rangle/R_x^2$ and $\Delta\langle y^2\rangle/R_y^2$, and curves indicating the superfluid hydrodynamics model, as fit simultaneously to all three sets of data. Open symbols indicate those points where vortices were likely to have entered the system. Adapted with permission from LeBlanc, L. J., et al. (2012), Observation of a superfluid Hall effect, *Proc. Nat. Acad. Sci. (USA)*, **109**, 10811 [31]. Copyright (2012) National Academy of Sciences, USA.

the modulated BEC, good agreement was found between the measurements and the calculations (Fig. 15.2e). The microscopic interaction properties of the BEC can be determined by comparing the model (which has as a key parameter the interaction strength between particles) and the experiment. While these interaction characteristics are well understood in the context of a harmonically trapped BEC, the development of a method that uses only a bulk measurement to determine microscopic properties is an important tool that can be extended to systems with more complicated internal dynamics.

Though on the edge of the available experimental resolution, one feature is evident in the measurements of the superfluid Hall effect: there is a deviation from the model at high artificial field values. This change indicates a change in the system's internal dynamics, which in this configuration suggests the emergence of vortices, whose current patterns are quite different from the uniform system. Even without directly observing the zero-density cores of the vortices (which would be difficult in that particular geometry), this method indicates the structural phase transition between the uniform system and one with phase defects.

15.3.2 Time-of-Fight Measurements and Vortex Nucleation

Complementing the *in situ* measurements, TOF measurements provide a direct probe of the system's momentum distributions. In the first demonstration of the Raman-induced artificial magnetic field [17], the emergence of directly imaged vortices was used as evidence of the magnetic field's existence. However, further inspection of these data shows a secondary signal: the bulk shape of each cloud is "sheared" after TOF, and the degree of shear increases with $\mathcal{B}$.

While the first measurements [17] were performed in a nearly round pancake-shaped trap (which abetted the imaging of vortex cores; Fig. 15.3a), the TOF shear in elongated surfboard-shaped traps (Fig. 15.3b) is more pronounced because the change in $\mathcal{A}$ along the length is greater. In these measurements, the shear parameter s_{xy} was measured for release from different initial field strengths where the spatially dependent part of changing vector potential was proportional to the Landau gauge. The shear parameter increases as the initial $\mathcal{B}$ increased, up to a critical value (Fig. 15.3d). After this point, the marginal increase in shear with $\mathcal{B}$ is smaller, or negative, and the density distribution within the clouds becomes fragmented (Fig. 15.3c). Both these (rather dramatic) signals point to the emergence of vortices and the structural phase transition from the superfluid hydrodynamic state to the one exhibiting vortices, again indicating a phase transition from a uniform phase distribution to the proliferation of vortices in a superfluid BEC. GPE calculations confirm that vortices nucleate at higher field strengths and are evident in both the *in situ* and TOF density distributions (Fig. 15.3b,c).

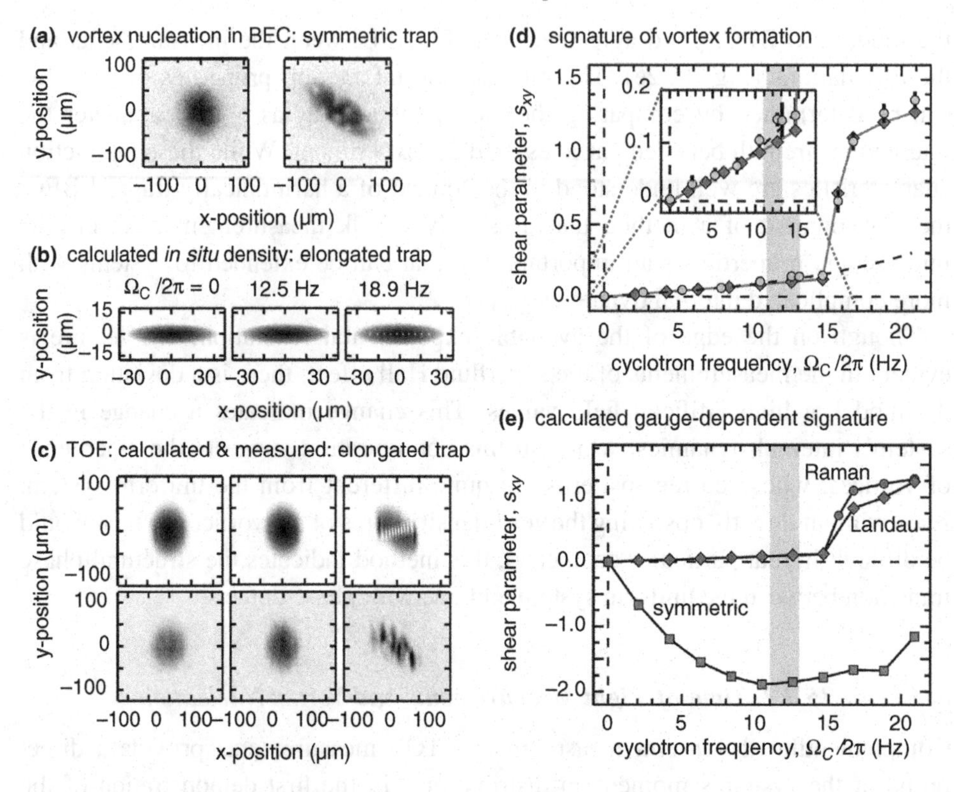

Figure 15.3 Indirect detection of vortex nucleation (a) TOF images of atoms released from a cylindrically symmetric ("pancake") trap below (left) and above (right) the critical field for vortex nucleation. Adapted with permission from Lin, Y.-J., et al. (2009), Synthetic magnetic fields for ultracold neutral atoms, *Nature* **462**, 628 [17]. Copyright (2009) by the Nature Publishing Group. (b) Calculated density profiles of an elongated BEC subject to three artificial magnetic field strengths. (c) Calculated (top) and measured (bottom) time-of-flight density profiles after release from the trap. Columns correspond to (b). (d) Signature of vortex nucleation as measured through the shear of TOF density distributions. Light points are experimental; dark diamonds are GPE calculated points, connected by dark lines; dashed line is the superfluid hydrodynamic model. The shaded region indicates the estimated critical field for vortex nucleation, including uncertainty (mainly due to atom number uncertainty). Inset shows the low-field points. (e) TOF density distribution shear, calculated for different "natural" gauge configurations with the same $\mathcal{B}$: circles for the Raman configuration, as used in experiment, diamonds for the Landau gauge, and squares for the symmetric gauge. Shaded region is as in (c). Plots (b) - (e) adapted with permission from LeBlanc, L. J. et al. (2015) *New J. Phys.* **17**, 065016, under the Creative Commons Attribution 3.0 Unported license.

As an interesting aside, these measurements reveal the existence of a "natural" gauge that accompanies each experimental configuration: there is a gauge for which $\mathcal{A} = 0$ when the control parameter is zero. In the case of conventional vector potentials that create electric and magnetic fields, this is the gauge for which $\mathcal{A}$ is zero when the current is zero: in parallel current sheets, the Landau gauge is the obvious choice, while for a solenoid, it is the symmetric gauge. In these experiments, the choice is revealed during the removal of the vector potential that accompanies the initiation of TOF: the change in vector potential will always be proportional to the Landau gauge, as described above, due to the geometry of this system, thereby making this the "natural" descriptor of this system. Calculations of the TOF shear show different behavior, depending on which gauge is natural in the system (Fig. 15.3e). In the case of these experiments, the Landau gauge configuration shows a much clearer signal, demonstrating the importance of designing experimental measurements with the natural gauge in mind.

15.4 BECs with Spin-Orbit Coupling

In this section, we turn our attention to systems with spin-orbit coupling, i.e., where linear momentum is coupled to the internal spin degrees of freedom. For example, we might study contributions to the atomic Hamiltonian such as $H_{\rm SOC} \propto k_x \check{\sigma}_3$, where motion along $\mathbf{e}_x$ couples to spin via the $\check{\sigma}_3$ Pauli matrix.

If adding artificial magnetic fields to cold atom systems was a clear path forward from the onset, the idea of spin-orbit coupled atomic gases came as a surprise. In this section, we will discuss the basic physics of spin-orbit coupled Bose-Einstein condensates (SOBECs), a new kind of quantum matter, unanticipated until its experimental realization. Because of the expected fractional quantum Hall effect physics of cold atoms in large magnetic fields, much initial work with artificial gauge fields – both experimental and theoretical – centered on conventional magnetic fields. In this case, the vector potential $\mathcal{A}$, or in a lattice the hopping phase ϕ, takes center stage.

Given this backdrop, the initial proposals for creating spin-orbit coupling in atomic gases were framed in the language of non-Abelian gauge fields in which each component of the vector potential $\check{\mathcal{A}}$ is an operator acting on the atomic spin degree of freedom. For example, the vector potential $\check{\mathcal{A}} = \mathcal{A}_0 \left(\check{\sigma}_2 \mathbf{e}_x - \check{\sigma}_1 \mathbf{e}_y\right)$ has noncommuting components proportional to the Pauli matrices, making this vector potential non-Abelian. This vector potential enters into the Hamiltonian as usual, but expanding the kinetic energy

$$H = \frac{\hbar^2}{2m}\left(\mathbf{k} - \frac{\check{\mathcal{A}}}{\hbar}\right)^2 = \frac{\hbar^2 \mathbf{k}^2}{2m} + \frac{\hbar^2 \mathcal{A}_0}{m}\left(-k_x \check{\sigma}_2 + k_y \check{\sigma}_1\right) + \mathcal{A}_0$$

shows that it is equivalent to SOC of the Rashba form.

Here we will describe the experimentally realized scheme in which the spin-orbit term in the Hamiltonian is $H_{\rm SOC} \propto k_x \check{\sigma}_3$, realized in the lab by the same pair of counterpropagating "Raman" lasers used to create artificial magnetic fields described above. This experimental geometry was first proposed in Ref. [8], where the authors used a language of periodic laser dressing rather than that of SOC. We will show the equivalence of these two pictures.

15.4.1 Formal Underpinning: Rotation

The first artificial gauge-field experiments with BECs used rotation as a means to introduce laboratory-tunable vector potentials. We understand these experiments in terms of the time-dependent Hamiltonian

$$\hat{H}(t) = \frac{\hbar^2\hat{\mathbf{k}}^2}{2m} + V\left(\hat{x}\cos\Omega t + \hat{y}\sin\Omega t, -\hat{x}\sin\Omega t + \hat{y}\cos\Omega t\right) \tag{15.12}$$

for particles in 2D, where the confining potential $V(\hat{\mathbf{r}})$ rotates with constant angular frequency Ω. We then consider the spatial rotation operator $\hat{R} = \exp(+i\hat{L}_z\Omega t/\hbar)$ that transforms into the frame co-rotating with the potential where

$$\hat{H}' = \hat{R}_z\hat{H}(t)\hat{R}_z^\dagger = \frac{\hbar^2\hat{\mathbf{k}}^2}{2m} + V\left(\hat{\mathbf{r}}\right) \quad \text{and} \quad \left|\psi'\right\rangle = \hat{R}_z\left|\psi\right\rangle. \tag{15.13}$$

This transformation seems nothing special, until we derive the rotating-frame time-dependent Schrödinger equation (TDSE)

$$\begin{aligned} i\hbar\frac{\partial}{\partial t}\left|\psi\right\rangle = \hat{H}(t)\left|\psi\right\rangle \quad &\rightarrow \quad i\hbar\hat{R}_z\frac{\partial}{\partial t}\left[\hat{R}_z^\dagger\left|\psi'\right\rangle\right] = \hat{H}'(t)\left|\psi'\right\rangle \\ &\rightarrow \quad i\hbar\frac{\partial}{\partial t}\left|\psi'\right\rangle = \left[\hat{H}'(t) + \hat{L}_z\Omega\right]\left|\psi'\right\rangle \end{aligned} \tag{15.14}$$

that contains a spatially varying vector potential $\hat{\mathcal{A}} = m\Omega\left(\hat{y}\mathbf{e}_x - \hat{x}\mathbf{e}_y\right)$ giving artificial field $\mathcal{B} = 2m\Omega$ along with an anticonfining potential $\hat{V}'(\hat{\mathbf{r}}) = -m\Omega^2\hat{\mathbf{r}}^2/2$. This method for creating artificial gauge fields requires a potential that is time independent in the rotating frame and a technique for bringing cold atomic gases into the rotating frame. These two technical issues together stalled progress in reaching the high-field limit with rotating gases; creating artificial gauge fields using laser light provided a new path forward that is now being followed by numerous experimental groups world wide. In the present context, an important lesson is that the gauge field appeared as a result of the unitary transformation between two frames: one with explicit time dependence and the other without. The gauge term resulted not directly from the transformation, but rather because the transformation depended explicitly on time so that $[\partial_t, \hat{R}_z] \neq 0$. Using this understanding, we will describe spin-orbit coupling in BECs as the outcome of a frame transformation.

15.4.2 Formal Underpinning: 1D Spin-Orbit Coupling

SOC with atomic gases can be understood in the language of dispersion relations formulated at the beginning of this chapter, but the connection to gauge fields is more clear if we begin with a description in real space, rather than focusing straight away on the momentum space description.

As was observed in Refs. [35, 36, 37], spin-independent scalar optical potentials U_s are accompanied by additional spin-dependent terms which become significant near atomic resonance: the rank-1 vector light shift U_v, and the rank-2 tensor light shifts [35] (negligible for the experiments described below). For the alkali atoms, adiabatic elimination of the $j = 1/2$ (D1) and $j = 3/2$ (D2) excited states yields a second-order Hamiltonian

$$H_L = \left[u_s(\mathbf{E}^* \cdot \mathbf{E}) + \frac{i u_v (\mathbf{E}^* \times \mathbf{E})}{\hbar} \cdot \check{\mathbf{J}} \right]$$

for ground-state atoms. Here $\mathbf{E}$ is the optical electric field; $u_v = -2u_s \Delta_{\rm FS}/3(\omega - \omega_0)$ determines the vector light shift; $\Delta_{\rm FS} = \omega_{3/2} - \omega_{1/2}$ is the fine-structure splitting; $\hbar\omega_{1/2}$ and $\hbar\omega_{3/2}$ are the D1 and D2 transition energies; and $\omega_0 = (2\omega_{1/2} + \omega_{3/2})/3$ is the oscillator-strength weighted average resonance frequency. u_s sets the scale of the light shift and is proportional to the atoms' ac polarizability.

The scalar and vector light shifts in H_L can be independently tuned using the laser frequency and electric field strength. The vector light shift is a contribution to the total Hamiltonian acting like an effective magnetic field

$$\mathbf{B}_{\rm eff} = \frac{i u_v (\mathbf{E}^* \times \mathbf{E})}{\mu_B \, g_J}$$

that effects $\check{\mathbf{J}} = \check{\mathbf{L}} + \check{\mathbf{S}}$, the combined orbital $\check{\mathbf{L}}$, and electron-spin $\check{\mathbf{S}}$ angular momentum, but not the nuclear spin $\check{\mathbf{I}}$. For small applied magnetic fields B_0, the effective Hamiltonian for a single manifold of total angular momentum $\check{\mathbf{F}} = \check{\mathbf{J}} + \check{\mathbf{I}}$ states is

$$H_0 + H_L = u_s(\mathbf{E}^* \cdot \mathbf{E}) + \frac{\mu_B g_F}{\hbar} (\mathbf{B} + \mathbf{B}_{\rm eff}) \cdot \check{\mathbf{F}}.$$

Notice that $\mathbf{B}_{\rm eff}$ acts as a true magnetic field and adds vectorially with $\mathbf{B}$.

Now, we consider an atom in a magnetic field $\mathbf{B} = B_0 \mathbf{e}_z$, illuminated by lasers with frequencies ω and $\omega + \delta\omega$, where $\delta\omega \approx |g_F \mu_B B_0/\hbar|$ differs by a small detuning $\delta = g_F \mu_B B_0/\hbar - \delta\omega$ from the linear Zeeman shift between m_F states (where $|\delta| \ll \delta\omega$). The electric field $\mathbf{E} = \mathbf{E}_{\omega_-} \exp(-i\omega t) + \mathbf{E}_{\omega_+} \exp[-i(\omega + \delta\omega)t]$ gives the effective magnetic field

$$\begin{aligned} \mathbf{B} + \mathbf{B}_{\rm eff} = & B_0 \mathbf{e}_z + \frac{i u_v}{\mu_B g_J} \Big[\left(\mathbf{E}^*_{\omega_-} \times \mathbf{E}_{\omega_-} \right) + \left(\mathbf{E}^*_{\omega_+} \times \mathbf{E}_{\omega_+} \right) \\ & + \left(\mathbf{E}^*_{\omega_-} \times \mathbf{E}_{\omega_+} \right) e^{-i\delta\omega t} + \left(\mathbf{E}^*_{\omega_+} \times \mathbf{E}_{\omega_-} \right) e^{i\delta\omega t} \Big]. \end{aligned}$$

The first two terms are time independent and add to the static bias field $B_0\mathbf{e}_z$. The remaining two time-dependent terms lead to transitions between different m_F levels. When $B_0 \gg |\mathbf{B}_{\text{eff}}|$ and $\delta\omega$ are large compared with the kinetic energy, the Hamiltonian can be simplified by time-averaging to zero the time-dependent terms in the scalar light shift. After making the rotating wave approximation (RWA) to eliminate the time dependence of the coupling fields, the Hamiltonian

$$\hat{H}_{\text{RWA}} = U(\mathbf{r})\check{1} + \boldsymbol{\Omega}\cdot\check{\mathbf{F}} \tag{15.15}$$

contains both a scalar potential

$$U(\mathbf{r}) = u_s\left(\mathbf{E}^*_{\omega_-}\cdot\mathbf{E}_{\omega_-} + \mathbf{E}^*_{\omega_+}\cdot\mathbf{E}_{\omega_+}\right) \tag{15.16}$$

and an RWA effective magnetic field

$$\begin{aligned}\boldsymbol{\Omega} = &\left[\delta + i\frac{u_v}{\hbar}\left(\mathbf{E}^*_{\omega_-}\times\mathbf{E}_{\omega_-} + \mathbf{E}^*_{\omega_+}\times\mathbf{E}_{\omega_+}\right)\cdot\mathbf{e}_z\right]\mathbf{e}_z \\ &-\frac{u_v}{\hbar}\text{Im}\left[\left(\mathbf{E}^*_{\omega_-}\times\mathbf{E}_{\omega_+}\right)\cdot\left(\mathbf{e}_x - i\mathbf{e}_y\right)\right]\mathbf{e}_x \\ &-\frac{u_v}{\hbar}\text{Re}\left[\left(\mathbf{E}^*_{\omega_-}\times\mathbf{E}_{\omega_+}\right)\cdot\left(\mathbf{e}_x - i\mathbf{e}_y\right)\right]\mathbf{e}_y.\end{aligned} \tag{15.17}$$

This expression is valid for $g_F > 0$ (for $g_F < 0$ the sign of the $i\mathbf{e}_y$ term would be positive, owing to selecting the opposite complex terms in the RWA).

Fig. 15.4a,b shows the typical experimental geometry [17, 19, 20, 22, 25] and level diagram for creating SOBECs with two counterpropagating Raman beams. In this simple case:

$$\mathbf{E}_{\omega_-} = Ee^{ik_{\text{L}}x}\mathbf{e}_y \qquad \text{and} \qquad \mathbf{E}_{\omega_+} = Ee^{-ik_{\text{L}}x}\mathbf{e}_z$$

describe the electric field of two lasers counterpropagating along $\mathbf{e}_x$ with equal intensities and crossed linear polarization. The resulting scalar light shift $U(\mathbf{r})$ and the effective magnetic field $\boldsymbol{\Omega}$ describing the vector light shift are

$$U(\mathbf{r}) = u_s\left(\mathbf{E}^*_{\omega_-}\cdot\mathbf{E}_{\omega_-} + \mathbf{E}^*_{\omega_+}\cdot\mathbf{E}_{\omega_+}\right) = 2u_sE^2$$

$$\boldsymbol{\Omega} = \delta\mathbf{e}_z + \Omega_R\left[\sin\left(2k_{\text{L}}x\right)\mathbf{e}_x - \cos\left(2k_{\text{L}}x\right)\mathbf{e}_y\right],$$

where $\Omega_L = u_vE^2/\hbar$. These describe a constant scalar light shift along with a spatially rotating effective magnetic field, as discussed in Ref. [19], which produced an artificial spin-orbit coupling and, in different notation, is equivalent to the proposal of Ref. [10]. Because this Hamiltonian is invariant under spatial translations with primitive vector $\mathbf{u} = \pi/k_{\text{L}}\mathbf{e}_x$, it would be expected to describe a periodic lattice.

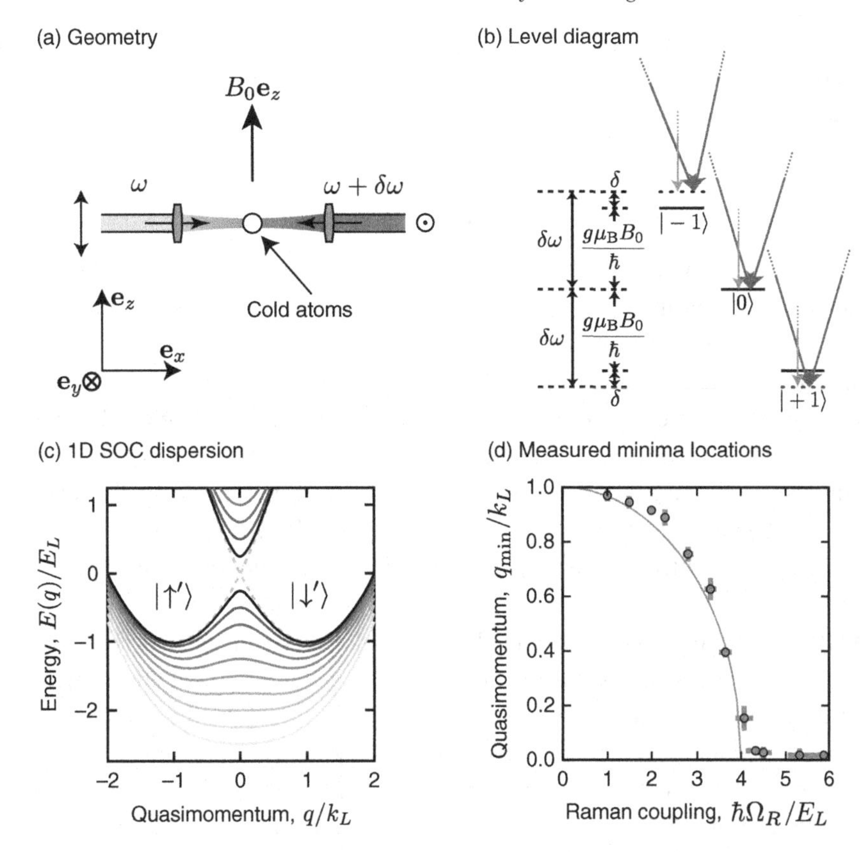

Figure 15.4 Experimental schematic. (a) Geometry with two counterpropagating lasers linearly polarized along $\mathbf{e}_y$ and $\mathbf{e}_z$ with frequencies ω and $\omega + \Delta\omega$. (b) Level diagram for coupling in an $f = 1$ manifold such as for ^{87}Rb. (c) Resulting spin-orbit dispersion for a range of coupling. (d) Measured quasimomentum of BECs adiabatically loaded into local minima of the SOC dispersion relation, clearly showing the prediction (solid curve) of the engineered Hamiltonian. Data shown here and the accompanying theory were provided by P. Engels, and image is adapted with permission from Hamner, C., et al. (2014), Dicke-type phase transition in a spin-orbit-coupled Bose-Einstein condensate. *Nat. Comm.*, **5** [38] (copyright (2014) by the Nature Publishing Group), where this behavior was interpreted in the context of a Dicke transition.

However, transforming the complete Hamiltonian according to the spatially dependent spin-rotation $\check{U}(x)\hat{H}\check{U}^\dagger(x)$, with $\check{U}(x) = \exp\left[i\check{F}_3\left(2k_{\mathrm{L}}x - \pi/2\right)/\hbar\right]$ gives

$$\hat{H} = \frac{\hbar^2}{2m}\left(\hat{k} - 2k_{\mathrm{L}}\check{F}_3/\hbar\right)^2 + U(\mathbf{r})\check{1} + \delta\check{F}_3 + \Omega_R\check{F}_1, \tag{15.18}$$

in which the position dependence has vanished from coupling vector $\delta\mathbf{e}_z + \Omega_R\mathbf{e}_y$, in exchange for a matrix valued gauge field, equivalent to SOC when the kinetic energy term is expanded. In analogy with the case for rotation, a unitary transformation (here a local spin rotation) is responsible for introducing the vector potential term into the Hamiltonian.

15.4.3 Experiment

Most experiments with SOC in atomic gases focused on just two levels – a pseudospin-1/2 system – in which any additional atomic levels are sufficiently far from two-photon resonance that they are neglected, as illustrated in Fig. 15.1d. The eigenenergies – spin-orbit coupled dispersion relations – for this two-level case are plotted in Fig. 15.4, depicting a pair of displaced parabolas (the displacement marks the presence of SOC) which become coupled as the laser intensity increases (this coupling results from the $\check{F}_1$ term in Eq. (15.18)). This characteristic dispersion relation features a double well in momentum space, where the minima shift together and finally merge as a function of increasing laser strength.

The first experimental evidence [19] of SOC in a BEC was provided by tracking the location of the minima in the double-minima structure, where a BEC was prepared in a superposition of both pseudospin states, and the laser intensity was adiabatically ramped on. More recent high-quality measurements from Ref. [38] provided by P. Engels show the measured location of the minima is plotted along with theory in Fig. 15.4d. This shift results directly from adding laser coupling to the atomic Hamiltonian and is single-particle physics.

The engineered single-particle SOC Hamiltonian was unexpectedly accompanied by new many-body physics; while the details are beyond the scope of this chapter, the basic physics is straightforward. In ^{87}Rb, the $|f = 1, m_F = -1\rangle$ and $|f = 1, m_F = 0\rangle$ hyperfine spin states used in the construction of the SOC Hamiltonian are miscible: a BEC formed from a mixture of these states will mix (an easy-plane ferromagnet) rather than phase separate (an easy-axis ferromagnet). This outcome depends on the specific atom under study; for example, in ^{23}Na these states are immiscible. In Ref. [19], the authors observed that as a function of Raman coupling strength Ω_R, this mixing behavior gave way to phase separation at a critical coupling strength. The physical interpretation of this is that with SOC the mixed phase acquires density modulations with amplitude $\propto \Omega_R^2$, increasing the interaction energy (for this reason, the mixed phase is often called the stripe phase). At the critical point, this increase overcomes the intrinsic energy gain from mixing, and the SOBEC undergoes a quantum phase transition into a spin-polarized ferromagnetic phase. Examples of this behavior are shown in Fig. 15.5a. The experimental observation of this phase transition marked SOBECs as a frontier system for realizing new forms of Bose-Einstein condensation.

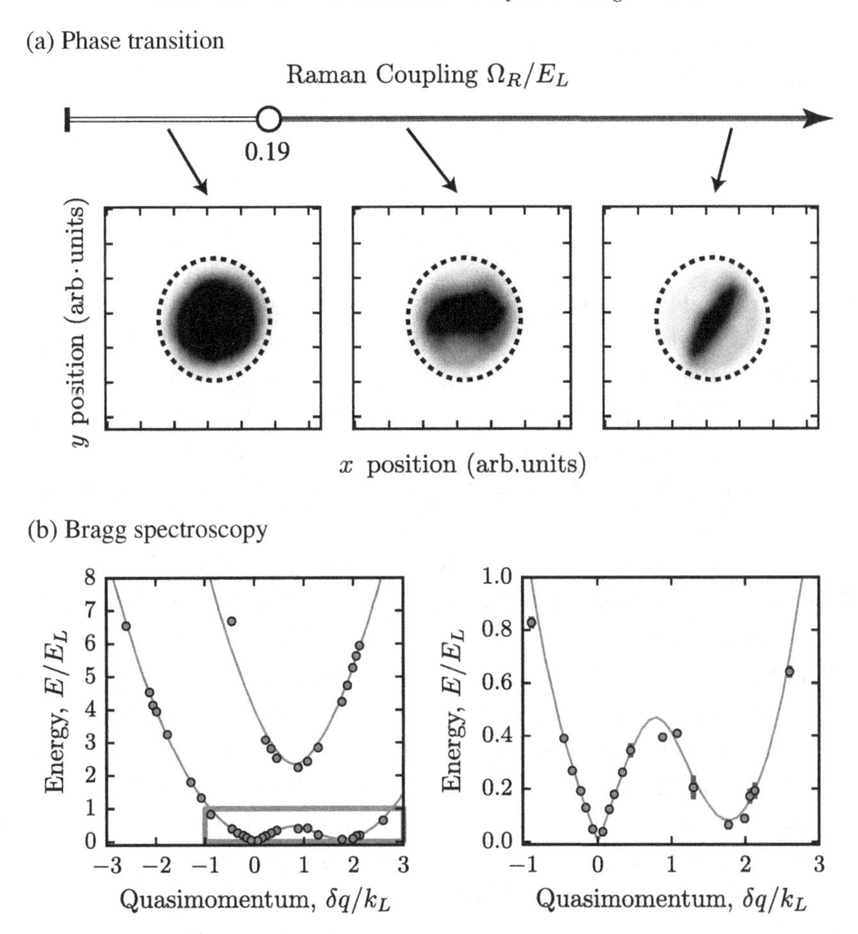

Figure 15.5 Experimental schematic. (a) Observed phase transition from miscible (striped) to immiscible, showing spin mixing (dark) for the whole BEC at small Ω_R and phase separation (white denotes regions of equal $|f = 1, m_F = -1\rangle$ and $|f = 1, m_F = 0\rangle$ within the dashed outline that indicates the extent of the BEC) for large Ω_R. Adapted with permission from Lin, Y.-J., et al. (2011), Spin-orbit-coupled Bose-Einstein condensates, *Nature*, **471**, 83 [19]. Copyright (2011) by the Nature Publishing Group. (b) Collective mode spectra showing roton-like mode at $\Omega_R = 2E_L$, which softens with decreasing Ω_R leading to the transition to the phase mixed phase. The right panel shows the inset marked by the gray box in the left panel. Adapted with permission from Ji, S.-C., et al. (2015), Softening of roton and phonon modes in a Bose-Einstein condensate with spin-orbit coupling, *Phys. Rev. Lett.*, **114**, 105301 [39]. Copyright (2015) by the American Physical Society, as provided by S. Chen.

This understanding can be turned about on itself by considering the Bogoliubov spectrum of the system (first computed by Higbie and Stamper-Kurn in Ref. [8]) starting in the phase separated phase: the double-well structure from the single particle Hamiltonian gives rise to a roton like feature[2] shown in Fig. 15.5, with roton-minimum lifted from zero by the "stripe energy." This feature falls with decreasing Ω_R and reaches zero at the critical coupling strength: a classic roton instability.

15.5 Conclusion

Artificial gauge fields offer the opportunity for studying BECs, and other ultracold quantum gases, under conditions that mimic those with applied electromagnetic fields or other effects like spin-orbit coupling. In this chapter, we discussed how artificial gauge fields change conventional BECs by explaining the modifications to the Gross-Pitaevskii formalism, especially in the case of an artificial magnetic field. We explored several experiments that demonstrated how these effects have made it possible to create and study analogues to the Hall effect, superfluid vortex nucleation in BECs, and a magnetic phase transition in an SOBEC.

While some of the original motivating experiments have yet to be realized – including cold-atom quantum-Hall-like systems [40, 41, 42] or the fractional quantum Hall effect [9] – these are increasingly within reach. The primary limitation to these methods is the heating associated with spontaneous emission, which ultimately limits the strength of the applied fields (especially in experiments with the alkali species). Apart from the goal of creating systems with very strong fields, the recognition that these methods could facilitate spin-orbit coupling in quantum gases has, perhaps, created an even greater opportunity for exploring new systems without conventional counterparts. Artificial SOC provides the path toward, for example, quantum gas analogues of the quantum spin Hall effect [43, 44] and exotic interacting topological insulators in Bose gases [16, 45]. In addition to the Raman-transition methods described here, several groups have recently demonstrated alternative methods for realizing artificial gauge fields by implementing light-assisted tunneling in optical lattices [46], including realizing strong magnetic fields [47] and simulating the Harper [48], Haldane [49], and Hofstadter [50] Hamiltonians. (See also Chapter 14.) Artificial magnetic fields, and their cousins, such as SOC, have shown us that the Hamiltonians governing BECs can be substantially re-engineered, giving us the freedom to explore both a variety of systems for which we have conventional analogues, as well as systems with unique properties that might only be possible in an ultracold quantum gas environment.

Acknowledgments: We appreciate enlightening conversations and collaborations with N. R. Cooper, G. Juzeliunas, J. V. Porto, and W. D. Phillips. This

[2] It is currently popular in cold-atom experiments to term these features rotons; however, as they are single-particle effects slightly altered by interactions, they are at best "roton-like."

work was partially supported by the Army Research Office (ARO) with funding from the Defense Advanced Research Projects Agency's (DARPA) Optical Lattice Emulator (OLE) program and the Atomtronics–Multidisciplinary University Research Initiative (MURI); and the National Science Foundation (NSF) through the Joint Quantum Institute (JQI) Physics Frontier Center. L. J. L. acknowledges support during the writing of this chapter from, in part, the Canada Research Chairs program and the Natural Sciences and Engineering Research Council (NSERC).

References

[1] Bloch, I., Dalibard, J., and Nascimbène, S. 2012. Quantum simulations with ultracold quantum gases. *Nat. Phys.*, **8**, 267–276.

[2] Matthews, M. R., Anderson, B. P., Haljan, P. C., Hall, D. S., Wieman, C. E., and Cornell, E A. 1999. Vortices in a Bose-Einstein condensate. *Phys. Rev. Lett.*, **83**, 2498–2501.

[3] Maragó, O. M., Hopkins, S. A., Arlt, J., Hodby, E., Hechenblaikner, G., and Foot, C J. 2000. Observation of the scissors mode and evidence for superfluidity of a trapped Bose-Einstein condensed gas. *Phys. Rev. Lett.*, **84**, 2056–2059.

[4] Madison, K. W., Chevy, F., Wohlleben, W., and Dalibard, J. 2000. Vortex formation in a stirred Bose-Einstein condensate. *Phys. Rev. Lett.*, **84**, 806.

[5] Haljan, P, Coddington, I., Engels, P., and Cornell, E. 2001. Driving Bose-Einstein-condensate vorticity with a rotating normal cloud. *Phys. Rev. Lett.*, **87**, 210403.

[6] Fetter, Alexander. 2009. Rotating trapped Bose-Einstein condensates. *Rev. Mod. Phys.*, **81**, 647–691.

[7] Dalfovo, F., Giorgini, S., Pitaevskii, L. P., and Stringari, S. 1999. Theory of Bose-Einstein condensation in trapped gases. *Rev. Mod. Phys.*, **71**, 463–512.

[8] Higbie, J., and Stamper-Kurn, D. 2002. Periodically dressed Bose-Einstein condensate: a superfluid with an anisotropic and variable critical velocity. *Phys. Rev. Lett.*, **88**, 090401.

[9] Sørensen, A., Demler, E., and Lukin, M. 2005. Fractional quantum Hall states of atoms in optical lattices. *Phys. Rev. Lett.*, **94**, 086803.

[10] Juzeliūnas, G., Ruseckas, J., Öhberg, P., and Fleischhauer, M. 2006. Light-induced effective magnetic fields for ultracold atoms in planar geometries. *Phys. Rev. A*, **73**.

[11] Zhu, Shi-Liang, Fu, Hao, Wu, C. J., Zhang, S. C., and Duan, L. M. 2006. Spin Hall effects for cold atoms in a light-induced gauge potential. *Phys. Rev. Lett.*, **97**, 240401.

[12] Cheneau, M., Rath, S. P., Yefsah, T., Günter, K. J., Juzeliūnas, G., and Dalibard, J. 2008. Geometric potentials in quantum optics: a semi-classical interpretation. *Europhys. Lett.*, **83**, 60001.

[13] Günter, K., Cheneau, M., Yefsah, T., Rath, S., and Dalibard, J. 2009. Practical scheme for a light-induced gauge field in an atomic Bose gas. *Phys. Rev. A*, **79**, 011604.

[14] Klein, A., and Jaksch, D. 2009. Phonon-induced artificial magnetic fields in optical lattices. *Europhys. Lett.*, **85**, 13001.

[15] Spielman, I. 2009. Raman processes and effective gauge potentials. *Phys. Rev. A*, **79**, 063613.

[16] Dalibard, J., Gerbier, F., Juzeliūnas, G., and Öhberg, P. 2011. Colloquium: artificial gauge potentials for neutral atoms. *Rev. Mod. Phys.*, **83**, 1523–1543.

[17] Lin, Y.-J., Compton, R. L., Jimenez-García, K., Porto, J. V., and Spielman, I. B. 2009. Synthetic magnetic fields for ultracold neutral atoms. *Nature*, **462**, 628–632.

[18] Lin, Y.-J., Compton, R. L., Jimenez-García, K., Phillips, W. D., Porto, J. V., and Spielman, I. B. 2011a. A synthetic electric force acting on neutral atoms. *Nat. Phys.*, **7**, 531.

[19] Lin, Y.-J., Jimenez-García, K., and Spielman, I. B. 2011b. Spin-orbit-coupled Bose-Einstein condensates. *Nature*, **471**, 83–86.

[20] Cheuk, L. W., Sommer, A. T., Hadzibabic, Z., Yefsah, T., Bakr, W. S., and Zwierlein, M. W. 2012. Spin-injection spectroscopy of a spin-orbit coupled Fermi gas. *Phys. Rev. Lett.*, **109**, 095302.

[21] Zhang, J.-Y., Ji, S.-C., Chen, Z., Zhang, L., Du, Z.-D., Yan, B., Pan, G.-S., Zhao, B., Deng, Y.-J., and Zhai, H. 2012. Collective dipole oscillations of a spin-orbit coupled Bose-Einstein condensate. *Phys. Rev. Lett.*, **109**, 115301.

[22] Wang, P., Yu, Z.-Q., Fu, Z., Miao, J., Huang, L., Chai, S., Zhai, H., and Zhang, J. 2012. Spin-orbit coupled degenerate Fermi gases. *Phys. Rev. Lett.*, **109**, 095301.

[23] Qu, C., Hamner, C., Gong, M., Zhang, C., and Engels, P. 2013. Observation of Zitterbewegung in a spin-orbit-coupled Bose-Einstein condensate. *Phys. Rev. A*, **88**, 021604(R).

[24] Olson, Abraham J., Wang, Su-Ju, Niffenegger, Robert J., Li, Chuan-Hsun, Greene, Chris H., and Chen, Yong P. 2014. Tunable Landau-Zener transitions in a spin-orbit-coupled Bose-Einstein condensate. *Phys. Rev. A*, **90**, 013616.

[25] Lin, Y.-J., Compton, R., Perry, A., Phillips, W., Porto, J., and Spielman, I. 2009. Bose-Einstein condensate in a uniform light-induced vector potential. *Phys. Rev. Lett.*, **102**, 130401.

[26] Pitaevskii, L. P. 1961. Vortex lines in an imperfect Bose gas. *Sov. Phys. JETP*, **13**, 451–454 [J. Exptl. Theoret. Phys. (U.S.S.R.) 40, 646–651 (1961).].

[27] Gross, Eugene P. 1963. Hydrodynamics of a superfluid condensate. *J. Math. Phys.*, **4**, 195.

[28] Stringari, S. 1996. Collective excitations of a trapped bose-condensed gas. *Phys. Rev. Lett.*, **77**, 2360.

[29] Zambelli, F., and Stringari, S. 2001. Moment of inertia and quadrupole response function of a trapped superfluid. *Phys. Rev. A*, **63**, 033602.

[30] Cozzini, M., and Stringari, S. 2003. Macroscopic dynamics of a Bose-Einstein condensate containing a vortex lattice. *Phys. Rev. A*, **67**, 041602.

[31] LeBlanc, L. J., Jiménez-García, K., Williams, R. A., Beeler, M. C., Perry, A. R., Phillips, W. D., and Spielman, I. B. 2012. Observation of a superfluid Hall effect. *Proc. Nat. Acad. Sci. (USA)*, **109**, 10811–10814.

[32] Castin, Y., and Dum, R. 1996. Bose-Einstein condensates in time dependent traps. *Phys. Rev. Lett.*, **77**, 5315–5319.

[33] Ashcroft, N. W., and Mermin, N. D. 1976. *Solid State Physics*. New York: Holt, Rinehart and Winston.

[34] Fetter, A. L. 1973. Electrodynamics of a layered electron gas. I. Single layer. *Annals of Physics*, **81**, 367–393.

[35] Deutsch, I. H., and Jessen, P. S. 1998. Quantum-state control in optical lattices. *Phys. Rev. A*, **57**, 1972–1986.

[36] Dudarev, A., Diener, R., Carusotto, I., and Niu, Q. 2004. Spin-orbit coupling and Berry phase with ultracold atoms in 2D optical lattices. *Phys. Rev. Lett.*, **92**, 153005.

[37] Sebby-Strabley, J., Anderlini, M., Jessen, P. S., and Porto, J. V. 2006. Lattice of double wells for manipulating pairs of cold atoms. *Phys. Rev. A*, **73**, 033605.

[38] Hamner, C., Qu, C., Zhang, Y., Chang, J., Gong, M., Zhang, C., and Engels, P. 2014. Dicke-type phase transition in a spin-orbit-coupled Bose-Einstein condensate. *Nature Comm.*, **5**, 4023.

[39] Ji, S.-C., Zhang, L., Xu, X.-T., Wu, Z., Deng, Y., Chen, S., and Pan, J.-W. 2015. Softening of roton and phonon modes in a Bose-Einstein condensate with spin-orbit coupling. *Phys. Rev. Lett.*, **114**, 105301.

[40] Cazalilla, M. A., Barberan, N., and Cooper, N. R. 2005. Edge excitations and topological order in a rotating Bose gas. *Phys. Rev. B*, **71**, 121303.

[41] Goldman, N., Dalibard, J., Dauphin, A., Gerbier, F., Lewenstein, M., Zoller, P., and Spielman, I. B. 2013. Direct imaging of topological edge states in cold-atom systems. *Proc. Nat. Acad. Sci. (USA)*, **110**, 6736–6741.

[42] Spielman, I. B. 2013. Detection of topological matter with quantum gases. *Ann. Phys. (Berlin)*, **525**, 797–807.

[43] van der Bijl, E., and Duine, R. A. 2011. Anomalous Hall conductivity from the dipole mode of spin-orbit-coupled cold-atom systems. *Phys. Rev. Lett.*, **107**, 195302.

[44] Beeler, M. C., Williams, R. A., Jiménez-García, Karina, LeBlanc, L. J., Perry, A. R., and Spielman, I. B. 2013. The spin Hall effect in a quantum gas. *Nature*, **498**, 201–204.

[45] Stanescu, T., Galitski, V., Vaishnav, J., Clark, C., and Das Sarma, S. 2009. Topological insulators and metals in atomic optical lattices. *Phys. Rev. A*, **79**, 053639.

[46] Jaksch, D., and Zoller, P. 2003. Creation of effective magnetic fields in optical lattices: the Hofstadter butterfly for cold neutral atoms. *New J. Phys.*, **5**, 56.

[47] Aidelsburger, M., Atala, M., Nascimbène, S., Trotzky, S., Chen, Y.-A., and Bloch, I. 2011. Experimental realization of strong effective magnetic fields in an optical lattice. *Phys. Rev. Lett.*, **107**, 255301.

[48] Miyake, H., Siviloglou, G. A., Kennedy, C. J., Burton, W. C., and Ketterle, W. 2013. Realizing the Harper Hamiltonian with laser-assisted tunneling in optical lattices. *Phys. Rev. Lett.*, **111**, 185302.

[49] Jotzu, G., Messer, M., Desbuquois, R., Lebrat, M., Uehlinger, T., Greif, D., and Esslinger, T. 2014. Experimental realization of the topological Haldane model with ultracold fermions. *Nature*, **515**, 237–240.

[50] Aidelsburger, M., Lohse, M., Schweizer, C., Atala, M., Barreiro, J. T., Nascimbène, S., Cooper, N. R., Bloch, I., and Goldman, N. 2014. Measuring the Chern number of Hofstadter bands with ultracold bosonic atoms. *Nat. Phys.*, **11**, 162–166.

16

Second Sound in Ultracold Atomic Gases

LEV PITAEVSKII

INO-CNR BEC Center and Dipartimento di Fisica, Università di Trento, Povo, Italy
Kapitza Institute for Physical Problems, Moscow, Russia

SANDRO STRINGARI

INO-CNR BEC Center and Dipartimento di Fisica, Università di Trento, Povo, Italy

We provide an overview of the recent theoretical and experimental advances in the study of second sound in ultracold atomic gases. Starting from Landau's two fluid hydrodynamic equations, we develop the theory of first and second sound in various configurations characterized by different geometries and quantum statistics. These include the weakly interacting three-dimensional (3D) Bose gas, the strongly interacting Fermi gas at unitarity in the presence of highly elongated traps, and the dilute two-dimensional (2D) Bose gas, characterized by the Berezinskii-Kosterlitz-Thouless transition. An explicit comparison with the propagation of second sound in liquid helium is carried out to elucidate the main analogies and differences. We also make an explicit comparison with the available experimental data and point out the crucial role played by the superfluid density in determining the temperature dependence of the second sound speed.

16.1 Introduction

Superfluidity is one of the most challenging features characterizing the behavior of ultracold atomic gases and provides an important interdisciplinary connection with the physics of other many-body systems in both condensed matter (superfluid helium, superconductivity) and high-energy physics (nuclear superfluidity, neutron stars). It is a phenomenon deeply related to Bose-Einstein condensation (BEC), which, in the case of Fermi superfluids, corresponds to the condensation of pairs. Several manifestations of superfluidity have already emerged in experiments on ultracold atomic gases, in the case of either bosons or fermions [1].

Important phenomena are the consequence of the irrotational nature of the velocity field associated with the superfluid flow. They include the occurrence of quantized vortices, which is actually a direct manifestation of the absence of shear viscosity, the quenching of the moment of inertia, and the behavior of the collective oscillations at zero temperature. Vortices in a superfluid are characterized by the quantization of the circulation $\oint \mathbf{v}_s d\mathbf{l}$ of the superfluid velocity around the vortex line. They have become experimentally available using different approaches. A first procedure consists of creating the vortical configuration with optical methods in two-component condensates [2]. A second method, which shares a closer resemblance with the rotating bucket experiment of superfluid helium, makes use of a suitable rotating modulation of the trap to stir the condensate [3]. Above a critical angular velocity, one observes the formation of vortices which are imaged after expansion. Configurations containing a large number of vortices can be created by stirring the condensate at angular velocities close to the centrifugal limit and have been realized in both Bose [4, 5], and Fermi [6] superfluid gases. The quenching of the moment of inertia due to superfluidity shows up in the peculiar behavior of the scissors mode [7, 8], and its temperature dependence has been detected directly by measuring the angular momentum of a trapped superfluid rotating at a given angular velocity [9]. The effects of irrotationality on the collective oscillations of both Bose and Fermi superfluid gases have been the object of extensive theoretical and experimental work, and are accurately described by superfluid hydrodynamics in the presence of harmonic traps [10].

An important consequence of superfluidity is the occurrence of a critical velocity below which the superfluid can move without dissipation (Landau's criterion). These experiments were carried out with moving focused laser beams ([11]) as well as with moving one-dimensional optical lattices [12]. Other manifestations of superfluidity are provided by the coherent macroscopic behavior responsible for quantum tunneling in the double-well geometry [13] and, in the case of Fermi superfluids, by the appearance of a gap in the single-particle excitation spectrum, revealed by radio frequency transitions [14]. Let us also mention the recent experimental investigation of the equation of state of a strongly interacting Fermi gas [15] (see also Fig. 9.2 and associated discussion). This measurement has revealed the typical λ-behavior of the specific heat at the superfluid transition and an accurate determination of the critical temperature.

A key property of a superfluid at finite temperature is its two-fluid nature, i.e., the simultaneous presence of two flows: a dissipationless superfluid flow and a normal flow. The propagation of second sound [16], which represents the topic of discussion of the present work, is a direct manifestation of this two-fluid nature. An important feature of second sound is that it can provide unique information on the temperature dependence of the superfluid density, as proven in the case of superfluid helium. For this reason, its investigation permits researchers to better clarify the conceptual difference between superfluidity and Bose-Einstein condensation. This distinction is particularly important in strongly interacting gases (such as the Fermi gas at unitarity), and it is even more dramatic in two dimensions, where long-range order, yielding Bose-Einstein condensation, is ruled out at finite temperature by the Hohenberg-Mermin-Wagner theorem.

In this chapter, we summarize some advances in the understanding of second sound in quantum gases that have been achieved theoretically and experimentally in the last few years. In Section 16.2, we discuss the general behavior of Landau's equations of two-fluid hydrodynamics. In Section 16.3, we discuss the solutions of these equations in the case of a 3D dilute Bose gas. In Section 16.4, we apply the equations to the case of a Fermi gas at unitarity in the presence of tight radial trapping conditions, giving rise to highly elongated configurations which are suitable for the experimental investigation of sound waves. Finally, in Section 16.5 we discuss the behavior of second sound in 2D Bose gases where its measurement could provide unique information on the behavior of the superfluid density.

16.2 Two-Fluid Hydrodynamics: First and Second Sound

In the present chapter, we will describe the macroscopic dynamic behavior of a superfluid at finite temperature. We will consider situations where the free path of the elementary excitation is small compared with the wavelength of the sound wave and local thermodynamic equilibrium is ensured by collisions. This condition is rather severe at low temperature, where collisions are rare. It permits us to define locally (in space and time) the thermodynamic properties of the fluid, like the temperature, the pressure, etc. The general system of equations describing the macroscopic dynamic behavior of the superfluid at finite temperature was obtained by Landau [17] (see also Ref. [18]).

In the following, we will limit ourselves to the description of small velocities and small amplitude oscillations, corresponding to the linearized solutions of the equations of motion. The propagation of such oscillations at finite temperature provides an important tool to investigate the consequences of superfluidity. The equation for the superfluid velocity $\mathbf{v}_s$ obeys the fundamental irrotational form

$$m\frac{\partial \mathbf{v}_s}{\partial t} + \nabla(\mu + V_{ext}) = 0 \tag{16.1}$$

fixed by the chemical potential $\mu(n,T)$ of uniform matter, which, in general, depends on the local value of the density and of the temperature.

The equation for the density n has the usual form of the continuity equation:

$$\frac{\partial n}{\partial t} + \text{div}\mathbf{j} = 0, \tag{16.2}$$

where the current density $\mathbf{j}$ can be separated into the normal and superfluid components $\mathbf{j} = n_s\mathbf{v}_s + n_n\mathbf{v}_n$ with $n_s + n_n = n$. The time derivative of the current density is equal to the force per unit volume which, in the linear approximation, is fixed by the gradient of the pressure and by the external potential (Euler equation):

$$\frac{\partial \mathbf{j}}{\partial t} + \frac{\nabla P}{m} + \frac{n}{m}\nabla V_{ext} = 0. \tag{16.3}$$

The last equation is the equation for the entropy density s. This can be derived from general arguments. In fact if dissipative processes such as viscosity and thermoconductivity are absent, the entropy is conserved and the equation for s takes the form of a continuity equation. Furthermore, only elementary excitations contribute to the entropy whose transport is hence fixed by the normal velocity $\mathbf{v}_n$ of the fluid. The equation for the entropy then takes the form

$$\frac{\partial s}{\partial t} + \mathrm{div}(s\mathbf{v}_n) = 0 \tag{16.4}$$

where, in the linear regime, the entropy entering the second term should be calculated at equilibrium.

The thermodynamic quantities entering the above equations are not independent and obey the Gibbs-Duhem thermodynamic equation $nd\mu = -sdT + dP$. It is worth mentioning that also this thermodynamic identity is valid only in the linear regime. In fact, in general the thermodynamic functions can exhibit a dependence also on the relative velocity $(\mathbf{v}_n - \mathbf{v}_s)$ between the normal and the superfluid velocities.

At $T = 0$, where $n_n = 0$, the entropy identically vanishes and the Euler equation (16.3), thanks to the Gibbs-Duhem relation, coincides with Eq. (16.1). The two-fluid hydrodynamic equations (16.1–16.4) then reduce to the equation of continuity and to the equation for the irrotational superfluid velocity. Vice versa, for temperatures larger than the critical value, where the superfluid density vanishes and the equation for the superfluid velocity can be ignored, the other three equations reduce to the classical equations of dissipationless hydrodynamics.

Let us for the moment ignore the external potential V_{ext}. By looking for plane wave solutions varying in space and time like $e^{-i\omega(t-x/c)}$, after a lengthy but straightforward calculation one obtains (see, for example, [1]) the Landau equation for the sound velocity [17],

$$c^4 - \left[\frac{1}{mn\kappa_s} + \frac{n_s T\bar{s}^2}{mn_n\bar{c}_v}\right]c^2 + \frac{n_s T\bar{s}^2}{mn_n\bar{c}_v}\frac{1}{mn\kappa_T} = 0\,, \tag{16.5}$$

where $\bar{c}_v = T(\partial\bar{s}/\partial T)_n$ is the specific heat at constant volume per particle; $\kappa_s = (\partial n/\partial P)_{\bar{s}}/n$ and $\kappa_T = (\partial n/\partial P)_T/n$ are, respectively, the adiabatic and isothermal compressibilities while $\bar{s} = s/n$ is the entropy per particle.

If $n_s \neq 0$, Eq. (16.5) gives rise to two distinct sound velocities, known as first and second sound. This is the consequence of the fact that in a superfluid there are two degrees of freedom associated with the normal and superfluid components. The existence of two types of sound waves in a Bose-Einstein condensed system was first noted by Tisza [19]. The second sound velocity was measured in superfluid

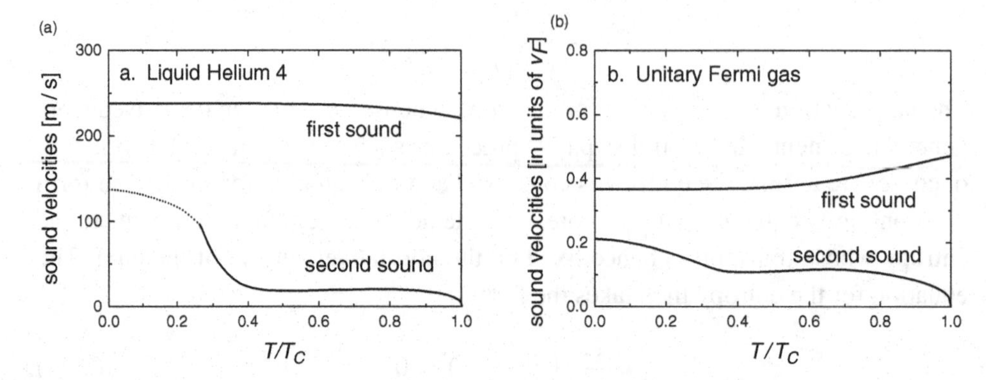

Figure 16.1 (a) Experimental values of first and second sound speeds in superfluid ^{4}He. Dots correspond to the theoretical calculation accounting for the contribution of the phonon–roton excitations to the thermodynamic functions. (b) Two-fluid sound speeds in a uniform Fermi gas at unitarity calculated using the Nozieres and Schmitt-Rink (NSR) thermodynamic functions. Figure adapted with permission from Taylor, E., et al. (2009), First and second sound in a strongly interacting Fermi gas, *Phys. Rev. A*, **80**, 053601 [21]. Copyright (2009) by the American Physical Society.

^{4}He by Peshkov [20]. The values of the two sound velocities, measured in superfluid ^{4}He, are reported in Fig. 16.1a as a function of T.

The knowledge of the two sound velocities can be used to evaluate the density strengths associated with the corresponding modes and the consequent possibility of exciting them with a density probe. For example, using a sudden laser perturbation, one excites both first and second sound with a relative weight given by the relative contribution of the two modes to the inverse energy weighted moment

$$\int_{-\infty}^{\infty} d\omega \frac{1}{\omega} S(\mathbf{q}, \omega) = \frac{1}{2} n \kappa_T \tag{16.6}$$

also known as the compressibility sum rule (see Chapter 7 in Ref. [1]), fixed by the isothermal compressibility and here derived in the limit of small momentum transfer $\mathbf{q}$. In Eq. (16.6), $S(\mathbf{q}, \omega)$ is the dynamical structure factor with momentum $\mathbf{q}$ and frequency ω. Taking into account the fact that at small wave vectors not only the inverse energy weighted sum rule but also the energy weighted moment $\int_{-\infty}^{\infty} \omega S(\mathbf{q}, \omega) = q^2/2m$, known as f-sum rule, is exhausted by the two sound modes [22], one straightforwardly finds that the relative contributions to the compressibility sum rule (16.6) from each sound are given by [23]

$$W_1 \equiv \frac{1 - mn\kappa_T c_2^2}{2m(c_1^2 - c_2^2)}, \qquad W_2 \equiv \frac{mn\kappa_T c_1^2 - 1}{2m(c_1^2 - c_2^2)}. \tag{16.7}$$

Landau's equation (16.5) is easily solved as $T \to T_c$, i.e., close to the transition temperature where $\rho_s \to 0$. In this case, the upper (first sound) and lower (second sound) solutions are given by

$$c_1^2 = \frac{1}{mn\kappa_s} = \frac{1}{m}\left(\frac{\partial P}{\partial n}\right)_{\bar{s}}, \qquad c_2^2 = \frac{n_s T \bar{s}^2}{mn_n \bar{c}_p}. \tag{16.8}$$

The first sound velocity is given by the isoentropic value and exhibits a continuous transition to the usual sound velocity above T_c. The second sound velocity is instead fixed by the superfluid density and vanishes at the transition where $n_s = 0$ in 3D systems. In the above equation, we have introduced the specific heat at constant pressure $\bar{c}_p$, related to $\bar{c}_v$ by the thermodynamic relation $\bar{c}_p/\bar{c}_v = \kappa_T/\kappa_s$.

Results (16.8) hold also in the $T \to 0$ limit, where one can set $\kappa_T = \kappa_s$ and $\bar{c}_p = \bar{c}_v$. In this limit, all the thermodynamic functions as well as the normal density ρ_n are fixed by the thermal excitation of phonons and one can easily prove that Eq. (16.8) yields result $c_2^2 = c_1^2/3$ for the second sound velocity.

For systems characterized by a small thermal expansion coefficient $\alpha = -(1/n)$ $\partial n/\partial T|_P = (\kappa_T/\kappa_s - 1)/T$, and hence very close values of the isothermal and adiabatic compressibilities, Eq. (16.8) provides an excellent approximation to the second sound velocity in the entire superfluid region $0 < T < T_c$. More precisely, the applicability of results (16.8) requires the condition

$$\frac{c_2^2}{c_1^2}\alpha T \ll 1. \tag{16.9}$$

Systems satisfying the condition (16.9) include not only the well-celebrated superfluid ^{4}He but also the interacting Fermi gas at unitarity (see Section 16.4). Furthermore, since in these cases $c_1^2 \sim (mn\kappa_s)^{-1} \sim (mn\kappa_T)^{-1}$, the second sound contribution to the compressibility sum rule (16.6) is negligible, and consequently second sound should be excited using a thermal rather than a density perturbation [24]. Eq. (16.8) for c_2 points out explicitly the crucial role played by the superfluid density and was actually employed to determine the temperature dependence of the superfluid density in liquid ^{4}He in a wide interval of temperatures [25].

Equations (16.8) are instead inadequate to describe the sound velocities in dilute Bose gases at intermediate values of T, due to the high isothermal compressibility exhibited by these systems, as we will discuss in Sections 16.3 and 16.5.

16.3 Second Sound in Weakly Interacting 3D Bose Gases

An important consequence of the high compressibililty exhibited by weakly interacting Bose gases is the occurrence of a hybridization phenomenon between first and second sound. This phenomenon, first investigated in the seminal paper by Lee and Yang [26] and later discussed by Griffin and co-workers (see, for example, [23, 27, 28]) has been recently investigated in detail in [29] where a suitable perturbative approach has been developed to explore the behavior of second sound below and above the hybridization point. The mechanism of hybridization is caused by the tendency of the velocity of the two modes to cross at very low temperatures, of the order of the zero temperature value of the chemical potential $\mu(T = 0) = gn$, where $g = 4\pi\hbar^2 a/m$ is the bosonic coupling constant. This phenomenon characterizes dilute Bose gases and is absent in strongly-interacting superfluids. It is natural to call the upper and lower branches as first and second sounds, respectively. For temperatures below the hybridization point, the velocity of the upper branch approaches the Bogoliubov value $c_B = \sqrt{gn/m}$, while the lower branch approaches Landau's result $c_2 = c_B/\sqrt{3}$. Above the hybridization point, the role of first and second sound is inverted, in the sense that the lower (second sound) branch is essentially an oscillation of the superfluid density, which practically coincides with the condensate density $n_0(T)$.

For temperatures higher than the hybridization temperature ($k_B T \gg gn$), a useful expression for the second sound velocity is obtained by evaluating all the quantities entering the quartic equation (16.5), except the isothermal compressibility and the superfluid density, using the ideal Bose gas model. One can actually show that the adiabatic compressibility, the specific heat at constant volume, and the entropy density of a weakly-interacting Bose gas deviate very little from the ideal Bose gas predictions in a wide interval of temperatures above the hybridization point. This simplifies significantly the solution of Eq. (16.5). In fact, in the ideal Bose gas model one finds

$$\frac{1}{mn\kappa_s} + \frac{n_s T\bar{s}^2}{mn_n\bar{c}_v} = \frac{nT\bar{s}^2}{mn_n\bar{c}_v} \tag{16.10}$$

for the coefficient of the c^2 term. It is now easy to derive the two solutions satisfying the condition $c_1 \gg c_2$. To obtain the larger velocity c_1, one can neglect the last term in (16.5). Using the thermodynamic relations of the ideal Bose gas model and identifying the normal density with the thermal density ($n_T = n - n_0$), one obtains the prediction (here and in the following, we take the Boltzmann constant $k_\mathrm{B} = 1$)

$$c_1^2 = \frac{5}{3}\frac{g_{5/2}}{g_{3/2}}\frac{T}{m} \tag{16.11}$$

for the first sound velocity [26], where $g_{5/2}$ and $g_{3/2}$ are integrals of the Bose distribution function (see, for example, Section 3.2 in Ref. [1]). To calculate c_2, we must instead neglect the c^4 term in (16.5) and using result (16.10) one finds the useful result

$$c_2^2 = \frac{n_s}{n}\frac{1}{mn\kappa_T} \tag{16.12}$$

revealing that the superfluid density and the isothermal compressibility are the crucial parameters determining the value of the second sound velocity of weakly interacting Bose gases for $T \gg gn$. Furthermore, in a weakly interacting 3D Bose gas, one can safely identify (except close to T_c) the inverse isothermal compressibility with its $T = 0$ value $(n\kappa_T)^{-1} = gn$, yielding the simple expression

$$c_2^2 = \frac{gn_s(T)}{m} \tag{16.13}$$

for the second sound velocity, revealing that second sound can be regarded, in this temperature range, as a finite temperature generalization of the Bogoliubov sound mode propagating at $T = 0$ at the velocity $\sqrt{gn/m}$ (at $T = 0$, one has $n_s = n$). Since in a weakly interacting 3D Bose gas the superfluid density can be safely approximated with the condensate density ($n_s(T) = n_0(T)$), the above result also permits us to understand that in the same regime of temperatures second sound corresponds to an oscillation of the condensate, which can be easily excited through a density perturbation of the gas and subsequently imaged. In Fig. 16.2, we report the sound velocity measured in the experiment of [30], where a density perturbation was applied to a Bose gas confined in a highly elongated trap. Due to the harmonic radial confinement, the density n entering the Bogoliubov formula should be actually replaced by $n/2$, where n is the central axial density [31] (a similar renormalization is expected to occur for the superfluid density and the condensate density at finite temperature). The good agreement between the measured sound velocity and the Bogoliubov value $\sqrt{gn_0(T)/2m}$, with the condensate fraction measured at the same value of T, permits us to conclude that the sound excitation investigated in [30] actually corresponds to the second sound mode described above. The remaining small deviations are likely due to the approximations made to derive result (16.13) and to the identification $n_s = n_0$.

In order to explore the behavior of the sound velocities in the whole range of temperatures, including the hybridization region at low T, a more accurate knowledge of the thermodynamic functions entering the two-fluid hydrodynamic equations is needed. At low temperature ($T \ll T_c$), one can use Bogoliubov theory where the elementary excitations of the gas, whose dispersion is given by the Bogoliubov law $\varepsilon(\mathbf{p}) = \sqrt{(p^2/2m)(p^2/2m + 2gn)}$, are thermally excited according to the bosonic rule $N_{\mathbf{p}}(\varepsilon) = (e^{\varepsilon(\mathbf{p})/kT} - 1)^{-1}$. Bogoliubov theory fails at temperatures of the order of the critical temperature, where a more reliable approach is provided by the perturbation theory developed in Ref. [32] based on the Beliaev diagrammatic technique at finite temperature [33]. In this latter approach, thermal effects in the Bogoliubov excitation spectrum are accounted for through a self-consistent procedure. At low temperatures, this approach coincides with Bogoliubov theory, except for temperatures smaller than $(na^3)^{1/4}gn$, i.e., at temperatures much smaller than the

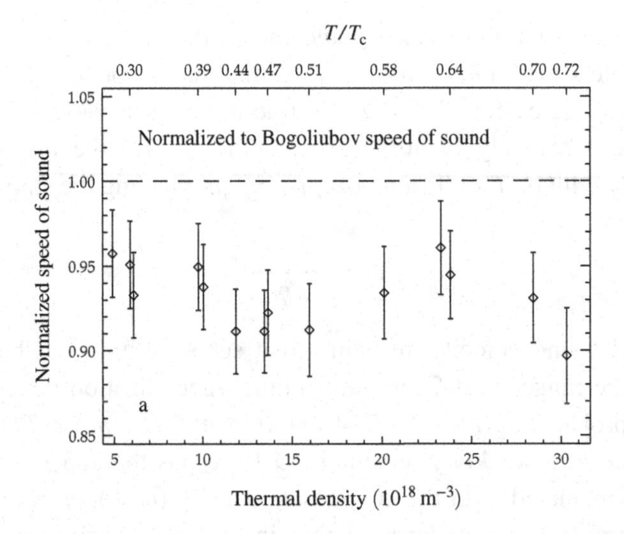

Figure 16.2 Speed of sound as a function of the thermal density. The upper axis gives the reduced temperature T/T_c for the corresponding data point. The speed of sound is normalized to the Bogoliubov sound velocity $\sqrt{gn_0(T)/2m}$ based on the central BEC density. Reprinted with permission from Meppelink, R., et al. (2009), Sound propagation in a Bose-Einstein condensate at finite temperatures, *Phys. Rev. A*, **80**, 043605 [30]. Copyright (2009) by the American Physical Society.

hybridization point. At higher temperatures, the theory of Ref. [32] turns out to be very accurate in dilute gases when compared with exact Monte Carlo simulations. It follows that, at least for small values of the gas parameter na^3, Bogoliubov theory and the diagrammatic approach of Ref. [32] match exactly in the hybridization region of temperatures $T \sim gn$, and that the thermodynamic behavior of the gas is consequently under control for all ranges of temperatures, both below and above gn.

The coefficients of the quartic equation (16.5) depend not only on the equilibrium thermodynamic functions, but also on the normal and superfluid densities. The normal density can be calculated using Landau's prescription

$$mn_n = -\frac{1}{3}\int \frac{dN_{\mathbf{p}}(\varepsilon)}{d\varepsilon} p^2 \frac{d\mathbf{p}}{(2\pi\hbar)^3} \tag{16.14}$$

in terms of the elementary excitations of the gas. Landau's prescription (16.14) ignores interaction effects among elementary excitations and in a dilute Bose gas is expected to be very accurate in the low-temperature regime characterizing the hybridization point and to hold also at higher temperatures, except close to the critical point.

A peculiar property of the weakly-interacting Bose gas is that all the thermodynamic functions entering Eq. (16.5), as well as the normal density ρ_n, can be written in a rescaled form as a function of the reduced temperature $\tilde{t} \equiv T/gn$ and of the reduced chemical potential $\eta \equiv gn/T_C^0$, where

$$T_c^0 = \frac{2\pi\hbar^2}{m}\left(\frac{n}{\zeta(3/2)}\right)^{2/3} \tag{16.15}$$

is the critical temperature of the ideal Bose gas. In weakly-interacting Bose gases, T_c^0 does not coincide with the actual critical temperature which contains a small correction fixed by the value of the gas parameter na^3: $T_c = T_c^0(1 + \gamma(na^3)^{1/3})$ with $\gamma \sim 1.3$ [34, 35, 36]. The reduced chemical potential can be also expressed in terms of the gas parameter as $\eta = 2\zeta(3/2)^{2/3}(na^3)^{1/3}$.

Using Bogoliubov theory, the free energy $F = U - TS$ can be written as

$$\frac{F}{gnN} = \frac{1}{2}\left[1 + \frac{128}{15\sqrt{\pi}}(na^3)^{1/2}\right] + \frac{2\tilde{t}}{\zeta(3/2)\sqrt{2\pi}}\eta^{3/2}\int_0^\infty \tilde{p}^2 \ln\left(1 - e^{-\frac{\tilde{p}}{2\tilde{t}}\sqrt{\tilde{p}^2+4}}\right) d\tilde{p}, \tag{16.16}$$

where we have defined $\tilde{p} = p/\sqrt{mgn}$ and, for completeness, we have included the Lee-Huang-Yang correction to the $t = 0$ value of the free energy (term proportional to $(na^3)^{1/2}$). Starting from expression (16.16) for the free energy [29], all the thermodynamic functions entering Eq. (16.5) are easily calculated using standard thermodynamic relations. The normal density (16.14) is instead evaluated using Eq. (16.14), written in terms of the energy of the elementary excitations.

The idea now is to calculate the two solutions of Eq. (16.5) for a fixed value of $\tilde{t}$ of the order of unity, taking the limit $\eta \to 0$. Physically this corresponds to considering very low temperatures (of the order of gn) and small values of the gas parameter na^3. For example, in the case of ^{87}Rb the value of the scattering length is $a = 100a_0$ (where a_0 is the Bohr radius) and typical values of the density correspond to $na^3 \sim 10^{-6}$. This yields $\eta \approx 0.04$.

Writing the solutions of Eq. (16.5) in terms of the $T = 0$ Bogoliubov velocity $c_B = \sqrt{gn/m}$, one finds that, as $\eta \to 0$, the two sound velocities only depend on the dimensionless parameter $\tilde{t}$ and are given by

$$c_+^2 = c_B^2\,, \qquad c_-^2 = c_B^2 f(\tilde{t}), \tag{16.17}$$

where $f(\tilde{t}) = \lim_{\eta\to 0} \frac{n}{n_n}\frac{T\bar{s}^2}{\bar{c}_v}\frac{1}{gn}$. One actually easily finds that $\bar{s} \propto \eta^{3/2}$, $\bar{c}_v \propto \eta^{3/2}$, and $n_n \propto \eta^{3/2}$ as $\eta \to 0$ and that f, in this limit, is consequently a function of $\tilde{t}$, independent of η. In the $\eta \to 0$ limit, the two velocities shown in Fig. 16.3a cross each other at the value $\tilde{t}_{\text{hyb}} \approx 0.6$. At lower temperatures, c_-^2 approaches, as expected, the zero-temperature value $c_B^2/3$. By considering finite, although small, values of η,

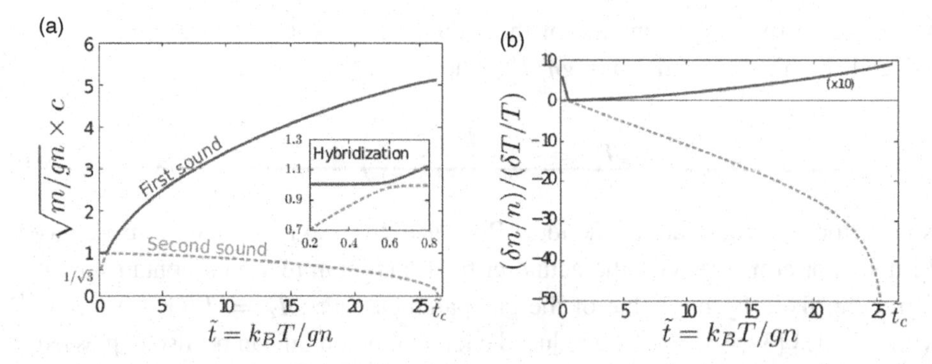

Figure 16.3 (a) Sound velocities (dashed gray and solid black lines) computed interpolating Bogoliubov theory and the diagrammatic approach of Ref. [32], over the whole range of temperatures $0 < \tilde{t} < \tilde{t}_c$. The inset shows the hybridization region. (b) Ratio $(\delta n/n)/(\delta T/T)$ for the lower branch (dashed gray) and the upper branch (solid black). The parameters are chosen as in Fig. 16.3a. For both graphs, the gas parameter is chosen to be $na^3 = 10^{-6}$, and the critical point corresponds to $\tilde{t}_c = T_c/gn = 26.7$. Reprinted with permission from Verney, L., et al. (2015), Hybridization of first and second sound in a weakly interacting Bose gas, *Europhys. Lett.*, **111** (4), 40005 [29]. Copyright (2015) by the Institute of Physics.

it is possible to show that, at the hybridization point, the two branches exhibit a gap proportional to $\eta^{3/4}$. The mechanism of hybridization, explicitly shown in the inset of Fig. 16.3a, permits us to identify an upper branch c_1 (which coincides with c_+ for $\tilde{t} < \tilde{t}_{\mathrm{hyb}}$ and with c_- for $\tilde{t} > \tilde{t}_{\mathrm{hyb}}$), called "first sound." The lower branch c_2 (called "second sound") instead coincides with c_- for $\tilde{t} < \tilde{t}_{\mathrm{hyb}}$ and with c_+ for $\tilde{t} > \tilde{t}_{\mathrm{hyb}}$.

The validity of Eqs. (16.17) is limited to very low temperatures where the thermal depletion of the condensate can be ignored and Bogoliubov theory can be safely applied. When the temperature is comparable to the critical temperature T_c (corresponding to $\tilde{t}_c = 26.7$ in the case of Fig. 16.3a, where we have chosen $\eta = 0.04$), Bogoliubov theory is no longer applicable. In fact, at such temperatures the thermal depletion of the condensate becomes important and the Bogoliubov expression for the dispersion law is inadequate. The superfluid density fraction, calculated according to Eq. (16.14) with the $T = 0$ value of the Bogoliubov dispersion law, vanishes at $T \sim 1.2\,T_c$, well above the critical temperature, further revealing the inadequacy of the theory at high temperatures. As anticipated above, a better approach to be used at temperatures of the order of the critical value is the diagrammatic approach of Ref. [32], whose predictions for the first and second sound velocities at temperatures higher than the hybridization temperature are reported in Fig. 16.3a and which properly interpolates with the predictions of Bogoliubov theory near the hybridization point.

In Fig. 16.3b, we show the ratio $(\delta n/n)/(\delta T/T)$ between the relative density and temperature variations calculated for the first and second sound solutions of Eq. (16.5) as a function of temperature. This quantity represents an important characterization of the two branches. It is in fact well known that sound in an ideal classical gas is an adiabatic oscillation characterized by the value $3/2$ for the ratio $(\delta n/n)/(\delta T/T)$. For the upper solution of the hydrodynamic equations (first sound), this ratio is positive over the full range of temperatures below the transition and its value increases with T, getting close to unity near the transition. The most important feature emerging from Fig. 16.3b is that the ratio between the relative density and temperature changes associated with the second sound solution has an opposite sign and a much larger value in modulus, reflecting that second sound, for temperatures larger than the hybridization value, is dominated by the fluctuations of the density rather than by the ones of the temperature. This important feature is also revealed by the fact that, in the same range of temperatures, second sound practically exhausts the compressibility sum rule (16.6). This result can be easily understood from Eq. (16.7). In fact, if $T \gg gn$, one has $mc_1^2 \gg (n\kappa_T)^{-1}$ and consequently $W_2 \gg W_1$. The fact that the compressibility sum rule (16.6) is strongly affected by second sound is a remarkable feature exhibited by the weakly interacting Bose gas above the hybridization point, which distinguishes in a profound way its behavior from the one of strongly interacting superfluids, such as ^{4}He or the unitary Fermi gas, where the density fluctuations associated with second sound are very small.

From an experimental point of view, the results discussed in this section, and in particular Fig. 16.3b, reveal that, in dilute Bose gases, second sound is more easily accessible than first sound, being very sensitive to the coupling with the density probe. This is confirmed by the experimental identification of second sound in the experiment of Ref. [30]. It is finally worth noticing that the fact that second sound in a Bose gas can be easily excited through a density probe is not specific to the 3D case. Indeed, a similar behavior takes place also in 2D Bose gases [37], where its measurement could provide an efficient determination of the superfluid density, including its discontinuity at the Berezinskii-Kosterlitz-Thouless transition (see Section 16.5).

16.4 Second Sound in the Unitary Fermi Gas

The possibility of determining the temperature dependence of the superfluid density from the measurement of second sound represents a major challenge in the physics of interacting Fermi gases where, differently from the case of weakly interacting Bose gases discussed in the previous section, the superfluid density n_s cannot be identified with the condensate fraction, and the theoretical determination of n_s at finite temperature remains a difficult problem from the many-body point of view. Recent progress in the experimental measurement of second sound of a Fermi gas [38] at unitarity has opened new perspectives in this direction and the first determination of the superfluid density in a Fermi superfluid.

First predictions for the temperature dependence of the first and second sound velocities in the 3D unitary Fermi gas were obtained in [21] by calculating the thermodynamic functions entering the two-fluid Landau's hydrodynamic equation (16.5) using the Nozieres and Schmitt-Rink (NSR) approach [39, 40] (see Fig. 16.1b). These results have provided a first estimate of the second sound velocity in this strongly interacting Fermi system. (See also Ref. [41].) It is remarkable to see that the qualitative behavior of the second sound velocity of the unitary Fermi gas looks very similar to the one of superfluid ^{4}He (Fig. 16.1a).

In the following, we will discuss the behavior of second sound developing the hydrodynamic formalism in the presence of a highly elongated trap, a configuration particularly suited to explore experimentally the propagation of sound. The effect of the radial confinement has the important consequence that the variations of the temperature $\delta T(z,t)$ and of the chemical potential $\delta\mu(z,T)$, as well as the axial velocity field $v_n^z(z,t)$, do not exhibit any dependence on the radial coordinate during the propagation of sound, which can then be considered one dimensional (1D) in nature, although the local equilibrium properties of the system can be described using the 3D themodynamic functions in the local density approximation. The validity of the 1D-like assumption is ensured by collisional effects, which restore a radial local thermodynamic equilibrium and require the condition that the viscous penetration depth $\sqrt{\eta_s/mn_{n1}\omega}$ be larger than the radial size of the system. Here n_{n_1} is the 1D normal density, obtained by radial integration of the normal density, η_s is the shear viscosity, and ω is the frequency of the sound wave. In terms of the radial trapping frequency $\omega_\perp$, the condition can be written in the form $\omega \ll \omega_\perp^2\tau$, where τ is a typical collisional time here assumed, for simplicity, to characterize both the effects of viscosity and thermal conductivity. The condition of radial local thermodynamic equilibrium would have a dramatic consequence in the presence of a tube geometry with hard walls. In fact, in this case the viscosity effect near the wall would cause the vanishing of the normal velocity field, with the consequent blocking of the normal fluid. The resulting motion, involving only the superfluid fraction with the normal component at rest, is called fourth sound and was observed in liquid helium confined in narrow capillaries. In the presence of radial harmonic trapping, this effect is absent and the normal component can propagate as well, allowing for the propagation of both first and second sound.

Under the above conditions of local radial equilibrium, one can easily integrate radially the hydrodynamic equations (16.1–16.4), which then keep the same form as for the 3D uniform gas [42]:

$$\frac{\partial n_1}{\partial t} + \frac{\partial j_z}{\partial z} = 0 \tag{16.18}$$

$$m\frac{\partial v_s^z}{\partial t} = -\frac{\partial}{\partial z}\left(\mu_1(z) + V_{ext}(z)\right) \tag{16.19}$$

$$\frac{\partial s_1}{\partial t} + \frac{\partial}{\partial z}(s_1 v_n^z) = 0 \tag{16.20}$$

$$\frac{\partial j_z}{\partial t} - \frac{1}{m}\frac{\partial P_1}{\partial z} - \frac{n}{m}\frac{\partial}{\partial z}V_{ext}(z) \ , \tag{16.21}$$

where $n_1(z,t) = \int dxdy\, n(\mathbf{r},t)$, $s_1(z,t) = \int dxdy\, s(\mathbf{r},t)$, and $P_1(z,t) = \int dxdy\, P(\mathbf{r},t)$ are the radial integrals of their 3D counterparts, namely the particle density, the entropy density, and the local pressure. The integration accounts for the inhomogeneity caused by the radial component of the trapping potential. In the above equations, $j_z = n_{n1}v_n^z + n_{s1}v_s^z$ is the current density, $n_{s1} = \int dxdy\, n_s$ and $n_{n1} = \int dxdy\, n_n$ are the superfluid and the normal 1D densities, respectively, with $n_1 = n_{n1} + n_{s1}$, while v_s^z and v_n^z are the corresponding velocity fields. In Eq. (16.19), $\mu_1(z,t) \equiv \mu(\mathbf{r}_\perp = 0, z, t)$ is the chemical potential calculated on the symmetry axis of the trapped gas. Its dependence on the 1D density and on the temperature is determined by the knowledge of the radial profile, which can be calculated employing the equation of state of uniform matter in the local density approximation.

By setting the axial trapping $V_{ext}(z)$ equal to zero, which corresponds to considering a cylindrical geometry, we can look for sound wave solutions propagating with a phase factor of the form $e^{i(qz-\omega t)}$. One then derives the same Landau's equation (16.5) obtained in uniform matter, with the thermodynamic functions replaced by the corresponding 1D expressions; in particular, the 1D entropy per particle is given by $\bar{s}_1 = s_1/n_1$ and the 1D specific heat at constant pressure by $\bar{c}_{p1} = T(\partial\bar{s}_1/\partial T)_{P_1}$.

Also in the highly elongated 1D cigar configuration, Landau's equations admit two different solutions, corresponding to the first (c_1) and second (c_2) sound velocities. Simple results for the sound velocities are obtained under the condition (16.9), yielding the results

$$mc_1^2 = \left(\frac{\partial P_1}{\partial n_1}\right)_{\bar{s}_1} \quad \text{and} \quad mc_2^2 = T\frac{n_{s1}\bar{s}_1^2}{n_{n1}\bar{c}_{p1}} \tag{16.22}$$

for the first and second sound velocity, respectively.

Some comments are in order here: (i) The assumption (16.9), yielding results (16.22), is well satisfied in strongly interacting superfluid Fermi gases in the whole temperature interval below T_c due to their small compressibility (from this point of view, the behavior of the solutions of the hydrodynamic equations deeply differs from the case of dilute Bose gases discussed in the previous section). (ii) The expression (16.22) for the second sound velocity contains the specific heat at constant pressure. This reflects the fact that second sound actually corresponds to a wave propagating at constant pressure rather than at constant density. This difference is very important from the experimental point of view. In fact, even a small value of the thermal expansion coefficient is crucial in order to give rise to measurable density fluctuations during the propagation of second sound [38]. (iii) The thermodynamic ingredients P_1, $\bar{s}_1$, and $\bar{c}_{P1}$ are known with good precision at unitarity in a useful range of temperatures. They can be easily determined starting from the 3D thermodynamic relations holding for the uniform Fermi gas at unitarity.

In the following, we will outline the calculation of the themodynamic functions in the case of the unitary Fermi gas [43]. At unitarity, the only length scales for uniform configurations are the interparticle distance, fixed by the density of the gas, and the thermal wavelength $\lambda_T = \sqrt{2\pi\hbar^2/mT}$, fixed by the temperature. Correspondingly, the energy scales are now the temperature and the Fermi temperature

$$T_F = \frac{\hbar^2}{2m}\left(3\pi^2 n\right)^{2/3} \tag{16.23}$$

or, alternatively, the chemical potential μ. It follows that at unitarity all the thermodynamic functions can be expressed [44] in terms of a universal function $f_p(x)$ depending on the dimensionless parameter $x \equiv \mu/T$. This function can be defined in terms of the pressure and the density of the gas, according to the relationships

$$P\frac{\lambda_T^3}{T} = f_p(x), \qquad n\lambda_T^3 = f_p'(x) \equiv f_n(x). \tag{16.24}$$

From Eq. (16.24), for the density one derives the useful relationship $T/T_F = 4\pi/[3\pi^2 f_n(x)]^{2/3}$ for the ratio between the temperature and the Fermi temperature in terms of the ratio x.

In terms of f_n and f_p, we can calculate all the thermodynamic functions of the unitary Fermi gas. For example, using the thermodynamic relation $s/n = -(\partial\mu/\partial T)_P$, we find the result $\bar{s} = s/n = (5/2)f_p/f_n - x$ for the entropy per unit mass. It is worth noting that the above equations for the thermodynamic functions of the unitary Fermi gas are formally identical to the ones of the ideal Fermi gas, the dimensionless functions f_p and f_n replacing, apart from a factor 2 caused by spin degeneracy, the usual $F_{5/2}$ and $F_{3/2}$ Fermi integrals (see Section 16.1 in [1]). Since the entropy $\bar{s}$ depends only on the dimensionless parameter x, from Eqs. (16.24) one finds that during an adiabatic transformation the quantity $P/n^{5/3}$ remains constant, which is the same condition characterizing a noninteracting monoatomic gas.

The scaling function $f_p(x)$ (and hence its derivative $f_n(x)$) can be determined through microscopic many-body calculations or extracted directly from experiments carried out in trapped configurations from which it is possible to build the equation of state of uniform matter [15]. The experimental analysis of the thermodynamic functions (in particular, the isothermal compressibility and the specific heat) has allowed for the identification of the critical temperature associated with the superfluid phase transition for which the authors of [15] have found the result

$$T_c = 0.167T_F, \tag{16.25}$$

in agreement with the most recent reliable theoretical calculations, based on quantum Monte Carlo [45, 46] and diagrammatic [47] techniques. The value of T_c corresponds to $x_c = \mu_c/T_c = 2.48$.

Using the local density approximation along the radial direction, which is obtained by replacing the chemical potential with $\mu = \mu_0 - (1/2)m\omega_\perp^2 r_\perp^2$, and integrating the thermodynamic functions P, n, and s along the radial directions, it is possible to construct the corresponding 1D thermodynamic functions in terms of the 1D chemical potential $\mu_1 = x_1 T$, which are needed to calculate the ingredients entering Landau's equation for the 1D two sound velocities. One finds [43]

$$P_1 = \frac{2\pi}{m\omega_\perp^2}\frac{T^2}{\lambda_T^3}f_q(x_1)\,, \qquad n_1(x_1, T) = \frac{2\pi}{m\omega_\perp^2}\frac{T}{\lambda_T^3}f_p(x_1), \tag{16.26}$$

with $f_q = \int_{-\infty}^{x} dx' f_p(x')$, allowing for the determination of the various thermodynamic functions. In particular, the 1D entropy per unit mass takes the form $\bar{s}_1 = s_1/n_1 = (7/2)f_q(x_1) - x_1 f_p(x_1)$. From the above equations, it follows immediately that $(\partial P_1/\partial n_1)_{\bar{s}_1} = (7/5)P_1/n_1$, which differs from the adiabatic result $(\partial P/\partial n)_{\bar{s}} = (5/3)P/n$ holding in the 3D case. Using Eq. (16.22), one then predicts the result

$$mc_1^2 = \frac{7}{5}\frac{P_1}{n_1} \tag{16.27}$$

for the first sound velocity in good agreement with the experimental findings (see Fig. 16.4).

Since the temperature dependence of the superfluid density is not known theoretically in the unitary Fermi gas with sufficient accuracy, for the discussion of second sound we will follow a different strategy: we will employ the measured value of the 1D second sound velocity (see discussion below) to extract $n_{s1}(T)$ from Eq. (16.22). By the way, in the low-temperature regime, the calculation of the thermodynamic functions in the highly elongated geometry predicts $c_2 \propto \sqrt{T} \to 0$, differently from the uniform 3D case where $c_2 \to c_1/\sqrt{3}$. Thus the 1D second sound velocity c_2 tends to zero both as $T \to T_c$ and $T \to 0$. As a consequence, in the cigar geometry, the condition (16.9) is rather well satisfied for all temperatures and Eq. (16.22) is expected to be particularly accurate.

Fig. 16.4 shows the measured sound velocities in the experiment of [38] carried out in a highly elongated Fermi gas at unitarity. The excitation of first and second

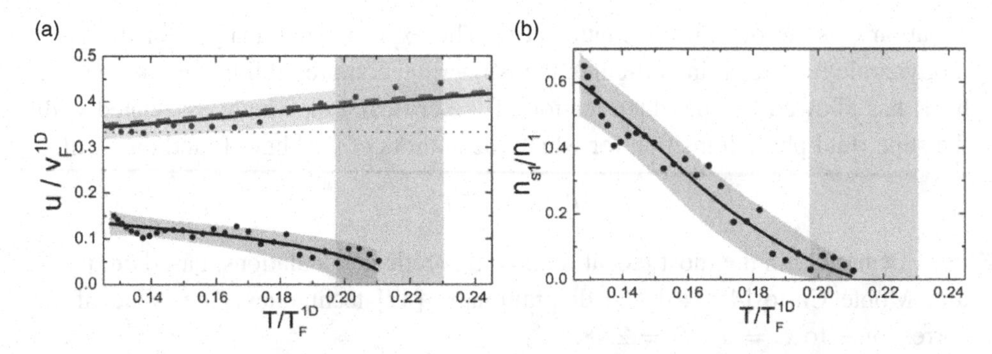

Figure 16.4 (a) Speeds of first and second sound of the unitary Fermi gas in a highly elongated trap, normalized to the local Fermi speed and plotted as a function of the reduced temperature. The dashed curve is a prediction based on Eq. (16.27) and the equation of state (EOS) from Ref. [15]. The dotted horizontal line is the zero-temperature limit for the speed of first sound. (b) Temperature dependence of the 1D superfluid fraction n_{s1}/n_1. Adapted with permission from Sidorenkov, L. A., et al. (2013), Second sound and the superfluid fraction in a Fermi gas with resonant interactions, *Nature*, **498** 7 [38]. Copyright (2013) by the Nature Publishing Group.

sounds was obtained by generating, respectively, a sudden local perturbation of the density and of the temperature in the center of the trapped gas. Due to the finite, although small, value of the thermal expansion coefficient of the unitary Fermi gas, also the thermal perturbation, generating the second sound wave, gives rise to a measurable density pulse. In this experiment, both the first and second sound velocities were obtained by measuring, for a fixed value of T, the time-dependence position of the density pulses generated by the perturbation. These measurements give access to the dependence of the sound velocity on the ratio T/T_F^{1D}, where $T_F^{1D} = (15\pi/8)^{2/5}(\hbar\omega_\perp)^{4/5}(\hbar^2 n_1^2/2m)^{1/5}$ is a natural definition for Fermi temperature in 1D cylindrically-trapped configurations [38, 43]. If n_1 is calculated for an ideal Fermi gas at zero temperature, T_F^{1D} coincides with the usual 3D definition of the Fermi temperature (16.23), with n calculated on the symmetry axis. In the presence of axial trapping, the value of T/T_F^{1D} actually increases as one moves from the center, because of the density decrease, and eventually the density pulse reaches the transition point where the superfluid vanishes.

From the measurement of the temperature dependence of the 1D superfluid density and recalling the definition $n_{s1} = \int dxdy\, n_s$, one can reconstruct the 3D superfluid density n_s as a function of the ratio T/T_c [38], which is reported in Fig. 16.5. The measurement of n_s reported in the figure represents the first experimental determination of the temperature dependence of the superfluid density in a superfluid Fermi gas and could be used as a benchmark for future many-body calculations.

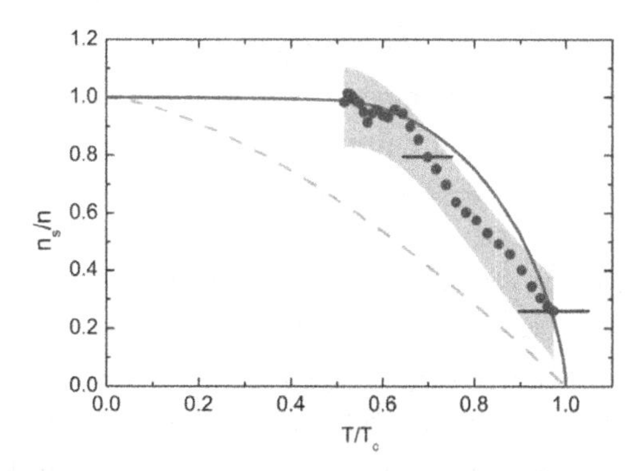

Figure 16.5 Superfluid fraction for the homogeneous case. The data points and the corresponding uncertainty range (shaded region) show the superfluid fraction for a uniform resonantly interacting Fermi gas versus T/T_c as reconstructed from its 1D counterpart in Fig. 16.4. The two horizontal error bars indicate the systematic uncertainties resulting from the limited knowledge of the critical temperature T_c. For comparison, we show the fraction for superfluid helium (solid line) as measured in Ref. [25] and the textbook expression $1 - (T/T_c)^{3/2}$ for the Bose-Einstein condensed fraction of the ideal Bose gas (dashed line). Reprinted with permission from Sidorenkov, L. A., et al. (2013), Second sound and the superfluid fraction in a Fermi gas with resonant interactions, *Nature*, **498** 7 [38]. Copyright (2013) by the Nature Publishing Group.

16.5 Second Sound in the 2D Bose Gas

Two-dimensional superfluids differ in a profound way from their 3D counterparts. In fact, the Hohenberg-Mermin-Wagner theorem [48, 49] rules out the occurrence of long-range order at finite temperature in 2D systems with continuous symmetry. Furthermore, the superfluid density approaches a finite value at the critical point of a 2D superfluid, known as the Berezinskii-Kosterlitz-Thouless (BKT) transition [50, 51, 52, 53], rather than vanishing, as happens in 3D. (A brief overview of BKT across different physical systems is given in Chapter 10.) With respect to the second-order phase transitions characterizing the onset of superfluidity in 3D, the nature of the BKT phase transition is deeply different, being associated with the emergence of a topological order, resulting from the pairing of vortices and antivortices. A peculiar property of these 2D systems is also the absence of discontinuities in the other thermodynamic functions at the critical temperature characterizing the transition to the superfluid phase. In order to identify the transition point, one has consequently to measure suitable transport properties. This is the case of the recent experiment of Ref. [54] on dilute two-dimensional Bose gases where the superfluid critical velocity was measured in a useful range of temperatures, pointing at the occurrence of a sudden jump at a critical temperature when one enters the superfluid regime.

In this section, we discuss the behavior of both first and second sound in 2D superfluid gases, with particular emphasis on the discontinuity of their velocities at the critical point, caused by the jump of the superfluid density. Although the thermodynamic behavior of 2D dilute Bose gases is now well understood both theoretically [55, 56, 57] and experimentally [58, 59], the measurement of the superfluid density remains one of the main open issues. Besides the prospect of measuring the superfluid density, the measurement of second sound itself would be important because second sound has never been measured in any 2D system so far. In helium films, the normal component of the liquid is in fact clamped to the substrate and cannot participate in the propagation of sound; only the superfluid is free to move (third sound). In helium films, the value of n_s became accessible via third sound [60] and torsional oscillator measurements [61], confirming the superfluid jump at the transition. Since the trapping of dilute atomic gases is provided by smooth potentials, second sound is expected to propagate in these systems also in 2D and to be properly described by the two-fluid Landau hydrodynamic equations. Its propagation in dilute Bose gases exhibits very peculiar features as compared with less compressible fluids such as helium or strongly interacting Fermi superfluid gases. In fact, in these latter systems, second sound can be identified as an entropy oscillation and corresponds with good accuracy to an isobaric oscillation. This is not the case of dilute Bose gases, which are highly compressible, giving rise to sizable coupling effects between density and entropy oscillations. Furthermore, with respect to the 3D case, in 2D, the superfluid density exhibits a jump at the transition, and this shows up as a discontinuity of both first and second sound velocities [37], as we will discuss in the following.

We start our investigation by considering Landau's two fluid hydrodynamic equations to describe the dynamics of the system in the superfluid phase of a 2D uniform configuration. The third direction is assumed to be blocked by a tight harmonic confinement, a condition well achieved in current experiments. We will focus on the dilute 2D Bose gas, where all the thermodynamic ingredients can be written in terms of dimensionless functions [56, 58, 59]. These depend only on the variable $x \equiv \mu_2/T$ and on the 2D coupling constant $g_2 = (\hbar^2/m)\sqrt{8\pi}a/l$, where μ is the chemical potential, a is the three-dimensional scattering length, and l is the oscillator length in the confined direction. Here, we assume $l \gg a$ so that the interaction is momentum independent [62]. We introduce the dimensionless reduced pressure $\mathcal{P}$ and the phase space density $\mathcal{D}$ by

$$P_2\lambda_T^2/T \equiv \mathcal{P}(x, mg_2/\hbar^2), \qquad n_2\lambda_T^2 \equiv \mathcal{D}(x, mg_2/\hbar^2), \tag{16.28}$$

where λ_T is the thermal de Broglie wavelength, and P_2 and n_2 are the 2D pressure and particle number density, respectively. The simple relation $\partial\mathcal{P}/\partial x = \mathcal{D}$ follows from thermodynamics. Starting from Eq. (16.28), one can evaluate the thermodynamic functions of the gas. In particular, the entropy per unit mass, for a fixed value of g_2, depends solely on the parameter x_2: $\bar{s}_2 = s_2/n_2 = 2\mathcal{P}/\mathcal{D} - x_2$. The above equations reflect the universal nature of the thermodynamic behavior of the 2D Bose gas, for a fixed value of the coupling constant g_2. The function $\mathcal{D}$ has been numerically calculated for values of g_2 much smaller than $\hbar^2/m$ in [56], and both $\mathcal{P}$ and $\mathcal{D}$ have been theoretically [57] and experimentally [58, 59] determined around the superfluid transition. The results available from different methods well agree with each other.

The superfluid density n_s cannot be calculated in terms of the universal functions introduced above, but can be nevertheless expressed in terms of another dimensionless function $\lambda_T^2 n_s \equiv \mathcal{D}_s(x, g_2)$, which is known close to the transition [56] as well as in the highly degenerate phonon regime (large and positive x). At the critical point, one has $\mathcal{D}_s = 4$, which follows from the universal Nelson-Kosterlitz result $n_{2s} = 2mT_c/\pi\hbar^2$ [53], where T_c is the BKT transition temperature, providing an important relationship between the jump of the superfluid density at the transition and the value of the critical temperature.

The superfluid transition of a weakly interacting gas is predicted to take place at the value [55] $x_c = (mg_2/\pi\hbar^2)\log(\xi_\mu/g_2)$ with $\xi_\mu \approx 13.2\hbar^2/m$. For example, for $g_2 = 0.1\hbar^2/m$, a value relevant for the experiments of [59, 63, 54], the critical point corresponds to $x_c \approx 0.16$ or, in terms of the density, to $T_c = \{2\pi/\mathcal{D}(x_c)\}n_2/m \approx 0.76n/m$.

In Fig. 16.6a, we show the superfluid fraction n_{2s}/n_2, calculated as a function of T/T_c for a fixed value of the total density, using the data from [56]. The figure corresponds to the value $g_2 = 0.1\hbar^2/m$ and point out the large value of the superfluid fraction at the transition and the consequent jump. As we will see below, both the jump of the superfluid density and the large value of the thermal expansion coefficient play an important role in characterizing the solutions of the Landau's equation near the transition.

In Fig. 16.6b, we show, for the same g_2 as above, values for the first and second sound velocities predicted by the solutions of Eq. (16.5). These values are expressed in units of the zero-temperature value of the Bogoliubov sound velocity $c_0 \equiv \sqrt{g_2 n/m}$ and are calculated at fixed total density. The most remarkable feature emerging from the figure is the discontinuity exhibited by both the first and second sound velocities at the transition. Changing the value of the coupling constant g_2 does not affect the qualitative behavior of the sound velocities.

Using the parameters from the experiment of [54] carried out on a gas of ^{87}Rb atoms ($mg_2/\hbar^2 = 0.093$, $n_2 = 50/\mu\text{m}^2$), we predict the value $c_2 \approx 0.88$ mm/s for the second sound velocity at the transition. This value is close to the critical velocity observed in [54], thereby suggesting that the excitation of second sound is a possible mechanism for the onset of dissipation in this experiment.

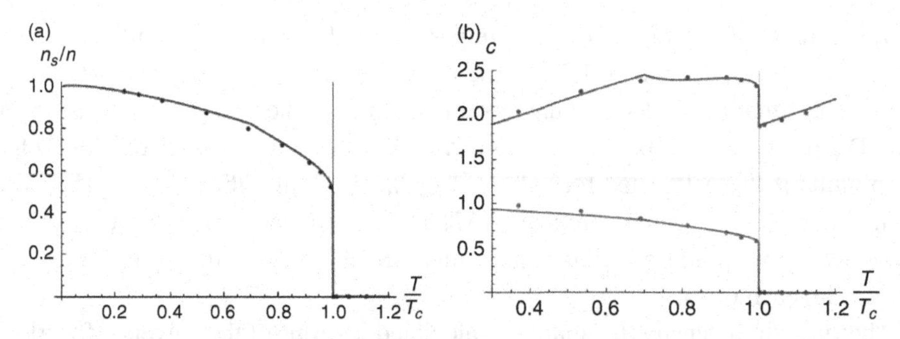

Figure 16.6 (a) Normalized superfluid density n_s/n for $mg_2/\hbar^2 = 0.1$. The line and the dots are calculated, respectively, from the approximate analytical and numerical results of [56]. Two analytical expressions valid at low and high temperatures are connected to give the curve, resulting in an unphysical kink at $T/T_c \sim 0.7$. (b) First and second sound velocities in units of the zero temperature Bogoliubov sound velocity $c_0 = \sqrt{g_2 n_2}/m$ with $mg_2/\hbar^2 = 0.1$, calculated by solving Eq. (16.5). Adapted from Ozawa, T., and Stringari, S. (2014), Discontinuities in the first and second sound velocities at the Berezinskii-Kosterlitz-Thouless transition, *Phys. Rev. Lett.*, **112**, 025302 [37]. Copyright (2014) by the American Physical Society.

Concerning the physical characterization of the two sounds, it is worth recalling that in dilute Bose gases, first and second sound cannot be interpreted, respectively, as isoentropic and isobaric oscillations (see Eq. (16.8)). In Section 16.3, we have already shown that in a weakly-interacting 3D Bose gas, the second sound speed c_2 is instead well approximated by the expression (16.12) [1] rather than by Eq. (16.8), in the relevant region of temperatures $T \gg \mu$, where the isoentropic compressibility, the entropy, and the specific heat at constant volume are very close to the predictions of the ideal Bose gas model. In two dimensions, Eq. (16.12) describes exactly the second sound velocity in the limit of small interactions ($g_2 \to 0$) [37]. For $g_2 = 0.1\hbar^2/m$, it is a good approximation (within $\sim 10\%$) to the second sound in the whole range of temperatures shown in Fig. 16.6b. On the other hand, the first sound velocity around the transition can be estimated by solving Eq. (16.5) for small values of g_2, which gives [37]

$$c_1^2 = c_{10}^2 + \alpha T c_{20}^2 \tag{16.29}$$

with c_{10}^2 and c_{20}^2 defined by Eq. (16.8), with the thermodynamic quantities calculated in 2D. This result shows that both the nonzero thermal expansion coefficient α and the discontinuity in the superfluid density are responsible for the jump of the first sound velocity at the BKT transition.

One finds that in the whole interval of temperatures, second sound in the 2D Bose gas corresponds to an oscillation where mainly the superfluid is moving, the normal part remaining practically at rest. First sound, on the other hand, corresponds to an oscillation involving mainly the normal component, similarly to the case of 3D Bose gases (see Section 16.2).

In order to measure first and second sound, a first important requirement is the reachability of the collisional hydrodynamic regime of fast collisions ($\omega\tau \ll 1$, where ω is the frequency of the sound and τ is a typical collisional time) in the normal part. This requirement is likely more problematic for first sound due to its higher velocity. The excitation of second sound should be more easily accessible not only because the velocity is lower but also because in dilute 2D Bose gases it can be naturally excited by density perturbations. For example, using a sudden laser perturbation, applied to the center of the trap, one excites both first and second sound with a relative weight given by Eq. (16.7). Similarly to the 3D case, also in the 2D Bose gas one finds that in the relevant temperature region $\mu \ll T < T_c$ one has $mc_1^2 \gg (n_2\kappa_T)^{-1}$ and hence $W_2 \gg W_1$. In this range of temperatures, second sound hence provides most of the contribution to the compressibility sum rule, thereby making its experimental excitation favorable through density perturbations. At lower temperatures, in the phonon regime, the situation is modified and a typical hybridization effect between the two sounds takes place [1], as discussed in the 3D case in Section 16.2. As $T \to 0$, the thermodynamics is governed by phonons and the second sound velocity approaches the value $c_0/\sqrt{2}$. Above T_c, the difference between the isoentropic and isothermal compressibilities is instead responsible for the occurrence of a diffusive mode at low frequency, as in 3D gases [23].

Typically, in experiments of dilute ultracold gases, atoms are harmonically trapped also along the radial direction. In such systems, T/T_c depends on the local density and thus, by exciting the sound modes through perturbing the center of the trap and tracing the propagation of the modes, one could reveal the T/T_c dependence of the sound velocities, as observed in the case of three-dimensional unitary Fermi gas [38] (see the previous section).

16.6 Conclusions

We have presented an overview of recent theoretical and experimental advances in the study of the second sound velocity in ultracold atomic gases. Concerning possible perspectives and open problems in the future studies on second sound, we would like to mention the experimental determination of the second sound velocity in 2D quantum gases, which would allow for the determination of the temperature dependence of the superfluid density in the presence of the Berezinskii-Kosterlitz-Thouless transition. Other topics of great relevance, and so far unexplored either from the theoretical and experimental perspective, are the study of second sound in superfluid mixtures of Fermi and Bose gases [64] and the role of spin-orbit coupling, causing the breaking of Galilean invariance [65, 66, 67].

Acknowledgments: We would like to thank stimulating collaborations with Gianluca Bertaina, Rudolf Grimm, Yanhua Hou, Tomoki Ozawa, Leonid Sidorenkov, and Meng Khoon Tey. We are also grateful to David Papoular for the final preparation of the chapter. This work has been supported by the European Research Council (ERC) through the Quantum Gases Beyond Equilibrium (QGBE) grant.

References

[1] Pitaevskii, L. P., and Stringari, S. 2016. *Bose-Einstein Condensation and Superfluidity*. Oxford University Press (New York).

[2] Matthews, M. R., Anderson, B. P., Haljan, P. C., Hall, D. S., Wieman, C. E., and Cornell, E. A. 1999. Vortices in a Bose-Einstein condensate. *Phys. Rev. Lett.*, **83**, 2498.

[3] Madison, K. W., Chevy, F., Wohlleben, W., and Dalibard, J. 2000. Vortex formation in a stirred Bose-Einstein condensate. *Phys. Rev. Lett.*, **84**, 806.

[4] Abo-Shaeer, J. R., Raman, C., Vogels, J. M., and Ketterle, W. 2001. Observation of vortex lattices in Bose-Einstein condensates. *Science*, **292**, 476.

[5] Coddington, I., Engels, P., Schweikhard, V., and Cornell, E. A. 2003. Observation of Tkachenko oscillations in rapidly rotating Bose-Einstein condensates. *Phys. Rev. Lett.*, **91**, 100402.

[6] Zwierlein, M. W., Abo-Shaeer, J. R., Schirotzek, A., Schunck, C. H., and Ketterle, W. 2005. Vortices and superfluidity in a strongly interacting Fermi gas. *Nature*, **435**, 1047.

[7] Guery-Odelin, D., and Stringari, S. 1999. Scissors mode and superfluidity of a trapped Bose-Einstein condensed gas. *Phys. Rev. Lett.*, **83**, 4452.

[8] Maragò, O. M., Hopkins, S. A., Arlt, J., Hodby, E., Hechenblaikner, G., and Foot, C. J. 2000. Observation of the scissors mode and evidence for superfluidity of a trapped Bose-Einstein condensed gas. *Phys. Rev. Lett.*, **84**, 2056.

[9] Riedl, S., Sanchez Guajardo, E. R., Kohstall, C., Hecker Denschlag, J., and Grimm, R. 2011. Superfluid quenching of the moment of inertia in a strongly interacting Fermi gas. *New J. Phys.*, **13**, 035003.

[10] Stringari, S. 1996. Collective excitations of a trapped Bose-condensed gas. *Phys. Rev. Lett.*, **77**, 2360.

[11] Onofrio, R., Raman, C., Vogels, J. M., Abo-Shaeer, J. R., Chikkatur, A. P., and Ketterle, W. 2000. Observation of superfluid flow in a Bose-Einstein condensed gas. *Phys. Rev. Lett.*, **85**, 2228.

[12] Miller, D. E., Chin, J. K., Stan, C. A., Liu, Y., Setiawan, W., Sanner, C., and Ketterle, W. 2007. Critical velocity for superfluid flow across the BEC-BCS crossover. *Phys. Rev. Lett.*, **99**, 070402.

[13] Albiez, M., Gati, R., Fölling, J., Hunsmann, S., Cristiani, M., and Oberthaler, M. K. 2005. Direct observation of tunneling and nonlinear self-trapping in a single bosonic Josephson junction. *Phys. Rev. Lett.*, **95**, 010402.

[14] Chin, C., Bartenstein, M., Altmeyer, A., Riedl, S., Jochim, S., Hecker Denschlag, J., and Grimm, R. 2004. Observation of the pairing gap in a strongly interacting Fermi gas. *Science*, **305**, 1128.

[15] Ku, M. J. H., Sommer, A. T., Cheuk, L. W., and Zwierlein, M. W. 2012. Revealing the superfluid lambda transition in the universal thermodynamics of a unitary Fermi gas. *Science*, **335**, 563.

[16] Donnelly, R. 2009. The two-fluid theory and second sound in liquid helium. *Physics Today*, **62**, 34.

[17] Landau, L. D. 1941. The theory of superfuidity of helium II. *J. Phys. USSR*, **5**, 71.

[18] Landau, L. D., and Lifshitz, E. M. 1987. *Fluid Mechanics*. Pergamon (Oxford).

[19] Tisza, L. 1940. Sur la théorie des liquides quantiques. Application a l'hélium liquid. *J. Phys. Radium*, **1**, 164.

[20] Peshkov, V. P. 1946. *J. Phys. USSR*, **10**, 389.

[21] Taylor, E., Hu, H., Liu, X.-J., Pitaevskii, L. P., Griffin, A., and Stringari, S. 2009. First and second sound in a strongly interacting Fermi gas. *Phys. Rev. A*, **80**, 053601.

[22] Pines, D., and Nozieres, P. 1990. *Theory of Quantum Liquids*. Addison-Wesley (Redwood City).

[23] Hu, H., Taylor, E., Liu, X.-J., Stringari, S., and Griffn, A. 2010. Second sound and the density response function in uniform superfluid atomic gases. *New J. Phys.*, **12**, 043040.

[24] Lifshitz, E. M. 1944. *J. Phys. USSR*, **8**, 110.

[25] Dash, J. G., and Taylor, R. D. 1957. Hydrodynamics of oscillating disks in viscous fluids: density and viscosity of normal fluid in pure He^4 from 1.2°K to the lambda point. *Phys. Rev.*, **105**, 7.

[26] Lee, T. D., and Yang, C. N. 1959. Low-temperature behavior of a dilute Bose system of hard spheres. II. Nonequilibrium properties. *Phys. Rev.*, **113**, 1406.

[27] Zaremba, E., Nikuni, T., and Griffin, A. 1999. Dynamics of trapped Bose gases at finite temperatures. *J. Low Temp. Phys.*, **116**, 277.

[28] Griffin, A., Nikuni, T., and Zaremba, E. 2009. *Bose-Condensed Gases at Finite Temperatures*. Cambridge University Press (New York).

[29] Verney, L., Pitaevskii, L., and Stringari, S. 2015. Hybridization of first and second sound in a weakly interacting Bose gas. *EPL (Europhysics Letters)*, **111**, 40005.

[30] Meppelink, R., Koller, S. B., and van der Straten, P. 2009. Sound propagation in a Bose-Einstein condensate at finite temperatures. *Phys. Rev. A*, **80**, 043605.

[31] Zaremba, E. 1998. Sound propagation in a cylindrical Bose-condensed gas. *Phys. Rev. A*, **57**, 518.

[32] Capogrosso-Sansone, B., Giorgini, S., Pilati, S., Pollet, L., Prokof'ev, N., Svistunov, B., and Troyer, M. 2010. The Beliaev technique for a weakly interacting Bose gas. *New J. Physics*, **12**, 043010.

[33] Abrikosov, A. A., Gorkov, L. P., and Dzyaloshinskii, I. E. 1975. *Methods of Quantum Field Theory in Statistical Physics*. Dover (Mineola, NY).

[34] Baym, G., Blaizot, J.-P., Holzmann, M., Laloë, F., and Vautherin, D. 1999. The transition temperature of the dilute interacting Bose gas. *Phys. Rev. Lett.*, **83**, 1703.

[35] Arnold, P., and Moore, G. 2001. BEC transition temperature of a dilute homogeneous imperfect Bose gas. *Phys. Rev. Lett.*, **87**, 120401.

[36] Kashurnikov, V. A., Prokof'ev, N. V., and Svistunov, B. V. 2001. Critical temperature shift in weakly interacting Bose gas. *Phys. Rev. Lett.*, **87**, 120402.

[37] Ozawa, T., and Stringari, S. 2014. Discontinuities in the first and second sound velocities at the Berezinskii-Kosterlitz-Thouless transition. *Phys. Rev. Lett.*, **112**, 025302.

[38] Sidorenkov, L. A., Tey, M. K., Grimm, R., Hou, Y.-H., Pitaevskii, L., and Stringari, S. 2013. Second sound and the superfluid fraction in a Fermi gas with resonant interactions. *Nature* (London), **498**, 78.

[39] Nozieres, P., and Schmitt-Rink, S. 1985. Bose condensation in an attractive fermion gas: from weak to strong coupling superconductivity. *J. Low Temp. Phys.*, **59**, 195.

[40] Hu, H., Liu, X.-J., and Drummond, P. D. 2006. Temperature of a trapped unitary Fermi gas at finite entropy. *Phys. Rev. A*, **73**, 023617.

[41] Salasnich, L. 2010. Low-temperature thermodynamics of the unitary Fermi gas: superfluid fraction, first sound, and second sound. *Phys. Rev. A*, **82**, 063619.

[42] Bertaina, G., Pitaevskii, L., and Stringari, S. 2010. First and second sound in cylindrically trapped gases. *Phys. Rev. Lett.*, **105**, 150402.

[43] Hou, Y.-H., Pitaevskii, L. P., and Stringari, S. 2013. First and second sound in a highly elongated Fermi gas at unitarity. *Phys. Rev. A*, **88**, 043630.

[44] Ho, T.-L. 2004. Universal thermodynamics of degenerate quantum gases in the unitarity limit. *Phys. Rev. Lett.*, **92**, 090402.

[45] Burovski, E., Prokof'ev, N., Svistunov, B., and Troyer, M. 2006. Critical temperature and thermodynamics of attractive fermions at unitarity. *Phys. Rev. Lett.*, **96**, 160402.

[46] Goulko, O., and Wingate, M. 2010. Thermodynamics of balanced and slightly spin-imbalanced Fermi gases at unitarity. *Phys. Rev. A*, **82**, 053621.

[47] Haussmann, R., Rantner, W., Cerrito, S., and Zwerger, W. 2007. Thermodynamics of the BCS-BEC crossover. *Phys. Rev. A*, **75**, 023610.

[48] Hohenberg, P. C. 1967. Existence of long-range order in one and two dimensions. *Phys. Rev.*, **158**, 383.

[49] Mermin, N. D., and Wagner, H. 1966. Absence of ferromagnetism or antiferromagnetism in one- or two-dimensional isotropic Heisenberg models *Phys. Rev. Lett.*, **17**, 1133.

[50] Berezinskii, V. L. 1972. Destruction of long-range order in one-dimensional and two-dimensional systems possessing a continuous symmetry group. II. Quantum systems. *Sov. Phys. JETP*, **34**, 610 [Zh. Eksp. Teor. Fiz. 61, 1144 (1971)].

[51] Kosterlitz, J. M., and Thouless, D. J. 1972. Long range order and metastability in two dimensional solids and superfluids (application of dislocation theory). *J. Phys. C*, **5**, L124.

[52] Kosterlitz, J. M., and Thouless, D. J. 1973. Ordering, metastability and phase transitions in two-dimensional systems. *J. Phys. C.*, **6**, 1181.

[53] Nelson, D. R., and Kosterlitz, J. M. 1977. Universal jump in the superfluid density of two-dimensional superfluids. *Phys. Rev. Lett.*, **39**, 1201.

[54] Desbuquois, R., Chomaz, L., Yefsah, T., Leonard, J., Beugnon, J., Weitenberg, C., and Dalibard, J. 2012. Superfluid behaviour of a two-dimensional Bose gas. *Nature Phys.*, **8**, 645.

[55] Prokof'ev, N., Ruebenacker, O., and Svistunov, B. 2001. Critical point of a weakly interacting two-dimensional Bose gas. *Phys. Rev. Lett.*, **87**, 270402.

[56] Prokof'ev, N., and Svistunov, B. 2002. Two-dimensional weakly interacting Bose gas in the fluctuation region. *Phys. Rev. A*, **66**, 043608.

[57] Rancon, A., and Dupuis, N. 2012. Universal thermodynamics of a two-dimensional Bose gas. *Phys. Rev. A*, **85**, 063607.

[58] Hung, C.-L., Zhang, X., Gemelke, N., and Chin, C. 2011. Observation of scale invariance and universality in two-dimensional Bose gases. *Nature* (London), **470**, 236.

[59] Yefsah, T., Desbuquois, R., Chomaz, L., Günther, K. J., and Dalibard, J. 2011. Exploring the thermodynamics of a two-dimensional Bose gas. *Phys. Rev. Lett.*, **107**, 130401.

[60] Rudnick, I. 1978. Critical surface density of the superfluid component in He 4 films. *Phys. Rev. Lett.*, **40**, 1454.

[61] Bishop, D. J., and Reppy, J. D. 1978. Study of the superfluid transition in two-dimensional He 4 films. *Phys. Rev. Lett.*, **40**, 1727.

[62] Petrov, D. S., Holzmann, M., and Shlyapnikov, G. V. 2000. Bose-Einstein condensation in quasi-2D trapped gases. *Phys. Rev. Lett.*, **84**, 2551.

[63] Tung, S., Lamporesi, G., Lobser, D., Xia, L., and Cornell, E. A. 2010. Observation of the presuperfluid regime in a two-dimensional Bose gas. *Phys. Rev. Lett.*, **105**, 230408.

[64] Ferrier-Barbut, I., Delehaye, M., Laurent, S., Grier, A. T., Pierce, M., Rem, B. S., Chevy, F., and Salomon, C. 2014. A mixture of Bose and Fermi superfluids. *Science*, **345**, 1035.

[65] Zhu, Q., Zhang, C., and Wu, B. 2012. Exotic superfluidity in spin-orbit coupled Bose-Einstein condensates. *Europhys. Lett.*, **100**, 50003.

[66] Zheng, W., Yu, Z.-Q., Cui, X., and Zhai, H. 2013. Properties of Bose gases with the Raman-induced spin–orbit coupling. *J. Phys. B*, **46**, 134007.

[67] Ozawa, T., Pitaevskii, L. P., and Stringari, S. 2013. Supercurrent and dynamical instability of spin-orbit-coupled ultracold Bose gases. *Phys. Rev. A*, **87**, 063610.

17

Quantum Turbulence in Atomic Bose-Einstein Condensates

NICK G. PARKER, A. JOY ALLEN, CARLO F. BARENGHI, AND NICK P. PROUKAKIS

Joint Quantum Centre (JQC) Durham-Newcastle, Newcastle University, UK

The past decade has seen atomic Bose-Einstein condensates emerge as a promising prototype system to explore the quantum mechanical form of turbulence, buoyed by a powerful experimental toolbox to control and manipulate the fluid, and the amenity to describe the system from first principles. This chapter presents an overview of this topic, from its history and fundamental motivations, its characteristics and key results to date, and finally to some promising future directions.

17.1 A Quantum Storm in a Teacup

A befitting title to this chapter could have been "a quantum storm in a teacup." The storm refers to a turbulent state of a fluid, teeming with swirls and waves. Quantum refers to the fact that the fluid is not the classical viscous fluid of conventional storms but rather a quantum fluid in which viscosity is absent and the swirls are quantized. The quantum fluid in our story is a quantum-degenerate gas of bosonic atoms, an atomic Bose-Einstein condensate (BEC), formed at less than a millionth of a degree above absolute zero. And finally the teacup refers to the bowl-like potential used to confine the gas; this makes the fluid inherently inhomogeneous and finite-sized. A typical image of our quantum storm in a teacup is shown in Fig. 17.1a.

This chapter reviews quantum turbulence in atomic condensates, tracing its history (Section 17.2), introducing the main theoretical approach (Section 17.3) and the underyling quantum vortices (Section 17.4). We then turn to describing physical characteristics (Section 17.5), the experimental observations to date (Section 17.6), methods of generating turbulence (Section 17.7), and some exciting research directions (Section 17.8) before presenting an outlook (Section 17.9).

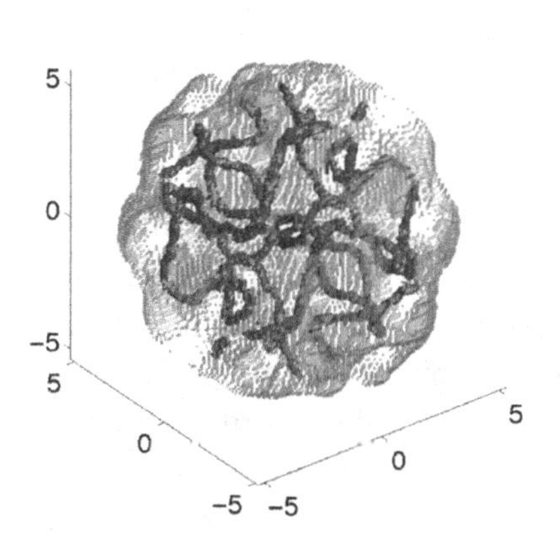

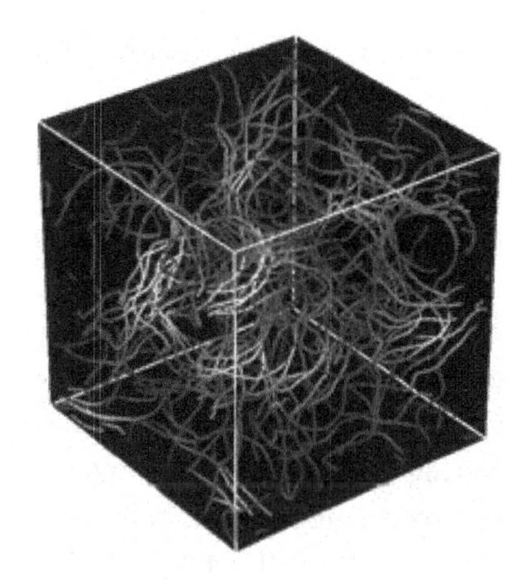

Figure 17.1 (a) A turbulent state in a trapped atomic Bose-Einstein condensate, highlighting the tangle of vortex cores (dark gray) and the condensate surface (light gray). Reprinted with permission from White, A. C., et al. (2010), Non-classical velocity statistics in a turbulent atomic Bose-Einstein condensate, *Phys. Rev. Lett.*, **104**, 075301 [1]. Copyright (2010) by the American Physical Society. (b) A vortex tangle in a large homogeneous box, applicable to turbulent helium-II. Vortex lines are colored according to the magnitude of the coarse-grained vorticity, with light (dark) regions corresponding to high (low) vorticity. Light regions correspond to bundles of parallel vortex lines. Adapted with permission from Baggaley, A. W., et al. (2012), Vortex-density fluctuations, energy spectra, and vortical regions in superfluid turbulence, *Phys. Rev. Lett.*, **109**, 205304 [2]. Copyright (2012) by the American Physical Society. Both images correspond to numerical simulations.

17.2 Origins

Turbulence refers to a highly agitated, disordered, and nonlinear fluid motion, characterized by the presence of eddies and energy across a range of length and time scales [3]. It occurs ubiquitously in nature, from blood flow and waterways to atmospheres and the interstellar medium, and is of practical importance in many industrial and engineering contexts. Since da Vinci's first scientific study of turbulent flow of water past obstacles, circa 1507, research into turbulence in classical viscous fluids continues with vigor; however, due to its rich complexities, the physical essence and mathematical description of turbulence remain a challenge.

The seeds for quantum turbulence were set by Feynman in 1955 [4] for the only known quantum fluid at that time, helium II (bosonic ^{4}He in its superfluid phase below 2.2 K). Working from Onsager's proposal that superfluid vortices carried quantized circulation,[1] he pictured the ensuing "quantum turbulence"[2] as a

[1] This seminal proposal was made as a comment to a paper by Gorter [5]; see comments therein.
[2] The term "quantum turbulence" was not coined until some thirty years later by Donnelly and Swanson [6].

disordered tangle of quantized vortex lines and vortex rings. A modern incarnation of Feynman's concept is depicted in Fig. 17.1b. Feynman may have seen quantum fluids as a prototypical perspective to approach classical turbulence, which he acknowledged as "the most important unsolved problem of classical physics" [7]. Soon after, Vinen pioneered the experimental study of turbulence in helium II [8], driven by thermally induced counterflow (whereby the normal and superfluid components are driven to flow in opposing directions). With no classical analog to thermal counterflow, these studies remained disparate from classical turbulence. However, when experiments in the 1990s turned to conventional methods of generating turbulence, e.g., moving grids and counterrotating discs, striking similarities began to emerge between quantum and classical turbulence, most notably the appearance of energy spectra which follow the classical Kolmogorov behavior [9, 10]. The current status of quantum turbulence in helium can be found in recent reviews [11, 12]. For example, questions are being asked about the role played in turbulence by vortex reconnections [13, 14], and about a new, nonclassical quantum turbulent regime (called "Vinen" or "ultraquantum" turbulence), which, according to experiments [15] and numerical simulations [16], is different from ordinary turbulence in terms of energy spectrum and decay. It is hoped that new methods of flow visualization will clarify this problem [17]. In the very low temperature limit of a pure superfluid, experiments are concerned with quantum turbulence in ^{3}He-B [18, 19], a fermionic superfluid, and with a new, non-classical turbulent energy cascade [20], similar to the Kolmogorov cascade, but arising from interacting Kelvin waves along vortices at very large wavenumbers [21].

The realization of gaseous atomic Bose-Einstein condensates in 1995 [22, 23] introduced a new and meritable system in which to analyze the dynamics of quantum fluids and vortices therein. Due to their weak interactions, condensate fractions of over 95% are achievable in the limit of zero temperature (which in practice corresponds to $T \ll T_C$, where T_C is the critical temperature for condensation). A practical implication of this is that the atomic density, readily imaged via absorption and phase contrast techniques, provides direct visualization of the condensate and vortices. Techniques from atomic physics enable vast control over the physical nature of the system. For example, the strength of the atomic interactions can be precisely tuned (by use of a molecular Feshbach resonance [24]); the potential $V(\mathbf{r}, t)$ used to confine the condensates can be varied almost arbitrarily in both time and space [25]; condensates of reduced dimensionality can be formed [26]; and additional condensate species can be introduced to create multicomponent quantum fluids [27].

In the past decade, atomic BECs have emerged as a promising new setting for studying quantum turbulence [28, 29, 30]. This began with theoretical proposals in the mid-2000s [31, 32, 33, 34], with recent landmark experimental reports of quantum turbulence in three [35] and two dimensions [36, 37]. The key motivation is their embodiment of a highly controllable, prototype system in which to elucidate quantum turbulence (and many-vortex dynamics in general) from the bottom up. However, as we will see, fundamental questions exist over the extent to which turbulence can be generated and observed in such small and inhomogeneous systems.

17.3 Theory

An attractive feature of atomic condensates is their amenity to first-principles theoretical modeling. Due to their high condensate fraction, diluteness, and weak interactions, the system is well described by the microscopic theory of the weakly interacting Bose gas [38]. In the zero temperature limit (often a suitable approximation in typical experiments which operate at much less than the critical temperature for condensation), the system can be parameterized by a single mean-field "macroscopic wavefunction" $\Psi(\mathbf{r}, t)$. This encapsulates both the atomic density distribution $n(\mathbf{r}, t) = |\Psi(\mathbf{r}, t)|^2$ and the coherent phase of the condensate $S(\mathbf{r}, t) = \arg[\Psi(\mathbf{r}, t)]$. The spatial and temporal behavior of Ψ is governed by the Gross-Pitaevskii equation (GPE):

$$i\hbar\frac{\partial\Psi}{\partial t} = \left(-\frac{\hbar^2}{2m}\nabla^2 + V(\mathbf{r}, t) + \frac{4\pi\hbar^2 a_s}{m}|\Psi|^2\right)\Psi,$$

where a_s is the s-wave scattering length which parameterizes the dominant elastic two-body scattering in the condensate. The potential V is typically harmonic, e.g., $V(\mathbf{r}) = \frac{1}{2}m\left[\omega_r^2(x^2 + y^2) + \omega_z^2 z^2\right]$ for an axisymmetric system. Taking $\omega_z \gg \omega_r$ causes the condensate to become flattened in z and provides access to two-dimensional physics [26].

The hydrodynamical interpretation of Ψ is completed with the definition of a condensate fluid velocity $v(\mathbf{r}, t) = (\hbar/m)\nabla S(\mathbf{r}, t)$; indeed, within this picture, the GPE corresponds to a classical continuity equation and the Euler equation for an inviscid compressible fluid. The compressible nature of the condensate is a key difference from helium and has important consequences for the turbulent dynamics.

The GPE has proven an accurate description of many aspects of the condensate, from its shape and collective modes to vortices and phonons [27, 39]. Departures from the GPE become significant, however, at raised temperatures, and a description of finite-temperature extensions to the GPE can be found elsewhere [40, 41, 42, 43].

In liquid helium, the assumption of a local effective interaction is not valid, and while the GPE may provide some qualitative insight of vortex dynamics, it is not physically accurate. A more common model is to treat the vortices as filaments whose mutually induced dynamics are governed by a Biot-Savart law [12]. Alternatively, incorporation of a nonlocal interaction term into the GPE provides a closer physical model of helium [44].

17.4 Quantized Vortices

Vortices are the "sinews and muscles of fluid motion" [45], and quantum turbulence is dominated by a distribution of quantized vortex lines. The quantization of circulation is a consequence of the coherent phase $S(\mathbf{r}, t)$ of the condensate. To preserve the single-valuedness of Ψ, the change in S around any closed path must be quantized as $2\pi q$, where $q = 0, \pm 1, \pm 2, \ldots$. Since a gradient in S is related to the fluid velocity, it immediately follows that the circulation $\Gamma = \oint_C \mathbf{v} \cdot \mathbf{dl}$ is quantized as $q(h/m)$. A vortex occurs for nonzero q. Note that $|q| > 1$ vortices are energetically unstable compared with multiple singly charged vortices, and rarely arise unless engineered [46].

A schematic of a straight vortex line through a harmonically trapped condensate is shown in Fig. 17.2. The vortex has a well-defined core, with zero density and the phase singularity at its center, relaxing to the background condensate density over lengthscale of the order of the healing length $\xi = 1/\sqrt{4\pi n a_s}$; this is typically of the order of $0.1 - 1$ μm but can be tuned by means of a Feshbach resonance (note that in helium $\xi \approx 10^{-10}$ m is fixed by nature). The condensate circulates azimuthally around the singularity with radial speed $v(r) = q\hbar/(mr)$. In contrast, in classical fluids (e.g., air, water), vorticity is a continuous field; circulation is unconstrained; and, in the case of filamentary structures (e.g., tornadoes), the core size is arbitrary.

In three dimensions (3D), the vortices may bend, e.g., into tangles and rings; carry helical Kelvin wave excitations; and undergo reconnections. However, under

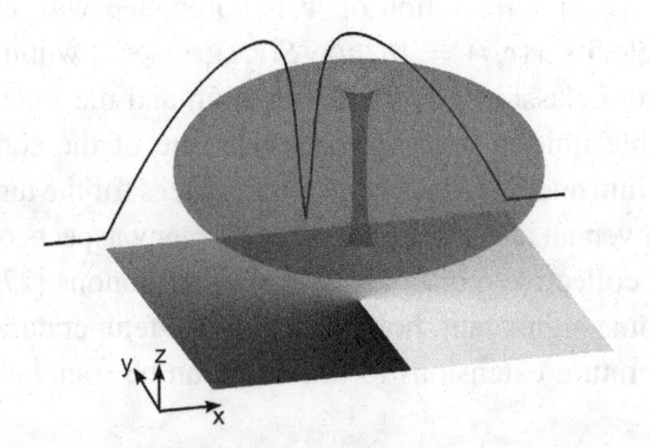

Figure 17.2 Three-dimensional density (isosurface plot) of a trapped condensate featuring a vortex line along the z-axis. The corresponding 2D phase profile and central 1D density profile are also depicted.

strong axial confinement of the condensate, the dynamics become effectively two-dimensional (2D); here the vortices approach the paradigm of point vortices [47, 48]. Being topological defects, vortices can only disappear via annihilation with an oppositely charged vortex or by exiting the fluid (at a boundary). The latter effect is promoted by thermal dissipation, which causes a vortex to spiral out of a trapped condensate [49, 50, 51, 52]. The role of quantum fluctuations has also been considered [53].

Vortical structures have been generated in the form of single vortices [54, 55], vortex–antivortex pairs [36, 56], vortex rings [57], and vortex lattices [58, 59], as well as the disordered vortex distributions of interest here [35, 37, 60]. Methods to generate vortices in condensates include optical imprinting of the phase [46], by a rapid quench through the transition temperature for the onset of Bose-Einstein condensation (i.e., the Kibble-Zurek mechanism [61, 62]) [55, 63] (see also Chapters 6 and 7 or Chapter 30 for ^{3}He), dragging of an obstacle [36, 64, 65], and mechanical rotation [54, 58, 59].

Optical absorption imaging of the vortices is typically preceded by expansion of the cloud to enlarge the cores [54, 64]. This method has been extended to provide real-time imaging of vortex dynamics [55]. While this imaging approach detects density only, the vortex circulation is detectable via a gyroscopic technique [66].

17.5 Characteristics

17.5.1 Scales

Under suitable continuous forcing and dissipation, turbulence reaches a statistical steady state. This is characterized by being statistically self-similar in time and space, with physical quantities behaving according to scaling laws. In ordinary turbulence, these ranges can be vast, e.g., the lengthscale of eddies in atmospheric turbulence span the range $10^{-4}-10^4$ m. In contrast, the accessible lengthscales in condensates are limited from approximately 10^{-6} m (the healing length ξ; smallest scale of density variations) to 10^{-4} m (the system size D), and the true range of self-similarity may be considerably narrower. This raises questions as to the nature of turbulence on such restricted scales, including whether it is justifiably "turbulent." Nonetheless, condensates provide a useful setting for analyzing vortex dynamics, from chaos and self-ordering to the quantum-classical crossover. Of great benefit is that all relevant fluidic scales in the system can be simulated in unison, something not possible for many classical scenarios.

Large-scale flow in quantum turbulence occurs through coherent vortex structures. For example, in 2D, clusters of like-sign vortices form a collective macroscopic flow which, when coarse-grained, mimics classical vorticity. In 3D, the equivalent is for localized bundling of like-sign vortex lines [2], as depicted in Fig. 17.1b. These structures provide the intuition of approaching classical behavior as the number of quanta is increased.

Another important lengthscale is the typical intervortex distance δ. In condensates, the lengthscales ξ, δ, and D are compact, with roughly one order of magnitude separating them (Fig. 17.1a). This should be compared with the many orders of magnitude which separate these distances in helium II (Fig. 17.1b); it is a result of this that vortex filament models are the workhorse for modeling turbulence therein, rather than the microscopic GPE. The interactions of vortices with each other and with the boundaries are magnified, which in turn promotes the occurrence of processes such as reconnections and nucleation/loss of vortices at the boundary.

17.5.2 3D Quantum Turbulence

Quantum turbulence is two-faced: on one side sharing common properties with ordinary turbulence; on another showing striking nonclassical phenomena. Steady-state turbulence in a bulk ordinary fluid, forced at some large scale, is characterized by the famous Kolmogorov behavior: energy is transferred from larger to smaller scales without dissipation over an inertial range of wavenumber k, leading to the Kolmogorov spectrum $k^{-5/3}$ of energy. This process is called the Richardson cascade: large-scale eddies, created at the forcing lengthscale, evolve into progressively smaller eddies. Remarkably, this classical Kolmogorov spectrum emerges in vortex filament calculations [12] which model helium and GPE simulations of 3D quantum turbulence [67, 68, 69] in the spectral range $k \gg 2\pi/\xi$. This classical-like behavior is believed to arise from bundling of vortex lines [2, 12]; that is, vortex lines of the same polarity come together, forming metastable structures. Note that, to link to classical incompressible fluids, the relevant energy in the compressible quantum fluid is the incompressible kinetic energy, i.e., the energy associated with the vortices (the remainder of the kinetic energy, the compressible part, is associated with phonons) [30]. Reconnections between vortex lines are believed to facilitate this cascade by producing small vortex rings. The vortex lines support helical Kelvin wave excitations, which become excited during reconnection events. These excitations carry energy at relatively small scales (high k) and lead to a k^{-3} scaling of the incompressible kinetic energy in the range $k \lesssim 2\pi/\xi$ [69]. These results were obtained in a large homogeneous system, and it remains to be established if this scaling behavior persists in trapped condensates and whether this can be experimentally detected.

Further differences emerge in the turbulent velocity field. In classical isotropic turbulence, the probability distribution of the fluid velocity components is a near-Gaussian. However, in quantum turbulence the velocity statistics have a power-law behavior, as confirmed experimentally in helium [70] and through GPE-based simulations of trapped condensates [1]. This distinction arises from the singular nature of quantized vorticity and the $1/r$ velocity profile. However, classical near-Gaussian behavior is recovered by coarse-graining over the typical intervortex distance [71, 72].

Due to their compressible nature, condensates also support self-similar cascades in acoustic energy, which can be analyzed in the context of weak-wave turbulence [73].

The typical harmonic trapping of condensates leads to a background density which is inhomogeneous throughout and soft boundaries. The recent achievement of an atomic condensate in a boxlike trap [74] allows for the study of quasi homogeneous turbulence and thus the crossover toward the homogeneous setting of helium.

17.5.3 2D Quantum Turbulence

Highly flattened condensates provide access to two-dimensional quantum turbulence. As known from classical fluids, 2D turbulence is vastly different from its 3D counterpart. The vortices become rectilinear, negating Kelvin waves, and reconnections become annihilation events. More importantly, however, 2D incompressible fluids possess an additional conserved quantity, the "enstrophy," which is the total squared vorticity [75, 76]. This has profound consequences: from some forcing lengthscale, energy is transferred to *larger* scales, associated with the aggregation of vorticity into larger structures. Jupiter's Great Red Spot is the natural exemplar of this *inverse cascade*. The inverse cascade leads to a $k^{-5/3}$ scaling of incompressible kinetic energy for $k < k_{\mathrm{f}}$ (where k_{f} is the forcing wavenumber); additionally, enstrophy is transported toward smaller scales, leading to a k^{-3} scaling for $k > k_{\mathrm{f}}$.

In the quantum fluid, enstrophy is simply proportional to the number of vortices. However, compressibility allows for vortices of opposite sign to annihilate, breaking the conservation of enstrophy, and also introducing interactions with phonons. Nonetheless, Reeves et al. [77] were able to suppress these compressibility effects (by suitable forcing and dissipation) and demonstrate the inverse cascade within GPE simulations. This arose through the clustering of vortices with like sign, and progressive increasing in their net charge and spatial scale (thereby mimicking rigid-body rotation [78]). The incompressible kinetic energy was confirmed to follow a $k^{-5/3}$ scaling, with accumulation at the largest scales of the system.

The role of the vortices in the condensate's kinetic energy spectrum has been studied theoretically [79]. For $k > \xi^{-1}$, the spectrum is k^{-3} and associated with the vortex core structure, while for $k < \xi^{-1}$ the spectrum follows the $k^{-5/3}$, which is determined by the distribution of the vortices. Furthermore, the system allows one to study the crossover to weak-wave turbulence [80, 81].

17.5.4 Vortex–Sound Interactions and the Acoustic Sink

In ordinary fluids, the sink at small scales is provided by viscosity, which dissipates small-scale eddies into heat. In finite temperature quantum fluids, an effective viscosity due to the interaction of the condensate with the thermal cloud provides a similar dissipation [43]. In the zero temperature limit, however, viscosity is absent; then dissipation is thought to occur at small lengthscales through the decay of vortical motion into phonon excitations (sound waves) [82].

One such contribution comes from vortex reconnections, where the abrupt change in the fluid topology and velocity field produces a burst of sound waves, as predicted in GPE simulations [83, 85] and depicted in Fig. 17.3a. The sudden snaplike dynamics of reconnections is so dominant that temperature has no significant effect [86, 87]. In 2D, the analog is the sound-generating annihilation of a vortex–antivortex pair. Another contribution occurs under acceleration of a vortex line segment, which leads to the emission of sound waves [84, 88], depicted in Fig. 17.3b for a precessing 2D vortex. This is analogous to the electromagnetic (Larmor) radiation from an accelerating charge [89]. However, this transfer of incompressible, vortex energy to compressible, sound energy is not one way: sound energy can be re-absorbed by the emitting vortex or another vortex [84, 90]. Evidence suggests that these sound-mediated interactions should not disrupt the inverse cascade in 2D quantum turbulence [91].

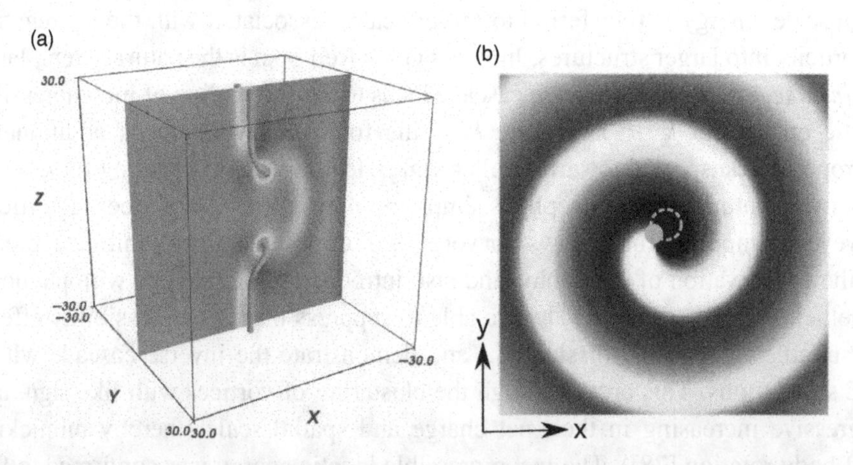

Figure 17.3 (a) Sound emission from a reconnection of two vortex lines (light gray). Reprinted with the permission of AIP Publishing from Zuccher, S., et al. (2012), Quantum vortex reconnections, *Phys. Fluids*, **24**, 125108 [83]. (b) A vortex moving in a circular path radiates sound waves in a quadrupolar pattern, distorted into a spiral by the vortex motion. Adapted with permission from Parker, N. G., et al. (2004), Controlled vortex–sound interactions in atomic Bose-Einstein condensates, *Phys. Rev. Lett.*, **92**, 160403 [84]. Copyright (2004) by the American Physical Society.

17.6 Experiments

17.6.1 3D Quantum Turbulence (São Paulo)

In 2009, Henn et al. [35] reported the generation of quantum turbulence in an elongated condensate via an oscillating trap perturbation. While weak agitation created collective modes of the condensate, i.e. dipole, quadrupole, and scissors modes, stronger agitation led to vortices becoming nucleated into the condensate (likely through a dynamical instability of the condensate surface, analogous to that arising under rotation [92]). Initially well separated, the vortices increased in density with agitation until a disordered tangle of vortex lines permeated the condensate (Fig. 17.3). Under 2D absorption images, the vortex tangle has low contrast, and it remains a challenge to image the 3D vortical structure, although recent experimental advances in simultaneously imaging in two directions may assist in vortex identification [93].

Following release from the trap, the condensate expanded with approximately constant aspect ratio, in contrast to the inversion of aspect ratio expected for ordinary condensates [38]. This anomalous expansion is consistent with the distribution of vorticity throughout the cloud [94], and thus provides an experimental signature of isotropic quantum turbulence. Later work has shown that the turbulent and non-turbulent condensates have very different momentum distributions [95].

17.6.2 2D Quantum Turbulence (Arizona)

Neely et al. [60] employed a flattened condensate with harmonic trapping ($\omega_z/\omega_r \sim 10$), pierced in the center by a localized repulsive Gaussian obstacle from a blue-detuned laser beam so as to form an annular net potential. The relative motion of an obstacle through a condensate is known to generate vortex–antivortex pairs when the motion exceeds a critical speed [96, 97]. Keeping the obstacle fixed, the harmonic trap was translated off-center and moved in a circular path for a short time, before returning to its original position. This generated a disordered distribution of approximately twenty vortices (Fig. 17.4). Note the high visibility of the vortices, due to their alignment along z as per the 2D geometry. The number of vortices decayed over time, until reaching a final state of a (vortex-free) persistent current in the annular trap.

These dynamics bear analogy to the inverse cascade of isotropic 2D compressible turbulence: small-scale stirring generated small-scale vortical excitations, which self-order over time into a large-scale flow (the persistent current). However, the stirring imparts angular momentum to the condensate, which may drive the evolution toward the persistent current.

Numerical simulations of the experiment reveal that stirring leads to the formation of a Kolmogorov-like $k^{-5/3}$ spectrum of the incompressible kinetic energy for

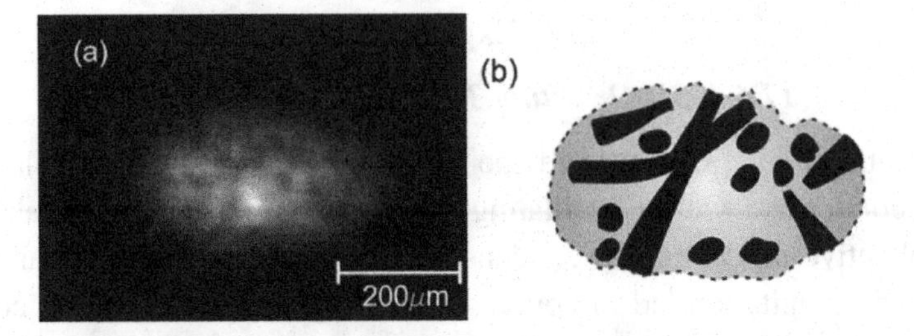

Figure 17.4 (a) Absorption image of the condensate (following expansion). The dark filamentary regions represent vortex lines. (b) An illustration of the inferred vortex distribution in (a). Figures adapted with permission from Henn, E., et al. (2009), Emergence of turbulence in an oscillating Bose-Einstein condensate, *Phys. Rev. Lett.*, **103**, 045301 [35]. Copyright (2009) by the American Physical Society.

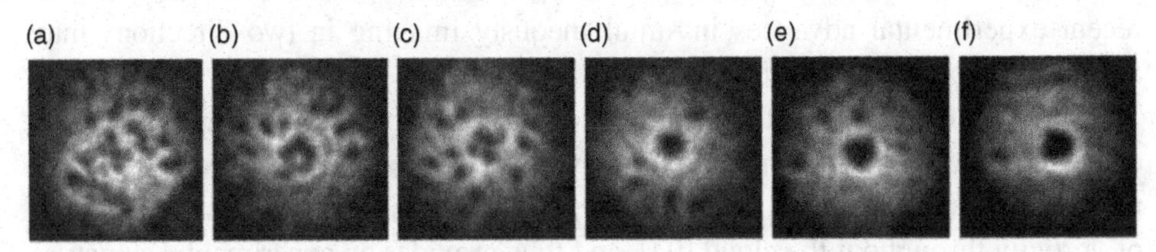

Figure 17.5 Absorption images of the condensate at various times following cessation of the stirring (from left to right: 0, 0.15, 0.33, 0.67, 1.17, and 8.17 s). An initial disordered vortex distribution evolves toward a persistent current. Adapted with permission from Neely, T. W., et al. (2013), Characteristics of two-dimensional quantum turbulence in a compressible superfluid, *Phys. Rev. Lett.*, **111**, 235301 [60]. Copyright (2013) by the American Physical Society.

$k < k_s$ and a k^{-3} spectrum for $k > \xi^{-1}$. Additionally, vortices of the same sign tend to form clusters. These features becomes gradually lost once the stirring ceases, as vortices decay toward the persistent current.

17.6.3 2D Quantum Turbulence (Seoul)

Kwon et al. [37] employed a similar setup as above, apart from the harmonic trap begin translated linearly (in x) relative to the obstacle and subsequently removed. When the translation speed v exceeded a critical value ($\sim$ 0.5 mm s^{-1}), vortices were nucleated into the condensate [56, 65]. The number of vortices N_v increased with v, saturating at around sixty vortices (Fig. 17.5a). The vortices were distributed in a disordered manner, characteristic of 2D quantum turbulence (Fig. 17.5b), and decayed over time.

The experiment focused on this decay of vorticity and suggested two contributions: (i) thermal dissipation (resulting in drifting of vortices to the edge of the condensate) and (ii) vortex–antivortex annihilations. To support this picture, it was found that the decay of N_v followed the form:

$$\frac{dN_v}{dt} = -\Gamma_1 N_v - \Gamma_2 N_v^2, \tag{17.1}$$

where the linear and quadratic terms, parameterized by the coefficients Γ_1 and Γ_2, model these two decay processes. The decay was examined at various temperatures, with both Γ_1 and Γ_2 increasing with temperature. Crescent-shaped, density-depleted regions occasionally appeared in the images and were associated with vortex–antivortex annihilation events.

Numerical simulations [98] have shed further light on these dynamics. As the condensate moves relative to the obstacle, vortices are nucleated, often in clusters of like sign, forming quasiclassical wakes. The condensate sloshes in the trap, generating further vortices and mixing the new with the old. Over time, the clusters disaggregate and the vortices become randomized. Here the condensate is characterized by a disordered distribution of vortices (with no net circulation), collective modes, and a tempestuous field of sound waves, indicative of two-dimensional quantum turbulence. Incorporation of phenomenological dissipation into the simulations (to mimic thermal dissipation) leads to a decay of N_v consistent with Eq. (17.1). The separate decay contributions from drifting and annihilations can be identified, and show that as $T \to 0$ drifting becomes negligible relative to annihilation.

17.7 Generation

The particular method used to generate turbulence can strongly bias the turbulent state which forms, and a key challenge is to establish experimentally feasible approaches which promote isotropic and homogeneous turbulence in condensates, despite their small size (relative to the vortex core size), soft boundaries, and sensitivity to heating. In turn, the generation method may allow control over the nature of the turbulence, e.g., the forcing scale, the transition from chaotic dynamics of few vortices to turbulence, and the ratio of compressible to incompressible kinetic energy. The experimental stirring methods above tend to impart momentum (linear and/or angular) to the condensate and excite collective modes. Simple modifications of these strategies may promote isotropy: stirring in a figure-of-eight path (rather than circular) leads to negligible momentum transfer and an isotropic velocity field [99], while making the Gaussian obstacle elliptical enhances vortex nucleation, which allows for reduced stirring speeds and in turn reduced surface modes and phonons [97, 98, 100].

While rotation of a condensate (in an anisotropic trap) leads to vortex lattices, Kobayashi et al. [33] showed that rotation about a second axis leads to a dense tangle of vortices with Kolmogorov behavior. An attractive feature of this setup is that it provides access to the crossover from isotropic turbulence to polarized turbulence to vortex lattices.

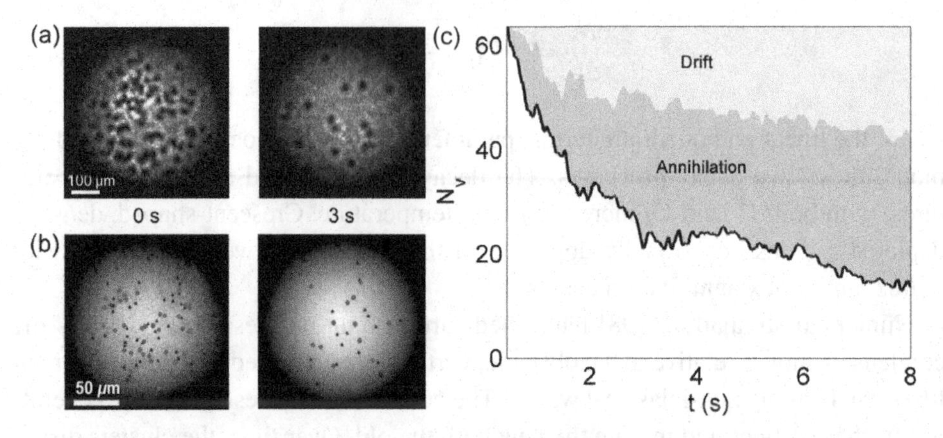

Figure 17.6 (a) Absorption images of the expanded condensate in Ref. [37] at various times following the removal of the obstacle. Images courtesy of Yong-il Shin. Adapted from Kwon, W. J., et al. (2014), Relaxation of superfluid turbulence in highly oblate Bose-Einstein condensates, *Phys. Rev. A*, **90**, 063627 [37]. Copyright (2014) by the American Physical Society. (b) Corresponding images of (unexpanded) condensate density from dissipative GPE simulations [98]. Vortices with positive (negative) circulation are highlighted by circles (triangles). The vortices appear much smaller since the condensate has not been expanded. (c) Decay of the vortex number N_v, with the contribution of drifting and annihilation depicted by the shaded regions. Panels (b) and (c) reprinted with permission from Stagg, G. W., et al. (2015), Generation and decay of two-dimensional quantum turbulence in a trapped Bose-Einstein condensate, *Phys. Rev. A*, **91**, 013612 [98]. Copyright (2015) by the American Physical Society.

Vortices may be imposed deterministically as an initial state by phase imprinting [46], and a recent variation of this, based on orbital angular momentum transfer from a holographic light beam [101], allows for creation of vortex configurations with almost arbitrary charge and geometry. What's more, relatively simple vortex structures initially present in the condensate may undergo an instability to produce quantum turbulence [102, 103]. A turbulence vortex tangle is also predicted to occur during the evolution of a Bose gas from highly nonequilibrium conditions [31]. (See Chapter 7 by Davis et al. for a more detailed discussion of this process and its relation to condensate formation.)

While turbulence is associated with no net vorticity, it is possible to create disordered distributions of vortices dominated by one sign of circulation, e.g., as transient states in rotating condensates [32, 104, 105], through stirring [106] and the decay of a giant vortex [107]. Although not strictly turbulent, these systems offer insight into vortex clustering and self-organization processes.

The generation of statistical steady states of quantum turbulence, with suitable forcing and dissipation, is more challenging in trapped condensates. However, a significant step has been achieved numerically [77]: with forcing provided by flow past a series of obstacles and phenomenological damping, a steady state of 2D quantum turbulence was achieved. Recent simulations have also shown that vortex nucleation and the formation of vortex tangles in three dimensions may be enhanced at finite temperature [108].

17.8 Exotica

Condensates offer access to exotic forms of quantum turbulence; here we list some topical examples.

17.8.1 Quantum Ferrofluid Turbulence

Condensates have been formed with ^{52}Cr [109], ^{164}Dy [110], and ^{168}Er [111] atoms. These possess significant magnetic dipole moments, such that the atomic interactions develop a long-range and anisotropic dipolar contribution (also discussed in Chapter 18), in contrast to the short-range and isotropic van der Waals interactions. The quantum fluid then acquires a global ferromagnetic behavior, forming a "quantum ferrofluid" [112]. Each vortex then behaves as a macroscopic dipole, introducing an additional interaction between vortices, which is long-range and anisotropic [113, 114]. This has significant consequences for the vortex dynamics, controlling the annihilation threshold for vortex–antivortex pairs and the anisotropy of the vortex–vortex dynamics (Fig. 17.7a,b). This can be expected to lead to rich behavior in the ensuing turbulence and provide a handle on the role of annihilations.

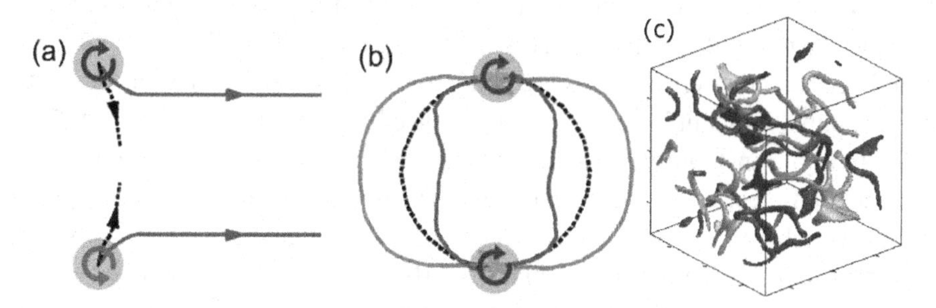

Figure 17.7 (a), (b) Vortex dynamics in a quantum ferrofluid. Dipolar interactions can (a) stabilize vortex-antivortex pairs against annihilation and (b) induce anisotropic co-rotation for vortex–vortex pairs. The nondipolar dynamics are shown by black dotted lines. Adapted with permission from Mulkerin, B. C. et al. (2013), Anisotropic and long-range vortex interactions in two-dimensional dipolar Bose gases, *Phys. Rev. Lett.*, **111**, 170402 [113]. Copyright (2013) by the American Physical Society. (c) Turbulence in a binary condensate can exist as a two inter coupled vortex tangles, shown here through a simulation of the binary GPE [115].

17.8.2 Onsager Vortices and Negative Temperatures

In 1949, Onsager considered turbulence in 2D classical fluids as a disordered collection of point vortices [116]. This model has a finite phase space, the consequence of which is that as energy is added the system it will become more ordered, i.e., lose entropy, a process associated with negative temperature. Remarkably, this phenomenon appears accessible in compressible 2D condensates [117, 118]. Out of an initially disordered distribution of $\pm$ vortices in an isolated (undriven, undamped) system, giant long-lived macroscopic clusters of like-signed vortices emerge [118], the "Onsager vortices." This self-ordering process is associated with the evaporation of vortices, which increases the average energy per vortex, a so-called evaporative heating. This provides a setting to understand a universal phenomena of vortex clustering and the counterintuitive realm of negative temperatures.

17.8.3 Multicomponent Turbulence

Several coexisting condensates can be formed, of different atoms, different isotopes, or different hyperfine states [27]. The components are coupled, and may be either miscible or immiscible, depending on a playoff between the interspecies and intraspecies interactions; in this manner, turbulence may exist as a multiple interpenetrating and coupled vortex tangles, as depicted for a two-component (binary) condensate in Fig. 17.7c.

The presence of intercomponent hydrodynamical instabilities opens up new mechanisms to generate quantum turbulence. For miscible binary condensates made to counterpropagate, a counterflow instability develops above a critical relative speed; this leads to vortex nucleation and ultimately two interpenetrating turbulent quantum fluids [119]. This is analogous to the thermal counterflow method of ^{4}He; however, vortices are generated intrinsically within the bulk rather than through "remnant" vortices. For immiscible binary condensates [120], a Rayleigh-Taylor instability can set in at the interfacial boundary, leading to mixing of the condensates; a Kelvin-Helmholtz instability subsequently produces vortex bundles, which can evolve into turbulence. Unlike other methods for generating quantum turbulence, this can take place at arbitrary small Mach number. Aspects of universal behavior in the dynamics of binary condensates far from equilibrium have also been considered [121].

Multicomponent condensates may possess spin degrees of freedom, termed "spinor" condensates (reviewed in Chapter 18), which allow for unconventional topological defects, including fractional vortices, monopoles and skyrmions, and rich superfluid dynamics [122, 123]. In particular, it is possible to generate spin turbulence, in both homogeneous and trapped condensates, whereby the spin density vectors are disordered [124, 125]. Spin turbulence obeys characterstic scaling laws which are distinct from conventional quantum turbulence, and depends on the form of the spin-dependent interaction. The spin configuration is also coupled to the density, and so coupled spin-density turbulence is a further possibility for unconventional turbulence.

17.9 Outlook

The study of turbulence in quantum gases is still in its infancy. The discrete nature of the vorticity and the scarcity of dissipation (particularly in the low temperature limit $T \ll T_c$, where T_c is the critical temperature) make quantum fluids ideal testing grounds to explore the emergence of classical behavior observed in ordinary fluids, such as the Kolmogorov spectrum [2, 9, 10] and Gaussian velocity statistics [71, 72], from many elementary quanta of circulation. In other words, under certain conditions quantum turbulence seems to capture the main properties of turbulence, thus representing, perhaps, its skeleton. On the other hand, the study of this new form of turbulence has also revealed new physics which is worth studying *per se*. The study of quantum turbulence in atomic BECs presents opportunities (such as the transition from 2D to 3D turbulence) as well as theoretical and experimental challenges (such as the finite-temperature generalization of the GPE and the visualization of 3D turbulence).

Further questions of interest relate to the possibility of generating such turbulent states in other recently realized 2D quantum gases reviewed in this book – most notably polariton BECs, which are an inherently driven-dissipative system [126] – and clarifying the connection between turbulent states emerging in both 2D and 3D dynamically upon quenching a system through the critical region and corresponding turbulent states generated in a more controlled manner, e.g., by shaking or stirring.

Acknowledgments: We acknowledge support from the Engineering and Physical Sciences Research Council (Grant Nos. EP/I019413/1; AJA, CFB, NPP and EP/M005127/1; NGP). We also thank Y. I. Shin for providing the experimental images in Fig. 17.6a.

References

[1] White, A. C., Barenghi, C. F., Proukakis, N. P., Youd, A. J., and Wacks, D. H. 2010. Nonclassical velocity statistics in a turbulent atomic Bose-Einstein condensate. *Phys. Rev. Lett.*, **104**, 075301.

[2] Baggaley, A. W., Laurie, J., and Barenghi, C. F. 2012. Vortex-density fluctuations, energy spectra, and vortical regions in superfluid turbulence. *Phys. Rev. Lett.*, **109**, 205304.

[3] Frisch, U. 1995. *Turbulence: The Legacy of Kolmogorov*. Cambridge, UK: Cambridge University Press.

[4] Feynman, R. P. 1955. Application of quantum mechanics to liquid helium. *Progress in Low Temperature Physics*, **1**, 17–53.

[5] Gorter, C. J. 1949. The two fluid model for helium II. *Il Nuovo Cimento Series 9*, 245–250.

[6] Donnelly, R. J., and Swanson, C. E. 2006. Quantum turbulence. *J. Fluid Mech.*, **173**, 387.

[7] Feynman, R., Leighton, R. B., and Sands, M. 1964. *The Feynman Lectures on Physics*. Boston, MA: Addison-Wesley.

[8] Vinen, W. F. 1957. Mutual friction in a heat current in liquid helium II. I. Experiments on steady heat currents. *Proc. R. Soc. A*, **240**, 114–127.

[9] Maurer, J., and Tabeling, P. 1998. Local investigation of superfluid turbulence. *Europhys. Lett.*, **43**, 29.

[10] Salort, J., Baudet, C., Castaing, B., Chabaud, B., Daviaud, F., Didelot, T., Diribarne, P., Dubrulle, B., Gagne, Y., Gauthier, F., Girard, A., Hbral, B., Rousset, B., Thibault, P., and Roche, P.-E. 2010. Turbulent velocity spectra in superfluid flows. *Phys. Fluids*, **22**, 125102.

[11] Skrbek, L. 2011. Quantum turbulence. *J. Phys. Conf. Ser.*, **318**, 012004.

[12] Barenghi, C. F., Skrbek, L., and Sreenivasan, K. R. 2014. Introduction to quantum turbulence. *Proc. Nat. Acad. Sci.*, **111**(Supplement 1), 4647–4652.

[13] Kerr, R. M. 2011. Vortex stretching as a mechanism for quantum kinetic energy decay. *Phys. Rev. Lett.*, **106**, 224501.

[14] Baggaley, A. W., Barenghi, C. F., and Sergeev, Y. A. 2014. Three-dimensional inverse energy transfer induced by vortex reconnections. *Phys. Rev. E*, **89**, 013002.

[15] Walmsley, P. M., and Golov, A. I. 2008. Quantum and quasiclassical types of superfluid turbulence. *Phys. Rev. Lett.*, **100**, 245301.

[16] Baggaley, A. W., Barenghi, C. F., and Sergeev, Y. A. 2012. Quasiclassical and ultraquantum decay of superfluid turbulence. *Phys. Rev. B*, **85**, 060501.

[17] Guo, W., Cahn, S. B., Nikkel, J. A., Vinen, W. F., and McKinsey, D. N. 2010. Visualization study of counterflow in superfluid ^{4}He using metastable helium molecules. *Phys. Rev. Lett.*, **105**, 045301.

[18] Bradley, D. I., Fisher, S. N., Guenault, A. M., Haley, R. P., Pickett, G. R., Potts, D., and Tsepelin, V. 2011. Direct measurement of the energy dissipated by quantum turbulence. *Nat. Phys.*, **7**, 473.

[19] Hosio, J. J., Eltsov, V. B., Heikkinen, P. J., Hänninen, R., Krusius, M., and L'vov, V. S. 2013. Energy and angular momentum balance in wall-bounded quantum turbulence at very low temperatures. *Nat. Comm.*, **4**, 1614.

[20] Krstulovic, G. 2012. Kelvin-wave cascade and dissipation in low-temperature superfluid vortices. *Phys. Rev. E*, **86**, 055301.

[21] Kozik, E., and Svistunov, B. 2004. Kelvin-wave cascade and decay of superfluid turbulence. *Phys. Rev. Lett.*, **92**, 035301.

[22] Anderson, M. H., Ensher, J. R., Matthews, M. R., Wieman, C. E., and Cornell, E. A. 1995. Observation of Bose-Einstein condensation in a dilute atomic vapor. *Science*, **269**, 198.

[23] Davis, K. B., Mewes, M. O., Andrews, M. R., van Druten, N. J., Durfee, D. S., Kurn, D. M., and Ketterle, W. 1995. Bose-Einstein condensation in a gas of sodium atoms. *Phys. Rev. Lett.*, **75**, 3969.

[24] Inouye, S., Andrews, M. R., Stenger, J., Miesner, H.-J., Stamper-Kurn, D. M., and Ketterle, W. 1998. Observation of Feshbach resonances in a Bose-Einstein condensate. *Nature*, **392**, 151.

[25] Henderson, K., Ryu, C., MacCormick, C., and Boshier, M. G. 2009. Experimental demonstration of painting arbitrary and dynamic potentials for Bose-Einstein condensates. *New J. Phys.*, **11**, 043030.

[26] Görlitz, A., Vogels, J. M., Leanhardt, A. E., Raman, C., Gustavson, T. L., Abo-Shaeer, J. R., Chikkatur, A. P., Gupta, S., Inouye, S., Rosenband, T., and Ketterle, W. 2001. Realization of Bose-Einstein condensates in lower dimensions. *Phys. Rev. Lett.*, **87**, 130402.

[27] Kevrekidis, P. G., Frantzeskakis, D. J., and Carretero-Gonzalez, R. (eds). 2008. *Emergent Nonlinear Phenomena in Bose-Einstein Condensates: Theory and Experiment.* Berlin, Germany: Springer.

[28] Tsubota, M., Kobayashi, M., and Takeuchi, H. 2013. Quantum hydrodynamics. *Phys. Rep.*, **522**, 192–238.

[29] Allen, A. J., Parker, N. G., Proukakis, N. P., and Barenghi, C. F. 2014. Quantum turbulence in atomic Bose-Einstein condensates. *J. Phys.: Conf. Ser.*, **544**, 012023.

[30] White, A. C., Anderson, B. P., and Bagnato, V. S. 2014. Vortices and turbulence in trapped atomic condensates. *Proc. Nat. Acad. Sci.*, **111**(Supplement 1), 4719–4726.

[31] Berloff, N. G., and Svistunov, B. V. 2002. Scenario of strongly nonequilibrated Bose-Einstein condensation. *Phys. Rev. A*, **66**, 013603.

[32] Parker, N. G., and Adams, C. S. 2005. Emergence and decay of turbulence in stirred atomic Bose-Einstein condensates. *Phys. Rev. Lett.*, **95**, 145301.

[33] Kobayashi, M., and Tsubota, M. 2007. Quantum turbulence in a trapped Bose-Einstein condensate. *Phys. Rev. A*, **76**, 045603.

[34] Tsubota, M., and Kobayashi, M. 2008. Quantum turbulence in trapped atomic Bose-Einstein condensates. *J. Low Temp. Phys.*, **150**, 402–409.

[35] Henn, E., Seman, J., Roati, G., Magalhães, K., and Bagnato, V. 2009. Emergence of turbulence in an oscillating Bose-Einstein condensate. *Phys. Rev. Lett.*, **103**, 045301.

[36] Neely, T. W., Samson, E. C., Bradley, A. S., Davis, M. J., and Anderson, B. P. 2010. Observation of vortex dipoles in an oblate Bose-Einstein condensate. *Phys. Rev. Lett.*, **104**, 160401.

[37] Kwon, W. J., Moon, G., Choi, J., Seo, S. W., and Shin, Y. 2014. Relaxation of superfluid turbulence in highly oblate Bose-Einstein condensates. *Phys. Rev. A*, **90**, 063627.

[38] Pethick, C. J., and Smith, H. 2002. *Bose-Einstein Condensation in Dilute Gases.* Cambridge, UK: Cambridge University Press.

[39] Dalfovo, Franco, Giorgini, Stefano, Pitaevskii, Lev P., and Stringari, Sandro. 1999. Theory of Bose-Einstein condensation in trapped gases. *Rev. Mod. Phys.*, **71**, 463–512.

[40] Proukakis, Nick P., and Jackson, Brian. 2008. Finite-temperature models of Bose-Einstein condensation. *J. Phys. B*, **41**, 203002.

[41] Blakie, P. B., Bradley, A. S., Davis, M. J., Ballagh, R. J., and Gardiner, C. W. 2008. Dynamics and statistical mechanics of ultra-cold Bose gases using c-field techniques. *Adv. Phys.*, **57**, 363–455.

[42] Proukakis, N. P., Gardiner, S. A., Davis, M. J., and Szymańska, M. H. (eds). 2013. *Quantum Gases: Finite Temperature and Non-Equilibrium Dynamics.* London, UK: Imperial College Press.

[43] Berloff, N. G., Brachet, M., and Proukakis, N. P. 2014. Modeling quantum fluid dynamics at nonzero temperatures. *Proc. Nat. Acad. Sci.*, **111**(Supplement 1), 4675–4682.

[44] Berloff, Natalia G., and Roberts, Paul H. 1999. Motions in a Bose condensate: VI. Vortices in a nonlocal model. *J. Phys. A*, **32**, 5611.

[45] Kuchemann, D. 1965. Report on the I.U.T.A.M symposium on concentrated vortex motions in fluids. *J. Fluid Mech.*, **21**, 1–20.
[46] Leanhardt, A. E., Görlitz, A., Chikkatur, A. P., Kielpinski, D., Shin, Y., Pritchard, D. E., and Ketterle, W. 2002. Imprinting vortices in a Bose-Einstein condensate using topological phases. *Phys. Rev. Lett.*, **89**, 190403.
[47] Middelkamp, S., Torres, P. J., Kevrekidis, P. G., Frantzeskakis, D. J., Carretero-González, R., Schmelcher, P., Freilich, D. V., and Hall, D. S. 2011. Guiding-center dynamics of vortex dipoles in Bose-Einstein condensates. *Phys. Rev. A*, **84**, 011605.
[48] Aref, H. 2007. Point vortex dynamics: a classical mathematics playground. *J. Math. Phys.*, **48**, 065401.
[49] Jackson, B., Proukakis, N. P., Barenghi, C. F., and Zaremba, E. 2009. Finite-temperature vortex dynamics in Bose-Einstein condensates. *Phys. Rev. A*, **79**, 053615.
[50] Rooney, S. J., Bradley, A. S., and Blakie, P. B. 2010. Decay of a quantum vortex: test of nonequilibrium theories for warm Bose-Einstein condensates. *Phys. Rev. A*, **81**, 023630.
[51] Allen, A. J., Zaremba, E., Barenghi, C. F., and Proukakis, N. P. 2013. Observable vortex properties in finite-temperature Bose gases. *Phys. Rev. A*, **87**, 013630.
[52] Gautam, S., Roy, Arko, and Mukerjee, Subroto. 2014. Finite-temperature dynamics of vortices in Bose-Einstein condensates. *Phys. Rev. A*, **89**, 013612.
[53] Thompson, L., and Stamp, P. C. E. 2012. Quantum dynamics of a Bose superfluid vortex. *Phys. Rev. Lett.*, **108**, 184501.
[54] Madison, K. W., Chevy, F., Wohlleben, W., and Dalibard, J. 2000. Vortex formation in a stirred Bose-Einstein condensate. *Phys. Rev. Lett.*, **84**, 806–809.
[55] Freilich, D. V., Bianchi, D. M., Kaufman, A. M., Langin, T. K., and Hall, D. S. 2010. Real-time dynamics of single vortex lines and vortex dipoles in a Bose-Einstein condensate. *Science*, **329**, 1182.
[56] Kwon, W. J., Seo, S. W., and Shin, Y. 2015. Periodic shedding of vortex dipoles from a moving penetrable obstacle in a Bose-Einstein condensate. *Phys. Rev. A*, **92**, 033613.
[57] Anderson, B. P., Haljan, P. C., Regal, C. A., Feder, D. L., Collins, L. A., Clark, C. W., and Cornell, E. A. 2001. Watching dark solitons decay into vortex rings in a Bose-Einstein condensate. *Phys. Rev. Lett.*, **86**, 2926–2929.
[58] Hodby, E., Hechenblaikner, G. A., Hopkins S. M., Marago O., and Foot, C. J. 2001. Vortex nucleation in Bose-Einstein condensates in an oblate, purely magnetic potential. *Phys. Rev. Lett.*, **88**, 010405.
[59] Abo-Shaeer, J. R., Raman, C., Vogels, J. M., and Ketterle, W. 2001. Observation of vortex lattices in Bose-Einstein condensates. *Science*, **292**, 476.
[60] Neely, T. W., Bradley, A. S., Samson, E. C., Rooney, S. J., Wright, E. M., Law, K. J. H., Carretero-González, R., Kevrekidis, P. G., Davis, M. J., and Anderson, B. P. 2013. Characteristics of two-dimensional quantum turbulence in a compressible superfluid. *Phys. Rev. Lett.*, **111**, 235301.
[61] Kibble, T. W. B. 1976. Topology of cosmic domains and strings. *J. Physics A*, **9**, 1387.
[62] Zurek, W. H. Cosmological experiments in superfluid helium? *Nature*, **317**.
[63] Weiler, C. N., Neely, T. W., Scherer, D. R., Bradley, A. S., Davis, M. J., and Anderson, B. P. 2008. Spontaneous vortices in the formation of Bose-Einstein condensates. *Nature*, **455**, 948.
[64] Raman, C., Abo-Shaeer, J. R., Vogels, J. M., Xu, K., and Ketterle, W. 2001. Vortex nucleation in a stirred Bose-Einstein condensate. *Phys. Rev. Lett.*, **87**, 210402.

[65] Kwon, W. J., Moon, G., Seo, S. W., and Shin, Y. 2015. Critical velocity for vortex shedding in a Bose-Einstein condensate. *Phys. Rev. A*, **91**, 053615.

[66] Powis, A. T., Sammut, S. J., and Simula, T. P. 2014. Vortex gyroscope imaging of planar superfluids. *Phys. Rev. Lett.*, **113**, 165303.

[67] Nore, C., Abid, M., and Brachet, M. E. 1997. Decaying Kolmogorov turbulence in a model of superflow. *Phys. Fluids*, **9**, 2644.

[68] Kobayashi, M., and Tsubota, M. 2005. Kolmogorov spectrum of superfluid turbulence: numerical analysis of the Gross-Pitaevskii equation with a small-scale dissipation. *Phys. Rev. Lett.*, **94**, 065302.

[69] Yepez, J., Vahala, G., Vahala, L., and Soe, M. 2009. Superfluid turbulence from quantum Kelvin wave to classical Kolmogorov cascades. *Phys. Rev. Lett.*, **103**, 084501.

[70] Paoletti, M. S., Fisher, Michael E., Sreenivasan, K. R., and Lathrop, D. P. 2008. Velocity statistics distinguish quantum turbulence from classical turbulence. *Phys. Rev. Lett.*, **101**, 154501.

[71] Baggaley, A. W., and Barenghi, C. F. 2011. Quantum turbulent velocity statistics and quasiclassical limit. *Phys. Rev. E*, **84**, 067301.

[72] Mantia, M. La, and Skrbek, L. 2014. Quantum, or classical turbulence? *Europhys. Lett.*, **105**, 46002.

[73] Proment, D., Nazarenko, S., and Onorato, M. 2009. Quantum turbulence cascades in the Gross-Pitaevskii model. *Phys. Rev. A*, **80**, 051603.

[74] Gaunt, A. L., Schmidutz, T. F., Gotlibovych, I., Smith, R. P., and Hadzibabic, Z. 2013. Bose-Einstein condensation of atoms in a uniform potential. *Phys. Rev. Lett.*, **110**, 200406.

[75] Kraichnan, R. H., and Montgomery, D. 1980. Two-dimensional turbulence. *Rep. Prog. Phys.*, **43**, 547.

[76] Boffetta, G., and Ecke, R. E. 2012. Two-dimensional turbulence. *Annu. Rev. Fluid Mech.*, **44**, 427–451.

[77] Reeves, M. T., Billam, T. P., Anderson, B. P., and Bradley, A. S. 2013. Inverse energy cascade in forced two-dimensional quantum turbulence. *Phys. Rev. Lett.*, **110**, 104501.

[78] Reeves, M. T., Billam, T. P., Anderson, B. P., and Bradley, A. S. 2014. Signatures of coherent vortex structures in a disordered two-dimensional quantum fluid. *Phys. Rev. A*, **89**, 053631.

[79] Bradley, A. S., and Anderson, B. P. 2012. Energy spectra of vortex distributions in two-dimensional quantum turbulence. *Phys. Rev. X*, **2**, 041001.

[80] Nazarenko, Sergey, and Onorato, Miguel. 2006. Wave turbulence and vortices in Bose-Einstein condensation. *Physica D*, **219**, 1–12.

[81] Reeves, M. T., Anderson, B. P., and Bradley, A. S. 2012. Classical and quantum regimes of two-dimensional turbulence in trapped Bose-Einstein condensates. *Phys. Rev. A*, **86**, 053621.

[82] Barenghi, C. F., Parker, N. G., Proukakis, N. P., and Adams, C. S. 2005. Decay of quantised vorticity by sound emission. *J. Low Temp. Phys.*, **138**, 629–634.

[83] Zuccher, S., Caliari, M., Baggaley, A. W., and Barenghi, C. F. 2012. Quantum vortex reconnections. *Phys. Fluids*, **24**, 125108.

[84] Parker, N. G., Proukakis, N. P., Barenghi, C. F., and Adams, C. S. 2004. Controlled vortex-sound interactions in atomic Bose-Einstein condensates. *Phys. Rev. Lett.*, **92**, 160403.

[85] Leadbeater, M., Winiecki, T., Samuels, D. C., Barenghi, C. F., and Adams, C. S. 2001. Sound emission due to superfluid vortex reconnections. *Phys. Rev. Lett.*, **86**, 1410–1413.

[86] Allen, A. J., Zuccher, S., Caliari, M., Proukakis, N. P., Parker, N. G., and Barenghi, C. F. 2014. Vortex reconnections in atomic condensates at finite temperature. *Phys. Rev. A*, **90**, 013601.

[87] Bewley, G. P., Paoletti, M. S., Sreenivasan, K. R., and Lathrop, D. P. 2008. Characterization of reconnecting vortices in superfluid helium. *Proc. Nat. Acad. Sci.*, **105**, 13707–13710.

[88] Vinen, W. F. 2001. Decay of superfluid turbulence at a very low temperature: the radiation of sound from a Kelvin wave on a quantized vortex. *Phys. Rev. B*, **64**, 134520.

[89] Arovas, D. P., and Freire, J. A. 1997. Dynamical vortices in superfluid films. *Phys. Rev. B*, **55**, 1068–1080.

[90] Parker, N. G., Allen, A. J., Barenghi, C. F., and Proukakis, N. P. 2012. Coherent cross talk and parametric driving of matter-wave vortices. *Phys. Rev. A*, **86**, 013631.

[91] Lucas, A., and Surówka, P. 2014. Sound-induced vortex interactions in a zero-temperature two-dimensional superfluid. *Phys. Rev. A*, **90**, 053617.

[92] Parker, N. G., van Bijnen, R. M. W., and Martin, A. M. 2006. Instabilities leading to vortex lattice formation in rotating Bose-Einstein condensates. *Phys. Rev. A*, **73**, 061603.

[93] Lamporesi, G., Donadello, S., Serafini, S., Dalfovo, F., and Ferrari, G. 2013. Spontaneous creation of Kibble-Zurek solitons in a Bose-Einstein condensate. *Nat. Phys.*, **9**, 656.

[94] Caracanhas, M., Fetter, A. L., Muniz, S. R., Magalhes, K. M. F., Roati, G., Bagnato, G., and Bagnato, V. S. 2012. Self-similar expansion of the density profile in a turbulent Bose-Einstein condensate. *J. Low Temp. Phys.*, **166**, 49–58.

[95] Thompson, K. J., Bagnato, G. G., Telles, G. D., Caracanhas, M. A., dos Santos, F. E. A., and Bagnato, V. S. 2014. Evidence of power law behavior in the momentum distribution of a turbulent trapped Bose-Einstein condensate. *Laser Phys. Lett.*, **11**, 015501.

[96] Frisch, T., Pomeau, Y., and Rica, S. 1992. Transition to dissipation in a model of superflow. *Phys. Rev. Lett.*, **69**, 1644–1647.

[97] Stagg, G. W., Allen, A. J., Barenghi, C. F., and Parker, N. G. 2015a. Classical-like wakes past elliptical obstacles in atomic Bose-Einstein condensates. *J. Phys. Conf. Ser.*, **594**, 012044.

[98] Stagg, G. W., Allen, A. J., Parker, N. G., and Barenghi, C. F. 2015b. Generation and decay of two-dimensional quantum turbulence in a trapped Bose-Einstein condensate. *Phys. Rev. A*, **91**, 013612.

[99] Allen, A. J., Parker, N. G., Proukakis, N. P., and Barenghi, C. F. 2014. Isotropic vortex tangles in trapped atomic Bose-Einstein condensates via laser stirring. *Phys. Rev.*, **89**, 025602.

[100] Stagg, G. W., Parker, N. G., and Barenghi, C. F. 2014. Quantum analogues of classical wakes in Bose-Einstein condensates. *J. Phy. B*, **47**, 095304.

[101] Brachmann, J. F. S., Bakr, W. S., Gillen, J., Peng, A., and Greiner, M. 2011. Inducing vortices in a Bose-Einstein condensate using holographically produced light beams. *Opt. Express*, **19**, 12984–12991.

[102] Horng, T.-L., Hsueh, C.-H., and Gou, S.-C. 2008. Transition to quantum turbulence in a Bose-Einstein condensate through the bending-wave instability of a single-vortex ring. *Phys. Rev. A*, **77**, 063625.

[103] Horng, T.-L., Hsueh, C.-H., Su, S.-W., Kao, Y.-M., and Gou, S.-C. 2009. Two-dimensional quantum turbulence in a nonuniform Bose-Einstein condensate. *Phys. Rev. A*, **80**, 023618.

[104] Schweikhard, V., Coddington, I., Engels, P., Tung, S., and Cornell, E. A. 2004. Vortex-lattice dynamics in rotating spinor Bose-Einstein condensates. *Phys. Rev. Lett.*, **93**, 210403.

[105] Wright, T. M., Ballagh, R. J., Bradley, A. S., Blakie, P. B., and Gardiner, C. W. 2008. Dynamical thermalization and vortex formation in stirred two-dimensional Bose-Einstein condensates. *Phys. Rev. A*, **78**, 063601.

[106] White, A. C., Barenghi, C. F., and Proukakis, N. P. 2012. Creation and characterization of vortex clusters in atomic Bose-Einstein condensates. *Phys. Rev. A*, **86**, 013635.

[107] Cidrim, A., dos Santos, F. E. A., Galantucci, L., Bagnato, V. S., and Barenghi, C. F. 2016. Controlled polarization of two-dimensional quantum turbulence in atomic Bose-Einstein condensates. *Phys. Rev. A*, **93**, 033651.

[108] Stagg, G. W., Pattinson, R. W., Barenghi, C. F., and Parker, N. G. 2016. Critical velocity for vortex nucleation in a finite-temperature Bose gas. *Phys. Rev. A*, **93**, 023640.

[109] Griesmaier, A., Werner, J., Hensler, S., Stuhler, J., and Pfau, T. 2005. Bose-Einstein condensation of chromium. *Phys. Rev. Lett.*, **94**, 160401.

[110] Lu, M., Burdick, N. Q., Youn, S. H., and Lev, B. L. 2011. Strongly dipolar Bose-Einstein condensate of dysprosium. *Phys. Rev. Lett.*, **107**, 190401.

[111] Aikawa, K., Frisch, A., Mark, M., Baier, S., Rietzler, A., Grimm, R., and Ferlaino, F. 2012. Bose-Einstein condensation of erbium. *Phys. Rev. Lett.*, **108**, 210401.

[112] Lahaye, T., Menotti, C., Santos, L., Lewenstein, M., and Pfau, T. 2009. The physics of dipolar bosonic quantum gases. *Rep. Prog. Phys.*, **72**, 126401.

[113] Mulkerin, B. C., van Bijnen, R. M. W., O'Dell, D. H. J., Martin, A. M., and Parker, N. G. 2013. Anisotropic and long-range vortex interactions in two-dimensional dipolar Bose gases. *Phys. Rev. Lett.*, **111**, 170402.

[114] Mulkerin, B. C., O'Dell, D. H. J., Martin, A. M., and Parker, N. G. 2014. Vortices in the two-dimensional dipolar Bose gas. *J. Phys. Conf. Ser.*, **497**, 012025.

[115] Pattinson, R. W., Proukakis, N. P., and Parker, N. G. In preparation. Quantum turbulence via repeated interspecies interaction quenches of a binary Bose gas at finite temperature.

[116] Eyink, Gregory L., and Sreenivasan, Katepalli R. 2006. Onsager and the theory of hydrodynamic turbulence. *Rev. Mod. Phys.*, **78**, 87–135.

[117] Billam, T. P., Reeves, M. T., Anderson, B. P., and Bradley, A. S. 2014. Onsager-Kraichnan condensation in decaying two-dimensional quantum turbulence. *Phys. Rev. Lett.*, **112**, 145301.

[118] Simula, T., Davis, M. J., and Helmerson, K. 2014. Emergence of order from turbulence in an isolated planar superfluid. *Phys. Rev. Lett.*, **113**, 165302.

[119] Takeuchi, H., Ishino, S., and Tsubota, M. 2010. Binary quantum turbulence arising from countersuperflow instability in two-component Bose-Einstein condensates. *Phys. Rev. Lett.*, **105**, 205301.

[120] Kobyakov, D., Bezett, A., Lundh, E., Marklund, M., and Bychkov, V. 2014. Turbulence in binary Bose-Einstein condensates generated by highly nonlinear Rayleigh-Taylor and Kelvin-Helmholtz instabilities. *Phys. Rev. A*, **89**, 013631.

[121] Karl, M., Nowak, B., and Gasenzer, T. 2013. Tuning universality far from equilibrium. *Sci. Rep.*, **3**, 2394.

[122] Kasamatsu, K., Tsubota, M., and Ueda, M. 2005. Vortices in multicomponent Bose-Einstein condensates. *Int. J. Mod. Phys. B*, **19**, 1835–1904.
[123] Stamper-Kurn, D. M., and Ueda, M. 2013. Spinor Bose gases: symmetries, magnetism, and quantum dynamics. *Rev. Mod. Phys.*, **85**, 1191–1244.
[124] Fujimoto, K., and Tsubota, M. 2012. Counterflow instability and turbulence in a spin-1 spinor Bose-Einstein condensate. *Phys. Rev. A*, **85**, 033642.
[125] Fujimoto, K., and Tsubota, M. 2014. Spin-superflow turbulence in spin-1 ferromagnetic spinor Bose-Einstein condensates. *Phys. Rev. A*, **90**, 013629.
[126] Berloff, N. G. 2010. Turbulence in exciton-polariton condensates. *arXiv:1010.5225*.

18

Spinor-Dipolar Aspects of Bose-Einstein Condensation

MASAHITO UEDA

Department of Physics, The University of Tokyo, Japan

Selected topics on spinor-dipolar aspects of Bose-Einstein condensation are overviewed. Spinor aspects include spin correlations, fragmentation, dynamical symmetries, and quantum mass acquisition. Dipolar aspects include magnetostriction, d-wave collapse, roton-maxon spectrum, supersolidity, and ferrofluidity. Finally, spinor-dipolar aspects concern the Einstein–de Haas effect and spontaneous mass and spin currents in the ground state.

18.1 Introduction

Bose-Einstein condensation is a phenomenon in which a macroscopic number of particles occupy the same single-particle state as a consequence of symmetrization of a many-body wave function and thereby quantum effects are amplified to the macroscopic level. The amplified single-particle state serves as a complex order parameter whose phase behaves as an emergent thermodynamic quantity, with its temporal and spatial variations giving the chemical potential and the superfluid velocity, respectively. If the interparticle interaction is repulsive, a Bose-Einstein condensate (BEC) acquires stability against excitations out of the condensate because they would cost the Fock exchange energy. This rigidity of the order parameter or the condensate wave function endows a BEC with several remarkable transport properties which are collectively known as superfluidity. In particular, the superfluid fraction can be 100% at zero temperature even though the condensate fraction is depleted due to interparticle interactions. Interparticle interactions also make excited particles phase-locked to the condensate, leading to Bogoliubov quasiparticles, which are the manifestation of phase-coherent particle–hole excitations.

When constituent particles have spin degrees of freedom, a BEC features magnetism and spin nematicity which conspire with superfluidity to produce a rich phase diagram. In the absence of an external magnetic field, the mean-field ground-state phase diagram includes two [1, 2], five [3, 4, 5], and eleven phases [6, 7] for spin-1, 2, and 3 BECs, respectively. BECs with spin degrees of freedom are called spinor condensates. Moreover, the magnetic dipole–dipole interaction (MDDI) between atoms lends distinct twists to spinor condensates because it is long-ranged and anisotropic in sharp contrast with other interactions, which are short-ranged and isotropic. In this chapter, we will present a brief overview on the spinor-dipolar aspects of BEC. An early experimental work of spinor BECs is reviewed in Ref. [8]. Later developments with an emphasis on theoretical aspects are described in Ref. [9]. Many-body physics of polarized dipolar gases is reviewed in Ref. [10]. A more recent review on general aspects of spinor gases is given in Refs. [11, 12].

Atomic-gas BECs are extremely dilute with the atomic density five orders of magnitude lower than that of the air. The typical interatomic distance is 1000 Å, which is much longer than the range of the interaction. Thus, the interatomic interaction can be modeled with a contact interaction except for the dipole–dipole interaction (DDI). To be precise, let us recall that the partial-wave expansion is applicable if the interaction potential decays sufficiently rapidly with the interatomic distance r. More specifically, the lth partial wave expansion is valid if the potential decays more rapidly than $1/r^{2l+3}$ [13]. For a power-law decaying potential $\propto r^{-n}$, the partial-wave expansion is applicable for $l < (n-3)/2$. This implies that the s-wave approximation is valid if the interaction potential decays faster than r^{-3}. However, the magnetic DDI decays with $n = 3$, where all partial waves contribute to the scattering amplitude even at absolute zero, and we must use the full DDI, which is invariant under simultaneous rotations of the spin and spatial coordinates by the same amount. Thus, the magnetic DDI couples the spin with orbital degrees of freedom and makes unique contributions to the fundamental properties of the spinor-dipolar condensates even though its energy scale is much smaller than that of the spin-exchange interaction.

18.2 Spinor Interaction

We first consider the case without the magnetic DDI. Let $\hat{\psi}_m(\mathbf{r})$ be the field operator of a spin-f boson at position $\mathbf{r}$ with magnetic sublevel $m = f, f-1, \ldots, -f$. The s-wave interaction depends on the total spin F of a pair of interacting bosons, which must be even due to Bose symmetry. Thus, there are $f+1$ distinct scattering lengths a_F $(F = 0, 2, \ldots, 2f)$ and the same number of coupling constants $c_f^{(i)}$ $(i = 0, 1, \ldots, f)$, which are described by a_F. The isotropy in spin space requires that the interaction depends only on scalar combinations of the number-density operator $\hat{n} = \sum_{m=-f}^{f} \hat{\psi}_m^\dagger \hat{\psi}_m$ $(\alpha = x, y, z)$, the spin-density operator

$\hat{f}_\alpha = \sum_{m,n=-f}^{f} \hat{\psi}_m^\dagger (f_\alpha)_{mn} \hat{\psi}_n$ ($\alpha = x, y, z$), and the spin-singlet-pair creation operator $\hat{\mathcal{A}}^\dagger = \sum_{m=-f}^{f} \hat{\psi}_m^\dagger T \hat{\psi}_m$, where $(f_\alpha)_{mn}$ are the matrix elements of the spin-f matrices and T is the time-reversal operator $[T\hat{\psi}_m = (-1)^m \hat{\psi}_{-m}^\dagger]$. It follows from these considerations that the interaction Hamiltonians for spin-1 [1, 2], spin-2 [3, 4, 5], and spin-3 [6, 14] bosons are given by

$$\hat{V}^{(f=1)} = \int d\mathbf{r} : \left(c_0^{(1)} \hat{n}^2 + c_1^{(1)} \hat{\mathbf{f}} \cdot \hat{\mathbf{f}} \right) :, \tag{18.1}$$

$$\hat{V}^{(f=2)} = \int d\mathbf{r} : \left(c_0^{(2)} \hat{n}^2 + c_1^{(2)} \hat{\mathbf{f}} \cdot \hat{\mathbf{f}} + c_2^{(2)} \hat{\mathcal{A}}^\dagger \hat{\mathcal{A}} \right) :, \tag{18.2}$$

$$\hat{V}^{(f=3)} = \int d\mathbf{r} : \left(c_0^{(3)} \hat{n}^2 + c_1^{(3)} \hat{\mathbf{f}} \cdot \hat{\mathbf{f}} + c_2^{(3)} \hat{\mathcal{A}}^\dagger \hat{\mathcal{A}} + c_3^{(3)} (\hat{\mathbf{f}} \cdot \hat{\mathbf{f}})^2 \right) :, \tag{18.3}$$

where :: denotes normal ordering which places annihilation operators to the right of creation operators; the last term in Eq. (18.3) describes spin nematicity; and $\hat{\mathbf{f}} = (\hat{f}_x, \hat{f}_y, \hat{f}_z)$ is the spin-vector operator for each spin f.

In Eqs. (18.1)–(18.3), the term $\hat{\mathbf{f}} \cdot \hat{\mathbf{f}}$ describes processes in which two interacting bosons change their magnetic quantum numbers by one while keeping their sum conserved. Typical spin-exchanging processes include $(m = 0) + (m = 0) \leftrightarrow (m = +1) + (m = -1)$. Such spin-exchange interactions lead to population oscillations of magnetic sublevels as observed in ^{87}Rb [15, 16, 17, 18, 19] and ^{23}Na [20] condensates. For $f \geq 2$, the spin-singlet term $\hat{\mathcal{A}}^\dagger \hat{\mathcal{A}}$ appears in the interaction Hamiltonian. Physically, this term projects a pair of atoms into the spin-singlet state. Such a process is not explicit for the $f = 1$ case because the corresponding projection operator $\hat{\mathcal{P}}_0$ can be expressed in terms of the Casimir operator of the SO($2f + 1$) group as $\hat{\mathcal{P}}_0 = [\hat{N}(\hat{N} + 2f - 1) - \hat{C}_2(\mathrm{SO}(2f + 1))]/(2f + 1)$ [21], where $\hat{N}$ is the total particle-number operator. Thus, for the case of $f = 1$, the spin-singlet term can expressed in terms of the quadratic Casimir operator $\hat{C}_2(\mathrm{SO}(3))$ of the SO(3) group, which is nothing but $\hat{\mathbf{f}} \cdot \hat{\mathbf{f}}$. For $f \geq 3$, the spin-nematic term $(\hat{\mathbf{f}} \cdot \hat{\mathbf{f}})^2$ describes processes in which two interacting bosons change their magnetic quantum numbers by up to two while keeping their sum conserved. At the mean-field level, it reduces to the trace of the squared nematic tensor, $\mathrm{tr}\hat{\mathcal{N}}^2$, where $(\hat{\mathcal{N}})_{ij} = \langle \hat{f}_i \hat{f}_j + \hat{f}_j \hat{f}_i \rangle / 2$ [6].

18.3 Mean-Field Phase Diagram

We consider the mean-field phase diagram of spinor condensates in the uniform system. In this case, the kinetic energy can be ignored and the Hartree term $: \hat{n}^2 :$ is constant, so the ground state can be found by minimizing the interaction energy.

18.3.1 Spin-1 Case

In the absence of an external magnetic field, the ground state is determined by the last term in Eq. (18.1). Thus the ground state is ferromagnetic if $c_1 < 0$ and antiferromagnetic if $c_1 > 0$. The normalized spinor of the ferromagnetic state is $(\zeta_1, \zeta_0, \zeta_{-1}) = (1, 0, 0)$ and that of the antiferromagnetic state is $(1/\sqrt{2}, 0, 1/\sqrt{2})$, which is obtained from (0,1,0) by a rotation about the x-axis through $\pi/2$. The state (0,1,0) is called polar because the spinor changes its sign upon inversion in spin space; that is, it has polarity. Both the ferromagnetic and antiferromagnetic states are inert states, where an inert state is the one whose order parameter does not depend on any of the interaction parameters such as the scattering length [22].

An external magnetic field B has distinct effects on the spinor and dipolar parts. The linear Zeeman term influences only the dipolar interaction but not the spinor interaction because the total spin and its projection onto the quantization axis are conserved. Therefore, the quadratic Zeeman effect makes the dominant contribution to the spinor interaction. A quadratic Zeeman term has the form $q\hat{f}_z^2$ with $q = (g\mu_B B)^2/\Delta A_{hf}$, where g, μ_B, and ΔA_{hf} are the g-factor, the Bohr magneton, and the hyperfine energy-level splitting, respectively. Thus magnetic sublevels with large (small) $|m|$ are energetically favored for $q < 0$ $(q > 0)$. For the $f = 1$ ^{87}Rb condensate, $q > 0$ and the $m = 0$ state will be energetically favored for sufficiently large B. Thus, there is an interesting competition between the ferromagnetic interaction and the quadratic Zeeman effect. For intermediate magnetic fields with $0 < q < 2|c_1^{(1)}|n$, all spin components are populated and a transverse magnetization emerges. This implies that the axial symmetry around an applied magnetic field is spontaneously broken [23] and hence this phase is called the broken axisymmetry phase. Since the magnetization depends on the ratio $q/(2|c_1^{(1)}|n)$, this state is not inert. For the $f = 1$ ^{23}Na condensate, the spinor interation is antiferromagnetic, so there is no such competition unlike the case of ^{87}Rb, and the role of the quadratic Zeeman effect is to favor (0,1,0) against $(1/\sqrt{2}, 0, 1/\sqrt{2})$.

18.3.2 Spin-2 Case

In the absence of an external magnetic field, the ground state is determined by minimizing the last two terms in Eq. (18.2), and we find three distinct phases [3, 4, 5]. For $c_1^{(2)} < 0$ and $c_2^{(2)} > 4c_1^{(2)}$, the ferromagnetic phase with $|\langle \hat{\mathbf{f}} \rangle| = 2$ and $|\langle \hat{\mathcal{A}} \rangle| = 0$ is realized. The standard spinor for this state is $(1, 0, 0, 0, 0)$.

For $c_2^{(2)} < 0$ and $c_2^{(2)} < 4c_1^{(2)}$, the antiferromagnetic (or nematic) phase with $|\langle \hat{\mathbf{f}} \rangle| = 0$ and $|\langle \hat{\mathcal{A}} \rangle| = 1$ is realized. The standard spinor for this case is $(\sin\eta/\sqrt{2}, 0, \cos\eta, 0, \sin\eta/\sqrt{2})$ [5], where η characterizes the profile of the spinor order parameter. This state has the inversion symmetry in spin space, unlike the polar phase. To be more specific, $\eta = 0$ and $\pi/2$ correspond to the uniaxial-nematic and biaxial-nematic phases, and other values describe the dihedral-2 phase. All of these phases are degenerate at the mean-field level due to the SO(5) symmetry of the spin-singlet state.

For $c_1^{(2)} > 0$ and $c_2^{(2)} > 0$, neither the ferromagnetic nor antiferromagnetic phase is energetically favorable; it turns out that in this case, three bosons form a spin-singlet trio, and the trio bosons undergo BEC. (Note that such a trio singlet cannot be constructed from spin-1 bosons.) This phase is called cyclic or tetrahedral because the Majorana representation of the mean-field ground state has the symmetry of a tetrahedron. The standard order parameter of the cyclic phase is $(1/2, 0, -i/\sqrt{2}, 0, 1/2)$ [5] which is degenerate with and related through a rotation in spin space to $(1/, \sqrt{3}, 0, 0, \sqrt{2/3}, 0)$.

Both the antiferromagnetic phase and the cyclic phase are nonmagnetic, and the time-reversal symmetry is broken (unbroken) in the cyclic (antiferromagnetic) phase. It is remarkable that in the absence of an external magnetic field, all ground states for both spin-1 and spin-2 cases are inert states.

In the presence of the quadratic Zeeman effect, the degeneracy in the nematic phases is lifted and the uniaxial-nematic (biaxial-nematic) phase is favored for $q > 0$ $(q < 0)$. The quadratic Zeeman effect also lifts the degeneracy between the two cyclic phases described above and distorts them into $\zeta^{\mathrm{C}} = (\sin\theta/\sqrt{2}, 0, -i\cos\theta, 0, \sin\theta/\sqrt{2})$ with $\cos^2\theta = 1/2 + 5q/(c_2^{(2)}n)$ and $\zeta^{\mathrm{M}} = (\cos\theta, 0, 0, \sin\theta, 0)$ with $\cos^2\theta = 1/3 - q/(3c_1^{(2)})$. The former phase ζ^{C} remains nonmagnetic, whereas the latter phase ζ^{M} is magnetized and termed the mixed phase after such magnetic hybridization [24]. All of these states are noninert, as they depend explicitly on the external magnetic field and the interaction parameters.

18.3.3 Spin-3 Case

The mean-field ground state of a uniform spin-3 BEC is determined by the competition between the last three terms in Eq. (18.3) and involves eleven phases. Four of them are inert states and the others are noninert. Some of the detailed properties of spin-3 BECs are reviewed in Ref. [7]. The only experimentally realized spin-3 BEC is that of ^{52}Cr [25], which has the electronic spin of three and zero nuclear spin. Here the magnetic DDI is significant and spinor-dipolar physics becomes important, as described in Section 18.6.

18.4 Many-Body Aspects of Spinor Bose-Einstein Condensates

18.4.1 Spin Correlations

Many-body spin correlations may be investigated analytically under situations in which the system is tightly confined so that the excitation energies of the spatial modes are much larger than the energy scales of the interaction and the temperature. In this case, the field operator can be expressed as $\hat{\psi}_m(\mathbf{r}) = \hat{a}_m\phi(\mathbf{r})$, where $\phi(\mathbf{r})$ is the spatial mode of the ground state and $\hat{a}_m$ is the annihilation operator of a boson in that spatial mode with magnetic quantum number m. In this single-mode approximation, the complete eigenspectrum of the many-body Hamiltonian is found for the spin-1 case in Ref. [26] and the spin-2 case in Refs. [3, 5], and a partial solution is obtained for the spin-3 case in Ref. [27].

Let us first discuss the spin-1 case. In this case, only the last term in Eq. (18.1) is relevant for many-body spin correlations. When $c_1^{(1)} < 0$, the ferromagnetic spin correlation is energetically favored, and atomic spins are likely to align in the same direction unless constraints such as the total magnetization are imposed. When $c_1^{(1)} > 0$, the spins of two colliding bosons are likely to be antiparallel to each other and we may expect the spin-singlet pair correlation to be formed. In fact, we may rewrite the last term in Eq. (18.1) as $: \hat{\mathbf{f}} \cdot \hat{\mathbf{f}} :=: \hat{n}^2 : -3\hat{\mathcal{A}}^\dagger\hat{\mathcal{A}}$. This implies that the many-body ground state is given by $(\hat{\mathcal{A}}^\dagger)^{N/2}|\text{vac}\rangle$, which is indeed an eigenstate of both $: \hat{n}^2 :$ and $\hat{\mathcal{A}}^\dagger\hat{\mathcal{A}}$. Here we assume for simplicity that the total number of bosons N is even. Since this state is isotropic in spin space, all magnetic sublevels are equally populated, in sharp contrast with the mean-field ground state in the antiferromagnetic phase. For the spin-2 case, similar arguments apply to the ferromagnetic and antiferromagnetic phases. An interesting feature arises in the cyclic phase. In this case, neither spin parallel nor antiparallel configuration is energetically favored because $c_1^{(2)} > 0$ and $c_2^{(2)} > 0$. It turns out that the cyclic phase circumvents such two-body spin frustration by forming a spin-singlet state of three bosons [5].

A striking feature of many-body spin correlations is that they drastically affect populations of magnetic sublevels due to Bose enhancement. For example, in the antiferromagnetic phase of a spin-2 BEC, the average population of the $m = 0$ level is given by [5]

$$\langle \hat{a}_0^\dagger \hat{a}_0 \rangle \sim N_\text{S} \left(\frac{1 + 2n_{22}}{n_{12}} \right), \tag{18.4}$$

where N_S is the number of spin-singlet pairs, n_{12} is the number of $m = 2$ bosons, and n_{22} is the number of pairs that have the total spin of two and hence characterizes the spin correlation. Depending on this number, the $m = 0$ population changes by a large factor.

18.4.2 Fragmentation

In most examples of BECs, the single-particle reduced density matrix has one and only one (macroscopic) extensive eigenvalue which is proportional to an effective volume of the condensate. However, there are cases in which more than one extensive eigenvalue exists. Such a BEC is said to be fragmented [28, 29]. A fragmented BEC must have two or more degenerate single-particle states that can be occupied macroscopically. The interaction between these degenerate states must be attractive to gain the Fock exchange energy due to fragmentation. Furthermore, the system must be mesoscopic and should not be in the thermodynamic limit; otherwise, even an infinitesimal symmetry-breaking perturbation would mix the degenerate state into a single coherent superposition state. The many-body ground states of spin-1 and spin-2 antiferromagnetic BECs and a spin-2 cycic BEC are all fragmented. For an antiferromagnetic BEC, a positive (negative) quadratic Zeeman term, together with bosonic enhancement, drives the system into the $m = 0$ state (the $m = \pm f$ states), thereby making the system a single coherent BEC. To stabilize a fragmented BEC, the system size must be smaller than the spin healing length and the quadratic Zeeman effect should be sufficiently small.

A fragmented BEC may be viewed as a coherent superposition of ordinary (non-fragmented) condensates. To see this, let us consider a polar state of the spin-1 BEC. By rotating the standard spinor (0,1,0) to an arbitrary direction specified by an Euler rotation $U(\alpha, \beta, \gamma) := e^{-i\alpha S_z} e^{-i\beta S_y} e^{-i\gamma S_z}$, we obtain a general spinor for the polar state: $(\zeta_1, \zeta_0, \zeta_{-1}) = (-e^{-i\alpha}/\sqrt{2}, \cos\beta, e^{i\alpha}\sin\beta/\sqrt{2})$, which is independent of γ. The corresponding creation operator is given by $\hat{a}_\zeta^\dagger = \sum_{m=-1}^{1} \zeta_m \hat{a}_m^\dagger$ and the state vector of the single condensate is $|\zeta\rangle = (\hat{a}_\zeta^\dagger)^N/\sqrt{N!}|\mathrm{vac}\rangle$. Averaging this state over the solid angle $d\Omega_\zeta = \sin\beta d\alpha d\beta/(4\pi)$, we obtain the many-body spin-singlet BEC which is fragmented:

$$\int d\Omega_\zeta |\zeta\rangle = \begin{cases} 0 & \text{if } N \text{ is odd;} \\ \dfrac{1}{(N+1)\sqrt{N!}}(a_0^{\dagger 2} - 2a_1^\dagger a_{-1}^\dagger)^{N/2}|\mathrm{vac}\rangle & \text{if } N \text{ is even.} \end{cases} \tag{18.5}$$

Thus, the fragmented BEC is given as an equal-weighted average of single coherent BECs. Similar arguments apply to other fragmented condensates. That is, a fragmented BEC is an equal-weighted superposition state over the entire degenerate mean-field states. This is the reason why the fragmented BEC is fragile against symmetry-breaking perturbations.

18.4.3 Dynamical Symmetries

In the single-mode approximation, the eigenspectra of spin-1 and spin-2 BECs can be obtained exactly, and one might wonder whether exact solutions can be found for higher-spin cases. In this respect, it is worthwhile to note that dynamical symmetry lies behind the exact solvability in spinor condensates. Dynamical symmetry is the symmetry inherent in the Hamiltonian and is to be distinguished from space-time symmetries. Examples include the U(n) symmetry in the n-dimensional harmonic-oscillator model and the O(4) symmetry in the three-dimensional Coulomb system [30]. In a spin-f BEC, the particle-number conservation leads to the U($2f+1$) symmetry. To solve the eigenvalue problem, we must find the complete set of quantum numbers. For a spin-f BEC, we need $2f+1$ eigenvalues to completely specify the eigenstate. Also, since a spin-f condensate has $f+1$ coupling constants, we need to find the same number of quadratic Casimir invariants to diagonalize the Hamiltonian.

Let us first consider the $f = 1$ case, where we need two quadratic Casimir invariants to diagonalize the Hamiltonian and three quantum numbers to completely characterize the eigenstate. We start from U(3) and consider the chain of groups U(3)$\supset$SO(3)$\supset$SO(2). This chain provides the desired two quadratic Casimir invariants $\hat{N}^2, \hat{\mathbf{F}}^2$ and one linear Casimir operator $\hat{F}_z$, where $\hat{N}$, $\hat{\mathbf{F}}$, and $\hat{F}_z$ are the total particle-number operator, the total spin operator, and the total magnetic quantum number operator, respectively. Thus, the Hamiltonian can be diagonalized by using two quadratic Casimir invariants $\hat{N}^2$ and $\hat{\mathbf{F}}^2$, and the eigenvalues N, F, and F_z completely characterize the eigenstate $|N, F, F_z\rangle$.

Let us next consider the $f = 2$ case, where we need three quadratic Casimir invariants to diagonalize the Hamiltonian and five quantum numbers to uniquely specify the eigenstate. We start from U(5) and consider a chain of groups U(5)$\supset$SO(5)$\supset$SO(3)$\supset$SO(2). The first three groups provide three quadratic Casimir invariants $\hat{N}^2, \hat{\mathcal{A}}^\dagger\hat{\mathcal{A}}, \hat{F}^2$ and the last group gives a linear Casimir operator $\hat{F}_z$, where $\hat{\mathcal{A}}^\dagger$ is the creation operator of the spin-singlet pair. Because three quadratic Casimir invariants are available, they can be used to diagonalize the Hamiltonian. However, only four quantum numbers are available to specify the eigenstate, and one quantum number is lacking, which is called a missing label. The four quantum numbers are N, F, F_z, and τ, where τ is the number of unpaired bosons and called the seniority quantum number [31]. We may choose the number of spin-singlet trios n_{30} as the missing label (we may also use the number of spin-singlet pairs $n_{20} := (N-\tau)/2$ instead of τ) and the eigenspectrum is degenerate with respect to the missing label. The eigenstate is thus given by $|N, \tau, n_{30}, F, F_z\rangle$, where the eigenstates are degenerate with respect to n_{30}.

Finally, let us discuss the case of $f = 3$. In this case, the chain of groups is U(7)$\supset$SO(7)$\supset$$G_2$ $\supset$SO(3)$\supset$SO(2). However, the quadratic Casimir invariant of the exceptional group G_2 is proportional to that of SO(7) [21], so that only three quadratic Casimir invariants are available and one quadratic Casimir invariant is lacking. Therefore, only for a special class of coupling constants can the Hamiltonian be diagonalized by dynamical symmetry alone. Also, only four quantum numbers, N, F, F_z, and τ, are available, so three missing labels are needed to uniquely specify eigenstates [32], and the eigenspectrum is degenerate with respect to these three missing labels.

18.4.4 Quantum Mass Acquisition

Quantum mass acquisition is a phenomenon in which a massless quasiparticle becomes massive due to quantum corrections. It has been discussed in several subfields of physics but experimentally remained elusive because the emergent energy gap is too small to be distinguished from other secondary effects. This type of quasiparticles are called quasi-Nambu-Goldstone (qNG) bosons, which, at the mean-field level, have gapless excitations that do not originate from spontaneous symmetry breaking and acquire energy gaps once quantum corrections are taken into account [33, 34]. It is generally held that the zero-point energy sets the energy scale for the qNG mode [35]. Remarkably, in the uniaxial-nematic phase of a spin-2 BEC, which is the most likely ground state of the spin-2 ^{87}Rb BEC at zero magnetic field, the zero-point energy does not set the energy scale of the emergent energy gap because the mean-field biaxial-nematic state becomes dynamically unstable once quantum corrections beyond the Bogoliubov level [36] are taken into account [37]. Because of this special situation, the zero-point energy plays no role and the energy gap turns out to be two orders of magnitude larger than the zero-point energy [38]. For the uniaxial-nematic phase, the number of spontaneously broken gauge and spin-rotation symmetries is three, leading to three Nambu-Goldstone excitations (one photon and two magnon modes). However, the Bogoliubov spectrum exhibits five gapless excitations [5]. The two extra gapless modes, which are quadrupolar (or nematic) modes, arise because the mean-field ground state of the nematic phase has an SO(5) symmetry [35], which is larger than the symmetry of the Hamiltonian SO(3)×U(1). It is these two extra modes that are expected to acquire mass once quantum fluctuations of the condensate beyond the Bogoliubov approximation are taken into account. Due to the time-reversal and spin-space inversion symmetries, the two quadrupolar excitations are degenerate and thus there is a single emergent mass gap. The magnitude of the emergent energy gap Δ depends strongly on the value of $c_2^{(2)}$ and provides precise information on the latter. For $c_2^{(2)} = c_1^{(2)}$, $\Delta/\hbar \simeq 2.3$ Hz. The qNG modes can be excited by transferring a fraction of atoms from the $|m_F = 0\rangle$ to $|m_F = \pm 2\rangle$ hyperfine states. Then the relative phase between these two states will oscillate at a frequency given by the energy gap of the qNG modes and can therefore be measured by phase-contrast imaging.

18.5 Magnetic Dipole–Dipole Interaction

As discussed above, there are contact-type interactions between two atoms. Furthermore, if two atoms have magnetic moments $\mathbf{d}_1, \mathbf{d}_2$, there will be an MDDI between them:

$$V_{\rm dd}(\mathbf{r}) = c_{\rm dd}\frac{\hat{\mathbf{d}}_1 \cdot \hat{\mathbf{d}}_2 - 3(\hat{\mathbf{d}}_1 \cdot \hat{\mathbf{r}})(\hat{\mathbf{d}}_2 \cdot \hat{\mathbf{r}})}{r^3}, \tag{18.6}$$

where $\hat{\mathbf{d}}_1 = \mathbf{d}_1/d_1$, $\hat{\mathbf{d}}_2 = \mathbf{d}_2/d_2$, and $\hat{\mathbf{r}} = \mathbf{r}/r$ are unit vectors, and $c_{\rm dd} = \mu_0 d_1 d_2/(4\pi)$ with μ_0 being the magnetic permeability. As can be seen from this expression, the MDDI is long-ranged and anisotropic, and can be repulsive or attractive depending on the relative directions of the two magnetic dipoles. In particular, if the two dipoles are polarized in the z direction, Eq. (18.6) reduces to

$$V_{\rm dd}(\mathbf{r}) = c_{\rm dd}\frac{1 - 3\cos^2\theta}{r^3}. \tag{18.7}$$

Thus, the DDI is attractive for small θ and repulsive for large $\theta = 0$ with its sign changing at the magic angle

$$\theta_{\rm m} = \cos^{-1}\frac{1}{\sqrt{3}} \simeq 54.7^\circ. \tag{18.8}$$

Thus, the DDI is repulsive when two dipoles are aligned side by side and attractive when they are placed head to tail. This leads to an instability of a condensate if it is confined in a sufficiently elongated trap and if the dipoles are polarized along the elongated direction [39]. As we discussed in this chapter's introduction, all partial waves contribute in the presence of the DDI. Moreover, since the interaction is anisotropic, different partial waves are coupled. These unique features of the MDDI endow BECs with several unique properties.

18.5.1 Magnetostriction

The anisotropic interaction causes magnetostriction of a BEC, resulting in a spontaneous deformation of the condensate due to the DDI as observed in ^{52}Cr [40]. By equating $c_{\rm dd}/r^3$ to $\hbar^2/(Mr^2)$, one obtains a characteristic length scale of the magnetic DDI: $a_{\rm dd} = c_{\rm dd}M/(3\hbar^2)$, where the numerical factor is introduced so that the collapsing instability due to the DDI occurs for $a_{\rm dd} > a$, with a being the s-wave scattering length. The ratio $\epsilon_{\rm dd} = a_{\rm dd}/a$ gives a relative strength of the DDI against the contact interaction. Specifically, we have $\epsilon_{\rm dd} \sim 0.007, 0.16, 0.67$, and 1.32 for ^{87}Rb, ^{52}Cr, ^{166}Er, and ^{162}Dy, respectively. It is noteworthy that this ratio can be enhanced by reducing a. Such a technique was used to create magnetic DDI-dominated condensates for ^{52}Cr [40] and ^{7}Li [41].

18.5.2 D-Wave Collapse

Yet another important feature of the DDI is the d-wave nature. In fact, the numerator of Eq. (18.7) is proportional to the zeroth component of the spherical harmonic function of rank 2, Y_2^0. Thus, the selection rule of the DDI is $(l,m) \to (l,m)$ or $(l,m) \to (l \pm 2, m)$, where l and m are the orbital angular momentum and the magnetic quantum number, respectively. In general, Eq. (18.6) involves Y_2^m and changes the orbital angular momentum l by 0, ± 2. A spectacular consequence of the d-wave interaction was displayed in the experiment of a collapsing dipolar BEC, where the collapse was followed by an explosion of the condensate showing a cloverleaf pattern characteristic of the d-wave interaction [42].

18.5.3 Roton-Maxon Spectrum

The dispersion relation of a polarized dipolar BEC that is harmonically confined in the direction of the polarization and free in the other two directions exhibits a characteristic roton-maxon spectrum [43]. When the wavenumber k of the excitation is smaller than the inverse length scale of the confinement L^{-1}, dipoles are aligned side by side and repulsively interact with each other, and the dispersion relation is phononic. However, as k increases beyond L^{-1}, the excitation becomes three dimensional and atoms can align head to tail and interact attractively. Thus, the slope of the dispersion relation becomes negative and reaches a minimum before it becomes free-boson-like for larger k.

18.5.4 Soliton, Supersolid, and Ferrofluidity

The Gross-Pitaevskii equation with a contact interaction does not admit a stable solution in two dimensions; the system will expand or collapse. However, with the DDI a stable solition can exist in a two-dimensional harmonic trap [44]. The basic idea is that the polarized DDI can be either attractive or repulsive, depending on the trap geometry. Thus, for an attractive (repulsive) contact interaction, a BEC can be stabilized by $\epsilon_{\rm dd} > 0$ ($\epsilon_{\rm dd} < 0$). Other interesting possibilities of polarized dipolar BECs include supersolidity in optical lattices [45, 46] and ferrofluidity which emerges at an interface between dipolar and nondipolar BECs [47].

18.6 Spinor-Dipolar Bose-Einstein Condensate

When a dipolar condensate is confined in an optical trap, the spin degrees of freedom are liberated and the system will have its own dynamics governed by both contact spinor interactions and anisotropic long-ranged dipolar interactions. We shall refer to such a condensate as a spinor-dipolar condensate. A spinor-dipolar condensate forms nontrivial spin textures over the length scale of the dipole healing length $\xi_{\rm dd} = \hbar/\sqrt{2Mc_{\rm dd}n}$, where M is the atomic mass and n is the particle-number density. In a spinor-dipolar BEC, even a weak DDI can significantly alter spin textures because changing the direction of a local spin costs little energy if its spatial variation extends over a distance much longer than $\xi_{\rm dd}$.

In the second-quantized form, the magnetic DDI of a spinor BEC is given by

$$\hat{V}_{\rm dd} = \frac{c_{\rm dd}}{2} \int d\mathbf{r} \int d\mathbf{r}' \sum_{\nu\nu'} : \hat{F}_\nu(\mathbf{r}) Q_{\nu\nu'}(\mathbf{r}-\mathbf{r}') \hat{F}_{\nu'}(\mathbf{r}') :, \tag{18.9}$$

where $\hat{F}_\nu$ is the spin-density operator, $c_{\rm dd} = \mu_0 (g\mu_{\rm B})^2/(4\pi)$, with g being the Landé g-factor for the atom, and $Q_{\nu\nu'}$ is the dipole interaction kernel which is given by $Q_{\nu\nu'}(\mathbf{r}) \equiv (\delta_{\nu\nu'} - 3\hat{r}_\nu \hat{r}_{\nu'})/r^3$. The nonlocal interaction (18.9) produces an effective magnetic field

$$B_\nu^{\rm eff}(\mathbf{r}) = \frac{c_{\rm dd}}{g\mu_B} \int d\mathbf{r}' \sum_{\nu\nu'} Q_{\nu\nu'}(\mathbf{r}-\mathbf{r}') F_{\nu'}(\mathbf{r}'). \tag{18.10}$$

Such a nonlocal DDI leads to spin waves and spin ordering in deep one- and two-dimensional optical lattices [48, 49, 50].

18.6.1 Einstein-de Haas Effect

The magnetic DDI couples the spin with the orbital degrees of freedom and is invariant under simultaneous rotations of the spin and orbital angular momenta. Thus, the interaction (18.9) conserves not the spin but the total (spin plus orbital) angular momentum. As a consequence, an initially polarized BEC will transfer its angular momentum from the spin to orbital sector, and the BEC begins to rotate spontaneously. This is the Einstein–de Haas (EdH) effect in a spinor-dipolar BEC [14, 51]. The EdH effect may be interpreted as a consequence of spin precession due to the effective magnetic field (18.10). Note, however, that for a fully polarized initial state, the integration on the right-hand side of Eq. (18.10) vanishes if the system is uniform. We thus conclude that the EdH effect is unique to nonuniform systems.

For a ferromagnetic spinor condensate such as a spin-1 ^{87}Rb BEC, the spin-polarized state should be rather stable. In this case, the EdH effect could be induced if we apply an external magnetic field opposite to the direction of the spin polarization and make the kinetic energy cost of the orbital angular momentum resonant with the linear Zeeman gain upon spin flip [52], and a further enhancement is predicted to occur by using a resonant ac magnetic field [53]. On the other hand, for nonferromagnetic condensates such as spin-1 ^{23}Na and spin-3 ^{52}Cr condensates, spin-polarized states are dynamically unstable against the EdH effect at a sufficiently low external magnetic field. Such spontaneous demagnetization has been observed in a thermal gas [54] and a condensate [55, 56] of ^{52}Cr atoms.

18.6.2 Ground-State Circulation

In a solid-state ferromagnet, the minimization of the MDDI in Eq. (18.9) is achieved if the magnetization $\mathbf{F}$ satisfies the flux-closure relation $\nabla \cdot \mathbf{F}(\mathbf{r})$. An interesting situation arises in a ferromagnetic spinor condensate due to the spin-gauge symmetry. In a trapped system, the particle density is not uniform, so the effective magnetic field (18.10) is caused by the DDI. Thus, the system spontaneously develops a spin texture. Then, due to the spin-gauge symmetry, a mass current should also flow spontaneously [57, 58]. Thus, we may expect a nonzero mass current in the ground state of a spinor BEC. The ground-state phase diagram of a spin-1 ferromagnetic BEC is predicted to have three phases: flower (FL), chiral-spin vortex (CSV), and polar-core vortex (PCV) phases [57]. The total angular momentum per atom is one for the former two phases and zero for the last one. The CSV phase has nonzero spin circulation (i.e, spin chirality) and a substantial mass circulation. In the PCV phase, the spin is completely quenched on the symmetry axis of the trap with a spin circulation around it (namely, the spin vortex) and no net mass current. All of these phases can be realized for spin-1 ^{87}Rb condensates as a function of $\xi_{\rm sp}/R_{\rm TF}$ and $\xi_{\rm dd}/R_{\rm TF}$, where $\xi_{\rm sp} = \hbar/\sqrt{2M|c_1^{(1)}n}$ is the spin healing length and $R_{\rm TF}$ is the Thomas-Fermi radius [57].

18.6.3 Dipole–Dipole Interaction Under a Magnetic Field

Under a magnetic field, the spin undergoes the Larmor precession at the Larmor frequency $\omega_{\rm L} = g\mu_{\rm B}B/\hbar$. When $\hbar\omega_{\rm L} \gg c_{\rm dd}n$, the time-dependent part of the MDDI is averaged out and only the part of the MDDI that is time independent on the rotating frame of reference at $\omega_{\rm L}$ remains nonvanishing. This part no longer conserves the total angular momentum but conserves the projected total (but not necessarily local) spin angular momentum and the projected orbital angular momentum separately. The remaining part still preserves the long-range and anisotropic nature of the MDDI, and the resulting time-averaged MDDI tends to quench magnetization along the magnetic field and help spontaneous formation of transverse magnetization [59]. Such MDDI-induced spin textures have been observed in spin-1 ^{87}Rb Bose-Eisntein condensates [60].

18.7 Conclusions and Outlook

This chapter overviews selected topics of spinor-dipolar aspects of BECs. It is by no means intended to be comprehensive. Among the important subjects that cannot be covered here due to lack of space are topological excitations [61] and spinor-dipolar physics in optical lattices [62, 12] and in the presence of synthetic gauge fields [63].

All of these topics are still developing at a rapid pace. If this chapter serves as an introduction to this fascinating field, it will have well served its intended purpose.

References

[1] Ohmi, T., and Machida, K. 1998. Bose-Einstein condensation with internal degrees of freedom in alkali atom gases. *J. Phys. Soc. Jpn.*, **67**, 1822.

[2] Ho, T.-L. 1998. Spinor Bose condensates in optical trap. *Phys. Rev. Lett.*, **81**, 742.

[3] Koashi, M., and Ueda, M. 2000. Exact eigenstates and magnetic response of spin-1 and spin-2 Bose-Einstein condensates. *Phys. Rev. Lett.*, **84**, 1066.

[4] Ciobanu, C. V., Yip, S.-K., and Ho, T.-L. 2000. Phase diagrams of F=2 spinor Bose-Einstein condensates. *Phys. Rev. A*, **61**, 033607.

[5] Ueda, M., and Koashi, M. 2002. Theory of spin-2 Bose-Einstein condensates: spin correlations, magnetic response, and excitation spectra. *Phys. Rev. A*, **65**, 063602.

[6] Diener, R. B., and Ho, T.-L. 2006. ^{52}Cr spinor condensate: a biaxial of uniaxial spin nematic. *Phys. Rev. Lett.*, **96**, 190405.

[7] Kawaguchi, Y., and Ueda, M. 2011. Symmetry classification of spinor Bose-Einstein condensates. *Phys. Rev. A*, **84**, 053616.

[8] Stamper-Kurn, D. M., and Ketterle, W. 2001. *Spinor Condensates and Light Scattering from Bose-Einstein Condensates*. New York: Springer-Verlag. Chap. 2, pages 137–218.

[9] Kawaguchi, Y., and Ueda, M. 2012. Spinor Bose Einstein condensates. *Phys. Rep.*, **520**, 253–381.

[10] Baranov, M. A. 2008. Theoretical progress in many-body physics with ultracold dipolar gases. *Phys. Rep.*, **464**, 71.

[11] Ueda, M. 2012. Bose gases with nonzero spin. *Annual Review of Condensed Matter Physics*, **3**, 263–283.

[12] Stamper-Kurn, D. M., and Ueda, M. 2013. Spinor Bose gases: symmetries, magnetism, and quantum dynamics. *Rev. Mod. Phys.*, **85**, 1191.

[13] Landau, L. D., and Lifshitz, E. M. 1981. *Quantum Mechanics (Non-Relativistic Theory)*, 3rd Edition. Butterworth-Heinemann.

[14] Santos, L., and Pfau, T. 2006. Spin-3 chromium Bose-Einstein condensates. *Phys. Rev. Lett.*, **96**, 190404.

[15] Schmaljohann, H., Erhard, M., Kronjäger, J., Kottke, M., van Staa, S., Cacciapuoti, L., Arlt, J. J., Bongs, K., and Sengstock, K. 2004. Dynamics of $F = 2$ spinor Bose-Einstein condensates. *Phys. Rev. Lett.*, **92**, 040402.

[16] Kuwamoto, T., Araki, K., Eno, T., and Hirano, T. 2004. Magnetic field dependence of the dynamics of Rb 87 spin-2 Bose-Einstein condensates. *Phys. Rev. A*, **69**, 063604.

[17] Chang, M.-S., Hamley, C. D., Barrett, M. D., Sauer, J. A., Fortier, K. M., Zhang, W., You, L., and Chapman, M. S. 2004. Observation of spinor dynamics in optically trapped ^{87}Rb Bose-Einstein condensates. *Phys. Rev. Lett.*, **92**, 140403.

[18] Kronjäger, J., Becker, C., Brinkmann, M., Walser, R., Navez, P., Bongs, K., and Sengstock, K. 2005. Evolution of a spinor condensate: coherent dynamics, dephasing, and revivals. *Phys. Rev. A*, **72**, 063619.

[19] Widera, A., Gerbier, F., Fölling, S., Gericke, T., Mandel, O., and Bloch, I. 2006. Precision measurement of spin-dependent interaction strengths for spin-1 and spin-2 ^{87}Rb atoms. *New J. Phys.*, **8**, 152.

[20] Black, A. T., Gomez, E., Turner, L. D., Jung, S., and Lett, P. D. 2007. Spinor dynamics in an antiferromagnetic spin-1 condensate. *Phys. Rev. Lett.*, **99**, 070403.

[21] Uchino, S., Otsuka, T., and Ueda, M. 2008. Dynamical symmetry in spinor Bose-Einstein condensates. *Phys. Rev. A*, **78**, 023609.

[22] Volovik, G. E., and Gorkov, L. P. 1985. Superconducting classes in heavy-fermion systems. *Sov. Phys. JETP*, **61**, 843.

[23] Murata, K., Saito, H., and Ueda, M. 2007. Broken-axisymmetry phase of a spin-1 ferromagnetic Bose-Einstein condensate. *Phys. Rev. A*, **75**, 013607.

[24] Saito, H., and Ueda, M. 2005. Diagnostics for the ground-state phase of a spin-2 Bose-Einstein condensate. *Phys. Rev. A*, **72**, 053628.

[25] Griesmaier, A., Werner, J., Hensler, S., Stuhler, J., and Pfau, T. 2005. Bose-Einstein condensation of chromium. *Phys. Rev. Lett.*, **94**, 160401.

[26] Law, C. K., Pu, H., and Bigelow, N. P. 1998. Quantum spins mixing in spinor Bose-Einstein condensates. *Phys. Rev. Lett.*, **81**, 5257.

[27] Uchino, S., Kobayashi, M., and Ueda, M. 2010. Bogoliubov theory and Lee-Huang-Yang corrections in spin-1 and spin-2 Bose-Einstein condensates in the presence of the quadratic Zeeman effect. *Phys. Rev. A*, **81**, 063632.

[28] Nozières, P., and Saint James, D. Particle vs. pair condensation in attractive Bose liquids. *J. Phys. (Paris)*, **43**, 1133.

[29] Mueller, E. J., Ho, T.-L., Ueda, M., and Baym, G. 2006. Fragmentation of Bose-Einstein condensates. *Phys. Rev. A*, **74**, 033612.

[30] Iachello, F. 2006. *Lie Algebras and Applications*. Berlin: Springer Verlag.

[31] Talmi, Igal. 2993. *Simple Models of Complex Nuclei: The Shell Model and Interacting Boson Model*. London: Harwood Acad. Publ.

[32] Rohoziński, S. G. 1978. The oscillator basis for octupole collective motion in nuclei. *J. Phys. G*, **4**, 1075.

[33] Weinberg, S. 1972. Approximate symmetries and pseudo-Goldstone bosons. *Phys. Rev. Lett.*, **29**, 1698.

[34] Georgi, H., and Pais, A. 1975. Vacuum symmetry and the pseudo-Goldstone phenomenon. *Phys. Rev. D*, **12**, 508.

[35] Uchino, S., Kobayashi, M., Nitta, M., and Ueda, M. 2010. Quasi-Nambu-Goldstone modes in Bose-Einstein condensates. *Phys. Rev. Lett.*, **105**, 230406.

[36] Phuc, N. T., Kawaguchi, Y, and Ueda, M. 2013a. Beliaev theory of spinor Bose-Einstein condensates. *Ann. Phys. (Berlin)*, **328**, 158.

[37] Phuc, N. T., Kawaguchi, Y, and Ueda, M. 2013b. Fluctuation-induced and symmetry-prohibited metastabilities in spinor Bose-Einstein condensates. *Phys. Rev. A*, **88**, 043629.

[38] Phuc, N. T., Kawaguchi, Y, and Ueda, M. 2014. Quantum mass acquisition in spinor Bose-Einstein condensates. *Phys. Rev. Lett.*, **113**, 230401.

[39] Santos, L., Shlyapnikov, G. V., Zoller, P., and Lewenstein, M. 2000. Bose-Einstein condensation in trapped dipolar gases. *Phys. Rev. Lett.*, **85**, 1791.

[40] Lahaye, T., Koch, T., Frohlich, B., Fattori, M., Metz, J., Griesmaier, A., Giovanazzi, S., and Pfau, T. 2007. Strong dipolar effects in a quantum ferrofluid. *Nature*, **448**, 672.

[41] Pollack, S. E., Dries, D., Junker, M., Chen, Y. P., Corcovilos, T. A., and Hulet, R. G. 2009. Extreme tunability of interactions in a ^{7}Li Bose-Einstein condensate. *Phys. Rev. Lett.*, **102**, 090402.

[42] Lahaye, T., Metz, J., Froehlich, B., Koch, T., Meister, M., Griesmaier, A., Pfau, T., Saito, H., Kawaguchi, Y., and Ueda, M. 2008. d-wave collapse and explosion of a dipolar Bose-Einstein condensate. *Phys. Rev. Lett.*, **101**, 080401.

[43] Santos, L., Shlyapnikov, G. V., and Lewenstein, M. 2003. Roton-maxon spectrum and stability of trapped dipolar Bose-Einstein condensates. *Phys. Rev. Lett.*, **90**, 250403.

[44] Pedri, P., and Santos, L. 2005. Two-dimensional bright solitons in dipolar Bose-Einstein condensates. *Phys. Rev. Lett.*, **95**, 200404.

[45] Góral, K., Santos, L., and Lewenstein, M. 2002. Quantum phases of dipolar bosons in optical lattices. *Phys. Rev. Lett.*, **88**, 170406.

[46] Yi, S., Li, T., and Sun, C. P. 2007. Novel quantum phases of dipolar Bose gases in optical lattices. *Phys. Rev. Lett.*, **98**, 260405.

[47] Saito, H., Kawaguchi, Y., and Ueda, M. 2009. Ferrofluidity in a two-component dipolar Bose-Einstein condensate. *Phys. Rev. Lett.*, **102**, 230403.

[48] Pu, H., Zhang, W., and Meystre, P. 2001. Ferromagnetism in a lattice of Bose-Einstein condensates. *Phys. Rev. Lett.*, **87**, 140405.

[49] Gross, K., Search, C. P., Pu, H., Zhang, W., and Meystre, P. 2002. Magnetism in a lattice of spinor Bose-Einstein condensates. *Phys. Rev. A*, **66**, 033603.

[50] Zhang, W., Pu, H., Search, C., and Meystre, P. 2002. Spin waves in a Bose-Einstein condensed atomic spin chain. *Phys. Rev. Lett.*, **88**, 060401.

[51] Kawaguchi, Y., Saito, H., and Ueda, M. 2006. Einstein-de Haas effect in dipolar Bose-Einstein condensates. *Phys. Rev. Lett.*, **96**, 080405.

[52] Gawryluk, K., Brewczyk, M., Bongs, K., and Gajda, M. 2007. Resonant Einstein–de Haas effect in a rubidium condensate. *Phys. Rev. Lett.*, **99**, 130401.

[53] Gawryluk, K., Bongs, K., and Brewczyk, M. 2011. How to observe dipolar effects in spinor Bose-Einstein condensates. *Phys. Rev. Lett.*, **106**, 140403.

[54] Hensler, S., Werner, J., Griesmaier, A., Schmidt, P. O., Gorlitz, A., Pfau, T., Giovanazzi, S., and Rzazewski, K. 2003. Dipolar relaxation in an ultra-cold gas of magnetically trapped chromium atoms. *Appl. Phys. B*, **77**, 765.

[55] Pasquiou, B., Bismut, G., Beaufils, Q., Crubellier, A., Maréchal, E., Pedri, P., Vernac, L., Gorceix, O., and Laburthe-Tolra, B. 2010. Control of dipolar relaxation in external fields. *Phys. Rev. A*, **81**, 042716.

[56] Pasquiou, B., Maréchal, E., Bismut, G., Pedri, P., Vernac, L., Gorceix, O., and Laburthe-Tolra, B. 2011. Spontaneous demagnetization of a dipolar spinor Bose gas in an ultralow magnetic field. *Phys. Rev. Lett.*, **106**, 255303.

[57] Kawaguchi, Y., Saito, H., and Ueda, M. 2006. Spontaneous circulation in ground-state spinor dipolar Bose-Einstein condensates. *Phys. Rev. Lett.*, **97**, 130404.

[58] Yi, S., and Pu, H. 2006. Spontaneous spin textures in dipolar spinor condensates. *Phys. Rev. Lett.*, **97**, 020401.

[59] Kawaguchi, Y., Saito, H., and Ueda, M. 2007. Can spinor dipolar effects be observed in Bose-Einstein condensates? *Phys. Rev. Lett.*, **98**, 110406.

[60] Yujiro E., Hiroki S., and Hirano, T. 2014. Observation of dipole-induced spin texture in a ^{87}Rb Bose-Einstein condensate. *Phys. Rev. Lett.*, **112**, 185301.

[61] Ueda, M. 2014. Topological aspects in spinor Bose-Einstein condensates (Key Issues Review). *Reports on Progress in Physics*, **77**, 122401.

[62] Lewenstein, M., Sanpera, A., Ahufinger, V., Damski, B., Sen, A., and Sen, U. 2007. Ultracold atomic gases in optical lattices: mimicking condensed matter physics and beyond. *Adv. Phys*, **56**, 243–379.

[63] Dalibard, J., Gerbier, F, Juzeliūnas, G., and Öhberg, P. 2011. Artificial gauge potentials for neutral atoms. *Rev. Mod. Phys.*, **83**, 1523.

Part IV

Condensates in Condensed Matter Physics

Editorial Notes

As discussed in Chapter 1, experimental studies of Bose-Einstein condensation (BEC) have moved well beyond liquid helium and atomic gases in the past decade. In particular, there are now many condensed matter systems which exhibit various features of BEC, including some which show effects that cannot easily be seen in liquid helium or gases.

One set of such condensates may be called "optical condensates." The photon is a good boson, and so it is natural to expect that one can make a Bose-Einstein condensate of photons. To make a system analogous to atomic condensates, photons can be given an effective mass by manipulation of their dispersion relation in a cavity. As discussed in Chapter 4, there are different limits in which macroscopic coherence of optical condensates can be seen. In the "pure photon" limit, discussed in Chapter 19, photons are absorbed and re-emitted from a medium at a fixed temperature. Number is not conserved, but average number is kept constant, analogous to the grand canonical ensemble of thermodynamics, with significant number fluctuations. This system can be viewed as a laser with a very large number, i.e., a near-continuum, of nearby modes, such that the occupations of these modes can thermalize.

In the polariton, or "strong coupling," limit, a sharp electronic resonance in the system is coupled to a cavity photon (at any given momentum, there is one relevant photon mode and one relevant extended electronic excitation state). The polaritons which are the eigenstates of this coupled system can be viewed as photons engineered to have a strong repulsive interaction between them, and their interaction with the rest of the system can be small enough that they can be viewed as approximately number-conserved. This system is then highly analogous to atomic condensates. The existence of those two limits naturally gives rise to another type of crossover, which is unique to the optical condensates: the lasing-BEC crossover, characterized by the level of thermal equilibrium of the condensate, as reviewed in Chapter 20.

Polariton condensates exhibit vortex dynamics, which can be studied directly by phase maps, as discussed in Chapter 21. Moreover, they are especially useful for studying Berezinskii-Kosterlitz-Thouless (BKT) transitions, as they are intrinsically two-dimensional (2D) systems. Some of this work was reviewed in Chapter 10. Much work has also been done in polariton systems on the crossover from dissipative to nondissipative condensates (see Chapter 11), the BEC-BCS crossover (Chapter 12), and the localization-delocalization transition, a.k.a. the phase locking of spatially separated condensates (Chapters 22 and 23.)

There is a third type of system, not reviewed in this volume, often also called an optical condensate. In that system, explored in depth by Jason Fleischer and coworkers[1] in recent years, a coherent optical field impinges on a nonlinear medium. Since the input light is already coherent and well described by a classical field, there is no spontaneous coherence. However, the system obeys a Gross-Pitaevskii equation that describes the evolution of a condensate. The evolution of this type of optical system is therefore exactly analogous to the evolution of the low-energy, highly-occupied states near a condensate, in what is known as the classical-field approximation, briefly discussed in Chapter 7.

In addition to these optical condensates, there are also condensation effects in spin systems. In one approach, microwave radiation is used to excite a steady-state magnon population, analogous to the way optical pumping is used to create a steady-state photon or polariton population. This work is reviewed in Chapters 25 and 26. In another approach, a permanent population of boson quasiparticles is created by tuning the magnetic field to a point where the ground state corresponds to a macroscopic number of bosonic degrees of freedom. Chapters 27 and 28 discuss this type of experiment.

Interestingly, polariton condensates can also be treated as spin systems, since polaritons have two allowed angular momentum states corresponding to left-circular and right-circular polarization, which can be mapped to a pseudospin. Chapter 24 reviews some of these effects; theorist Alexey Kavokin and coworkers have explored pseudospin effects in many other versions, such as the optical spin-Hall effect.[2]

The general study of Bose-Einstein condensation in condensed matter systems has now expanded to such a degree that it is not possible to survey all of the different types in this book. In particular, coherent transport has been seen in bilayer excitonic condensates, analogous in many ways to BCS superconductors, but with Coulombic pairing instead of phonon-mediated pairing.[3] Work also continues apace on optically pumped excitonic condensates, which were among the earliest theoretically predicted condensates in solid state systems.[4]

[1] C. Sun et al., Observation of kinetic condensation of classical waves, *Nature Physics*, **8**, 470 (2012).
[2] C. Leyder et al., Observation of the optical spin-hall effect, *Nature Physics*, **3**, 628 (2007).
[3] D. Nandi et al., Exciton condensation and perfect Coulomb drag, *Nature*, **488**, 481 (2012).
[4] For a review of early theory, see S.A. Moskalenko and D. W. Snoke, *Bose-Einstein Condensation of Excitons and Biexcitons* (Cambridge University Press, 2000).

19

Bose-Einstein Condensation of Photons and Grand-Canonical Condensate Fluctuations

JAN KLAERS

Institut für Angewandte Physik, Universität Bonn, Germany
Present address:
Institute of Quantum Electronics, ETH Zürich, Switzerland

MARTIN WEITZ

Institut für Angewandte Physik, Universität Bonn, Germany

We review recent experiments on the Bose-Einstein condensation of photons in a dye-filled optical microresonator. The most well-known example of a photon gas, photons in blackbody radiation, does not show Bose-Einstein condensation. Instead of massively populating the cavity ground mode, photons vanish in the cavity walls when they are cooled down. The situation is different in an ultrashort optical cavity imprinting a low-frequency cutoff on the photon energy spectrum that is well above the thermal energy. The latter allows for a thermalization process in which both temperature and photon number can be tuned independently of each other or, correspondingly, for a nonvanishing photon chemical potential. We here describe experiments demonstrating the fluorescence-induced thermalization and Bose-Einstein condensation of a two-dimensional photon gas in the dye microcavity. Moreover, recent measurements on the photon statistics of the condensate, showing Bose-Einstein condensation in the grand-canonical ensemble limit, will be reviewed.

19.1 Introduction

Quantum statistical effects become relevant when a gas of particles is cooled, or its density is increased, to the point where the associated de Broglie wavepackets spatially overlap. For particles with integer spin (bosons), the phenomenon of Bose-Einstein condensation (BEC) then leads to macroscopic occupation of a single quantum state at finite temperatures [1]. Bose-Einstein condensation in the gaseous case was first achieved in 1995 by laser and subsequent evaporative cooling of a dilute cloud of alkali atoms [2, 3, 4], as detailed in Chapter 3 of this volume. The condensate atoms can be described by a macroscopic single-particle wavefunction, similar to the case of liquid helium [1]. Bose-Einstein condensation has also been observed for exciton-polaritons, which are hybrid states of matter and light [5, 6, 7] (see Chapter 4), magnons [8] (see Chapter 25), and other physical systems. Other than material particles, photons usually do not show Bose-Einstein condensation [9]. In blackbody radiation, the most common Bose gas, photons at low temperature disappear, instead of condensing to a macroscopically occupied ground-state mode. In this system, photons have a vanishing chemical potential, meaning that the number of photons is determined by the available thermal energy and cannot be tuned independently from temperature. Clearly, a precondition for a Bose-Einstein condensation of photons is a thermalization process that allows for an independent adjustment of both photon number and temperature. Early theoretical work has proposed a thermalization mechanism by Compton scattering in plasmas [10]. Chiao et al. proposed a two-dimensional photonic quantum fluid in a nonlinear resonator [11]. Thermal equilibrium here was sought from photon–photon scattering, in analogy to atom–atom scattering in cold atom experiments, but the limited photon–photon interaction in available nonlinear materials has yet prevented a thermalization [12]. In the strong coupling regime, (quasi-)equilibrium Bose-Einstein condensation of exciton-polaritons, mixed states of matter and light, has been achieved [5, 6, 7]. Here interparticle interactions of the excitons drive the system into or near thermal equilibrium. More recently, evidence for superfluidity of polaritons has been reported [13, 14]. Other experimental work has observed the kinetics of condensation of classical optical waves [15].

Bose-Einstein condensation of photons in a dye-filled microresonator has been realized in 2010 in our group at the University of Bonn and in 2014 at Imperial College London [16, 17]. Thermalization of the photon gas with the dye solution is achieved by repeated absorption and re-emission processes. For liquid dye solutions at room-temperature conditions, it is known that rapid decoherence from frequent collisions (10^{-14} s time scale) with solvent molecules prevents a coherent excitation exchange between photonic and electronic degrees of freedom [18, 19], so that the condition of strong light-matter coupling is not met. It is therefore justified to regard the bare photonic and electronic excitations of the system as the true energy eigenstates. The separation between the two curved resonator mirrors (see Fig. 19.1a) is of order of the photon wavelength. The small cavity spacing causes a large frequency spacing between the longitudinal resonator modes, which is of order of the emission width of the dye molecules; see Fig. 19.1b. Under these conditions, only photons of a fixed longitudinal mode are observed to populate the resonator, which effectively makes the photon gas two-dimensional as only the two transversal motional degrees of freedom remain. The lowest lying mode of this manifold ($q = 7$), the TEM_{00} transverse ground mode, acts as a low-frequency cutoff at an energy of $\hbar\omega_c \simeq 2.1\,\text{eV}$ in the Bonn experiment [16, 20]. This restricts the photon spectrum to energies $\hbar\omega$ well above the thermal energy

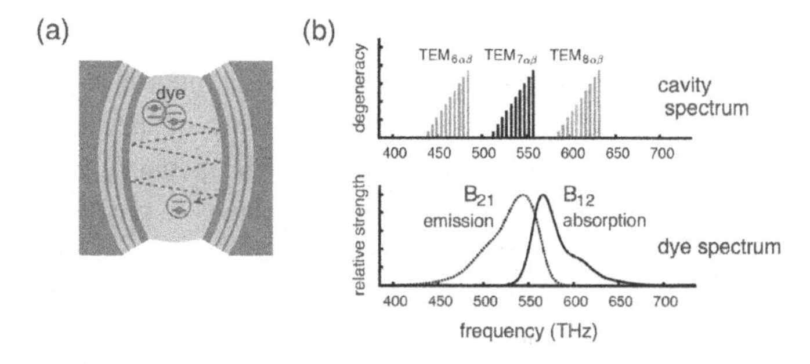

Figure 19.1 (a) Scheme of the experimental setup used in [16]. (b) Schematic spectrum of cavity modes. Transverse modes belonging to the manifold of longitudinal mode number $q = 7$ are shown by black lines, those of other longitudinal mode numbers in gray. The bottom graph indicates the relative absorption coefficient and fluorescence strength of rhodamine 6G dye versus frequency. Figure adapted with permission from Klaers, J., et al. (2010), Bose-Einstein condensation of photons in an optical microcavity, *Nature*, **468**, 545 [16]. Copyright (2010) by the Nature Publishing Group.

$k_B T \simeq 1/40\,\text{eV}$, i.e., $\hbar\omega \geq \hbar\omega_c \gg k_B T$, which to good approximation decouples the number of photons from the heat content of the system (nonvanishing chemical potential). In this situation, the photon number becomes tunable by (initial) optical pumping, which can be regarded as fully analogous to the loading of cold atoms into a magnetic or optical dipole trap.

In the microcavity, the energy-momentum relation moreover becomes quadratic, as for a massive particle, and the mirror curvature induces an effective trapping potential in the transverse plane. In general, significant population of high transverse modes ($\text{TEM}_{\alpha\beta}$ with high transversal mode numbers α and β and, correspondingly, high eigenfrequencies) is expected at high temperatures, while the population concentrates to the lowest transverse modes when the system is cold. One can show that the photon gas in the resonator is formally equivalent to a harmonically trapped two-dimensional gas of massive bosons with effective mass $m_{\text{ph}} = \hbar\omega_c(n/c)^2$. Here c denotes the vacuum speed of light and n the index of refraction of the resonator medium. In thermal equilibrium, such a system is known to undergo Bose-Einstein condensation at a finite temperature [21]. Both the thermalization of the photon gas to room temperature [20] and the Bose-Einstein condensation [16] has been verified experimentally.

Several theoretical publications have discussed different aspects of photon Bose-Einstein condensation using a variety of approaches [22, 23, 24, 25, 26, 27, 28, 29, 30, 31, 32], including work based on a superstatistical approach [23], on a Schwinger-Keldysh theory [26], and on a master equation approach [24]. Investigated topics include first-order coherence properties such as the dynamics of phase coherence onset [25] and equilibrium phase fluctuations of the photon condensate [30]. Moreover, second-order coherence properties of photon condensates have been studied in some detail [22, 23, 32]. The coupling of the photon gas to the dye medium, which allows for both energy and particle exchange, can be described by a grand-canonical ensemble representation. This leads to physically observable consequences in the condensed phase regime, in which the condensate performs anomalously large intensity fluctuations [22]. Another key topic is the relation between lasing and condensation. These different regimes have been studied in a theory model accounting for photon loss leading to partial thermal equilibrium of the photon gas [24].

In the following, Section 19.2 gives a theoretical description of the fluorescence-induced thermalization mechanism, as well as the expected thermodynamic behavior of the two-dimensional photon gas in the dye-filled microcavity system. Further, Section 19.3 describes experiments observing the thermalization and Bose-Einstein condensation of the photon gas at room temperature. Section 19.4 reviews theory and experimental results regarding the grand-canonical nature of the condensate fluctuations. Finally, Section 19.5 concludes this contribution.

19.2 Thermodynamics of a Two-Dimensional Photon Gas

19.2.1 Thermal and Chemical Equilibrium

In the dye-filled microcavity system, the photon gas in the resonator is thermally coupled to the dye medium. This thermalization mechanism relies on two preconditions. First, the dye medium itself has to be in thermal equilibrium. Consider an idealized dye molecule with an electronic ground state and an electronically excited state separated by the energy $\hbar\omega_{\mathrm{ZPL}}$ (zero-phonon-line), each subject to additional rotational and vibrational level splitting [33]. Frequent collisions of solvent molecules with the dye, on the time scale of a few femtoseconds at room temperature, rapidly alter the rovibrational state of the dye molecules. These collisions are many orders of magnitude faster than the electronic processes (the upper state natural lifetime of, e.g., rhodamine 6G dye is 4 ns), so that both absorption and emission processes will take place from an equilibrated internal state. One can show that the Einstein coefficients for absorption and emission $B_{12,21}(\omega)$ then will be linked by a Boltzmann factor

$$\frac{B_{21}(\omega)}{B_{12}(\omega)} = \frac{w_\downarrow}{w_\uparrow} e^{-\frac{\hbar(\omega-\omega_{\mathrm{ZPL}})}{k_{\mathrm{B}}T}}, \tag{19.1}$$

where $w_{\downarrow,\uparrow}$ are statistical weights related to the rovibrational density of states [22]. This relation is known as the Kennard-Stepanov law [34, 35, 36, 37, 38, 33]. Experimentally, the Kennard-Stepanov relation is well fulfilled for many dye molecules.

Deviations from this law can either arise from imperfect rovibrational relaxation or a reduced fluorescence quantum yield [39].

The second precondition for the light-matter thermalization process is the chemical equilibrium between photon gas and dye medium. Absorption and emission processes can be regarded as a photochemical reaction of the type $\gamma + \downarrow \rightleftharpoons \uparrow$ between photons (γ), excited ($\uparrow$), and ground-state ($\downarrow$) molecules. This reaction reaches chemical equilibrium if the rates of competing processes (such as pump and loss) are negligible and there is no net change in the densities of one of the species anymore. The corresponding chemical potentials then satisfy $\mu_\gamma + \mu_\downarrow = \mu_\uparrow$, which can also be expressed as [22]

$$e^{\frac{\mu_\gamma}{k_B T}} = \frac{w_\downarrow}{w_\uparrow} \frac{\rho_\uparrow}{\rho_\downarrow} e^{-\frac{\hbar\omega_{\mathrm{ZPL}}}{k_B T}}, \tag{19.2}$$

where $\rho_\uparrow$ ($\rho_\downarrow$) denotes the density of excited (ground-state) molecules. In equilibrium, the photon chemical potential is thus determined by the excitation ratio $\rho_\uparrow/\rho_\downarrow$ of the medium. Assuming both the Kennard-Stepanov law, Eq. (19.1), and chemical equilibrium, as expressed by Eq. (19.2), one can show that multiple absorption-emission cycles drive the photon gas into thermal equilibrium with the dye solution at temperature T, and with a photon chemical potential μ_γ determined by the molecular excitation ratio [22].

19.2.2 Cavity Photon Dispersion and BEC Criticality

The energy of a cavity photon is determined by its longitudinal (k_z) and transversal wavenumber (k_r) as $E = (\hbar c/n)\sqrt{k_z^2 + k_r^2}$, where n again denotes the index of refraction of the medium. Owing to the curvature of the mirrors, the boundary conditions for the photon field depend on the distance to the optical axis $r = |\mathbf{r}|$. For the longitudinal component, we set $k_z(\mathbf{r}) = q\pi/D(r)$, where q denotes the longitudinal mode number and $D(r)$ describes the mirror separation as a function of r. For a symmetric resonator consisting of two spherically curved mirrors with separation D_0 and radius of curvature R, in a paraxial approximation ($r \ll R$, $k_r \ll k_z$), the photon energy is given by [20]

$$E \simeq m_{\mathrm{ph}}(c/n)^2 + \frac{(\hbar k_r)^2}{2m_{\mathrm{ph}}} + \frac{1}{2} m_{\mathrm{ph}} \Omega^2 r^2 , \tag{19.3}$$

with an effective photon mass $m_{\mathrm{ph}} = \pi\hbar n q/cD_0$ and trapping frequency $\Omega = c/n\sqrt{D_0 R/2}$. This describes a particle moving in the two-dimensional transversal plane with nonvanishing (effective) mass subject to a harmonic trapping potential with trapping frequency Ω. Such a system is known to undergo Bose-Einstein condensation at finite temperature [21]. If we account for the twofold polarization degeneracy of photons, condensation is expected when the particle number exceeds the critical particle number

$$N_c = \frac{\pi^2}{3}\left(\frac{k_B T}{\hbar\Omega}\right)^2 . \tag{19.4}$$

The typical trapping frequency in our setup is $\Omega/2\pi \simeq 41\,\text{GHz}$, and at room temperature ($T = 300\,\text{K}$) one obtains a critical photon number of $N_c \simeq 77.000$, which is experimentally feasible. The physical reason for the possibility to observe Bose-Einstein condensation at room temperature conditions is the small effective photon mass $m_{ph} = \hbar\omega_c(n/c)^2 \simeq 7 \cdot 10^{-36}\,\text{kg}$, which is ten orders of magnitude smaller than, e.g., the mass of the rubidium atom.

19.2.3 Equilibrium versus Nonequilibrium

In general, particle loss can drive a system out of equilibrium, if the time scale associated with loss is not well separated from the time scale for the equilibration of the system. Separated time scales clearly can be achieved for the case of dilute atomic gases. The true ground state for, e.g., an atomic rubidium gas is a cloud of molecular dimers. However, researchers have learned in the 1980s to the early 1990s that the recombination rate from three-body collisions to the molecular state can be kept sufficiently small by the use of very dilute atomic clouds for which the corresponding rates are sufficiently small [40, 41]. High-phase space densities can nevertheless be achieved by cooling to ultralow temperatures in the nano-Kelvin regime. Correspondingly, quantum degeneracy of a cloud of bosonic atoms can be reached under conditions that are close to equilibrium.

In the case of photons, nonequilibrium conditions can arise either from a violation of the Kennard-Stepanov law (Eq. (19.1)) or from a violation of chemical equilibrium (Eq. (19.2)), if, e.g., the photon loss rate is not negligible compared with the photon absorption and emission rate. Clearly, the latter situation is well known from typical laser operation. Both laser operation and Bose-Einstein condensation, either of photons or atoms [42], rely on Bose enhancement. However, to achieve lasing at the desired wavelength, it is usually necessary to break the chemical equilibrium between photons and molecules, allowing for a departure from Bose-Einstein statistics and for a photon energy distribution independent of energetics. For this purpose, gain and loss are deliberately engineered, for example, by frequency-selective components. In the field of exciton-polaritons, the question whether a system that is pumped and exhibits losses should be regarded as polariton laser or polariton Bose-Einstein condensate has been extensively discussed [43, 44, 45]; see also following chapters in this volume. For the case of photonic Bose-Einstein condensation, the role of losses and pumping has been theoretically investigated by

Kirton and Keeling [24]. Experimentally, the crossover between equilibrium and nonequilibrium photon gases has been studied both in the nondegenerate and in the quantum degenerate regime [17, 20, 46].

19.3 Experiments on Photon Condensation

A scheme of the setup used in the Bonn photon condensation experiment is shown in Fig. 19.1a. The optical resonator consists of two highly reflecting spherically curved mirrors ($\simeq 0.999985$ reflectivity in the relevant wavelength region) with radius of curvature $R = 1$ m. One of mirrors is cut to $\simeq 1$ mm surface diameter to allow for a cavity length in the micrometer range ($D_0 \simeq 1.46\,\mu$m), as measured by the cavity-free spectral range, despite the mirror curvature. The resonator contains a drop of liquid dye, typically rhodamine 6G or perylenedimide (PDI), dissolved in an organic solvent. Both of these dyes have high quantum efficiencies between 0.95 and 0.97 and fulfill the Kennard-Stepanov relation in good approximation. Fig. 19.1b shows the cavity spectrum (top) along with the absorption and fluorescence spectrum for rhodamine dye (bottom). The resonator setup is off-resonantly pumped with a laser beam near 532 nm wavelength derived from a frequency-doubled Nd:YAG laser inclined at less than a 45° angle to the cavity axis.

In initial experiments, the thermalization of the two-dimensional photon gas in the dye-filled microresonator was carefully tested [20]. Fig. 19.2a shows experimental spectra of the light transmitted through one cavity mirror for two different temperatures of the setup (top: $T \simeq 300$ K, room temperature; bottom $T \simeq 365$ K). In these experiments, the average photon number inside the cavity ($N \simeq 50$) is three orders of magnitude below the critical particle number. The experimental data (dots) in both cases are well described by a Boltzmann distribution of photon energies at the corresponding temperature (solid line). In other experiments, the pump spot was transversely displaced by a variable amount and the position where the maximum of the observed fluorescence occurs was monitored; see Fig. 19.2b. As expected in the presence of a trapping potential, a spatial relaxation of the photons toward regions of low potential energy near the optical axis was observed.

In subsequent experiments, the dye-filled microcavity was operated at photon numbers sufficiently high to reach quantum degeneracy. To avoid excessive population of dye molecules in triplet states and heat deposition, the optical pump beam was acousto-optically chopped to $0.5\,\mu$s long pulses, with an 8 ms repetition time. Fig. 19.3a shows typical spectra of the photon gas at different photon numbers [16]. While the observed spectrum resembles a Boltzmann distribution at small intracavity optical powers, near the phase transition a shift of the maximum toward the cutoff frequency is observed, and the spectrum more resembles a Bose-Einstein distribution. At intracavity powers above the critical value, the

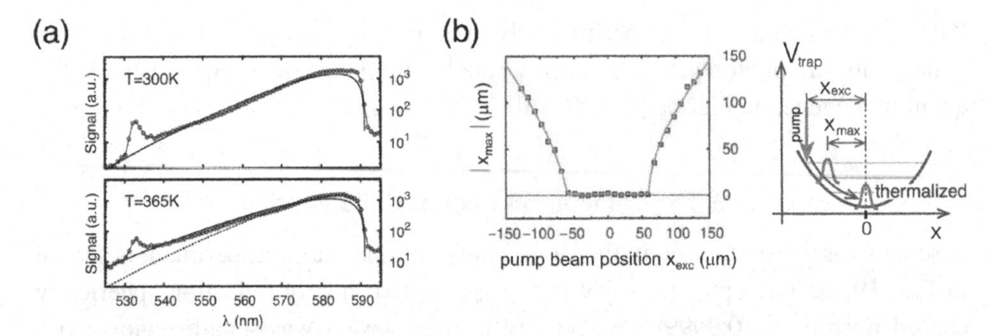

Figure 19.2 (a) Measured spectral intensity distributions (connected dots) of the cavity emission for temperatures of the resonator setup of 300 K (top) and 365 K (bottom) at an average photon number of 60 ± 10 inside the cavity, i.e., far below the onset of a BEC. The solid lines are theoretical spectra based on a Bose-Einstein distribution. For illustration, a $T = 300$ K distribution is also inserted in the bottom graph (dashed line). (b) Distance of the fluorescence intensity maximum from the optical axis $|x_{max}|$ versus transverse position of the pump spot, x_{exc}. Due to the thermalization, the photon gas accumulates in the trap center, where the potential exhibits a minimum value. This holds as long as the excitation spot is closer than approximately 60 μm distance. Figure reprinted with permission from Klaers, J., et al. (2010), Thermalization of a two-dimensional photonic gas in a white-wall photon box, *Nat. Phys.*, **6**, 512 [20]. Copyright (2010) by the Nature Publishing Group.

Bose-Einstein condensate occurs as a spectrally sharp peak at the position of the cutoff. The observed spectral width of the condensate peak is limited by the resolution of the used spectrometer. The experimental results are in good agreement with theoretical expectations (see the inset of the figure). At the phase transition, the optical intracavity power is $P_{c,exp} = (1.55 \pm 0.6)$ W, which corresponds to a photon number of $(6.3 \pm 2.4) \cdot 10^4$.

Fig. 19.3b shows spatial images of the light transmitted through one of the cavity mirrors (real image onto a color charge-coupled device [CCD] camera) both below (top) and above (bottom) the condensate threshold. Both images show a shift from the yellow spectral regime for the transversally low excited cavity modes located near the trap center to the green for transversally higher excited modes appearing at the outer trap regions. In the lower image, a bright spot is visible in the center with a measured full width at half maximum (FWHM) diameter of (14 ± 2) μm. Within the quoted experimental uncertainties, this corresponds well to the expected diameter of the TEM_{00} transverse ground-state mode of 12.2 μm, yielding clear evidence for a single-mode macroscopic population of the ground state. Fig. 19.3c gives normalized intensity profiles (cuts along one axis through the trap center) for different powers. One observes that not only the height of the condensate peak

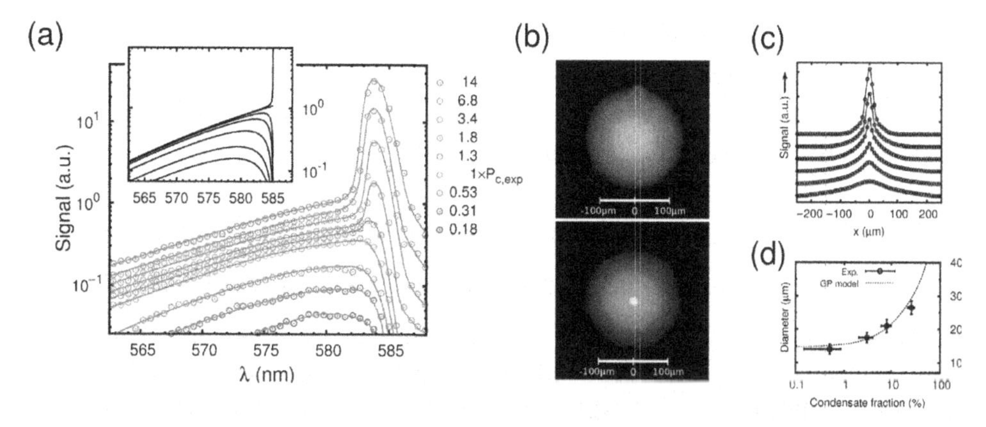

Figure 19.3 (a) The connected circles show measured spectral intensity distributions for different pump powers. The legend gives the optical intracavity power, determining the photon number. On top of a broad thermal wing, a spectrally sharp condensate peak at the position of the cavity cutoff is visible above a critical power. The observed peak width is limited by the spectrometer resolution. The inset gives theoretical spectra based on Bose-Einstein distributed transversal excitations. (b) Images of the radiation emitted along the cavity axis, below (top) and above (bottom) the critical power. In the latter case, a condensate peak is visible in the center. (c) Cuts through the center of the observed intensity distribution for increasing optical pump powers. (d) The data points give the measured width of the condensate peak versus condensate fraction, and the dotted line is the result of a theoretical model based on the Gross-Pitaevskii equation. Figure reprinted with permission from Klaers, J., et al. (2010), Bose-Einstein condensation of photons in an optical microcavity, *Nature*, **468**, 545 [16]. Copyright (2010) by the Nature Publishing Group.

increases for larger condensate fractions, but also its width; see also Fig. 19.3d. This effect is not expected for an ideal photon gas, and suggests a weak repulsive self-interaction mediated by the dye solution. The origin of the self-interaction is thermal lensing, which under steady-state conditions can be described by a nonlinear term analogous to the Gross-Pitaevskii equation (see [16]). In general, the interplay between optical and heat flow equations can lead to nonlocal interactions; see [29]. By comparing the observed increase of the mode diameter with numerical solutions of the two-dimensional Gross-Pitaevskii equation, a dimensionless interaction parameter of was estimated [16]. This interaction parameter is found to be significantly smaller than the values reported for two-dimensional atomic physics quantum gas experiments [47, 48] and also below the values at which Kosterlitz-Thouless physics can be expected to become important in the harmonically trapped case [49]. Experimentally, when directing the condensate through a Michelson-type sheering interferometry, no signatures of phase blurring (which occur in two-dimensional atomic gas experiments) were observed [50].

Further signatures consistent within the framework of Bose-Einstein condensation include the expected scaling of the critical photon number with resonator geometry, and a spatial relaxation process that leads to a strongly populated ground mode even for a spatially displaced pump spot [16].

19.4 Fluctuations of Photon Condensates

19.4.1 Photon Condensates Coupled to a Particle Reservoir

In this section, we discuss quantum statistical properties of photon condensates, in particular the photon number distribution and particle number fluctuations. The main result is that photon Bose-Einstein condensates in the dye microcavity system, owing to the grand-canonical nature of the light-matter thermalization process, can show unusually large particle number fluctuations, which are not observed in present atomic Bose-Einstein condensates.

In statistical physics, different statistical ensembles reflect different laws of conservation that can be realized in experiments. The microcanonical ensemble corresponds to a physical system with energy and particle number strictly fixed at all times, while in the canonical ensemble energy fluctuates around a mean value determined by the temperature of a heat reservoir. Under grand-canonical conditions, an exchange of both energy and particles with a large reservoir is allowed, leading to fluctuations in both quantities. The photon gas in the dye microcavity investigated here, with photons being frequently absorbed and emitted by dye molecules, belongs to the latter class of experiments. As discussed in Section 19.2, absorption and emission can be regarded as the two directions of a photochemical reaction $\gamma + \downarrow \rightleftarrows \uparrow$, where photons ($\gamma$), ground state ($\downarrow$), and excited dye molecules ($\uparrow$) are repeatedly converted into each other, and the dye molecules act as a "reservoir species" for the photon gas.

A common assumption is that the ensemble conditions realized in a physical system are not essential for its physical behavior. The various statistical approaches are correspondingly expected to become interchangeable in the thermodynamic limit [9, 51], in the sense that relative fluctuations vanish in all of them, i.e., $\delta N/N \to 0$ for the average total particle number N and its root mean square deviation δN. This assumption is, however, violated in the grand-canonical treatment of the ideal Bose gas, where the occupation of any single particle state undergoes relative fluctuations of 100% of the average value [52, 53]. For a macroscopically occupied ground state of a Bose-Einstein condensed gas, this implies fluctuations of order of the total particle number, i.e., $\delta N \simeq N$. While one usually expects fluctuations to freeze out at low temperatures, here the reverse situation is encountered: the total particle number starts to strongly fluctuate as the condensate fraction approaches unity, a behavior that has been recognized early in BEC theory [54] and later has been termed "grand-canonical fluctuation catastrophe" [53, 55, 56]. In experiments with cold atoms, this anomaly has not been observed so far, as sufficiently large particle reservoirs are usually not experimentally realizable. For those systems, much theoretical work has been performed to obtain the particle number fluctuations in a (micro-)canonical description [55, 57, 58, 59] and to account for trapping potentials [60, 61]. A review can be found in Ref. [53]. Noteworthy, the microcanonical ensemble description of the ideal Bose gas shows interesting connections to the partitioning and factorizing problem of integer numbers [62].

For a photon Bose-Einstein condensate, grand-canonical ensemble conditions can be an inherent feature of the thermalization process and can therefore influence the second-order coherence properties [22]. We consider a situation in which the photon condensate is coupled to the electronic transitions of M dye molecules (located in the volume of the electromagnetic ground mode) by absorption and emissions processes. In this way, the condensate exchanges excitations with a reservoir of a given (finite) size. Using a master equation approach, one can show that the probability $\mathcal{P}_n$ to find n photons in the ground state follows

$$\frac{\mathcal{P}_n}{\mathcal{P}_0} = \frac{(M-X)!\,X!}{(M-X+n)!\,(X-n)!} e^{-n\hbar(\omega_c-\omega_{ZPL})/k_B T}, \tag{19.5}$$

where the excitation number X is defined as the sum of ground mode photon number and electronically excited molecules in the reservoir. As before, ω_c and ω_{ZPL} denote the frequencies of the condensate mode and zero-phonon-line of the medium, respectively. In this calculation, X is constant; i.e., it is not expected to perform large fluctuations on its own. The photon number distribution, which can also be derived in a superstatistical approach [23], in general interpolates between Bose-Einstein statistics and Poisson statistics. Assuming that the excitation level $\rho_\uparrow/\rho_\downarrow \simeq X/(M-X)$ of the medium stays fixed, which conserves the chemical potential μ (Eq. (19.2)), the average condensate number $\bar{n}_0$, and total particle number $\bar{N}$, one finds that large reservoirs M lead to Bose-Einstein–like statistics with an exponentially decaying photon number distribution starting at a maximum value for zero photon number $n = 0$. For small reservoirs, $\mathcal{P}_n$ becomes poissonian with a maximum value at a nonzero photon number. The distinction between these two statistical regimes is not unambiguous due to the smooth crossover behavior between them. A natural choice for a borderline is the point at which "finding zero photons" ceases to be the most probable event, which occurs at $\mathcal{P}_0 = \mathcal{P}_1$ and resembles a common laser threshold definition [63]. For the temperature T_x at which this condition is reached, given a certain system size $\bar{N}$ (average photon number) and the reservoir size M, one obtains the equation

$$\bar{N} - \frac{\pi}{6}\left(\frac{k_B T_x}{\hbar\Omega}\right)^2 \simeq \sqrt{\frac{M/2}{1+\cosh\frac{\hbar\Delta}{k_B T_x}}}\,. \tag{19.6}$$

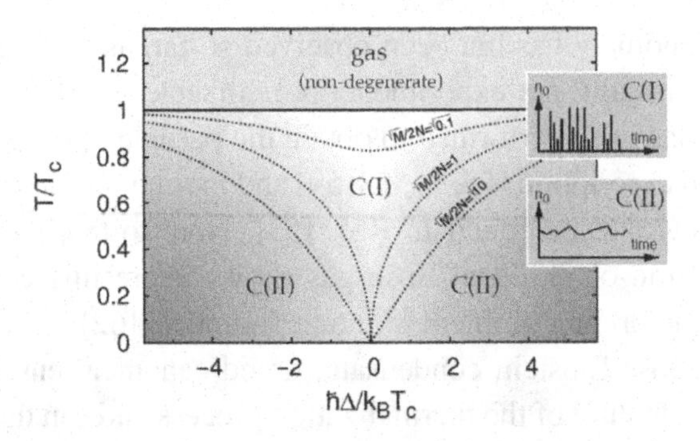

Figure 19.4 Phase diagram of the two-dimensional photon gas for fixed average photon number $\bar{N}$ in the plane spanned by the reduced temperature T/T_c and the dye-cavity detuning $\hbar\Delta/k_B T_c$. The solid line marks the BEC phase transition. The dotted lines (three cases are shown) separate two regimes: a condensate regime with large number fluctuations and a Bose-Einstein–like photon number distribution C(I), and a regime of non-fluctuating condensates obeying Poisson statistics C(II). The temperature of the crossover C(I)–C(II) depends on the ratio $\sqrt{M}/\bar{N}$, where the reservoir size M denotes the number of dye molecules in the mode volume of the ground state. The insets give a sketch of the corresponding temporal evolution of the condensate photon number $n_0(t)$. Reprinted with permission from Klaers, J., et al. (2012), Statistical physics of Bose-condensed light in a dye microcavity, *Phys. Rev. Lett*, **108**, 160403 [22]. Copyright (2012) by the American Physical Society.

Here Δ denotes the detuning between condensate mode and zero-phonon-line of the dye, defined as $\Delta = \omega_c - \omega_{ZPL}$. For zero dye-cavity detuning $\Delta = 0$, one finds the analytic solution $T_{x,\Delta=0} \simeq T_c\sqrt{1 - \sqrt{M}/2\bar{N}}$, provided that $\sqrt{M}/2\bar{N} < 1$. For general detunings Δ, Eq. (19.6) has to be solved numerically. Fig. 19.4 gives a phase diagram, where solutions for three different cases $\sqrt{M}/2\bar{N} = \sqrt{0.1}, 1$, and $\sqrt{10}$ are marked as dotted lines, which separate two different regimes of the photon condensate, denoted by C(I) with Bose-Einstein–like photon statistics and C(II) with Poisson statistics, respectively. In terms of second-order correlations, the dotted lines correspond to $g^{(2)}(0) \simeq \pi/2$, or relative fluctuations of $\delta n/\bar{n}_0 = \sqrt{g^{(2)}(0) - 1} = 0.75$. Note that both T_c and T_x are conserved in a thermodynamic limit $\bar{N}$, M, $R \to \infty$ in which $\bar{N}/R = \text{const}$, and $\sqrt{M}/\bar{N} = \text{const}$. Recent work has investigated the possible effects of fast photon–photon interactions on the photon number statistics [32].

19.4.2 Observation of Anomalous Condensate Fluctuations

The intensity correlations and fluctuations of the condensate have been measured using a Hanbury Brown–Twiss setup [64, 65]. In this experiment, the condensate

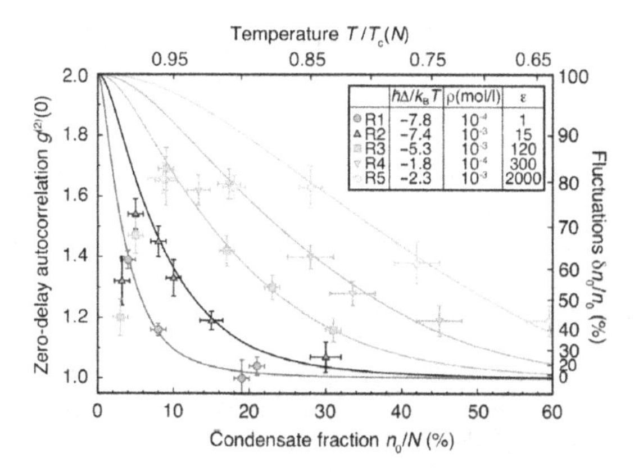

Figure 19.5 Zero-delay autocorrelations $g^{(2)}(0)$ and condensate fluctuations $\delta n_0/\bar{n}_0$ versus condensate fraction $\bar{n}_0/\bar{N}$ (or equivalently the corresponding reduced temperature $T/T_c(\bar{N})$ at $T = 300\,\mathrm{K}$) for five different reservoirs R1–R5. The increase of the effective molecular reservoir size from R1 to R5 is quantified by the parameter ϵ (third column), defined in Eq. (19.8). Condensate fluctuations extend deep into the condensed phase for high dye concentration ρ and small dye-cavity detuning Δ (R5). Results of a theoretical model based on Eq. (19.5) are shown as solid lines. The error bars indicate statistical uncertainties. Experimental parameters: condensate wavelength $\lambda_0 = \{598, 595, 580, 598, 602\}\,\mathrm{nm}$ for data sets R1–R5; dye concentration $\rho = \{10^{-4}, 10^{-3}, 10^{-3}\}\,\mathrm{mol/l}$ for R1–R3 (rhodamine 6G), and $\rho = \{10^{-4}, 10^{-3}\}\,\mathrm{mol/l}$ for R4–R5 (perylene red). For the theory curves, we find effective reservoir sized of $M = \{5.5 \pm 2.2,\ 20 \pm 7,\ 16 \pm 6,\ 2.1 \pm 0.4,\ 11 \pm 4\} \times 10^9$ for R1–R5. Reprinted with permission from Schmitt, J., et al. (2014), Observation of grand-canonical number statistics in a photon Bose-Einstein condensate, *Phys. Rev. Lett.*, **112**, 030401 [64]. Copyright (2014) by the American Physical Society.

mode is separated from the higher transversal modes by spatial filtering in the far field, which corresponds to a transverse momentum filter. The beam is split into two paths, each of which is directed onto single-photon avalanche photodiodes. Time correlations of the condensate population can be determined with a temporal resolution of 60 ps with this setup. The second-order correlation function $g^{(2)}(t_1, t_2) = \langle n_0(t_1) n_0(t_2)\rangle / \langle n_0(t_1)\rangle \langle n_0(t_2)\rangle$ to good approximation is found to depend only on the time delay $\tau = t_2 - t_1$. Further analysis is thus performed with the time-averaged function $g^{(2)}(\tau) = \left\langle g^{(2)}(t_1, t_2)\right\rangle_{\tau = t_2 - t_1}$.

We have varied the reservoir size systematically to test for the grand-canonical nature of the system. Fig. 19.5 shows the zero-delay correlations $g^{(2)}(0)$ as a function of the condensate fraction $\bar{n}_0/\bar{N}$ for five combinations of dye concentration ρ and dye-cavity detuning Δ. The data sets labeled R1–R3 have been obtained with rhodamine 6G dye ($\omega_{\mathrm{ZPL}} = 2\pi c/545$ nm). For measurements R4 and R5, we have used perylene red ($\omega_{\mathrm{ZPL}} = 2\pi c/585$ nm) as the dye species, which allows us to reduce the detuning between condensate and dye reservoir and to effectively increase the reservoir size. Following Eq. (19.6), this effective reservoir size can be quantified as

$$M_{\mathrm{eff}} = \frac{M/2}{1 + \cosh\left(\hbar\Delta/k_{\mathrm{B}}T\right)} . \tag{19.7}$$

Furthermore, a relative reservoir size is obtained by normalizing to the reservoir size in measurement R1

$$\epsilon = \frac{M_{\mathrm{eff},R_i}}{M_{\mathrm{eff},R1}} = \frac{\rho_{R_i}}{\rho_{R1}} \times \frac{1 + \cosh(\hbar\Delta_{R1}/k_{\mathrm{B}}T)}{1 + \cosh(\hbar\Delta_{R_i}/k_{\mathrm{B}}T)} . \tag{19.8}$$

For the lowest dye concentration and largest detuning (R1, $\epsilon = 1$), the particle reservoir is so small that the condensate fluctuations are damped almost directly above the condensation threshold ($\bar{N} \geq N_{\mathrm{c}}$). By increasing dye concentration and decreasing the dye-cavity detuning, one can systematically extend the regime of large fluctuations to higher condensate fractions (R1–R5). Our experimental results are recovered by a theoretical modeling shown as solid lines in Fig. 19.5, except for small condensate fractions below 5%. Here the visible drop-off in the correlation signal is attributed to imperfect mode filtering, which does not fully preclude photons in higher transversal cavity modes that are statistically uncorrelated to the ground mode photons from reaching the avalanche photo detectors. The maximum observed zero-delay autocorrelation is $g^{(2)}(0) \simeq 1.67$, corresponding to relative fluctuations of $\delta n_0/\bar{n}_0 = 82\%$, which is slightly less than theoretically expected. For the largest reservoir realized (R5, $\epsilon = 2000$), we observe zero-delay correlations of $g^{(2)}(0) \simeq 1.2$ at a condensate fraction of $\bar{n}_0/\bar{N} \simeq 0.6$. At this point, the condensate still performs large relative fluctuations of $\delta n_0/\bar{n}_0 = \sqrt{g^{(2)}(0) - 1} \simeq 45\%$, although its occupation number is comparable to the total photon number. This clearly demonstrates that the observed super-Poissonian photon statistics are determined by the grand-canonical particle exchange between condensate and dye reservoir.

19.5 Conclusions

We have described recent experiments on photon Bose-Einstein condensation in a dye-filled optical microresonator. Thermalization of the photon gas is achieved by a fluorescence-induced thermalization mechanism which establishes a thermal contact to the room-temperature dye medium and allows for a freely adjustable chemical potential. The photons here act like a gas of material particles with a phase transition temperature that is many orders of magnitude higher than for dilute atomic Bose-Einstein condensates. A further notable system property is a regime with unconventional fluctuation properties in which statistical fluctuations of the condensate number comparable to the total particle number occur. This is a yet unexplored regime of Bose-Einstein condensation that originates from the grand-canonical nature of the light-matter thermalization process and breaks the usual assumption of ensemble equivalence in statistical physics. Moreover, the unconventional second-order coherence properties of a photon condensate can draw a further borderline (in addition to the equilibrated system state) to laser-like behavior, if one follows the usual definition of a laser as a both first- and second-order coherent light source.

For the future, it will be interesting to test for the first-order coherence of photon condensation in the grand-canonical limit and to verify whether such a condensate exhibits superfluidity. A further fascinating perspective is the exploration of periodic potentials for the photon gas, which may allow researchers to tailor novel quantum many-body states of light.

Acknowledgments: We acknowledge funding from the European Research Council (ERC) (Interacting Photon Bose-Einstein Condensates in Variable Potentials [INPEC]) and the German Research Council (DFG) (We 1748-17).

References

[1] Leggett, A. J. 2006. *Quantum Liquids.* Oxford University Press, Oxford.

[2] Anderson, M. H., Ensher, J. R., Matthews, M. R., Wieman, C. E., and Cornell, E. A. 1995. Observation of Bose-Einstein condensation in a dilute atomic vapor. *Science*, **269**, 198.

[3] Davis, K. B., Mewes, M.-O., Andrews, M. R., van Druten, N. J., Durfee, D. M., Kurn, D. M., and Ketterle, W. 1995. Bose-Einstein condensation in a gas of sodium atoms. *Phys. Rev. Lett.*, **75**, 3969.

[4] Bradley, C. C., Sackett, C. A., and Hulet, R. G. 1997. Bose-Einstein condensation of lithium: observation of limited condensate number. *Phys. Rev. Lett.*, **78**, 985.

[5] Deng, H., Weihs, G., Santori, C., Bloch, J., and Yamamoto, Y. 2002. Condensation of semiconductor microcavity exciton polaritons. *Science*, **298**, 199.

[6] Kasprzak, J., Richard, M., Kundermann, S., Baas, A., Jeambrun, P., Keeling, J. M. J., Marchetti, F. M., Szymanska, M. H., Andre, R., Staehli, J. L., Savona, V., Littlewood, P. B., Deveaud, B., and Dang, L. S. 2006. Bose-Einstein condensation of exciton polaritons. *Nature*, **443**, 409.

[7] Balili, R., Hartwell, V., Snoke, D., and Pfeiffer, L. 2007. Bose-Einstein condensation of microcavity polaritons in a trap. *Science*, **316**, 1007.

[8] Demokritov, S. O., Demidov, V. E., Dzyapko, O., Melkov, G. A., Serga, A. A., Hillebrands, B., and Slavin, A N. 2006. Bose-Einstein condensation of quasi-equilibrium magnons at room temperature under pumping. *Nature*, **443**, 430.

[9] Huang, K. 1987. *Statistical Mechanics*. Wiley, New York.

[10] Zel'dovich, Y. B., and Levich, E. V. 1969. Bose condensation and shock waves in photon spectra. *Sov. Phys. JETP*, **28**, 1287.

[11] Chiao, R.Y. 2000. Bogoliubov dispersion relation for a "photon fluid": is this a superfluid? *Opt. Commun.*, **179**, 157.

[12] Mitchell, M. W., Hancox, C. I., and Chiao, R. Y. 2000. Dynamics of atom-mediated photon–photon scattering. *Phys. Rev. A*, **62**, 043819.

[13] Amo, A., Sanvitto, D., Laussy, F. P., Ballarini, D., Del Valle, E., Martin, M. D., Lemaitre, A., Bloch, J., Krizhanovskii, D. N., Skolnick, M. S., C., Tejedor, and L., Vina. 2009a. Collective fluid dynamics of a polariton condensate in a semiconductor microcavity. *Nature*, **457**, 291.

[14] Amo, A., Lefrère, J., Pigeon, S., Adrados, C., Ciuti, C., Carusotto, I., Houdré, R., Giacobino, E., and Bramati, A. 2009b. Superfluidity of polaritons in semiconductor microcavities. *Nat. Phys.*, **5**, 805.

[15] Sun, C., Jia, S., Barsi, C., Rica, S., Picozzi, A., and Fleischer, J. W. 2012. Observation of the kinetic condensation of classical waves. *Nat. Phys.*, **8**, 470.

[16] Klaers, J., Schmitt, J., Vewinger, F., and Weitz, M. 2010. Bose-Einstein condensation of photons in an optical microcavity. *Nature*, **468**, 545.

[17] Marelic, J., and Nyman, R. A. 2015. Experimental evidence for inhomogeneous pumping and energy-dependent effects in photon Bose-Einstein condensation. *Phys. Rev. A*, **91**, 033813.

[18] Angelis, E. De, Martini, F. De, and Mataloni, P. 2000. Microcavity quantum superradiance. *J. Opt. B – Quantum S. O.*, **2**, 149.

[19] Yokoyama, H., and Brorson, S. D. 1989. Rate equation analysis of microcavity lasers. *J. Appl. Phys.*, **66**, 4801.

[20] Klaers, J., Vewinger, F., and Weitz, M. 2010. Thermalization of a two-dimensional photonic gas in a "white-wall" photon box. *Nat. Phys.*, **6**, 512.

[21] Bagnato, V., and Kleppner, D. 1991. Bose-Einstein condensation in low-dimensional traps. *Phys. Rev. A*, **44**, 7439.

[22] Klaers, J., Schmitt, J., Damm, T., Vewinger, F., and Weitz, M. 2012. Statistical physics of Bose-condensed light in a dye microcavity. *Phys. Rev. Lett*, **108**, 160403.

[23] Sob'yanin, D. N. 2012. Hierarchical maximum entropy principle for generalized superstatistical systems and Bose-Einstein condensation of light. *Phys. Rev. E*, **85**, 061120.

[24] Kirton, P., and Keeling, J. 2013. Nonequilibrium model of photon condensation. *Phys. Rev. Lett.*, **111**, 100404.

[25] Snoke, D. W., and Girvin, S. M. 2013. Dynamics of phase coherence onset in Bose condensates of photons by incoherent phonon emission. *J. Low Temp. Phys.*, **171**, 1.

[26] de Leeuw, A.-W., and Duine, H. T. C., and Stoof R. A. 2013. Schwinger-Keldysh theory for Bose-Einstein condensation of photons in a dye-filled optical microcavity. *Phys. Rev. A*, **88**, 033829.

[27] Nyman, R. A., and Szymańska, M. H. 2014. Interactions in dye-microcavity photon condensates and the prospects for their observation. *Phys. Rev. A*, **89**, 033844.

[28] Kruchkov, A. 2014. Bose-Einstein condensation of light in a cavity. *Phys. Rev. A*, **89**, 033862.

[29] Strinati, M. C., and Conti, C. 2014. Bose-Einstein condensation of photons with nonlocal nonlinearity in a dye-doped graded-index microcavity. *Phys. Rev. A*, **90**, 043853.

[30] de Leeuw, A.-W., van der Wurff, E. C. I., Duine, R. A., and Stoof, H. T. C. 2014. Phase diffusion in a Bose-Einstein condensate of light. *Phys. Rev. A*, **90**, 043627.

[31] Sela, E., Rosch, A., and Fleurov, V. 2014. Condensation of photons coupled to a Dicke field in an optical microcavity. *Phys. Rev. A*, **89**, 043844.

[32] van der Wurff, E.C.I., de Leeuw, A.-W., Duine, R.A., and Stoof, H.T.C. 2014. Interaction effects on number fluctuations in a Bose-Einstein condensate of light. *Phys. Rev. Lett.*, **113**, 135301.

[33] Lakowicz, J. R. 1999. *Principles of Fluorescence Spectroscopy*. Kluwer Academic, New York.

[34] Kennard, E. H. 1918. On the thermodynamics of fluorescence. *Phys. Rev.*, **11**, 29.

[35] Kennard, E. H. 1927. The excitation of fluorescence in fluorescein. *Phys. Rev.*, **29**, 466.

[36] Stepanov, B. I. 1957. Universal relation between the absorption spectra and luminescence spectra of complex molecules. *Dokl. Akad. Nauk. SSSR+*, **112**, 839.

[37] Kazachenko, L. P., and Stepanov, B. I. 1957. Mirror symmetry and the shape of absorption and luminescence bands of complex molecules. *Opt. Spektrosk.*, **2**, 339.

[38] McCumber, D. E. 1964. Einstein relations connecting broadband emission and absorption spectra. *Phys. Rev.*, **136**, A954.

[39] Klaers, J. 2014. The thermalization, condensation and flickering of photons. *J. Phys. B: At. Mol. Opt.*, **47**, 243001.

[40] Cornell, E. A., and Wieman, C. E. 2002. Nobel lecture: Bose-Einstein condensation in a dilute gas, the first 70 years and some recent experiments. *Rev. Mod. Phys.*, **74**, 875.

[41] Ketterle, W. 2002. Nobel lecture: when atoms behave as waves: Bose-Einstein condensation and the atom laser. *Rev. Mod. Phys*, **74**, 1131.

[42] Lee, M. D., and Gardiner, C. W. 2000. Quantum kinetic theory. VI. The growth of a Bose-Einstein condensate. *Phys. Rev. A*, **62**, 033606.

[43] Deng, H., Press, D., Götzinger, S., Solomon, G. S., Hey, R., Ploog, K. H., and Yamamoto, Y. 2006. Quantum degenerate exciton-polaritons in thermal equilibrium. *Phys. Rev. Lett.*, **97**, 146402.

[44] Kasprzak, J., Solnyshkov, D. D., Andre, R., Dang, L. S., and Malpuech, G. 2008. Formation of an exciton polariton condensate: thermodynamic versus kinetic regimes. *Phys. Rev. Lett.*, **101**, 146404.

[45] Wouters, M., Carusotto, I., and Ciuti, C. 2008. Spatial and spectral shape of inhomogeneous nonequilibrium exciton-polariton condensates. *Phys. Rev. B*, **77**, 115340.

[46] Schmitt, Julian, Damm, Tobias, Dung, David, Vewinger, Frank, Klaers, Jan, and Weitz, Martin. 2015. Thermalization kinetics of light: from laser dynamics to equilibrium condensation of photons. *Phys. Rev. A*, **92**, 011602.

[47] Hadzibabic, Z., Krüger, P., Cheneau, M., Battelier, B., and Dalibard, J. 2006. Berezinskii-Kosterlitz-Thouless crossover in a trapped atomic gas. *Nature*, **441**, 1111.

[48] Clade, P., Ryu, C., Ramanathan, A., Helmerson, K., and Phillips, W. D. 2009. Observation of a 2D Bose gas: from thermal to quasicondensate to superfluid. *Phys. Rev. Lett.*, **102**, 170401.

[49] Hadzibabic, Z., and Dalibard, J. 209. Two-dimensional Bose fluids: an atomic physics perspective. *Lecture Notes from the Varenna Summer School 23 June–3 July 2009.* Published in Proceedings of the International School of Physics "Enrico Fermi," Course CLXXIII, "Nano optics and atomics: transport of light and matter waves." Editors R. Kaiser, D. Wiersma, and L. Fallani. IOS Press, Amsterdam, and SIF, Bologna, 273.

[50] Klaers, J., Schmitt, J., Damm, T., Vewinger, F., and Weitz, M. 2011. Bose-Einstein condensation of paraxial light. *Appl. Phys. B*, **105**, 17.

[51] Huang, K. 2001. *Introduction to Statistical Physics*. CRC Press, Boca Raton.

[52] Ziff, R. M., Uhlenbeck, G. E., and Kac, M. 1977. Ideal Bose-Einstein gas, revisited. *Phys. Rep.*, **32**, 169.

[53] Kocharovsky, V. V., Kocharovsky, V. V., Holthaus, M., Ooi, C. H. Raymond, Svidzinsky, A. A., Ketterle, W., and Scully, M. O. 2006. Fluctuations in ideal and interacting Bose-Einstein condensates: from the laser phase transition analogy to squeezed states and Bogoliubov quasiparticles. *Adv. Atom. Mol. Opt. Phy.*, **53**, 291.
[54] Fierz, M. 1955. Über die statistischen Schwankungen in einem kondensierenden System. *Helv. Phys. Acta*, **29**, 47.
[55] Grossmann, S., and Holthaus, M. 1996. Microcanonical fluctuations of a Bose system's ground state occupation number. *Phys. Rev. E*, **54**, 3495.
[56] Holthaus, M., Kalinowski, E., and Kirsten, K. 1998. Condensate fluctuations in trapped Bose gases: canonical vs. microcanonical ensemble. *Ann. Phys. (NY)*, **270**, 198.
[57] Fujiwara, I., ter Haar, D., and Wergeland, H. 1970. Fluctuations in the population of the ground state of Bose systems. *J. Stat. Phys.*, **2**, 329.
[58] Politzer, H. D. 1996. Condensate fluctuations of a trapped, ideal Bose gas. *Phys. Rev. A*, **54**, 5048.
[59] Navez, P., Bitouk, D., Gajda, M., Idziaszek, Z., and Rzazewski, K. 1997. Fourth statistical ensemble for the Bose-Einstein condensate. *Phys. Rev. Lett.*, **79**, 1789.
[60] Grossmann, S., and Holthaus, M. 1997. Maxwell's demon at work: two types of Bose condensate fluctuations in power-law traps. *Opt. Express*, **1**, 262.
[61] Weiss, C., and Wilkens, M. 1997. Particle number counting statistics in ideal Bose gases. *Opt. Express*, **1**, 272.
[62] Weiss, C., Page, S., and Holthaus, M. 2004. Factorising numbers with a Bose-Einstein condensate. *Physica A*, **341**, 586.
[63] Scully, M. O., and Lamb, W. E. 1967. Quantum theory of an optical maser. I. General theory. *Phys. Rev.*, **159**, 208.
[64] Schmitt, J., Damm, T., Dung, D., Vewinger, F., Klaers, J., and Weitz, M. 2014. Observation of grand-canonical number statistics in a photon Bose-Einstein condensate. *Phys. Rev. Lett.*, **112**, 030401.
[65] Ciuti, C. 2014. Statistical flickers in a Bose-Einstein condensate of photons. *Physics*, **7**, 7.

20

Laser Operation and Bose-Einstein Condensation: Analogies and Differences

ALESSIO CHIOCCHETTA AND ANDREA GAMBASSI
Scuola Internazionale Superiore di Studi Avanzati (SISSA) – International School for Advanced Studies and
Istituto Nazionale di Fisica Nucleare (INFN), Trieste, Italy

IACOPO CARUSOTTO
Istituto Nazionale di Ottica – National Research Council of Italy (INO-CNR)
Bose-Einstein Condensation (BEC) Center and
Dipartimento di Fisica, Università di Trento, Povo, Italy

After reviewing the interpretation of laser operation as a nonequilibrium Bose-Einstein condensation phase transition, we illustrate the novel features arising from the nonequilibrium nature of photon and polariton Bose-Einstein condensates recently observed in experiments. We then propose a quantitative criterion to experimentally assess the equilibrium versus nonequilibrium nature of a specific condensation process, based on fluctuation-dissipation relations. The power of this criterion is illustrated on two models which show very different behaviors.

20.1 Historical and Conceptual Introduction

The first introduction of concepts of nonequilibrium statistical mechanics into the realm of optics dates back to the early 1970s with pioneering works by Graham and Haken [1] and by DeGiorgio and Scully [2], who proposed a very insightful interpretation of the laser threshold in terms of a spontaneous breaking of the $U(1)$ symmetry associated with the phase of the emitted light. Similar to what happens to the order parameter at a second-order phase transition, such an optical phase is randomly chosen every time the device is switched on and remains constant for macroscopic times. Moreover, a long-range spatial order is established, as light emitted by a laser device above threshold is phase-coherent on macroscopic distances.

While textbooks typically discuss this interpretation of laser operation in terms of a phase transition for the simplest case of a single-mode laser cavity, rigorously speaking this is valid only in spatially infinite systems. In fact, only in this case one can observe nonanalytic behaviors of the physical quantities at the transition point. In particular, the long-range order is typically assessed by looking at the long-distance behavior of the correlation function of the order parameter, which, for a laser, corresponds to the first-order spatial coherence of the emitted electric field $\hat{E}(\mathbf{r})$,

$$\lim_{|\mathbf{r}-\mathbf{r}'|\to\infty} \left\langle \hat{E}^\dagger(\mathbf{r})\,\hat{E}(\mathbf{r}')\right\rangle : \tag{20.1}$$

the spontaneous symmetry breaking is signaled by this quantity becoming nonzero (see Chapter 5). The average $\langle\ldots\rangle$ is taken on the stationary density matrix of the system. In order to be able to probe this long-distance behavior, experimental studies need devices with a spatially extended active region. The so-called vertical cavity surface-emitting lasers (VCSELs) are perhaps the most studied examples in this class [3, 4]: by using an active medium sandwiched between a pair of plane-parallel semiconductor mirrors, one can realize devices of arbitrarily large size, the only limitation coming from extrinsic effects such as the difficulty of having a spatially homogeneous pumping of the active material and of avoiding disorder of the semiconductor microstructure.

The most celebrated phase transition breaking a $U(1)$ symmetry in statistical physics is perhaps the Bose-Einstein condensation (BEC). While in textbooks [5, 6] BEC is typically described in terms of the emergence of a macroscopic occupation of a single quantum level, which gives a finite condensate density n_{BEC}, an alternative, mathematically equivalent condition – the so-called Penrose-Onsager criterion [7] – involves the long-distance limit of the coherence function of the matter field $\hat{\Psi}(\mathbf{r})$ describing the Bose particles undergoing condensation, i.e.,

$$\lim_{|\mathbf{r}-\mathbf{r}'|\to\infty} \left\langle \hat{\Psi}^\dagger(\mathbf{r})\,\hat{\Psi}(\mathbf{r}')\right\rangle = n_{BEC} > 0, \tag{20.2}$$

where the average $\langle\ldots\rangle$ is taken on the thermal density matrix. On this basis, it is natural to see BEC as a phase transition which spontaneously breaks the global $U(1)$ gauge symmetry of the quantum matter field $\hat{\Psi}(\mathbf{r}) \to e^{i\theta}\hat{\Psi}(\mathbf{r})$.

There is, however, an important difference between textbook BEC and laser operation: in the former, the system is assumed to be in thermal equilibrium at temperature T, and therefore its density matrix $\rho_{\rm eq}$ is given by the Boltzmann ensemble of equilibrium statistical mechanics $\rho_{\rm eq} \propto \exp(-H/k_BT)$. A laser is, instead, an intrinsically nonequilibrium device, whose steady state is determined by a dynamical balance of pumping and losses, the latter being essential to generate the output laser beam used in any application. As a result, one can think of laser operation in spatially extended devices as an example of *nonequilibrium BEC*. In between the two extreme limits of equilibrium BEC and laser operation, experiments with gases of exciton-polaritons in microcavities [8], of magnons [9], and of photons [10] have explored a full range of partially thermalized regimes depending on the ratio between the loss and the thermalization rates, the latter being typically due to interparticle collisions within the gas and/or to interactions with the host material.

The first part of this chapter will be devoted to a brief review of the novel features of the nonequilibrium BEC as compared to its equilibrium counterpart. In the second part we will dwell on the equilibrium versus nonequilibrium character of these phenomena by proposing a quantitative criterion to experimentally probe the nature of the stationary state in specific cases, based on the fluctuation-dissipation theorems. In order to illustrate the practical utility of this approach, two toy models of condensation will be discussed.

20.2 Textbook Equilibrium Bose-Einstein Condensation

The textbook discussion of the equilibrium BEC in statistical physics [5, 6] is based on a grand-canonical description of an ideal three-dimensional Bose gas at thermal equilibrium with an inverse temperature $\beta = 1/k_B T$ and a chemical potential $\mu \leq 0$ in terms of the Bose distribution, $n_{\mathbf{k}} = [e^{\beta(\epsilon_{\mathbf{k}} - \mu)} - 1]^{-1}$, where $\epsilon_{\mathbf{k}} = \hbar^2 k^2/2m$ is the energy of the state of momentum $\hbar\mathbf{k}$, and m the mass of the particles.

For any given temperature T, the maximum density of particles that can be accommodated in the excited states at $\mathbf{k} \neq 0$ grows with μ and then saturates at $n_{\max}(T)$ for $\mu = 0$. If the actual density of particles n exceeds this threshold, the extra particles must accumulate into the lowest state at $\mathbf{k} = 0$, forming the so-called condensate. As the coherence in Eq. (20.2) is the Fourier transform of the momentum distribution $n_{\mathbf{k}}$, it is immediate to see that the two different definitions of condensate fraction n_{BEC} in terms of a macroscopic occupation $n_{\mathbf{k}=0}$ of the lowest mode and of the long-distance behavior of the coherence actually coincide.

An alternative but equivalent description of this physics was presented in Ref. [11]: in order to highlight the spontaneous symmetry-breaking mechanism it is convenient to introduce a fictitious external field η coupling to the quantum matter field via the Hamiltonian

$$H_{\text{ext}} = -\int \mathrm{d}^3\mathbf{r} \left[\eta \hat{\Psi}^\dagger(\mathbf{r}) + \eta^* \hat{\Psi}(\mathbf{r})\right], \tag{20.3}$$

which explicitly breaks the $U(1)$ symmetry. As in a ferromagnet, the direction of the magnetization is selected by an external magnetic field B; in the presence of an η field the quantum matter field $\hat{\Psi}(\mathbf{r})$ acquires a finite expectation value. For $T < T_{BEC}$, this expectation value

$$\Psi_0 = \lim_{\eta \to 0} \lim_{V \to \infty} \left\langle \hat{\Psi}(\mathbf{r}) \right\rangle \tag{20.4}$$

remains finite even for vanishing η in the thermodynamic limit $V \to \infty$, and is related to the condensate density by $n_{BEC} = |\Psi_0|^2$. As usual for phase transitions, the order in which the zero-field and the infinite-volume limits are taken in Eq. (20.4) is crucial [5]. Alternative descriptions of the condensate have been investigated, in terms of either Fock or coherent states [12], unveiling subtle mechanisms leading to interference fringes even with Fock states [13] and illustrating the roles of spontaneous amplification of quantum coherence toward coherent states [14] and, conversely, of the phase decoherence mechanisms [15].

In the presence of interparticle pair interactions, all condensation criteria based on the lowest mode occupation, on the long-distance coherence, and on the spontaneous coherent matter field remain valid. However, the underlying physics is much more complex and we refer the reader to the specialized literature on the subject, e.g., Ref. [6]. For our purposes, we only need to mention that at equilibrium at $T = 0$, the condensate order parameter $\Psi_0(\mathbf{r})$ in the presence of an external potential $V(\mathbf{r})$ can be obtained in the dilute gas regime by minimizing the so-called Gross-Pitaevskii (GP) energy functional,

$$E[\Psi_0] = \int \mathrm{d}^3\mathbf{r} \left\{ \frac{\hbar^2}{2m} |\nabla\Psi_0(\mathbf{r})|^2 + V(\mathbf{r})|\Psi_0(\mathbf{r})|^2 + \frac{g}{2}|\Psi_0(\mathbf{r})|^4 \right\}, \tag{20.5}$$

where the normalization of Ψ_0 is set to the total number N of particles in the system, $N = \int d^3\mathbf{r}\, |\Psi_0(\mathbf{r})|^2$, and $g = 4\pi\hbar^2 a_0/m$ quantifies the strength of the (local) interactions proportional to the s-wave interparticle scattering length a_0. In terms of a_0, the dilute gas regime corresponds to $na_0^3 \ll 1$. The mathematical form of the GP energy functional (20.5) guarantees that the condensate wavefunction $\Psi_0(\mathbf{r})$ keeps a constant phase throughout the whole system even in the presence of the external potential $V(\mathbf{r})$, while the density profile $|\Psi_0(\mathbf{r})|^2$ develops large variations in space.

Finally, at this same level of approximation, the condensate dynamics is ruled by the time-dependent Gross-Pitaevskii equation,

$$i\hbar\frac{\partial\Psi_0(\mathbf{r},t)}{\partial t} = -\frac{\hbar^2}{2m}\nabla^2\Psi_0(\mathbf{r},t) + V(\mathbf{r})\Psi_0(\mathbf{r},t) + g|\Psi_0(\mathbf{r},t)|^2\,\Psi_0(\mathbf{r},t), \tag{20.6}$$

which has the mathematical form of a nonlinear Schrödinger equation.

20.3 Mean-Field Theory of Nonequilibrium BEC

While the shape of an equilibrium condensate is obtained by minimizing the energy functional (20.5) (of the Ginzburg-Landau type), nonequilibrium BECs appear to be significantly less universal as their theoretical description typically requires some microscopic modeling of the specific dissipation and pumping mechanisms present in a given experimental setup. For a comprehensive discussion of the main configurations, we refer the interested reader to the recent review article [16]. Here we summarize the simplest and most transparent of such descriptions, which is based on the following complex Ginzburg-Landau evolution equation for the order parameter,

$$i\hbar\frac{\partial\Psi_0}{\partial t} = -\frac{\hbar^2}{2m}\nabla^2\Psi_0 + V(\mathbf{r})\Psi_0 + g|\Psi_0|^2\,\Psi_0 - \frac{i\hbar}{2}\left[\gamma - \frac{P}{1+\frac{|\Psi_0|^2}{n_{\text{sat}}}}\right]\Psi_0, \quad (20.7)$$

inspired by the so-called semiclassical theory of the laser. In addition to the terms already present in the equilibrium description (20.6), Eq. (20.7) accounts for the losses at a rate γ and for the stimulated pumping of new particles into the condensate at a bare rate P, which then saturates once the condensate density exceeds the saturation density n_{sat}. Given the driven-dissipative nature of this evolution equation, the steady state has to be determined as the long-time limit of the dynamical evolution.

This apparently minor difference has profound implications, as the breaking of time-reversal symmetry by the pumping and loss terms in (20.7) allows for steady-state configurations with a spatially varying phase of the order parameter, which physically corresponds to finite particle currents through the condensate. This feature was first observed in Ref. [17] as a ring-shaped condensate emission in the wavevector $\mathbf{k}$-space and, in the presence of disorder, as an asymmetry of the emission pattern under reflections, $n(\mathbf{k}) \neq n(-\mathbf{k})$ [18]. A theoretical interpretation was proposed in Ref. [19] and soon confirmed by the more detailed experiments in Ref. [20].

Another feature of Eq. (20.7) which clearly distinguishes nonequilibrium systems from their equilibrium counterparts is the dispersion of the collective excitations on top of a condensate. As first predicted in Refs. [21, 22, 23], the usual Bogoliubov dispersion of spatially homogeneous condensates $\hbar\omega_{\text{eq},\mathbf{k}} = [\epsilon_{\mathbf{k}}(\epsilon_{\mathbf{k}} + 2g|\Psi_0|^2)]^{1/2}$ is modified by pumping and dissipation to $\omega_{\text{neq},\mathbf{k}} = -i\Gamma/2 + [\omega^2_{\text{eq},\mathbf{k}} - \Gamma^2/4]^{1/2}$, where the dissipation parameter $\Gamma = \gamma(1 - P_c/P)$ depends on the pumping power P in units of its threshold value P_c. In particular, the usual sonic (linear) dispersion of low-wavevector excitations in equilibrium condensates is strongly modified into a flat, diffusive region [24].

20.4 Noncondensed Cloud and Quasicondensation

In an equilibrium system at low temperature, the collective excitation modes discussed in the previous section are thermally populated according to a Bose distribution with zero chemical potential. In the noninteracting limit, this Bogoliubov approach recovers the textbook prediction for the noncondensed density discussed in Section 20.2. In the interacting case, in addition to these thermal fluctuations of the matter Bose field, a further contribution to the noncondensed fraction comes from the so-called quantum depletion of the condensate, i.e., quantum fluctuations due to virtual scattering of condensed particles into the noncondensed modes.

In spatial dimension $d \leq 2$, the long-range order of the condensate is not stable against thermal fluctuations and is replaced by a so-called quasi-condensate, in agreement with the Hohenberg-Mermin-Wagner theorem of statistical physics [5]: while order is present up to intermediate length scales, upon increasing the distance it decays (at finite T) with an exponential (in $d = 1$) or algebraic (in $d = 2$) law. While this result was first discussed in the late 1960s for equilibrium systems [25, 26], a first mention in the non-equilibrium context was already present in the above-cited seminal work by Graham and Haken [1]: in this latter case, however, fluctuations did not have a thermal origin, but were an unavoidable consequence of the quantum nature of the field undergoing condensation. Independently of this pioneering work, this result was rediscovered later on in Refs. [21, 23] and then extended to the critical region using renormalization-group techniques. Within this approach, several new features have been anticipated: in $d = 3$ novel critical exponents appear [27], while the $d = 2$ algebraic long-range decay of correlations is destroyed and replaced by a stretched exponential [28].

20.5 How to Quantitatively Assess Equilibrium?

In most experiments on BEC in polariton and photon gases so far, a special effort was made in order to understand whether the system was thermalized or not. One can in fact expect that some effective thermalization should occur even in nonequilibrium regimes as soon as the thermalization time is shorter than the lifetime of the particles. In order to experimentally assess this *quasi-equilibrium* condition, the measured momentum and/or energy distribution of the noncondensed particle was typically compared with a Bose distribution.

The observation of such thermal distributions is to be expected in polariton and photon gases showing frequent collisions [8, 10], as illustrated in Fig. 20.1a. On the other hand, it was quite a surprise when the experiment in Ref. [29] reported a momentum distribution with a thermal-like tail even in a lasing regime where the photons should not be thermalized — see Fig. 20.1b. This observation cast some doubts on the interpretation of similar available experimental results; in particular, several authors have tried to develop alternative models to justify the observed thermal tail in terms of generalized, strongly nonequilibrium laser theories [30, 31].

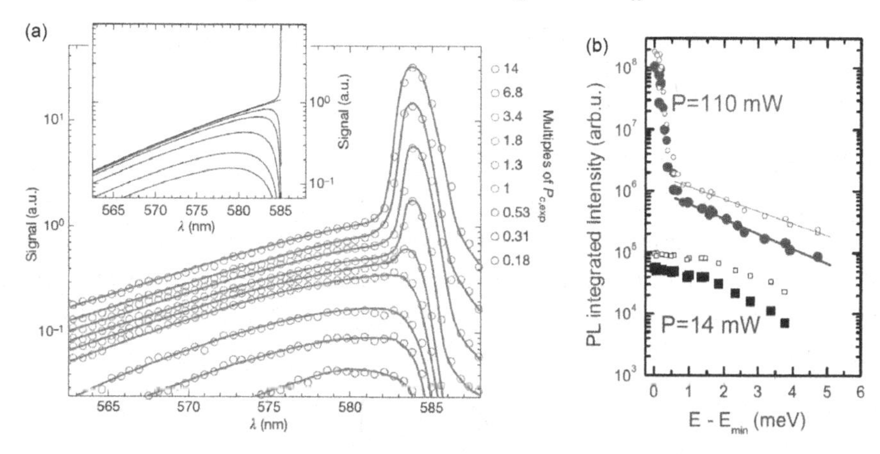

Figure 20.1 (a) Wavelength distribution of the emission from a photon condensate across the condensation threshold. Adapted with permission from Klaers, J., et al. (2010), Bose-Einstein condensation of photons in an optical microcavity. *Nature*, **468**, 545 [10]. Copyright (2010) by the Nature Publishing Group. (b) Energy distribution of the emission from a microcavity for pump values P respectively below (squares) and above (circles) the condensation threshold. In the latter case, photons are expected to be almost noninteracting particles. Reprinted with permission from Bajoni, D., et al. (2007), Photon lasing in GaAs microcavity: similarities with a polariton condensate. *Phys. Rev. B*, **76**, 201305 [29]. Copyright (2007) by the American Physical Society.

This ongoing debate calls for the identification of novel criteria to quantitatively assess the equilibrium versus nonequilibrium nature of a system. A possible approach to this problem will be the subject of the next sections.

20.5.1 Fluctuation-Dissipation Theorems

Thermodynamical equilibrium is not only a property of the *state* of a system, but also of its *dynamics*. A remarkable consequence of equilibrium which involves dynamical quantities is the so-called fluctuation-dissipation theorem (FDT) [32], which provides a relationship between the linear response of a system to an external perturbation of frequency ω and the thermal fluctuations of the same system at the same frequency ω. While FDT relations hold for any pair of operators, in the following sections of this work we focus on the annihilation and creation operators $b_{\mathbf{k}}$ and $b_{\mathbf{k}}^{\dagger}$ of a $\mathbf{k} \neq 0$ noncondensed mode of the intracavity photon/polariton field undergoing BEC in a spatially homogeneous geometry. To account for the particle number variation, it is convenient to include a chemical potential in the Hamiltonian, $H = H_0 - \mu N$, so that frequencies ω are measured from the chemical potential. In the presence of a condensate, μ coincides with the oscillation frequency of the condensate mode, which in optical systems is observable as the condensate emission frequency ω_{BEC}.

To state the FDT, it is convenient to introduce the two functions

$$C_{bb^\dagger}(\mathbf{k}, t-s) = \frac{1}{2}\left\langle\left\{\hat{b}_{\mathbf{k}}(t), \hat{b}^\dagger_{\mathbf{k}}(s)\right\}\right\rangle, \quad \chi''_{bb^\dagger}(\mathbf{k}, t-s) = \frac{1}{2}\left\langle\left[\hat{b}_{\mathbf{k}}(t), \hat{b}^\dagger_{\mathbf{k}}(s)\right]\right\rangle, \tag{20.8}$$

where the time dependence of the operator corresponds to their Heisenberg evolution under the system Hamiltonian H and the average $\langle \ldots \rangle$ is taken in a thermal equilibrium state at temperature T with density matrix $\rho \propto \exp(-H/k_B T)$. In such a state, these correlations only depend on the time difference $t - s$ and we can define their Fourier transforms $C_{bb^\dagger}(\mathbf{k}, \omega)$ and $\chi''_{bb^\dagger}(\mathbf{k}, \omega)$. The explicit form of the FDT then reads

$$C_{bb^\dagger}(\mathbf{k}, \omega) = \coth\left(\frac{\omega}{2k_B T}\right)\chi''_{bb^\dagger}(\mathbf{k}, \omega). \tag{20.9}$$

In order to understand the physical content of this relation, we note that $\chi''_{bb^\dagger}(\mathbf{k}, \omega)$ appearing in this relation is directly related to the imaginary part of the response function $\chi_{bb^\dagger}(\mathbf{k}, \omega)$ of the system, which quantifies the energy it absorbs from the weak perturbation [33], i.e., $\chi''_{bb^\dagger}(\mathbf{k}, \omega) = -\mathrm{Im}[\chi_{bb^\dagger}(\mathbf{k}, \omega)]$. As usual, $\chi_{bb^\dagger}(\mathbf{k}, \omega)$ is defined as the Fourier transform of the linear response susceptibility $\chi_{bb^\dagger}(\mathbf{k}, t) = -2i\theta(t)\,\chi''_{bb^\dagger}(\mathbf{k}, t)$.

An alternative, fully equivalent formulation of the FDT (20.9) is the Kubo-Martin-Schwinger (KMS) condition [34, 35], which in our example reads

$$S_{b^\dagger b}(\mathbf{k}, -\omega) = \mathrm{e}^{-\beta\omega} S_{bb^\dagger}(\mathbf{k}, \omega), \tag{20.10}$$

where $S_{bb^\dagger}(\mathbf{k}, t) = \langle \hat{b}_{\mathbf{k}}(t)\hat{b}^\dagger_{\mathbf{k}}\rangle$ and $S_{b^\dagger b}(\mathbf{k}, t) = \langle \hat{b}^\dagger_{\mathbf{k}}(t)\hat{b}_{\mathbf{k}}\rangle$.

The FDT has quite often been used to probe the effective thermalization of a system and to characterize its possible departure from equilibrium [36, 37, 38]. In particular, given a pair of correlation functions, one can always define from (20.10) an effective temperature T_{eff} such that the functions satisfy an FDT: if the system is really at equilibrium, T_{eff} has a constant value independently of $\mathbf{k}$ and ω and equal to the thermodynamic temperature. On the other hand, if the system is out of equilibrium T_{eff} will generically develop a nontrivial dependence on $\mathbf{k}$ and ω.

20.5.2 Application to Photon/Polariton Condensates

Applying these ideas to the photon/polariton condensates discussed in the previous sections provides a quantitative criterion to assess the equilibrium or nonequilibrium nature of the condensate: the protocol we propose consists in measuring different correlation and/or response functions and in checking if they satisfy the FDT.

On the one hand, the correlation function $S_{b^\dagger b}$ can be related to the angle- and frequency-resolved photoluminescence intensity $S(\mathbf{k}, \omega)$ coming from the noncondensed particles via $\mathcal{S}(\mathbf{k}, \omega + \omega_{BEC}) = S_{b^\dagger b}(\mathbf{k}, -\omega)$, where the condensate emission frequency ω_{BEC} plays the role of the chemical potential μ in the nonequilibrium context. On the other hand, $\chi''_{bb^\dagger}(\mathbf{k}, \omega)$ is related to the imaginary part of the linear response to an external monochromatic field with momentum $\mathbf{k}$ and frequency $\omega + \omega_{BEC}$. To be more specific, let us assume a two-sided cavity illuminated by an external classical field $E_{\mathbf{k}}^{\text{inc}}(t)$ incident from the left. This field couples to the intracavity bosons through the Hamiltonian [16, 39]

$$H_{\text{pump}} = i \int \frac{d^d k}{(2\pi)^d} \left[\eta_{\mathbf{k}}^{\text{l}} E_{\mathbf{k}}^{\text{inc}}(t) b_{\mathbf{k}}^\dagger - \eta_{\mathbf{k}}^{\text{l}*} E_{\mathbf{k}}^{\text{inc}*}(t) b_{\mathbf{k}} \right], \tag{20.11}$$

where $\eta_{\mathbf{k}}^{\text{l}}$ ($\eta_{\mathbf{k}}^{\text{r}}$) is the transmission amplitude of the left (right) mirror of the cavity. In the steady state, the intracavity field vanishes $\langle b_{\mathbf{k}} \rangle_{\text{eq}} = 0$ and the incident classical field induces a perturbation $\langle b_{\mathbf{k}}(\omega) \rangle = i\eta_{\mathbf{k}}^{\text{l}} \chi_{bb^\dagger}(\mathbf{k}, \omega) E_{\mathbf{k}}^{\text{inc}}(\omega)$, where $b_{\mathbf{k}}(\omega) = \int dt\, b_{\mathbf{k}}(t) e^{i\omega t}$. From the boundary conditions at the two mirrors, the reflected and transmitted fields can be related to the intracavity field [16, 39] as

$$E_{\mathbf{k}}^{\text{refl}}(\omega) = \left[1 - i|\eta_{\mathbf{k}}^{\text{l}}|^2 \chi_{bb^\dagger}(\mathbf{k}, \omega)\right] E_{\mathbf{k}}^{\text{inc}}(\omega), \tag{20.12}$$

$$E_{\mathbf{k}}^{\text{tr}}(\omega) = -i\eta_{\mathbf{k}}^{\text{r}*} \eta_{\mathbf{k}}^{\text{l}} \chi_{bb^\dagger}(\mathbf{k}, \omega) E_{\mathbf{k}}^{\text{inc}}(\omega). \tag{20.13}$$

Accordingly, from a measurement of the reflected or transmitted fields it is possible to reconstruct the response function $\chi''_{bb^\dagger}(k, \omega)$ through the formulas

$$\chi''_{bb^\dagger}(\mathbf{k}, \omega) = \frac{1}{|\eta_{\mathbf{k}}^{\text{l}}|^2} \left(1 - \text{Re}\left[\frac{E_{\mathbf{k}}^{\text{refl}}(\omega)}{E_{\mathbf{k}}^{\text{inc}}(\omega)}\right]\right) = -\frac{1}{|\eta_{\mathbf{k}}^{\text{l}}|^2} \text{Re}\left[\frac{\eta_{\mathbf{k}}^{\text{l}*}}{\eta_{\mathbf{k}}^{\text{r}*}} \frac{E_{\mathbf{k}}^{\text{tr}}(\omega)}{E_{\mathbf{k}}^{\text{inc}}(\omega)}\right], \tag{20.14}$$

where both the amplitude and the phase of $E_{\mathbf{k}}^{\text{refl}}$ and $E_{\mathbf{k}}^{\text{tr}}$ can be measured with standard optical tools and the $\eta_{\mathbf{k}}^{\text{l,r}}$ coefficients can be extracted from reflection and transmission measurements on the unloaded cavity.

At equilibrium, these quantities are related to the angle- and frequency-resolved luminescence spectrum $\mathcal{S}$ by the FDT

$$\chi''_{bb^\dagger}(\mathbf{k}, \omega) = \frac{1 - e^{-\beta\omega}}{2} S_{bb^\dagger}(\mathbf{k}, \omega) = \frac{e^{\beta\omega} - 1}{2} \mathcal{S}(\omega_{BEC} + \omega, \mathbf{k}); \tag{20.15}$$

as both sides of this equation are experimentally measurable, any discrepancy is a signature of a nonequilibrium condition.

As a further verification, the interested reader may check that this FDT is satisfied for an empty cavity which is illuminated from both sides by thermal radiation at the same temperature.

20.5.3 Application to Some Models of Photon/Polariton BEC

As a final point, we will illustrate the behavior of the FDT for two simple models of photon/polariton BEC. In doing this, one has to keep in mind that some of the approaches usually used to describe open quantum systems are intrinsically unable to correctly reproduce the FDT, so one must be careful not to mistake an equilibrium system for a nonequilibrium one just because of the approximations made in the theoretical model. In particular, every quantum master equation governed by a Lindblad super-operator always violates the FDT [40, 41], even if the stationary solution has the form of a thermal density matrix: the reason of this pathology lies in the full Markovian approximation, which is inherent in the master equations [42].

Quantum Langevin Model

In Ref. [31], two of us proposed a simple model of nonequilibrium condensation based on a generalized laser model. The idea is to model the complex scattering processes responsible for condensation in terms of a spatially uniform distribution of population-inverted two-level atoms, which can emit light into the cavity mode. Once the population inversion is large enough, laser operation will occur into the cavity. The dynamics of quantum fluctuations on top of the coherent laser emission are then described by means of quantum Langevin equations for the noncondensed mode amplitudes.

To check the effective lack of thermalization of the system, in Fig. 20.2 we study the ω and $\mathbf{k}$ dependence of the effective inverse temperature $\beta_{\rm eff}$, as extracted from

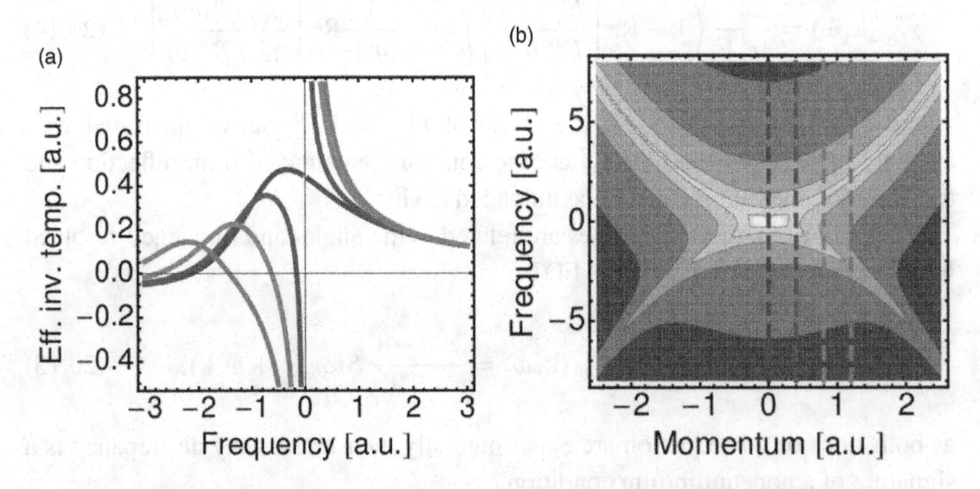

Figure 20.2 (a) Effective inverse temperatures versus frequency for the model of Ref. [31]. The various curves correspond to different values of the in-plane momenta $\mathbf{k}$. (b) Angle- and frequency-resolved photoluminescence intensity $\mathcal{S}(\mathbf{k}, \omega_{BEC} + \omega)$ (originally reported in Ref. [31]), with dashed vertical lines corresponding to the in-plane momenta considered in panel (a).

the KMS relation (20.10) using the quantum Langevin prediction for the correlation functions. The resulting $\beta_{\text{eff}}(\mathbf{k}, \omega)$ strongly depends on both ω and $\mathbf{k}$ and becomes even negative in some regions: all these features are a clear signature of a very nonequilibrium condition.

Non-Markovian Toy Model

The situation is much more intriguing for the model recently proposed in Refs. [43, 44] for studying the photon BEC experiments of Ref. [10]: clear signatures of a thermal distribution of the noncondensed cloud were observed as soon as the thermalization rate under the effect of repeated absorption and emission cycles by the dye molecules becomes comparable to the photon loss rate. On the other hand, when thermalization is too slow, the thermal-like features disappear and the system reproduces the nonequilibrium physics of a laser.

In order to investigate how this crossover affects the FDT, we introduce a non-Markovian toy model which extends the theory in Refs. [43, 44] in order to avoid spurious effects due to the Markov approximation. For simplicity, we consider a single noncondensed mode of frequency ω_c described by operators $\hat{b}, \hat{b}^\dagger$. The frequency-dependent absorption and amplification by the dye molecules are modeled in terms of two distinct baths of harmonic oscillators $\hat{a}_n, \hat{a}_n^\dagger$ and $\hat{c}_n, \hat{c}_n^\dagger$ with frequencies ω_n^- and ω_n^+, respectively:

$$H = \omega_c \hat{b}^\dagger \hat{b} + \sum_n \left(\omega_n^- \hat{a}_n^\dagger \hat{a}_n + \omega_n^+ \hat{c}_n^\dagger \hat{c}_n\right) + \sum_n \left(\eta_n^- \hat{b}\hat{a}_n^\dagger + \eta_n^+ \hat{b}^\dagger \hat{c}_n^\dagger + \text{h.c.}\right). \quad (20.16)$$

The baths take into account the two processes pictorially represented in Fig. 20.3a, in which the photon absorption and emission processes are associated with

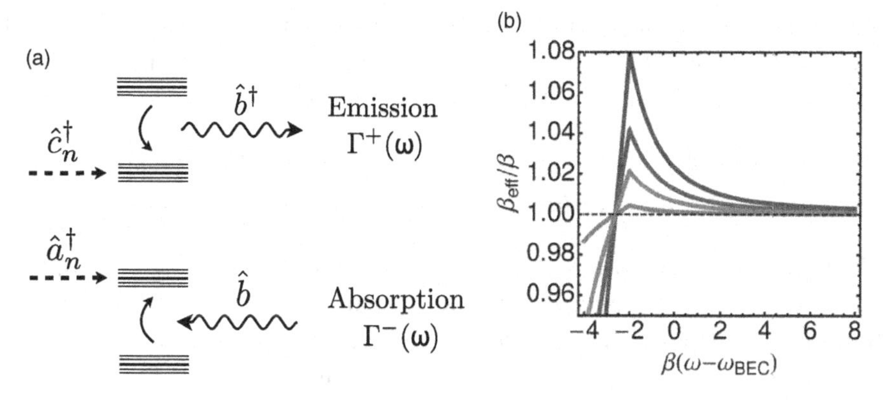

Figure 20.3 (a) Sketch of the energy levels under consideration. (b) Effective inverse temperatures β_{eff} for $\beta(\omega_{BEC} - \bar{\omega}) = 2$ and for different values of $\kappa/\gamma^- = 0.2, 0.1, 0.05, 0.01$ (from top to bottom).

the creation/destruction of rovibrational phonons. An analogous absorbing bath is used to model cavity losses due to the imperfect mirrors.

According to the usual quantum-Langevin theory [42], we solve the Heisenberg equations of motions for $\hat{a}_n, \hat{a}_n^\dagger$ and $\hat{c}_n, \hat{c}_n^\dagger$ and replace the formal solution into the Heisenberg equation for $\hat{b}$, which takes the simple form

$$\frac{\mathrm{d}\hat{b}}{\mathrm{d}t} = -\left(i\omega_c + \frac{\kappa}{2}\right)\hat{b} - \int_{-\infty}^{+\infty} \mathrm{d}t' \, [\Gamma^-(t-t') - \Gamma^+(t-t')]\hat{b}(t') + F, \quad (20.17)$$

where κ is the decay rate of the cavity photon, $F = F^\kappa + F^- + F^+$ is the total noise operator, and the memory kernels $\Gamma^\pm$ are defined as $\Gamma^\pm(t) = \theta(t)\int_{-\infty}^{+\infty} \mathrm{d}\omega \, \rho^\pm(\omega)\mathrm{e}^{\pm i\omega t}/(2\pi)$, in terms of the absorption and emission spectral functions $\rho^\pm(\omega) = 2\pi \sum_n |\eta_n^\pm|^2 \delta(\omega - \omega_n^\pm)$.

Given the form of the bath-system coupling (20.16) and of the memory kernels $\Gamma^\pm(t)$, absorption (viz. amplification) of a photon at ω_c is proportional to $\rho^-(\omega)$ (viz. $\rho^+(-\omega)$). The Kennard-Stepanov (KS) relation between the absorption and emission spectra from molecules in thermal contact with an environment at an inverse temperature β then translates into

$$\rho^+(-\omega) = \mathcal{C}\,\mathrm{e}^{-\beta\omega}\rho^-(\omega), \quad (20.18)$$

the constant $\mathcal{C}$ depending on the pumping conditions, e.g., the fraction of molecules in the ground and excited electronic states. In what follows, we assume all baths to be initially in their vacuum state, so as to model irreversible absorption and emission processes. In this regime, the structure factors read

$$S_{b^\dagger b}(-\omega) = \frac{\rho^+(-\omega)}{[\omega - \Sigma(\omega)]^2 + \Gamma_T^2(\omega)}, \quad S_{bb^\dagger}(\omega) = \frac{\kappa + \rho^-(\omega)}{[\omega - \Sigma(\omega)]^2 + \Gamma_T^2(\omega)}, \quad (20.19)$$

where $\Sigma(\omega) = \omega_c - \mathrm{Im}[\Gamma^-(\omega)] + \mathrm{Im}[\Gamma^+(\omega)]$ takes into account the Lamb-shift induced by the baths and the total relaxation rate is $\Gamma_T(\omega) = \kappa/2 + \mathrm{Re}[\Gamma^-(\omega)] - \mathrm{Re}[\Gamma^+(\omega)] > 0$. For concreteness, we choose the forms

$$\rho^-(\omega) = \gamma^-\big\{[n(\omega-\bar{\omega}) + 1]\,\theta(\omega-\bar{\omega}) + n(\bar{\omega}-\omega)\,\theta(\bar{\omega}-\omega)\big\}, \quad (20.20)$$

$$\rho^+(-\omega) = \gamma^+\big\{n(\omega-\bar{\omega})\,\theta(\omega-\bar{\omega}) + [n(\bar{\omega}-\omega) + 1]\,\theta(\bar{\omega}-\omega)\big\}, \quad (20.21)$$

modeling phonon-assisted absorption on a molecular line at $\bar{\omega}$, with $\gamma^\pm$ proportional to the molecular population in the ground and excited states and $n(\omega) = [\exp(\beta\omega) - 1]^{-1}$. It is straightforward to check that these forms for $\rho^\pm$ indeed satisfy the KS relation (20.18) with $\mathcal{C} = e^{\beta\bar{\omega}}\gamma^+/\gamma^-$. Dynamical stability of the condensate imposes the further condition $\kappa + \rho^-(\omega_{BEC}) = \rho^+(-\omega_{BEC})$, which translates into $\mathcal{C} = e^{\beta\omega_{BEC}}[1 + \kappa/\rho^-(\omega_{BEC})]$.

As discussed in Section 20.5.1, the frequencies appearing in the KMS condition (20.10) are measured from the chemical potential. Even with this rescaling, it can immediately be seen that the structure factors do not generally satisfy the KMS condition (20.10) signaling a nonequilibrium behavior. However, in the plots shown in Fig. 20.3b of the KMS effective inverse temperature

$$\beta_{\text{eff}}(\omega) = \frac{\log\left(\frac{\kappa+\rho^{-}(\omega)}{\rho^{+}(-\omega)}\right)}{\omega - \omega_{BEC}} = \beta + \frac{1}{\omega - \omega_{BEC}} \log\left[\frac{1+\kappa/\rho^{-}(\omega)}{1+\kappa/\rho^{-}(\omega_{BEC})}\right], \qquad (20.22)$$

one easily sees that an effective equilibrium at the inverse temperature β is recovered in the $\kappa \to 0$ limit where the KS condition (20.18) makes the KMS condition to be trivially fulfilled. Physically, if the repeated absorption and emission cycles by the molecules are much faster than cavity losses, the KS condition imposes a full thermal equilibrium condition in the photon gas.

20.6 Conclusions

After reviewing the most intriguing novel features of nonequilibrium BEC and laser operation as compared to textbook BEC, we have proposed and characterized a quantitative criterion to experimentally assess the equilibrium versus nonequilibrium nature of a condensate. This criterion has been applied to a strongly nonequilibrium model of condensation inspired by the semiclassical theory of laser and to a simple non-Markovian model of the photon BEC: provided photons undergo repeated absorption-emission cycles before being lost, the photon gas can inherit the thermal condition of the dye molecules. With respect to static properties, such as the momentum distribution, considered so far in experiments, our criterion based on fluctuation-dissipation relations imposes stringent conditions also on the dynamical properties of the gas: its experimental implementation appears feasible with state-of-the-art technology and would give conclusive evidence of thermal equilibrium in the gas.

References

[1] Graham, R., and Haken, H. 1970. Laserlight – first example of a second-order phase transition far away from thermal equilibrium. *Z. Phys. A*, **237**, 31–46.

[2] DeGiorgio, V., and Scully, M. O. 1970. Analogy between the laser threshold region and a second-order phase transition. *Phys. Rev. A*, **2**, 1170–1177.

[3] Chang-Hasnain, C. J. 2000. Tunable VCSEL. *IEEE J. Sel. Top. Quantum Electron.*, **6**(6), 978–987.

[4] Lundeberg, L. D. A., Lousberg, G. P., Boiko, D. L., and Kapon, E. 2007. Spatial coherence measurements in arrays of coupled vertical cavity surface emitting lasers. *Appl. Phys. Lett.*, **90**, 021103.

[5] Huang, K. 1987. *Statistical Mechanics*. Wiley.

[6] Pitaevskii, L. P., and Stringari, S. 2016. *Bose-Einstein Condensation and Superfluidity*. Oxford University Press.

[7] Penrose, O., and Onsager, L. 1956. Bose-Einstein condensation and liquid helium. *Phys. Rev.*, **104**, 576–584.

[8] Kasprzak, J., Richard, M., Kundermann, S., Baas, A., Jeambrun, P., Keeling, J. M. J., Marchetti, F. M., Szymanska, M. H., Andre, R., Staehli, J. L., Savona, V., Littlewood, P. B., Deveaud, B., and Dang, L. S. 2006. Bose-Einstein condensation of exciton polaritons. *Nature*, **443**, 409–414.

[9] Demokritov, S. O., Demidov, V. E., Dzyapko, O., Melkov, G. A., Serga, A. A., Hillebrands, B., and Slavin, A. N. 2006. Bose–Einstein condensation of quasi-equilibrium magnons at room temperature under pumping. *Nature*, **443**, 430–433.

[10] Klaers, J., Schmitt, J., Vewinger, F., and Weitz, M. 2010. Bose-Einstein condensation of photons in an optical microcavity. *Nature*, **468**, 545–548.

[11] Gunton, J. D., and Buckingham, M. J. 1968. Condensation of the ideal Bose gas as a cooperative transition. *Phys. Rev.*, **166**, 152–158.

[12] Castin, Y. 2001. Bose-Einstein condensates in atomic gases: simple theoretical results. Pages 1–136 of: *Coherent Atomic Matter Waves*. Springer.

[13] Castin, Y., and Dalibard, J. 1997. Relative phase of two Bose-Einstein condensates. *Phys. Rev. A*, **55**, 4330–4337.

[14] Snoke, D. W., and Girvin, S. M. 2013. Dynamics of phase coherence onset in Bose condensates of photons by incoherent phonon emission. *Journal of Low Temperature Physics*, **171**, 1–12.

[15] Lewenstein, M., and You, L. 1996. Quantum phase diffusion of a Bose-Einstein condensate. *Phys. Rev. Lett.*, **77**, 3489–3493.

[16] Carusotto, I., and Ciuti, C. 2013. Quantum fluids of light. *Rev. Mod. Phys.*, **85**, 299–366.

[17] Richard, M., Kasprzak, J., Romestain, R., André, R., and Dang, L. S. 2005. Spontaneous coherent phase transition of polaritons in CdTe microcavities. *Phys. Rev. Lett.*, **94**, 187401.

[18] Richard, M., Kasprzak, J., André, R., Romestain, R., Dang, L. S., Malpuech, G., and Kavokin, A. 2005. Experimental evidence for nonequilibrium Bose condensation of exciton polaritons. *Phys. Rev. B*, **72**, 201301.

[19] Wouters, M., Carusotto, I., and Ciuti, C. 2008. Spatial and spectral shape of inhomogeneous nonequilibrium exciton-polariton condensates. *Phys. Rev. B*, **77**, 115340.

[20] Wertz, E., Ferrier, L., Solnyshkov, D. D., Johne, R., Sanvitto, D., Lemaître, A., Sagnes, I., Grousson, R., Kavokin, A. V., Senellart, P., Malpuech, G., and Bloch, J. 2010. Spontaneous formation and optical manipulation of extended polariton condensates. *Nat. Phys.*, **6**, 860–864.

[21] Wouters, M., and Carusotto, I. 2006. Absence of long-range coherence in the parametric emission of photonic wires. *Phys. Rev. B*, **74**, 245316.

[22] Wouters, M., and Carusotto, I. 2007. Goldstone mode of optical parametric oscillators in planar semiconductor microcavities in the strong-coupling regime. *Phys. Rev. A*, **76**, 043807.

[23] Szymańska, M. H., Keeling, J., and Littlewood, P. B. 2006. Nonequilibrium quantum condensation in an incoherently pumped dissipative system. *Phys. Rev. Lett.*, **96**, 230602.

[24] Wouters, M., and Carusotto, I. 2007. Excitations in a nonequilibrium Bose-Einstein condensate of exciton polaritons. *Phys. Rev. Lett.*, **99**, 140402.

[25] Reatto, L., and Chester, G. V. 1967. Phonons and the properties of a Bose system. *Phys. Rev.*, **155**, 88–100.

[26] Popov, V. N. 1972. On the theory of the superfluidity of two- and one-dimensional Bose systems. *Theor. Math. Phys.*, **11**, 565–573.
[27] Sieberer, L. M., Huber, S. D., Altman, E., and Diehl, S. 2013. Dynamical critical phenomena in driven-dissipative systems. *Phys. Rev. Lett.*, **110**, 195301.
[28] Altman, E., Sieberer, L. M., Chen, L., Diehl, S., and Toner, J. 2015. Two-dimensional superfluidity of exciton polaritons requires strong anisotropy. *Phys. Rev. X*, **5**, 011017.
[29] Bajoni, D., Senellart, P., Lemaître, A., and Bloch, J. 2007. Photon lasing in GaAs microcavity: similarities with a polariton condensate. *Phys. Rev. B*, **76**, 201305.
[30] Fischer, B., and Weill, R. 2012. When does single-mode lasing become a condensation phenomenon? *Opt. Express*, **20**, 26704–26713.
[31] Chiocchetta, A., and Carusotto, I. 2014. Quantum Langevin model for nonequilibrium condensation. *Phys. Rev. A*, **90**, 023633.
[32] Kubo, R. 1966. The fluctuation-dissipation theorem. *Rep. Prog. Phys.*, **29**, 255.
[33] Mahan, G. D. 2000. *Many-Particle Physics*. Springer Science & Business Media.
[34] Kubo, R. 1957. Statistical-mechanical theory of irreversible processes. I. General theory and simple applications to magnetic and conduction problems. *J. Phys. Soc. Jpn.*, **12**, 570–586.
[35] Martin, P. C., and Schwinger, J. 1959. Theory of many-particle systems. I. *Phys. Rev.*, **115**, 1342–1373.
[36] Cugliandolo, L. F. 2011. The effective temperature. *J. Phys. A: Math. Theor.*, **44**, 483001.
[37] Foini, L., Cugliandolo, L. F., and Gambassi, A. 2011. Fluctuation-dissipation relations and critical quenches in the transverse field Ising chain. *Phys. Rev. B*, **84**, 212404.
[38] Foini, L., Cugliandolo, L. F., and Gambassi, A. 2012. Dynamic correlations, fluctuation-dissipation relations, and effective temperatures after a quantum quench of the transverse field Ising chain. *J. Stat. Mech.*, **2012**, P09011.
[39] Walls, D. F., and Milburn, G. J. 2007. *Quantum Optics*. Springer Science & Business Media.
[40] Talkner, P. 1986. The failure of the quantum regression hypothesis. *Annals of Physics*, **167**, 390–436.
[41] Ford, G. W., and O'Connell, R. F. 1996. There is no quantum regression theorem. *Phys. Rev. Lett.*, **77**, 798–801.
[42] Gardiner, C., and Zoller, P. 2004. *Quantum Noise: A Handbook of Markovian and Non-Markovian Quantum Stochastic Methods with Applications to Quantum Optics*. Vol. 56. Springer Science & Business Media.
[43] Kirton, P., and Keeling, J. 2013. Nonequilibrium model of photon condensation. *Phys. Rev. Lett.*, **111**, 100404.
[44] Kirton, P., and Keeling, J. 2015. Thermalization and breakdown of thermalization in photon condensates. *Phys. Rev. A*, **91**, 033826.

21

Vortices in Resonant Polariton Condensates in Semiconductor Microcavities

DMITRY N. KRIZHANOVSKII, KURUMURTHY GUDA, MAKSYM SICH, MAURICE S. SKOLNICK

Department of Physics and Astronomy, University of Sheffield, UK

L. DOMINICI, AND D. SANVITTO

Istituto di Nanotecnologia (NANOTEC) – Consiglio Nazionale delle Ricerche (CNR), Lecce, Italy

We review studies of quantised vortices in polariton condensates in three main configurations, namely off-resonant, optical parametric oscillator (OPO), and direct resonant excitation. A brief introduction is given on the typical interferometric detection and spinor nature of polaritons. Specific experiments are described in detail, highlighting the dynamics of spontaneous and imprinted vortices in OPO and resonant polariton condensate and the role of nonlinearities. Time-resolved measurements reveal metastable rotational polariton flow indicating superfluidlike behaviour. In the case of a ring-shaped pump, a transition from the vortex state with angular momentum $M = 1$ to $M = 2$ is observed due to interplay between gain and polariton–polariton interactions. Finally, we demonstrate the direct pulsed initialization of a condensate carrying a half-vortex, and the spontaneous creation of vortices when starting from ring-shaped condensates. These are created in vortex–antivortex pairs due to the interplay between breaking of $y \mapsto -y$ reflection symmetry in the system and conservation of orbital angular momentum.

21.1 Introduction

Coherent macroscopically occupied states (condensates) attract major interest since they exhibit a number of interesting phenomena, such as superfluidity, vortices, and solitons. As well as coherent condensates of liquid helium and dilute atomic gases, condensates of semiconductor microcavity polaritons constitute an appealing testbed allowing easy optical manipulation and direct imaging of the condensate due to the photonic part of their hybrid light-matter wavefunction [1, 2]. In contrast to the atomic Bose-Einstein condensate (BEC), the polariton system is a nonequilibrium system [3], where a dynamic balance between loss and gain from an external pump field is formed. Furthermore, polaritons also have strong nonlinearities while their spinor nature allows for half-integer vorticity.

In this chapter, we review the main observations of polariton vortices and also describe some specific experiments in more detail. In the first paragraphs, we give an introduction to the quantised vortices in condensates and how they can be recognised using interferometric detection, and to the spinor nature of polaritons. We briefly discuss spontaneous generation of vortices under nonresonant excitation. Then we consider nonequilibrium polariton condensates arising from polariton–polariton parametric scattering into macroscopically occupied signal at $k_s \sim 0$ and idler states at $k_i \sim 2k_p$ for resonant pump excitation into the lower polariton branch at $k = k_p$. This regime corresponds to the so-called optical parametric oscillator (OPO; see Fig. 22.1), which has attracted significant interest [4, 5]. Importantly, the phase of the signal is not imprinted by the pump but forms spontaneously due to $U1$ symmetry breaking as in the case of incoherently pumped condensates [6]. The OPO system consists of three macroscopically occupied coherently coupled signal, pump, and idler states and hence is more complicated than a single resonant fluid or nonresonantly pumped condensate. In this chapter, we mainly focus on the properties of vortices in the signal condensate and how these are affected by the pump and idler. Finally, a description of some main experiments in the case of resonant excitation at $k = 0$ is presented, comprising the case of a direct imprint of a polariton condensate with the topological charge of a full- or half-vortex.

The phenomenology of spontaneous pattern formation is much richer in nonequilibrium than in equilibrium systems [7], with vortices being a typical example. Vortices are topological defects occurring in optics and condensed matter as well as in particle physics and cosmology. In vortices in quantum fluids, the phase of a field winds around a vortex core where the density is almost zero with a change for a complete loop being an integer multiple of 2π. Therefore, a vortex can be described by a state with quantised orbital angular momentum (OAM), also called phase winding. In optics, this degree of freedom can be further used in photonic quantum information applications, and also polariton vortices have been proposed for information processing [8].

Atomic Bose-Einstein condensates [9], liquid helium [10], and semiconductor lasers [11] (vertical cavity surface emitting lasers [VCSELs]) exhibit formation of vortices (see Chapters 10 and 17 for reviews of various vortex effects). In equilibrium systems such as a BEC of cold atoms or liquid helium, the existence of stable vortex is a manifestation of superfluidity. In nonequilibrium systems, like nonresonant incoherently pumped polariton condensates in CdTe microcavities, spontaneous vortices may form due to the interplay between polariton flows and natural photonic potential disorder, e.g., localised defects [12]. Half-vortices, where only one spin (or circular polarisation, which is associated to spin angular momentum [SAM]) component of the condensate exhibits a vortex, can be also observed [13]. It is important that the excitonic component in the polariton wavefunction leads to very strong polariton–polariton interactions and consequently to a giant $\chi^{(3)}$ optical nonlinearity, which affects the vortex formation as discussed in this volume. These very strong interactions also enable observation of effects such as bright solitons [14, 15] or polariton superfluidity at very moderate particle densities. Vortices can be generated in polariton superfluids flowing against an obstacle as a result of oblique dark solitons breaking into streets or trains of vortex–antivortex pair at subsonic critical velocities of the quantum fluid [16, 17].

The aim of this chapter is to describe several experimental results showing the effect of interactions and of the nonequilibrium polariton nature on the formation and dynamics of polariton vortices in OPO and resonant condensates. First, we demonstrate that vortices in a polariton condensate can be created using a weak external imprinting beam (which we term the *im*-beam). The vortex core radius is determined by polariton–polariton interactions leading to a decrease of the vortex radius with increasing particle density [18]. It is shown that OAM is conserved during pair polariton–polariton scattering and hence the imprinting of vortices on the signal state is accompanied by formation of an antivortex in the corresponding idler state [18]. The measurements of temporal evolution of an imprinted vortex reveal metastable polariton flow consistent with a superfluid-like behaviour of the interacting polariton system. Second, we show that vortices may arise spontaneously in an artificially created potential landscape [19]. The optically induced potential can be created by a ring-shaped pump beam, carrying zero OAM. The formation of a stable vortex with OAM $M = 1$ is observed due to breaking of the $y \mapsto -y$ reflection symmetry of the system. At higher excitation density, the interplay between the kinetic energy of the vortex due to localisation, the potential interaction energy within the condensate, and the spatial distribution of the gain results in spontaneous formation of a vortex with OAM $M = 2$. Finally, we investigated vortices in an artificial condensate injected directly at $k = 0$ by a pulsed pump beam [19]. One powerful tool of the resonant scheme is that it allows one to create the condensate directly with an initial full- or half-vortex topological charge and to follow its evolution in different regimes. Also, it is possible to start with a zero-winding state but with a space inhomogeneous profile, in order to induce and observe the formation of a vortex–antivortex pair, dictated by the conservation of OAM. All the experiments were performed at 4 K. The detailed description of the GaAs-based microcavity samples and experimental techniques can be found in Refs. [18, 19].

21.2 Signature of Vortex in Interference and Phase Maps

One useful characteristic of polariton condensates is their hybrid photonic-electronic nature, which continuously emits photons coherently with the condensate wavefunction. In this sense, not only imaging of the fluid is possible, but also of the phase of their complex wavefunction. This can be retrieved when letting the emission from the fluid interfere with a coherent wave: two major cases are possible. First, the emission from a fluid which is created by nonresonant excitation (and hence becomes noncoherent with the excitation beam) has to be allowed to auto-interfere with itself, either with a centre- or axis-symmetric copy of the emission image or with an expanded and homogeneous portion of the same field. This also allows time-resolved measures of the emission provided the coherence time is long enough, while the time resolution is limited by the emission duration itself. Second, the emission from a resonantly excited polariton fluid can be allowed to interfere instead with the original exciting pulse, which is possible and indeed typical only for the resonant excitation scheme and not for the off-resonant one. Here the time resolution is given by the pulse width. In both cases, signature of a vortex is seen by typical patterns in the interferograms: a forklike dislocation with an excess number of fringes on one side equal to the quantised charge of the vortex M. Here we show in Fig. 21.1 an example case of the vortex recognition scheme for a resonant OPO vortex, where panel (a) is the intensity map and (b) the interferogram (extracted from [20]).

Typically such images are used, together, to report on the observation, formation, and motion of quantised vortices. A further possible step is to use digital

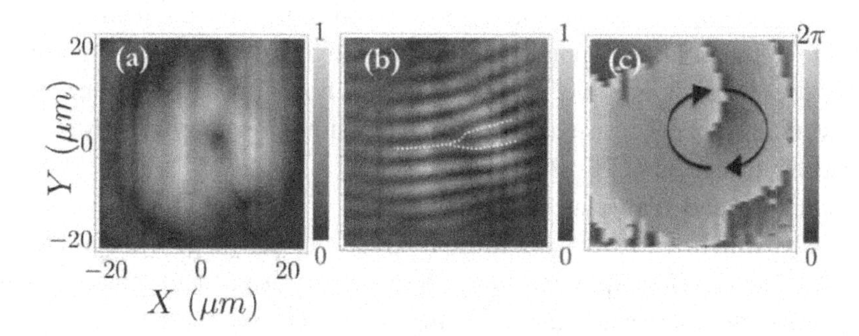

Figure 21.1 (a) Space distribution of the intensity associated to the vortex state. When overlapped to a homogeneous wavefront of a coherent beam, the resulting interferogram shows a forklike pattern as that shown in (b). The associated phase map retrieved by means of Fourier filtering is reported in (c), displaying a phase singularity with an $M = +1$ winding of the phase around the core. Adapted with permission from Antón, C., et al. (2012), Role of supercurrents on vortices formation in polariton condensates, *Opt. Express*, **20**, 16366 [20]. Copyright (2012) by the Optical Society of America.

holographic technique, i.e., Fourier analysis, to retrieve maps of the phase distribution of the fluid, as in panel (c). These show for each excess fringe in the original interferograms a phase singularity, which is the very centre of the vortex around which the phase winds M times. Note that a fork-like interference slit or a thickness varying plate can also both be used for the shaping of a photonic Gaussian LG_{00} beam into an LG_{0M} Laguerre Gauss vortex beam, to be used for direct resonant imprint. Alternatively, space light modulators (SLM; cf. Ref. [21]) or q-plate devices (patterned liquid crystal phase-retarder; see [22, 23, 24]) can be used with more versatility to such an extent. For a more detailed explanation of time-resolved digital holography imaging in polariton fluids, see [20, 25].

21.3 Polariton Spinor Nature and Half-Vortices

The polarisation degree of freedom within photons coupled to the spin of excitons makes the polariton condensate a spinor quantum fluid, analogous to a two-component atomic BEC. This property allows one to obtain a rich variety of topological excitations featuring different polarisation patterns, as described in [13, 26, 27, 28]. All these states are analogous to purely photonic vectorial fields, apart from three main features typical for polaritons: strong nonlinearity, sample disorder landscape, and symmetry breaking terms such as transverse electric (TE)–transverse magnetic (TM) or x-y splitting. Two main interesting states are possible when looking at a full basis of, say, right and left (R, L) polarisation components. In the first case, usually referred to as a full-vortex (FV), a vortex with the same winding exists in both polarisations. This results in a fixed linear polarisation with an integer winding of the phase. The second case is a vortex in only one spin population coupled to a Gaussian or homogeneous profile in the opposite spin population, and is usually referred to as a half-vortex (HV). Such a term derives from the half-integer winding of both phase and linear polarization direction which is seen in an HV state, as illustrated in Fig. 21.2.

Here the left and right panels show the cases of lemon and star half-vortices, respectively, which are associated to homologue or heterologue OAM and SAM directions. Nonresonant spontaneous generation or resonant imprinting of half-vortices are both possible: see the following and the last sections.

21.4 Nonresonant Excitation and Spontaneous Vortices

Several different observations of spontaneous vortices have been made in nonresonantly created polariton condensates. Initially, the first detection of spontaneous formation of pinned quantised vortices in the Bose-condensed phase of a polariton fluid allowed parallels to be drawn between polariton systems and conventional

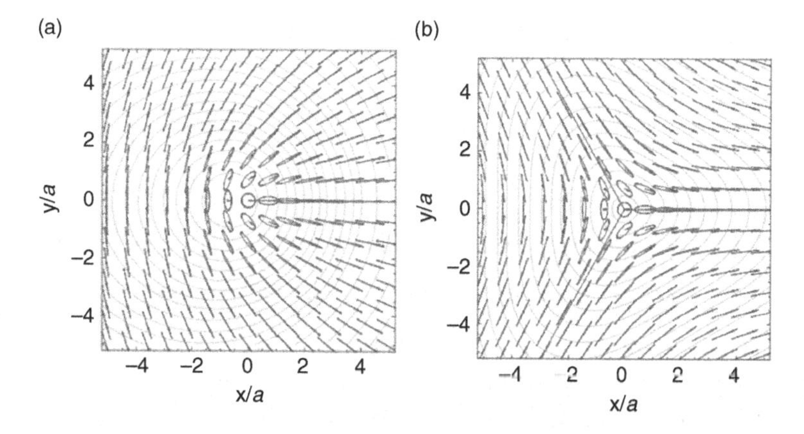

Figure 21.2 The two kind of half-vortices, lemon (a) and star (b), with their associated polarisation patterns. At large distance from the centre, the polarisation is linear and its direction makes a half-rotation when one moves around a space circumference. Reprinted with permission from Toledo-Solano, M., et al. (2014), Warping and interactions of vortices in exciton-polariton condensates, *Phys. Rev. B*, **89**, 035308 [28]. Copyright (2014) by the American Physical Society.

BEC [12]. Subsequently, a coexistence of half-quantum vortices (HV) and single-quantum vortices (FV) was shown in microcavity polaritons, as expected for a spinor quantum fluid [13, 29]. In these cases, Lagoudakis et al. showed the role of the landscape disorder in pinning one or the other of the two elementary states. This role was later confirmed by the dynamical imaging of the dissociation of a full-vortex into a pair of half-vortices [26], which also put into evidence the creation of a different combination of elementary charges giving rise to patterns such as that of a hyperbolic vortex [27].

Recently, starting from a ring-shaped potential, the formation of a different kind of more generic half-vortex with the polarisation winding spanning both linear and circular states [30] was observed. Inhomogeneous spot profiles can be used to induce formation of a single vortex–antivortex pair [31] or of a lattice consisting of a few vortices [32], due to an interplay of phase fluctuations and disorder pinning. Other nonhomogeneous optical pump beams such as multiple spots can generate a lattice of many vortices, due to the locking of propagating flows over a long area and their mutual interference [33]. A structured optical pump beam could also be used to induce the breaking of chiral symmetry, inducing transfer of orbital angular momentum and creation of a single exciton–polariton vortex (chiral polaritonic lenses; see [34]). In either case, the role of nonlinearities is fundamental, in both driving the polaritonic flows out of the originally excited area and in locking their mutual phase.

The effect of spontaneous free-single or bound-pair vortices on the long-range-order coherence of a superfluid and their link to the Berezinskii-Kosterlitz-Thouless

(BKT) transition are discussed in [35]. Spontaneous vortices under off-resonant pumping have been observed also in organic polariton condensates in a disordered landscape. Here fluctuating formation of vortices was highlighted thanks to single- or few-shot interferograms [36, 37].

21.5 OPO Condensates with Imprinted and Spontaneous Vortices

21.5.1 Vortices Imprinted on OPO Condensate

In this section, we describe a method to excite a vortex in an OPO system (a schematic of which is shown in Fig. 21.3a). Fig. 21.3b shows the real space image of the OPO signal recorded at pump power three times above the condensation

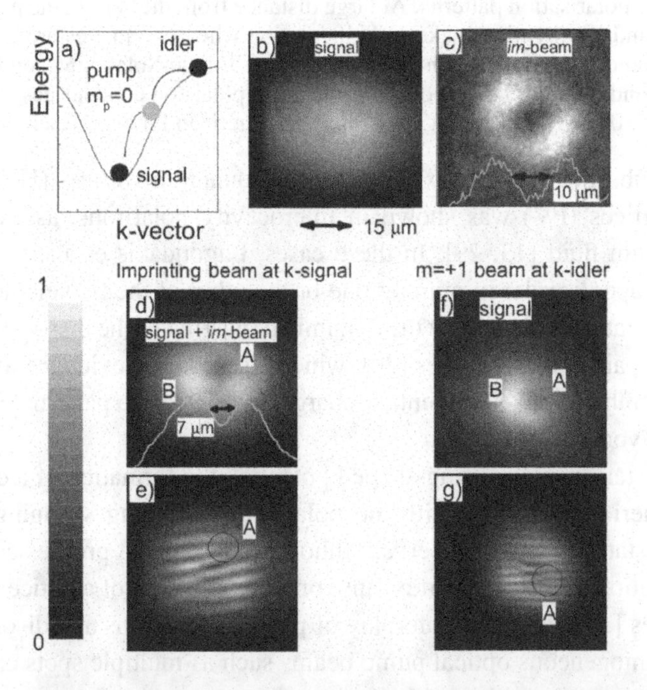

Figure 21.3 (a) Schematic diagram of OPO. (b) Real space image of the signal with no imprinting beam. (c) Image of Gauss-Laguerre imprinting beam (*im*-beam). (d) Signal with weak *im*-beam of (c), showing an imprinted vortex labelled A. (e) Interferogram revealing the 2π phase variation around the vortex of (d). (f) Vortex in signal, labelled A, created by excitation at the idler position with Gauss-Laguerre beam. (g) Interferogram revealing the vortex with OAM $M_s = -1$ created in (f). Plots (c) and (d) also display (noisy grey lines) cross-sections through the vortex cores, with the sizes of the cores (FWHM) as indicated. Adapted with permission from Krizhanovskii, D. N., et al. (2010), Effect of interactions on vortices in a nonequilibrium polariton condensate, *Phys. Rev. Lett.*, **104**, 126402 [18]. Copyright (2010) by the American Physical Society.

threshold, P_{th}. The uniform spatial distribution of the emission indicates the high quality of the studied microcavity sample with very weak disorder [38]. In order to create a vortex state in the OPO signal emission, a weak continuous wave *im*-beam carrying OAM $M = +1$ is introduced (Fig. 21.3c). The power of the *im*-beam is $\sim$ 40 times less than that of the signal, and its frequency is tuned in resonance with the signal frequency. As shown in Fig. 21.3d under the application of the *im*-beam, the spatial distribution of the signal is modified with a resultant well-defined dip of diameter (FWHM) $\approx 7\,\mu$m, (radius $\approx 3.5\,\mu$m (HWHM)) labelled A. The dip arises from a vortex imprinted on the condensate profile. This is demonstrated by measuring the spatial phase variation of the OPO condensate [12]. For this purpose, the interference pattern between the signal image and the image inverted around a central point of symmetry was recorded, so that the region of a vortex core (labelled A) interferes with the region B (see Fig. 21.3d) where the phase is nearly constant. The resultant interference pattern shown in Fig. 21.3e reveals the two forklike dislocations, demonstrating formation of a single vortex in OPO signal state with OAM $M_s = 1$ around region A.

Since the *im*-beam is very weak, it does not modify substantially the OPO signal population. However, it changes spatial distribution of the phase of the OPO emission. Since the phase of the signal is undetermined in the OPO, the few polaritons injected by the *im*-beam lock the signal phase to their own. That is why the process above is described as 'imprinting'. We note that experimentally, a minimum ratio of the *im*-beam to signal power density of about $\sim 1/45$ is required to imprint a vortex in the OPO condensate. Decoherence processes occurring in polariton condensates due to fluctuations probably determine this limit [39, 40].

21.5.2 Temporal Dynamics of a Vortex Imprinted on OPO Polariton Condensate

In another experiment [41], a picosecond laser pulse carrying OAM was employed to imprint a vortex state into the OPO signal steady state driven by a cw laser pump. The vortex behaviour with time was then tracked using a streak camera. The time evolution of an $M_s = 1$ vortex excited on top of the OPO signal and its interference pattern, which characterises unequivocally the vortex state, is shown in Fig. 21.4. The external pulsed *im*-beam was set to have an intensity smaller than the signal state. In the sequence of Fig. 21.4, it is possible to follow the vortex dynamics after the probe has generated it within the first few picoseconds. The vorticity, which was imprinted into the steady state of the signal, remains in the condensate for a time which is at least one order of magnitude longer than the polariton lifetime and is only limited by the time it takes to get out from the pump spot area. Such observation is clear evidence that the quantum fluid of the polariton is behaving as

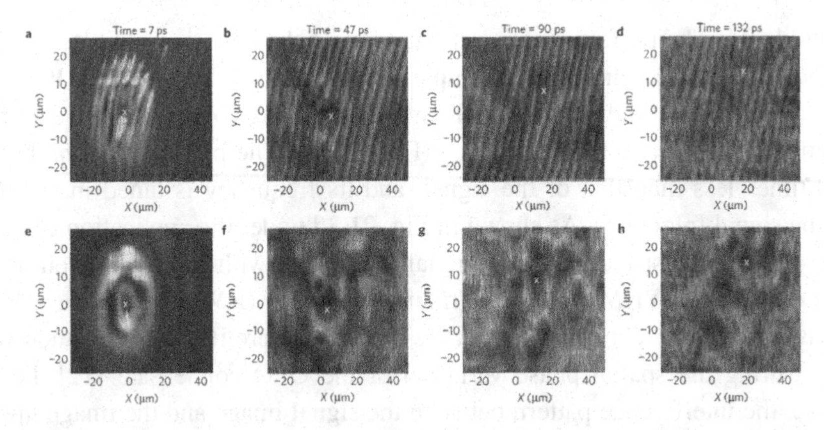

Figure 21.4 Dynamics of a vortex of angular momentum $M_s = 1$ imprinted into the signal steady state of a cw OPO signal. Four snapshots of the evolution of the vortex core at 7 ps, 47 ps, 90 ps, and 132 ps after the probe pulse has arrived are shown. In (a)–(d) the vortex core can be identified by the fork dislocation, which is formed when the signal is made to interfere with a constant wavefront taken from a spatial expanded part of the emission. In (e)–(h), the same images are shown in the intensity map. Reprinted with permission from Sanvitto, D., et al. (2010), Persistent currents and quantized vortices in a polariton superfluid, *Nat. Phys.*, **6**, 527533 [41]. Copyright (2010) by the Nature Publishing Group.

a superfluid, showing frictionless rotation for times longer than the coherence time of the state: fringes, which show the typical forklike dislocation, disappear while the vortex core is still detectable in the intensity maps.

The formation of metastable rotating currents, in polariton fluids, showing the absence of scattering with defects (always present in microcavity samples), is consistent with the theoretical model of excitations in non-equilibrium polariton condensates [42]. This model predicts that perturbations created by defects are not able to propagate over long distances due to finite damping rates of the excitations. As a result, it is much harder to break the topological stability of the supercurrents.

The sample inhomogeneities, however, still play a significant role and are very useful in this experiment since they provide a deterministic path for the injected vortex that would otherwise undergo a random drift out of the injection spot. Such effect allows for the detection of the vortex dynamics, which contrarily would be washed out by averaging over many experimental realisations.

21.5.3 The Effect of Polariton–Polariton Interactions on Vortex Size

In Ref. [18], the effect of polariton–polariton interactions on the vortex size has been investigated. We note that orbital angular momentum of light has been investigated in parametric down-conversion experiments [43], where the optical nonlinearity has a $\chi^{(2)}$ form. By contrast, polariton–polariton interactions lead to the $\chi^{(3)}$ nonlinearity, which gives rise to OPO processes as well as to the blueshifts of polariton modes. As a result of these interactions, the vortex size is determined by the healing length [44], a characteristic length scale within which a locally perturbed condensate wavefunction returns to its unperturbed value. In the case of a vortex, the expression for the healing length of the OPO condensate $\xi \approx M_s(2m_{\text{eff}}\kappa n_s)^{-\frac{1}{2}}$ can be obtained by equating a typical kinetic energy associated with a vortex in the condensate, $\approx M_s^2(2m_{\text{eff}}\xi^2)^{-1}$, to the interaction energy (blueshift), which is $\approx \kappa n_s$ (all energies are scaled to $\hbar^2$). Here m_{eff} is the polariton effective mass, κ is the strength of the nonlinearity, M_s is the value of OAM in the vortex, and n_s is the signal population density. Therefore, the healing length should scale with the number of particles as $n_s^{-0.5}$.

In order to investigate this population dependence the vortex radius in the OPO, the condensate was measured at different pump powers, which determine the density of particles in the signal. The real space images of the OPO condensate with an imprinted vortex are shown in Fig. 21.5a,b for pump powers P $\cong 1.5P_{th}$ and P $\cong 4.5P_{th}$, respectively. The vortex radius (healing length)

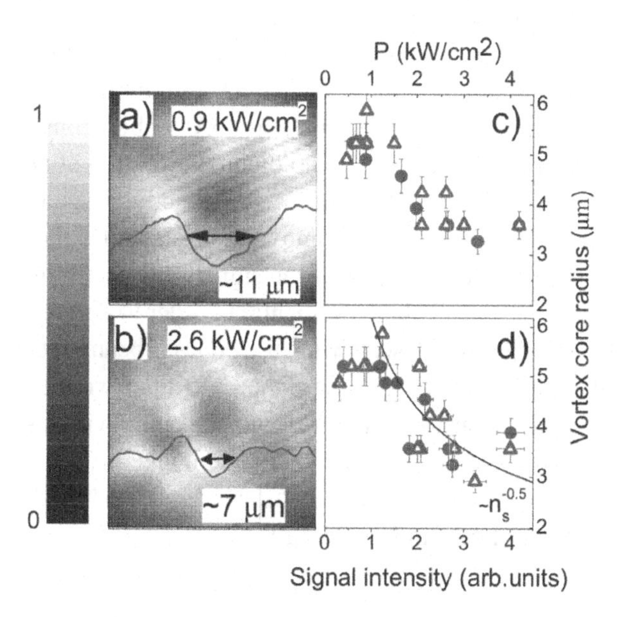

Figure 21.5 (a), (b) Images of a signal vortex at (a) 0.9 kWcm^{-2} and (b) 2.6 kWcm^{-2}. Experimental vortex core radius of the OPO signal as a function of pump power (c) and as a function of the signal intensity (d) for an *im*-beam core radius of $\sim$ 4 μm (circles) and $\sim$ 7 μm (triangles), respectively. Adapted with permission from Krizhanovskii, D. N., et al. (2010), Effect of interactions on vortices in a nonequilibrium polariton condensate, *Phys. Rev. Lett.*, **104**, 126402 [18]. Copyright (2010) by the American Physical Society.

apparently reduces with increasing signal particle density. The vortex core radius (Fig. 21.5c) decreases from $\sim$ 5.5 μm at threshold down to $\sim$ 3 μm at excitation density five times above threshold. Moreover, very similar variations of the vortex size with polariton density were observed for two different sizes of the *im*-beam, indicating that the profile of the imprinted vortex is an intrinsic property of the interacting polariton condensate. As is shown in Fig. 21.5d the dependence of the vortex radius versus signal intensity scales approximately as $n_s^{-\frac{1}{2}}$(solid line) in agreement with the expression for the healing length above. However, at small signal intensities the vortex radius does not diverge as expected from the $n_s^{-\frac{1}{2}}$, but is probably limited by the finite signal size.

Let's compare the magnitudes of the vortex radius in the polariton and atom systems: for polaritons, $m_{\rm eff} \sim 10^{-5} m_e$, and $\kappa n_s \sim 10^{-1}$ meV. By contrast, in a rubidium BEC with a density $n \sim 10^{14}$ cm^{-3}, the effective mass of an atom is much larger, $m_{\rm eff} \sim 10^5 m_e$, than that of a polariton, which is compensated by the much smaller interaction energy per single particle, $gn = 4\pi a_s n\hbar^2/M \sim 10^{-7}$ meV, where the atom scattering length $a_s \approx 5$ nm [44] and g is the atom interaction strength. As a result, the healing length for the polariton and the atom systems are of the order of 10 μm and 0.1 μm, respectively.

In the OPO system, the vortex is imprinted onto the signal condensate, which arises due to polariton–polariton scattering from the pump into the signal and idler. If the pump state is driven by a laser with zero OAM, then conservation of OAM in the polariton–polariton scattering should dictate formation of an antivortex state with OAM $M_i = -1$ in the idler. It is difficult to directly image the idler state and hence to demonstrate the creation of an antivortex state, since the idler intensity is typically $>$100 times weaker than that of the signal [40] due to the small photonic component of polariton states at high momenta. In order to demonstrate the conservation of OAM in the parametric scattering, a seed laser beam with OAM $M_i = +1$ was applied at the angle and energy where the OPO idler would appear. The seed laser stimulates the pair polariton scattering from the pump with resultant formation of the 'signal' forming at $k \sim 0$. The signal state imaged in Fig. 21.3f carries an antivortex state with OAM $M_s = -1$ as is further confirmed by the corresponding self-interference image shown in Fig. 21.3g [18].

21.5.4 Spontaneous Vortices with OAM $M_s = 1$ in an OPO Condensate Using Ring-Shaped Excitation

In this section, we show that spontaneous vortices may also appear in the OPO condensates subject to optically shaped external potential due to polariton flow from a high to low potential energy (and high to low gain) region. The OPO is

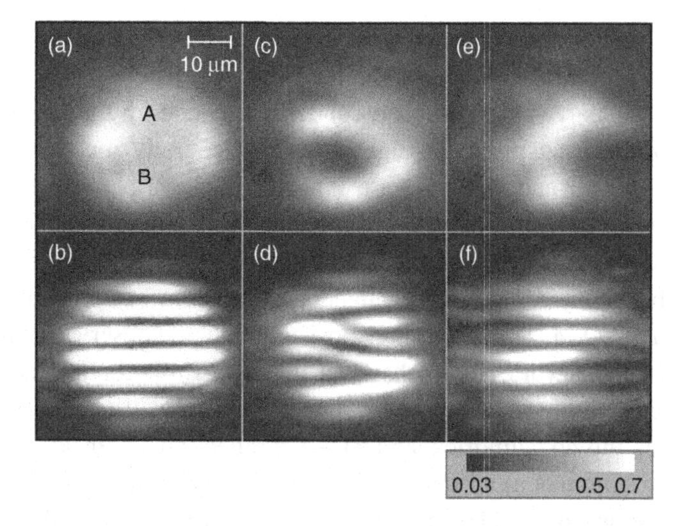

Figure 21.6 (a) Spatial image of a uniform OPO condensate. (b) Interference pattern obtained by interfering the image with its inverted image and overlapping positions A and B. No fork, i.e., no vortex, is observed. (c) Image of a ring-shaped condensate, induced by a ring-shaped pump beam. The corresponding interference image (d) reveals a single spontaneous vortex inside the ring. Panels (e) and (f) show the case where the mask was moved towards the edge of the beam and no vortex forms. Reprinted with permission from Guda, K., et al. (2013), Spontaneous vortices in optically shaped potential profiles in semiconductor microcavities, *Phys. Rev. B*, **87**, 081309 [19]. Copyright (2013) by the American Physical Society.

excited using the ring-shaped pump, which is prepared by placing an opaque mask in the way of the Gaussian excitation laser beam [19]. As a result of such excitation, the OPO condensate also has a ring-shaped profile, as shown in Fig. 21.6c. The corresponding self-interference pattern of the signal emission is shown in Fig. 21.6d, which exhibits clearly the forklike dislocations indicating vortex formation with OAM $M_s = 1$, as in the case of imprinted vortex.

Polariton–polariton interactions, which are stronger in the high- than in low-density region, form an optically induced potential trap due to density-induced polariton blueshifts. As a result, polaritons flow from the high- to low-density region in the center of the pump spot acquiring kinetic energy and rotary motion with the resulant formation of a vortex state. The radius of the optical potential trap coincides with the typical vortex size (healing length) $\xi \sim 5$ μm [18] described by OAM $M_s = +1$, as discussed above. In this case, the potential energy associated with repulsive interactions in the high-density region ($\sim 0.1-0.2$ meV) is similar to the kinetic energy associated with the vortex in the low-density region, making it more favourable for the condensate to form a stable single vortex. We note that an optical trap with well-defined boundaries is required for the spontaneous vortex to be formed: no vortices are observed if the opaque mask is moved close to the edge of the excitation spot, as shown in Fig. 21.6e,f.

It is not very clear why a polariton condensate excited by a ring-shaped laser favours the vortex with a particular sign of OAM. Generally, the vortex states with OAM of both $M_s = 1$ and $M_s = -1$ are expected to be formed if the OPO system has $y \mapsto -y$ reflection symmetry perpendicular to the pump wave-vector $(k_p, 0)$. An irregular doughnut shape of the pump spot can easily break this reflection symmetry, which may result in a spontaneous formation of a vortex state with a particular sign of OAM. The results are consistent with the observation in Ref. [34], where a spontaneous vortex was observed in a polariton condensate excited by the spatially shaped chiral pump beam, which also created an optically induced potential possessing no cylindrical and $y \mapsto -y$ reflection symmetry.

We note that Manni et al. [45] studied nonresonantly pumped condensates using ring-shaped excitation in CdTe microcavities and in contrast to the results here shown, observed a multilobe standing wave pattern, which is a coherent superposition of wavefunctions with OAMs $M_s = +1$ and $M_s = -1$. Their observation indicates the breaking of cylindrical but not of reflection symmetry. On the other hand, the reflection symmetry can be easily broken by some disorder potential across the sample, and spontaneous vortices only with OAM $M = 1$ can be observed in the same CdTe-based microcavities [12].

21.5.5 Observation of Spontaneous Vortex with OAM $M_s = 2$ in an OPO Condensate

In this section, we explore the effect of the density-dependent polariton interactions on the healing length in condensates using the ring-shaped pump beam excitation. Fig. 21.7a–f shows real space density images of the OPO condensate and the corresponding interference patterns for increasing pump powers from 73 to 180 mW. The images in Fig. 21.7a, d correspond to a power of $1.1P_{th} = 73$ mW, just above condensation threshold P_{th}. The formation of a vortex with OAM $M_s = 1$ can be observed. For increasing pump powers up to 150 mW, the condensate keeps its ringlike shape and the vortex survives maintaining its charge of $M_s = 1$ (Fig. 21.7b, e). However, a surprising change takes place for $P = 180$ mW (Fig. 21.7c, f), when, besides an apparent increase of the core diameter, the forklike dislocation develops an additional arm corresponding to the next quantised vortex state with OAM $M_s = 2$.

Fig. 21.7g shows intensity profiles across the center of the vortex for varying pump intensities. The vortex core diameter (FWHM of the intensity profiles) versus pump power is shown in Fig. 21.7h. For increasing pump powers in the range of 70 to 150 mW, the core diameter of the vortex changes slightly from about 13.5 μm

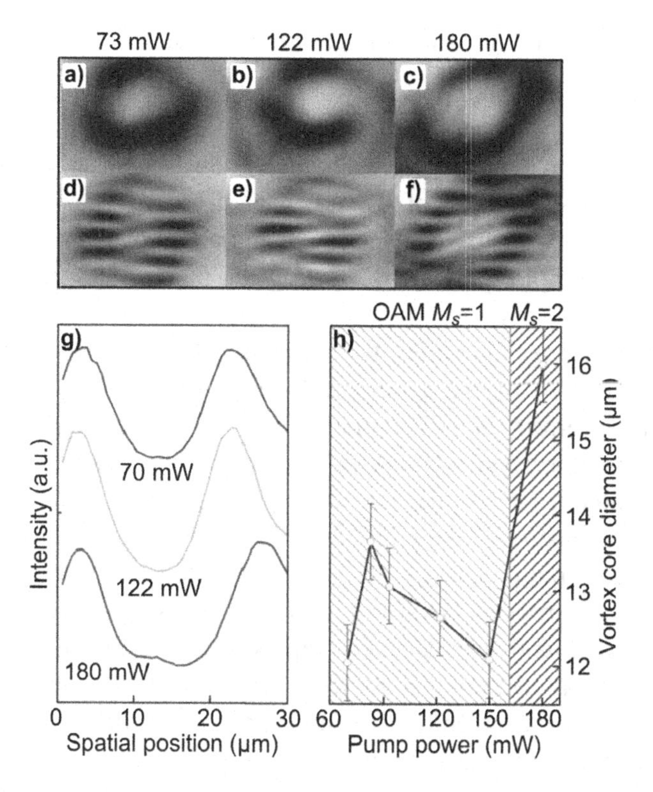

Figure 21.7 Spatial images of vortex with OAM $M_s = 1$ (a, b) and OAM $M_s = 2$ (c). (d)–(f) The interference patterns revealing vortex states with OAM $M_s = 1$ (d, e) and $M_s = 2$ (f). (g) Intensity profiles across the centre of the vortex for varying pump intensities. (h) The vortex core diameter versus pump power.

down to 12 μm. However, a prominent jump of the core diameter to 16 μm occurs at 180 mW pump power and coincides with the transition from the vortex state with OAM $M_s = 1$ to the state with OAM $M_s = 2$. With increasing polariton population density, the interaction energy of polaritons away from the vortex core also increases, which is, as discussed above, compensated by the increased kinetic energy of polaritons in the vortex core with resultant slight shrinking of the vortex size (healing length). However, in this experimental arrangement the OPO condensate is supported by the gain from the ring-shaped pump, which hence places a lower bound on the vortex core size (the pump density is almost zero in the middle of the pump spot). Therefore, it appears to be more favourable for the system to increase the kinetic energy by forming a higher-order quantised vortex state with $M_s = 2$. As a result of such a transformation, the healing length $\xi = M_s(2m_{\text{eff}}\kappa n_s)^{-\frac{1}{2}}$ also should increase abruptly in agreement with the experimental observation in Fig. 21.7.

21.6 Dynamics of Vortices in Resonant Polariton Condensates

In this section, we see that the versatility of the resonant excitation allows the direct imprint of a polariton condensate carrying a single-vortex, half-vortex, or multiple vortices, in other terms, directly created with an integer nonzero OAM. Naturally, spontaneous couple-generation of secondary vortices is also possible within the resonant scheme, for example, induced by using an inhomogeneous pump beam such as ring or multiple spots or by creating the fluid with a given finite-k and sending it against a defect.

In a fundamental configuration, the resonant scheme allows one to generate a condensate directly carrying a single full- or half-vortex state as initial conditions [22]. The wavefronts of the photonic excitation pulse are shaped by a patterned phase retarder, in order to allow Laguerre Gauss beams to be sent on the microcavity sample. A time-resolved imaging of the polariton fluid is obtained thanks to delay line interferometry of the emission with a coherent and spatially homogeneous pulse. Each spin population of a full-vortex, and only one spin component in the half-vortex, carries an OAM = +1. Every phase singularity is then digitally tracked in time and its dynamics can be represented as 2D+t vortex strings/lines. The primary singularities (original vortices) evolve due to an interplay of disorder landscape and nonlinear potential, which changes in time due to dissipation. An example case shows that the twin cores of a full-vortex can undergo an erratic movement at low exciting densities (or at long times) and eventually separate each from the other under the action of a symmetry-breaking term such as xy or TE-TM splitting. This is similar to what is observed in the case of spontaneously formed vortices under a nonresonant excitation scheme where disorder played a pivotal role [26]. At larger densities instead, the two main cores are seen to move together for longer times, while the nonlinear potential somehow screens out the disorder potential. Interspin interactions which are weak and attractive can supposedly help keep the cores together. Another manifestation of them is that the singularity in a half-vortex undergoes a spiralling trajectory around the density maximum of the opposite spin population, which acts as a Gaussian trap in the case of a metastable rotating vortex, as predicted in [46] and experimentally reported in the example of Fig. 21.8.

Instead, when a single polariton condensate is directly injected at $k = 0$ with an initial zero OAM, an inhomogeneous shape can be used to induce a fluid redistribution associated to radial currents. For example, in [19] the condensate was excited with a pulsed laser beam focussed into a large spot. A dip in the condensate background intensity was created by passing the beam through an opaque mask as above, and the temporal dynamics of vortex formation due to polariton flows have been measured in such an optically induced potential. The detailed experimental

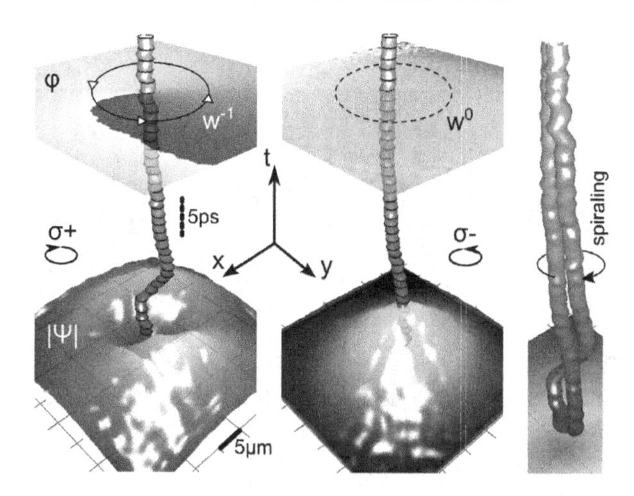

Figure 21.8 A half-vortex state is represented by means of its two spin components σ_+ and σ_-, on left and middle columns, respectively. The initial space distributions of the amplitude are reported in the bottom of the figure, while the phase maps appear in the top. The 2D+t strings represent the evolution of the phase singularity for the vortex component and of the maximum of the density for the Gaussian one, in a 40 ps time range. The right-most panel shows a spiralling of the vortex core around the centroid of the opposite spin population. Figure adapted from data previously reported in Ref. [22].

arrangements are very similar to that in Ref. [16]. The real spatial image of the injected condensate is shown in Fig. 21.9a with a dark region (encircled) created by the opaque mask. Fig. 21.9b–d shows the resultant interference patterns recorded as a function of time after the arrival of the excitation pulse. At time $t = 0$, the interference pattern indicates a nearly homogeneous phase distribution of the polariton field: no forklike dislocations and hence no vortices are observed in the system. At later times, the phase pattern becomes disturbed and then a vortex (V)–antivortex (AV) pair is clearly observed at about $t = 16$ ps (Fig. 21.9d). These vortices arise from polariton flow to the low-density region with small potential energy. Polaritons propagate at a typical speed of about ~ 1 μm/ps, which probably determines the time scale of V-AV formation. In this experimental arrangement, spontaneous vortices are always observed in V-AV pairs, but never as a single vortex, which is a consequence of the OAM conservation law. Since the polariton condensate was initially prepared in a state with OAM of $M = 0$ at $t = 0$, any spontaneous vortex with $M = 1$ arising from polariton flow should be accompanied by an antivortex [16, 17] with $M = -1$. In the OPO case studied in previous sections, only single vortex states are observed in the signal, which is consistent with the fact that an antivortex state is created in the idler. Both vortex and antivortex states

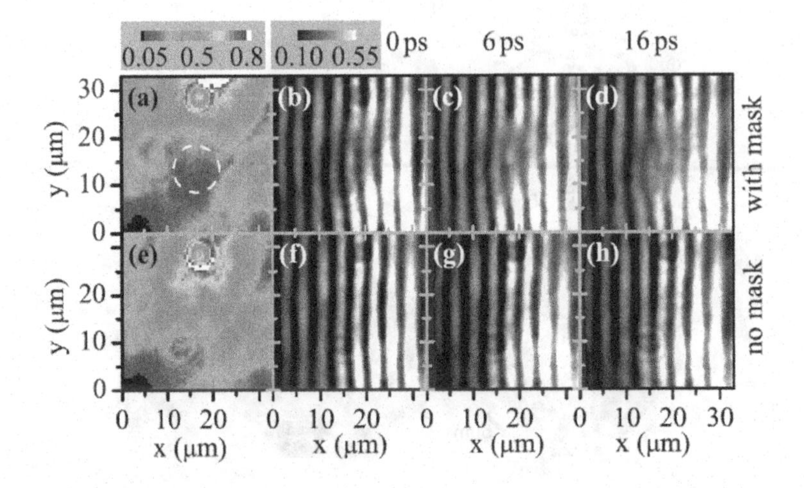

Figure 21.9 Images of the transmitted beam with and without a 10 micron mask in the pump beam path are shown in (a) and (e), respectively. A potential trap due to a lower polariton density is formed when the mask is present in the encircled area. (b)–(d) Interference images obtained with a reference pulse at $t = 0$, 6, and 16 ps showing the dynamics of the formation of a vortex–antivortex pair. (f)–(h) Without the potential trap, no vortex pair is formed. Reprinted with permission from Guda, K., et al. (2013), Spontaneous vortices in optically shaped potential profiles in semiconductor microcavities, *Phys. Rev. B*, **87**, 081309 [19]. Copyright (2013) by the American Physical Society.

are possible in the OPO signal despite the presence of the idler, although this is not observed in our experiments. For example, a moving A-AV pair was observed in a spatially extended OPO signal perturbed with a probe pulse beam carrying an OAM of $M_s = 1$ [47].

In a different configuration, when the fluid is resonantly injected with a cw finite-k and sent to hit a localised defect (e.g., photonic pointlike defect), the hydrodynamics regime changes upon density passing from a superfluid state without turbulence to the formation of oblique dark solitons and vortex streets in the wake of the potential barrier [17]. The similar emission of vortex–antivortex pairs from a localised defect could be observed also under pulsed pumping, by ultrafast imaging of their transient dynamics [16]. Also, lattices of vortices induced by resonant multiple-spot interference were theoretically studied, together with the stabilising effects of nonlinearities [48, 49], and experimentally realised by use of a mask-shaped potential [50], somehow analogously to what was reported for the lattices described in the nonresonant section. Recently, even an annular chain of co-winding vortices was resonantly injected, and their evolution in the linear and nonlinear regime was experimentally explored [51].

21.7 Conclusions

To conclude, the polariton OPO supports a novel topological excitation consisting of a vortex in the signal and an antivortex in the idler. We have shown that the core radius of vortices in the polariton condensate is determined by polariton–polariton interactions. With ultrafast measurement techniques, the imprinting method also provides a means to investigate vortex dynamics on timescales inaccessible in other systems. Metastable rotational polariton flow is observed indicating superfluidlike behaviour of the nonequilibrium condensate. The spontaneous formation of quantised vortices in OPO polariton condensates is also observed in an optically induced ring-shaped potential trap. A transition from a vortex state with OAM $M_s = 1$ to one with $M_s = 2$ is observed in OPO condensates driven by a ring-shaped pump, which inhibits shrinking of the vortex size with power density and makes it more favourable for the system to switch to the state with higher OAM. In the case of a polariton population directly injected resonantly at $k = 0$ by a pump pulse, a pair of OAM of opposite sign can develop, which preserves the initial injected OAM of $M = 0$. This excludes the observation of single vortices for such configuration. Neverthless, the versatility of coherent scheme allows also the direct excitation of both a single full- or half-vortex spinor condensate, and enables one to follow their dynamics during the lifetime of polaritons, devising the interplay of nonlinearity, disorder, symmetry splitting terms and dissipation rates.

Acknowledgements: We acknowledge support by Engineering and Physical Sciences Research Council (EPSRC) grants EP/G001642 and EP/J007544. DS and LD acknowledge support from the European Research Council (ERC) project POLAFLOW.

References

[1] Timofeev, V., and Sanvitto, D. 2012. *Exciton Polariton in Microcavities: New Frontiers, Springer Series in Solid-State.* Berlin, Germany: Springer.

[2] Kavokin, A., Baumberg, J., Malpuech, G., and F. Laussy, F. *Microcavities.* Oxford, UK: Oxford University Press.

[3] Wouters, M., and Carusotto, I. 2007. Excitations in a nonequilibrium Bose-Einstein condensate of exciton polaritons. *Phys. Rev. Lett.*, **99**, 140402.

[4] Stevenson, R. M., Astratov, V. N., Skolnick, M. S., Whittaker, D. M., Emam-Ismail, M., Tartakovskii, A. I., Savvidis, P. G., Baumberg, J. J., and Roberts, J. S. 2000. Continuous wave observation of massive polariton redistribution by stimulated scattering in semiconductor microcavities. *Phys. Rev. Lett.*, **85**, 3680–3683.

[5] Tartakovskii, A. I., Krizhanovskii, D. N., and Kulakovskii, V. D. 2000. Polariton polariton scattering in semiconductor microcavities: distinctive features and similarities to the three-dimensional case. *Phys. Rev. B*, **62**, R13298–R13301.

[6] Wouters, M., and Carusotto, I. 2007. Goldstone mode of optical parametric oscillators in planar semiconductor microcavities in the strong-coupling regime. *Phys. Rev. A*, **76**, 043807.

[7] Cross, M. C., and Hohenberg, P. C. 1993. Pattern formation outside of equilibrium. *Rev. Mod. Phys.*, **65**, 851–1112.

[8] Sigurdsson, H., Egorov, O. A., Ma, X., Shelykh, I. A., and Liew, T. C. H. 2014. Information processing with topologically protected vortex memories in exciton-polariton condensates. *Phys. Rev. B*, **90**, 014504.

[9] Matthews, M. R., Anderson, B. P., Haljan, P. C., Hall, D. S., Wieman, C. E., and Cornell, E. A. 1999. Vortices in a Bose-Einstein condensate. *Phys. Rev. Lett.*, **83**, 2498–2501.

[10] Vinen, W. F. 1961. The detection of single quanta of circulation in liquid helium II. *Proc. R. Soc. Lond*, **260**, 218.

[11] Scheuer, J., and Orenstein, M. 1999. Optical vortices crystals: spontaneous generation in nonlinear semiconductor microcavities. *Science*, **285**, 230–233.

[12] Lagoudakis, K. G., Wouters, M., Richard, M. Baas, A., Carusotto, I. Andre, R., Dang, L. S., and Deveaud-Pledran, B. 2008. Quantized vortices in an exciton-polariton condensate. *Nat. Phys.*, **4**, 706.

[13] Lagoudakis, K. G., Ostatnický, T., Kavokin, A. V., Rubo, Y. G., André, R., and Deveaud-Plédran, B. 2009. Observation of half-quantum vortices in an exciton-polariton condensate. *Science*, **326**, 974–976.

[14] Sich, M., Krizhanovskii, D. N., Skolnick, M. S., Gorbach, A. V., Hartley, R., Skryabin, D. V., Cerda-Méndez, E. A., Biermann, K., Hey, R., and Santos, P. V. 2012. Observation of bright polariton solitons in a semiconductor microcavity. *Nat. Photonics*, **6**, 50–55.

[15] Sich, M., Fras, F., Chana, J. K., Skolnick, M. S., Krizhanovskii, D. N., Gorbach, A. V., Hartley, R., Skryabin, D. V., Gavrilov, S. S., Cerda-Méndez, E. A., Biermann, K., Hey, R., and Santos, P. V. 2014. Effects of spin-dependent interactions on polarization of bright polariton solitons. *Phys. Rev. Lett.*, **112**, 046403.

[16] Nardin, G., Grosso, G., Leger, Y., Petka, B., Morier-Genoud, F., and Deveaud-Plédran, B. 2011. Hydrodynamic nucleation of quantized vortex pairs in a polariton quantum fluid. *Nat. Phys.*, **7**, 635.

[17] Amo, A., Pigeon, S., Sanvitto, D., Sala, V. G., Hivet, R., Carusotto, I., Pisanello, F., Leménager, G., Houdré, R., Giacobino, E., Ciuti, C., and Bramati, A. 2011. Polariton superfluids reveal quantum hydrodynamic solitons. *Science*, **332**, 1167–1170.

[18] Krizhanovskii, D. N., Whittaker, D. M., Bradley, R. A., Guda, K., Sarkar, D., Sanvitto, D., Vina, L., Cerda, E., Santos, P., Biermann, K., Hey, R., and Skolnick, M. S. 2010. Effect of interactions on vortices in a nonequilibrium polariton condensate. *Phys. Rev. Lett.*, **104**, 126402.

[19] Guda, K., Sich, M., Sarkar, D., Walker, P. M., Durska, M., Bradley, R. A., Whittaker, D. M., Skolnick, M. S., Cerda-Méndez, E. A., Santos, P. V., Biermann, K., Hey, R., and Krizhanovskii, D. N. 2013. Spontaneous vortices in optically shaped potential profiles in semiconductor microcavities. *Phys. Rev. B*, **87**, 081309.

[20] Antón, C., Tosi, G., Martín, M. D., Viña, L., Lemaître, A., and Bloch, J. 2012. Role of supercurrents on vortices formation in polariton condensates. *Opt. Express*, **20**, 16366.

[21] Dreismann, Alexander, Cristofolini, Peter, Balili, Ryan, Christmann, Gabriel, Pinsker, Florian, Berloff, Natasha G., Hatzopoulos, Zacharias, Savvidis, Pavlos G., and Baumberg, Jeremy J. 2014. Coupled counterrotating polariton condensates in optically defined annular potentials. *Proc. Natl. Acad. Sci.*, **111**, 8770–8775.

[22] Dominici, Lorenzo, Dagvadorj, Galbadrakh, Fellows, Jonathan M., Ballarini, Dario, De Giorgi, Milena, Marchetti, Francesca M., Piccirillo, Bruno, Marrucci, Lorenzo, Bramati, Alberto, Gigli, Giuseppe, Szymańska, Marzena H., and Sanvitto, Daniele.

2015. Vortex and half-vortex dynamics in a nonlinear spinor quantum fluid. *Science Advances*, **1**, E1500807.

[23] D'Ambrosio, Vincenzo, Baccari, Flavio, Slussarenko, Sergei, Marrucci, Lorenzo, and Sciarrino, Fabio. 2015. Arbitrary, direct and deterministic manipulation of vector beams via electrically-tuned q-plates. *Sci. Rep.*, **5**, 7840.

[24] Cardano, Filippo, Karimi, Ebrahim, Marrucci, Lorenzo, de Lisio, Corrado, and Santamato, Enrico. Generation and dynamics of optical beams with polarization singularities. *Opt. Express*, **21**, 8815–8820.

[25] Dominici, L., Colas, D., Donati, S., Cuartas, J. P., Restrepo, Giorgi, M. De, Ballarini, D., Guirales, G., Carreño, J. C. Lopez, Bramati, A., Gigli, G., del Valle, E., Laussy, F. P., and Sanvitto, D. 2014. Ultrafast control and Rabi oscillations of polaritons. *Phys. Rev. Lett*, **113**, 226401.

[26] Manni, F, Lagoudakis, K. G., and Liew, T. C. H. 2012. Dissociation dynamics of singly charged vortices into half-quantum vortex pairs. *Nat. Commun.*, **3**, 1309.

[27] Manni, F., Léger, Y., Rubo, Y. G., André, R., and Deveaud, B. 2013. Hyperbolic spin vortices and textures in exciton-polariton condensates. *Nat. Commun.*, **4**, 2590.

[28] Toledo-Solano, M., Mora-Ramos, M. E., Figueroa, A., and Rubo, Y. G. 2014. Warping and interactions of vortices in exciton-polariton condensates. *Phys. Rev. B*, **89**, 035308.

[29] Lagoudakis, K. G., Manni, F., Pietka, B., Wouters, M., Liew, T. C. H., Savona, V., Kavokin, A. V., André, R., and Deveaud-Plédran, B. 2011. Probing the dynamics of spontaneous quantum vortices in polariton superfluids. *Phys. Rev. Lett.*, **106**, 115301.

[30] Liu, G., Snoke, D. W., Daley, A., Pfeiffer, L. N., and West, K. 2015. A new type of half-quantum circulation in a macroscopic polariton spinor ring condensate. *Proc. Natl. Acad. Sci.*, **112**, 2676–2681.

[31] Roumpos, Georgios, Fraser, Michael D., Loffler, Andreas, Hofling, Sven, Forchel, Alfred, and Yamamoto, Yoshihisa. 2011. Single vortex–antivortex pair in an exciton-polariton condensate. *Nat. Phys.*, **7**, 129–133.

[32] Manni, F., Liew, T. C. H., Lagoudakis, K. G., Ouellet-Plamondon, C., André, R., Savona, V., and Deveaud, B. 2013. Spontaneous self-ordered states of vortex–antivortex pairs in a polariton condensate. *Phys. Rev. B*, **88**, 201303.

[33] Tosi, G., Christmann, G., Berloff, N. G., Tsotsis, P., Gao, T., Hatzopoulos, Z., Savvidis, P. G., and Baumberg, J. J. 2012. Geometrically locked–vortex lattices in semiconductor quantum fluids. *Nat. Commun.*, **3**, 1243.

[34] Dall, Robert, Fraser, Michael D., Desyatnikov, Anton S., Li, Guangyao, Brodbeck, Sebastian, Kamp, Martin, Schneider, Christian, Höfling, Sven, and Ostrovskaya, Elena A. 2014. Creation of orbital angular momentum states with chiral polaritonic lenses. *Phys. Rev. Lett.*, **113**, 200404.

[35] Nitsche, W. H., Kim, Na Y., Roumpos, G., Schneider, C., Kamp, M., Höfling, S., Forchel, A., and Yamamoto, Y. 2014. Algebraic order and the Berezinskii-Kosterlitz-Thouless transition in an exciton-polariton gas. *Phys. Rev. B*, **90**, 205430.

[36] Plumhof, J. D., Stöferle, T., Mai, L., Scherf, U., and Mahrt, R. F. 2014. Room-temperature Bose-Einstein condensation of cavity exciton-polaritons in a polymer. *Nat. Mater.*, **13**, 247–252.

[37] Daskalakis, K. S., Maier, S. A., and Kéna-Cohen, S. 2015. Spatial coherence and stability in a disordered organic polariton condensate. *Phys. Rev. Lett.*, **115**, 035301.

[38] Tinkler, L., Walker, P. M., Clarke, E., Krizhanovskii, D. N., Bastiman, F., Durska, M., and Skolnick, M. S. 2015. Design and characterization of high optical quality InGaAs/GaAs/AlGaAs-based polariton microcavities. *Appl. Phys. Lett.*, **106**, 021109.

[39] Love, A. P. D., Krizhanovskii, D. N., Whittaker, D. M., Bouchekioua, R., Sanvitto, D., Rizeiqi, S. Al, Bradley, R., Skolnick, M. S., Eastham, P. R., André, R., and Dang, Le Si. 2008. Intrinsic decoherence mechanisms in the microcavity polariton condensate. *Phys. Rev. Lett.*, **101**, 067404.

[40] Krizhanovskii, D. N., Sanvitto, D., Love, A. P. D., Skolnick, M. S., Whittaker, D. M., and Roberts, J. S. 2006. Dominant effect of polariton–polariton interactions on the coherence of the microcavity optical parametric oscillator. *Phys. Rev. Lett.*, **97**, 097402.

[41] Sanvitto, D., Marchetti, F. M., Szymaska, M. H., Tosi, G., Baudisch, M., Laussy, F. P., Krizhanovskii, D. N., Skolnick, M. S., Marrucci, L., Lemaitre, A., Bloch, J., Tejedor, C., and Vina, L. 2010. Persistent currents and quantized vortices in a polariton superfluid. *Nat. Phys.*, **6**, 527533.

[42] Wouters, M., and Carusotto, I. 2010. Superfluidity and critical velocities in nonequilibrium Bose-Einstein condensates. *Phys. Rev. Lett.*, **105**, 020602.

[43] Martinelli, M., Huguenin, J. A. O., Nussenzveig, P., and Khoury, A. Z. 2004. Orbital angular momentum exchange in an optical parametric oscillator. *Phys. Rev. A*, **70**, 013812.

[44] Pitaevskii, L. P., and Stringari, S. 2003. *Bose–Einstein Condensation*. Oxford, UK: Clarendon Press.

[45] Manni, F., Lagoudakis, K. G., Liew, T. C. H., André, R., and Deveaud-Plédran, B. 2011. Spontaneous pattern formation in a polariton condensate. *Phys. Rev. Lett.*, **107**, 106401.

[46] Ostrovskaya, E. A., Abdullaev, J., Desyatnikov, A. S., Fraser, M. D., and Kivshar, Y. S. 2012. Dissipative solitons and vortices in polariton Bose-Einstein condensates. *Phys. Rev. A*, **86**, 013636.

[47] Tosi, G., Marchetti, F. M., Sanvitto, D., Anton, C., Szymanska, M. H., Berceanu, A., Tejedor, C., Marrucci, L., Lemaitre, A., Bloch, J., and Vina, L. 2011. Onset and dynamics of vortex–antivortex pairs in polariton optical parametric oscillator superfluids. *Phys. Rev. Lett.*, **107**, 036401.

[48] Liew, T. C. H., Rubo, Yuri G., and Kavokin, A. V. 2008. Generation and dynamics of vortex lattices in coherent exciton-polariton fields. *Phys. Rev. Lett.*, **101**, 187401.

[49] Gorbach, A. V., Hartley, R., and Skryabin, D. V. 2010. Vortex lattices in coherently pumped polariton microcavities. *Phys. Rev. Lett.*, **104**, 213903.

[50] Hivet, R., Cancellieri, E., Boulier, T., Ballarini, D., Sanvitto, D., Marchetti, F. M., Szymanska, M. H., Ciuti, C., Giacobino, E., and Bramati, A. 2014. Interaction-shaped vortex–antivortex lattices in polariton fluids. *Phys. Rev. B*, **89**, 134501.

[51] Boulier, T., Terças, H., Solnyshkov, D. D., Glorieux, Q., Giacobino, E., Malpuech, G., and Bramati, A. 2015. Vortex chain in a resonantly pumped polariton superfluid. *Sci. Rep.*, **5**, 9230.

22

Optical Control of Polariton Condensates

GABRIEL CHRISTMANN
Foundation for Research and Technology-Hellas, Heraklion, Crete, Greece

PAVLOS G. SAVVIDIS
Foundation for Research and Technology-Hellas, Heraklion, Crete, Greece
Department of Materials Science and Technology, University of Crete, Greece
Cavendish Laboratory, University of Cambridge, UK

JEREMY J. BAUMBERG
Cavendish Laboratory, University of Cambridge, UK

Microcavity polaritons, the bosonic quasiparticles resulting from the strong coupling between a cavity photon and a quantum well exciton, offer unique opportunities to study quantum fluids on a semiconductor chip. Their excitonic part leads to strong repulsive polariton–polariton interactions, and their photonic part allows one to probe their properties using conventional imaging and spectroscopy techniques. In this chapter, we report on recent results on the optical manipulation and control of polariton condensates. Using spatially engineered excitation profiles, it is possible to create potential landscapes for the polaritons. This leads to the observation of effects such as long distance spontaneous polariton propagation; confined states in a parabolic potential, in a configuration similar to a quantum harmonic oscillator; and vortex lattices.

22.1 Introduction

Wave-particle duality is one of the most striking features of quantum physics and has led to numerous discussions spreading far beyond the field of physics. The fact that the properties of a particle are described by a wavefunction redefined physics between the 19th and 20th centuries. When technological progress started to allow experimental access to microscopic particles, wave effects could be observed. Around the same time, the observation of the photoelectric effect eventually explained by Einstein, introduced the concept of photons, as quanta of electromagnetic radiation [1]. This also forced a reconsideration of the wave theory of light, which at that time was well established thanks to interferometry experiments and Maxwell's equations. Such quantisation in fact linked back to the ideas of light corpuscles as introduced by Newton.

In the 1920s, Einstein, on the basis of Bose's work on the statistics of photons [2], proposed the idea that an atomic gas of noninteracting bosons should exhibit, below a finite temperature, a macroscopic occupation of the lowest energy quantum state [3]. This is what is now called Bose-Einstein condensation and is the main topic of this book. This phenomenon extends the wave properties of matter to an ensemble of particles and therefore to the macroscopic scale. At first, this purely theoretical prediction was first rejected by the scientific community. However, when superfluidity of ^{4}He was observed [4, 5], London proposed that this observation was in fact linked to Bose-Einstein condensation [6]. The situation of liquid helium was however quite far from the picture of a gas of noninteracting particles. Not until the 1990s was Bose-Einstein condensation observed in a system that was similar to the theoretical picture: dilute atomic gases [7, 8]. This enabled observations of macroscopic wavefunctions on a large ensemble of particles and striking effects such as interference between two Bose-Einstein condensates [9].

Since atoms are fairly heavy, the effects were observed at very low temperature only using sophisticated laser cooling techniques. For this reason, there has been a drive to observe Bose-Einstein condensations in other systems. Using quasiparticles with low effective mass should allow condensation at much higher temperatures. In this context, significant achievements have been recently obtained with excitons [10, 11], magnons [12], cavity polaritons [13], and cavity photons [14]. In particular, Bose-Einstein condensation of polaritons, which is the topic of this chapter, is now a well-established fact and has been observed even at room temperature in several systems [15, 16, 17]. Cavity polaritons are the quasiparticles resulting from a coherent superposition between semiconductor excitons and the light field inside a micron-sized Fabry-Pérot interferometer [18]. (See Chapter 4 for a general introduction to polaritons.) As a result, a key advantage in these systems is that they can be studied by measuring the light leaking from the optical cavity. Polariton condensates can therefore be imaged directly from the surface of a semiconductor chip. This microcavity system has enabled many results reminiscent of what has been obtained with cold atoms, such as quantised vortices [19], superfluidity [20, 21], and solitons [22]. In parallel, properties more reminiscent of the field of nonlinear optics and lasers were demonstrated using microcavities in the strong coupling regime, the most striking being the demonstration of polariton optical parametric amplification [23, 24, 25], illustrating the connection between Bose-Einstein condensation and nonlinear optics.

There is now a drive to manipulate these polariton condensates in advanced configurations where confinement effects come into play. This can be achieved, for example, by processing the semiconductor chip into mesas [26]. In this way, further properties are accessible: condensation in pillars [27], propagating condensates in waveguides [28], polariton molecules [29] or lattices [30]. Confinement has also been obtained using strain traps [31] and metallic structures [32]. In this chapter, a different way to control polariton condensates will be presented using spatially structured illumination. By controlling the excitation pattern of light, localised populations of reservoir excitons are created which, through repulsive interactions with polaritons, create a spatial potential able to confine polaritons. The outline of this chapter is the following. First a general presentation of polariton condensation will be given. Then optical control of the polariton flow will be discussed. Then results on the locking and confinement of a polariton condensates will be presented, first in the case where a harmonic trap is created by two excitation spots and finally for more complicated excitation patterns, ultimately leading to full confinement of polariton condensates.

22.2 General Features of Bose-Einstein Condensation of Polaritons

Polaritons in semiconductor microcavities are the quasiparticles resulting from coherent coupling between a cavity photon and a quantum well exciton. Building from the picture introduced by Hopfield in the case of bulk materials where the translation invariance leads to coherent oscillation between the light field and the exciton field [33], here in one direction this translation invariance is broken by the cavity. If the cavity has a sufficiently large quality factor ($Q = \lambda/\Delta\lambda$), then a photon in the cavity will interact with the quantum well before decaying by escaping through the non-100% reflectivity mirrors. Due to the two-dimensional (2D) nature of the quantum well and of the cavity confinement, the polariton dispersion is also two-dimensional and is shown in Fig. 22.1, resulting from the mixed combination of the exciton and the cavity photon dispersions. Within the in-plane wavevectors ($k_{\|}$) generally considered, which correspond to optically accessible cavity modes, the exciton dispersion is mostly flat. The curvature of the polariton dispersion curve is therefore dominated by that of the cavity mode, which is nearly parabolic (Fig. 22.1). Lower cavity polaritons thus experience a near-harmonic trap around $k_{\|} = 0$ associated with a very small effective mass, typically 10^{-4} that of excitons, and thus 10^{-8} that of atoms. To sum up, the polaritons therefore combine a very small effective mass thanks to their photonic component with the possibility to interact and thus thermalise thanks to their excitonic components through dipole–dipole repulsive interactions [13]. This is what allows the observation of Bose-Einstein condensation at elevated temperatures [13, 15].

The other very important thing about the lower polariton dispersion curve is its inflection point at higher $k_{\|}$. This leads to a separation in the density of state between the low density in the harmonic trap near $k_{\|} = 0$ and the high density at high $k_{\|}$ where the lower polariton becomes mostly exciton-like. This results in a relaxation bottleneck for excitons in this area (Fig. 22.1, [34]) set by the inability

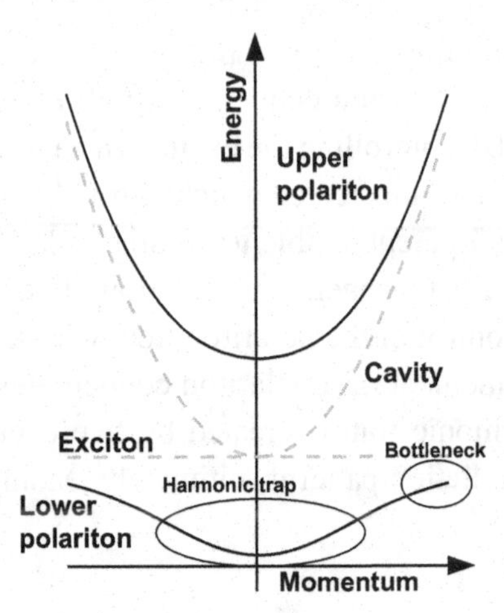

Figure 22.1 Typical in-plane dispersion curve of polaritons.

of phonons to remove sufficient energy for the small momentum changes involved. The consequence is that when a condensate is formed by relaxation from a hot population of photocreated excitons, although there will be a sufficient population of thermalized polaritons in the trap for Bose-Einstein condensation, there will be an even larger population of reservoir excitons. These reservoir excitons will interact with the polaritons through repulsive dipole–dipole interactions, creating a potential hill in the location where the hot excitons are injected. This effect originally caused difficulties for clear demonstrations of Bose-Einstein condensation of polaritons where it seemed that the *condensate* was not forming at $k_{\|} = 0$ [35]. By using a larger spot size, the reservoir exciton effect was mitigated enough to finally allow unambiguous demonstration of polariton Bose-Einstein condensation [13, 36].

Fig. 22.2 shows this polariton condensation effect in the case of a tightly focussed pump spot. The sample from which the results presented here are obtained is a high quality factor $5\lambda/2$ AlGaAs microcavity with four sets of three GaAs/AlGaAs quantum wells. It has been studied in detail in Ref. [37]. For increasing pump power, the emitted intensity exhibits a clear threshold associated with a change of the emission pattern (Fig. 22.2a). Below threshold the emission more or less corresponds to the pump spot, while above threshold the emission spreads over tens of microns and shows a characteristic sunflower pattern [38]. Fig. 22.2b shows the energy versus real-space emission below threshold where most of the emission comes from the pump spot area itself while Fig. 22.2c shows emission from the extended condensate above threshold. This emission is at a single energy, and significantly blue-shifted from the lower polariton at

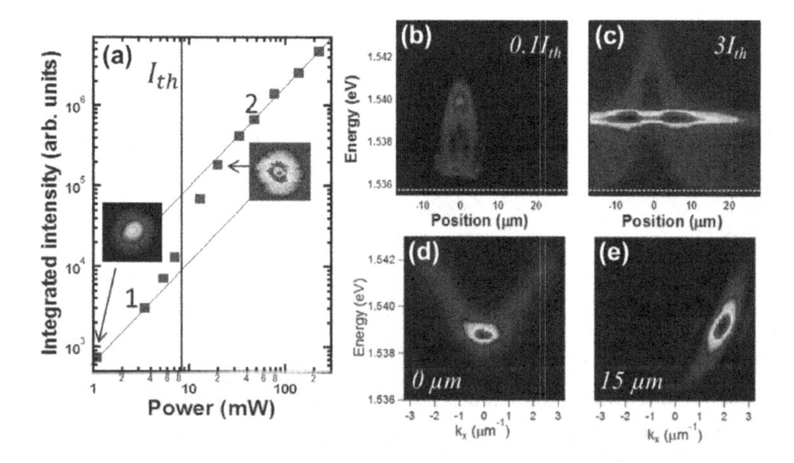

Figure 22.2 (a) Integrated emission intensity versus pump power. Vertical line marks the position of the threshold power I_{th}. Insets shows the 2D emission real-space image at the corresponding pump powers. Labels 1 and 2 correspond to b and c, respectively. (b, c) Energy versus real-space emission image for a line going through the centre of the pump spot, for pump powers of $0.1I_{th}$ and $3I_{th}$, respectively. White dashed line corresponds to non-blueshifted lower polariton energy at $k_{\|} = 0$. (d, e) Energy vs $k_{\|}$ emission images for emission areas selected at 0 and 15 μm from the pump spot, respectively.

$k_{\|} = 0$ (white dashed line). At the pump spot location, at lower energies than the condensate, no residual emission from incoherent polaritons is visible, while away from this spot some weak emission is seen. This dark area at the pump spot is explained by the presence at the pump spot of an significant population of reservoir excitons which, through repulsive dipole–dipole interactions, creates a potential for the polaritons. At the pump spot, the condensate is therefore created at the lowest energy state available for polaritons, consistent with the Bose-Einstein condensation mechanism. However, as polaritons move away from this spot, the reservoir exciton population diminishes rapidly, and thus the repulsive potential decreases. Therefore these polaritons, in order to keep their energy, have to gain in-plane momentum and accelerate away from the spot, creating a radially expanding figure. This is confirmed by analysis of the $k_{\|}$ space densities. In Fig. 22.2d, measured at the centre, the polaritons from the condensate are clearly at $k_{\|} = 0$ while when the same measurement is taken away from the pump spot (Fig. 22.2e), these polaritons are at nonzero $k_{\|}$, confirming their propagation. Finally, the characteristic sunflower pattern observed on the expanding polariton condensate has been shown to be due to coherent resonant Rayleigh scattering [38] of a small fraction of the polaritons.

To sum up, nonresonant excitation in semiconductor microcavities creates a strong blueshift for polaritons. In the case of exciting condensates with a tight spot, this creates a propagating coherent population of polaritons both in planar microcavities [38, 39] and in fabricated wire structures [28]. Furthermore, by locally creating a population of reservoir excitons, this will create a potential for polaritons.

22.3 Optical Control of Polariton Flow

The first experiments on the optical control of the polariton flow used resonant excitation configurations. This means that the coherent population of polaritons is not formed from a spontaneous condensation of incoherent polaritons but created instead by resonant injection of photons into the corresponding polariton state using a coherent source (a laser). This technique is more versatile to create a coherent polariton population in the desired configuration, but all coherence is imposed from outside. This can be done either with the pump directly coupled to the desired condensate state or through a parametric scattering process to reach the desired state [23]. With such techniques, it is possible to produce well-controlled polariton flows, and this allowed demonstrations of polariton superfluidity [20, 21], dark [22] and bright [40] solitons, as well as vortex stability [41]. It has also been proposed to control the flow of polaritons using a resonant beam [42], thanks to polariton–polariton interactions so that the resonant beam creates an additional potential for polaritons. Using this technique, the authors were able to create point and line barriers for the polariton flow. Further work showed the possibility to create even more complex potentials [43]. Compared to the scheme with reservoir excitons presented in the introduction, the height of the potential created by resonantly imprinting an optical pattern typically creates a smaller potential for polaritons. This is because polaritons have a much smaller density of states than reservoir excitons so it is harder to create a population as significant as in the case of reservoir excitons. However, this configuration with resonant excitation can combine control of polariton flow with the bistability effects that can be obtained in strongly coupled microcavities [44]. Hence it has been possible to realize an all-optical transistor [45]. Creating such optical circuits is currently an emerging hot topic and several schemes have been proposed [46, 47].

It has also been possible to use non-resonant excitation to control the flow of propagating polariton condensates. This is achieved by both confining the polaritons in mesa structures and affecting their flow using a nonresonant control pump beam. By etching the microcavity in a one-dimensional structure, it is possible to have long-range propagation of polariton condensates [28]. A polariton condensate created at one extremity of the waveguide will propagate to the other end and

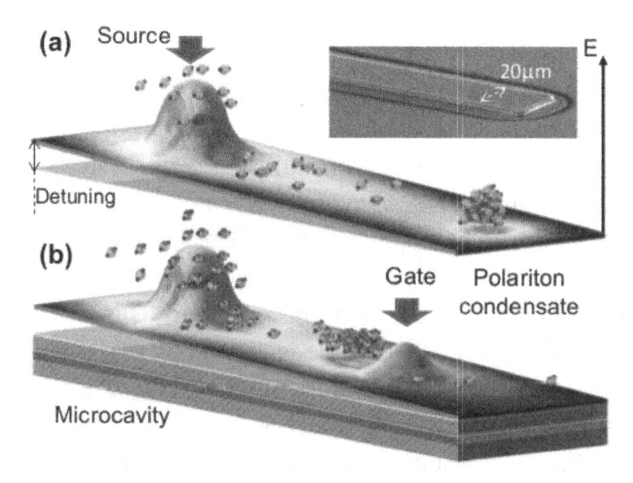

Figure 22.3 Schematic of the polariton condensate transistor based on a microcavity ridge, (a) without and (b) with the gate. Scanning electron microscopy image of a 20 μm ridge.

accumulate there. Such accumulation of polaritons at the end of the wire is favoured by progressively increasing negative detunings towards one end of the wire. Furthermore, strain relaxation at the ridge end lowers further the polariton energy, providing a strong trap for polaritons. By sending a control pump spot between the pump and the end of the wire as shown in Fig. 22.3, it is possible to completely stop the polariton flow and thus quench the polaritons accumulating at the end of the wire [48]. In this way, a transistor switch is realised with switching speed of about 200 *ps* limited by exciton lifetimes [49]. A very similar result has been obtained using a resonant tunnelling diode structure etched on a microcavity [50]. There, two polariton waveguides are separated by a fully zero-dimensional structure, allowing transmission only when one of the quantised levels in this structure is aligned with the propagating polariton condensate energy. Thanks to reservoir excitons created by a control pump, it is possible to tune these levels and switch on and off the transmission. A similar scheme was employed to induce phase shifts in Mach-Zehnder and Sagnac interferometers [51]. The devices were created by etching into the microcavity structure different interferometers. The interferometers are fed by a resonantly pumped propagating polariton condensate, and then a control pump is applied on one arm, locally creating a blueshift thanks to the population of reservoir excitons. Contrary to the transistor case, the control pump is kept low enough in order to avoid creating a complete barrier for polaritons. The result is that the polaritons travelling there will reduce their in-plane wavevector, thus adding a phase shift which is detected by the interference pattern with polaritons travelling along the other arm.

22.4 Complex Optical Imprinted Potentials for Polariton Condensates

By using beam shaping techniques, it is possible to produce more complex excitation patterns. Moderately complex shapes can be produced using classical optical elements: beam splitters and mirrors for multispot configurations [52, 53], axicon prisms for ring excitation [53, 54], or cylindrical lenses for line excitation. For more complex patterns, spatial light modulators have been successfully used [55, 56, 57], virtually enabling any excitation profile to be reconfigured on the fly.

The first rather simple configuration presented here (Fig. 22.4a) is a two-spot excitation [52]. In this regime, the sample is excited by two tightly focussed non-resonant laser spots separated by a distance of a few tens of microns. Two radially expanding condensates are created by reservoir excitons at the pump spot locations and interact together. The resulting pattern is shown in Fig. 22.4b, and exhibits clear mutual interference fringes. Note that this emission pattern is associated to a single energy. This observation indicates that both independently created condensates build a stable common phase thanks to polariton–polariton interactions and finally form a single coherent state. This is in complete contradiction to what would be expected from two coherent photonic modes with almost identical energies

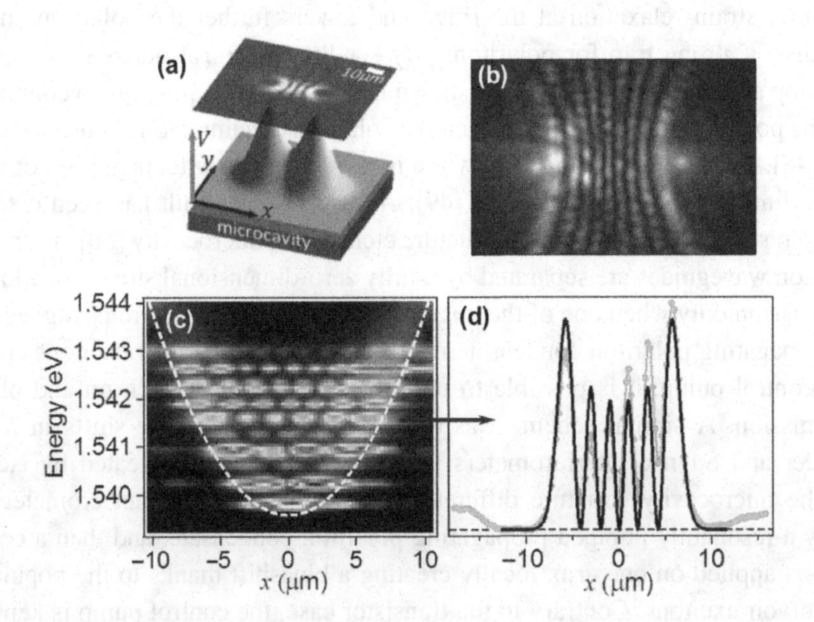

Figure 22.4 (a) Schematic description of the two-spot excitation pattern. (b) Two-dimensional emission pattern of the microcavity under this excitation scheme. (c) Energy versus real-space emission image under two-spot pumping with spatial cut taken along the two pump spots. (d) Spatial emission profile for the sixth state in (c) (light grey) compared to a Hermite polynomial (black).

meeting each other. In this case, because photons do not interact, no fringes would be observed on a time-integrated image contrary to the one shown in Fig. 22.4b.

The coherence of a polariton Bose-Einstein condensate was clearly demonstrated interferometrically [13]. Locking between different localised polariton condensates was first observed by measuring the energies of these localised modes. While they have slightly different energies below the condensation threshold, they lock at a single energy value above threshold, showing that due to polariton–polariton interactions, the system attempts to occupy a single macroscopic coherent state [58]. In the case of the two-spot excitation discussed above, phase locking between two radially expanding condensates (the case shown in Fig. 22.4) can be directly observed [59]. By utilising a Michelson interferometer with a retroreflector in one arm, it is possible to interfere the emission of one condensate with that of the other one and then further study dynamically the locking utilising time-resolved Streak camera measurements (Fig. 22.5a). It is very clearly seen (using interferometric contrast) that before coming into contact, the two expanding polariton condensates have no stable phase relationship as no interference fringes are observed (Fig. 22.5b,c). After that, once they come into contact (Fig. 22.5d), they rapidly build a stable common phase, highlighted by the appearance of fringes with rapidly increasing contrast (Fig. 22.5e,f), and finally produce a fully coherent state whose spatial profile is very similar to what is obtained under continuous wave excitation (Fig. 22.5g). These experiments show that as soon as condensates ballistically interact, nonlinear scattering rapidly builds up, allowing population transfer and

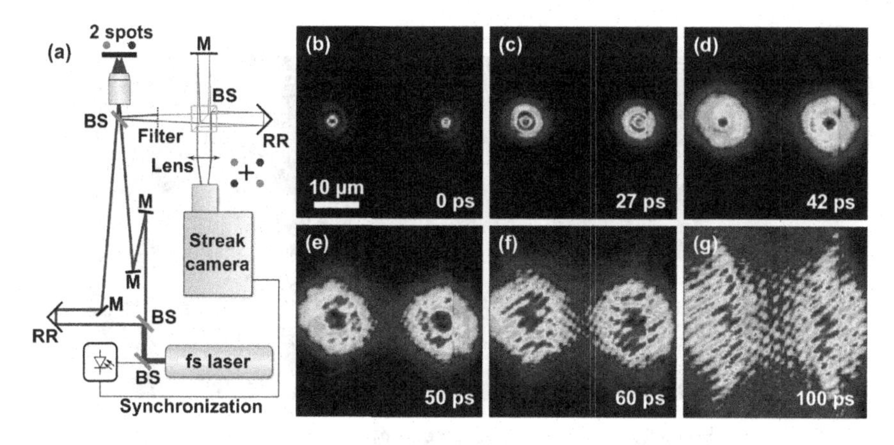

Figure 22.5 (a) Schematic of the experimental setup, showing two spatially separated radially expanding condensates which are interferometrically overlapped and then scanned tomographically with a Streak camera to give temporal images. (b–g) Interferometric images of the sample emission for increasing time delays reconstructed by tomography.

phase locking at a rate given by the polariton gain and loss. (See Chapter 23 for a discussion of the theoretical issues of phase locking of condensates.)

Interestingly, when raising the excitation power from the case shown in Fig. 22.4, the emission pattern changes and several states are observed between the two spots (Fig. 22.4c). These states are confined in the harmonic potential profile created by reservoir excitons at the two pumping spots and are equally spaced in energy. This is consistent with the different quantum states inside a harmonic potential and is further confirmed by looking at the emission profile of such states (Fig. 22.4d). The photoluminescence spatial distribution is indeed well fit by a Hermite polynomial, which is the solution of the quantum harmonic oscillator. The population of these different states can be explained by parametric scattering which is very compatible with equally spaced levels, similarly to what is observed with frequency combs in resonators [60]. Note that the two-dimensional emission pattern of each state can be recovered by tomography [52], and is very similar to Fig. 22.4b. It has also been shown that all these modes have a stable phase relationship interferometrically [52]. This suggests that all these modes are mode locked and form a coherent oscillating wavepacket, a scheme which is consistent with complex Ginzburg-Landau simulations that predict dark soliton oscillations [52]. Similar dark soliton oscillations have also been observed in cold atom systems [61, 62], and mode locking of lasers into dark pulses has also been achieved [63].

These oscillations can be time resolved with picosecond resolutions using a streak camera [59]. In Fig. 22.6, time versus real-space emission images taken along a line joining the two pump spots are reported for increasing pump power. In Fig. 22.6a, which is taken just above the threshold for condensation, at short times

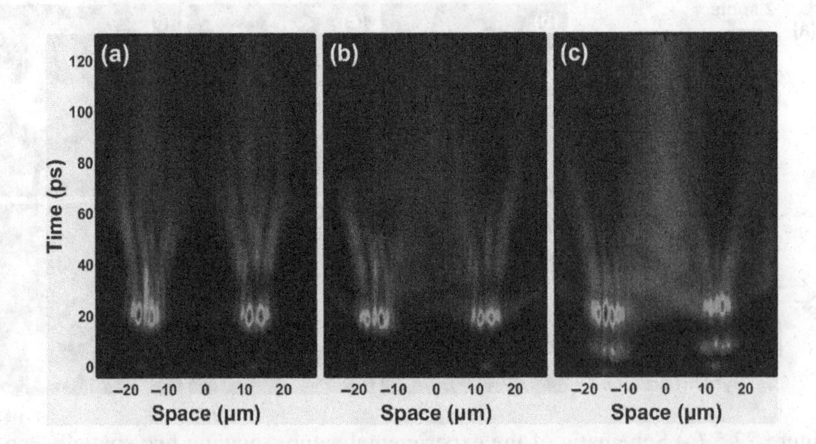

Figure 22.6 Time versus real-space Streak camera images of the microcavity emission for increasing pump powers. (a) Low pump power (just above threshold, 4.5 *mW* per spot). (b) Intermediate pump power (11 *mW* per spot) showing dark oscillation. (c) High pump power (28 *mW* per spot) showing bright oscillation.

expansion of the condensates is observed and for longer times mutual interference fringes are observed. This image does not show any oscillatory behaviour and corresponds to the formation dynamics of the single energy pattern as seen in Fig. 22.4b. The pump power is too small to trigger polariton relaxation in the discrete states of the harmonic potential. By raising the power (Fig. 22.6b), a dark oscillation is now clearly observed between the two spots. The power is now sufficient to trigger polariton relaxation into the discrete states and their mode locking leads to a dark oscillation as predicted by theory [52]. Interestingly, increasing the power even more leads to a drastic change of the oscillatory wavepacket, which becomes now bright. This transition from a dark to a bright soliton oscillation has been explained by a change of the polariton effective mass from positive to negative values, as a result of the inflection point in the dispersion curve [59]. Polaritons with positive effective masses are expected to support dark solitons [64] while those with a negative mass are expected to support bright ones [65].

Condensation under more complex excitation patterns have also been studied. The phase locking observed in the case of two-spot excitation Fig. 22.4b can be extended to three spots arranged in a regular triangle. The corresponding emission pattern is reported in Fig. 22.7a showing that phase-locked emission from

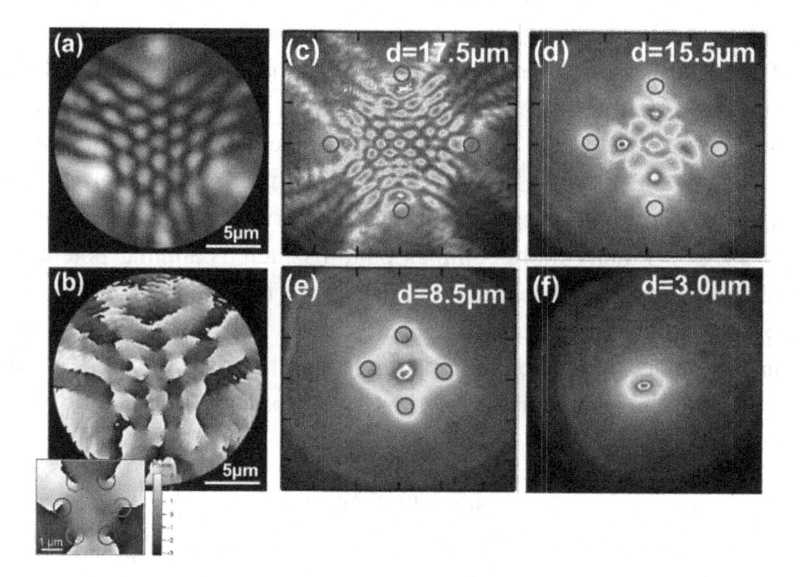

Figure 22.7 (a) Emission from a polariton condensate under three-spot excitation. (b) Corresponding phase image extracted from interferometry measurements. Inset: zoom of the central area of the image. Red and green crosses and arrows mark vortices of opposite circulation. (c–f) Emission from a polariton condensate under four-spot excitation for reducing excitation pattern size. The distance is measured from the centre to one of the spots. Red circles indicate the pump spot positions.

the three spots produces bright spots arranged in a triangular lattice. The phase profile extracted from interferometry measurements is reported in Fig. 22.7b. It exhibits a honeycomb lattice of vortices which forms a triangle lattice of vortex pairs of opposite circulation, as expected from the interference from three expanding coherent waves [66]. Note that similar vortex lattices were obtained previously inside nonlinear crystals [67], and subsequently also in polaritons with a variety of resonant and nonresonant excitation schemes [68, 69].

A development of previous experiments is exploring similar regular patterns but with an increasing number of spots, which is easily done using a spatial light modulator system [56]. More and more complex self-interference patterns are observed, and an example with four spots is displayed in Fig. 22.7c. The other very interesting consequence of increasing spot numbers is that the blueshift potentials created by reservoir excitons from the pump spots tend to create an increasingly sharply confined 2D potential. Note that the size of the pattern is also critical for the definition of this confinement potential. This is exemplified by Fig. 22.7c–e, where by reducing the pattern size the emission moves from a self-interference pattern (c) to a confined state (d, e). The order of these modes also decreases with the size of the excitation profile, with condensation in the ground state only seen in (e). Finally when the excitation spots are too close, the blueshift potentials merge into a single hill, and the emission pattern becomes similar to single spot excitation (f). This transition from locked states to trapped states has been studied in detail, and preferential condensation in a trapped state is associated with a dramatic threshold reduction [56]. Note that increasing the number of spots favours a trapped state and that in the limiting case of a ring excitation, only trapped states are observed, as the confining potential is never lost at holes in widely separated pump spots [53, 54, 56]. Finally more complex patterns such as annular potentials have been used, producing ring-shaped stationary patterns indicating the presence of two counter propagating condensates [57]. These can be potentially be utilised in ring interferometer devices and show a remarkable stability over hundreds of microns and large areas of the sample unhindered by any disorder.

22.5 Conclusion

In conclusion, numerous recent results on the optical control of polariton condensates are opening up enhanced ways to understand and control such superfluids. Such control is achieved thanks to polariton–polariton repulsive dipolar interactions in the resonant case, and polariton–reservoir exciton interactions in the nonresonant case. Control of polariton flows both in the non resonant and resonant cases show the great potential of polaritons to realise a wide variety of optical devices. On the other hand, polariton condensation in complex potential landscapes have also been reported, showing either patterns from phase-locked states or fully confined polariton condensates. Such a complex nonlinear superfluid offers unusual configurations that have not until now been open to study in other systems, because of the precise way that condensates can be excited and confined in arbitrary locations. These results illustrate how semiconductor microcavities offer a versatile system for the observation of interacting Bose gases in a wide variety of configurations and suggest how many experimental approaches are now open.

Acknowledgements: We would like to acknowledge greatly all our collaborators at the University of Cambridge, the University of Crete and the Foundation for Research and Technology (FORTH)–Institute of Electronic Structure and Laser (IESL), who contributed to the work presented in this chapter, as well as funding from the Engineering and Physical Sciences Research Council (EPSRC) EP/G060649/1, the European Union (EU) CLERMONT4 235114, EU INDEX 289968 and Greek General Secretariat for Research and Technology (GSRT) programs ARISTEIA and Irakleitos II.

References

[1] Einstein, A. (1905) 'Über einen die Erzeugung und Verwandlung des Lichtes betreffenden heuristischen Gesichtspunkt', *Ann. Phys.* **322**, 132–148.

[2] Bose, S. N. (1924) 'Plancks gesetz und lichtquantenhypothese', *Z. Phys. A* **26**, 178–181.

[3] Einstein, A. (1925) 'Quantum theory of mono-atomic ideal gas. Second paper', *Sitzungsber. K. Preuss. Akad. Wiss.*, 3–14.

[4] Allen, J. F. & Misener, A. D. (1938) 'Flow of liquid helium II', *Nature* **141**, 75.

[5] Kapitza, P. (1938) 'Viscosity of liquid helium below the λ-Point', *Nature* **141**, 74.

[6] London, F. (1938) 'On the Bose-Einstein condensation', *Phys. Rev.* **54**, 947–954.

[7] Anderson, M. H., Ensher, J. R., Matthews, M. R., Wieman, C. E. & Cornell, E. A. (1995) 'Observation of Bose-Einstein condensation in a dilute atomic vapor', *Science* **269**, 198–201.

[8] Davis, K. B., Mewes, M.-O., Andrews, M. R., van Druten, N. J., Durfee, D. S., Kurn, D. M. & Ketterle, W. (1996) 'Bose-Einstein condensation in a gas of sodium atoms', *Phys. Rev. Lett.* **75**, 3969–3973.

[9] Andrews, M. R., Townsend, C. G., Miesner, H. J., Durfee, D. S., Kurn, D. M. & Ketterle, W. (1997) 'Observation of interference between two Bose condensates', *Science* **275**, 637–641.

[10] High, A. A., Leonard, J. R., Remeika, M., Butov, L. V., Hanson, M. & Gossard, A. C. (2012) 'Condensation of excitons in a trap', *Nano Lett.* **12**, 2605–2609.

[11] Alloing, M., Beian, M., Lewenstein, M., Fuster, D., González, Y., González, L., Combescot, R., Combescot, M. & Dubin, F. (2014) 'Evidence for a Bose-Einstein condensate of excitons', *Europhys. Lett.* **107**, 10012.

[12] Demokritov, S. O., Demidov, V. E., Dzyapko, O., Melkov, G. A., Serga, A. A., Hillebrands, B. & Slavin, A. N. (2006) 'Bose-Einstein condensation of quasi-equilibrium magnons at room temperature under pumping', *Nature* **443**, 430–433.

[13] Kasprzak, J., Richard, M., Kundermann, S., Baas, A., Jeambrun, P., Keeling, J. M. J., Marchetti, F. M., Szymanska, M. H., André, R., Staehli, J. L., Savona, V., Littlewood, P. B., Deveaud, B. & Le Si Dang (2006) 'Bose-Einstein condensation of exciton polaritons', *Nature* **443**, 409–414.

[14] Klaers, J., Schmitt, J., Vewinger, F. & Weitz M. (2010) 'BoseEinstein condensation of photons in an optical microcavity', *Nature* **468**, 545–548.

[15] Christopoulos, S., von Hgersthal, G. B. H., Grundy, A. J. D., Lagoudakis, P. G., Kavokin, A. V., Baumberg, J. J., Christmann, G., Butté, R., Feltin, E., Carlin, J.-F. & Grandjean, N. (2007) 'Room-temperature polariton lasing in semiconductor microcavities', *Phys. Rev. Lett.* **98**, 126405.

[16] Kéna-Cohen, S. & Forrest, S. R. (2010) 'Room-temperature polariton lasing in an organic single-crystal microcavity', *Nat. Photon.* **4**, 371–375.

[17] Guillet, T., Mexis, M., Levrat, J., Rossbach, G., Brimont, C., Bretagnon, T., Gil, B., Butté, R., Grandjean, N., Orosz, L., Réveret, F., Leymarie, J., Zúñiga-Pérez, J., Leroux, M., Semond, F. & Bouchoule, S. (2011) 'Polariton lasing in a hybrid bulk ZnO microcavity', *Appl. Phys. Lett.* **99**, 161104.

[18] Weisbuch, C., Nishioka, M., Ishikawa, A. & Arakawa, Y. (1992) 'Observation of the coupled exciton–photon mode splitting in a semiconductor quantum microcavity', *Phys. Rev. Lett.* **69**, 3314.

[19] Lagoudakis, K. G., Wouters, M., Richard, M., Baas, A., Carusotto, I., André, R., Le Si Dang & Deveaud-Plédran, B. (2008) 'Quantized vortices in an exciton–polariton condensate', *Nat. Phys.* **4**, 706–710.

[20] Amo, A., Sanvitto, D., Laussy, F. P., Ballarini, D., del Valle, E., Martin, M. D., Lemaître, A., Bloch, J., Krizhanovskii, D. N., Skolnick, M. S., Tejedor, C. & Viña, L. (2009) 'Collective fluid dynamics of a polariton condensate in a semiconductor microcavity', *Nature* **457**, 291–295.

[21] Amo, A., Lefrère, J., Pigeon, S., Adrados, C., Ciuti, C., Carusotto, I., Houdré, R., Giacobino, E. & Bramati, A. (2009) 'Superfluidity of polaritons in semiconductor microcavities', *Nat. Phys.* **5**, 805–810.

[22] Amo, A., Pigeon, S., Sanvitto, D., Sala, V. G., Hivet, R., Carusotto, I., Pisanello, F., Leménager, G., Houdré, R., Giacobino, E., Ciuti, C. & Bramati, A. (2011) 'Polariton superfluids reveal quantum hydrodynamic solitons', *Science* **332**, 1167–1170.

[23] Savvidis, P. G., Baumberg, J. J., Stevenson, R. M., Skolnick, M. S., Whittaker, D. M. & Roberts J. S. (2000) 'Angle-resonant stimulated polariton amplifier', *Phys. Rev. Lett.* **84**, 1547.

[24] Stevenson, R. M., Astratov, V. N., Skolnick, M. S., Whittaker, D. M., Emam-Ismail, M., Tartakovskii, A. I., Savvidis, P. G., Baumberg, J. J., & Roberts J. S. (2000) 'Continuous wave observation of massive polariton redistribution by stimulated scattering in semiconductor microcavities', *Phys. Rev. Lett.* **85**, 3680.

[25] Diederichs, C., Tignon, J., Dasbach, G., Ciuti, C., Lemaître, A., Bloch, J., Roussignol, Ph. & Delalande, C. (2006) 'Parametric oscillation in vertical triple microcavities', *Nature* **440**, 904–907.

[26] El Daïf, O., Baas, A., Guillet, T., Brantut, J.-P., Idrissi Kaitouni, R., Staehli, J. L., Morier-Genoud, F. & Deveaud, B. (2006) 'Polariton quantum boxes in semiconductor microcavities', *Appl. Phys. Lett.* **88**, 061105.

[27] Bajoni, D., Senellart, P., Wertz, E., Sagnes, I., Miard, A., Lemaître, A. & Bloch, J. (2008) 'Polariton laser using single micropillar GaAsGaAlAs semiconductor cavities', *Phys. Rev. Lett.* **100**, 047401.

[28] Wertz, E., Ferrier, L., Solnyshkov, D. D., Johne, R., Sanvitto, D., Lemaître, A., Sagnes, I., Grousson, R., Kavokin, A. V., Senellart, P., Malpuech, G. & Bloch, J. (2010)

'Spontaneous formation and optical manipulation of extended polariton condensates', *Nat. Phys.* **6**, 860–864.

[29] Galbiati, M., Ferrier, L., Solnyshkov, D. D., Tanese, D., Wertz, E., Amo, A., Abbarchi, M., Senellart, P., Sagnes, I., Lemaître, A., Galopin, E., Malpuech, G. & Bloch, J. (2012) 'Polariton condensation in photonic molecules', *Phys. Rev. Lett.* **108**, 126403.

[30] Jacqmin, T., Carusotto, I., Sagnes, I., Abbarchi, M., Solnyshkov, D. D., Malpuech, G., Galopin, E., Lemaître, A., Bloch, J. & Amo, A. (2014) 'Direct observation of dirac cones and a flatband in a honeycomb lattice for polaritons', *Phys. Rev. Lett.* **112**, 116402.

[31] Balili, R., Hartwell, V., Snoke, D., Pfeiffer, L. & West, K. (2007) 'Bose-Einstein condensation of microcavity polaritons in a trap', *Science* **316**, 1007–1010.

[32] Lai, C. W., Kim, N. Y., Utsunomiya, S., Roumpos, G., Deng, H., Fraser, M. D., Byrnes, T., Recher, P., Kumada, N., Fujisawa, T. & Yamamoto, Y. (2007) 'Coherent zero-state and π-state in an excitonpolariton condensate array', *Nature* **450**, 529–532.

[33] Hopfield, J. J. (1958) 'Theory of the contribution of excitons to the complex dielectric constant of crystals', *Phys. Rev.* **112**, 1555–1567.

[34] Tassone, F., Piermarocchi, C., Savona, V., Quattropani, A. & Schwendimann, P. (1997) 'Bottleneck effects in the relaxation and photoluminescence of microcavity polaritons', *Phys. Rev. B* **56**, 7554.

[35] Richard, M., Kasprzak, J., Romestain, R., André, R. & Le Si Dang (2005) 'Spontaneous coherent phase transition of polaritons in CdTe microcavities', *Phys. Rev. Lett.* **94**, 187401.

[36] Richard, M., Kasprzak, J., André, R., Romestain, R., Le Si Dang, Malpuech, G. & Kavokin, A. (2005) 'Experimental evidence for nonequilibrium Bose condensation of exciton polaritons', *Phys. Rev. B* **72**, 201301.

[37] Tsotsis, P., Eldridge, P. S., Gao, T., Tsintzos, S. I., Hatzopoulos, Z. & Savvidis, P. G. (2012) 'Lasing threshold doubling at the crossover from strong to weak coupling regime in GaAs microcavity', *New J. Phys.* **14**, 023060.

[38] Christmann, G., Tosi, G., Berloff, N. G., Tsotsis, P., Eldridge, P. S., Hatzopoulos, Z., Savvidis, P. G. & Baumberg, J. J. (2012) 'Polariton ring condensates and sunflower ripples in an expanding quantum liquid', *Phys. Rev. B* **85**, 235303.

[39] Wertz, E., Ferrier, L., Solnyshkov, D. D., Senellart, P., Bajoni, D., Miard, A., Lemaître, A., Malpuech, G. & Bloch, J. (2009) 'Spontaneous formation of a polariton condensate in a planar GaAs microcavity', *Appl. Phys. Lett.* **95**, 051108.

[40] Sich, M., Krizhanovskii, D. N., Skolnick, M. S., Gorbach, A. V., Hartley, R., Skryabin, D. V., Cerda-Méndez, E. A., Biermann, K., Hey, R. & Santos, P. V. (2012) 'Observation of bright polariton solitons in a semiconductor microcavity', *Nat. Photon.* **6**, 50–55.

[41] Sanvitto, D., Marchetti, F. M., Szymańska, M. H., Tosi, G., Baudisch, M., Laussy, F. P., Krizhanovskii, D. N., Skolnick, M. S., Marrucci, L., Lemaître, A., Bloch, J., Tejedor, C. & L. Viña (2010) 'Persistent currents and quantized vortices in a polariton superfluid', *Nat. Phys.* **6**, 527–533.

[42] Amo, A., Pigeon, S., Adrados, C., Houdré, R., Giacobino, E., Ciuti, C. & Bramati A. (2010) 'Light engineering of the polariton landscape in semiconductor microcavities' *Phys. Rev. B* **82**, 081301.

[43] Sanvitto, D., Pigeon, S., Amo, A., Ballarini, D., De Giorgi, M., Carusotto, I., Hivet, R., Pisanello, F., Sala, V. G., Guimaraes, P. S. S., Houdré, R., Giacobino, E., Ciuti, C., Bramati, A. & Gigli, G. (2011) 'All-optical control of the quantum flow of a polariton condensate', *Nat. Photon.* **5**, 610–614.

[44] Baas, A., Karr, J. Ph., Eleuch, H. & Giacobino, E. (2004) 'Optical bistability in semiconductor microcavities', *Phys. Rev. A* **69**, 023809.

[45] Ballarini, D., De Giorgi, M., Cancellieri, E., Houdré, R., Giacobino, E., Cingolani, R., Bramati, A., Gigli, G. & Sanvitto, D. (2012) 'All-optical polariton transistor', *Nat. Comms.* **4**, 1778.

[46] Liew, T. C. H., Kavokin, A. V. & Shelykh, I. A. (2008) 'Optical circuits based on polariton neurons in semiconductor microcavities', *Phys. Rev. Lett.* **101**, 016402.

[47] Liew, T. C. H., Shelykh, I. A. & Malpuech, G. (2011) 'Polaritonic devices', *Physica E* **43**, 1543–1568.

[48] Gao, T., Eldridge, P. S., Liew, T. C. H., Tsintzos, S. I., Stavrinidis, G., Deligeorgis, G., Hatzopoulos, Z. & Savvidis, P. G. (2012) 'Polariton condensate transistor switch', *Phys. Rev. B* **85**, 235102.

[49] Antón, C., Liew, T. C. H., Sarkar, D., Martín, M. D., Hatzopoulos, Z., Eldridge, P. S., Savvidis, P. G. & Viña, L. (2014) 'Operation speed of polariton condensate switches gated by excitons', *Phys. Rev. B* **89**, 235312.

[50] Nguyen, H. S., Vishnevsky, D., Sturm, C., Tanese, D., Solnyshkov, D., Galopin, E., Lemaître, A., Sagnes, I., Amo, A., Malpuech, G. & Bloch, J. (2013) 'Realization of a double-barrier resonant tunneling diode for cavity polaritons', *Phys. Rev. Lett.* **110**, 236601.

[51] Sturm, C., Tanese, D., Nguyen, H. S., Flayac, H., Galopin, E., Lemaître, A., Sagnes, I., Solnyshkov, D., Amo, A., Malpuech, G. & Bloch, J. (2014) 'All-optical phase modulation in a cavity-polariton Mach-Zehnder interferometer', *Nat. Comms.* **5**, 3278.

[52] Tosi, G., Christmann, G., Berloff, N. G., Tsotsis, P., Gao, T., Hatzopoulos, Z., Savvidis, P. G. & Baumberg, J. J. (2012a) 'Sculpting oscillators with light within a nonlinear quantum fluid', *Nat. Phys.* **8**, 190–194.

[53] Manni, F., Lagoudakis, K. G., Liew, T. C. H., André, R. & Deveaud-Plédran, B. (2011) 'Spontaneous pattern formation in a polariton condensate', *Phys. Rev. Lett.* **107**, 106401.

[54] Askitopoulos, A., Ohadi, H., Kavokin, A. V., Hatzopoulos, Z., Savvidis, P. G. & Lagoudakis, P. G. (2013) 'Polariton condensation in an optically induced two-dimensional potential', *Phys. Rev. B* **88**, 041308.

[55] Aßmann, M., Veit, F., Bayer, M., Löffler, A., Höfling, S., Kamp, M. & Forchel, A. (2012) 'All-optical control of quantized momenta on a polariton staircase', *Phys. Rev. B* **85**, 155320.

[56] Cristofolini, P., Dreismann, A., Christmann, G., Franchetti, G., Berloff, N. G., Tsotsis, P., Hatzopoulos, Z., Savvidis, P. G. & Baumberg, J. J. (2013) 'Optical superfluid phase transitions and trapping of polariton condensates', *Phys. Rev. Lett.* **110**, 186403.

[57] Dreismann, A., Cristofolini, P., Balili, R., Christmann, G., Pinsker, F., Berloff, N. G., Hatzopoulos, Z., Savvidis, P. G. & Baumberg, J. J. (2014) 'Coupled counterrotating polariton condensates in optically defined annular potentials', *Proc. Natl. Acad. Sci. U.S.A.* **111**, 8770–8775.

[58] Baas, A., Lagoudakis, K. G., Richard, M., André, R., Le Si Dang & Deveaud-Plédran, B. (2008) 'Synchronized and desynchronized phases of exciton-polariton condensates in the presence of disorder', *Phys. Rev. Lett.* **100**, 170401.

[59] Christmann, G., Tosi, G., Berloff, N. G., Tsotsis, P., Eldridge, P. S., Hatzopoulos, Z., Savvidis, P. G. & Baumberg, J. J. (2014) 'Oscillatory solitons and time-resolved phase locking of two polariton condensates', *New J. Phys.* **16**, 103039.

[60] Kippenberg, T. J., Holzwarth, R. & Diddams, S. A. (2011) 'Microresonator-based optical frequency combs', *Science*, **332**, 555–559.

[61] Burger, S., Bongs, K., Dettmer, S., Ertmer, W., Sengstock, K., Sanpera, A., Shlyapnikov, G. V. & Lewenstein, M. (1999) 'Dark solitons in Bose-Einstein condensates', *Phys. Rev. Lett.* **83**, 5198.

[62] Weller, A., Ronzheimer, J. P., Gross, C., Esteve, J., Oberthaler, M. K., Frantzeskakis, D. J., Theocharis, G. & Kevrekidis, P. G. (2008) 'Experimental observation of oscillating and interacting matter wave dark solitons', *Phys. Rev. Lett.* **101**, 130401.

[63] Feng, M., Silverman, K. L., Mirin, R. P. & Cundiff, S. T. (2010) 'Dark pulse quantum dot diode laser', *Opt. Express* **18**, 13385.

[64] Yulin, A. V., Egorov, O. A., Lederer, F. & Skryabin, D. V. (2008) 'Dark polariton solitons in semiconductor microcavities', *Phys. Rev. A* **78**, 061801.

[65] Egorov, O. A., Skryabin, D. V., Yulin, A. V. & Lederer, F. (2009) 'Bright cavity polariton solitons', *Phys. Rev. Lett.* **102**, 153904.

[66] Tosi, G., Christmann, G., Berloff, N. G., Tsotsis, P., Gao, T., Hatzopoulos, Z., Savvidis, P. G. & Baumberg, J. J. (2012b) 'Geometrically locked vortex lattices in semiconductor quantum fluids', *Nat. Comms.* **3**, 1243.

[67] Tikhonenko, V., Christou, J., Luther-Davies, B. & Kivshar, Y. S. (1996) 'Observation of vortex solitons created by the instability of dark soliton stripes', *Opt. Lett.* **21**, 1129–1131.

[68] Manni, F., Liew, T. C. H., Lagoudakis, K. G., Ouellet-Plamondon, C., André, R., Savona, V. & Deveaud, B. (2013) 'Spontaneous self-ordered states of vortex-antivortex pairs in a polariton condensate', *Phys. Rev. B* **88**, 201303.

[69] Hivet, R., Cancellieri, E., Boulier, T., Ballarini, D., Sanvitto, D., Marchetti, F. M., Szymanska, M. H., Ciuti, C., Giacobino, E. & Bramati, A. (2014) *Phys. Rev. B* **89**, 134501.

23

Disorder, Synchronization, and Phase-locking in Nonequilibrium Bose-Einstein Condensates

PAUL R. EASTHAM

School of Physics and Centre for Research on Adaptive Nanostructures and Nanodevices (CRANN), Trinity College Dublin, Ireland

BERND ROSENOW

Institut für Theoretische Physik, Universität Leipzig, Germany

We review some theories of nonequilibrium Bose-Einstein condensates in potentials, in particular of the Bose-Einstein condensate of polaritons. We discuss such condensates, which are steady states established through a balance of gain and loss, in the complementary limits of a double-well potential and a random disorder potential. For equilibrium condensates, the former corresponds to a Josephson junction, whereas the latter is the setting for the superfluid/Bose glass transition. We explore the nonequilibrium generalization of these phenomena and highlight connections with mode selection and synchronization.

23.1 Introduction

It is twenty years since Bose-Einstein condensation (BEC) was achieved, in its ideal setting of a weakly interacting ultracold gas. In other settings, namely superconductivity (which we understand in terms of a Bose-Einstein condensate of Cooper pairs), Bose-Einstein condensates have been available in laboratories for over a century. Yet their behaviour is still startling. Because the many particles of the condensate occupy the same quantum state, collective properties become described by a macroscopic wavefunction, with an interpretation parallel to that of the single-particle wavefunction of Schrödinger's equation. Thus many of the phenomena of single-particle quantum mechanics appear as behaviours of the condensate.

At the mean-field level, a BEC is described by an order parameter Ψ, which is a complex field $\Psi(r,t) = \sqrt{n}e^{i\phi}$. Its square modulus is the local condensate or superfluid density, and it obeys the Schrödinger-like Gross-Pitaevskii equation. Because the order is described by a complex field, i.e., there is a spontaneous breaking of a U(1) symmetry (see Chapter 5), there is a new conserved current, given by the usual probability current of a wavefunction. This describes the condensate contribution to the macroscopic current flow.

The wavelike behaviour of condensates leads to interesting effects in a potential. For the simple double-well potential, and related two-state problems, one obtains the Josephson effects [1]. In particular, the double-well supports a d.c. Josephson state, with a current flowing due to a difference in the phases ϕ of the two wells. Because the phase is compact, there is a maximum current supported by such a state; attempting to impose a larger current by external bias typically leads to an a.c. Josephson state, where the relative phase oscillates or winds. The other extreme is a complex disorder potential, where it is natural to ask whether the ordered state survives, i.e., whether a global phase is established, and hence whether the system supports superfluidity [2]. This problem is closely related to wave localization, and the result is that superfluidity persists up to a critical disorder strength where the order is destroyed, leading to a glasslike state.

The aim of this chapter is to review some theories of how these phenomena generalize to nonequilibrium Bose-Einstein condensates. We have in mind, primarily, the Bose-Einstein condensate of polaritons. Here there is a continuous gain and loss of particles in the condensate, due to pumping and decay. However, the concepts are also relevant to other topical nonequilibrium condensates, including those of magnons and photons, and are linked to aspects of laser physics. Our aim is not a comprehensive review. Rather, we hope to indicate a unifying framework for understanding nonequilibrium condensates in inhomogeneous settings, from Josephson-like double-well systems to complex disorder potentials. We think that these problems can be understood in terms of the synchronization and phase-locking of coupled oscillators, as well as the related phenomenology of mode selection in lasers. The connection between synchronization and the physics of equilibrium Josephson junctions is well known, and reviewed, for example, in Ref. [3].

23.2 Models of Nonequilibrium Condensation

As the basis for discussing nonequilibrium BEC, we will use a generalization of the Gross-Pitaevskii equation (GPE) in the form [4]

$$i\hbar\frac{\partial\Psi}{\partial t} = \left[-\frac{\hbar^2}{2m}\nabla^2 + V(\mathbf{x})\right]\Psi + U|\Psi|^2\Psi + i\big(\gamma - \Gamma|\Psi|^2\big)\Psi, \qquad (23.1)$$

for definiteness supposing two space dimensions, as appropriate to microcavity polaritons. $\Psi(\mathbf{x}, t)$ is the macroscopic wavefunction for the condensate. The first three terms on the right-hand side comprise the usual GPE [1], with contributions from the kinetic energy, potential energy, and repulsive interactions. The terms in the final bracket model, in a phenomenological way, a continual gain and loss of particles in the condensate, due to scattering into and out of external incoherent reservoirs. Noting that $|\Psi|^2$ is the condensate density, we see that the term proportional to $i\gamma$ generates an exponential growth or decay of the condensate. In general, it models both stimulated scattering into the condensate from one reservoir and spontaneous emission out of it into another. If the former exceeds the latter, then $\gamma > 0$ and the net effect is an exponential growth, so that $\gamma = 0$ marks the threshold for condensation. Above threshold, the condensate density builds up and the growth rate is reduced by the nonlinear term proportional to Γ, reaching zero at a steady-state density, which, in the homogeneous case, is $n_0 = \gamma/\Gamma$. In the language of laser physics, this final term is the lowest-order nonlinear gain [5], describing the depletion of the gain by the buildup of the condensate. The scattering of particles into the condensate causes, in addition to its growth, a reduction of the occupation in the gain medium, and hence a reduction in the linear gain. The linear growth rate, γ, can of course vary with position, for example, where the external pumping, and so the reservoir population, is inhomogeneous.

The generalized Gross-Pitaevskii model (23.1) was introduced for polariton Bose-Einstein condensation by Keeling and Berloff [4]. It is closely related to the GPE introduced by Wouters and Carusotto [6], in which the gain depends explicitly on a reservoir population, which in turn obeys a related first-order rate equation. Such a theory reduces to (23.1) if the reservoir population can be adiabatically eliminated, and the gain expanded in powers of the condensate density. Whether this is correct will depend on the scattering rates in the reservoir and hence the relaxation time for its population.

There are many other interesting extensions of the model (23.1) that may be considered. In particular, it is a mean-field equation that ignores the stochastic nature of the gain and loss process. In reality, these lead to fluctuations in the condensate density, which have an observable signature in the finite linewidth of the light emitted from the polariton condensate [7] (i.e., a finite correlation time for the U(1) phase). Both single-mode and multimode theories including such fluctuations have been developed within the density matrix formalism [8]. They can be treated within the GPE by introducing stochastic terms, related by the fluctuation-dissipation theorem to the gain and loss [9, 10]. Such stochastic GPEs have been derived from the truncated Wigner approximation [9] and used to study the coherence properties of polariton condensates. Another potentially important extension is to allow some degree of thermalization with the reservoirs, which corresponds to a frequency-dependent gain.

In the following, we shall focus on two specific applications of (23.1). Firstly, we consider a double well with a single relevant orbital on each side. In the usual way [11], we may expand the wavefunction in terms of the amplitudes for the left and right wells as $\Psi(\mathbf{x}, t) = \Psi_l(t)\phi_l(\mathbf{x}) + \Psi_r(t)\phi_r(\mathbf{x})$, where $\phi_{l,r}$ are wavefunctions localized on the left and right. Inserting this into (23.1) gives the equations for the amplitudes $\Psi_{l,r}$:

$$\begin{aligned} i\hbar\frac{d\Psi_l}{dt} &= \frac{\epsilon}{2}\Psi_l - J\Psi_r + U_l|\Psi_l|^2\Psi_l + i[g_l - \Gamma_l|\Psi_l|^2]\Psi_l, \\ i\hbar\frac{d\Psi_r}{dt} &= -\frac{\epsilon}{2}\Psi_r - J\Psi_l + U_r|\Psi_r|^2\Psi_r + i[g_r - \Gamma_r|\Psi_l|^2]\Psi_r, \end{aligned} \tag{23.2}$$

assuming the overlap of ϕ_l and ϕ_r is small. Here ϵ is the energy difference between the wells, and J is the tunneling matrix element. $g_{l,r}$ corresponds to the gain/loss of each well, $g_{l,r} = \int \phi^*_{l,r}\gamma\phi_{l,r}\, d^2\mathbf{x}$. If the pumping is uniform, γ is independent of position and $g_l = g_r = \gamma$. $U_{l,r}$ and $\Gamma_{l,r}$ are the nonlinearities for each well, $\Gamma_{l,r} = \int \left|\phi_{l,r}\right|^4 \Gamma\, d^2\mathbf{x}$. Secondly, we shall consider a nonequilibrium condensate in a random disorder potential. Thus we will consider Eq. (23.1) with V being a Gaussian random potential, whose correlation function is characterized by its first two moments, which we take to be $\langle\langle V(\mathbf{x})\rangle\rangle = 0$ and $\langle\langle V(\mathbf{x})V(\mathbf{y})\rangle\rangle = V_0^2\delta^{(2)}(\mathbf{x}-\mathbf{y})$, where angle brackets denote an average over disorder realizations.

23.3 Josephson Effects and Synchronization

The physics of synchronization and phase locking is well described elsewhere, for example, in Ref. [3], so we summarize it only briefly. The starting point is the idea that self-sustained oscillators, which oscillate at their own frequencies when isolated, can be coupled together. We will say that oscillators are synchronized if they oscillate at a common frequency. Synchronization is the phenomenon that oscillators become synchronized when coupled. This occurs above a critical coupling which increases with the detuning, i.e., the difference in frequencies when uncoupled. Two oscillators with the same intrinsic frequency are of course synchronized, in our sense, even for zero coupling, but for detuned oscillators a nonzero coupling is required if they are to establish a common frequency. We will also use the term phase-locked, by which we mean that the coupling establishes some definite relation between the phases of the oscillations. This is stronger than our notion of synchronized. Note that the definitions of these terms are not standardized, and some other authors use them somewhat differently.

The relevance of this physics to the nonequilibrium Josephson junction, Eq. (23.2), is immediate. When the tunneling $J = 0$, the equations decouple. Each well is a self-sustained oscillator with its own frequency. The steady-state amplitude of the left well, for example, is $\Psi_l = \sqrt{n_{0,l}}e^{-i\omega_l t + i\theta_l}$, with occupation $n_{0,l} = g_l/\Gamma_l$. $\omega_l = (\epsilon/2 + U_l n_{0,l})/\hbar$ is the frequency of this oscillator, with a corresponding expression, with $\epsilon \to -\epsilon$, $l \to r$, for the other. θ_l and θ_r are arbitrary, and independent, phase offsets.

The Josephson coupling term, proportional to J, allows these oscillators to drive one another. Because the oscillators are nonlinear, as described by the terms proportional to both U and Γ, their phases become coupled. A physical picture of this is that as the oscillators force one another, their amplitudes change, which changes their frequency difference through the nonlinearity and hence shifts their relative phase. This can establish a steady state with a constant relative phase and a single frequency.

As a simple model of synchronization, one might suggest that a suitably defined relative phase δ should obey an equation of the form [3]

$$\frac{d\delta}{dt} = -(\omega_l - \omega_r) + c\sin(\delta), \tag{23.3}$$

on the grounds that the first term generates the appropriate winding when the oscillators are uncoupled, and the second is the simplest coupling one can write consistent with the 2π periodicity. This is the Adler equation, which can be seen to have solutions of both constant relative phase and continuously increasing relative phase. The case of a constant relative phase corresponds to a steady-state solution to Eq. (23.2) of the form

$$\Psi_{l,r} = \sqrt{n_{l,r}}e^{-i\omega t \pm i\theta/2}e^{i\theta_0}, \tag{23.4}$$

which contains a single frequency ω and a single undetermined phase θ_0.

The conditions for a synchronized solution for the dissipative double well can be established by inserting Eq. (23.4) into Eq. (23.2) and examining whether there are physical solutions to the resulting equations. This approach was taken by Wouters [12], using a slightly more complex model. He obtains the conditions on the detuning and tunneling required for the synchronized solution and predicts properties of the states. The dynamics of the two-mode problem is also treated in this way in Ref. [13], where the two modes correspond to two polarizations. A recent numerical analysis of that problem can be found in Ref. [10].

We summarize this type of steady-state analysis of the two-mode model using Eq. (23.2). For simplicity, we take the two wells to be identical, so that $g_l = g_r = g$, etc. It is convenient to choose $n_0 = g/\Gamma$ to be the density scale, by replacing $\Psi_l \to \Psi_l\sqrt{n_0}$, so that $n_{l,r} = 1$ in the uncoupled steady state. We also set $\hbar = 1$ and take as the energy scale Un_0. From (23.2) and (23.4) we then find

$$\alpha(1 - n_l)n_l = -J\sqrt{n_l n_r}\sin(\theta) = -\alpha(1 - n_r)n_r, \tag{23.5}$$

$$E_l n_l = J\sqrt{n_l n_r}\cos(\theta) = E_r n_r, \tag{23.6}$$

where α, J, E_l and E_r are energies measured in units of Un_0. $\alpha = g/(Un_0) = \Gamma/U$ is a dimensionless measure of the gain, and $E_{l,r} = \pm\frac{\epsilon}{2} - \omega + n_{l,r}$ the energies of each well, including the mean-field shifts, relative to ω. Note that $\alpha \to 0$ corresponds to the unpumped Josephson junction, whereas $\alpha \to \infty$ is the limit where the interaction U is negligible as, for example, in a laser.

Eq. (23.5) describes current flows in the nonequilibrium double well. The term on the far left corresponds to the net current flowing between the reservoirs and the left well. If the density there deviates from $n_l = 1$, the gain will no longer be reduced to zero by the nonlinear term, and there will be a source ($n_l < 1$) or sink ($n_l > 1$) of particles. In a steady state, this current must flow into the other well, as the Josephson current visible in the center of the chain of equalities. It must then match the current flowing between the right well and the reservoirs, which is the quantity on the right. Eq. (23.6) is a related condition, stating that the two wells must be in mechanical equilibrium through the Josephson coupling.

The presence of trigonometric functions in Eqs. (23.5) and (23.6), which have magnitude less than one, is why the synchronized solution only exists over limited parameter regimes. In particular, Eq. (23.5) limits the range of J, in which there is a synchronized solution, and Eq. (23.6) limits the range of detunings ϵ. Consider, for example, starting in a synchronized solution with $\epsilon = 0$. Then as ϵ increases, the real part of the steady-state equation, which is essentially Schrödinger's equation for a double well, will concentrate the wavefunction to one side or another. For such a wavefunction, the pumping will generate a net interwell current. If this exceeds the Josephson critical current $J\sqrt{n_l n_r}$, then the synchronized steady state breaks down. The transition is thus analogous to that between the d.c. and a.c. Josephson effects, but with currents generated by gain and loss rather than external bias.

A complementary route to understanding the physics of nonequilibrium condensates in potentials, and particularly the presence of both synchronized and desynchronized states, is to relate it to that of mode selection in lasers. For polariton condensates, this was done by one of us [14] using the model of Eq. (23.1). We outline it here for comparison.

The general approach is to take those parts of Eqs. (23.1) or (23.2) that form the Schrödinger equation as an unperturbed problem. The remainder can then be dealt with using a form of degenerate perturbation theory. To do this, we expand the solutions in terms of the orbitals, which are eigenfunctions of the first bracket in Eq. (23.1). We assume two states, for simplicity, and write $\Psi(\mathbf{x}, t) = \Psi_1(t)\phi_1(\mathbf{x}) + \Psi_2(t)\phi_2(\mathbf{x})$. The time dependence of the unperturbed amplitudes will be $\Psi_{1,2}(t) = e^{-iE_{1,2}t/\hbar}$, where $E_{1,2}$ are the energies of $\phi_{1,2}$. For Eqs. (23.2), the Schrödinger part is explicitly diagonalized by a rotation, writing $\Psi_{l,r} = \cos(\theta)\Psi_1 \mp \sin(\theta)\Psi_2$ and choosing $\tan(2\theta) = -2J/\epsilon$.

Such a unitary transformation leaves the equations-of-motion for the amplitudes Ψ_1 and Ψ_2 coupled, because it neither diagonalizes the interactions, nor the linear gain terms (unless $g_l = g_r$). However, these off-diagonal terms can be neglected if their magnitudes are small compared with the unperturbed level spacing $E_1 - E_2$. For the nonlinear couplings, this requires that the nonlinearities (both the mean-field shift Un_0 and the corresponding scale from the gain depletion, Γn_0) are small compared with the level spacing.

In the equations-of-motion, the off-diagonal couplings correspond to terms which oscillate at frequencies of order $E_1 - E_2$ and therefore average to zero. In the energy functional, they are terms such as $\Psi_1^*\Psi_1^*\Psi_2\Psi_2$ (from the interactions) and $\Psi_1^*\Psi_2$ (from the linear gain), which, in a quantized theory, describe scattering processes that do not conserve energy. Retaining only the resonant terms gives

$$i\hbar\frac{d\Psi_{1,2}}{dt} = \left[E_{1,2} + ig_{1,2} + (U - i\Gamma)\left(\eta_{1,2}|\Psi_{1,2}|^2 + 2\beta|\Psi_{2,1}|^2\right)\right]\Psi_{1,2}, \qquad (23.7)$$

where we suppose only two states, and nonlinearities which do not depend on position, for simplicity. Here $g_{1,2}$ are diagonal matrix elements of the linear gain for orbitals ϕ_1, ϕ_2, cf., the discussion after Eq. (23.2). $\eta_{1,2} = \int |\phi_{1,2}|^4 d^2\mathbf{x}$ and $\beta = \int |\phi_1|^2|\phi_2|^2 d^2\mathbf{x}$ are matrix elements of the nonlinearities within and between the single-particle orbitals, respectively. It follows from Eq. (23.7) that the occupations obey

$$\hbar\frac{\partial n_{1,2}}{\partial t} = 2\left[g_{1,2} - \Gamma\left(\eta_{1,2}n_{1,2} + 2\beta n_{2,1}\right)\right]n_{1,2}. \qquad (23.8)$$

Equations such as (23.8) describe mode competition in lasers with local gain [5]. The linear term gives an exponential growth of each mode, which is controlled by gain depletion effects. The buildup of one mode of course reduces its own gain, as described by the term proportional to η, but it also reduces the gain for any other mode which shares the same gain medium, i.e., overlaps in space. This cross-gain depletion is the term proportional to β.

The steady-state structure of Eq. (23.8) is straightforward to determine and reflects the density profiles of the orbitals [14]. For the general two-mode case, there are states in which only one of n_1, n_2 is nonzero. These are synchronized states, as they have only a single oscillation frequency, which corresponds to the energy of the occupied orbital, shifted by interactions. Note that although the orbitals involved are linear eigenstates, the synchronization itself is due to nonlinearities: it is the nonlinear gain which selects an eigenstate in which to form the condensate. Furthermore, there are also states in which $n_1 \neq 0, n_2 \neq 0$. These are the desynchronized states, with oscillations at two frequencies, analogous to the a.c. Josephson state.

We conclude this discussion by commenting on a few of the many experiments addressing these issues with microcavity polaritons. For polaritons, the difference between synchronized and desynchronized states is immediate, because Ψ is the amplitude of the macroscopic electric field in the microcavity. The spectrum of Ψ thus corresponds to the spectrum of the light emitted from the microcavity. Thus in a potential with two relevant orbitals, the synchronized state has one single narrow emission line and the desynchronized state two. In the language of laser physics, this is the distinction between single-mode and multimode lasing.

The double well was studied experimentally in Ref. [15], and oscillations in the intensity observed in the time domain. Such oscillations would be expected where two linear eigenmodes of a double well are both highly occupied, i.e., in a multimode condensate. The experiment, however, is not completely consistent with that picture. The density oscillations have a deterministic phase, implying that there are processes which fix the relative phase of the two macroscopically occupied orbitals. These can be found among the terms neglected above. Simulations including them can be found alongside the experiments, and show good agreement.

Particularly in extended geometries, where there is a potential due to in-plane disorder, polariton condensates do emit at many distinct frequencies [16]. A detailed study of the spectra was performed by Baas et al. [17], who identified pairs of modes which, while having independent frequencies at low densities, locked to a single frequency above a critical density. The low density state thus appears consistent with Eq. (23.7). The transition to a synchronized state in the few-mode problem occurs due to the neglected nonlinear coupling terms. In particular, increasing density in a multimode solution leads first to nonlinear mixing effects, which finally drive the formation of a synchronized state. This is shown numerically in Ref. [14]. On the other hand, Thunert et al. [18] find statistical evidence that disorder effects persist at high densities, for a nonequilibrium condensate in a random potential.

23.4 The Bose Glass and Phase Locking

We now turn to consider the complementary problem of a nonequilibrium condensate in a random potential, reviewing first the interplay of disorder and BEC in two dimensions, as discussed in [2]. Due to the localization effects of randomness, one expects that sufficiently strong disorder causes a destruction of superfluidity. The resulting phase is called a Bose glass and is characterized by a finite compressibility $\kappa = \partial n/\partial \mu$, has gapless excitations with a finite density of states at zero energy, and as a consequence has infinite superfluid susceptibility. The susceptibility is determined by the ensemble averaged retarded correlation function

$$G^R(\mathbf{x},t) = -i\theta(t)\langle\langle[\Psi(\mathbf{x},t),\Psi^\dagger(0,0)]\rangle\rangle, \tag{23.9}$$

where $[,]$ denotes the commutator, $\langle\langle\cdots\rangle\rangle$ the combined average over disorder and quantum fluctuations, and $\theta(t)$ the Heaviside step function. Due to the localizing effects of disorder, G^R decays exponentially as a function of distance. The local Green function can be represented in terms of the single-particle density of states $\rho(\omega)$ as

$$G^R(0,t) = 2\pi i\theta(t)\int_0^\infty d\omega\rho(\omega)\, e^{-i\omega t} . \tag{23.10}$$

Here, the quasi-particle energy is measured with respect to the chemical potential. When there is a finite density of states at zero energy, $\rho(0) \neq 0$, then the long time asymptotics of the Green function are given by $G^R(t) \sim 2\pi\rho(0)\theta(t)/t$, giving rise to a divergent uniform superfluid susceptibility $\chi = \int d^d r dt G^R(\mathbf{r},t)$. The susceptibility is dominated by rare localized regions which have anomalously low quasiparticle excitation energies. While much of the focus in [2] is on the scaling behavior of the superfluid Bose glass and also the superfluid insulator transition, the competition between a disorder potential and its screening by a weak repulsive interaction in the presence of a harmonic trap was discussed in [19].

An analysis of the superfluid to Bose-glass transition with a focus on poariton condensates was presented in [20]. There, the excitation spectrum of the system was obtained by computing stationary solutions of the GPE, $\Psi_j(\mathbf{r},t) = \Psi_j(\mathbf{r})\exp(-i\omega_j t)$. From these solutions the retarded Green function can be obtained as

$$G^R(\mathbf{x},\mathbf{x}';\omega) = \sum_j \frac{\Psi_j(\mathbf{x})\Psi_j^\star(\mathbf{x}')}{\omega - \omega_j + i\eta}. \tag{23.11}$$

Fourier transformation with regard to $\mathbf{r}-\mathbf{r}'$ leads to an excitation spectrum comparable to experimental observations [20]. For a chemical potential below the bottom of the disordered parabolic band, the system is almost empty, there are no excitations at the chemical potential, and the superfluid susceptibility does not diverge. Introducing a finite density of bosons, first the potential minima (traps) with the lowest energy are filled. Due to the density-dependent blue shift, the density in each trap adjusts itself in such a way that it is filled up to the chemical potential, and there are many low-energy excitations, giving rise to a diverging superfluid susceptibility as discussed above. The compressibility of this Bose glass state is finite since in the absence of a periodic potential there are many states available to be filled. Increasing the density further, the local condensates increase in size, until the different condensate puddles connect with each other to form a percolating cluster, which then represents a global superfluid. The order parameter for this transition is the superfluid density or superfluid fraction as discussed below in Eq. (23.17). In [20], the superfluid fraction is computed numerically as a function of the condensate density, and good agreement is found with the static analytical calculation discussed below.

Insight about the destruction of superfluidity by disorder can be gained from an Imry-Ma type argument for the pinning by weak disorder [19, 21, 22] of a fragmented condensate cloud of radius R. Localizing the condensate within a spatial region of radius R costs kinetic energy, but allows the condensate to lower its potential energy by taking advantage of local minima of the disorder potential, giving rise to a total energy

$$\epsilon(R) = \frac{\hbar^2}{2m}\frac{1}{R^2} - \frac{V_0}{\sqrt{\pi}R} . \tag{23.12}$$

The disorder energy decreases inversely proportional to the square root of the cloud area due to averaging over independent local fluctuations of the random potential. Minimizing this energy yields the density Larkin length $\mathcal{L}_{\mathrm{n}} = \sqrt{\pi}\hbar^2/mV_0$. Superfluidity is destroyed when the density Larkin length is equal to the healing length $\xi = \sqrt{\hbar^2/2mn_0U}$, giving rise to a critical density for the onset of superfluidity $n_c = mV_0^2/2\pi\hbar^2U$.

In the following, we provide a more quantitative discussion of the Bose glass superfluid transition and include the influence of nonequilibrium. We analyze the dimensionless form of Eq. (23.1),

$$i\frac{\partial\Psi}{\partial t} = \left(-\nabla^2 + \vartheta + |\Psi|^2\right)\Psi + i\alpha\left(1 - |\Psi|^2\right)\Psi, \tag{23.13}$$

where we measure length in units of the healing length ξ, energy in units of the blueshift n_0U, and time in units of $\hbar/n_0U$. The strength of the non-equilibrium fluctuations is controlled by the parameter $\alpha = \Gamma/U$. We assume that the correlation length of the disorder potential is the shortest length scale in the problem, such that the Gaussian random potential can be characterized by its average values $\langle\langle\vartheta(\mathbf{x}) = 0\rangle\rangle$ and $\langle\langle\vartheta(\mathbf{x})\vartheta(\mathbf{y})\rangle\rangle = \kappa^2\delta(\mathbf{x}-\mathbf{y})$. The dimensionless disorder strength is related to the dimensionful parameters via $\kappa = V_0/\xi^{d/2}n_0U$.

In the synchronized regime, the polariton condensate emits coherent light at a frequency ω, which can be described by the ansatz

$$\Psi(\mathbf{x}, t) = \sqrt{n(\mathbf{x})}\, e^{i\phi(\mathbf{x}) - i\omega t}. \tag{23.14}$$

Following the discussion in [23], the condensate frequency ω can be computed by inserting this ansatz into Eq. (23.13). The real part gives

$$\omega = (\nabla\phi)^2 + \frac{1}{4}\frac{(\nabla n)^2}{n^2} - \frac{1}{2}\frac{\nabla^2 n}{n} + n + \vartheta, \tag{23.15}$$

which is a pressure balance equation analogous to Eq. (23.6). The imaginary part gives the analog of Eq. (23.5),

$$\nabla \cdot (n\nabla\phi) = \alpha n(1-n), \tag{23.16}$$

which is a continuity equation for the supercurrent, including the sources and sinks generated, via the gain depletion, by density fluctuations.

In the absence of disorder, the Bogolubov excitation mode is diffusive out of equilibrium [6], and a naive application of the Landau stability criterion yields a vanishing superfluid velocity. However, taking into account the imaginary part of the excitation energies, the drag force on a small moving object and the onset of fringes in the density profile are found to have a sharp threshold as a function of the velocity [24]. Similarly, superfluidity is found to survive [25] if the superfluid density is defined via the irrotational response at long wavelengths. To establish the behavior in the presence of disorder, we calculate the superfluid stiffness, which characterizes the superfluid Bose-glass transition. To do this, we apply twisted boundary conditions $\phi_\theta(\mathbf{x}+L\mathbf{e}_\theta)-\phi_\theta(\mathbf{x})=\theta$. For the actual calculation, we apply a local transformation $\nabla\phi_\theta = \nabla\phi + \mathbf{A}_\theta$ with a twist current $\mathbf{A}_\theta = (\theta/L)\mathbf{e}_\theta$ with periodic boundary conditions imposed on $\phi(\mathbf{x})$. The superfluid stiffness [26, 27] is then obtained from the frequency shift as

$$f_s = \lim_{\theta\to 0}\frac{L^2}{\theta^2}[\omega(\theta)-\omega(0)]. \tag{23.17}$$

In the presence of weak disorder, we expand both the density and the condensate phase in powers of the disorder strength κ: $n = 1+\eta_{(1)}+\mathcal{O}(\kappa^2)$ and $\nabla\phi = \nabla\phi_{(1)}+\mathcal{O}(\kappa^2)$ with $\eta_{(1)},\nabla\phi_{(1)} \sim \mathcal{O}(\kappa)$. From Eq. (23.15), we then obtain the expansion of the frequency shift. As the frequency shift is self-averaging, only even powers in κ contribute. Odd powers of κ in Eqs. (23.15, 23.16) are used to compute the solutions for density and condensate phase according to $\eta_1(\mathbf{k}) = G_\eta(\mathbf{k},\mathbf{A}_\theta)\vartheta_\mathbf{k}$, $\phi_1(\mathbf{k}) = G_\phi(\mathbf{k},\mathbf{A}_\theta)\vartheta_\mathbf{k}$ with the Green functions [23]

$$G_\eta(\mathbf{k},\mathbf{A}_\theta) = \frac{-k^2\chi_k}{k^2+2\,\mathbf{ik}\cdot\mathbf{A}_\theta(\mathbf{ik}\cdot\mathbf{A}_\theta+\alpha)\chi_k}, \tag{23.18}$$

$$G_\phi(\mathbf{k},\mathbf{A}_\theta) = \frac{-(\mathbf{ik}\cdot\mathbf{A}_\theta+\alpha)\chi_k}{k^2+2\,\mathbf{ik}\cdot\mathbf{A}_\theta(\mathbf{ik}\cdot\mathbf{A}_\theta+\alpha)\chi_k}, \tag{23.19}$$

with $\chi_k = (k^2/2+1)^{-1}$. These Green functions give the correlation functions of density and phase fluctuations in the ground state with $\mathbf{A}_\theta = 0$. The correlation function for density fluctuations decays exponentially on the scale of the healing length, allowing density fluctuations to screen the disorder potential on short length scales. As discussed above, a weak disorder potential becomes important only on the scale of the density Larkin length $\mathcal{L}_\mathrm{n} \sim 1/\kappa$. The driven nature of the polariton condensate becomes apparent when considering phase fluctuations, which are imprinted onto the condensate by random sources and sinks as described by Eq. (23.16). Long-range order is destroyed by these phase fluctuations on the scale of the phase Larkin length $\mathcal{L}_\phi \sim 1/\alpha\kappa$, which is also the decay length of the phase correlation function.

In the next step, we calculate the condensate stiffness using Eq. (23.17), perturbatively to order κ^2, finding

$$f_s \approx 1 - \left\{c_1 + g_1(L)\,\alpha^2 + \left(g_2(L) + c_2 L^2\right)\alpha^4\right\}\kappa^2\,, \tag{23.20}$$

where we have omitted finite-size corrections vanishing for $L \to \infty$. The coefficients in this expansion are [23]

$$\begin{aligned} c_1 &= \tfrac{1}{2\pi}\,, \qquad c_2 = \tfrac{1}{(2\pi)^3}\,, \\ g_1(L) &= -\tfrac{1}{\pi}\left(\log\tfrac{2L^2}{(2\pi)^2} - \tfrac{19}{12}\right), \\ g_2(L) &= -\tfrac{1}{\pi}\left(\log\tfrac{2L^2}{(2\pi)^2} - \tfrac{13}{12}\right). \end{aligned}$$

In the equilibrium limit, $\alpha \to 0$, which reproduces previous findings [28, 29, 30]: the stiffness is zero above a critical $\kappa = \sqrt{2\pi}$. For finite α, however, the stiffness is driven to zero for any $\kappa \neq 0$ beyond the length scale $\mathcal{L}_s \sim 1/\alpha^2\kappa$. This indicates that for a driven system, superfluidity is always destroyed in the thermodynamic limit.

One expects, from this perturbative result, that the stiffness reduction will scale with the parameter $\alpha^4\kappa^2L^2$. This is confirmed numerically [23], as shown in Fig. 23.1. The numerically computed stiffness decays exponentially for large values of the parameter $c_2\alpha^4\kappa^2L^2 \sim L^2/\mathcal{L}_s{}^2$. Using this insight, we propose a scaling form for the superfluid stiffness which reproduces both the perturbative results for small values of the scaling parameter, and which displays an exponential decay for large values

$$f_s = e^{-c_2\,\alpha^4\kappa^2L^2}\left(1 - g\left(\alpha, \kappa, \log L\right)\right). \tag{23.21}$$

In Fig. 23.1, the numerically computed stiffness is plotted as a function of the scaling variable $c_2\alpha^4\kappa^2L^2 \sim L^2/\mathcal{L}_s{}^2$ to demonstrate data collapse, confirming the validity of scaling and the specific form of the scaling function above.

What is the interpretation of the length scale $\mathcal{L}_s \propto 1/\alpha^2\kappa$ for the decay of superfluidity? Imposing twisted boundary conditions and following the difference $\nabla\phi_\theta - \nabla\phi$ for individual disorder realizations, one observes that the phase twist does not relax in a uniform manner from one boundary of the sample to the other, but rather relaxes over domain walls of width $\mathcal{L}_s$. Since the relaxation of the phase twist θ imposes the existence of a local supercurrent $j = n(\nabla\phi_\theta)$, it is energetically

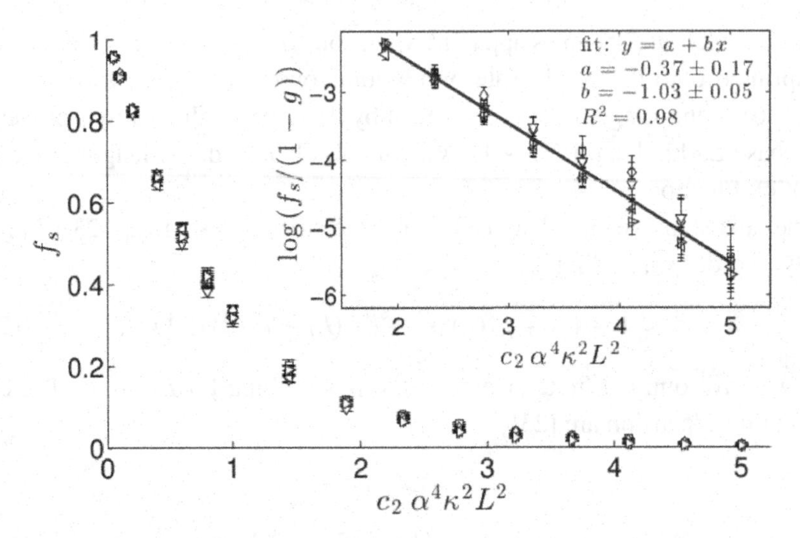

Figure 23.1 Test of the scaling form for the stiffness of a nonequilibrium condensate, Eq. (23.21). A clear data collapse is observed when plotting the numerically obtained superfluid stiffness as a function of $c_2\alpha^4\kappa^2L^2 \sim L^2/\mathcal{L}_s{}^2$. Inset: exponentially small tail of f_s compared to the scaling form Eq. (23.21), using the perturbative results for c_2 and g. Data points for $L \times L$ lattices with $L = 64$ and 96; $\alpha = 0.9, 1$ and 1.2, and for up to 1320 disorder realizations. Reprinted with permission from Janot, A., et al. (2013), Superfluid stiffness of a driven dissipative condensate with disorder, *Phys. Rev. Lett.*, **111**, 230403 [23]. Copyright (2013) by the American Physical Society.

favorable for the relaxation to occur in spatial regions where the local distribution of current sources and sinks leads to a pinning of such a domain wall.

23.5 Summary

We have outlined how, within a generalized Gross-Pitaevskii model, gain and loss affect the physics of condensates in potentials. For the double-well potential, there are transitions between synchronized and desynchronized states. These states can be understood in a similar way to the d.c. and a.c. Josephson effects, with the transitions caused by currents associated with gain and loss, rather than external bias. A complementary perspective is that of mode selection in lasers, where the form of the nonlinear gain can select single or multimode behavior. The key phenomena of Josephson oscillations and multimode polariton condensation have been observed experimentally. More recent experiments are further expanding the phenomenology of nonequilibrium condensation in few-mode systems.

Interest in the behavior of Bose Einstein condensates in a disordered environment has been reinvigorated by experiments on Anderson localization of cold atomic gases. Polariton condensates naturally allow the observation of such physics, enriched by their nonequilibrium nature. Theory predicts it has a dramatic effect. The presence of condensate currents in the steady state converts potential disorder to symmetry-breaking disorder [23, 31], which can destroy long-range order. Moreover, the absence of a true continuity equation allows a localized response to an imposed long-wavelength phase twist, so that the superfluid stiffness is driven to zero. The resulting disordered phase has a vanishing stiffness, like the Bose glass, and appears as soon as one departs from the equilibrium limit. This allows for the observation of phase-coherent phenomena and superfluidity as finite size effects only.

It would be interesting to attempt to observe this physics experimentally in more detail, as there are several open issues. One is whether the disordered phase really only has short-range order as predicted by perturbation theory, or whether this result is modified by nonperturbative effects. More broadly, it remains to establish the full phase diagram of the nonequilibrium problem, in terms of the parameters κ and α. At present, we know there is an ordered phase for $\alpha = 0, \kappa < \kappa_c$, and a disordered one for $\alpha \neq 0, \kappa \neq 0$. The disordered phase we describe above is, in the terminology of Section 23.3, synchronized (it has a single frequency) but not phase-locked (it has no stiffness). Experimentally and numerically, however, there is also a disordered phase which is neither synchronized nor phase-locked. This corresponds to the a.c. Josephson state of the double well or more generally to multimode condensation. It remains to determine whether this is a distinct disordered phase, and if so, where in parameter space it occurs.

References

[1] Leggett, A. J. 2001. Bose-Einstein condensation in the alkali gases: some fundamental concepts. *Rev. Mod. Phys.*, **73**, 307–356.

[2] Fisher, M. P. A., Weichman, P. B., Grinstein, G., and Fisher, D. S. 1989. Boson localization and the superfluid-insulator transition. *Phys. Rev. B*, **40**, 546–570.

[3] Pikovsky, A., Rosenblum, M., and Kurths, J. 2001. *Synchronization*. Cambridge, UK: Cambridge University Press.

[4] Keeling, J., and Berloff, N. G. 2008. Spontaneous rotating vortex lattices in a pumped decaying condensate. *Phys. Rev. Lett.*, **100**, 250401.

[5] Siegman, A. E. 1986. *Lasers*. Oxford, UK: Oxford University Press.

[6] Wouters, M., and Carusotto, I. 2007. Excitations in a nonequilibrium Bose-Einstein condensate of exciton polaritons. *Phys. Rev. Lett.*, **99**, 140402.

[7] Love, A. P. D., Krizhanovskii, D. N., Whittaker, D. M., Bouchekioua, R., Sanvitto, D., Rizeiqi, S. Al, Bradley, R., Skolnick, M. S., Eastham, P. R., André, R., and Dang, Le Si. 2008. Intrinsic decoherence mechanisms in the microcavity polariton condensate. *Phys. Rev. Lett.*, **101**, 067404.

[8] Racine, D., and Eastham, P. R. 2014. Quantum theory of multimode polariton condensation. *Phys. Rev. B*, **90**, 085308.

[9] Wouters, M., and Savona, V. 2009. Stochastic classical field model for polariton condensates. *Phys. Rev. B*, **79**, 165302.

[10] Read, D., Rubo, Y. G., and Kavokin, A. V. 2010. Josephson coupling of Bose-Einstein condensates of exciton–polaritons in semiconductor microcavities. *Phys. Rev. B*, **81**, 235315.

[11] Zapata, I., Sols, F., and Leggett, A. J. 1998. Josephson effect between trapped Bose-Einstein condensates. *Phys. Rev. A*, **57**, R28–R31.

[12] Wouters, M. 2008. Synchronized and desynchronized phases of coupled nonequilibrium exciton–polariton condensates. *Phys. Rev. B*, **77**, 121302(R).

[13] Borgh, M. O., Keeling, J., and Berloff, N. G. 2010. Spatial pattern formation and polarization dynamics of a nonequilibrium spinor polariton condensate. *Phys. Rev. B*, **81**, 235302.

[14] Eastham, P. R. 2008. Mode locking and mode competition in a nonequilibrium solid-state condensate. *Phys. Rev. B*, **78**, 035319.

[15] Lagoudakis, K. G., Pietka, B., Wouters, M., André, R., and Deveaud-Plédran, B. 2010. Coherent oscillations in an exciton–polariton Josephson junction. *Phys. Rev. Lett.*, **105**, 120403.

[16] Krizhanovskii, D. N., Lagoudakis, K. G., Wouters, M., Pietka, B., Bradley, R. A., Guda, K., Whittaker, D. M., Skolnick, M. S., Deveaud-Plédran, B., Richard, M, André, R, and Dang, Le Si. 2009. Coexisting nonequilibrium condensates with long-range spatial coherence in semiconductor microcavities. *Phys. Rev. B*, **80**, 045317.

[17] Baas, A., Lagoudakis, K. G., Richard, M., André, R., Dang, Le Si, and Deveaud-Plédran, B. 2008. Synchronized and desynchronized phases of exciton–polariton condensates in the presence of disorder. *Phys. Rev. Lett.*, **100**, 170401.

[18] Thunert, M., Janot, A., Franke, H., Sturm, C., Michalsky, T., Martín, M. D., Viña, L., Rosenow, B., Grundmann, M., and Schmidt-Grund, R. 2016. Cavity polariton condensate in a disordered environment. *Phys. Rev. B*, **93**, 064203.

[19] Nattermann, T., and Pokrovsky, V. L. 2008. Bose-Einstein condensates in strongly disordered traps. *Phys. Rev. Lett.*, **100**, 060402.

[20] Malpuech, G., Solnyshkov, D. D., Ouerdane, H., Glazov, M. M., and Shelykh, I. 2007. Bose glass and superfluid phases of cavity polaritons. *Phys. Rev. Lett.*, **98**, 206402.

[21] Larkin, A. I. 1970. Effect of inhomogeneities on the structure of the mixed state of superconductors. *Sov. Phys. JETP*, **31**, 784–786.

[22] Imry, Y., and Ma, S.-K. 1975. Random-field instability of the ordered state of continuous symmetry. *Phys. Rev. Lett.*, **35**, 1399–1401.

[23] Janot, A., Hyart, T., Eastham, P. R., and Rosenow, B. 2013. Superfluid stiffness of a driven dissipative condensate with disorder. *Phys. Rev. Lett.*, **111**, 230403.

[24] Wouters, M., and Carusotto, I. 2010. Superfluidity and critical velocities in nonequilibrium Bose-Einstein condensates. *Phys. Rev. Lett.*, **105**, 020602.

[25] Keeling, J. 2011. Superfluid density of an open dissipative condensate. *Phys. Rev. Lett.*, **107**, 080402.

[26] Fisher, M. E., Barber, M. N., and Jasnow, D. 1973. Helicity modulus, superfluidity, and scaling in isotropic systems. *Phys. Rev. A*, **8**, 1111–1124.

[27] Leggett, A. J. 1970. Can a solid be "superfluid"? *Phys. Rev. Lett.*, **25**, 1543–1546.

[28] Huang, K., and Meng, H.-F. 1992. Hard-sphere Bose gas in random external potentials. *Phys. Rev. Lett.*, **69**, 644–647.

[29] Meng, H.-F. 1994. Quantum theory of the two-dimensional interacting-boson system. *Phys. Rev. B*, **49**, 1205–1210.

[30] Giorgini, S., Pitaevskii, L., and Stringari, S. 1994. Effects of disorder in a dilute Bose gas. *Phys. Rev. B*, **49**, 12938–12944.

[31] Kulaitis, G., Krüger, F., Nissen, F., and Keeling, J. 2013. Disordered driven coupled cavity arrays: nonequilibrium stochastic mean-field theory. *Phys. Rev. A*, **87**, 013840.

24

Collective Topological Excitations in 1D Polariton Quantum Fluids

HUGO TERÇAS

Institute for Theoretical Physics, University of Innsbruck, Austria
Present Address: Physics of Information and Quantum Technologies Group, Instituto de Telecomunicações, Lisboa, Portugal

DMITRY D. SOLNYSHKOV AND GUILLAUME MALPUECH

Institut Pascal, PHOTON-N2, Clermont Université, Blaise Pascal University, CNRS, Aubière Cedex, France

We discuss some recent advances in the spin dynamics in photonic systems and polariton superfluids. In particular, we describe how the spin degree of freedom affects the collective behaviour of the half-soliton gas. First, we demonstrate that the anisotropy in the intra- and interspin interaction leads to the formation of a one-dimensional ordered phase: the topological Wigner crystal. Second, we show that half-solitons behave as magnetic monopoles in effective magnetic fields. We study the transport properties and demonstrate a deviation from the usual Ohm's law for moderate values of the magnetic field.

24.1 Introduction

Photonic systems offer great opportunities for the study of quantum fluids, due to the possibility of creation of macroscopically occupied states with well-controlled properties by coherent excitation with lasers, and the full access to the wavefunction of the quantum fluid by well-established optical methods [1]. The main distinctive feature of quantum fluids as compared with classical ones are the topological defects, which, once created, cannot be removed by a continuous transformation. The most well-known example of such defect is a quantum vortex, which can appear in two-dimensional (2D) and three-dimensional (3D) systems. Its analog in one-dimensional (1D) Bose-Einstein condensates (BECs) is a *soliton* [2]. In fact, solitons are ubiquitous in systems described by the self-defocussing nonlinear Schrödinger equations. Specifically, for Bose gases, they are associated with the excitations of type II of the Lieb and Liniger theory. Spinor BECs (particularly with two pseudospin projections) offer a plethora of nonlinear spin effects, including half-integer topological defects possible in BECs with spin-anisotropic interactions [3]. Recent experimental work reports on the emergent monopole behaviour of half-solitons in the presence of an effective magnetic field [4]. In this chapter, we highlight some theoretical advances concerning not only the dynamics but also the many-body aspects of the physics of half-solitons.

24.2 A Topological Wigner Crystal

It is commonly accepted that the Wigner crystal is one of the most simple yet dramatic many-body effects. In the seminal work published in 1931 [5], Wigner showed that as a result of the competition between the long-ranged potential and kinetic energies, electrons spontaneously form a self-organized crystal at low densities, in a state that strongly differs from the Fermi gas. Experimental observations of this effect have been reported in carbon nanotubes [6]. The phenomenon has been discussed in ultracold Fermi gases with dipolar interactions [7, 8] and also investigated in systems with short-range interactions [9]. The concept of Wigner crystal is used for the description of more exotic systems, such as holons (charged solitons) in the vicinity of a metal-Mott insulator transition [10], and the ground-state properties of nuclear matter [11]. We here show how a half-soliton gas can form an ordered state exhibiting the same correlation properties as the Wigner crystal.

The half-soliton appears as a solitary excitation, being a solution of the 1D spinor Gross-Pitaevskii equation [2]

$$i\hbar\frac{\partial\psi_\sigma}{\partial t} = -\frac{\hbar^2}{2m}\nabla^2\psi_\sigma + \left(\alpha_1|\psi_\sigma|^2 + \alpha_2|\psi_{-\sigma}|^2\right)\psi_\sigma. \tag{24.1}$$

Here, m represents the boson mass, α_1 the interaction constant between particles having the same spin, and α_2 the interaction constant between particles of opposite projections. When one soliton is present in a spin component and absent in the other, the corresponding object is called a *half-soliton*. The stability of such objects is ensured in the case of spin-anisotropic interactions, when $|\alpha_1| \gg |\alpha_2|$, which is typically realised in exciton–polariton BEC [12, 13], because the singlet interaction is a second-order process. Let us consider that a soliton is formed in the σ_+ spin projection, for definiteness. In that case, the single-soliton solution travelling with speed v is given by

$$\psi_+(x) = \sqrt{n_0/2}\left[i\beta + \sqrt{1-\beta^2}\tanh\left(\sqrt{1-\beta^2}\frac{x}{\sqrt{2}\xi}\right)\right], \tag{24.2}$$

where $\beta = v/c_s$, $c_s \simeq \sqrt{\alpha_1 m/c_s}$ is the sound velocity and $\xi = \hbar/\sqrt{2m\alpha_1 n_0}$ denotes the healing length. Due to the small interspin interaction α_2, the other component contains a small density minimum. From mechanical arguments, we can derive variationally the pseudopotential associated with the interaction between the two σ_+ solitons [14]

$$U(x) = \frac{M_*}{2} c_s^2 \frac{1 - \beta^2}{\sinh^2\left(\frac{\sqrt{2}\sqrt{1-\beta^2}x}{\xi}\right)}, \tag{24.3}$$

where M_* represents the effective mass of the solitons. In fact, it reduces to an effective potential for the case of almost black ($v \sim 0$) collisions,

$$U(x) \simeq \frac{M_*}{2} c_s^2 \text{cosech}^2\left(\frac{\sqrt{2}x}{\xi}\right). \tag{24.4}$$

In order to incorporate the interaction with the σ_- spin projection, we have to take into account the spinor nature of the ground state of the condensate. In that case, the function of two HSs in different components located at positions x_1 and x_2, travelling with opposite speeds v and $-v$, can be written using the center-of-mass and relative coordinates $\zeta = (x_1 + x_2)/2$ and $\eta = (x_1 - x_2)/2$ respectively [15]. Integrating over the soliton centroid ζ to compute the energy $E = \int \mathcal{E}\, d\zeta$, with $\mathcal{E}$ denoting the energy density, we can obtain the following potential between almost black σ_+–σ_- solitons [16]

$$V(x) = \frac{M_*}{2} |\Lambda| c_s^2 \left[\frac{\sqrt{2}x \cosh(\sqrt{2}x/\xi) - \sinh(\sqrt{2}x/\xi)}{\sinh^3(\sqrt{2}x/\xi)}\right], \tag{24.5}$$

where $|\Lambda| = |\alpha_2|/\alpha_1$ is the measure of the spin-anisotropy. This potential is repulsive for attractive interspin interactions ($\alpha_2 < 0$), which is the case considered here.

24.2.1 Kinetic Description

To describe the dynamics of a gas of HSs, we postulate that the phase-space distributions $f^\pm(x, v, t)$ are governed by a Vlasov kinetic equation

$$\frac{df^\pm}{dt} \equiv \frac{\partial f^\pm}{\partial t} + v\frac{\partial f^\pm}{\partial x} + \dot{v}^\pm \frac{\partial f^\pm}{\partial v} = 0, \tag{24.6}$$

where the collision integral is neglected, assuming only elastic processes. The acceleration term can be simply given by $\dot{v}^\pm = -1/M_* \partial U_{\text{eff}}^\pm/\partial x$, where $U_{\text{eff}}^\pm(x)$ is the coarse-grained mean-field potential in terms of Eqs. (24.4, 24.5)

$$\begin{aligned} U_{\text{eff}}^\pm(x) = &\int_{-\infty}^{\infty}\int_{-\infty}^{\infty} U(x - x')f^\pm(x', v)dx'dv \\ &+ \int_{-\infty}^{\infty}\int_{-\infty}^{\infty} V(x - x')f^\mp(x', v)dx'dv \end{aligned}. \tag{24.7}$$

To get some insight in the stability, we linearise the system around its equilibrium, $f^{\pm} = f_0^{\pm} + \delta f^{\pm}$ and obtain the kinetic dispersion relation for the excitations of the HS gas

$$(1 - I_U^{+})(1 - I_U^{-}) - I_V^{-} I_V^{+} = 0, \tag{24.8}$$

where the integrals $I_U^{\pm}$ and $I_V^{\pm}$ have the form

$$I_X^{\pm}(k, \omega) = \frac{k^2 \tilde{X}(k)}{M_*} \int_{-\infty}^{\infty} \frac{f_0^{\pm}}{(\omega - kv)^2} dv, \tag{24.9}$$

with $\tilde{X}(k)$ representing the Fourier transform of the potentials U and V. To close the mean-field analysis, we use the fact that the excitations in 1D Bose gases below the critical velocity c_s follow a fermionic statistics, as established in the famous Lieb-Liniger theory [17, 18]. Actually, this result is quite easy to understand by simply looking at the phase of the two-soliton wave function in Eq. (24.2): exchanging two solitons located at positions x_1 and x_2, we obtain an overall phase shift of π in agreement with fermionic statistics [16]. As a consequence, the equilibrium configuration is that of a 1D Fermi gas $f_0(x, v)^{\pm} = N_0^{\pm}/2v_F^{\pm}\Theta(v_F^{\pm} - |v|)$, where $v_F^{\pm} = \pi\hbar N_0^{\pm}/M_*$ is the 1D Fermi velocity and $N_0^{\pm}$ is the soliton density. In that case, the excitation spectrum around the equilibrium configuration contains two branches $\omega_{\pm}(k)$ displaying acoustic behaviour in the long-wavelength limit $k \to 0$ as $\omega_{\pm} \approx v_{\pm}k$ [16]. In the special case of a spin-balanced gas $N_0^{+} = N_0^{-} = N_0$, we obtain

$$v_{\pm} = \frac{v_F}{2\sqrt{2}\pi\gamma} \sqrt{8\pi^2\gamma^2 - \sqrt{2}\gamma(4 \pm |\Lambda|)}, \tag{24.10}$$

with $\gamma = N_0\xi$ representing the dimensionless soliton concentration, an analogue of the Wigner-Seitz parameter. The relation between the Fermi velocity of the soliton gas and the sound velocity of the condensate $v_F/c_s = \sqrt{2}\pi\gamma$ has the meaning of a dimensionless interaction parameter.

Due to the competition between the statistical pressure and the interactions between the solitons, the dispersion relation (24.8) encodes very interesting features, as illustrated in Fig. 24.1a. If the soliton gas is dilute enough so that the concentration parameter lies below the critical value $\gamma^* \simeq 0.07$, the system does not exhibit any ordering, corresponding to a gaseous state. Instead, for $\gamma > \gamma^*$, the solitons start to perform periodic oscillations up to a critical value of the wavevector $k^* = k^*(\gamma)$. At this point, the frequency softens towards zero and the system undergoes crystallisation with the lattice constant given by $d = 2\pi/k^*$. The onset of instability is expected at higher k^* for higher values of γ, therefore corresponding to tighter lattices. When the interactions dominate and the distance becomes comparable with the characteristic scale of these interactions, the system

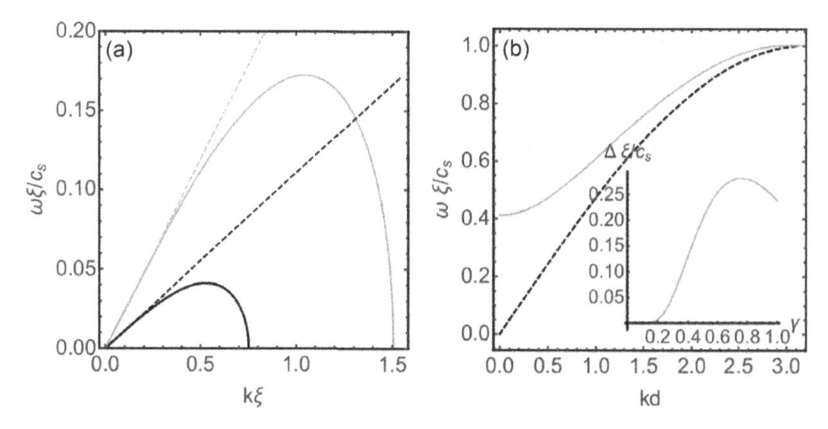

Figure 24.1 (a) Dispersion relation of a HS gas exhibiting dynamical instability, obtained for $\gamma_+ = 0.1$, $\gamma_- = 0.08$ and $|\Lambda| = 0.2$. The black (grey) full lines corresponds to ω_- (ω_+) modes in Eq. (24.9). The dashed lines illustrate to the low-wavelength limit $\omega_\pm \simeq v_\pm k$. (b) Spectrum of a symmetric crystal with $\gamma_+ = \gamma_- = 0.5$, corresponding to the half-filling configuration. Acoustic (black) and optic (grey) phonon modes. The inset represents the gap frequency Δ as a function of the parameter γ.

forms a regular lattice. The ratio of the Fermi velocity to the interaction potential (24.3) provides the following magnitude estimate $\left|v_F^2/U\right| \sim v_F^2/c_s^2 \sim \gamma^2$. For the case of an asymmetric mixture, a similar analysis allows us to conclude about the existence of different phases: a gaseous phase ($v_- = v_+ = 0$), defined by the region $\gamma_+ < (16 - |\Lambda|^2 - 16\sqrt{2}\pi^2\gamma_-)/[16\pi^2(\sqrt{2}\pi^2\gamma_-)]$, and two ordered phases, sustaining single-mode ($v_- = 0, v_+ > 0$) and a two-mode ($v_-, v_+ > 0$) oscillations.

What this mean-field analysis simply tells us is that the gas configuration is not dynamically stable and gives place to crystalline order. Let us start with the analysis of a hypothetically crystalline phase and conclude about its stability a posteriori. The configuration of minimum potential energy is obtained for a chain of alternating $\sigma_+ - \sigma_-$ solitons with lattice constant d, and quantum fluctuations are expected to lead only to low-amplitude oscillations around the equilibrium distance $d = 1/N_0$. In the harmonic approximation, the corresponding Hamiltonian reads

$$H = \sum_\ell \left(\frac{p_{u,\ell}^2}{2M_*} + \frac{p_{v,\ell}^2}{2M_*} \right) + \frac{1}{4} \sum_{\ell,\ell'} U''(\ell d)(u_\ell - u_{\ell+2\ell'})^2 + \frac{1}{4} \sum_{\ell,\ell'} V''(\ell d)(u_\ell - v_{\ell+\ell'})^2, \tag{24.11}$$

where u_ℓ (v_ℓ) is the deviation of the $\sigma_+(\sigma_-)$-soliton at site ℓ from its equilibrium position. The diagonalisation of (24.11) with the help of Hamilton equations leads to two modes, an acoustic mode ω_1 and a gapped optical mode ω_2. In the long-wavelength limit, they are given by $\omega_1 \approx u_1 k$ and $\omega_2^2 = \Delta^2 + u_2^2 k^2$, where the velocities u_1 and u_2 are functions of the ratio $d/\xi = 1/\gamma$ [16]. Within the first neighbour approximation, we obtain

$$u_{1,2} \simeq d\sqrt{\frac{U''(d)}{M_*} \pm \frac{V''(d)}{4M_*}}, \quad \Delta \simeq 2\sqrt{\frac{V''(d)}{M_*}}. \tag{24.12}$$

The *plasma* frequency Δ is a feature that clearly distinguishes the crystal from the gaseous configuration, and is a function of the spin-anisotropy $|\Lambda|$, $\Delta \sim \sqrt{|\Lambda|}$. In Fig. 24.1b, we plot the phonon modes $\omega_{1,2}$.

24.2.2 Luttinger Liquid Theory

In order to investigate the quantum features of the crystalline phase, we compute the density correlations. Due to the short-range character of the interactions, the Luttinger liquid theory can accurately describe the physical properties of the system [19]. For a model with $SU(2)$ spin symmetry such as the second-quantised version of the Hamiltonian (24.11), bosonisation predicts the decay of the density correlation function as [20]

$$\langle n_x n_0 \rangle \simeq -\frac{K}{\pi x^2} + \frac{A_2}{x^{1+K}}\cos(2k_F x) + \frac{A_4}{x^{4K}}\cos(4k_F x), \tag{24.13}$$

where A_2 and A_4 are some constants. The Luttinger parameter [19] $K = \sqrt{K_U^3 K_V / (K_U K_V + |\Lambda|)}$ can be given in terms of the single-spin parameters

$$K_X^{-1} = 1 + \frac{2}{\pi \hbar M_* v_F} \sum_{\ell} \xi X(\ell d)\left[1 - \cos\left(2k_F \ell d\right)\right], \tag{24.14}$$

with $X = U, V$. Apart from the x^{-2} dependence, which is familiar from the Fermi liquid theory, the properties of the system are universally defined in terms of K. Therefore, for $K > 1/2$, the $2k_F$ Friedel-like oscillations dominate, typical for a Luttinger liquid (LL). On the contrary, for $K < 1/3$, $4k_F$ quantum fluctuations dominate the system, leading to a modulation at the average distance $d = 1/N_0$ between the solitons. This corresponds to a quasi-Wigner crystal (qWC) state, where the (quasi-)long-range order is due to quantum fluctuations, being favoured in the low-density limit (notice that the interaction energy scales as $c_s/v_F \sim 1/\gamma$). In the intermediary region $1/3 < K < 1/2$, there is a mixture (M) between these two phases. An interesting feature distinguishing the qWC state from the Wigner crystal of electrons is the topological nature of the effective potentials, with an effective range defined by the BEC healing length ξ. This suggests that this state can

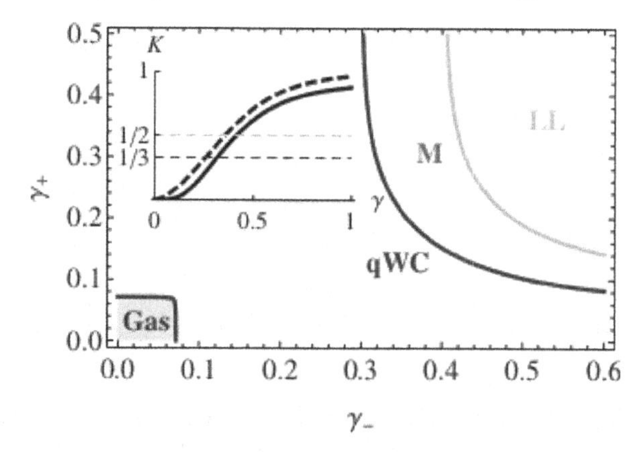

Figure 24.2 Phase diagram of the ordered phase described by the Hamiltonian (24.11). The contours $K = 1/2$ (top, grey line) and $K = 1/3$ (bottom, black line) delimit the Luttinger liquid (LL), the topological quasi-Wigner crystal (qWC) and the mixture (M) phases. Inset: Cut the Luttinger parameter K along $\gamma_+ = \gamma_- = \gamma$ for single (dashed) and two-species (solid) crystal. The small dashed region represents a zoom-out of the gas phase. Reprinted with permission from Terças, H., et al. (2013), Topological Wigner crystal of half-solitons in a spinor Bose-Einstein condensate, *Phys. Rev. Lett.*, **110**, 035303 [16]. Copyright (2013) by the American Physical Society.

be regarded as a topological Wigner crystal. Yet another difference is related with the absence of the logarithmic divergence of the Coulomb potential $\sim 1/x$ leading to a very slow decay in the $4k_F$ correlation (much slower than any power law [20]). Excitations about the qWC equilibrium correspond to the topological equivalent of the so-called charge density wave, occurring for $K < 1/4$ and characterised by commensurate condition of the filling factor $\gamma = 1/\nu$, where ν is an integer. In Fig. 24.2, we illustrate the relevant phases occurring in the system. We can observe that it exhibits the same phases as in electronic systems, except for the gaseous phase occurring for very dilute systems. This is a consequence of the short-range nature of the interactions. The existence of the stability of such phases can be checked by direct simulations of the Gross-Pitaevskii equation (24.1). In these simulations, the initial configuration was a condensate containing a gas of half-solitons. A noise term is introduced in order to mimic the quantum fluctuations. In Fig. 24.3, we depict the circular polarisation degree $\rho_c = (n_+ - n_-)/(n_+ + n_-)$. In panel a, the regular pattern corresponding to an ordered phase (Wigner crystal) is stable, with the solitons undergoing small oscillations around their equilibrium positions; in the other case (panel b), the pattern changes dramatically and the ordering melts down, as it is typical for a liquidlike state.

At this point, it may be important to notice that our analysis is valid in the zero temperature limit and in the absence of disorder. Similar to what happens in

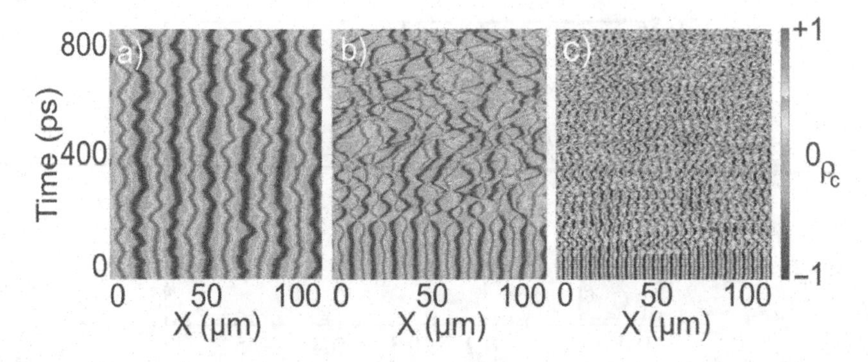

Figure 24.3 Circular polarisation degree of the condensate as a function of time showing (a) a stable Wigner crystal ($\gamma_+ = \gamma_- \approx 0.25$), and (b) 'melting' of such crystal into a liquidlike state, ($\gamma_+ = \gamma_- \approx 0.5$). In both situations, we took $|\Lambda| = 0.1$. Reprinted with permission from Terças, H., et al. (2013), Topological Wigner crystal of half-solitons in a spinor Bose-Einstein condensate, *Phys. Rev. Lett.*, **110**, 035303 [16]. Copyright (2013) by the American Physical Society.

other systems, we expect that both the temperature and the disorder can melt the crystal state [21]. We can estimate the robustness of the ordered state: in analogy with the criteria used in one-component plasmas [21], the melting occurs for $\Gamma \equiv E_{\text{int}}/E_{\text{kin}} \sim 170$; thus, we should expect the crystal to be robust for $\hbar\Delta \gg E_{\text{int}}, E_{\text{dis}}$, where E_{dis} is the disorder amplitude.

We believe that other interesting many-body effects with solitons may emerge as a consequence of the fermionic statistics of the half-solitons. For example, by taking into account the coupling between the condensate phonons (Bogoliubov excitations) and the solitons, we could expect a mecanism similar to that of Cooper pairing also to occur here. In the next section, we explore another important feature: the collective response to an effective magnetic field, putting in evidence the fact that half-solitons may indeed behave as magnetic monopoles, as previously suggested by some recent experiments [4].

24.3 Half-Solitons as Magnetic Monopoles

Magnetic monopoles have been one of the most important physical questions in quantum mechanics since the first ideas put forward by Dirac [22]. In fact, 'real' elementary magnetic charges have not been observed up to now, despite long efforts to detect them [23]. Recently, magnetically frustrated materials, or spin ices [24, 25], offered the possibility of investigating magnetic charges. Besides the substantial experimental evidence to support the existence of spin-ice magnetic monopoles [26, 27, 28], the measurement of the charge and current of magnetic monopoles has become possible [29].

Although photons do not interact with a real magnetic field, their spin can be affected by effective magnetic fields. The half-solitons in polariton BECs are good candidates for magnetic monopoles, due to their nontrivial spin texture, while their very small mass ($\sim 10^{-5} m_e$) allows expecting ultrafast 'magnetricity' [29]. Before digging into the collective behaviour of the gas, let us highlight some basic kinematic properties of individual half-solitons:

$$\psi_+ = \sqrt{\frac{n_0}{2}}\left[i\frac{v}{c} + \frac{1}{\gamma}\tanh\left(\frac{x-y}{\sqrt{2}\xi\gamma}\right)\right], \quad \psi_- = \sqrt{\frac{n_0}{2}}. \tag{24.15}$$

Here (see also Eq. 24.2), $y = vt + x_0$ is the soliton centroid and $\gamma = (1 - v^2/c^2)^{-1/2}$ is the relativistic factor. This solution is characterised by a divergent in-plane pseudospin pattern $S_x = \mathrm{Re}(\psi_+\psi_-^*)/2 \simeq (n_0/2\gamma)\mathrm{sign}(y - x)$. Fig. 5.1a shows the density and the pseudospin for two HS in opposite spin components. The magnetic charge is defined by analogy with Maxwell's equation $\rho = \nabla \cdot \mathbf{S}$, and the charge of a single HS is $q = \pm n/2 = \pm n_0/2\gamma$ (shown by the symbols $+$ and $-$ in Fig. 24.4. Since the charge is defined by the in-plane pseudospin texture, it does not depend on the $\sigma_\pm$ component in which the HS appears. The dynamics of each spin in a magnetic field is governed by the precession equation $\partial\mathbf{S}/\partial t = \mathbf{S} \times \mathbf{H}$. The monopole dynamics of Eq. (24.15) can be obtained by calculating the magnetic force $F_m = -n_0 H/2\gamma$ and the acceleration $a = n_0 H/2M_0\gamma^2$, where $M_0 = 2\sqrt{2}n_0\xi m$ is the absolute value of the HS rest mass.[1] Integrating once, the velocity is $v(t) = c\tanh(t/\tau_0) < c$ [30], where $\tau_0 = 2M_0 c/n_0 H$.

24.3.1 Relativistic Dynamics

Let us now consider *two* HSs located at the positions $\pm y/2$

$$\psi_\pm = \sqrt{\frac{n_0}{2}}\left[\pm i\frac{\dot{y}}{c} + \frac{1}{\gamma}\tanh\left(\frac{x \mp y/2}{\sqrt{2}\xi\gamma}\right)\right]. \tag{24.16}$$

The pseudospin texture is invariant with respect to the exchange of the two HSs, $y \to -y$: for this particular solution, the spin field is divergent for the soliton on the right. Moreover, it is impossible to have two solitons of the same type next to each other. Fig. 24.4a, therefore, is the most general spin texture. When two solitons cross each other, the 'sign' of each monopole is inverted, i.e., the one located in the σ_- projection, initially with a convergent texture, becomes divergent after crossing and vice versa. In Fig. 24.4b, c, and d, we depict the temporal evolution of $\rho_c = (n_+ - n_-)/(n_+ - n_-)$ of the HS by solving Eq. (24.1). Panel b illustrates

[1] The mass of each soliton is defined by the number of polaritons depleted from the condensate, $M = m\int \psi_\pm^* \psi_\pm dx = -2\sqrt{2}n_0\xi m/\gamma$, being therefore negative. M_0 is here defined as the magnitude of the rest mass.

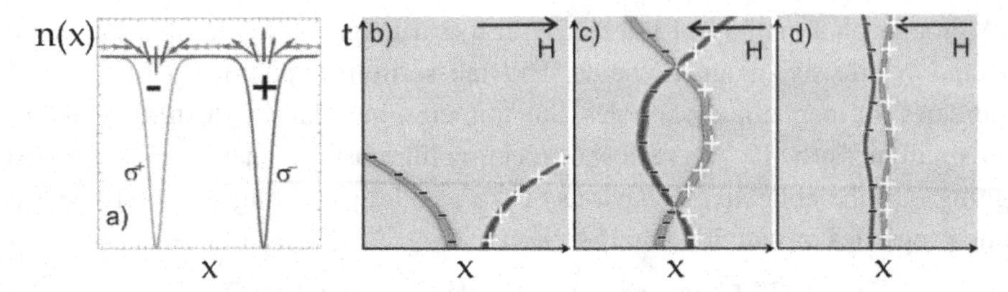

Figure 24.4 Polarization degree ρ_c for a pair of half-solitons. (a) The dark grey (light grey) line depicts the density profile of half-soliton in the $\sigma = -$ ($\sigma = +$) component. The arrows indicate the pseudospin S_x. Numerical time evolution of ρ_c as extracted from Eq. (24.1) in the presence of a constant magnetic field $\mathbf{H} = \pm 10\mathbf{e_x}$ μeV (black arrows), showing the trajectories of two HSs: acceleration for $H > 0$ (panel b), and oscillations (panel c) and bouncing (panel d) for $H < 0$. The symbols + and − indicate the sign of the magnetic charges. Adapted with permission from Terças, H., et al. (2014), High-speed DC transport of emergent monopoles in spinor photonic fluids, *Phys. Rev. Lett.*, **113**, 036403 [31]. Copyright (2014) by the American Physical Society.

the simplest behaviour: acceleration without crossing for $H > 0$; panel c illustrates the inversion of the charge (the 'red', σ_--soliton is initially accelerated to the left and then to the right). In this case, the two solitons undergo dipolar oscillations, forming a 'molecule', due to the inversion of the spin texture (charge). Panel d depicts the bouncing of the two HSs without the charge inversion, due to the interactions between spin components. Details on the variational analysis can be found in Ref. [31].

We consider now the situation of a dense soliton gas, for which the multiple-soliton solution with the Inverse Scattering Transform [32] provides a continuum of eigenvalues. In that case, the solitons are uncorrelated and therefore ergodic enough to justify a statistical treatment: kinetic equations have been used to describe wave turbulence in optical fibres [33], for example. We describe the evolution of the soliton distributions $f^{\pm}(x, v, t)$ with the following kinetic equations of the Boltzmann type

$$\frac{\partial f^{\pm}}{\partial t} + v\frac{\partial f^{\pm}}{\partial x} + \frac{q(v)}{M(v)}H\frac{\partial f^{\pm}}{\partial v} = I[f^{\pm}], \tag{24.17}$$

where $q(v)/M(v) = n_0/2M_0\gamma^2$ is the relativistic charge/mass ratio. In order to estimate the transport properties of the system, in analogy with the Drude model for electrons in the presence of an electric field [34, 35], we assume small deviations from equilibrium, allowing the collision integral $I[f^{\pm}]$ to be written in the relaxation-time approximation [36], $I[f^{\pm}] \simeq -(f^{\pm} - f_0^{\pm})/\tau^{\pm}$, where $\tau^{\pm}$ is the relaxation time and $f_0^{\pm}$ is the phase-space equilibrium distribution. We define the total magnetic current as $j = j^+ - j^-$, where $j^{\sigma} = \langle q(v)N_c v\rangle = \int q(v)N_c v f^{\sigma} dv$

and N_c is the concentration of magnetic charges. For symmetry, the total current is given as $j = 2j^+ = -2j^-$, so we calculate the current associated with the σ_+ component only. From Eq. (24.17), the DC magnetic current can be written as

$$j = \frac{N_c n_0^2 \tau}{2M_0} H \int v \left(1 - \frac{v^2}{c^2}\right) \frac{\partial f_0}{\partial v} dv. \tag{24.18}$$

Eq. (24.18) incorporates the relativistic behaviour of HS, which implies a vanishing current near the sound speed $v \simeq c$. To estimate the collision time τ, we make use of the Matthiessen's rule [37]: $1/\tau = 1/\tau_H + 1/\tau_{\sigma,\sigma} + 1/\tau_{\sigma,-\sigma}$, where τ_H is the collision rate induced by the field H; $\tau_{\sigma,\sigma}$ ($\tau_{\sigma,-\sigma}$) represents the collision rate due to the short-range potentials in Eqs. (24.4) and (24.5).

The two-body dynamics is in competition with the collective behaviour of the system. Thus, the concentration of unbound carriers is not necessarily the same as that of the gas. To estimate the concentration of carriers, we extend Onsager's theory for the conduction of weak electrolytes [38]. Using the fermionic statistics of solitons, as argued above, Eq. (24.18) yields

$$j = \frac{N_0 n_0^2}{2M_0} \tau H \eta \left(1 - \frac{v_F^2}{c^2}\right), \tag{24.19}$$

with η standing for the fraction of dissociated monopoles [31]. We note that the Fermi velocity of the gas, $v_F = \pi \hbar N_0 / M_0$, is small compared to the sound speed for the case of polariton condensates, but it is not necessarily the case for cold atomic condensates, for which we may easily have $n_0 \xi \sim 1$. The features of Eq. (24.19) are summarised in Fig. 24.5c, d. For small values of the field, η does not vary with H and the DC current satisfies the Ohm's law $j \propto H$ (see Fig. 24.5c). For moderate values of H, the system enters a non-Ohmic regime, characterised by $\partial j / \partial H < 0$. This behaviour is qualitatively different from the deviation from the Ohmic response observed in spin-ices, where the conductivity monotonically increases with the applied field [25]. The reason for such a difference resides in the fact that the soliton-pair dissociation energy depends on the density of the HS gas; besides, our system is 1D and the jamming of carriers is more important. In of Fig. 24.5d, we plot the conductivity against the HS gas density. For very low densities, the transport is dominated by two-particle dynamics and the DC current is strongly suppressed. For higher densities, the response of the system is dictated by collisions.

24.3.2 Numerical Checking of the Analytical Results

We have also performed numerical simulations using Eq. (24.1) with a gas of HS taken as initial condition. In Fig. 24.6, we illustrate the most relevant regimes of

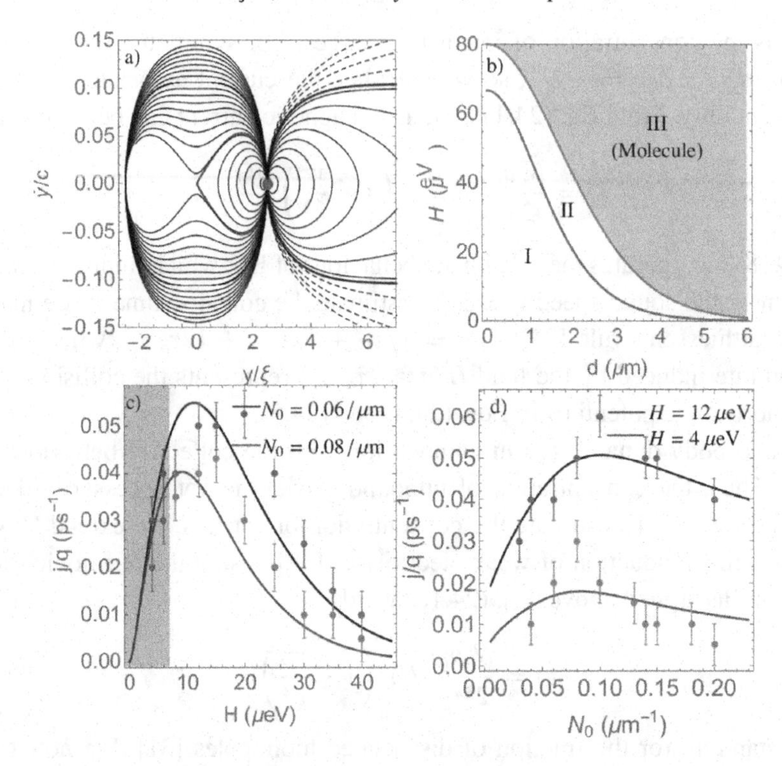

Figure 24.5 (a) Phase-space map for a pair of half-solitons initially separated by $d = 2.5\xi$. Full (dashed) lines are obtained for $H > 0$ ($H < 0$). The thick line is obtained for $H = 0$. The grey line is the separatrix between modes II and III. (b) Magnitude of the critical fields H_1 (grey line) and H_2 (dark grey line) as a function of the initial separation d. The DC monopole current as a function of the applied field (gas density) is shown in panels c., d. The shadow limits the Ohmic region. Solid lines (c, d) – theoretical predictions, dots with error bars – the numerical results. Other parameters: $m = 5 \times 10^{-5} m_e$, $\xi = 1\mu$m and $n_0 \sim 500\mu\text{m}^{-1}$. Reprinted with permission from Terças, H., et al. (2014), High-speed DC transport of emergent monopoles in spinor photonic fluids, *Phys. Rev. Lett.*, **113**, 036403 [31]. Copyright (2014) by the American Physical Society.

the magnetic current. In Fig. 24.6a, we observe the breaking of dipolar oscillations (or molecule dissociation) due to collisions between the solitons within the same component. For moderate values of density (Fig. 24.6b), such collisions, similarly to the Drude model for electron conduction, lead to the appearance of a net current of magnetic charges (Ohmic response). Finally, for higher values of density, the conductivity is suppressed (Fig. 24.6c), and the small-amplitude oscillations become the dominant mechanism. All these features are in qualitative agreement with the analytical estimates, as illustrated in Fig. 24.5. The deviation between the

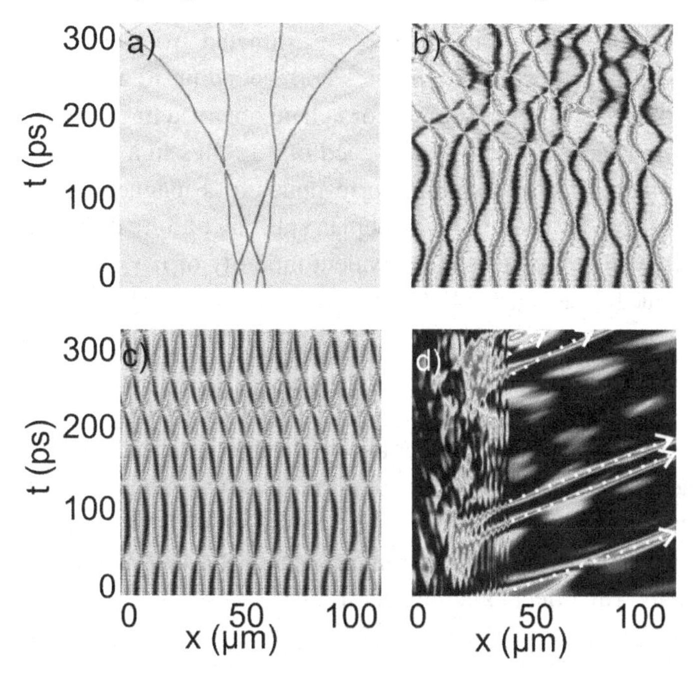

Figure 24.6 (a) Breakdown of oscillations due to the interactions between HS in the same component. (b) The onset of magnetricity ($N_0 \sim 0.08\ \mu m^{-1}$). (c) Suppression of conductivity due to short-range interaction ($N_0 \sim 0.15\ \mu m^{-1}$). (d) Extraction of half-solitons (light grey traces) from the trapping region ($\lambda = 0.5\ \mu m$, $L = 45 \mu m$, $U_0 = 2$ meV and $\tau = 30$ ps) by a field $H = 5\ \mu eV$. The white arrows are a guide for the eyes. Reprinted with permission from Terças, H., et al. (2014), High-speed DC transport of emergent monopoles in spinor photonic fluids, *Phys. Rev. Lett.*, **113**, 036403 [31]. Copyright (2014) by the American Physical Society.

analytic theory and the numerical results may stem in the fact that the transport model neglects the acoustic radiation of solitons as they accelerate [39].

To describe a more realistic experimental configuration, we simulate Eq. (24.1) for polaritons propagating in a one-dimensional cavity by adding (i) a narrow Gaussian barrier described by the potential $U_{\pm}\psi_{\pm} = U_0 \exp[-(x - L)^2/\lambda^2]\psi_{\pm}$, (ii) a coherent, linearly polarised pump $P_{\pm} = P_0(x)e^{i(kx-\omega t)}$ and (iii) the finite life-time term $-i\hbar\psi_{\pm}/2\tau$. The barrier, located at the position $x = L$, is strong enough to confine the HS gas. Initially, the magnetic field is absent, and the solitons remain trapped without escaping. Then, the effective magnetic field is switched on (it can be controlled externally [40]), and we observe the extraction of HS from the confined region (red traces propagating to the right in Fig. 24.6d), showing the linear polarisation degree of the condensate. An alternative way of generating soliton trains has been proposed recently [41].

The monopole mobility can be directly estimated from our calculations, for which we compare the potential energies corresponding to a fixed displacement. Indeed, a mobility of 10^6 cm^2/Vs (a record value obtained in graphene [42]) means that an electron is accelerated up to a speed of 10^6 cm/s in a field of 1 V/cm. The same displacement of a half-soliton corresponds to a magnetic energy of 10 eV (assuming $n_0 \sim 2 \times 10^2\ \mu\text{m}^{-1}$ and a Zeeman splitting of 5 μeV), while the velocity is $\sim 10^8$ cm/s, which provides an equivalent mobility of $\mu = 10^7$ cm^2/Vs. Such a high value is due to the extremely low polariton mass ($m < 10^{-4} m_e$).

24.4 Conclusions

In this chapter, we have studied the collective behaviour of half-integer topological defects – the half-solitons – in 1D polariton quantum fluids, demonstrating that a gas of half-solitons can form a Wigner crystal or melt into a Luttinger liquid, depending on the density. Under an effective magnetic field, a DC current of half-solitons appears – an effect called magnetricity.

References

[1] Carusotto, I., and Ciuti, C. 2013. Quantum fluids of light. *Rev. Mod. Phys.*, **85**, 299.

[2] Pitaevskii, L. P., and Stringari, S. 2003. *Bose-Einstein Condensation*. Oxford, UK: Clarendon Press.

[3] Salomaa, M. M., and Volovik, G. E. 1987. Quantized vortices in superfluid He3. *Rev. Mod. Phys.*, **59**, 533.

[4] Hivet, R., Flayac, H., Solnyshkov, D. D., Tanese, D., Boulier, T., Andreoli, D., Giacobino, E., Bloch, J., Bramati, A., Malpuech, G., and Amo, A. 2012. Half-solitons in a polariton quantum fluid behave like magnetic monopoles. *Nat. Phys.*, **8**, 724.

[5] Wigner, E. 1934. On the interaction of electrons in metals. *Physical Review*, **46**, 1002–1011.

[6] Deshpande, V. V., and Bockrath, M. 2007. The one-dimensional wigner crystal in carbon nanotubes. *Nat. Phys.*, **4**, 314.

[7] Xu, Zhihao, Li, Linhu, Xianlong, Gao, and Chen, Shu. 2013. Wigner crystal versus Fermionization for one-dimensional Hubbard models with and without long-range interactions. *J. Phys: Condens. Matter*, **25**, 055601.

[8] Łakomy, K., Nath, R., and Santos, L. 2012. Spontaneous crystallization and filamentation of solitons in dipolar condensates. *Phys. Rev. A*, **85**, 033618.

[9] Bloch, I., Dalibard, J., and Zwerger, W. 2008. Many-body physics with ultracold gases. *Rev. Mod. Phys.*, **80**, 885–964.

[10] Krive, I. V., Nersesyan, A. A., Jonson, M., and Shekhter, R. I. 1995. Influence of long-range Coulomb interaction on the metal-insulator transition in one-dimensional strongly correlated electron systems. *Phys. Rev. B*, **52**, 10865–10871.

[11] Weber, U., and McGovern, J. A. 1997. A self-consistent approach to the Wigner-Seitz treatment of soliton matter. *Phys. Rev. C*, **57**, 3376.

[12] Shelykh, I. A., Kavokin, A. V., Rubo, Y. G., Liew, T. C. H., and Malpuech, G. 2009. Polarization-sensitive phenomena in planar semiconductor microcavities. *Semiconductor Science and Technology*, **25**, 013001.

[13] Flayac, H., Solnyshkov, D. D., and Malpuech, G. 2011. Oblique half-solitons and their generation in exciton–polariton condensates. *Phys. Rev. B*, **83**, 193305.
[14] Frantzeskakis, D. J. 2010. Dark solitons in atomic Bose-Einstein condensates: from theory to experiments. *J. Phys. A*, **43**, 82.
[15] Öhberg, P., and Santos, L. 2001. Dark solitons in a two-component Bose-Einstein condensate. *Phys. Rev. Lett.*, **86**, 2918.
[16] Terças, H., Solnyshkov, D. D., and Malpuech, G. 2013. Topological Wigner crystal of half-solitons in a spinor Bose-Einstein condensate. *Phys. Rev. Lett.*, **110**, 035303.
[17] Lieb, E. H., and Liniger, W. 1963. Exact analysis of an interacting Bose gas. I. The general solution and the ground state. *Phys. Rev.*, **130**, 1605.
[18] Lieb, E. H. 1963. Exact analysis of an interacting bose gas. II. The excitation spectrum. *Phys. Rev.*, **130**, 1616.
[19] Haldane, F. D. M. 2000. 'Luttinger liquid theory' of one-dimensional quantum fluids. I. Properties of the Luttinger model and their extension to the general 1D interacting spinless Fermi gas. *J. Phys. C*, **14**, 2585.
[20] Schulz, H. 1990. Correlation exponents and the metal-insulator transition in the one-dimensional Hubbard model. *Phys. Rev. Lett.*, **64**, 2831.
[21] March, N. H. 1988. Comment on melting of the Wigner crystal at finite temperature. *Phys. Rev. A*, **37**, 4526.
[22] Dirac, P. A. M. 1931. Quantised singularities in the electromagnetic field. *Proc. Roy. Soc. London Ser. A*, **133**, 60–72.
[23] Pinfold, James L. 2010. Dirac's dream – the search for the magnetic monopole. *AIP Conference Proceedings*, **1304**, 234.
[24] Castelnovo, C., Moessner, R., and Sondhi, S. L. 2008. Magnetic monopoles in spin ice. *Nature*, **451**, 42.
[25] Bramwell, S. T., Giblin, S. R., Calder, S., Aldus, R., Prabhakaran, D., and Fennell, T. 2009. Measurement of the charge and current of magnetic monopoles in spin ice. *Nature*, **461**, 956.
[26] Jaubert, L. D. C., and Holdsworth, P. C. W. 2009. Signature of magnetic monopole and Dirac string dynamics in spin ice. *Nat. Phys.*, **5**, 258–261.
[27] Fennell, T., Deen, P. P., Wildes, A. R., Schmalzl, K., Prabhakaran, D., Boothroyd, A. T., Aldus, R. J., McMorrow, D. F., and Bramwell, S. T. 2009. Magnetic Coulomb phase in the spin ice Ho2Ti2O7. *Science*, **326**, 415.
[28] Morris, D. J. P., Tennant, D. A., Grigera, S. A., Klemke, B., Castelnovo, C., Moessner, R., Czternasty, C., Meissner, M., Rule, K. C., Hoffmann, J. U., Kiefer, K., Gerischer, S., Slobinsky, D., and Perry, R. S. 2009. Dirac strings and magnetic monopoles. *Science*, **326**, 411.
[29] Bramwell, S. T. 2012. Magnetic monopoles: magnetricity near the speed of light. *Nat. Phys.*, **8**, 703.
[30] Solnyshkov, D. D., Flayac, H., and Malpuech, G. 2012. Stable magnetic monopoles in spinor polariton condensates. *Phys. Rev. B*, **85**, 073105.
[31] Terças, H., Solnyshkov, D. D., and Malpuech, G. 2014. High-speed DC transport of emergent monopoles in spinor photonic fluids. *Phys. Rev. Lett.*, **113**, 036403.
[32] Zakharov, V. E. 1971. Kinetic equation for solitons. *Sov. Phys. JETP* **33**, 538 [*Zh. Eksp. Teor. Fiz.* **60**, 993 (1971)].
[33] El, G. A., and Kamchatnov, A. M. 2005. Kinetic equation for a dense soliton gas. *Phys. Rev. Lett.*, **95**, 204101.
[34] Drude, P. 1900. Zur Elektronentheorie der Metalle. *Ann. Phys. Ser.*, **4**, 566.
[35] Sommerfeld, A. 1928. Zur elektronentheorie der metalle auf grund der Fermischen Statistik. *Zeits. and Physik*, **47**, 1.

[36] Gantmaker, V. F., and Levinson, Y. B. 1987. *Carrier Scattering in Metal and Semiconductors*. North-Holland.

[37] Beaulac, T. P., Allen, P. B., and Pinski, F. J. 1982. Electron–phonon effects in copper. II. Electrical and thermal resistivities and Hall coefficient. *Phys. Rev. B*, **26**, 1549.

[38] Onsager, L. 1969. The motion of ions: principles and concepts. *Science*, **166**, 1359.

[39] Parker, N. G., Proukakis, N. P., Barenghi, C. F., and Adams, C. S. 2003. Dynamical instability of a dark soliton in a quasi-one-dimensional Bose-Einstein condensate perturbed by an optical lattice. *J. Phys. B*, **37**, 12.

[40] Malpuech, G., Glazov, M. M., Shelykh, I. A., Bigenwald, P., and Kavokin, K. V. 2006. Electronic control of the polarization of light emitted by polariton lasers. *Appl. Phys. Lett.*, **88**, 111118.

[41] Pinsker, F., and Flayac, H. 2014. On-demand dark soliton train manipulation in a spinor polariton condensate. *Phys. Rev. Lett.*, **112**, 140405.

[42] Bolotin, K. I., Sikes, K. J., Jiang, Z., Klima, M., Fudenberg, G., Hone, J., Kim, P., and Stormer, H. L. 2008. Ultrahigh electron mobility in suspended graphene. *Solid State Communications*, **146**, 351.

25

Microscopic Theory of Bose-Einstein Condensation of Magnons at Room Temperature

HAYDER SALMAN
School of Mathematics, University of East Anglia, Norwich Research Park, Norwich, UK

NATALIA G. BERLOFF
Skolkovo Institute of Science and Technology, Russian Federation
DAMTP, Centre for Mathematical Sciences, Cambridge, UK

SERGEJ O. DEMOKRITOV
Institute for Applied Physics and Center for Nonlinear Science, University of Muenster, Germany;
Institute of Metal Physics, Ural Division of the Russian Academy of Sciences (RAS), Yekaterinburg, Russia

A quantised spin wave – magnon – in magnetic films can undergo Bose-Einstein condensation (BEC) into two energetically degenerate lowest-energy quantum states with nonzero wave vectors $\pm \boldsymbol{k}_{\mathrm{BEC}}$. This corresponds to two interfering condensates forming spontaneously in momentum space. Brillouin Light Scattering studies for a microwave-pumped film with submicrometer spatial resolution experimentally confirm the existence of the two wavefunctions and show that their interference results in a nonuniform ground state of the condensate with the density oscillating in space. Moreover, fork dislocations in the density fringes provide direct experimental evidence for the formation of pinned half-quantum vortices in the magnon condensate. The measured amplitude of the density oscillation implies the formation of a nonsymmetric state that corresponds to nonequal occupation of two energy minima. We discuss the experimental findings and consider the theory of magnon condensates, which includes, to leading order, the contribution from the noncondensed magnons. The effect of the noncondensed magnon cloud is to increase the contrast of the asymmetric state and to bring about the experimental measurements.

25.1 Introduction

Magnons are quasiparticles corresponding to quantised spin waves that describe the collective motion of spins [1]. In recent years, it has become possible to realise a BEC of magnons in two remarkably different systems: in superfluid phases of an isotope of helium – Helium-3 at ultralow temperatures [2, 3] and in ferrimagnetic insulators [4, 5, 6, 7]. In the normal, noncondensed state, these magnetic materials exhibit a magnetically ordered state with a gas of magnons whose phases are not correlated. However, once condensation occurs, spins develop phase coherence resulting in a common global frequency and phase of precession. At room temperature, ferrimagnetic insulators, such as yttrium-iron garnet (YIG) films, together with a combination of an in-plane magnetic field and microwave radiation, are required for the magnon condensation [4, 5, 6, 7]. Therefore, in analogy with exciton–polariton condensates, this provides yet another example of a BEC of quasiparticle excitations in a solid-state system.

25.2 Experimental Realisation of a Bose-Einstein Condensate of Magnons in an Epitaxial YIG-film

A room-temperature Bose-Einstein condensate of magnons was created in an epitaxial YIG-film using an experimental setup shown in Fig. 25.1a, which, in general, is similar to that used in our previous studies [4, 5]. Given that magnons are quasiparticle excitations, their number is not conserved and, therefore, the chemical potential is zero. To reach the critical value of the chemical potential necessary for BEC formation, we inject additional magnons using microwave parametric pumping. After thermalisation, the injected magnons gather in two energetically degenerate minima of the energy spectrum located at $\pm \boldsymbol{k}_{\mathrm{BEC}} = \pm \mathbf{Q} = (0, \pm Q)$. In fact, the energy spectrum is anisotropic and arises from the combined effects of the long-range magnetic dipole interactions that break the isotropy of the spectrum and the short-range exchange interactions. The additional effects of pumping together with dissipation, caused by the spin–lattice interactions, results in a system that is driven out of equilibrium. However, for magnons, a quasi-equilibrium state with a nonzero chemical potential can be realised because the magnon lifetimes, set by the spin–lattice relaxation times ($\sim$ 1 ms), are long compared to the magnon–magnon thermalisation time ($\sim$ 100 ns). Therefore, in contrast to BECs of other quasiparticle excitations such as exciton-polariton condensates, a magnon BEC can be considered to be in quasi-equilibrium. This situation implies that an equilibrium statistical mechanical approach is expected to provide a reasonably accurate description of this system.

Condensation into two nonzero values of the wave vector $\pm \boldsymbol{k}_{\mathrm{BEC}}$ leads to an anisotropy of the condensate ground state and coexistence of two spatially overlapping condensates with the order parameters $\psi_{\pm Q}$. In [8], Brilloun Light Scattering (BLS) was used to measure the total magnon density $|\psi|^2$, where $\psi = \psi_Q \exp[iQz] + \psi_{-Q} \exp[-iQz]$, and we have assumed that the in-plane magnetic field is along the z-direction. Therefore, by scanning the probing laser spot in the

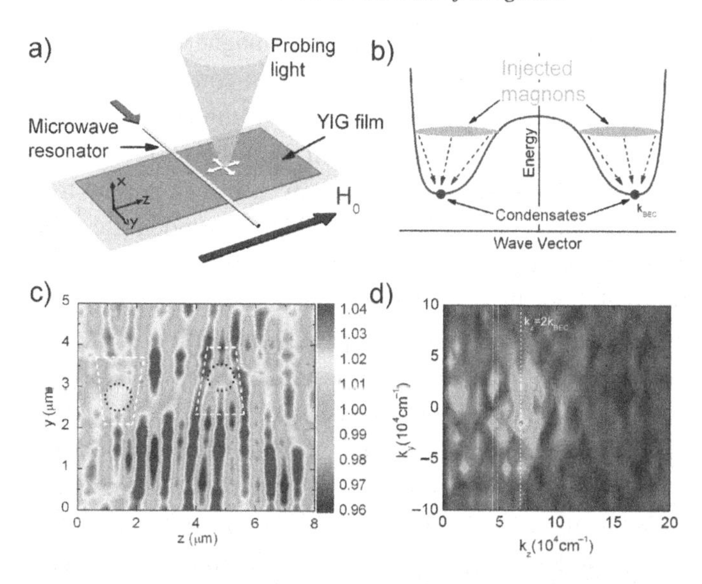

Figure 25.1 Schematic of the experiment and results of two-dimensional imaging of the condensate density. (a) Experimental setup. Magnons are injected into the YIG film using a microwave resonator. After thermalisation, they create a Bose-Einstein condensate, which is imaged by scanning the probing laser light in two lateral directions. (b) Qualitative picture of the magnon spectrum in a ferromagnetic film. Injected magnons thermalise and create two Bose-Einstein condensates at two degenerate spectral minima with nonzero wave vectors $\pm \boldsymbol{k}_{\mathrm{BEC}}$. (c) Measured two-dimensional spatial map of the BLS intensity proportional to the condensate density, obtained at the maximum used pumping power. Dashed circles show the positions of topological defects in the standing-wave pattern corresponding to a nonuniform ground state of the condensate. (d) Two-dimensional Fourier transform of the measured spatial map. The dashed line marks the value of the wave vector equal to $2k_{\mathrm{BEC}}$. The spread of the spectral peak and its slight displacement with respect to $k_y = 0$ are caused by the presence of topological defects resulting in a nonzero slope of the real-space stripe structure, as well as by slight misalignment between the static magnetic field and the scanning axis. Reprinted with permission from Nowik-Boltyk, P., et al. (2013), Spatially non-uniform ground state and quantized vortices in a two-component Bose-Einstein condensate of magnons, *Scientific Reports*, **2**, 482 [8]. Copyright (2013) by the Nature Publishing Group.

two lateral directions and recording the BLS intensity, the spatial distribution of the condensate density can be visualised. Fig. 25.1c shows the results of a two-dimensional mapping of the condensate density across an $8 \times 5\ \mu\mathrm{m}^2$ area of the YIG film adjacent to the pumping resonator, within which the field created by the resonator can be considered as being approximately uniform. The mapping was performed by repetitive scanning of the spatial area followed by the averaging of the recorded data to improve the signal-to-noise ratio. The map clearly demonstrates a periodic pattern along the direction of the static magnetic field created because of interference of the two components of the magnon condensate. The spatial period of the pattern 0.9 ± 0.1 μm obtained from a two-dimensional Fourier transform of the recorded map (Fig. 25.1d) agrees well with the period 0.92 μm calculated based on the known value $k_{\mathrm{BEC}} = 3.4 \times 10^4 \mathrm{cm}^{-1}$. It is important to recognise that in the absence of the phase locking between the two condensates, the evolution of the phase difference between the two condensate order parameters would lead to changes in the spatial positions of the maxima and minima and would not show in the time-averaged results. The observed spatial modulation in the BLS intensity is therefore clear evidence of the locking of the coherent phases between the two components that must be explained by any successful theory of magnon condensation.

The interference experiments not only provide direct evidence of the spatial coherence of the magnon condensate in YIG films, but also indicate that a strong asymmetry exists in the number density of the two condensates. To see this, we recall that the average BLS intensity is proportional to the total density of the condensate. It grows with pumping power, increasing from the BEC-transition threshold of 6 mW to 100 mW and then saturates due to the reduction of the parametric pumping efficiency as shown in Fig. 25.2. At the same time, the modulation depth

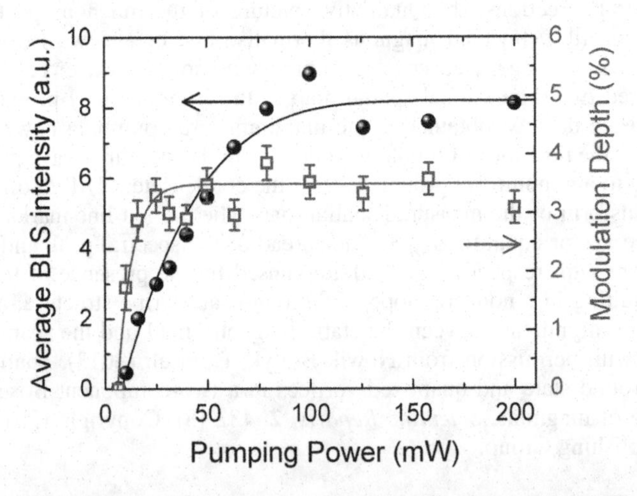

Figure 25.2 Effect of the pumping power on the condensate density and on the amplitude of the standing wave. Filled symbols show the average BLS intensity proportional to the total density of the condensate. Open symbols show the relative depth of the spatial modulation of the BLS intensity, which characterises the strength of the phase locking between the condensate components. The data were obtained from one-dimensional scans parallel to the direction of the static field. Reprinted with permission from Nowik-Boltyk, P., et al. (2013), Spatially non-uniform ground state and quantized vortices in a two-component Bose-Einstein condensate of magnons, *Scientific Reports*, **2**, 482 [8]. Copyright (2013) by the Nature Publishing Group.

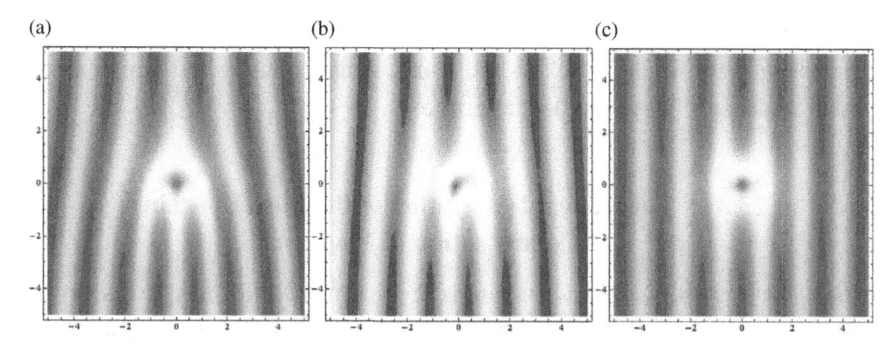

Figure 25.3 Density field in the presence of vortices: (a) Vortex in first component and antivortex in component two; (b) vortex in component one only; (c) vortices in both components.

increases quickly above the threshold and then stays nearly constant. As shown in the figure, the measured contrast defined as

$$\beta = \frac{|\psi|^2_{\mathrm{max}} - |\psi|^2_{\mathrm{min}}}{|\psi|^2_{\mathrm{max}} + |\psi|^2_{\mathrm{min}}} \tag{25.1}$$

is seen to be equal to $\beta = 0.035$. However, by accounting for the spatial resolution of the measuring probe, we estimate the actual contrast in the experiment to be $\beta = 0.12–0.15$.

The above observations provide direct evidence for the coexistence of a two-component BEC of magnons which are interlocked through their phase coherence. The production of interference fringes associated with the densities of the two components also allows us to identify vortices in our magnon condensate. A clear signature of the formation of vortices is the appearance of a forklike dislocation in the interference pattern as seen in Fig. 25.1c. These quantised vortices, which spontaneously appear during the condensation process [9], are pinned by the crystalline defects present in the sample. The number of prongs present in the interference fringes leads us to conclude that the vortices we observe are analogous to half-quantum vortices (fractional vortices), which have also been observed in exciton-polariton condensates [10]. To illustrate this, we present in Fig. 25.3 three density fields produced by assuming a condensate wavefunction of the form $\psi = \psi_Q \exp[iQz] + \psi_{-Q} \exp[-iQz]$. We prescribe the background values of $|\psi_Q|^2 = n_{+Q}$ and $|\psi_{-Q}|^2 = n_{-Q}$ to reproduce the measured contrast of 15%, but in each case[1]

[1] (a) $\psi_Q = \sqrt{n_{+Q}}(z+iy)/\sqrt{z^2+y^2+n_{+Q}^{-1}}$, $\psi_{-Q} = \sqrt{n_{-Q}}(z-iy)/\sqrt{z^2+y^2+n_{-Q}^{-1}}$;
(b) $\psi_Q = \sqrt{n_{+Q}}(z+iy)/\sqrt{z^2+y^2+n_{+Q}^{-1}}$, $\psi_{-Q} = \sqrt{n_{-Q}}$;
(c) $\psi_Q = \sqrt{n_{+Q}}(z+iy)/\sqrt{z^2+y^2+n_{+Q}^{-1}}$, $\psi_{-Q} = \sqrt{n_{-Q}}(z+iy)/\sqrt{z^2+y^2+n_{-Q}^{-1}}$,
with $n_{-Q} = 0.15 n_{+Q}$.

we make different assumptions on what vortices exist in each component. As can be seen, assuming a single vortex in one component produces best agreement with the observed density fringes in the experiments.

25.3 A Theoretical Description of the Bose-Einstein Condensation of Magnons

Recently, a description of BEC in microwave-pumped YIG films was proposed [11] that accounted for the fourth order interactions and magnon-nonconserving terms of the dipolar interactions. The theory was able to explain the appearance of asymmetric states but the contrast obtained was much smaller ($\sim 1\%$) than the actual experimentally observed ($\sim 12-15\%$). In fact, the value of 3.5% presented in Fig. 25.2 does not account for the spatial resolution of the probing laser light, which is about 0.55 μm and therefore of the same order as the wavelength 0.92 μm associated with the interference pattern created by the two condensates. This results in significant averaging over the oscillation and leads to an underestimation of the contrast. It can be shown that the experimental measurement needs to be multiplied by a correction factor of 3.43, producing this relatively higher value of contrast. This discrepancy has been remedied in [12] by including the effect of the non-condensed thermal cloud of magnons on the condensate within a Hartree-Fock approximation. It turns out that, apart from depleting the condensate, the thermally excited noncondensed magnons increase the contrast of the asymmetric state and bring about the experimental measurements.

To adopt a microscopic description of the problem, we recognise that on energy scales relevant to the experiments, only the lowest ferromagnetic magnon band is important. The magnetic properties of YIG can then be described in terms of a quantum Heisenberg ferromagnet on a cubic lattice given by the Hamiltonian

$$\hat{\mathcal{H}} = -g\mu_B H_0 \sum_j \hat{S}_j^z - J \sum_{j,\delta_L} \hat{\mathbf{S}}_j \cdot \hat{\mathbf{S}}_{j+\delta_L} + U_d \sum_{i \neq j} \frac{\hat{\mathbf{S}}_i \cdot \hat{\mathbf{S}}_j - 3(\hat{\mathbf{S}}_i \cdot \mathbf{n}_{ij})(\hat{\mathbf{S}}_i \cdot \mathbf{n}_{ij})}{|r_{ij}|^3},$$

for the spin operators $\hat{\mathbf{S}}_j = (\hat{S}_j^x, \hat{S}_j^y, \hat{S}_j^z)$. The various terms account for the Zeeman energy, the exchange interactions and the dipolar interactions, respectively. Here, μ_B is the magnetic moment (Bohr magneton), $g = 2$ is the Landé factor, H_0 is the externally applied magnetic field, and J and U_d are the exchange and dipolar interaction constants, respectively. The sums i, j are taken over lattice sites at positions $\mathbf{r}_i$, δ_L represents a vector to one of the nearest neighbours of i, $\mathbf{r}_{ij} = (\mathbf{r}_i - \mathbf{r}_j)/a_0$, $\mathbf{n}_{ij} = \mathbf{r}_{ij}/|\mathbf{r}_{ij}|$ and $a_0 = 12.376 \AA$ is the lattice constant.

Since we are interested in describing spin deviations from the direction of the in-plane externally applied magnetic field, we will consider the action of spin operators on spin states $|s_j^z\rangle$, where s_j^z denotes the z-quantum number on site j. The total spin operator at j then satisfies $\hat{\mathbf{S}}_j^2|s_j^z\rangle = (\hat{S}_j^{x^2} + \hat{S}_j^{y^2} + \hat{S}_j^{z2})|s_j^z\rangle = S(S+1)|s_j^z\rangle$, where S is the effective spin. This means that we can effectively describe the action of $\hat{S}_j^z$ in terms of the other two components of spin at j. We will therefore recast the Hamiltonian into a form that directly incorporates this constraint by eliminating the spin operator $\hat{S}_j^z$. The remaining operators are then encoded into spin raising and lowering operators defined as

$$\hat{S}_j^+ = \hat{S}_j^x + i\hat{S}_j^y, \qquad \hat{S}_j^+ = \hat{S}_j^x - i\hat{S}_j^y. \tag{25.2}$$

We can now relate these spin operators to creation and annihilation bosonic operators $\hat{a}_j^\dagger$ and $\hat{a}_j$ that act on the occupation number basis of spin deviations at site j. This is given by the Holstein-Primakoff transformation [14]

$$\hat{S}_j^+ = \sqrt{2S}\left(1 - \frac{\hat{a}_j^\dagger \hat{a}_j}{2S}\right)^{1/2} \hat{a}_j, \qquad \hat{S}_j^- = \sqrt{2S}\hat{a}_j^\dagger \left(1 - \frac{\hat{a}_j^\dagger \hat{a}_j}{2S}\right)^{1/2} \tag{25.3}$$

The bosonic operators can be expressed in terms of normal modes such that

$$\hat{a}_j = \sum_{\mathbf{k}} \hat{a}_k(x_j) e^{i\mathbf{k}\cdot\mathbf{r}}, \qquad \hat{a}_j^\dagger = \sum_{\mathbf{k}} \hat{a}_k^\dagger(x_j) e^{i\mathbf{k}\cdot\mathbf{r}}, \tag{25.4}$$

where $\mathbf{r} = (y, z)$ and $\boldsymbol{k} = (k_y, k_z)$. Given that the thickness of the film is much smaller than the lateral dimensions, we have performed only a partial Fourier summation over the wavenumbers lying within the (y, z) plane of the film. In contrast, we have retained the spatial dependence on the x-coordinate that is aligned along the thickness of the film. This is because the transverse modes along x are expected to be sensitive to the effect of the boundaries and cannot be approximated using plane waves. However, if we focus on formulating an effective theory for quantities averaged over the thickness of the film, we can drop the dependence on the x spatial coordinate. This results in the so-called uniform mode approximation. To proceed, we now assume that the number of spin deviations in the system is small and exploit the fact that the effective spin $S \simeq 14.2$ is quite large to perform an expansion in powers of $1/S$. Hence, by adopting the uniform mode approximation [13] together with a Holstein-Primakoff transformation [14], our Hamiltonian can be expanded up to fourth-order terms in $\hat{a}_k$ and $\hat{a}_k^\dagger$ which can be written as

$$\hat{\mathcal{H}} = \hat{\mathcal{H}}_0 + \hat{\mathcal{H}}_2 + \hat{\mathcal{H}}_4 + \cdots,$$

where the index of the corresponding part of the Hamiltonian corresponds to the order of the interactions. The quadratic terms given by $\hat{\mathcal{H}}_2$ takes the form

$$\hat{\mathcal{H}}_2 = \hbar \sum_{\mathbf{k}} \left(A_k \hat{a}_k^\dagger \hat{a}_k + \frac{1}{2} B_k \hat{a}_k \hat{a}_{-k} + \frac{1}{2} B_k^* \hat{a}_k^\dagger \hat{a}_{-k}^\dagger \right), \tag{25.5}$$

where

$$A_k = \gamma H_0 + \mathcal{D}k^2 + \gamma 2\pi M(1 - F_k)\sin^2\theta + \gamma 2\pi M F_k,$$
$$B_k = \gamma 2\pi M(1 - F_k)\sin^2\theta - \gamma 2\pi M F_k. \tag{25.6}$$

Here, $F_k = \big(1 - \exp(-|kd|)\big)/|kd|$, d is the film thickness, $k^2 = k_y^2 + k_z^2$, θ is the angle between the wavevector $\boldsymbol{k}$ and the in-plane applied magnetic field, $M = 0.14$kG is the magnetisation, $\gamma = 12\mu eV/kOe$ is the gyromagnetic ratio and $\mathcal{D} = 0.24eV\text{Å}^2$ is the coefficient of the exchange interactions [11]. These quadratic terms can be diagonalised by using the Bogoliubov transformation

$$\hat{a}_k = u_k\hat{c}_k + v_k\hat{c}^\dagger_{-k}, \quad \hat{a}^\dagger_k = u_k\hat{c}^\dagger_k + v_k^*\hat{c}_{-k},$$
$$u_k = \left(\frac{A_k + \hbar\omega_k}{2\hbar\omega_k}\right)^{1/2}, \quad v_k = \mathrm{sgn}(B_k)\left(\frac{A_k - \hbar\omega_k}{2\hbar\omega_k}\right)^{1/2}, \tag{25.7}$$

where sgn(·) is the sign function, $|u_k|^2 - |v_k|^2 = 1$, and the dispersion relation is given by $\omega_k = (A_k^2 - |B_k|^2)^{1/2}$. The quadratic term then reduces to $\hat{\mathcal{H}}_2 = \sum_k \hbar\omega_k\hat{c}^\dagger_k\hat{c}_k$. Since a magnon BEC consists of two condensates, we write the quadratic term for condensed magnons in terms of the annihilation and creation operators, $\hat{c}_{\pm\mathbf{Q}}$ and $\hat{c}^\dagger_{\pm\mathbf{Q}}$ for two energetically degenerate lowest energy quantum states with nonzero wave vectors $\mathbf{k}_{\mathrm{BEC}} \equiv \pm\mathbf{Q} = (0, \pm Q)$ in two-dimensional (2D) momentum space. This gives $\mathcal{H}_2 = \hbar\omega_{\mathbf{Q}}\big(\hat{c}^\dagger_{\mathbf{Q}}\hat{c}_{\mathbf{Q}} + \hat{c}^\dagger_{-\mathbf{Q}}\hat{c}_{-\mathbf{Q}}\big)$. The next order terms corresponding to $\mathcal{H}_4$ are given by

$$\hat{\mathcal{H}}_4 = A\left[\hat{c}^\dagger_{\mathbf{Q}}\hat{c}^\dagger_{\mathbf{Q}}\hat{c}_{\mathbf{Q}}\hat{c}_{\mathbf{Q}} + \hat{c}^\dagger_{-\mathbf{Q}}\hat{c}^\dagger_{-\mathbf{Q}}\hat{c}_{-\mathbf{Q}}\hat{c}_{-\mathbf{Q}}\right] + 2B\hat{c}^\dagger_{\mathbf{Q}}\hat{c}^\dagger_{-\mathbf{Q}}\hat{c}_{-\mathbf{Q}}\hat{c}_{\mathbf{Q}}$$
$$+ C\left[\hat{c}^\dagger_{\mathbf{Q}}\hat{c}_{\mathbf{Q}}\hat{c}_{\mathbf{Q}}\hat{c}_{-\mathbf{Q}} + \hat{c}^\dagger_{-\mathbf{Q}}\hat{c}_{-\mathbf{Q}}\hat{c}_{-\mathbf{Q}}\hat{c}_{\mathbf{Q}} + \mathrm{h.c}\right] + D\left[\hat{c}_{\mathbf{Q}}\hat{c}_{\mathbf{Q}}\hat{c}_{-\mathbf{Q}}\hat{c}_{-\mathbf{Q}} + \mathrm{h.c}\right], \tag{25.8}$$

where h.c denotes the hermitian conjugate.

The first two terms conserve the number of magnons and give rise to self- and mutual interaction of the condensates and were obtained in [15], the third term was introduced in [11] and the fourth term in [12]. The last two terms do not conserve the number of magnons and lead to two condensates locking their total phases. The expressions for the coefficients of Eq. (25.9) are

$$A = -\frac{\hbar\omega_M}{4SN}\Big[(\alpha_1 - \alpha_3)F_Q - 2\alpha_2(1 - F_{2Q})\Big] - \frac{\mathcal{D}Q^2}{2SN}(\alpha_1 - 4\alpha_2),$$
$$B = \frac{\hbar\omega_M}{2SN}\Big[(\alpha_1 - \alpha_2)(1 - F_{2Q}) - (\alpha_1 - \alpha_3)F_Q\Big] + \frac{\mathcal{D}Q^2}{SN}(\alpha_1 - 2\alpha_2),$$
$$C = \frac{\hbar\omega_M}{8SN}\left[(3\alpha_1 + 3\alpha_2 - 4\alpha_3)F_Q + \frac{8}{3}\alpha_3(1 - F_{2Q})\right] + \frac{\mathcal{D}Q^2}{SN}\frac{\alpha_3}{3},$$
$$D = \frac{\hbar\omega_M}{4SN}\Big[(\alpha_3 - 3\alpha_2)F_Q + 2\alpha_2(1 - F_{2Q})\Big] + \frac{\mathcal{D}Q^2}{2SN}\alpha_2, \tag{25.9}$$

where $\hbar\omega_M = 4\pi M\gamma$, $\alpha_1 = u_Q^4 + 4u_Q^2v_Q^2 + v_Q^4$, $\alpha_2 = 2u_Q^2v_Q^2$ and $\alpha_3 = 3u_Qv_Q(u_Q^2 + v_Q^2)$. Expressions for A and B were obtained in [15], C (correcting the coefficients in front of α_2 and α_3 [12]) was presented in [11] and, D in [12].

We now adopt a classical fields approximation and replace operators with complex numbers. Furthermore, we use the Madelung transformation for $c_{\pm\mathbf{Q}} = \sqrt{N_{\pm Q}}\exp[i\phi_\pm]$ and introduce the total number of condensed magnons $N_c = N_Q + N_{-Q}$, the occupation imbalance given by $\delta = N_Q - N_{-Q}$ and the total phase $\Phi = \phi_+ + \phi_-$ to obtain

$$\mathcal{H}_4 = \frac{1}{2}N_c^2\Big[(A+B) - (B + D\cos 2\Phi \quad \Lambda)(\delta/N_c)^2 + 2C\cos\Phi\sqrt{1-(\delta/N_c)^2}\Big]. \tag{25.10}$$

The total number of condensed magnons N_c is set by a balance established between pumping and relaxation that we will not model explicitly. We will, therefore, prescribe N_c based on the experimentally estimated values reported in [4, 5]. We then proceed to calculate the ground state by minimising the total energy of the system subject to fixed N_c. In this case, the minimum energy is determined by minimising the interaction term given by Eq. (25.10) with respect to δ and Φ. Differentiating (25.10) with respect to Φ and δ gives the fixed points (a) $\Phi = 0, \delta = 0$, (b) $\Phi = \pi, \delta = 0$, (c) $\Phi = 0, (\delta/N_c)^2 = 1 - C^2/(B + D - A)^2$, (d) $\Phi = \pi, (\delta/N_c)^2 = 1 - C^2/(B + D - A)^2$ or (e) $\Phi = \cos^{-1}(-C/2D), \delta = 0$. We discard (e) in the view of the smallness of D in comparison with $|C|$. $\mathcal{H}_4$ is minimised for $\Phi = \pi$ if $C > 0$ and for $\Phi = 0$ if $C < 0$. The minima are determined by the sign of the parameter $\Delta = A - B + |C| - D$. When $\Delta > 0$, $\delta = 0$, which gives rise to the symmetric case $N_Q = N_{-Q}$. When $\Delta < 0$, $\delta/N_c = 1 - C^2/(B + D - A)^2$ corresponds to an asymmetric case.

Substituting the parameters into Eqs. (25.9), we find $A = -0.1685$ mK/N, $B = 8.3395$ mk/N, $C = -0.0138$ mK/N and $D = -0.0017$ mK/N, so that $\Delta < 0$. This gives an asymmetric state with the contrast $\beta = 2\sqrt{N_QN_{-Q}}/N_c$ of the order of 1%. This is in disagreement with the experiment of [8], where the contrast is around 12–15% once corrections for the resolution of the probing laser light are taken into account. This discrepancy can be explained by accounting for the effect induced by the presence of the thermal cloud of noncondensed magnons that has been neglected in our discussion thus far in deriving the Hamiltonian of the system. By including the effect of the thermal cloud within a Hartree-Fock approximation, we find that the main effect on the condensate is to renormalise the coefficient C to $\tilde{C}$ such that

$$\tilde{C} = C + \frac{E}{N_c}, \tag{25.11}$$

where E is given by

$$E = \frac{1}{2}\Bigg\{-\sum_{\mathbf{k}}{}' (D_k/2+f_{1,k})/N\left[8(u_Q^2+v_Q^2)u_k v_k\right]$$

$$-(D_Q/2+f_{1,Q})/N\left[\sum_{\mathbf{k}}{}' 16u_Q v_Q(u_k^2+v_k^2)+\sum_{\mathbf{k}}{}' 8(u_Q^2+v_Q^2)u_k v_k\right]$$

$$-\sum_{\mathbf{k}}{}' (f_{2,k}+3f_{2,Q})/N\left[4(u_Q^2+v_Q^2)(u_k^2+v_k^2)\right]$$

$$+\sum_{\mathbf{k}}{}' (D_{k-Q}+f_{3,k-Q}+D_{k+Q}+f_{3,k+Q})/N\left[4(u_Q^2+v_Q^2)u_k v_k\right]$$

$$+\,(D_0+f_{3,0})/N\sum_{\mathbf{k}}{}'\left[8u_Q v_Q(u_k^2+v_k^2)\right]\Bigg\}\langle c_k^{\dagger}c_k\rangle, \qquad (25.12)$$

$\sum'$ implies that $\mathbf{k} \neq (0, \pm\mathbf{Q})$, and

$$f_1 = \frac{\hbar\gamma 2\pi M}{S}\left[(1-F_k)\sin^2\theta + F_k\right]/4,$$

$$f_2 = \frac{\hbar\gamma 2\pi M}{S}\left[(1-F_k)\sin^2\theta - F_k\right]/4,$$

$$f_3 = \frac{\hbar\gamma 2\pi M}{S}(1-F_k)\cos^2\theta,$$

$$D_k = -J\sum_{\boldsymbol{\delta}} e^{i\mathbf{k}\cdot\boldsymbol{\delta}} \approx \frac{\mathcal{D}k^2}{2S} - 2J. \qquad (25.13)$$

To evaluate E, we must determine the population of the thermal cloud in mode $\boldsymbol{k}$ denoted by $N_k = \langle c_k^{\dagger}c_k\rangle$. This must be interpreted as the occupation averaged over the sample thickness d since we are only considering in-plane wavenumbers. We can calculate this by summing over all the energy bands in the sample corresponding to the quantisation of the dispersion relation in the direction normal to the plane of the sample to obtain

$$n_{\mathbf{k}}(T,\mu) = \frac{1}{dL_xL_y}\sum_n \frac{1}{e^{(\hbar\omega_{\mathbf{k},n}-\mu)/k_B T_{\mathrm{eff}}}-1}. \qquad (25.14)$$

In the above, L_x and L_z denote the extent of the sample in the y and z-coordinate directions respectively and the effective temperature can be taken to be room temperature given by $T_{\mathrm{eff}} = 300K$. The above expression requires knowledge of the full energy spectrum in our sample. Here, we have used the form given by [16]. For the experiment of [8], this yields $E/N_c = -1.3043$ mK/N and for $n_c = N_c/V = 3.5 \times 10^{18}$ cm^{-3} gives $\widetilde{C} = -1.3181$ mK/N. This sets the contrast to the value

$\beta = |\widetilde{C}|/|(B+E-A)| = 15.5\%$, which is in much better agreement with experiments. We note that the value of n_c used here is in good agreement with experimentally quoted values. Moreover, it corresponds to a condensate number density to total number density ratio of 5.7%, which is also in reasonable agreement with experiments.

Finally, we comment on the mean-field equations for the two component magnon condensate. Starting with the full Hamiltonian of the system, we can now write [17] the Gross-Pitaevskii system of equations describing the collective excitations of the two condensates as

$$i\hbar\frac{\partial \psi_{\pm Q}}{\partial t} = -\frac{\delta \mathcal{H}}{\delta \psi^*_{\pm Q}}. \tag{25.15}$$

In analogy with approximations used in atomic condensates [18], we will assume a static thermal cloud of magnons. The system of equations is then given by

$$\begin{aligned} i\hbar\frac{\partial \psi_{\pm Q}}{\partial t} = & -\frac{\hbar^2}{2m}\nabla^2\psi_{\pm Q} + 2AV|\psi_{\pm Q}|^2\psi_{\pm Q} + 2BV|\psi_{\mp Q}|^2\psi_{\pm Q} \\ & +CV\big[(\psi_{\pm Q}\psi_{\mp Q} + \psi^*_{\pm Q}\psi^*_{\mp Q})\psi_{\pm Q} \\ & +(|\psi_{\pm Q}|^2 + |\psi_{\mp Q}|^2)\psi^*_{\mp Q}\big] + 2DV\psi^*_{\pm Q}\psi^{*2}_{\mp Q} \\ & \qquad +E\psi^*_{\mp Q} + F\psi_{\pm Q}, \end{aligned} \tag{25.16}$$

where F is an interaction coefficient that, similarly to E, is a function of the number of noncondensed magnons.

25.4 Conclusions

We have shown that condensation of magnons in ferrimagnetic insulators such as yttrium-iron garnet films exhibit a number of unique features. The condensation can be realised at room temperature and occurs at two nonzero momenta corresponding to the lowest energy states. Consequently, the ground state of the condensate appears as a real-space standing wave of the total condensate density attenable to the BLS measurements. The occupation of the states is asymmetric as indicated by contrast measurements associated with the standing wave pattern. Moreover, in analogy with exciton-polariton condensates [10], a magnon condensate supports the formation of half-quantised vortices between the two components that are pinned at the cristallographic defects. By modelling this two-component condensate, including the effects of noncondensed thermal cloud of magnons, it is possible to quantitatively explain the observed asymmetry of the occupation of the two condensates and predict a phase transition between symmetric and asymmetric states, as the film thickness and the applied magnetic field are varied. Future directions will require the development of models to study nonequilibrium phenomena, such as the process of condensate formation of magnons in YIG films.

Acknowledgements: SOD acknowledges the financial support by the Deutsche Forschungsgemeinschaft via the priority program SPP1538 "Spin Caloric Transport (SpinCaT)" and by the Russian Ministry of Education and Science via the program Megagrant No. 14.Z50.31.0025.

References

[1] V. L. Safonov, *Nonequilibrium Magnons: Theory, Experiment, and Applications* (2013). Wiley-VCH.
[2] G. Volovik, Twenty years of magnon Bose condensation and spin current superfluidity in 3HeB, *J. Low Temp. Phys.,* 153 (2008), 266.
[3] Y. M. Bunkov and G. V. Volovik, Magnon Bose Einstein condensation and spin superfluidity, J. *Phys.: Condens. Matter*, 22 (2010), 1.
[4] S. O. Demokritov, V. E. Demidov, O. Dzyapko, G. A. Melkov, A. A. Serga, B. Hillebrands, and A. N. Slavin, Bose-Einstein condensation of quasi-equilibrium magnons at room temperature under pumping, *Nature*, 443 (2006), pp. 430–433.
[5] V. E. Demidov, O. Dzyapko, S. O. Demokritov, G. A. Melkov, and A. N. Slavin, Observation of spontaneous coherence in BoseEinstein condensate of magnons, *Phys. Rev. Lett.*, 100 (2008), 047205.
[6] A. V. Chumak, G. A. Melkov, V. E. Demidov, O. Dzyapko, V. L. Safonov, and S. O. Demokritov, Bose Einstein condensation of magnons under incoherent pumping, *Phys. Rev. Lett.*, 102 (2009), 187205.
[7] O. Dzyapko, V. E. Demidov, M. Buchmeier, T. Stockhoff, G. Schmitz, G. A. Melkov, and S. O. Demokritov, Excitation of two spatially separated Bose-Einstein condensates of magnons, *Phys. Rev. B*, 80 (2009), 060401.
[8] P. Nowik-Boltyk O. Dzyapko, V. E. Demidov, N. G. Berloff and S. O. Demokritov, Spatially non-uniform ground state and quantized vortices in a two-component Bose-Einstein condensate of magnons, *Scientific Reports*, 2 (2013), 482.
[9] N. G. Berloff and B. V. Svistunov. Scenario of strongly nonequilibrated Bose Einstein condensation, *Phys. Rev. A*, 66 (2002), 013603.
[10] K. G. Lagoudakis, T. Ostatnický, A. V. Kavokin, Y. G. Rubo, R. André, and B. Deveaud-Plédran, *Science*, 326 (2009), 974.
[11] F. Li, W. M. Saslow and V.L. Pokrovsky, Phase diagram for magnon condensate in yttrium iron Garnet film, *Scientific Reports*, 3 (2013), 1372.
[12] H. Salman, N. G. Berloff and S. O. Demokritov, in preparation (2015).
[13] P. Krivosik and C. E. Patton, *Phys. Rev. B*, 82 (2010), 184428.
[14] T. Holstein and H. Primakoff, *Phys. Rev.*, 58, (1940) 1098.
[15] I. S. Tupitsyn et al., *Phys. Rev. Lett.*, 100 (2008), 257202.
[16] B. A. Kalinikos and A. N. Slavin, *J. Phys. C: Solid State Phys.*, 19 (1986), 7013.
[17] L. Pitaevskii, S. Stringari, *Bose-Einstein condensation* (2003). Clarendon Press.
[18] A. Griffin, T. Nikuni, and E. Zaremba, *Bose-Condensed Gases at Finite Temperatures* (2009). Cambridge University Press.

26

Spintronics and Magnon Bose-Einstein Condensation

REMBERT A. DUINE
Institute for Theoretical Physics and Center for Extreme Matter and Emergent Phenomena, Utrecht, The Netherlands

ARNE BRATAAS
Department of Physics, Norwegian University of Science and Technology, Trondheim, Norway

SCOTT A. BENDER AND YAROSLAV TSERKOVNYAK
Department of Physics and Astronomy, University of California, Los Angeles, USA

Spintronics is the science and technology of electric control over spin currents in solid-state-based devices. Recent advances have demonstrated a coupling between electronic spin currents in nonmagnetic metals and magnons in magnetic insulators. The coupling is due to spin transfer and spin pumping at interfaces between the normal metals and magnetic insulators. In this chapter, we review these developments and the prospects they raise for electric control of quasi-equilibrium magnon Bose-Einstein condensates and spin superfluidity in ferromagnets.

26.1 Introduction

In undergraduate texts, Bose-Einstein condensation is a phase transition that occurs in an ideal Bose gas at large enough densities. In developing setups to observe Bose-Einstein condensation in a way analogous to this textbook treatment, one faces several challenges. First of all, conserved bosons in condensed-matter systems are composite particles. For example, Cooper pairs of electrons are bosons, but the Bardeen-Cooper-Schieffer regime of condensation (in which the electrons condense as a result of a weak and attractive effective interaction) is rather different from the physics of noninteracting pointlike bosons [1]. The size of the Cooper pairs, determined in the weakly interacting limit by the coherence length $\xi \sim \epsilon_F/k_F\Delta$ with k_F the Fermi wavenumber, ϵ_F the Fermi energy, and Δ the superconducting gap, is much larger than the distance between the pairs. This latter distance is estimated from the fraction Δ/ϵ_F of the electrons that form Cooper pairs, so that the pair density is $n_p \sim k_F^3\Delta/\epsilon_F \sim k_F^2/\xi$, and we have that $n_p\xi^3 \sim (k_F\xi)^2 \gg 1$.

Cold atoms [2], on the other hand, are composite bosons where the internal degrees of freedom are typically at much higher energies than their temperature, so that they can be considered as point particles. The bosonic atoms in magnetically trapped ultracold atomic vapors are, however, not conserved and have a finite lifetime because atoms may escape from the trap (both as single atoms or after molecule-forming collision processes). The same mechanism also prevents the system from reaching its true thermodynamic ground state (a state that is most likely a solid). Relaxing the requirement of strict boson conservation, one may consider Bose-Einstein condensation as a quasi-equilibrium phenomenon. Prime examples of quasi-equilibrium Bose-Einstein condensation of nonconserved particles are condensates of photons [3] and (exciton-)polaritons [4, 5] (see Chapters 19 and 4 respectively). In true equilibrium, these particles would not be described by a Bose-Einstein distribution function with nonzero chemical potential, and for both photons and polaritons the system is maintained in quasi-equilibrium at nonzero chemical potential by external pumping. The mechanism of reaching quasi-equilibrium is very different for each of these systems.

In this chapter, we focus on magnons, quasiparticles corresponding to the quantized excitations of the magnetic order parameter, in magnetic insulators. These can undergo Bose-Einstein condensation as an equilibrium [6, 7] or a quasi-equilibrium [8] phenomenon (see Chapters 25, 27, and 28). Some examples of large, equilibrium magnon condensates include the easy-plane ferromagnets and antiferromagnets discussed in Chapter 27. Here, we focus on quasi-equilibrium condensation of magnons in an easy-plane ferromagnet subjected to a large out-of-plane magnetic field. A slight canting of the spin density order away from the field direction corresponds to a small, phase-coherent magnon condensate. This is the result of excited magnons coming to quasi-equilibrium by magnon-conserving scattering processes, before they dissipate energy and angular momentum to the lattice and relax. In experiments [8, 9] on quasi-equilibrium magnon condensation in solid-state magnetic insulators, the excitation of magnons is achieved by microwave pumping. Recent advances in spintronics, however, have demonstrated interactions between magnons and electrons at interfaces between metals and magnetic insulators [10]. As we will discuss in detail in this chapter, this opens the possibility of direct-current (DC) pumping of the magnetic system (see Fig. 26.1) and of achieving quasi-equilibrium magnon Bose-Einstein condensation in a solid-state DC transport experiment. On top of this, the integration of quasi-equilibrium magnon Bose-Einstein condensation with electronics opens the possibility for electronic transport probes of the associated spin superfluidity.

In the remainder of this chapter, we first discuss the recent developments in spintronics that make possible the manipulation of the magnetization in magnetic insulators by means of electrical current. Hereafter, we discuss the phase diagram

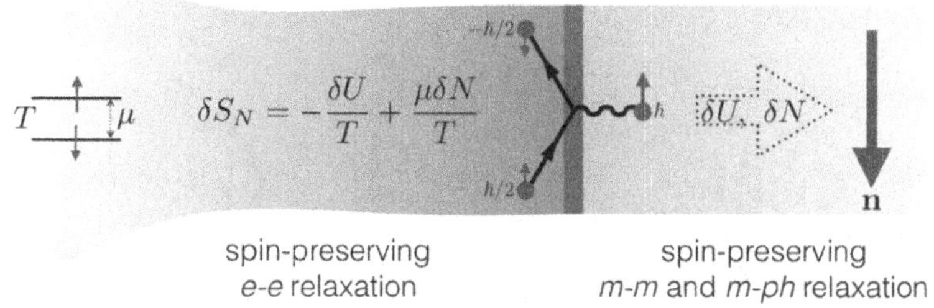

Figure 26.1 Schematic of a normal-metal/magnetic-insulator heterostructure, in which a finite chemical potential of magnons can be induced by a nonequilibrium electron-spin accumulation μ. In the idealized case with no magnon-number relaxation in the insulator, the metallic reservoir acts as a bath supplying magnons and energy into the ferromagnet, with temperature T and chemical potential μ. δS_N is the entropic change in the metal bath associated with the transfer of energy δU and creation of δN magnons in the ferromagnet (relative to the ordered spin orientation $\mathbf{n}$).

of the DC-pumped quasi-equilibrium magnon gas. Finally, we briefly discuss signatures of the resulting spin superfluidity and give future prospects.

26.2 Spintronics

In this section, we outline recent discoveries in spintronics that enable the integration of quasi-equilibrium magnon Bose-Einstein condensation with the control of spin currents. The first of these phenomena is the existence of spin currents across an interface between a normal metal and a magnetic insulator. The second concerns the generation of spin currents from charge current (or vice versa) via the (inverse) spin Hall effect. In the first two subsections below, we discuss the general theory for spin transport through the bulk of and across interfaces between metals and magnetic insulators. Hereafter, we discuss the (inverse) spin Hall effect and the spin Seebeck effect as applications.

26.2.1 Interfacial Spin Currents

We consider a magnetic insulator sandwiched between two normal metals, as depicted in Fig. 26.2. This geometry is chosen for convenience as it makes the theoretical analysis of the spin transport effectively one-dimensional. In an experiment, it is easiest to put the metals on top of a thin film of a magnetic insulator. In terms of materials, an often studied system is the magnetic insulator Yttrium-Iron-Garnett (YIG) interfaced with the nonmagnetic normal metal

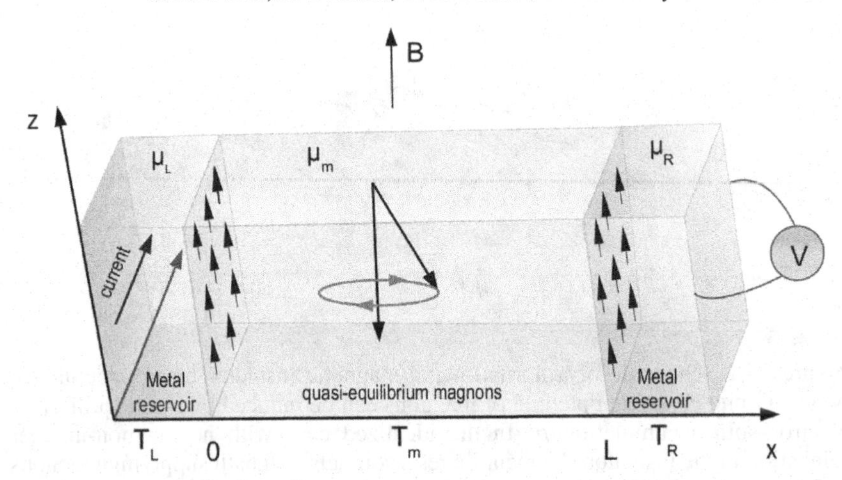

Figure 26.2 A set up that we consider in this chapter. A quasi-equilibrated magnetic insulator at temperature T_m and chemical potential μ_m is connected to left and right by metallic reservoirs with spin accumulations μ_L and μ_R, and temperatures T_L and T_R, respectively. The spin accumulation in the reservoirs results from the spin Hall effect in response to currents in the y-direction (shown for the left reservoir) and/or injected spin current from the magnetic insulator. Any spin current injected from the insulator to the reservoirs results in an inverse spin Hall voltage which can be measured (shown for the right reservoir). The spin quantization axis is defined by the external field that we choose in the z-direction so that the equilibrium spin density points in the $-z$-direction.

platinum. We assume that the electrons inside the normal metals on the left and right have respective spin accumulations μ_L and μ_R at each interface. This spin accumulation is an out-of-equilibrium electrochemical potential imbalance between up and down electrons, $\mu_L = \mu_\uparrow - \mu_\downarrow$ (with a similar definition of μ_R). While in a more general setup the spin accumulation would be a vector, here we assume that only its z-component is finite, as illustrated by the electron spins in Fig. 26.2 in the normal metal. How the spin accumulation is established is discussed below. We further assume that magnons in the magnetic insulator are in a quasi-equilibrium characterized by a magnon chemical potential μ_m and temperature T_m. The metallic leads are thermally biased, with T_L and T_R for the left and right reservoirs, respectively.

We first focus on the left interface. At the interface, there exists an interfacial exchange coupling between the localized spins in the magnetic insulator and the itinerant spins in the metal. Assuming this coupling is isotropic, a phenomenological expression is

$$\hat{V}_{int} = \int d\mathbf{x}d\mathbf{x}' V(\mathbf{x},\mathbf{x}')\hat{\mathbf{S}}(\mathbf{x}) \cdot \hat{\mathbf{s}}(\mathbf{x}') ,$$

where $\hat{\mathbf{S}}(\mathbf{x})$ corresponds to the spin density of the localized spins in the insulator and $\hat{\mathbf{s}}(\mathbf{x})$ the spin density of the electrons in the metal. The latter is given by

$$\hat{\mathbf{s}}(\mathbf{x}) = \frac{\hbar}{2} \sum_{\sigma,\sigma' \in \{\uparrow,\downarrow\}} \psi_{\sigma}^{\dagger}(\mathbf{x}) \boldsymbol{\tau}_{\sigma\sigma'} \psi_{\sigma'}(\mathbf{x})$$

in terms of the Pauli matrices $\boldsymbol{\tau}$ and electron creation and annihilation operators $\hat{\psi}_{\sigma}^{\dagger}(\mathbf{x})$ and $\hat{\psi}_{\sigma}(\mathbf{x})$. Finally, the matrix elements $V(\mathbf{x}, \mathbf{x}')$ decay rapidly away from the interface.

We first assume that the spin density in the magnetic insulator can be treated classically and is homogeneous, and, moreover, that the temperature is relatively low so that we can approximate $\langle \mathbf{S} \rangle \simeq \hbar s \mathbf{n}$, with the density $s = S/v$, where S is the total spin per unit cell, v the volume of a unit cell in the magnetic insulator, and $\mathbf{n}$ a unit vector in the direction of spin density. The z-component of the spin current from the magnetic insulator into the left reservoir, per unit area, is then given by [11] (we define positive spin current as flowing to the right)

$$j_s^{int} = -\frac{\hbar g_{\uparrow\downarrow}}{4\pi} \mathbf{n} \times \left. \frac{d\mathbf{n}}{dt} \right|_z , \tag{26.1}$$

with $g_{\uparrow\downarrow}$ the so-called spin-mixing conductance (in units of m^{-2} and disregarding its imaginary component). As we shall see, this latter conductance characterizes the efficiency of spin transport across the interface. It can be straightforwardly calculated, e.g., by using perturbation theory in the coupling $V(\mathbf{x}, \mathbf{x}')$. A microscopic expression is not needed at this point. The mixing conductance can also be determined from *ab initio* calculations [12] or from experiments [10]. For interfaces between YIG and Pt, the mixing conductance is estimated to be up to 5 nm^{-2} depending on interface quality [12, 13]. We note that the above expression captures the spin-pumping contribution and not the spin-transfer contribution that results from the spin accumulation in the normal metal.

We now consider the above expression for magnons. Using a linearized Holstein-Primakoff transformation, we have for the (circular) magnon annihilation operator that

$$\hat{b} = \sqrt{\frac{s}{2}} \left(\delta\hat{n}_x - i\delta\hat{n}_y \right) ,$$

where we assumed, as in Fig. 26.2, the magnetic order to be in the $-z$-direction, so that $\mathbf{n} \simeq (\delta\hat{n}_x, \delta\hat{n}_y, -1)$. Inserting this in the expression for the spin current leads to the replacement

$$\mathbf{n} \times \left. \frac{d\mathbf{n}}{dt} \right|_z \rightarrow \frac{4}{s} \frac{1}{V} \sum_{\mathbf{k}} n_{\mathbf{k}} \omega_{\mathbf{k}} ,$$

with $\hbar\omega_{\mathbf{k}}$ the magnon dispersion, and where we inserted an additional factor of two to incorporate constructive interference of magnon modes at the interface [14]. Here, we have made a Fourier transform so that the number of magnons at momentum $\mathbf{k}$ is $n_{\mathbf{k}} = \langle \hat{b}^{\dagger}_{\mathbf{k}} \hat{b}_{\mathbf{k}} \rangle$, and V is the volume of the magnetic insulator. Furthermore, we normal-ordered the magnon creation and annihilation operators and have kept only the expectation values of $\hat{b}^{\dagger}\hat{b}$ as we are interested in thermal magnons. Inserting this result in Eq. (26.1) yields in the first instance

$$j_s^{int} = -\frac{g_{\uparrow\downarrow}}{\pi s} \frac{1}{V} \sum_{\mathbf{k}} n_{\mathbf{k}} \hbar\omega_{\mathbf{k}} \, .$$

We now consider the situation of quasi-equilbrium magnons at chemical potential μ_m and temperature T_m so that

$$n_{\mathbf{k}} = n_B \left(\frac{\hbar\omega_{\mathbf{k}} - \mu_m}{k_B T_m} \right) ,$$

with $n_B(x) = [e^x - 1]^{-1}$ the Bose-Einstein distribution function. In equilibrium, there is no spin current across the interface. To account for this, we generalize our treatment with the replacement (see Ref. [15])

$$n_B \left(\frac{\hbar\omega_{\mathbf{k}} - \mu_m}{k_B T_m} \right) \to n_B \left(\frac{\hbar\omega_{\mathbf{k}} - \mu_m}{k_B T_m} \right) - n_B \left(\frac{\hbar\omega_{\mathbf{k}} - \mu_L}{k_B T_L} \right) .$$

The first contribution corresponds to spin pumping, the emission of spin current from an excited magnet. The second term reflects spin transfer, the injection and absorption of spin current into a magnet. Over the last decade, both these phenomena have been investigated extensively for a classical magnetization, i.e., not including contributions of magnons (see, e.g., Refs. [16] and [17]). Putting the above together, and rewriting momentum summations as energy integrations involving the magnon density of states $D(\epsilon)$, we finally find the result

$$j_s^{int} = -\frac{g_{\uparrow\downarrow}}{\pi s} \int d\epsilon D(\epsilon) \, (\epsilon - \mu_L) \left[n_B \left(\frac{\epsilon - \mu_m}{k_B T_m} \right) - n_B \left(\frac{\epsilon - \mu_L}{k_B T_L} \right) \right] . \tag{26.2}$$

In linear response, this yields

$$j_s^{int} = \frac{\sigma_s^{int}}{\hbar\Lambda} (\mu_L - \mu_m) + \frac{L_{SSE}^{int}}{\Lambda} (T_L - T_m) \, , \tag{26.3}$$

which defines an interface spin conductivity $\sigma_s^{int} = 3\zeta(3/2)\hbar g_{\uparrow\downarrow}/2\pi s \Lambda^2$ and the coefficient $L_{SSE}^{int} = 15\zeta(5/2) k_B g_{\uparrow\downarrow}/4\pi s \Lambda^2$. Here, we assumed a quadratic magnon dispersion $\hbar\omega_{\mathbf{k}} = J_s \mathbf{k}^2 + B$ in terms of the spin stiffness J_s (with B as the magnetic field in the z direction), so that the magnon density of states is $D(\epsilon) = \sqrt{\epsilon - B}/4\pi^2 J_s^{3/2}$. Furthermore, we have introduced the length scale $\Lambda = \sqrt{4\pi J_s / k_B T_m}$, the thermal de Broglie wavelength for the magnons. Eqs. (26.2) and

(26.3) are the main expressions for the magnon spin current across the interface that we will use below. The interface between the magnetic insulator and normal metal on the right can be treated analogously. The contribution proportional to the temperature difference in Eq. (26.3) and determined by the coefficient L_{SSE}^{int} gives rise to an interface contribution to the so-called spin Seebeck effect, i.e., a spin current as a result of a temperature gradient [18, 19]. This effect and its reciprocal (dubbed the spin Peltier effect [20]) are the subject of intense investigation [21].

26.2.2 Bulk Spin Transport in Normal Metals and Magnetic Insulators

Having discussed the spin transport through the interfaces between metals and insulators, we now turn to the bulk. We focus again on the left metallic reservoir and in the first instance assume transport is diffusive with weak spin-orbit interactions (so that the scattering mean free path is shorter than the spin-flip diffusion length). Even though the diffusive approach is not generally applicable to experimental situations, as detailed below, it for now serves our pedagogical purpose. The set of equations that describes the coupled spin and charge dynamics in the normal metal is then given by (within our quasi-one-dimensional geometry) [22]

$$
\begin{aligned}
j_c &= \sigma E + \frac{\sigma_{SH}}{2e}\frac{\partial \mu_L}{\partial x}\,, \\
\frac{2e}{\hbar} j_s &= -\frac{\sigma}{2e}\frac{\partial \mu_L}{\partial x} - \sigma_{SH} E\,.
\end{aligned}
\tag{26.4}
$$

In the above, the electric field E and charge current j_c are in the y-direction, j_s is in the x-direction, and the electron charge is e. The charge conductivity σ and spin Hall conductivity σ_{SH} are both in units of $\Omega^{-1}\ \mathrm{m}^{-1}$. The second term in the first equation is the charge current that results from a gradient in the spin accumulation via the inverse spin Hall effect [23]. This is the Onsager reciprocal of the spin Hall effect and is thus governed by the same coefficient σ_{SH}. The inverse spin Hall effect is a powerful means to detect spin current electrically, as discussed in more detail below. Writing $\sigma_{SH} = \theta_{SH}\sigma$, the spin Hall effect is quantified in terms of the dimensionless quantity (dubbed spin Hall angle) θ_{SH}. For Pt, $\theta_{SH} \sim 0.05$ and for Ta, its magnitude is similar, but the sign is opposite [24].

The above equations have to be complemented with a continuity like equation for spin,

$$
\frac{\partial j_s}{\partial x} = -\Gamma \mu_L\,,
$$

where the rate per unit volume Γ phenomenologically expresses spin-flip relaxation in the metal on the left of the insulator. Insertion of the expression for the spin current into the latter equation yields the spin-diffusion equation

$$\frac{\partial^2 \mu_L}{\partial x^2} = \frac{\mu_L}{\ell^2}, \tag{26.5}$$

with the spin-flip diffusion length of the left lead $\ell = \sqrt{\sigma\hbar/4e^2\Gamma}$. We assume a constant electronic temperature inside each of the normal metals so that bulk thermoelectic transport is neglected.

In the diffusive limit of magnon transport, the magnetic insulator is described by similar equations as the diffusive normal metal, i.e.,

$$j_s = -\frac{\sigma_s}{\hbar}\frac{\partial \mu_m}{\partial x} - L_{SSE}\frac{\partial T_m}{\partial x},$$
$$\frac{\partial^2 \mu_m}{\partial x^2} = \frac{\mu_m}{\ell_m^2}, \tag{26.6}$$

where σ_s is the spin conductivity of the magnetic insulator, L_{SSE} its bulk spin Seebeck coefficient, and ℓ_m the thermal-magnon propagation length. Contrary to the case of Pt normal metals, a diffusive approach is expected to be appropriate for thermal magnons, albeit that the various microscopic details (such as the role of complicated spin-wave dispersions and strong magnon-phonon and magnon-magnon interactions) are not well understood and may lead to a nontrivial dependence of the transport coefficients on magnetic field and temperature. The transport coefficients L_{SSE} and σ_s and the length scale ℓ_m are therefore neither experimentally nor theoretically very well understood for YIG at present. This, together with the complication of determining magnon temperature profiles from experimentally applied temperature differences [25, 26], hinders a full and quantitative understanding of the spin Seebeck effect (discussed in more detail below) in YIG. Here our purpose is to discuss a simple example of diffusive spin transport that will be contrasted with the case of superfluid magnons later on.

26.2.3 Spin Hall Effect

We now turn to the question of what establishes the spin accumulations μ_L and μ_R in the normal metals on the left and right. These are the result of spin-current injection from the magnetic insulator into the normal metals, combined with spin current due to the spin Hall effect. The spin Hall effect arises due to spin-orbit coupling and manifests as a spin current with a spin polarization and a spatial direction transverse to an applied electric field in a metal [27, 28]. In the geometry of Fig. 26.2, the electric field (or charge current) is oriented in the y-direction and gives rise to a spin current that flows in the x-direction with spin polarization in the z-direction.

We now assume the interface between the left metallic reservoir and the magnetic insulator is at position $x = 0$ and that the thickness of the metal is L_x in the

x-direction and L_z in the z-direction. The equations for the spin current and the spin accumulation are solved with the boundary conditions $j_s(x = -L_x) = 0$ and $j_s(x = 0) = j_s^{int}$. Using the linear-response expression in Eq. (26.3), we find that the spin accumulation at the interface in the left reservoir is given by (assuming $\theta_{SH} \ll 1$)

$$\mu_L = \frac{\theta_{SH} I_c \hbar^2 \Lambda \left[1 - \cosh\left(\frac{L_x}{\ell}\right)\right] + 2eLL_z\sigma_s^{int}\mu_m \cosh\left(\frac{L_x}{\ell}\right)}{2eL_xL_z\sigma_s^{int}\left[\cosh\left(\frac{L_x}{\ell}\right) + \left(\frac{\hbar}{2e}\right)^2 \frac{\Lambda}{\ell}\frac{\sigma}{\sigma_s^{int}} \sinh\left(\frac{L_x}{\ell}\right)\right]},$$

where $I_c = j_c L_x L_z$ is the total current through the normal metal (with the current density assumed to be homogeneous). Note that the above result shows that in the limit of no spin relaxation, i.e., $L_x \to \infty$, no net spin current is flowing across the interface, as j_s is constant and zero (according to the boundary condition at $-L_x$). Again we mention for completeness that an analogous derivation holds for the right metallic reservoir.

26.2.4 Inverse Spin Hall Voltage and the Spin Seebeck Effect

Spin current can be detected electrically via the inverse spin Hall effect as spin current injected into the metallic reservoirs gives an inverse spin Hall voltage in the y-direction across the reservoirs (in an open circuit geometry). As a further illustration, we extend the drift-diffusion treatment to capture the spin Seebeck effect. That is, we consider the inverse spin Hall voltage generated across the reservoirs as a result of a temperature gradient across the magnetic insulator. We assume, for simplicity, the temperature drops linearly $T_m(x) = (T_R - T_L)x/L_m + T_L$, with $T_L > T_R$ (see Fig. 26.3). We do not consider a temperature gradient across the metallic reservoirs, nor do we consider the effect of the temperature difference across the interface (governed by the Kapitza resistance).

We solve the spin diffusion equations in the magnetic insulator subject to the boundary conditions that the spin current vanishes at the left and right boundary of the system. For both interfaces, we use the boundary condition in Eq. (26.3) and take zero charge current in both reservoirs. The properties of the metallic reservoirs, as well as their interfaces with the magnetic insulator, are taken to be equal. Fig. 26.3 shows a schematic plot of the spin accumulation and magnon chemical potential that build up in the metallic reservoirs and magnetic insulator as a result of the thermal gradient. Within a distance $\sim \ell_m$ from the interfaces, the magnon chemical potential is nonzero. Similarly, the spin accumulation in the metallic reservoirs is nonzero within a distance $\sim \ell$. The jump from spin accumulation to magnon chemical potential across the interface is inversely proportional to the interface spin conductivity σ_s^{int}.

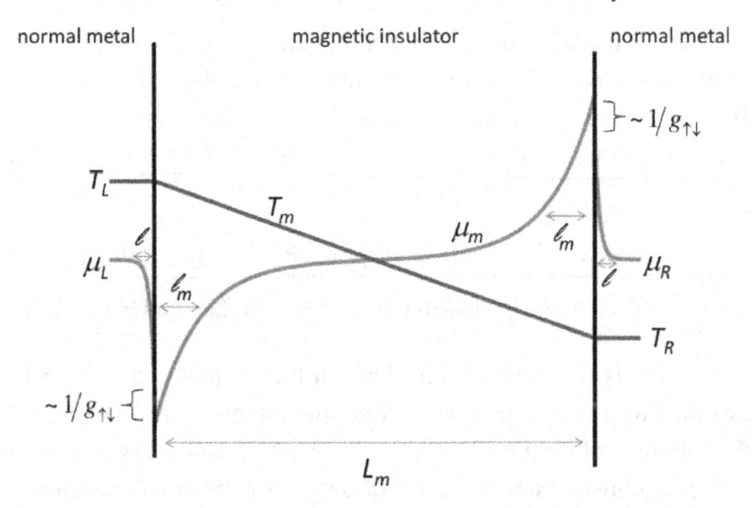

Figure 26.3 Spin accumulations μ_L and μ_R, and magnon chemical potential μ_m, that build up in the metallic reservoirs and magnetic insulator in response to a thermal gradient across the magnetic insulator. The spin accumulations and magnon chemical potential are nonzero over distances $\sim \ell$ and $\sim \ell_m$ (in metals and insulator, respectively) away from the interfaces. The jump across the interfaces is determined by the interface spin conductivity σ_s^{int} that is proportional to the spin mixing conductance $g_{\uparrow\downarrow}$.

The spin Seebeck coefficient is defined as $S_{SSE} = V_{ISHE}/(T_L - T_R)$, with the inverse spin Hall voltage V_{ISHE} found by computing the electric field via Eq. (26.4) from the injected spin current from magnetic insulator to metallic reservoir, and averaging over the x-direction. We consider now the limit $L_m \gg \ell_m$, which allows us to focus on one interface, which we choose to be the right. Within the drift-diffusion theory, the spin Seebeck coefficient is ultimately found as

$$S_{SSE} = -\frac{\theta_{SH}\hbar L_y \ell_m L_{SSE}}{2eL_x L_m \left\{\sigma_s^{int} + \left(\frac{\hbar}{2e}\right)^2 \left[\frac{\ell_m}{\ell}\sigma + \frac{\Lambda}{\ell}\frac{\sigma\sigma_s}{\sigma_s^{int}}\right]\right\}}, \tag{26.7}$$

where L_y is the length of the normal metals in the y-direction and where we also took the limit $L_x \gg \ell$. Within our drift-diffusion theory, the spin Seebeck coefficient (times L_m) saturates as a function of L_m/ℓ_m to the value determined by the above result. This saturation results from the fact that only magnons within a length ℓ_m from the interface contribute to the spin Seebeck voltage [29, 30].

In the above treatment, the properties of the insulator are characterized by the phenomenological constants ℓ_m, L_{SSE} and σ_s. The discussion is made more quantitative by computing these in the relaxation-time approximation in which $\sigma_s \sim J_s\tau/\Lambda^3$ and $L_{SSE} \sim J_s k_B \tau/\hbar\Lambda^3$, where τ is the magnon transport mean free time. Furthermore, the magnon spin propagation length is then given by $\ell_m \sim v_m\sqrt{\tau\tau_{sr}}$, where $v_m = 2\sqrt{J_s k_B T}/\hbar$ is the magnon thermal velocity and τ_{sr} is the magnon spin-relaxation time. Here, τ is the result of various magnon conserving and non-conserving relaxation mechanisms such as magnon–phonon scattering, magnon–magnon (Umklapp) scattering, and scattering of magnons with impurities. The spin-relaxation time τ_{sr} acquires contributions only from processes that do not conserve the magnon number. As we mentioned before, such relaxation and scattering processes are at present not fully understood. We remark that, by assuming Gilbert damping as the only relxation mechanism (which would be appropriate in clean systems at low temperatures), however, both time scales are on the order of $\hbar/\alpha k_B T$, with α the Gilbert damping constant (which for YIG can be as small as 10^{-4}).

Because the spin diffusion length of Pt (which is on the order of a few nm at room temperature [24]) is not large compared to the mean free path, and because in experiments the Pt layers may be rather thin and the interface disordered, a diffusive bulk treatment of the transport in the normal metal is not generally applicable, and the spin accumulation in the normal metal is in those situations not well defined. The appropriate variables are then the torque on the ferromagnet and the (electrical) current in the normal metal. The relations between current and torques are then found from symmetry considerations and microscopic calculations [31, 32, 33], and one finds that the general phenomenology remains the same. Moreover, the system is described phenomenologically in terms of an effective spin Hall angle and effective mixing conductance that characterize coupling between magnetic dynamics and current and loss of angular momentum at the interface, respectively [33]. The drift-diffusion theory discussed here gives an expression for these parameters in terms of the bulk spin Hall angle θ_{SH}, spin-mixing conductance $g_{\uparrow\downarrow}$, and other parameters that is appropriate if the diffusive treatment applies [33].

In summary, in this section we have established that temperature differences and differences between magnon chemical potential and electron spin accumulation drive spin transport across interfaces between normal metals and magnetic insulators. Furthermore, we have demonstrated how an electric current through a conductor that is parallel to its interface with a magnetic insulator sets up a nonzero spin accumulation via the spin Hall effect, and how spin current injected from a magnetic insulator to a normal metal through an interface can be detected via the inverse spin Hall effect in the normal metal.

We emphasize that the above analysis shows that the metallic reservoirs effectively act as grand-canonical baths for the magnons in the magnetic insulator (see Fig. 26.1), and that in this way the system we consider is rather close to the textbook grand-canonical treatment of Bose-Einstein condensation. In the next section, we discuss how by tuning the driving forces, i.e., electrical currents and temperature differences, a quasi-equilibrium Bose-Einstein condensate of magnons can be achieved and maintained.

26.3 Pumping of Quasi-equilibrium Magnon Condensation by Spin Current

In the previous section, we have reviewed recent developments in spintronics concerning the linearized interaction between electrons and thermal magnons at interfaces between magnetic insulators and normal metals. In this section, we consider the more general situation of a magnetic insulator that is partially condensed as a result of interactions with the metallic reservoirs.

To make the discussion concrete, we consider a magnetic insulator that is in the bulk described by a Heisenberg-model Hamiltonian,

$$\hat{H}[\hat{\mathbf{S}}] = -\frac{J}{2\hbar^2}\sum_{<i,j>}\hat{\mathbf{S}}_i \cdot \hat{\mathbf{S}}_j + \sum_i \left[\frac{K}{2\hbar^2}\hat{S}_{z,i}^2 + \frac{B}{\hbar}\hat{S}_{i,z}\right], \tag{26.8}$$

with J the nearest-neighbor exchange energy, $K > 0$ the easy-plane anisotropy constant, and $B > 0$ the external field in units of energy. At zero temperature and without dissipation, the dynamics of the average spin $\langle \hat{\mathbf{S}}_i \rangle \simeq \hbar S \mathbf{n}_i$ is governed by the Landau-Lifshitz equation

$$\frac{\partial \langle \mathbf{S}_i \rangle}{\partial t} = -\frac{1}{\hbar}\langle \mathbf{S}_i \rangle \times \frac{\partial \hat{H}[\langle \hat{\mathbf{S}} \rangle]}{\partial \langle \mathbf{S}_i \rangle}, \tag{26.9}$$

which describes precessional dynamics around the effective field.

The presence of the Bose-Einstein condensate is signaled by a nonzero expectation value $\Psi = \langle \hat{b} \rangle$ of the long-wavelength annihilation operator. Employing the same linearized Holstein-Primakoff transformation as in the previous section, we have that the magnetization direction for the condensed phase is at zero temperature given by $\mathbf{n} \simeq (\sqrt{2/s}\mathrm{Re}\Psi, -\sqrt{2/s}\mathrm{Im}\Psi, n_0/s - 1)$, where $\Psi = \sqrt{n_0}e^{-i\varphi}$ with n_0 the condensate spin density and φ the azimuthal angle of the magnetization with the x-axis. In the homogeneous situation, the Landau-Lifshitz equation then results in

$$\hbar\frac{d\varphi}{dt} = \hbar\omega = B - KS + KSn_0/s, \tag{26.10}$$

which plays the role of the condensate chemical potential. The easy-plane anisotropy leads to a mean-field self-interaction for the condensate. Within the Landau-Lifshitz description, the condensate density is "time-independent". We now incorporate spin injection at the interface and magnetization relaxation to derive a rate equation for the condensate density.

26.3.1 Condensate Rate Equation

For simplicity, we consider the situation as depicted in Fig 26.2, and focus only on the right reservoir. The interface spin current, determined by Eq. (26.1), is then given by

$$\frac{\hbar g_{\uparrow\downarrow} n_0}{2\pi s} \frac{d\varphi}{dt} .$$

Using Eq. (26.10), and substituting $\hbar\omega \rightarrow \hbar\omega - \mu_R$ to account for the nonzero spin accumulation in the right reservoir, this result is rewritten to yield the condensate contribution to the spin current across the interface between metal and insulator [15],

$$j_{s,c}^{int} = \frac{g_{\uparrow\downarrow} n_0}{2\pi s} \left(\hbar\omega - \mu_R\right) . \tag{26.11}$$

The condensate spin current is accompanied by the spin current carried by thermal magnons, which, using Eq. (26.2), is found to be

$$j_{s,x}^{int}(\mu_m, \mu_R, T_m, T_R) = \frac{g_{\uparrow\downarrow}}{\pi s} \int d\epsilon D(\epsilon) \left(\epsilon - \mu_R\right) \times \left[n_B\left(\frac{\epsilon - \mu_m}{k_B T_m}\right) - n_B\left(\frac{\epsilon - \mu_R}{k_B T_R}\right)\right] . \tag{26.12}$$

In the above expressions for both condensate and thermal magnon spin currents, the first term corresponds to spin pumping while the second accounts for spin transfer.

Relaxation processes are at small energies accurately described by the Landau-Lifshitz-Gilbert phenomenology [34]. This implies that the equation for the dynamics of the magnetization direction in Eq. (26.9) acquires a Gilbert damping term parameterized by a dimensionless Gilbert damping constant α, given by

$$\left.\frac{d\mathbf{n}}{dt}\right|_{rel} = -\alpha \mathbf{n} \times \frac{d\mathbf{n}}{dt} .$$

For the condensate, this results in a loss of condensed magnons according to

$$\frac{dn_0}{dt} = -\frac{2\alpha n_0 \hbar\omega}{\hbar} . \tag{26.13}$$

This loss term is similar in form to the contribution from spin pumping, as spin pumping corresponds to loss of angular momentum from the magnetic insulator to the normal metal across the interface.

When the magnons are partially condensed, the density of thermal magnons is fixed and any spin current entering the magnetic insulator through its interface with the right reservoir is eventually absorbed by the condensate [15]. We now assume that this absorption is instantaneous. Physically, this corresponds to the regime where the scattering rate due to interactions between the thermal cloud and condensate is fast compared to the rate of magnon absorption/excitation at the interface and to the damping rate. We denote $j_x \equiv -j_{s,x}^{int}(\hbar\omega, \mu_R, T_m, T_R)$, which is (minus) the interface spin current of thermal magnons in the partially condensed phase. The density of condensed magnons is then determined by the equation

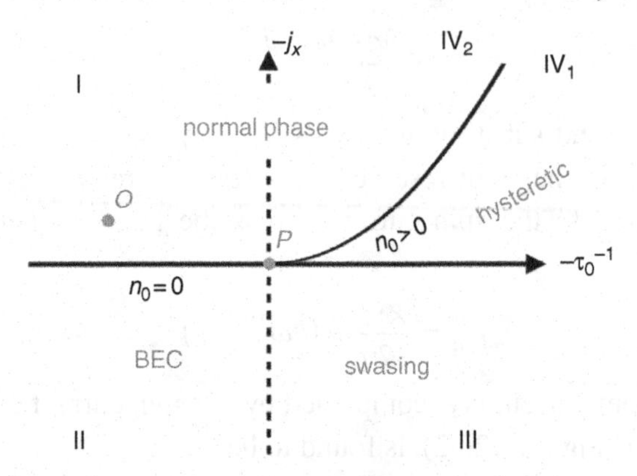

Figure 26.4 Phase diagram of a quasi-equilibrium Bose-Einstein condensate maintained by spin-current injection. The point O corresponds to thermal equilibrium, i.e., no driving forces. The possible phases that meet at the critical point P are separared by transitions (solid lines) or crossovers (dashed lines). For further details, see the main text.

$$\frac{dn_0}{dt} = \frac{j_x}{\hbar L_m} - \frac{g_{\uparrow\downarrow} n_0}{2\pi s L_m} \frac{(\hbar\omega - \mu_R)}{\hbar} - \frac{2\alpha n_0 \hbar\omega}{\hbar} \equiv \frac{j_x}{\hbar L_m} - \frac{n_0}{\tau_c}, \tag{26.14}$$

having defined

$$\frac{1}{\tau_c} = \frac{1}{\tau_0} - \frac{g_{\uparrow\downarrow} K S n_0}{2\pi s^2 \hbar L_m} - \frac{2\alpha K S n_0}{\hbar s},$$

where τ_0 denotes τ_c in the limit $n_0 \to 0$. For temperatures $k_B T \gg K$ the thermal spin current j_x is to a good approximation independent of n_0 [14].

The four possible choices of absolute and relative signs of τ_0 and j_x give rise to four distinct quadrants in the steady-state ($dn_0/dt = 0$) phase diagram (see Fig. 26.4): In region I, both condensate and thermal spin current lead to the loss of magnons and prevent formation of the condensate. In region II, the condensate spin current leads to the decay of condensate magnons, which are replenished from the thermal cloud by injection of thermal magnons. This results in a steady state in which a condensate exists. Region II crosses over to region III, which is the swasing regime [35], defined by the injection of spin directly into the condensate as the spin accumulation is then larger than the magnon ground-state energy. In this crossover, the number of condensed magnons becomes larger, more rapidly so at lower temperatures. Finally, in region IV$_2$, the condensate spin current is not large enough to overcome the losses due to spin current by thermal magnons. In region IV$_1$, on the other hand, two steady-state solutions exist. Only one of these is stable, however, and hence the number of condensate magnons needs to be above a critical number in order for the condensate to be maintained. This leads to hysteretic behavior in the first-order transitions from regions IV$_2$ and III to region IV$_1$. The phase diagram in terms of the spin accumulation μ_R and the magnon and electron temperatures is determined by calculating j_x and τ_0 in terms of them and reported in Ref. [14]. We conclude that quasi-equilibrium magnon Bose-Einstein condensation can be achieved by appropriate tuning of these driving forces.

26.3.2 Spin Superfluidity

One of the most striking consequences of Bose-Einstein condensation of magnons is the resulting spin superfluidity [36, 37]. Developing a theory for spin transport across the pumped magnon system is not straightforward, as the phase diagram itself will change in an inhomogeneous situation with respect to the homogeneous case discussed so far. Here, we ignore such issues and discuss spin superfluidity for the case of an equilibrium condensate. We also ignore the dynamic dipole–dipole interaction that can dramatically change the spin transport properties [38]. For the model Hamiltonian in Eq. (26.8), an equilibrium condensate forms at low temperatures when the external field is small enough so that the easy-plane anisotropy tilts the magnetization away from the z-axis. The general phenomenology of the superfluid spin transport is expected to be similar when comparing equilibrium and pumped condensates.

The spin superfluid transport is, for a small precession angle and low temperatures, conveniently described by the continuity equation for the density of condensed magnons (that includes a loss term due to Gilbert damping), and the Josephson equation for the condensate phase. These are found from the Landau-Lifschitz-Gilbert equation (Eq. 26.9 with the Gilbert damping added) and in the long-wavelength limit given by (at zero temperature)

$$\begin{aligned} \hbar\frac{dn_0}{dt} &= -\nabla\cdot\mathbf{j}_s - 2\alpha n_0\hbar\omega\,; \\ \frac{d\mathbf{v}_s}{dt} &= -\frac{2J_s\nabla\hbar\omega}{\hbar^2}\,, \end{aligned} \tag{26.15}$$

with $\mathbf{v}_s = -2J_s\nabla\varphi/\hbar$ the superfluid velocity and $\mathbf{j}_s = n_0\hbar\mathbf{v}_s$ the condensate (superfluid) spin current. In the above, we assume the condensate to be spatially homogeneous, and that $n_0 \ll s$ and $\alpha \ll 1$. The stiffness $J_s = JSa^2/2$, with a the lattice constant. For the configuration in Fig. 26.2, we find, by applying the boundary condition in Eq. (26.11) at both reservoirs, that the spin current injected into the right reservoir is given by

$$j_s = \frac{n_0}{4\pi s}\left(\frac{g_{\uparrow\downarrow}^2\mu_L}{g_{\uparrow\downarrow} + 2\pi\alpha s L_m}\right),$$

where we took $\mu_R = 0$. Consequently, spin superfluidity is signaled by an algebraic decay of the spin current, and the ensuing inverse spin Hall voltage, as a function of the size of the magnetic insulator in the spin current direction [37, 39, 40]. This is in contrast to the exponential decay (with the length scale ℓ_m) that is found in the normal state [41, 42, 43]. The algebraic decay is a consequence of the small but nonzero Gilbert damping. In fact, the above shows that in the case when damping is absent ($\alpha = 0$), the spin current through a spin superfluid magnetic insulator is limited only by the interface (via the mixing conductance $g_{\uparrow\downarrow}$).

For the geometry of our example, spin superfluidity may be pinned by anisotropies that break the rotation symmetry around the z-axis [37, 40] (which ultimately gives rise to the $U(1)$ symmetry that is spontaneously broken by the condensation). As an example, we consider the addition of a term $-K_x \sum_i \hat{S}_{i,x}^2/2\hbar^2$ to the Hamiltonian in Eq. (26.8). With this addition, the spin Hamiltonian for constant density reduces to the energy

$$E = \int d\mathbf{x} \left[J_s n_0 \left(\nabla\varphi\right)^2 + (B - KS)n_0 + \frac{KSn_0^2}{2s} - K_x S n_0 \cos^2\varphi \right], \qquad (26.16)$$

where we again took the long-wavelength limit. The spin-current-carrying state requires gradients in the phase. These are penalized by the term $\sim K_x$, which pins the phase at $\varphi = 0$ or π. The competition between exchange and in-plane anisotropy in the above energy thus defines a lower critical current

$$j_{c,low} = 2n_0\sqrt{J_s S K_x}/\hbar\,,$$

below which spin superfluid flow is pinned. Physically, this lower critical current follows from comparing the energy of the current-carrying state to the energy of a domain wall in φ from, e.g., 0 to π. Once the energy of the current-carrying state exceeds this domain wall energy, domain walls, and therefore gradients in the phase, are created, thus allowing finite superfluid spin currents. Below this lower critical current, the anisotropy makes it energetically favorable for the phase to remain homogeneous away from the interface [37, 44].

The upper critical current is found by realizing that the condensate density is actually $n_0 \approx s(1 + n_z)$ in terms of the z-component of the spin density. From the above energy, we then find that the superfluid flow is unstable toward decreasing n_z for $|\nabla\varphi| \sim B - KS$, which defines an upper critical current. When the current approaches this critical value, the superfluid-carrying state is relaxed by spontaneous vortex-induced phase slips transverse to our quasi-one-dimensional geometry [37].

26.4 Perspectives

In this chapter, we have discussed how spin currents, flowing across the interface between a magnetic insulator and normal metal, can be used to achieve quasi-equilibrium Bose-Einstein condensation of magnons in the magnetic insulator, and, moreover, to probe its spin transport properties. Future theoretical works should improve our treatment of interactions, in particular regarding the interactions between the condensate and thermal magnons and the role of phonons. Moreover, the description of the condensed phase may require inclusion of the dipolar interactions. Another direction for study is inhomogeneous situations and, in particular, the coupled spin-heat transport properties of the partially condensed magnon system.

On the experimental side, the best-studied systems are devices with only one metallic reservoir and where the magnetic insulator is YIG and the metal is Pt. More complicated setups involving more than one reservoir have not yet been seriously addressed. Moreover, from a materials science perspective, there is ample room for exploring novel materials and optimizing interface spin transport properties, e.g., by considering antiferromagnets [45]. In conclusion, we expect that the interplay between magnonic many-body physics and spintronics will be a source of new physics in the years to come.

Acknowledgments: This work was supported by the Stichting voor Fundamenteel Onderzoek der Materie (FOM) and is part of the Delta Institute for Theoretical Physics (D-ITP) consortium, a program of the Netherlands Organization for Scientific Research (NWO) that is funded by the Dutch Ministry of Education, Culture and Science (OCW) (R.A.D.), by the European Research Council (ERC) (R.A.D and A.B.), by the Eurpoean Union Future and Emerging Technologies (EU-FET) grant InSpin 612759 (A.B.), and by the US Department of Energy (DOE) Basic Energy Sciences (BES) under Award No. DE-SC0012190 (S.A.B. and Y.T.).

References

[1] Bardeen, J., Cooper, L. N., and Schrieffer, J. R. 1957. Microscopic theory of superconductivity. *Phys. Rev.*, **106**, 162–164.

[2] Pethick, C. J., and Smith, H. 2002. *Bose-Einstein Condensation in Dilute Gases*. Cambridge, UK: Cambridge University Press.

[3] Klaers, Jan, Schmitt, Julian, Vewinger, Frank, and Weitz, Martin. 2010. Bose-Einstein condensation of photons in an optical microcavity. *Nature*, **468**, 545–548.

[4] Balili, R., Hartwell, V., Snoke, D., Pfeiffer, L., and West, K. 2007. Bose-Einstein condensation of microcavity polaritons in a trap. *Science*, **316**, 1007–1010.

[5] Kasprzak, J., Richard, M., Kundermann, S., Baas A., Jeambrun, P., Keeling, J. M. J., Marchetti, F. M., Szymanska, M. H., André, R., Staehli, J. L., Savona, V., Littlewood,

P. B., Deveaud, B., and Dang, Le Si. 2010. Bose-Einstein condensation of exciton polaritons. *Nature*, **443**, 409–414.

[6] Nikuni, T., Oshikawa, M., Oosawa, A., and Tanaka, H. 2000. Bose-Einstein condensation of dilute magnons in $TlCuCl_3$. *Phys. Rev. Lett.*, **84**, 5868–5871.

[7] Rüegg, Ch., Cavadini, N., Furrer, A., Güdel, H.-U., Krämer, K., Mutka, H., Wildes, A., Habicht, K., and Vorderwisch, P. 2003. Bose-Einstein condensation of the triplet states in the magnetic insulator $TlCuCl_3$. *Nature*, **423**, 62–65.

[8] Demokritov, S. O., Demidov, V. E., Dzyapko, O., Melkov, G. A., Serga, A. A., Hillebrands, B., and Slavin, A. N. 2006. Bose-Einstein condensation of quasi-equilibrium magnons at room temperature under pumping. *Nature*, **443**, 430–433.

[9] Serga, A. A., Tiberkevich, V. S., Sandweg, C. W., Vasyuchka, V. I., Bozhko, D. A., Chumak, A. V., Neumann, T., Obry, B., Melkov, G. A., Slavin, A. N., and Hillebrands, B. 2014. Bose-Einstein condensation in an ultra-hot gas of pumped magnons. *Nature communications*, **5**, 3452.

[10] Weiler, M., Althammer, M., Schreier, M., Lotze, J., Pernpeintner, M., Meyer, S., Huebl, H., Gross, R., Kamra, A., Xiao, J., Chen, Y.-T., Jiao, H., Bauer, G. E. W., and Goennenwein, S. T. B. 2013. Experimental test of the spin mixing interface conductivity concept. *Phys. Rev. Lett.*, **111**, 176601.

[11] Tserkovnyak, Y., Brataas, A., and Bauer, G. E. W. 2002. Spin pumping and magnetization dynamics in metallic multilayers. *Phys. Rev. B*, **66**, 224403.

[12] Jia, X., Liu, K., Xia, K., and Bauer, G. E. W. 2011. Spin transfer torque on magnetic insulators. *Europhys. Lett.*, **96**, 17005.

[13] Burrowes, C., Heinrich, B., Kardasz, B., Montoya, E. A., Girt, E., Sun, Yiyan, Song, Young-Yeal, and Wu, Mingzhong. 2012. Enhanced spin pumping at yttrium iron garnet/Au interfaces. *Appl. Phys. Lett.*, **100**, 092403.

[14] Bender, S. A., Duine, R. A., Brataas, A., and Tserkovnyak, Y. 2014. Dynamic phase diagram of dc-pumped magnon condensates. *Phys. Rev. B*, **90**, 094409.

[15] Bender, S. A., Duine, R. A., and Tserkovnyak, Y. 2012. Electronic pumping of quasiequilibrium Bose-Einstein-condensed magnons. *Phys. Rev. Lett.*, **108**, 246601.

[16] Tserkovnyak, Y., Brataas, A., Bauer, G. E. W., and Halperin, B. I. 2005. Nonlocal magnetization dynamics in ferromagnetic heterostructures. *Rev. Mod. Phys.*, **77**, 1375–1421.

[17] Brataas, A., Kent, A. D., and Ohno, H. 2012. Current-induced torques in magnetic materials. *Nature Mat.*, **11**, 372–381.

[18] Uchida, K., Xiao, J., Adachi, H., Ohe, J., Takahashi, S., Ieda, J., Ota, T., Kajiwara, Y., Umezawa, H., Kawai, H., Bauer, G. E. W., Maekawa, S., and Saitoh, E. 2010. Spin Seebeck insulator. *Nature Mat.*, **9**, 894–897.

[19] Jaworski, C. M., Yang, J., Mack, S., Awschalom, D. D., Myers, R. C., and Heremans, J. P. 2011. Spin-Seebeck effect: a phonon driven spin distribution. *Phys. Rev. Lett.*, **106**, 186601.

[20] Flipse, J., Dejene, F. K., Wagenaar, D., Bauer, G. E. W., Youssef, J. Ben, and van Wees, B. J. 2014. Observation of the spin Peltier effect for magnetic insulators. *Phys. Rev. Lett.*, **113**, 027601.

[21] Bauer, G. E. W., Saitoh, E., and van Wees, B. J. 2012. Spin caloritronics. *Nature Mat.*, **11**, 391–399.

[22] D'yakonov, M. I., and Perel', V. I. 1971. Possibility of orienting electron spins with current. *Sov. Phys. JETP*, **13**, 467–469.

[23] Saitoh, E., Ueda, M., Miyajima, H., and Tatara, G. 2006. Conversion of spin current into charge current at room temperature: inverse spin-Hall effect. *Applied Physics Letters*, **88**, 182589.

[24] Liu, L., Buhrman, R. A., and Ralph, D. C. 2011. Review and analysis of measurements of the spin Hall effect in platinum. *arXiv:1111.3702*. Nov.

[25] Xiao, J., Bauer, G. E. W., Uchida, K.-C., Saitoh, E., and Maekawa, S. 2010. Theory of magnon-driven spin Seebeck effect. *Phys. Rev. B*, **81**, 214418.

[26] Schreier, M., Kamra, A., Weiler, M., Xiao, J., Bauer, G. E. W., Gross, R., and Goennenwein, S. T. B. 2013. Magnon, phonon, and electron temperature profiles and the spin Seebeck effect in magnetic insulator/normal metal hybrid structures. *Phys. Rev. B*, **88**, 094410.

[27] Sinova, J., Culcer, D., Niu, Q., Sinitsyn, N. A., Jungwirth, T., and MacDonald, A. H. 2004. Universal intrinsic spin Hall effect. *Phys. Rev. Lett.*, **92**, 126603.

[28] Murakami, S., Nagaosa, N., and Zhang, S.-C. 2003. Dissipationless quantum spin current at room temperature. *Science*, **301**, 1348.

[29] Hoffman, S., Sato, K., and Tserkovnyak, Y. 2013. Landau-Lifshitz theory of the longitudinal spin Seebeck effect. *Phys. Rev. B*, **88**, 064408.

[30] Kehlberger, A., Ritzmann, U., Hinzke, D., Guo, E.-J., Cramer, J., Jakob, G., Onbasli, M. C., Kim, D. H., Ross, C. A., Jungfleisch, M. B., Hillebrands, B., Nowak, U., and Kläui, M. 2015. Length scale of the spin Seebeck effect. *Phys. Rev. Lett.*, **115**, 096602.

[31] Knoester, M. E., Sinova, J., and Duine, R. A. 2014. Phenomenology of current-skyrmion interactions in thin films with perpendicular magnetic anisotropy. *Phys. Rev. B*, **89**, 064425.

[32] Hals, K. M. D., and Brataas, A. 2013. Phenomenology of current-induced spin-orbit torques. *Phys. Rev. B*, **88**, 085423.

[33] Tserkovnyak, Y., and Bender, S. A. 2014. Spin Hall phenomenology of magnetic dynamics. *Phys. Rev. B*, **90**, 014428.

[34] Gilbert, T. L. 2004. A phenomenological theory of damping in ferromagnetic materials. *Magnetics, IEEE Transactions on*, **40**, 3443–3449.

[35] Berger, L. 1996. Emission of spin waves by a magnetic multilayer traversed by a current. *Phys. Rev. B*, **54**, 9353–9358.

[36] Halperin, B. I., and Hohenberg, P. C. 1969. Hydrodynamic theory of spin waves. *Phys. Rev.*, **188**, 898–918.

[37] Sonin, E. B. 2010. Spin currents and spin superfluidity. *Advances in Physics*, **59**, 181–255.

[38] Skarsvåg, H., Holmqvist, C., and Brataas, A. 2015. Spin superfluidity and long-range transport in thin-film ferromagnets. *Phys. Rev. Lett.*, **115**, 237201.

[39] Takei, S., and Tserkovnyak, Y. 2014. Superfluid spin transport through easy-plane ferromagnetic insulators. *Phys. Rev. Lett.*, **112**, 227201.

[40] Chen, H., Kent, A. D., MacDonald, A. H., and Sodemann, I. 2014. Nonlocal transport mediated by spin supercurrents. *Phys. Rev. B*, **90**, 220401.

[41] Zhang, S. S.-L., and Zhang, S. 2012. Magnon mediated electric current drag across a ferromagnetic insulator layer. *Phys. Rev. Lett.*, **109**, 096603.

[42] Cornelissen, L. J., Liu, J., Duine, R. A., Ben Youssef, J., and Van Wees, B. J. 2015. Long distance transport of magnon spin information in a magnetic insulator at room temperature. *Nature Physics*, **11**, 1022.

[43] Cornelissen, L. J., Peters, K. J. H., Bauer, G. E. W., Duine, R. A., and van Wees, B. J. 2016. Magnon spin transport driven by the magnon chemical potential in a magnetic insulator. *Phys. Rev. B* **94**, 014412.
[44] König, J., Bønsager, M. C., and MacDonald, A. H. 2001. Dissipationless spin transport in thin film ferromagnets. *Phys. Rev. Lett.*, **87**, 187202.
[45] Takei, S., Halperin, B. I., Yacoby, A., and Tserkovnyak, Y. 2014. Superfluid spin transport through antiferromagnetic insulators. *Phys. Rev. B*, **90**, 094408.

27

Spin-Superfluidity and Spin-Current Mediated Nonlocal Transport

HUA CHEN AND ALLAN H. MACDONALD
Department of Physics, University of Texas at Austin, USA

Some strategies for reducing energy consumption in information processing devices involve the use of spin rather than charge to carry information. This idea is especially attractive when the spin current is a collective one carried by the condensate of a magnetically ordered state rather than a quasiparticle current carried by electrons or magnons. In this chapter, we explain how easy-plane magnets can be viewed as Bose-Einstein condensates (BECs) of magnons, defined in terms of quanta of the spin-component perpendicular to the easy plane, and how they can carry dissipationless spin-currents that induce nonlocal interactions between electrically isolated conducting channels. We comment specifically on important differences between superconductivity in normal/superconducting/normal circuits and spin-superfluidity in normal/magnetic/normal circuits.

27.1 Introduction

Spintronics, the study of the interplay between the electrical transport and magnetic properties of magnetically ordered solids, has made steady progress over the past few decades. Spintronics involves both phenomena such as giant magnetoresistance, in which transport properties are influenced by magnetic order configurations, and phenomena such as spin-transfer torques in which transport currents can be used to modify magnetic configurations. Pure spin currents, which do not involve charge flow, are routinely detected via the spin-transfer torques they exert on magnetic condensates and the electrical signals they give rise to when spins accumulate near sample boundaries or at electrodes. There are hopes that spin currents have advantages over charge currents that can be exploited to enable faster or lower-power electronic devices. In this chapter, we discuss the notion of spin-superfluidity in thin film magnetic systems, either ferromagnetic or antiferromagnetic and either metallic or insulating, that have approximate easy-plane magnetic order [1, 2, 3, 4, 5, 6, 7]. In spintronics, spin-superfluidity refers to the capacity for spin currents to be carried without dissipation by a metastable configuration of a magnetic condensate rather than by an electron or magnon quasiparticle current.

Our chapter is organized as follows. In Section 27.2, we introduce the concept of spin superfluidity using the common language of magnetism researchers by applying Landau-Lifshitz equations to easy-plane magnets. To motivate the spin-superfluidity concept, we compare the spin-transport properties of easy-plane magnets to the matter transport properties of an ideal classical fluid. At the end of the section, we discuss some similarities and differences between easy-plane ferromagnets and BEC systems. In Section 27.3, we discuss perpendicular spin injection in finite-size easy-plane magnetic systems. We then show that spin superfluids can exhibit Josephson-like I-V characteristics that arise ultimately from the topological stability of easy-plane magnetic order in thin films. Finally, we discuss potential applications of this behavior, and also the influence in realistic materials of magnetostatic interactions, magneto-crystalline anisotropy, and damped magnetization dynamics. We conclude in Section 27.4 with a discussion of the relationship between spin-superfluidity in easy-plane magnetic systems and superconductivity in metals.

This chapter is closely related to the previous chapter on spintronics and magnon Bose-Einstein condensation by Duine et al. (Chapter 26). Both chapters are motivated by advances in spintronics that allow spin currents to be routinely passed between different materials, including between metals and insulators. The phenomena that are addressed in Chapter 26 occur in easy-axis magnetic systems that are driven electrically into a quasi-equilibrium steady state.

27.2 Spin Superfluidity in Ideal Easy-Plane Magnets

To simplify the following discussion, we represent an ideal easy-plane magnet by the Ginzburg-Landau free energy functional [8, 2]

$$\mathcal{F} = \int dV \left[-|\alpha|\mathbf{M}\cdot\mathbf{M} + \frac{\beta}{2}(\mathbf{M}\cdot\mathbf{M})^2 + A|\nabla\mathbf{M}|^2 + KM_z^2 \right]. \tag{27.1}$$

In Eq. (27.1), the first two terms account for the magnetization magnitude and the ground-state free energy. The third term is a magnetic stiffness energy $A|\nabla\mathbf{M}|^2 \equiv A(|\nabla M_x|^2+|\nabla M_y|^2+|\nabla M_z|^2)$ that parameterizes the free energy cost of magnetization nonuniformity. In the easy-plane case of interest, K is positive and the last term characterizes the free energy cost of magnetization that is not oriented in the easy-plane. This expression ignores anisotropy within the easy plane, which we restore later, and also the complex term with long-range nonlocality that accounts for magnetostatic interactions [9]. The dynamics of the magnetization $\mathbf{M}$ is described by the Landau-Lifshitz equation [8]

$$\frac{d\mathbf{M}}{dt} = -\gamma \mathbf{M} \times \frac{\delta \mathcal{F}}{\delta \mathbf{M}}, \tag{27.2}$$

where $\gamma = g\mu_B/\hbar$ is the gyromagnetic ratio and we have assumed g to be negative for electrons. The Landau-Lifshitz equation are valid when the magnetization varies slowly in space and time and can be derived in a variety of different ways, for example, starting from a density-functional theory of the magnetically ordered state [10, 11]. Using the free energy expression in Eq. (27.1), the effective magnetic field that appears on the right-hand side of the Landau-Lifshitz equations and drives magnetization precession is

$$\frac{\delta \mathcal{F}}{\delta \mathbf{M}} \equiv -\mathbf{H}_{\text{eff}} = -2A\nabla^2\mathbf{M} + 2KM_z\hat{z}. \tag{27.3}$$

It is sometimes stated that the Landau-Lifshitz equation is a classical equation which describes spin-angular momentum precession. However, we prefer to view it as a quantum equation which describes the collective quantum dynamics of a magnetic order parameter; certainly its derivation is always quantum. It can be viewed as a classical equation only because the quantum spin dynamics of a macroscopic magnetic condensate is classical. In modern spintronics, the quantum nature of this equation is revealed by the appearance of $\hbar$ in the relationship between classical precession frequencies and spin electromotive forces [12, 13].

The classical ground state of the easy-plane ferromagnet has uniform in-plane magnetization. For small deviations from this classical ground state, we parameterize $\mathbf{M}$ as $M_0(\cos\phi, \sin\phi, m_z)$ with $m_z = M_z/M_0 \ll 1$. In this limit, which we assume below, the Landau-Lifshitz equations take the form

$$\begin{aligned} \dot{\phi} &= 2\gamma K M_0 m_z, \\ \dot{m}_z &= 2\gamma A M_0 \nabla^2 \phi. \end{aligned} \tag{27.4}$$

(We have ignored terms that are higher order in the small quantities $\nabla\phi$ and m_z.) Note that the second equation can be recognized as a continuity equation for m_z. In this interpretation, the current corresponding to m_z is the collective spin current

$$\mathbf{j}_z = -2\gamma A M_0 \nabla\phi. \tag{27.5}$$

The continuity equation is a direct consequence of the conservation of m_z in an ideal easy-plane ferromagnet, i.e., of the property that the Ginzburg-Landau energy is invariant under rotations around the $\hat{z}$-axis in magnetization space. As we discuss further below, a dissipationless spin current described by Eq. (27.5) flows through the system when the system has nonzero $\nabla\phi$ [3].

Sonin [1] has proposed a helpful analogy between a magnetic system carrying a dissipationless spin current and a rod that is twisted around its axis. The rod will rotate globally when a torque is applied at one end unless an opposite torque is applied at the other end. Although the net force on every individual atom in a twisted rod with balanced torques vanishes, the two torques can be viewed as giving rise to a uniform angular momentum flux, an angular momentum supercurrent, which passes through the cross section of the rod and transmits a torque applied at one end to the other end, where it is compensated. The nonlocal relationship between remote ends of the rod is supported by the rigidity of the rod, just as the nonlocal relationship between spin currents injected at opposite ends of an easy-plane magnet on which we focus is supported by the magnetic order parameter rigidity.

It is important to observe that all the spin-supercurrent phenomena in equilibrium easy-plane magnets that we comment on in this chapter have an alternate description solely in terms of the spin-torque language commonly used in spintronics, which applies to any magnetic system and is therefore more general. The analogy with superfluid phenomena is restricted to magnetic systems with easy-plane order, but is interesting nevertheless because of the properties it suggests, and because of the light it sheds on the relationship between the collective phenomena studied in superfluids and superconductors and those studied in modern magnetism research, in particular in spintronics. The conversion between normal metal currents and Cooper pair currents via Andreev scattering [14, 15], which is important in mesoscopic superconductivity, is simply the easy-plane limit of the spin-transfer torque concept so central in modern spintronics [16, 17, 18, 19, 20, 21, 22, 23]. To better explain the relationship of spin superfluidity to other superfluid phenomena, we now briefly summarize some key properties of fluids and superfluids.

27.2.1 Classical Superfluids

Part of the reason why easy-plane magnetic systems are usefully viewed as being *super* is that their properties are in compliance with the conventional definition of ideal fluids – fluids without viscosity and thermal conductivity (adiabatic). Ideal fluids can be simply described by Newton's second law, which is known in fluid dynamics as Euler's equation:

$$\frac{\partial \mathbf{v}}{\partial t} + \mathbf{v} \cdot \nabla \mathbf{v} = -\frac{1}{\rho} \nabla p, \tag{27.6}$$

where $\mathbf{v}$ is the velocity of an elemental volume of a fluid, ρ is the density of the fluid, and p is the pressure. Note that the left-hand side is simply $d\mathbf{v}/dt$.

An ideal fluid has an important property, referred to as Kelvin's theorem [24], that the velocity circulation is time independent. (The velocity circulation is defined as the line integral of the velocity around any closed loop in the fluid.) We emphasize later that a related property is essential to the stability of supercurrent states in superfluids. For now, we consider the case when the velocity circulation is zero, which means that the vorticity

$$\omega \equiv \nabla \times \mathbf{v} \tag{27.7}$$

vanishes identically everywhere in the fluid. Then one can always find a scalar function ϕ whose gradient is equal to the velocity, i.e.

$$\mathbf{v} = \nabla\phi. \tag{27.8}$$

Eq. (27.8) is similar to Eq. (27.5). Moreover, one can derive from Eq. (27.4) an equation for $\mathbf{j}$ that looks similar to Eq. (27.6), with the pressure term in the latter replaced by a term proportional to m_z. (We will return to this point in the next subsection.) Thus an easy-plane ferromagnet can indeed be viewed as an ideal fluid with density proportional to the perpendicular component of the magnetization.

In the following, we focus our attention on thin film systems in which the magnetization direction depends only on two spatial coordinates, since this is normally the case of greatest practical interest. The analogies we make will therefore be between thin film magnets and two-dimensional fluids. What is different between easy-plane magnet quantum superfluids and the classical ideal fluid is that the velocity potential ϕ is identified as a phase or azimuthal orientation angle in the quantum case. The line integral of the phase or angle gradient over any closed loop must then be an integer multiple of 2π. This circulation quantization leads to vortices, topological defects carrying nonzero circulation quanta. Since circulation is conserved in the bulk of an ideal fluid, a vortex will remain stable unless it reaches the boundary of the fluid where circulation is not well defined or it annihilates with another vortex with opposite circulation. In circular coordinates (r, θ), a vortex with circulation κ can be represented by the velocity field

$$\mathbf{v} = \frac{\kappa}{2\pi}\frac{\hat{\theta}}{r}. \tag{27.9}$$

One can then estimate the kinetic energy associated with a vortex by integrating $(1/2)\rho v^2$ over the whole fluid. It is easy to see that the energy of the vortex increases logarithmically with the system size. It follows that creation or annihilation of a vortex is associated with an unbounded energy change. Moreover, under the assumption of zero viscosity and adiabaticity, creation and annihilation of vortices is the only way for a superfluid to relax from a metastable state with nonzero

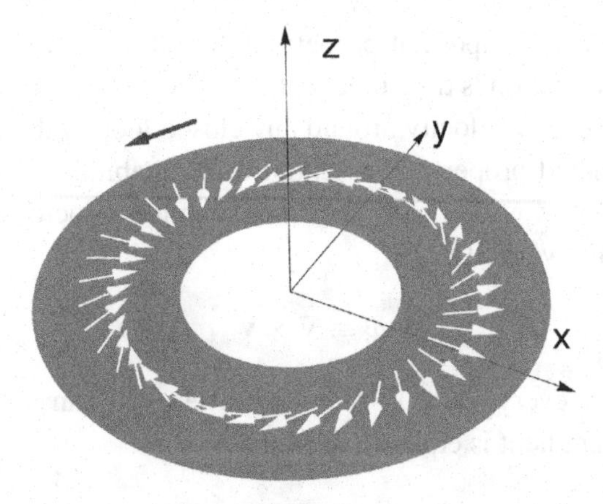

Figure 27.1 Metastable magnetization configuration formed by an easy-plane ferromagnet in a ring. This configuration carries a dissipationless spin supercurrent. The magnetization in this illustration changes by 4π upon enclosing the ring.

supercurrent to the zero current ground state. Because the creation of these topological defects requires that large energy barriers be overcome, the current state of a superfluid is extraordinarily stable.

It is instructive to consider the example in which we connect the two ends of a long, thin ferromagnetic wire to form a ring, as shown in Fig. 27.1. The in-plane magnetization angle must then rotate by an integer multiple of 2π as one moves around the ring to complete a cycle. Provided that the total rotation angle is not zero, there is according to Eq. (27.5) a persistent spin supercurrent in the ring because of the nonzero azimuthal angle gradient. The topological stability of this spin supercurrent state is then obvious since it is not possible to change the angle winding number by locally perturbing the magnetization. This geometry is similar to the twisted rod example given earlier in this chapter and is related to the celebrated rotating cylinder experiment in superfluid He^4.

Now imagine that a vortex with an angle winding of 2π is nucleated at one boundary of the ring and moves across the width of the ring. The azimuthal angle change from one end of the sample to the other, measured along the direction perpendicular to the path of the vortex, changes by 2π for every vortex which is nucleated on one edge of the ring, moves across, and is then annihilated at the other edge to restore a uniform superfluid. This *phase slip* is accompanied by a lower free energy when it reduces $|\nabla \mathbf{M}|^2$ and also by a smaller spin supercurrent. The barrier for supercurrent relaxation is thus proportional to the vortex nucleation energy and can greatly exceed $k_B T$ because it is a collective barrier involving many electronic spins. A similar argument explains the metastability of currents in superconductors. In the interior of a magnetic vortex, the magnetization is rotated out of the easy plane, allowing the in-plane magnetization to vanish and ϕ to change discontinuously. The energy cost of creating vortices is therefore related in part to the strength of the easy-plane ansiotropy. A nonzero uniaxial anisotropy energy is essential for the stability of the spin supercurrent [2], as we emphasize again in the next subsection.

As we have explained, the stability of supercurrent states in general superfluids can be understood in terms of the conservation of circulation, whether quantum or classical. However, we have not yet addressed the reason why the superfluids act like ideal fluids, i.e., why viscosity (or dissipation) is absent. This issue will be discussed in the next subsection.

27.2.2 Spin Superfluidity and Bose-Einstein Condensation

The prototypical superfluid, liquid ^{4}He, is also a Bose-Einstein condensate. Although the two concepts, superfluidity and BEC, are not equivalent, nor is one necessarily the consequence of the other, they are intimately related. In this subsection, we will discuss the relationship between BEC and superfluidity, while at the same time making a comparison between BEC and easy-plane magnetism.

Briefly, to avoid repeating material presented in earlier chapters, we define a BEC as a state of matter in which a macroscopic number of bosonic particles share the same single-particle wavefunction. For simplicity we assume here that *all* particles are in the same state. One can then write the wavefunction of this state Ψ as a direct product of the single-particle states ψ:

$$\Psi(\{\mathbf{r}_j\}, t) = \prod_{j=1}^{N} \psi(\mathbf{r}_j, t) \exp(-i\mu t/\hbar) \qquad (27.10)$$

where ψ satisfies a mean-field Schrödinger equation:

$$i\hbar\frac{\partial \psi}{\partial t} = \left(-\frac{\hbar^2}{2m}\nabla^2 - \mu\right)\psi + \psi \int |\psi(\mathbf{r}')|^2 N U(\mathbf{r} - \mathbf{r}') d\mathbf{r}'. \qquad (27.11)$$

Here μ is the chemical potential, and the last term is due to a weak interaction between particles. The factor N in the last term reflects the fact that the effective interaction strength scales with the number of particles in the condensate. Eq. (27.11) is called the (time-dependent) Gross-Pitaevskii (GP) equation [25, 26] and is is central to the earliest and also the most widely used microscopic theory of BECs formed by weakly interacting bosonic particles. Below we will absorb a $\sqrt{N}$ factor into ψ. The integral of $|\psi|^2$ over space is then the total number of particles in the condensate. We can therefore regard ψ as the order parameter of the condensate and the GP equation as an equation for order parameter dynamics.

Below we emphasize its similarity to the Landau-Lifshitz equation for the order parameter dynamics of an easy-plane magnet. The close relationship between these two equations is of course not coincidental.

Assuming the BEC order parameter is a complex scalar function of position and time $\Psi = \sqrt{n(\mathbf{r},t)}\, e^{i\phi(\mathbf{r},t)}$, where n is the density of the condensed particles, and the time-dependent GP equation can be rewritten as coupled equations for n and ϕ:

$$\hbar\dot{\phi} = \frac{\hbar^2}{2m}\left(\frac{1}{2n}\nabla^2 n - \frac{1}{4n^2}|\nabla n|^2\right) - \frac{\hbar^2}{2m}|\nabla\phi|^2 + \mu - U_0 n, \tag{27.12}$$

$$\dot{n} = -\frac{\hbar}{m}\left(n\nabla^2\phi + \nabla n\cdot\nabla\phi\right),$$

where $U_0(\mathbf{r}) \equiv \int U(\mathbf{r}-\mathbf{r}')d\mathbf{r}'$. The second equation has the form of a continuity equation if the current is

$$\mathbf{j} = n\frac{\hbar}{m}\nabla\phi \equiv n\mathbf{v}_s. \tag{27.13}$$

One can check that such a definition indeed agrees with that calculated from the standard formula $\mathbf{j} = -(i\hbar/2m)(\Psi^*\nabla\Psi - \Psi\nabla\Psi^*)$. Moreover, by taking the time derivative of $\mathbf{v}_s$ and making use of the first equation in Eq. (27.12), we obtain

$$\frac{\partial\mathbf{v}_s}{\partial t} + \mathbf{v}_s\cdot\nabla\mathbf{v}_s = -\nabla\left[\frac{U_0 n}{m} - \frac{\mu}{m} - \frac{\hbar^2}{2m^2}\left(\frac{1}{2n}\nabla^2 n - \frac{1}{4n^2}|\nabla n|^2\right)\right], \tag{27.14}$$

which coincides with Euler's equation for an ideal fluid, Eq. (27.6). Thus the condensate is an ideal fluid, conserves velocity circulation, and is irrotational ($\nabla\times\mathbf{v}_s = 0$). Moreover, its angular momentum must be carried by quantized vortices as we discussed in the previous subsection. (These conclusions apply only when the BEC order parameter is a complex scalar function and do not apply to spinor BECs.) The Landau-Lifshitz equations of easy-plane magnets, Eq. (27.4), correspond to the GP equations of BECs if we associate the term $\propto m_z$ in the first equation of the former with $\mu - U_0 n$. (The counterpart of the kinetic energy term in Eq. (27.12) has been ignored in Eq. (27.4), which considered the spatially constant order parameter case.) It is then interesting to ask if this means that an easy-plane ferromagnet can also be viewed as a BEC. Below we show that this is indeed the case.

Let us start from a single macrospin with angular momentum $s\hbar$, where s is a real number much larger than $1/2$. Taking the z direction to be the quantization axis, the raising and lowering operators for the z-spin are written as

$$S_+ = S_x + iS_y,\ S_+|s,s_z\rangle = \sqrt{s(s+1) - s_z(s_z+1)}\hbar|s,s_z+1\rangle, \tag{27.15}$$

$$S_- = S_x - iS_y,\ S_-|s,s_z\rangle = \sqrt{s(s+1) - s_z(s_z-1)}\hbar|s,s_z-1\rangle$$

where $|s, s_z\rangle$ is the eigenstate of S_z with the eigenvalue $s_z\hbar$. Letting S_- act repeatedly on eigenstates of S_z generates a set of states $|n\rangle$, which are eigenstates of S_z with eigenvalues $(s-n)\hbar$. One can then define a set of bosonic creation and annihilation operators acting on these Fock states, a^+ and a, which decrease (a^+) or increase (a) the spin projection in z direction by $\hbar$, i.e.,

$$\begin{aligned} [a, a^+] &= 1, \\ a|n\rangle &= \sqrt{n}|n-1\rangle, \\ a^+|n\rangle &= \sqrt{n+1}|n+1\rangle. \end{aligned} \tag{27.16}$$

a^+ and a are related to $S_\pm$ and S_z through the Holstein-Primakoff transformation [27]

$$\begin{aligned} S_+ &= \hbar\sqrt{2s - a^+ a}\, a, \\ S_- &= \hbar\sqrt{2s - a^+ a}\, a^+, \\ S_z &= \hbar(s - a^+ a). \end{aligned} \tag{27.17}$$

We first ignore magnetic anisotropy altogether by assuming for the moment that the Hamiltonian commutes not only with the total spin component S_z, as it does in ideal easy-plane ferromagnets, but also with S_x and S_y. In this case, all eigenstates occur in spin-multiplets and in the case of ferromagnets, the ground-state multiplet has a macroscopic value of s, proportional to the size of the system. Now consider the ground state of an easy-plane ferromagnet, which should be an eigenstate of $S_x^2 + S_y^2$. We define this state as $|XY\rangle$, which must have the property that

$$S_z^2|XY\rangle = \left[S^2 - (S_x^2 + S_y^2)\right]|XY\rangle = 0. \tag{27.18}$$

Therefore $|XY\rangle$ can be constructed using the bosonic operator a^+ acting on the vacuum–the eigenstate of S_z with eigenvalue $s\hbar$:

$$|XY\rangle = |s, s_z = 0\rangle = \frac{1}{\sqrt{s!}}\left(a^+ e^{i\phi}\right)^s |0\rangle, \tag{27.19}$$

where ϕ is the azimuthal orientation angle of the macrospin. Therefore, the ground state of an easy-plane ferromagnet can be viewed as a condensate of $N = M_{\text{tot}}/(|g|\mu_B\hbar)$ z-spin Holstein-Primakoff bosons (magnons). When magnetic anisotropy is included, the ground state weakly mixes states with slightly different values of s, but this picture still applies. For an easy-spin magnet, the Landau-Lifshitz equation can therefore be viewed as the counterpart of the GP equation for the z-spin magnon condensate. Quantum fluctuations in the local value of S_z correspond to quantum fluctuations in boson density and, quantum fluctuations in the azimuthal angle ϕ correspond to quantum fluctuations in the condensate phase. The correspondence between the m_z term in the $\dot{\phi}$ equation in Eq. (27.4) and $(\mu - U_0 n)$ in Eq. (27.12) is also clear since both express the energy change associated with changing the particle number by one. We note that another way to understand the condensate nature of an easy-plane ferromagnet is through its analogy with the pseudospin description of superconductivity by Anderson [28], with electron–electron pairing in superconductivity replaced by electron-hole pairing in easy–plane ferromagnetism [2].

It is now time to discuss the origin of vanishing viscosity in superfluids in relationship to analogous properties of easy-plane ferromagnets. First we discuss the analog of the Landau's criterion for superfluidity, namely that the system be in a metastable state that cannot relax to the ground state via elementary excitations, which we now briefly summarize. (Vortex nucleation requires an unbounded energy and is not an elementary excitation.) A fluid flowing with velocity $\mathbf{v}$ has kinetic energy $E = (1/2)Mv^2$. Consider the possibility of energy dissipation through creation of an elementary excitation that has energy ϵ and momentum $\mathbf{p}$ in the reference frame moving with the fluid. One can find that in the rest frame the energy of the excitation is $\epsilon + \mathbf{p} \cdot \mathbf{v}$. The moving fluid is metastable (ignoring thermal excitations at finite temperature) only if all excitations have positive energy in the rest frame, i.e., only if the velocity of the fluid

$$v < \min\left(\frac{\epsilon}{p}\right). \tag{27.20}$$

If the elementary excitations of the fluid have linear dispersion, this criterion can be satisfied below a critical velocity. Indeed, the elementary excitations in weakly interacting boson systems (as in superfluid He^4) are sound waves with linear dispersion as can be derived by linearizing the GP equation around its ground-state solution Ψ. Easy-plane ferromagnets are also superfluids in the same sense that their finite spin-current states can decay to the ground states only via vortex-nucleation processes and not via elementary excitations. As in a BEC, an easy-plane ferromagnet has linearly dispersing spin waves as elementary excitations. This result can be established by taking the second-order time derivative of ϕ and making use of the Landau-Lifshitz equation Eq. (27.4) to obtain

$$\ddot{\phi} = 4\gamma^2 M_0^2 AK\nabla^2\phi. \tag{27.21}$$

The spin-wave velocity

$$c = 2|\gamma|M_0\sqrt{AK} \tag{27.22}$$

is identical to the upper critical value of the spin supercurrent [2]. The linearly dispersive elementary excitations in both BEC and easy-plane ferromagnets are Goldstone modes related in the magnet case to spontaneous rotational symmetry breaking and in the BEC case to gauge symmetry breaking. Isotropic ferromagnets are not spin superfluids because their magnon dispersion is quadratic rather than linear at long wavelengths [1]. Landau's criterion is, however, not a sufficient condition for superfluidity, since it says nothing about the topological stability of the metastable superfluid states.

We end this section by noting that a discussion of superfluidity normally starts from the identification of a well-defined current. In other words, the discussion starts from a continuity equation that can be written down for the physical quantity that is transported without dissipation and whose total number is conserved. This is not a problem with the mass superfluidity in BEC or the charge superfluidity in superconductors, since the particle number is a good quantum number in both cases. However, no component of spin is ever really a good quantum number due to inevitable spin-orbit coupling and magnetostatic interaction processes. The concept of spin currents has nevertheless been useful in spintronics, because spin is *nearly* conserved. The use of this concept does, however, sometimes lead to debate and confusion [1, 29], especially in cases where spin-orbit coupling plays a dominant role [30]. In fact, the easy-plane anisotropy required for a finite critical current in our spin superfluid obviously requires spin-orbit coupling. If there is no other anisotropy, the z component of total spin is still a good quantum number, which means the z-spin supercurrent is well defined. In reality, however, there is always some anisotropy in the easy plane. The fact that S_z is not conserved leads to both dissipative and reactive effects which must both be taken into account in analyzing spin-transport phenomena. When we invoke the concept of spin-superfluidity, we have in mind the metastability of magnetic configurations that carry spin currents through a system collectively through the magnetic condensate, and not via nonequilibrium magnon or electron quasiparticles. In the next section, we will discuss realistic situations and show how the concept of spin superfluidity is useful even though S_z is not a good quantum number.

27.3 Dynamics of Spin Superfluids with Spin Injection

The central idea of spintronics is that spin can be used instead of or as a complement to charge to carry information through circuits and to store information. When spin-orbit coupling is negligible, total spin is a good quantum number. One can then define the spin current by multiplying spin with the probability current operator $\mathbf{j}$; for example, $S_z\mathbf{j}$ is the spin current operator for the $\hat{z}$-spin projection (see below). One therefore needs to trace over the spin degree of freedom to get the expectation value of the spin current. It is possible to have a spin current that is not accompanied by net charge transport, a pure spin current, when the charges carried by states with opposite spins cancel. Since spins couple to lattice vibrations much more weakly than charges, the Joule heating problem associated with electronics-based circuits could be mitigated if charge and spin transport could be decoupled.

In the absence of spin-orbit coupling and magnetostatic interactions,

$$\dot{\mathbf{S}} = \frac{i}{\hbar}[H, \mathbf{S}] = 0; \tag{27.23}$$

in other words, spin is a good quantum number. For an individual independent electron,

$$\frac{\partial \langle \mathbf{S}(\mathbf{r}, t)\rangle}{\partial t} = -\langle \mathbf{S} \otimes (\nabla \cdot \mathbf{j})\rangle \equiv -\nabla \cdot \langle \hat{j}_S \rangle, \tag{27.24}$$

where $\mathbf{j}$ is the usual probability current operator in quantum mechanics, and $\hat{j}_S$ is the spin current operator, which is a rank 2 tensor. When $\mathbf{S}$ is not a good quantum number,

$$\dot{\mathbf{S}} = \frac{i}{\hbar}[H, \mathbf{S}] \equiv \Pi \neq 0, \tag{27.25}$$

where Π is the spin torque operator, and

$$\frac{\partial \langle \mathbf{S}\rangle(\mathbf{r}, t)}{\partial t} = \frac{i}{\hbar}\left[(H\psi)^{\dagger}\mathbf{S}\psi - \psi^{\dagger}\mathbf{S}(H\psi)\right] + \psi^{\dagger}\Pi\psi. \tag{27.26}$$

One cannot isolate a current from the right-hand side of Eq. (27.26) in any unambiguous way. Even in the case that the first term on the right-hand side of Eq. (27.26) can be approximately identified as the divergence of the spin current defined in Eq. (27.24), the torque term can still change the spin density locally even with a uniform steady effective spin current. If one insists on maintaining the same definition of spin current, this torque term accounts for additional sources and sinks of spins.

It should be acknowledged that spin currents are in fact normally accompanied by dissipation. We distinguish two classes of mechanisms. (i) Dissipation associated with diffusive motion of magnon or electron quasiparticles: Quasiparticle scattering tends to relax the quasiparticles toward a state that is at rest with respect to the lattice, and in the process to transfer energy to phonons or magnons. In this case, the dissipation can be described by classical Boltzmann theory. There is little difference, particularly if spin is carried by electronic quasiparticles, between the dissipation associated with quasiparticle charge currents and spin currents. (ii) Dissipation due to relaxation of the magnetic condensate toward its minimum energy configuration. This type of dissipation is captured by the Gilbert damping terms which appear in the Landau-Lifshitz equations for collective dynamics. No analogous terms appear in the GP equations for an equilibrium BEC. Similar terms do appear, however, in phenomenological descriptions of magnon condensates, which are always nonequilibrium steady states that are not true thermal equilibrium. By exploiting spin supercurrents in an easy-plane ferromagnet, one can largely get rid of the dissipation due to the first mechanism. If S_z is conserved, the spin supercurrent is well defined and one can use the easy-plane ferromagnet as a dissipationless link to efficiently transport spin between remote spintronics devices.

To understand the role of Gilbert damping and magnetic anisotropy within the easy plane, we need to study the dynamics of spin superfluids subject to injection or extraction of normal spin currents, which is discussed in the next subsection. The spin spiral states of Fig. 27.1, which carried a spin supercurrent in the ideal case, are slightly distorted by weak in-plane magnetic anisotropy, but their metastability is largely unaffected. In Section 27.3.2, we discuss a possible spintronic device based on easy-plane ferromagnets that is conceptually similar to an N-S-N circuit containing normal metal leads connected to a superconducting wire.

27.3.1 Dynamics of Spin Superfluids with Spin Injection

In this subsection, we describe the basic ideas needed to understand spin supercurrents in a finite easy-plane ferromagnet coupled to external sources/drains of quasiparticle spin. The spintronics toolkit contains a variety of possible sources of spin currents with spin polarization perpendicular to the easy plane, including ones based on the spin Hall effect, ferromagnetic resonance, or electron tunneling from perpendicular anisotropy magnetic films. Note that electrical generation of spin currents always requires a charge bias potential. A normal spin current in an easy-plane ferromagnet can be supported by electronic quasiparticles only close to the current source. Assuming that S_z is a good quantum number for now, the continuity equation for S_z in this boundary layer guarantees that this current will be converted into a collective spin supercurrent:

$$\mathbf{j}_{nz} = 2\gamma A M_0 \nabla\phi\big|_B, \tag{27.27}$$

where $\mathbf{j}_{nz}$ is the z-spin current injected from the source, and the subscript B indicates that the spatial derivative of the azimuthal magnetization orientation ϕ should be evaluated at a position close to the source or drain.

By eliminating m_z in the Landau-Lifshitz equation Eq. (27.4), the dynamics of ϕ in the bulk of the easy-plane ferromagnet is described by Eq. (27.21). For simplicity, we consider a one-dimensional (1D) problem. In the steady state, $\phi(x,t) = \phi(x) - \omega t$, and $\phi(x)$ is the solution of

$$\begin{aligned} \partial_x^2\phi &= 0, \\ \mathbf{j}_{nz} &= 2\gamma A M_0 \nabla\phi\big|_B, \end{aligned} \tag{27.28}$$

where the boundary condition must be satisfied at both ends of the 1D system. These conditions yield

$$\phi(x,t) = \frac{j_{nz}}{2\gamma A M_0} x - \omega t, \tag{27.29}$$
$$\mathbf{j}_{nz,L} = \mathbf{j}_{nz,R}.$$

The easy-plane ferromagnet is driven to a spiral state with wave vector

$$q = \frac{j_{nz}}{2\gamma A M_0}, \tag{27.30}$$

and the net spin current injected into the system must be zero or the system will not be able to find a steady state. j_{nz} also has to be smaller than the critical value given in Eq. (27.22) in order for the supercurrent state to be sustained.

To understand the significance of the spin-precession frequency, we transform the spin part of the system into a rotating frame synchronized with the precession of the order parameter. The unitary operator which achieves this transformation is

$$U = e^{-i\frac{\omega t}{2}\sigma_z}. \tag{27.31}$$

In the mean-field Hamiltonian of the easy-plane ferromagnet, the time-dependent order parameter leads to a term proportional to $\cos(qx - \omega t)\sigma_x + \sin(qx - \omega t)\sigma_y$. Applying the unitary transformation on this operator yields

$$U\left[\cos(qx - \omega t)\sigma_x + \sin(qx - \omega t)\sigma_y\right] U^{\dagger} = \cos(qx)\sigma_x + \sin(qx)\sigma_y, \tag{27.32}$$

i.e., the precession is removed. The trade-off is that the Hamiltonian acquires a spin-dependent chemical potential shift, which can be seen from the modification to the time-evolution operator

$$|\psi(t)\rangle_R = U|\psi(t)\rangle = e^{-i\frac{\omega t}{2}\sigma_z} e^{-i\frac{Ht}{\hbar}}|\psi(0)\rangle = e^{-i\frac{t}{\hbar}(H + \frac{\hbar\omega}{2}\sigma_z)}|\psi(0)\rangle_R. \tag{27.33}$$

Note that the last equality requires S_z to be conserved. This equivalence between dynamics and spin-dependent chemical potential is well known in spintronics, where is it responsible for spin-pumping [31] and spin electromotive forces [12, 13].

We now consider the effect of adding in-plane uniaxial anisotropy along the x direction to the magnet's energy functional:

$$-K'M_x^2 = -\frac{1}{2}KM_0^2\cos(2\phi) + \text{const}, \tag{27.34}$$

where the constant term can be ignored. The discussion below can be easily generalized to other forms of anisotropy. A Hamiltonian contribution which gives rise to this anisotropy obviously does not commute with the z component of spin in the microscopic Hamiltonian. As a result, the z−spin current is rigorously speaking not a well-defined quantity. Nevertheless, as we have discussed earlier in the approximation that the spin density varies slowly in space, we can still use the spin-current language and separate the contribution to $\dot{m}_z$ into a current term and a torque term. This can be seen in the Landau-Lifshitz equations modified by this anisotropy:

$$\begin{aligned} \dot{\phi} &= 2\gamma K M_0 m_z, \\ \dot{m}_z &= 2\gamma A M_0 \nabla^2 \phi - \gamma K' M_0 \sin(2\phi), \end{aligned} \tag{27.35}$$

where we have assumed $K' \ll K$ and on this basis ignored its modification to the $\dot{\phi}$ equation. The second term on the right-hand side of the $\dot{m}_z$ equation is the extra torque from anisotropy within the easy plane. Eliminating m_z from Eq. (27.35), we obtain the sine-Gordon equation

$$\ddot{\phi} - c^2 \left[\nabla^2 \phi - \frac{\sin(2\phi)}{l^2} \right] = 0, \tag{27.36}$$

where c is given in Eq. (27.21), and $l = \sqrt{2A/K'}$. The simplest time-independent solution of Eq. (27.36) contains a single soliton (domain wall):

$$\phi(x) = 2 \arctan \left[\exp \left(\sqrt{2} \frac{x-a}{l} \right) \right], \tag{27.37}$$

where a is the arbitrary soliton position. The homogeneous spiral state in the absence of the easy-axis anisotropy is thus not a stable state of the system; for any given phase gradient, the system can lower its energy by locally rotating the in-plane polarization toward its easy axis, thereby distorting the simple spiral state. The collective spin supercurrent is nonuniform in space, with its divergence matching the rate of transverse spin creation or annihilation by the torque from the in-plane anisotropy. It is often still possible, however, to find metastable distorted spiral states which satisfy the boundary conditions imposed by spin currents injected or absorbed at sample boundaries by solving a boundary value problem with Neumann boundary conditions:

$$\begin{aligned} \partial_x^2 \phi - \frac{\sin(2\phi)}{l^2} &= 0, \\ \mathbf{j}_{nz}\big|_{L,R} &= 2\gamma A M_0 \partial_x \phi \big|_{L,R}. \end{aligned} \tag{27.38}$$

Strictly speaking, the boundary conditions should include a spin torque term due to the easy-axis anisotropy at the boundary. However, since the torque contribution is an integral over the volume of the boundary layer, we can always ignore this term provided that the boundary layer is thin enough. An example of the solution of Eq. (27.38) is shown in Fig. 27.2.

Since static solutions balance spatial variation in spin currents against the in-plane anisotropy torque, it is clear that when the net current injection exceeds a

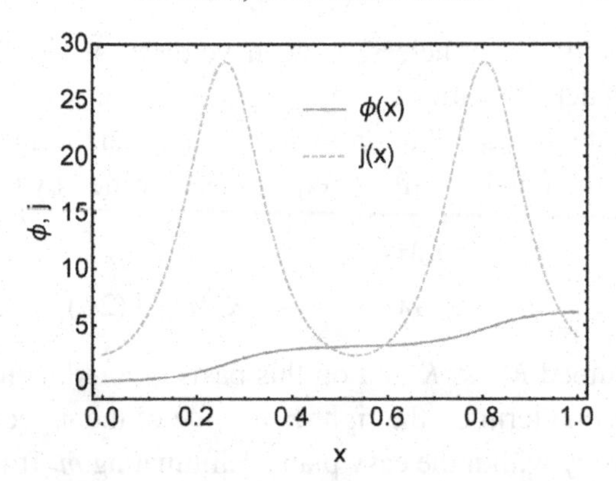

Figure 27.2 Distorted supercurrent spiral in a finite 1D system with spin injection at the sample ends and uniaxial easy-axis anisotropy along $\hat{x}$.

value determined by the easy-axis anisotropy, a static solution may not be found. An estimate of the critical current imbalance can be made by assuming the stiffness A is very large, so that both the wavelength of the spiral (Eq. 27.29) and the width of the domain wall l greatly exceed the system size. In this macrospin limit, Eq. (27.38) reduces to

$$-\gamma K' M_0 V \sin(2\phi) = \mathbf{I}_L + \mathbf{I}_R, \tag{27.39}$$

where V is the volume of the easy-plane ferromagnet and $\mathbf{I}_{L,R}$ are the normal spin currents injected. In this limit, the critical current imbalance is

$$\max(|\mathbf{I}_L + \mathbf{I}_R|) = |\gamma| K' M_0 V \equiv I_c. \tag{27.40}$$

A discussion of the opposite limit that $l \ll L$ where L is the length of a long easy-plane ferromagnet can be found in [6]. Note that static solutions are always available when the spin-current injected at one end of the sample is equal to the spin current removed at the other end of the sample.

The order parameter is not static when there are no metastable magnetic configurations that satisfy spin-injection boundary conditions. Under such circumstances, it is necessary to consider its damping. Collective magnetization dynamics, including damping, is accurately described by the Landau-Lifshitz-Gilbert (LLG) equation when magnetic order is well developed:

$$\frac{d\mathbf{M}}{dt} = -\gamma \mathbf{M} \times \frac{\delta \mathcal{F}}{\delta \mathbf{M}} + \frac{\alpha}{M_0} \mathbf{M} \times \frac{d\mathbf{M}}{dt}, \tag{27.41}$$

where α is the Gilbert damping parameter. Taking the in-plane easy-axis anisotropy into account, the LLG equation in terms of ϕ and m_z is

$$\dot{\phi} = 2\gamma K M_0 m_z - \alpha \dot{m}_z, \tag{27.42}$$

$$\dot{m}_z = 2\gamma A M_0 \nabla^2 \phi - \gamma K' M_0 \sin(2\phi) + \alpha\dot{\phi}.$$

Solving Eq. (27.42) in a finite system is challenging in general. Here we only consider the macrospin limit and assume a steady state in which $\dot{\phi}$ is spatially constant. For large easy-plane anisotropy, this means that $\dot{m}_z = 0$ according to the $\dot{\phi}$ equation in Eq. (27.42). We thus arrive at a single equation for ϕ:

$$-\gamma K' M_0 V \sin(2\phi) + \alpha V\dot{\phi} = \mathbf{I}_{\text{net}}. \tag{27.43}$$

For $|I_{\text{net}}| \gg I_c$., where $\mathbf{I}_{\text{net}} = \mathbf{I}_L + \mathbf{I}_R$, the solution is approximated by $\phi(t) = \phi_0 + (I_{\text{net}}/\alpha V)\, t$. When $|I_{\text{net}}| \sim I_c$, $\phi(t)$ has an additional oscillatory contribution. (*Cf.* Fig. 1b in [6].)

An important consequence of having both in-plane anisotropy and Gilbert damping in the easy-plane ferromagnet is that it is possible to drive the easy-plane ferromagnet across the transition between two very different spin-transport regimes. Specifically, recall that the precession of the in-plane magnetization is equivalent to a spin-dependent chemical potential shift $\delta\mu = -(\hbar\dot{\phi}/2)\sigma_z$ (Eq. (27.33)). When the magnetization is static, $\delta\mu = 0$ even for finite $I_{\text{net}} < I_c$ because of the easy-axis anisotropy within the easy plane, whereas in the steady precessing state $\delta\mu \approx -(\hbar I_{\text{net}}/2\alpha V)\,\sigma_z$, which can be very large when damping is small. The current dependence of the spin voltage in the system is thus highly nonlinear. In the next subsection, we will study this behavior in more detail and explore its potential use.

27.3.2 Device Based on an N-S-N Junction

In this subsection, we study a structure formed by an easy-plane ferromagnet sandwiched between two perpendicular anisotropy ferromagnetic tunnel junctions, as schematically illustrated in Fig. 27.3a. A ferromagnetic tunnel junction is formed by two easy-axis ferromagnets with opposite magnetizations, separated by dielectrics. When a tunneling current is established in the junction, z-spin conservation dictates that there must be pure spin currents injected into the easy-plane magnetic system at the position of the tunnel junction stack. These spin currents can be carried collectively from one stack to the other, even when the easy-plane system is not metallic. Because the quasiparticle spin currents in the ferromagnetic tunnel junctions are converted into spin supercurrents in the easy-plane ferromagnet, this geometry provides a magnetic analog of an N-S-N circuit.

The spin N-S-N junction can also be described using a microscopic model suitable for nonequilibrium Green's function calculations, which we briefly introduce here. The left and right metal stacks can be represented by nearest neighbor tight-binding models with no spin-orbit coupling and a difference between the up and down spin chemical potentials. To model the easy-plane magnet, we add to the tight-binding model a mean-field onsite anisotropic interaction

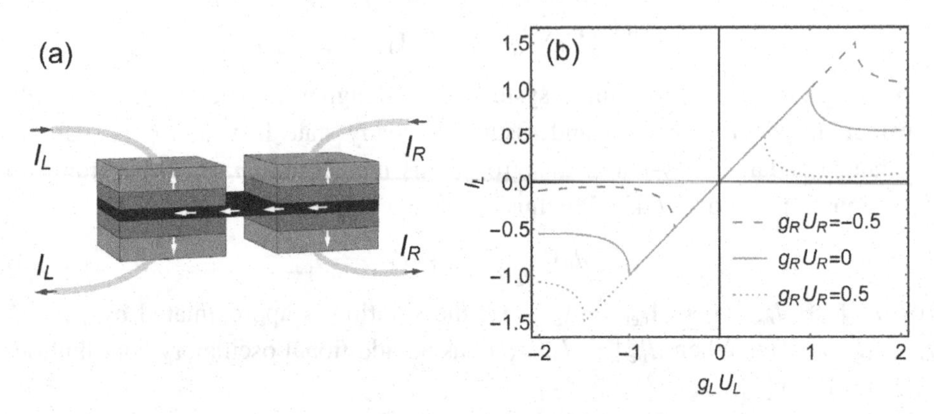

Figure 27.3 (a) A schematic illustration of the bistack magnetic transistor concept. (b) Corresponding DC current-voltage relationship of the device.

$$H_A = \sum_i \sum_{\alpha=x,y,z} U_\alpha S_{i\alpha} \langle S_{i\alpha} \rangle, \tag{27.44}$$

and set $U_x = U_y < U_z$ to account for the easy-plane or hard-axis anisotropy. H_A is also responsible for spontaneous magnetic ordering. This microscopic model complements the macroscopic Landau-Lifshitz description in the previous subsection by providing information on, e.g., the dependence of the magnitude of the in-plane magnetization on the potential biases in the leads, the decay length of normal spin currents injected into the easy-plane ferromagnet, and the difference in behavior between insulating and metallic easy-plane ferromagnets. The model can also be used to study spin superfluidity in antiferromagnets since the onsite interaction Eq. (27.44) is more likely to lead to antiferromagnetic ground state in equilibrium.

A benefit of using the ferromagnetic tunnel junctions to inject spin currents into the easy-plane ferromagnet is that the size of the spin current is directly determined by the electric voltages applied across the junctions. The magnetization dynamics of the easy-plane ferromagnet influences transport through the perpendicular magnetic anisotropy stacks through the effect we mentioned at the end of the last subsection. By contacting two ferromagnetic tunnel junctions to the same easy-plane ferromagnet, it is possible to realize highly nonlinear and nonlocal current-voltage characteristics, particularly when the easy-plane ferromagnet is driven across the transition between static and precessing states. Such a device has potential application as a field-effect transistor. Similar proposals have been made using other condensed matter systems, e.g., spatially indirect exciton condensates [32, 33].

To continue the analysis, we stay with the large easy-plane anisotropy and macrospin limit, for which Eq. (27.43) applies. It is, however, more relevant to use electric voltages across the ferromagnetic tunnel junctions instead of spin currents to characterize circuit characteristics. From the continuity equation of z-spin in the region of the ferromagnetic tunnel junctions, it follows that the spin current is proportional to the tunneling charge current, i.e., that

$$I_{L,R} = \frac{F_{L,R}}{e} g_{L,R} U_{L,R}, \tag{27.45}$$

where $g_{L,R}$ is the tunnel conductance, $U_{L,R}$ is the bias voltage across the tunnel junction, and $F_{L,R} \leq 1$ is a system-dependent parameter characterizing the conversion efficiency between charge (number) current and spin (m_z) current. When the in-plane magnetization of the easy-plane ferromagnet starts to precess, $U_{L,R}$ will be shifted by $-\hbar\dot{\phi}/e$ in the rotating frame of the easy-plane ferromagnet. It follows that in this case

$$I_{L,R} = \frac{F_{L,R}}{e} g_{L,R} \left(U_{L,R} - \frac{\hbar\dot{\phi}}{e} \right). \tag{27.46}$$

Eq. (27.43) then becomes

$$I_c \sin(2\phi) + g_i \frac{\hbar\dot{\phi}}{e^2} = \frac{F_L}{e} g_L \left(U_L - \frac{\hbar\dot{\phi}}{e} \right) + \frac{F_R}{e} g_R \left(U_R - \frac{\hbar\dot{\phi}}{e} \right), \tag{27.47}$$

where I_c is given in Eq. (27.40), and

$$g_i \equiv \alpha V e^2/\hbar \tag{27.48}$$

characterizes the Gilbert damping induced dissipation. When one increases $U_L + U_R$ so that I_{net} greatly exceeds I_c, the first term in Eq. (27.47) vanishes after time averaging. Combining Eqs. 27.47 and 27.46 yields

$$I_L^e = \frac{g_i + F_R g_R}{g_i + F_R g_R + F_L g_L} g_L U_L - \frac{F_R g_L}{g_i + F_R g_R + F_L g_L} g_R U_R, \tag{27.49}$$

where I_L^e means the tunneling electron current at the left tunnel junction. A similar equation for I_R^e can be obtained by interchanging L and R labels. A nonlocal correlation between the *charge* currents and voltages at the two tunnel junctions is thus established through the easy-plane ferromagnet, even when no charge paths connect the tunnel junction stacks.

The static and precessing regimes discussed above are partly analogous to the DC and AC Josephson effects in superconductors [34, 35]. The essence of the DC Josephson effect is that when the order parameter is static, the supercurrent is dependent on the position dependence of the condensate phase. A current can flow even when the voltage drop measured along the superconductor vanishes. In the AC Josephson effect, the order parameter phase is linearly increasing on time with a constant rate of change proportional to the voltage applied across the superconductor.

Comparing Eq. (27.49) to the static case in which I_L^e is simply equal to $g_L U_L$, we find that the effective conductance (with U_R fixed) is reduced by a factor of

$$r = \frac{g_i + F_R g_R}{g_i + F_R g_R + F_L g_L}. \tag{27.50}$$

The conductance reduction factor r reflects the property that when the critical current is exceeded, electrons can no longer flip their spins by scattering off the easy-plane magnetic condensate and must instead take advantage of the incoherent processes that contribute to Gilbert damping in order to make their way through the stack. r can in principle be much smaller than 1 if $g_i + F_R g_R \ll F_L g_L$, providing two states distinguished by very different DC resistances. Note that g_i is proportional to nano-particle volume whereas g_R and g_L are proportional to stack areas, so that large conductance reduction can be achieved only in high-quality thin film nanomagnets. Moreover, the transition between these two states can be controlled by U_R since it is determined by I_{net} (Eq. (27.40)) or $U_L + U_R$. The device behaves very much like a field effect transistor (FET) and can be used as a switch. The typical current-voltage characteristics of the device are shown in Fig. 27.3b. We note that when $|I_{\text{net}}|$ increases slightly above I_c from below, the charge current will have a large AC component while the DC component has a sudden drop.

The performance of a switch is evaluated based mainly on the following three considerations: the on/off ratio, the switch voltage (voltage around which the switching occurs), and the stability of the switching behavior against thermal fluctuations. We already see that small Gilbert damping αV (cf. Eq. (27.48)) and effective spin current injection (large $F_{L,R} g_{L,R}$) are necessary for a large on/off ratio. Permalloy is likely a suitable candidate for the easy-plane ferromagnetic junction because of its weak damping, and also because of the small anisotropy to which the switch voltage is proportional (cf. Eq. (27.40)). Since $g_i \propto V$, it is ideal if the cross-sectional area is dominated by ferromagnetic stacks rather than by the easy-plane link part. The thermal stability of the switch is determined by the energy barrier between the static and the precessing states [36, 37, 38], which is the in-plane anisotropy energy $\sim K' M_0^2 V$. Therefore, for the device to be operational, the minimal voltage difference between the on and the off states is $\delta U \sim (k_B T / K' M_0^2 V) \times (e I_c)/(F_L g_L)$ and should satisfy

$$\frac{e\delta U}{k_B T} \sim \frac{1}{M_0 F_L g_L} \ll 1, \tag{27.51}$$

where M_0 and g_L are in units of μ_B and $e^2/\hbar$, respectively. This relation means that because of the collective nature of the switching phenomena, fundamental limits on conventional devices based on single-electron behavior, can be circumvented.

27.4 Discussion and Conclusions

In this chapter, we have explained that ideal easy-plane magnets can be viewed as equilibrium magnon Bose-Einstein condensates. Magnon condensation in equilibrium differs qualitatively from condensation in systems with steady-state nonequilibrium populations of magnons, even when these partially thermalize. Just as Bose-Einstein condensation occurs in systems with conserved particle numbers, ideal equilibrium magnon condensation occurs in easy-plane magnetic systems in which the perpendicular $\hat{z}$ component of spin is a good quantum number. In these ideal systems, a spiral magnetization configuration is metastable and carries a spin current without dissipation.

In realistic cases, no component of spin is conserved. The concept of spin currents is nevertheless useful in both paramagnetic and ferromagnetic metals, even though it is necessary to be cautious in using the spin-current concept, which is sometimes ambiguous. This is also true for easy-plane magnetic systems regarded as spin superfluids. The spin-current contribution to collective spin dynamics which is readily identified in ideal systems is still present in the Landau-Liftshitz equations, which are the magnetic analog of the GP equations. There are still metastable magnetization configurations which carry spin currents without dissipation, although the spin current is not spatially constant because of the influence of torques associated with weak anisotropy within the easy plane. The dissipationless spin supercurrents are responsible for nonlocal relationships between the I-V characteristics of remote magnetic circuits, which are coupled only by interacting with the same magnetic condensate.

It is instructive to compare the properties of a system in which a bias voltage is applied across a superconductor by normal metal leads connected to a power supply, an N-S-N system, with the properties of a system in which an easy-plane magnet is connected to perpendicular magnetic anisotropy leads. In the superconductor case, the two normal metal leads do not normally have spin accumulation, i.e., they don't have well-defined chemical potentials for $\uparrow$ and $\downarrow$ spins that are different. In the magnetic case, spin accumulation is a common mechanism for the creation of spin currents. A spin accumulation can be established either by illumination at a magnetic resonance frequency or by application of a charge bias voltage.

The current which flows across an N-S interface is proportional to the chemical potential difference between the lead and the superconductor. The chemical potential of the superconductor is proportional to the time derivative of the condensate phase. In the steady state, its value is adjusted so that the current flowing into the superconductor across one N-S interface is exactly equal to the current flowing out of the superconductor across the other N-S interface. In the macrospin limit, the corresponding equation for the magnetic system is (cf. Eq. (27.47))

$$\hbar\dot{\phi} = \frac{g_L F_L e U_L + g_R F_R e U_R}{g_i + g_L F_L + g_R F_R}. \tag{27.52}$$

The left-hand side of this equation is effectively a chemical potential for magnons, measured from the ground-state chemical potential. If g_i in this equation is set to zero, the magnon chemical potential will adjust to guarantee that the spin current injected at one end is emitted from the following end. The total spin current which flows through the system will then depend only on the spin-accumulation difference between one end of the magnet and the other. In spintronics language, $\hbar\dot{\phi}$ is viewed as generating a spin-pumping contribution to the spin currents at each end of the system. The properties of the N-S-N junction and the easy-plane magnetic system are therefore quite similar when g_i is smaller than the electrode conductances.

There is another route which allows spin-supercurrent behavior to be revealed. In the N-S-N circuit, only the chemical potential difference between the two N electrodes influences transport. In the magnetic case, we have the ability to separately control the spin accumulations U_L and U_R and can, for example, choose their values such that the total injected current is below its critical value even when the individual injected currents are large in value. In this case, the large spin currents injected at one contact do not excite magnetization dynamics only because of the large compensated spin supercurrent injected at the other contact. The large spin supercurrent is carried along the sample without dissipation.

In conclusion, we point out that a number of considerations that are known to be important have not been extensively discussed in this brief chapter, and in some cases are only now being addressed in the literature. Among these, we mention in particular the role of long-range magnetostatic interactions, which are a serious complication in samples that are beyond the macrospin limit in size, and the possibility of using easy-plane antiferromagnetic materials [7] instead of ferromagnetic materials. In ferromagnets, magnetostatic interactions tend to destabilize the homogeneous magnetic configurations from which the spiral configurations arise, in favor of configurations containing domains with different orientations. This problem is interesting but perhaps mainly academic since magnetostatic interactions are less important in smaller systems, and the largest interest is in exploiting spin superfluid properties in nanoscale spintronic devices. Most of the observations made in this chapter apply equally well to ferromagnets and antiferromagnets, which have the advantages that magnetostatic interactions are absent and that dynamics are faster – possibly enabling spintronic devices that can be switched very rapidly.

Acknowledgments: This work was supported as part of Spins and Heat in Nanoscale Electronic Systems (SHINES), an Energy Frontier Research Center funded by the U.S. Department of Energy, Office of Science, Basic Energy Sciences under Award # SC0012670.

References

[1] Sonin, E. B. 2010. Spin currents and spin superfluidity. *Advances in Physics*, **59**, 181–255.

[2] König, J., Bønsager, M. C., and MacDonald, A. H. 2001. Dissipationless spin transport in thin film ferromagnets. *Phys. Rev. Lett.*, **87**, 187202.

[3] Heurich, J., König, J., and MacDonald, A. H. 2003. Persistent spin currents in helimagnets. *Phys. Rev. B*, **68**, 064406.

[4] Nogueira, F. S., and Bennemann, K.-H. 2004. Spin Josephson effect in ferromagnet/ferromagnet tunnel junctions. *EPL (Europhysics Letters)*, **67**, 620.

[5] Takei, S., and Tserkovnyak, Y. 2014. Superfluid spin transport through easy-plane ferromagnetic insulators. *Phys. Rev. Lett.*, **112**, 227201.

[6] Chen, H., Kent, A. D., MacDonald, A. H., and Sodemann, I. 2014. Nonlocal transport mediated by spin supercurrents. *Phys. Rev. B*, **90**, 220401.

[7] Takei, S., Halperin, B. I., Yacoby, A., and Tserkovnyak, Y. 2014. Superfluid spin transport through antiferromagnetic insulators. *Phys. Rev. B*, **90**, 094408.

[8] Landau, L. D., and Lifshitz, E. M. 1995. *Course of Theoretical Physics, Vol. 9*. Oxford: Butterworth-Heinemann.

[9] Skarsvåg, H., Holmqvist, C., and Brataas, A. 2015. Spin superfluidity and long-range transport in thin-film ferromagnets. *Phys. Rev. Lett.*, **115**, 237201.

[10] Garate, I., and MacDonald, A. 2009. Gilbert damping in conducting ferromagnets. I. Kohn-Sham theory and atomic-scale inhomogeneity. *Phys. Rev. B*, **79**, 064403.

[11] Garate, I., and MacDonald, A. 2009. Gilbert damping in conducting ferromagnets. II. Model tests of the torque-correlation formula. *Phys. Rev. B*, **79**, 064404.

[12] Berger, L. 1984. Exchange interaction between ferromagnetic domain wall and electric current in very thin metallic films. *Journal of Applied Physics*, **55**, 1954.

[13] Yang, S. A., Beach, G. S. D., Knutson, C., Xiao, D., Niu, Q., Tsoi, M., and Erskine, J. L. 2009. Universal electromotive force induced by domain wall motion. *Phys. Rev. Lett.*, **102**, 067201.

[14] Andreev, A. F. 1964. The thermal conductivity of the intermediate state in superconductors. *Sov. Phys. JETP*, **19**, 1228.

[15] Blonder, G. E., Tinkham, M., and Klapwijk, T. M. 1982. Transition from metallic to tunneling regimes in superconducting microconstrictions: excess current, charge imbalance, and supercurrent conversion. *Phys. Rev. B*, **25**, 4515–4532.

[16] Slonczewski, J. C. 1996. Current-driven excitation of magnetic multilayers. *Journal of Magnetism and Magnetic Materials*, **159**, L1–L7.

[17] Slonczewski, J. C. 1999. Excitation of spin waves by an electric current. *Journal of Magnetism and Magnetic Materials*, **195**, L261–L268.

[18] Berger, L. 1996. Emission of spin waves by a magnetic multilayer traversed by a current. *Phys. Rev. B*, **54**, 9353–9358.

[19] Berger, L. 2001. Effect of interfaces on Gilbert damping and ferromagnetic resonance linewidth in magnetic multilayers. *Journal of Applied Physics*, **90**, 4632.

[20] Tsoi, M., Jansen, A. G. M., Bass, J., Chiang, W.-C., Seck, M., Tsoi, V., and Wyder, P. 1998. Excitation of a magnetic multilayer by an electric current. *Phys. Rev. Lett.*, **80**, 4281–4284.

[21] Myers, E. B., Ralph, D. C., Katine, J. A., Louie, R. N., and Buhrman, R. A. 1999. Current-induced switching of domains in magnetic multilayer devices. *Science*, **285**, 867–870.

[22] Sun, J. Z. 1999. Current-driven magnetic switching in manganite trilayer junctions. *Journal of Magnetism and Magnetic Materials*, **202**, 157–162.

[23] Ralph, D. C., and Stiles, M. D. 2008. Spin transfer torques. *Journal of Magnetism and Magnetic Materials*, **320**, 1190–1216.
[24] Landau, L. D., and Lifshitz, E. M. 1995. *Course of Theoretical Physics, Vol. 6*. Oxford: Butterworth-Heinemann.
[25] Gross, E. P. 1961. Structure of a quantized vortex in boson systems. *Il Nuovo Cimento*, **20**, 454–477.
[26] Pitaevskii, L. P. 1961. Vortex lines in an imperfect Bose gas. *Soviet Physics JETP-USSR*, **13**, 451–454.
[27] Holstein, T., and Primakoff, H. 1940. Field dependence of the intrinsic domain magnetization of a ferromagnet. *Phys. Rev.*, **58**, 1098–1113.
[28] Anderson, P. W. 1958. Random-phase approximation in the theory of superconductivity. *Phys. Rev.*, **112**, 1900–1916.
[29] Shi, J., Zhang, P., Xiao, D., and Niu, Q. 2006. Proper definition of spin current in spin-orbit coupled systems. *Phys. Rev. Lett.*, **96**, 076604.
[30] Brataas, A., and Hals, K. M. D. 2014. Spin-orbit torques in action. *Nature Nanotechnology*, **9**, 86–88.
[31] Tserkovnyak, Y., Brataas, A., Bauer, G. E. W., and Halperin, B. I. 2005. Nonlocal magnetization dynamics in ferromagnetic heterostructures. *Rev. Mod. Phys.*, **77**, 1375–1421.
[32] Min, H., Bistritzer, R., Su, J.-J., and MacDonald, A. H. 2008. Room-temperature superfluidity in graphene bilayers. *Phys. Rev. B*, **78**, 121401.
[33] Banerjee, S. K., Register, L. F., Tutuc, E., Reddy, D., and MacDonald, A. H. 2009. Bilayer PseudoSpin Field-Effect Transistor (BiSFET): a proposed new logic device. *Electron Device Letters, IEEE*, **30**, 158–160.
[34] Josephson, B. D. 1962. Possible new effects in superconductive tunnelling. *Physics Letters*, **1**, 251–253.
[35] Tinkham, M. 1996. *Introduction to Superconductivity*. 2nd edn. McGraw-Hill, Inc.
[36] Koch, R. H., Katine, J. A., and Sun, J. Z. 2004. Time-resolved reversal of spin-transfer switching in a nanomagnet. *Phys. Rev. Lett.*, **92**, 088302.
[37] Krivorotov, I. N., Emley, N. C., Garcia, A. G. F., Sankey, J. C., Kiselev, S. I., Ralph, D. C., and Buhrman, R. A. 2004. Temperature dependence of spin-transfer-induced switching of nanomagnets. *Phys. Rev. Lett.*, **93**, 166603.
[38] Ambegaokar, V., and Halperin, B. I. 1969. Voltage due to thermal noise in the dc Josephson effect. *Phys. Rev. Lett.*, **22**, 1364–1366.

28

Bose-Einstein Condensation in Quantum Magnets

CORINNA KOLLATH
HISKP, University of Bonn, Germany

THIERRY GIAMARCHI
Department of Quantum Matter Physics, University of Geneva, Switzerland

CHRISTIAN RÜEGG
Paul Scherrer Institute, Laboratory for Neutron Scattering and Imaging, Switzerland
Department of Quantum Matter Physics, University of Geneva, Switzerland

The recent experimental advances in the field of quantum spin materials have led to well-controlled systems which can be used in order to simulate complex quantum models in a clean fashion. Here we discuss how fascinating effects normally known from bosonic systems such as the Bose-Einstein condensation can be investigated with the help of such materials. We discuss as an example how a crossover between an effectively one-dimensional geometry described by a Luttinger liquid to a three-dimensional Bose-Einstein condensation can be induced. Additionally, we point out how more complex phases as the Bose-glass phase could be realized in these quantum magnets.

28.1 Introduction

Although it can seem paradoxical, the quantum magnetic states found in insulators offer an interesting route to study Bose-Einstein condensation (BEC) and more generally phenomena related to itinerant, interacting bosonic systems.

In many materials, the repulsion between the electrons causes the ground state to be a Mott insulator with one electron localized per site. In such a case, the charge degrees of freedom are completely locked by large charge gaps (usually of the order of tens of electron volts). However, one can study the fascinating physics of the spin degrees of freedom. As was shown a long time ago, the localized spins are coupled by a so-called superexchange which may be ferro- or antiferromagnetic. The properties of the exchange coupling between the spins depend mainly on the underlying atomic structure of the material. The resulting physics is extremely rich and such materials can have a host of phases ranging from ordered magnetic phases to so-called spin-liquid phases in which spin–spin correlations decay extremely fast with distance. The field of research concerning these materials, dubbed quantum magnetism, is thus an extremely active field in its own right [1].

Additionally, one can exploit the formal mapping that exists between spin operators and bosonic ones, as will be described in the next section, to connect the magnetic materials to *itinerant* bosonic systems with interactions between the bosons. This opens the way to use such quantum spin systems as quantum simulators. A quantum simulator means that the experimental system can be employed in order to "solve" in a controlled way a problem/Hamiltonian known from a different less controlled physical realization, e.g., inspired by fundamental many-body physics, quantum information or computational applications. The enormous recent progress on the materials science side with the creation of new materials, on the experimental side with the development and improvement of experimental techniques to explore their complex properties, and on the theoretical side to describe these systems allowed for a very rich and rapidly expanding direction of research along those lines.

We here give a short overview of the recent progress in this field. In Section 28.2, we explain the mapping between spin systems and bosonic ones. In Section 28.3, we show how the spins can be used as quantum simulators, and in Section 28.4, we give an overview of the various theoretical techniques used. In Section 28.5, we discuss the Bose-Einstein condensate and Tomonaga-Luttinger liquid in three- and effectively one-dimensional systems and crossovers and transitions between them. In Section 28.6, we show additional examples for which the simulator enabled researchers to reach results that are eluding or difficult to obtain in a direct theoretical approach. Finally, we discuss some conclusions and perspectives in Section 28.7.

28.2 Spins and Bosons

The connection between localized spins and bosons goes back to Holstein-Primakoff and Matsubara-Matsuda [2, 3]. For spin $1/2$, the connection is quite simple. A spin down state maps onto an empty bosonic site and a spin up state onto a site occupied by one boson. To have this direct correspondence, one should limit the total number of states per site to a maximum of two by introducing a hardcore constraint for the bosons. This constraint prevents more than one boson on one site. Then the correspondence between the operators can be expressed as

$$\begin{aligned} S_i^+ &\to b_i^\dagger \\ S_i^z &\to b_i^\dagger b_i - \frac{1}{2}. \end{aligned} \tag{28.1}$$

Together with the hardcore constraint, the operators (28.1) satisfy all the proper commutation relations.

The standard spin *XXZ*-Hamiltonian is given by

$$H = \sum_{\langle i,j \rangle} \left(\frac{J_{XY}}{2} \left[S_i^+ S_j^- + \text{h.c.} \right] + J_Z S_i^z S_j^z \right) - h \sum_i S_i^z, \tag{28.2}$$

where $\langle i,j \rangle$ denote the sites that are coupled (usually the nearest neighbors), J_{XY} and J_Z are the magnetic exchange ($J_{XY} = J_Z$ is the Heisenberg model), and h is the externally imposed magnetic field. Employing the defined mapping between the spins and bosons, the Hamiltonian becomes

$$H_b = t \sum_{\langle i,j \rangle} \left[b_i^\dagger b_j + \text{h.c.} \right] + J_Z (n_i - 1/2)(n_j - 1/2) - \mu \sum_i (n_i - 1/2). \tag{28.3}$$

On a bipartite lattice, the gauge transformation

$$S_i^+ \to (-1)^i \tilde{S}_i^+ \quad , \quad S_i^z \to \tilde{S}_i^z \tag{28.4}$$

changes the sign of the kinetic energy from $t \to -t$ and thus the J_{XY} term corresponds to the kinetic energy of bosons. Additionally, there is an infinite onsite repulsion which takes into account the hardcore constraint, and a (typically) nearest-neighbor interaction that corresponds to the J_Z term of the spin system. Quite interestingly, the magnetic field h for the spins corresponds to a chemical potential μ for the bosons.

Although spin $1/2$ can be mapped exactly onto bosons, the mapping presents several disadvantages for a perfect simulation of bosonic systems. In particular, a filled or empty band of bosons corresponds to a fully polarized spin system. In that case, parasitic interactions such as dipolar interactions can play a role and complicate the Hamiltonian. Often a mapping starting from dimers formed by two spin $1/2$ instead of the spin $1/2$ themselves can be favorable; see Fig. 28.1a and b. For an isolated dimer, the ground state consists of a singlet $|S\rangle = |S, S^z\rangle = |0,0\rangle$ and a threefold degenerate triplet $|T_0\rangle = |1,0\rangle$, $|T_\pm\rangle = |1,\pm 1\rangle$. Under the application of a magnetic field, the lowest energy states are the singlet $|S\rangle$ and the triplet aligned with the field $|T_+\rangle$. A mapping to bosons can be performed by retaining only those two states, leading to the relation

$$\begin{aligned} S_{i,k}^+ &\to \frac{1}{\sqrt{2}} (-1)^{i+k} b_i^\dagger \\ S_{i,k}^z &\to \frac{1}{4} \left[1 + 2 \left(b_i^\dagger b_i - \frac{1}{2} \right) \right], \end{aligned} \tag{28.5}$$

where $k = 1,2$ labels the spins of a dimer and i the position of the dimers. A detailed derivation can be found, e.g., in [4] and the characteristics of the mappings are shown in Fig. 28.1.

One of the fascinating phenomena well known from three-dimensional bosonic systems is the phase transition below a critical temperature to a Bose-Einstein

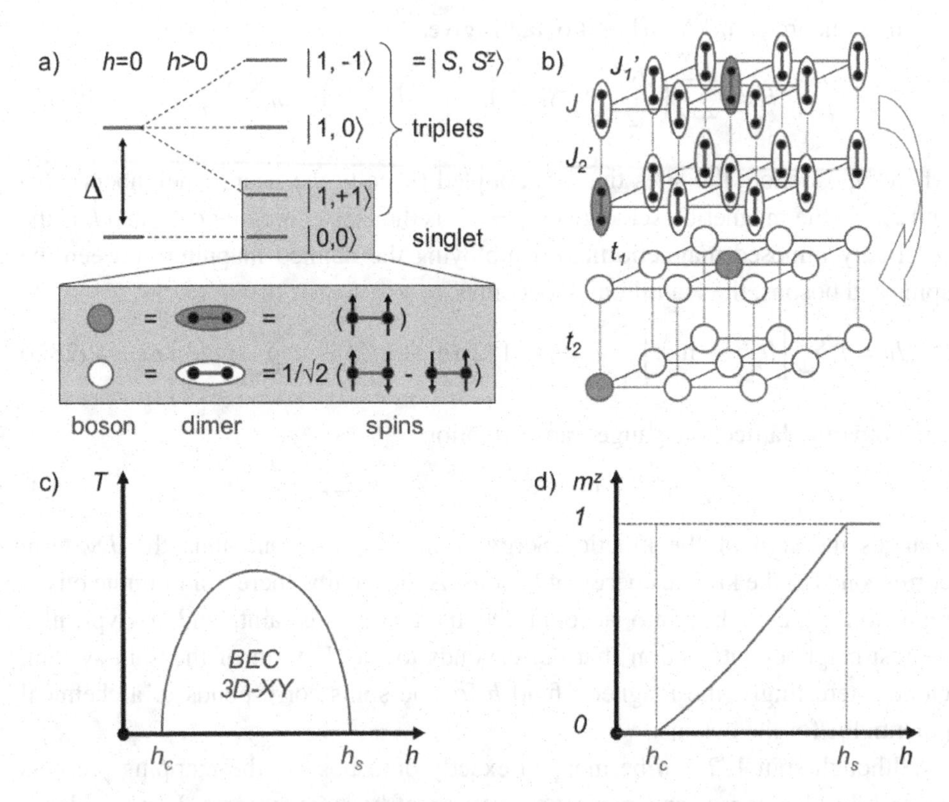

Figure 28.1 Spin to boson mapping and BEC. (a) For antiferromagnetic dimers, the ground state is a singlet, and there are three degenerate excited triplet states, separated by an energy gap Δ. In the presence of a magnetic field h, the triplet states split and for a critical field $h_c = \Delta$, one crosses the energy of the singlet state (not shown). Around this field, one can retain in the low-energy description only the singlet and the lowest triplet. In that case, one can map the singlet and lowest triplet to the absence (empty circles) and presence (gray shaded circles) of a boson, respectively. (b) If in addition to the dominant intradimer exchange (thick "internal" black links) there are weaker interdimer exchange interactions (thin black lines), then a dimer lattice (top) can be mapped onto a lattice of itinerant bosons (bottom) for which the kinetic energy of the bosons is related to the weak interdimer exchange. (c) These systems have a temperature (T) – magnetic field (h) phase diagram which contains a three-dimensional (3D) antiferromagnetic order in the direction perpendicular (XY) to the applied magnetic field h (Z). This phase and the corresponding transitions are identical in the boson language to the BEC phase and its transition. The angle of the staggered magnetization in the plane perpendicular to the magnetic field corresponds to the phase of the boson wavefunction. (d) In the regime between h_c and h_s, some of the dimers have been polarized into triplets, and thus in the boson language there is a finite density of bosons. The density of bosons corresponds to the uniform magnetization m_z along the magnetic field h. The linear behavior close to the critical points is characteristic of a three-dimensional (3D) BEC state. Such behavior will change crucially with the effective dimensionality of the problem (see text).

condensate with long-range order in the single-particle correlations $\langle b_i^\dagger b_{i+d} \rangle$. In the dimerized spin model, this translates via the mapping (28.5) to the appearance of long-range order in the staggered *transverse* correlations $\langle S_{i,k}^{x/y} S_{i+d,k}^{x/y} \rangle$. A typical phase diagram of a three dimensional dimer structure is given in Fig. 28.1c. At low magnetic field, the state of the system is a quantum-disordered spin-dimer liquid. In a certain range of magnetic fields $h_c < h < h_s$ and below a critical temperature T_c, the ordered phase with staggered transverse order (corresponding to the Bose-Einstein condensation) occurs. The number of bosons (longitudinal magnetization m_z in Fig. 28.1d) is controlled directly by the magnetic field h.

Some reference materials and their exchange geometries with various dimer motives are given in Fig. 28.2. More complete summaries are available in the literature (see, e.g., [5, 6] for reviews and further references). Excellent realizations of low-dimensional systems are, for example, the metal-organic materials $(C_5H_{12}N)_2CuBr_4$ and $Cu(NO_3)_2 \cdot 2.5H_2O$, in which the dimer lattices form one-dimensional ladders or alternating chains, respectively. These materials have energy gaps Δ below 1 meV and saturation fields h_s below 15 T, which can be reached conveniently by laboratory magnets with superconducting coils. Two-dimensional

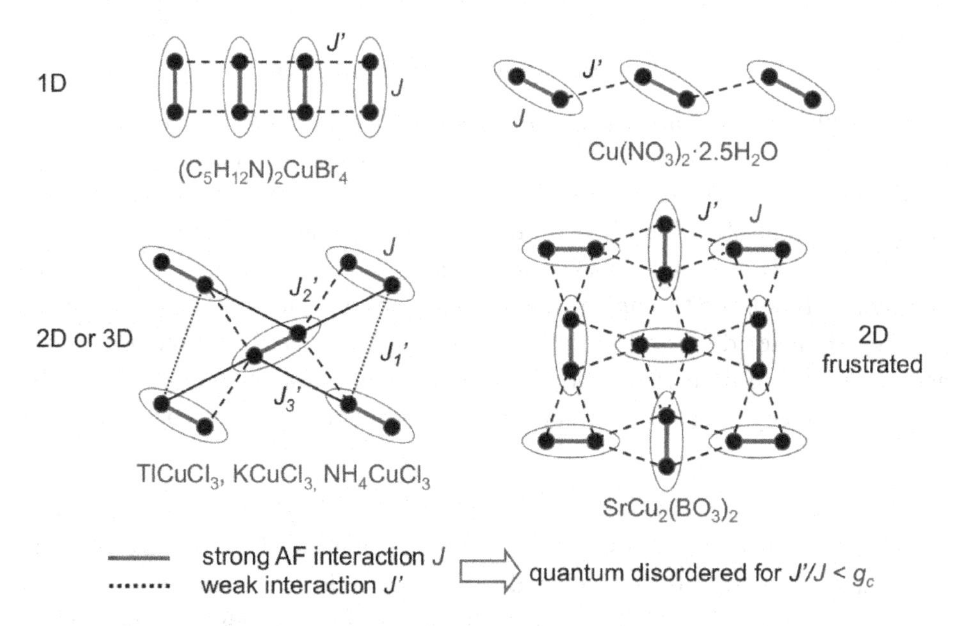

Figure 28.2 Materials with one-, two-, and three-dimensional frustrated and unfrustrated dimer lattices. Intradimer exchange J and interdimer exchange parameters J_i'. For $J'/J < g_c$, the ground state is quantum disordered with spin singlets $|S\rangle$ formed on each dimer. The critical ratio g_c marks a quantum critical point to a phase with long-range staggered transverse order and depends on the geometry and dimensionality of the lattice.

dimer lattices are found, e.g., in the layered materials $BaCuSi_2O_6$ and $SrCu_2(BO_3)_2$. These materials in addition show frustration effects between dimer layers or due to the Shastry-Sutherland geometry, respectively. Three-dimensional dimer lattices are realized, e.g., in the inorganic compounds $TlCuCl_3$, $KCuCl_3$, and NH_4CuCl_3 with dominant intradimer exchange J and a minimum of three interdimer exchange parameters J'_i, which can be determined directly by neutron spectroscopy from the measured triplet (or triplon) dispersions. The application of a magnetic field closes the gap Δ and thus allows one to see a quantum phase transition between the quantum-disordered singlet spin-dimer state and the phase with the staggered transverse order analog of the BEC transition, as we discuss in more detail in Section 28.5. Alternatively, the ordered phase can be reached via a quantum phase transition by increasing the ratio J'/J between the inter- and intradimer coupling, e.g., by applying pressure. This transition breaks $O(3)$ spin symmetry and is different from the BEC ($U(1)$ of the staggered transverse order). For materials with spin 1, e.g., Ni^{2+} ions, dominant spin anisotropy or the Haldane gap can lead to similar energy schemes with Zeeman splitting of the excited triplet states and eventually BEC. Some details of this transition are different [7, 8], but the general physics described here applies to all these microscopically diverse systems.

28.3 Spins as Quantum Simulators

The mapping discussed in the previous section renders the spin systems particularly interesting to serve as quantum simulators of itinerant bosons. A quantum simulator means that the experimental system is a faithful realization of a relatively simple or at least well-controlled Hamiltonian. In that case, reading the experimental results can be viewed as "solving" (in an analogic way) this Hamiltonian. Hence, quantum spin materials represent complementary realizations of quantum systems compared to other examples discussed in this book. We will not detail here the advantages and drawbacks of each class (for a comparison with ultracold atomic gases, see, e.g., [5]), but only focus on a certain number of advantageous properties of the spin materials:

I. Since the actual particles are localized in an insulating Mott state, the superexchange between the spins is dominating the coupling. Typically the superexchange is reasonably short ranged and reducible to few coupling constants J and J'_i. Their values can be extracted from macroscopic measurements such as magnetization or can be directly measured by, e.g., neutron scattering. Thus, the microscopic Hamiltonian is usually well known.
II. Due to progress in the design and synthesis of new spin materials, chemical compounds can be obtained with exchange parameters that are small enough

such that experimentally reachable magnetic fields allow for a sizable magnetization. In the bosonic language, this means that the density of bosons, which is directly related to the magnetization, can be tuned essentially from zero density to one boson per site with a high degree of control. This control of the density is, thus, excellent and certainly vastly superior to the couple of percent change which typically can be obtained for normal (mostly two-dimensional) electronic materials via gate doping.

III. Many well-established experimental techniques exist for magnetic materials giving considerable access to several properties of the equivalent bosonic system. For example, neutron scattering gives direct access to both the single-particle correlation function

$$\langle S_j^-(t)S_0^+(0)\rangle \to (-1)^j\langle b_j(t)b_0^\dagger(0)\rangle \tag{28.6}$$

as well as the density–density correlation function

$$\langle S_j^z(t)S_0^z(0)\rangle \to \langle \rho_j(t)\rho_0(0)\rangle, \tag{28.7}$$

where $\rho_j(t) = b_j^\dagger(t)b_j(t)$ is the density operator of the bosons. Note that the exact expressions of the measured correlations depend on whether the direct mapping (28.1) or the mapping of the dimerized system (28.5) is applicable.

IV. Since the sublattice carrying the spins lives in a crystalline environment of a solid, the exchange parameters can be influenced, for example, by chemical or pressure tuning or by substituting some of the elements out of the spin sites. This enables the investigation of the effects of doping, or disorder or of some modulation of the exchange on the bosonic system, as discussed below in Section 28.6.

28.4 Theoretical Techniques

In order to validate (benchmark) the simulator, it is important to be able to have an independent analytical or numerical solution of the same Hamiltonian. We will thus briefly review the techniques to tackle these spin systems theoretically.

28.4.1 Analytical Techniques

A large number of analytical techniques have been applied to the spin systems, and their efficiency clearly depends on the nature and dimensionality of the considered structure.

The simplest one is a mean-field approach in which one considers that there is some degree of order that is finite, such as the staggered spin expectation value along j on each site $(-1)^i\langle S_i^j\rangle$ for spin systems or the single-particle expectation value for bosons $\langle b_i\rangle$. Neglecting fluctuations around this average value, one obtains simplified equations which can be solved taking the self-consistent determination of the average value into account. Typically, such an approach works reasonably well in three dimensions and allows one to get an idea of which types of ordered phases can exist. For example, the mean-field approach predicts a transition to a Bose-Einstein condensate for the bosonic model (28.3) [9, 10]. As already mentioned above, in the spin models this phase transition to the BEC corresponds to the formation of a staggered *transverse* magnetic order, i.e. $(-1)^i\langle S_i^{x,y}\rangle \neq 0$, where i labels the site of the spin and of the dimer for the direct and the dimer mapping, respectively.

Although quite efficient in determining the standard types of order, the mean-field method fails in predicting more subtle types of order, and it is usually quite limited in determining quantitatively physical properties. For the bosonic operators, improvements of the mean-field theory exist which include fluctuations such as the Bogoliubov approximation [9, 10, 11]. The correspondence between the spin and the boson operators opens the way to carry over these improvements also to the spin operators. These improved techniques take part of the quantum fluctuations into account and can lead to much more precise physical answers. Other mean-field methods such as the bond-operator representation have also been developed for dimerized systems [12, 13].

Mean-field approximations become worse when the dimensionality of the problem is reduced. In particular, mean-field approximations completely fail for one-dimensional geometries for which quantum fluctuations essentially prevent the breaking of continuous symmetries even at zero temperature. Fortunately, a whole set of specific techniques can be used in that particular case, and we refer the reader to Refs. [14, 15, 16] for details. Two main classes of one-dimensional techniques are the so-called Bethe-ansatz and field theories. The first class, the Bethe-ansatz, is employed in order to treat the restricted set of exactly soluble models. The Bethe-ansatz provides a faultless solution of these models. However, one has to be clear by what a "solution" means. In many cases, one can obtain the ground state and in some cases the spectrum of excitations and the thermodynamics. Due to the extreme complexity of the solution, the calculation of more complex quantities such as the correlation functions is still a challenge. Since the original introduction of the method, a lot of time passed before correlation functions in different models at zero and finite temperatures could be computed recently. Typically, such a calculation demands a large numerical effort [17].

The second class is the field theoretical representation of the problem. These field theories (sometimes going under the name of bosonization) erase some of the microscopic details of the problem and give a faithful and relatively simple-to-compute description of the asymptotic properties of the system. The advantage is that it immediately shows the universal features. The drawback is that it is not usable at intermediate and large energies, e.g., comparable to the exchange energy scale.

28.4.2 Numerical Techniques

Several different quasi-exact numerical techniques can be used in order to treat spin or bosonic systems. Quantum Monte-Carlo techniques [18] relying on stochastic sampling procedures are best suited to tackle high-dimensional spin systems at finite temperature without frustration. Many physical properties can be determined, such as thermodynamic quantities or correlation functions. However, the calculation of dynamical correlation functions typically requires analytic continuation from imaginary to real time, which makes it often badly controlled.

In contrast, the density matrix renormalization group methods, also called matrix product state methods [19], and their tensor variants are quasi-exact variational methods and are very efficient in one or, more recently, two dimensions. These methods are very flexible and, in particular, also applicable to frustrated systems or the direct calculation of real-time and frequency-dependent correlations at zero or finite temperature. However, the limitations of this class of method show up, for example, in the resolution of the frequency or the time reachable in the dynamics and depend on the considered problem.

Another method which has been proven to be powerful especially for frustrated systems is exact diagonalization. The advantage of this method is that the entire spectrum and thus all physical quantities are accessible. However, this method is strongly limited in the number of lattice sites which can be dealt with, and a careful extrapolation to larger system sizes has to be performed.

It is important to note that none of these techniques can by itself give access to the full physical information on the broad range of problems. Numerical methods are at their best for short/intermediate times and distances. On the converse, analytical techniques such as field theories are at their best for asymptotically large distances and times. However, by combining the complementary sets of methods when available, one can get the best of both worlds and cover a large range of energy scales to obtain information on the physical states. This is particularly true in one dimension, where the combination of Bethe-ansatz or density-matrix renormalization group (DMRG) technique, and field theory (bosonization) essentially covers the entire range of reachable energy scales. This allows us in simple situations to validate the corresponding quantum simulators and compare directly to experiments. We will now examine some examples of successful implementations of this strategy.

28.5 BEC and Beyond – Dimensionality

We examine in this section several examples for which the magnetic insulators could be used as quantum simulators. We will examine three big classes of problems. The first one is the BEC of the "fake" bosons. This can be realized in materials for which the motion of bosons is high dimensional (two or three dimensional). Then we look at the case of one-dimensional bosonic systems, for which the universality of the Luttinger liquid behavior has been observed in quantum spin materials with very different exchange parameters along different spatial directions. Finally, we will comment on the effects of the effective dimensionality of materials and how one crosses over between these two extreme situations.

28.5.1 BEC in Three-Dimensional Magnetic Insulators

The first experimental observation of BEC in dimer insulators was reported by Nikuni and coworkers in the halide $TlCuCl_3$ [20]. In this material, spins 1/2 from Cu^{2+} ions are arranged in structural pairs on a three-dimensional lattice of low monoclinic symmetry, as shown as one of the examples in Fig. 28.2, with exchange parameters $J = 5.5$ meV, $J'_1 = -0.43$ meV, $J'_2 = 3.16$ meV, and $J'_3 = 0.91$ meV [21]. The resulting energy gap Δ is of the order of 0.7 meV, the triplet band extends up to 7 meV, and the critical field h_c lies in between 5.4 and 5.7 T depending on the direction of the magnetic field. For magnetic fields in excess of h_c, a quantum phase transition was observed to a phase with long-range magnetic order. This ordered phase shows the following characteristic properties of a BEC:

I. The longitudinal magnetization m_z is zero, at zero temperature, in the gapped phase. It becomes nonzero above h_c and increases continuously for $h > h_c$ up to saturation at h_s (indicating in the boson language that bosons are present in the system). Its magnetic field and especially its temperature dependence are consistent with the predictions [4] coming from the mapping to bosons and the existence of a BEC transition. In particular, the uniform magnetization shows a nonmonotonous temperature behavior with a kink when hitting the critical temperature at which the transverse order appears [20, 22].
II. The phase boundary $(h_c(T) - h_c(0)) \propto T^{\phi}$ shows [20, 23] a critical exponent ϕ that is consistent with the expected dependence between T_{BEC} and the chemical potential [4, 11, 24] with a mean-field exponent of $\phi = d/2$, where d is the spatial dimension.

III. The thermodynamic transition to the BEC phase with long-range transverse order shows a clear λ-anomaly [25].
IV. The transverse magnetic order as measured directly by, e.g., neutron diffraction has the characteristic field and temperature dependence of a BEC [26].
V. The excitation spectrum in the BEC phase measured by neutron spectroscopy consists of linear Goldstone modes at low energies and gapped quadratic modes at higher energies. The dispersions of these modes and their coexistence prove directly the nature of the condensate ground state [27].

For a large and ever-increasing number of dimer compounds, these characteristic properties of a BEC have been reported while specific details of the materials such as the energy scale of the exchange parameters and anisotropies may affect their observability [5, 6].

28.5.2 One-Dimensional Bosons – Tomonaga-Luttinger Liquids

In the previous section, one was considering a situation for which an order parameter for the BEC exists (three or quasi-two-dimensional lattices). In these situations, the system behaves essentially as nearly free bosons, in particular, close to the quantum critical point for which the density of bosons tends to zero. One can, however, use the quantum simulator provided by quantum magnets for more subtle situations. In particular, one can simulate effectively one-dimensional systems of bosons. One-dimensional bosons show quite different physics than their three-dimensional counterparts and behave much more akin to spinless fermions.

In many one-dimensional situations, the state of the system forms a so-called Tomonaga-Luttinger liquid (TLL) [15]. In such a Tomonaga-Luttinger liquid, the physics is characterized by correlation functions which at zero temperature decay as power laws with an exponent depending on a single parameter (called the Luttinger parameter). For example, the single-particle correlation function at long distances is dominated by

$$\langle b(x)b^\dagger(0)\rangle \propto \left(\frac{\alpha}{x}\right)^{\frac{1}{2K}}, \tag{28.8}$$

where $b(x)$ denotes the bosonic field at position x in the continuum. The parameter K is a nonuniversal number which depends on the precise details of the microscopic model and α is a short-distance cutoff parameter. The same behavior occurs also for fermions and spins in one-dimensional geometries. Testing the algebraic decay and its universality, however, is not easy. Various experimental systems have been analyzed in the past [15, 28], ranging from organic superconductors to cold atomic gases. However, in general a *quantitative* test is difficult and mostly the qualitative power law decay had been observed. Quite remarkably, the quantum magnet simulator offers the possibility for such a quantitative test. This is due to the following advantages of the spin materials over different realizations of quantum systems:

I. The microscopic Hamiltonian of the spin systems is perfectly known. This is typically not the case, for example, for fermions for which a complicated screened Coulomb interaction will be important. One can thus not only check power laws but *quantitatively* compare the theoretical calculations with the experiments *without* adjustable parameters.
II. The universality of the predictions can be tested by measuring different physical quantities.
III. The advances in the preparation of novel materials made the entire magnetization range – corresponding in the bosonic model to densities from zero to one boson per site – accessible. This fantastic tunability allows for a full check of the dependence of the physical properties on this control parameter and is very difficult to reach in alternative realizations of the TLL.
IV. Given the fact that the TLL physics depends crucially on the density (in particular, the exponents), it is especially important that the material is very homogeneous. The quantum spin materials are almost perfectly homogeneous, which ensures the high accuracy of the measurements of power law quantities.

As a consequence, the first quantitative tests of TLL physics were realized in such one-dimensional spin systems [29, 30, 31]. We will give additional details and the link between TLL and BEC physics in the next section. Since the first tests, several additional properties of one-dimensional systems can now be emulated such as $t-J$ model physics and fractionalization of the excitations, which are bound states of the elementary magnetic excitations in one dimension (1D) (spinons) [30, 32, 33]. Temperature-frequency scaling of a TLL could also be observed [34].

28.5.3 Dimensional Crossover and Quantum Critical Points

The quantum spin materials considered in the previous section are – even though spatially strongly anisotropic in the exchange parameters – not really one dimensional, but three dimensional from their structure. Therefore, the regime of validity of the description within the effectively reduced dimensionality has to be taken into account. In particular, due to strongly anisotropic exchange, the effective dimensionality of materials may vary between three-dimensional structures and effectively one-dimensional structures, depending on the temperature. Assume three distinct exchange parameters $J \gg J' \gg J''$ along the different lattice axes – for simplicity, say along the directions x, y, and z. Then the system behaves as one dimensional for temperature scales larger than J' and J'', since the thermal fluctuations along y and z will destroy any coherent effect of the exchange along these directions. Thus, the system behaves effectively as an array of decoupled one-dimensional chains with exchange J along x. Similarly, for a temperature $J, J' \gg k_B T \gg J''$, both the exchange parameters along x and y are important and the system is effectively two dimensional. Thus a dimensional crossover can be controlled by varying the temperature.

Quite remarkably, such dimensional crossover also becomes unavoidable even at the lowest temperatures when the energy of the excitations becomes very small, for example, close to one of the critical fields h_c or h_s. Indeed, as shown in Fig. 28.3, if the chemical potential for the excitations is far from the top or bottom of the band, then a small enough interchain coupling $J'' = J'$ will act on essentially a one-dimensional dispersion. The chains can then be seen as a quasi-one-dimensional system in the presence of a weak interchain coupling as described above. However, when the field becomes close to, e.g., h_c, regardless of the strength of the interchain coupling, such a coupling will become larger than the chemical potential. The full

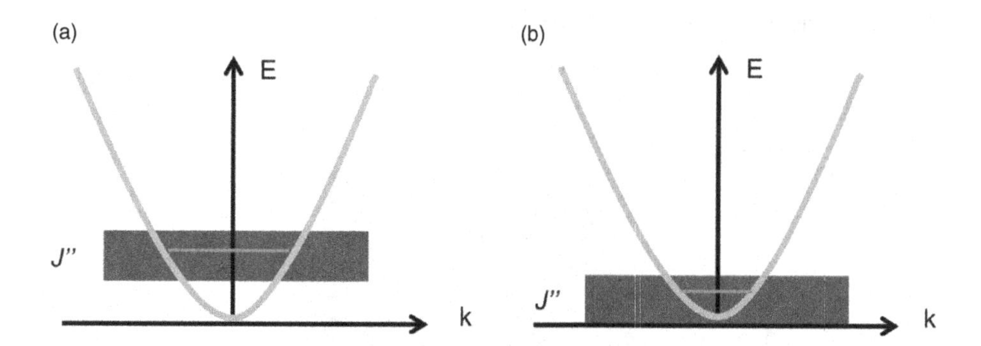

Figure 28.3 Dimensional crossover and different physical behaviors as a function of the interchain/interladder coupling. The energy E of excitations – for example, the triplon excitations – in a one-dimensional structure (chain or ladder) as a function of their quantum number (e.g., the momentum k) is schematically represented (curved gray line). The chemical potential around which excitations will take place is represented by the horizontal (gray) line. (a) The interchain/interladder exchange J'' is much smaller than the chemical potential. In that case, it affects only excitations around the chemical potential (schematically represented by the dark box), and one can treat these excitations by the usual one-dimensional approximation (e.g., bosonization). In that case, the system can be viewed as weakly coupled one-dimensional structures. (b) If J'' becomes larger *or* if one is close enough to a critical point so that the chemical potential is small, then all the excitations down to the bottom of the band are affected by the interchain/interladder exchange and the full (two- or three-dimensional) dispersion of the energy $E(k)$ must be taken into account. In that case, one cannot even start from a one-dimensional point of view, and the system is much better described by an anisotropic 3D material. These two limits correspond to quite different physical behaviors, and the crossover between the two is a dimensional crossover.

curvature of the band must then be taken into account and not just the excitations around the chemical potential. In that case, the problem must be considered as a (spatially anisotropic) two-dimensional (or three-dimensional) problem from the start [4]. More details on dimensional crossovers can be found in Ref. [35]. Examples of such systems are various recently discovered low-dimensional spin-ladder materials in the strong and weak (rung) coupling limit. The effective dimensionality of the systems can change their properties drastically. As an example, we discuss here the material $(C_5H_{12}N)_2CuBr_4$ or BPCB; see also Section 28.5.2 on TLL. The anisotropic exchange structure of the material, which consists of weakly coupled ladders with dominant rung, $J \gg J' \gg J''$, is sketched and the corresponding phase diagram shown in Fig. 28.4. For temperatures lower than the interladder exchange J'', a BEC with three-dimensional character as discussed in the previous sections has been observed [29, 31]. However, a TLL phase with one-dimensional character arises at a temperature, which is large compared to J'', but small compared to the intraladder exchange J and J' [29, 31, 36]. In BPCB, the universality of the TLL could be observed in a variety of properties ranging from specific heat [32, 36] to the nuclear magnetic resonance (NMR) relaxation rates [29, 32].

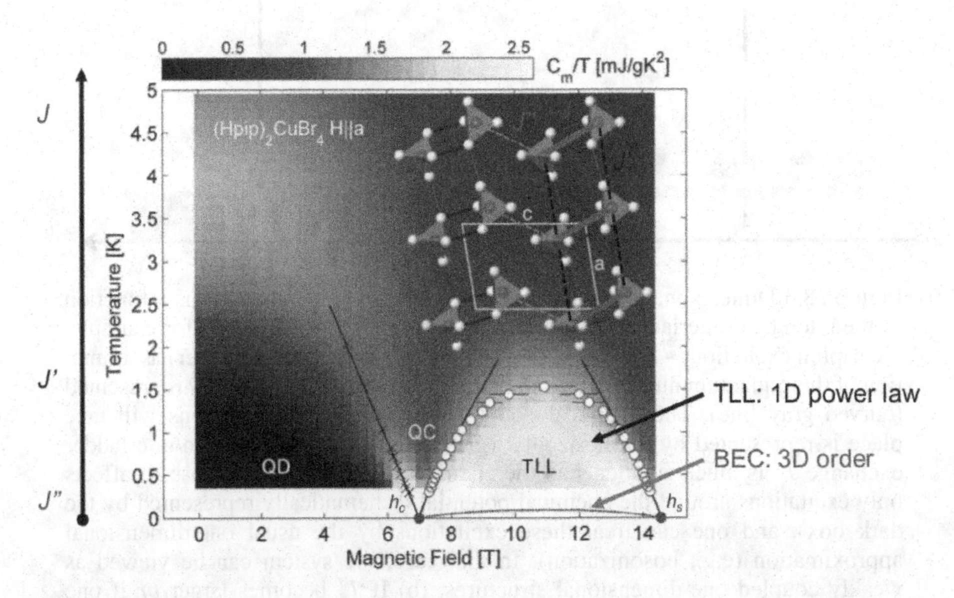

Figure 28.4 Phase diagram of BPCB. The quantum critical points are marked by h_c and h_s. QD: quantum disorder phase with excitation gap Δ, QC: quantum critical regime, TLL: spin Tomonaga-Luttinger liquid. The 3D BEC phase appears at intermediate fields below a temperature of 100 mK. Adapted with permission from Rüegg, Ch., et al. (2008), Thermodynamics of the Spin Luttinger liquid in a Model Ladder Material, *Phys. Rev. Lett.*, **101**, 247202 [36]. Copyright (2008) by the American Physical Society.

28.6 Some Things Worthy to Simulate – Dirty Bosons

In the previous section, we showed how these materials are efficient quantum simulators. The first two examples chosen, the regime of three-dimensional BEC and one-dimensional TLL character, had fortunately also an excellent handle on the theoretical side. This allowed us to actually benchmark such a simulator and to show its efficiency. The third example of the dimensional crossover has already been more problematic to tackle theoretically, and open questions can be addressed in the experiment. In this section, another example, a bosonic system in the presence of disorder and interactions, is presented for which our theoretical handle is even more limited. The existence of interesting phases such as the elusive Bose glass phase has been predicted [37, 38]. However, many properties of these systems are still far from being understood. In that respect, the utility of having an experimental quantum simulator takes all its importance. We examine how the quantum spin materials can be used to study the combined effects of disorder and interactions on bosonic systems.

The hopping parameter t, interaction J_Z, and the chemical potential μ in the boson Hamiltonian Eq. 28.3 are linked directly to microscopic parameters based on atomic superexchange paths in the specific material. For example, in BPCB two copper ions with spin 1/2 are connected via two bromine atoms to form a Cu-Br-Br-Cu dimer (see inset Fig. 28.4). Chemistry allows us to design and synthesize materials which have different values of these parameters. For example, the two spin ladder materials $(C_5H_{12}N)_2CuBr_4$ and $(C_5H_{12}N)_2CuCl_4$ share the same crystal structure and dimer lattice, but the latter chloride has an energy scale that is approximately factor 2 lower than the bromide compound, and its dimer (ladder rung) exchange is slightly less dominant. For $TlCuCl_3$ and its isostructural sister compound $KCuCl_3$, substitution of the cation Tl^+ with K^+, which participates indirectly in the superexchange, has a very similar effect.

Very interesting new directions arise when Br/Cl or Tl/K are partially substituted in these or related materials; see Fig. 28.5 for this and further examples. Since the substitution takes place at random places, this can cause a random variation of the exchange parameters, which translates in a random variation of the parameters in the bosonic model. The resulting local variation of the gap and chemical potential by this random substitution may then result in a Bose glass phase near a slightly shifted h_c with finite compressibility but absence of long-range order (which in contrast is characteristic for the homogeneous, coherent BEC). First experiments on quantum spin materials with random exchange from chemical substitution show evidence for the formation of this elusive phase [39, 40, 41, 42]. The design of disordered models is one of the current research directions. However, this is not the only one possible, and in Fig. 28.5 we summarize some of the exciting current

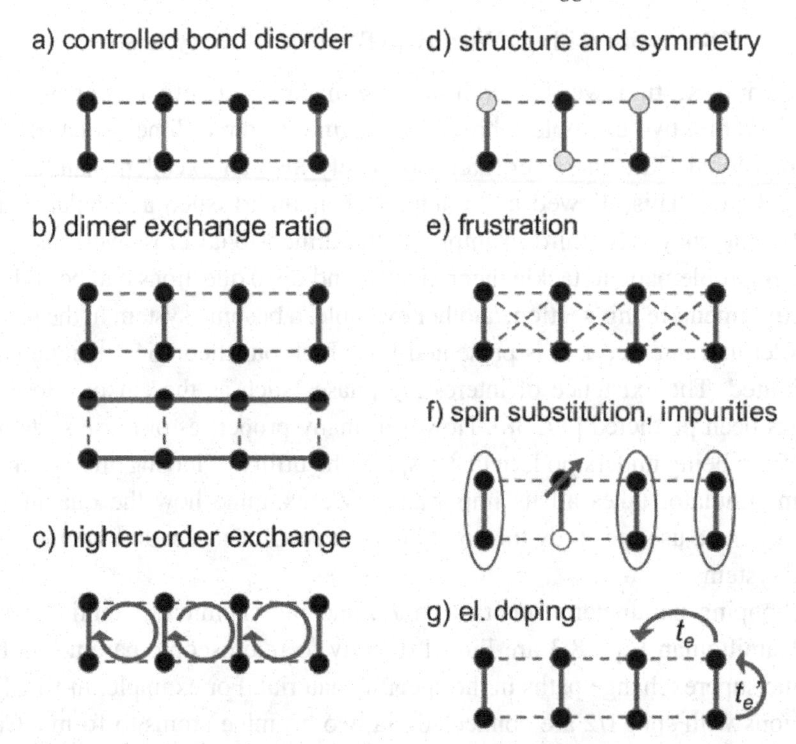

Figure 28.5 Control of interactions, disorder, and impurities in quantum spin (ladder) materials. (a) Bond disorder, for which Bose glass physics may be observed in a magnetic field [39, 40, 41, 42]; see text in Section 28.6. (b) Strong and weak (rung) limit of the ladder, in which attractive and repulsive interactions can be studied for the Tomonaga-Luttinger liquid [29, 30, 33]. (c) Higher-order spin (ring) exchange, which can tune the system to criticality with a different universality class and with interesting interplay with TLL and BEC physics. (d) Ladder system with reduced symmetry and protection of symmetric and antisymmetric (one- and two-particle) spectra. (e) Ladder with frustration and potentially incommensurate triplet excitations, BEC criticality, and order. (f) Doping with spin 0 ions, or more generally with ions with different spin, on the magnetic site. Substitution directly on the magnetic site leads to interesting quantum impurity problems that can be addressed both experimentally and in theory. (g) Metallic ladder or a system close to a Mott-insulator transition, where the charge gap disappears and electronic degrees of freedom become important.

directions in designing new Hamiltonians and materials for the simple case of the ladder geometry.

28.7 Conclusions and Perspectives

We have discussed in this chapter how magnetic insulators can be used to study physical properties of interacting itinerant (hardcore) bosons. One advantage of the realization in magnetic materials is the excellent control which one has over the design of the atomic structures. Many different lattice structures with a wide range of parameters can now be engineered reaching from effectively low-dimensional to disordered models. A large variety of effects ranging from Bose-Einstein condensation to more exotic properties of bosonic fluids such as the Tomonaga-Luttinger liquid for one-dimensional bosons have been observed. At the moment, many studies have focused on thermodynamic properties of such systems, with detailed investigations of the phase diagram in three, two, and one dimension. The excellent degree of control over external parameters such as the magnetic field or temperature enabled high-precision studies of quantum critical points. Further, dynamical response properties could be observed by measuring several correlation functions directly by the powerful probes that are available for spin systems (the two best known being clearly neutrons and NMR techniques). Quantum spin materials thus constitute excellent complementary systems for tackling boson physics such as cold atomic gases.

Several interesting perspectives are offered by solid-state quantum simulators based on quantum spin materials. For example, in dimer systems using the fact that additional states, besides the singlet and the lowest triplet, exist allows one to simulate more complex models such as the t-J model, which combines spin physics with itinerant models. As another example, disorder has already been mentioned and partly explored. However, this area clearly deserves much more investigations, for example, in the direction of quantum localization and the interplay with interaction. Thinking about how to further extend the simulators to additional systems is certainly a fruitful direction.

Last but not least, these systems should in principle allow one also to explore out-of-equilibrium situations. Although in condensed matter context one has always to be wary of relaxation processes, one has in principle now access to ultrafast pumps and probes that should allow exploring such physics as well as to physically isolated systems, such as the cold atoms. There are thus exciting times ahead where several different fields will join to shed light on the new physics that bosonic systems allow to obtain.

Acknowledgments: The works presented in this chapter were supported in part by the Swiss National Science Foundation under Division II, by the Royal Society (U.K.) and by the Deutsche Forschungsgemeinschaft (DFG).

References

[1] Auerbach, A. 1998. *Interacting Electrons and Quantum Magnetism.* Berlin: Springer.

[2] Holstein, T., and Primakoff, H. 1940. Field dependence of the intrinsic domain magnetization of a ferromagnet. *Phys. Rev.*, **58**, 1098.

[3] Matsubara, T., and Matsuda, H. 1956. A lattice model of liquid helium. *Prog. Theor. Phys.*, **16**, 569.

[4] Giamarchi, T., and Tsvelik, A. M. 1999. Coupled ladders in a magnetic field. *Phys. Rev. B*, **59**, 11398.

[5] Giamarchi, T., Rüegg, Ch., and Tchernyshyov, O. 2008. Bose-Einstein condensation in magnetic insulators. *Nature Physics*, **4**, 198.

[6] Zapf, V., Marcelo, J., and Batista, C. D. 2014. Bose-Einstein condensation in quantum magnets. *Rev. Mod. Phys.*, **86**, 563.

[7] Affleck, I. 1991. Bose condensation in quasi-one-dimensional antiferromagnets in strong fields. *Phys. Rev. B*, **43**, 3215.

[8] Chitra, R., and Giamarchi, T. 1997. Critical properties of gapped spin-chains and ladders in a magnetic field. *Phys. Rev. B*, **55**, 5816.

[9] Pitaevskii, L., and Stringari, S. 2003. *Bose-Einstein Condensation*. Oxford: Clarendon Press.

[10] Pethick, C. J., and Smith, H. 2002. *Bose-Einstein Condensation in Dilute Gases*. Cambridge: Cambridge University Press.

[11] Popov, V. N. 1987. *Functional Integrals and Collective Excitations*. Cambridge: Cambridge University Press.

[12] Sachdev, S., and Bhatt, R. N. 1990. Bond-operator representation of quantum spins: mean-field theory of frustrated quantum Heisenberg antiferromagnets. *Phys. Rev. B*, **41**, 9323.

[13] Normand, B., and Rüegg, Ch. 2011. Complete bond-operator theory of the two-chain spin ladder. *Phys. Rev. B*, **83**, 054415.

[14] Gogolin, A. O., Nersesyan, A. A., and Tsvelik, A. M. 1999. *Bosonization and Strongly Correlated Systems*. Cambridge: Cambridge University Press.

[15] Giamarchi, T. 2004. *Quantum Physics in One Dimension*. Vol. 121. Oxford: Oxford University Press.

[16] Cazallila, M. A., Citro, R., Giamarchi, T., Orignac, E., and Rigol, M. 2011. One dimensional bosons: from condensed matter systems to ultracold gases. *Rev. Mod. Phys.*, **83**, 1405.

[17] Caux, J.-S., Calabrese, P., and Slavnov, N. A. 2007. One-particle dynamical correlations in the one-dimensional Bose gas. *J. Stat. Mech.: Theor. Exp.*, **2007**, P01008.

[18] Pollet, L. 2012. Recent developments in quantum Monte Carlo simulations with applications for cold gases. *Reports on Progress in Physics*, **75**, 094501.

[19] Schollwöck, U. 2011. The density-matrix renormalization group in the age of matrix product states. *Annals of Physics*, **326**, 96.

[20] Nikuni, T., Oshikawa, M., Oosawa, A., and Tanaka, H. 2000. Bose-Einstein condensation of dilute magnons in $TlCuCl_3$. *Phys. Rev. Lett.*, **84**, 5868.

[21] Cavadini, N., Heigold, G., Henggeler, W., Furrer, A., Güdel, H.-U., Krämer, K., and Mutka, H. 2001. Magnetic excitations in the quantum spin system $TlCuCl_3$. *Phys. Rev. B*, **63**, 172414.

[22] Sirker, J., Weisse, A., and Sushkov, O. P. 2004. Consequences of spin-orbit coupling for the Bose-Einstein condensation of magnons. *Europhys. Lett.*, **68**, 275.

[23] Yamada, F., Ono, T., Tanaka, H., Misguich, G., Oshikawa, M., and Sakakibara, T. 2008. Magnetic-field induced Bose-Einstein condensation of magnons and critical behavior in interacting spin dimer system $TlCuCl_3$. *Journal of the Physical Society of Japan*, **77**, 013701.

[24] Batyev, E. G., and Braginskii, L. S. 1984. Antiferrornagnet in a strong magnetic field: analogy with Bose gas. *Journal of Experimental and Theoretical Physics*, **60**, 781.

[25] Oosawa, A., Katori, H. Aruga, and Tanaka, H. 2001. Specific heat study of the field-induced magnetic ordering in the spin-gap system $TlCuCl_3$. *Phys. Rev. B*, **63**, 134416.

[26] Tanaka, H., Oosawa, A., Kato, T., Uekusa, H., Ohashi, Y., Kakurai, K., and Hoser, A. 2001. Observation of field-induced transverse neel ordering in the spin gap system $TlCuCl_3$. *Journal of the Physical Society of Japan*, **70**, 939.

[27] Rüegg, Ch., Cavadini, N., Furrer, A., Güdel, H.-U., Krämer, K., Mutka, H., Habicht, A. K., and Vorderwisch, P. 2003. Bose-Einstein condensation of the triplet states in the magnetic insulator $TlCuCl_3$. *Nature*, **423**, 62.

[28] Giamarchi, T. 2012. Some experimental tests of Tomonaga-Luttinger liquids. *Int. J. Mod. Phys. B*, **26**, 1244004.

[29] Klanjšek, M., Mayaffre, H., Berthier, C., Horvatić, M., Chiari, B., Piovesana, O., Bouillot, P., Kollath, C., Orignac, E., Citro, R., and Giamarchi, T. 2008. Controlling Luttinger liquid physics in spin ladders under a magnetic field. *Phys. Rev. Lett.*, **101**, 137207.

[30] Thielemann, B., Rüegg, Ch., Rønnow, H. M., Läuchli, A. M., Caux, J.-S., Normand, B., Biner, D., Krämer, K. W., Güdel, H.-U., Stahn, J., Habicht, K., Kiefer, K., Boehm, M., McMorrow, D. F., and J., Mesot. 2009a. Direct observation of magnon fractionalization in the quantum spin ladder. *Phys. Rev. Lett.*, **102**, 107204.

[31] Thielemann, B., Rüegg, Ch., Kiefer, K., Rønnow, H. M., Normand, B., Bouillot, P., Kollath, C., Orignac, E., Citro, R., Giamarchi, T., Läuchli, A. M., Biner, D., Krämer, K. W., Wolff-Fabris, F., Zapf, V. S., Jaime, M., Stahn, J., Christensen, N. B., Grenier, B., McMorrow, D. F., and Mesot, J. 2009b. Field-controlled magnetic order in the quantum spin-ladder system $(Hpip)_2CuBr_4$. *Phys. Rev. B*, **79**, 020408(R).

[32] Bouillot, P., Kollath, C., Läuchli, A. M., Zvonarev, M., Thielemann, B., Rüegg, Ch., Orignac, E., Citro, R., Horvatić, M., Berthier, C., Klanjšek, M., and Giamarchi, T. 2011. Statics and dynamics of weakly coupled antiferromagnetic spin-$1/2$ ladders in a magnetic field. *Phys. Rev. B*, **83**, 054407.

[33] Schmidiger, D., Bouillot, P., Guidi, T., Bewley, R., Kollath, C., Giamarchi, T., and Zheludev, A. 2013. Spectrum of a magnetized strong-leg quantum spin ladder. *Phys. Rev. Lett.*, **111**, 107202.

[34] Povarov, K. Yu., Schmidiger, D., Reynolds, N., Bewley, R., and Zheludev, A. 2015. Scaling of temporal correlations in an attractive Tomonaga-Luttinger spin liquid. *Phys. Rev. B*, **91**, 020406.

[35] Giamarchi, T. 2010. Quantum phase transitions in quasi-one dimensional systems. Page 291 of: Carr, Lincoln D. (ed), *Understanding Quantum Phase Transitions*. CRC Press / Taylor & Francis.

[36] Rüegg, Ch., Kiefer, K., Thielemann, B., McMorrow, D. F., Zapf, V., Normand, B., Zvonarev, M. B., Bouillot, P., Kollath, C., Giamarchi, T., Capponi, S., Poilblanc, D., Biner, D., and Krämer, K. W. 2008. Thermodynamics of the spin Luttinger liquid in a model ladder material. *Phys. Rev. Lett.*, **101**, 247202.

[37] Giamarchi, T., and Schulz, H. J. 1988. Anderson localization and interactions in one-dimensional metals. *Phys. Rev. B*, **37**, 325.

[38] Fisher, M. P. A., Weichman, P. B., Grinstein, G., and Fisher, D. S. 1989. Boson localization and the superfluid-insulator transition. *Phys. Rev. B*, **40**, 546.

[39] Hong, T., Zheludev, A., Manaka, H., and Regnault, L.-P. 2010. Evidence of a magnetic Bose glass in $(CH_3)_2CHNH_3Cu(Cl_{0.95}Br_{0.05})_3$ from neutron diffraction. *Phys. Rev. B*, **81**, 060410.

[40] Yamada, F., Tanaka, H., Ono, T., and Nojiri, H. 2011. Transition from Bose Gass to a condensate of triplons in $Tl_{1-x}K_xCuCl_3$. *Phys. Rev. B*, **83**, 020409.

[41] Yu, R., Yin, L., Sullivan, N. S., Xia, J. S., Huan, C., Paduan-Filho, A., Oliveira Jr., N. F., Haas, S., Steppke, A., Miclea, C. F., Weickert, F., Movshovich, R., Mun, E.-D., Scott, B. S., Zapf, V. S., and Roscilde, T. 2012. Bose glass and Mott glass of quasiparticles in a doped quantum magnet. *Nature*, **489**, 379.

[42] Ward, S., Bouillot, P., Ryll, H., Kiefer, K., Kramer, K. W., Ruegg, C., Kollath, C., and Giamarchi, T. 2013. Spin ladders and quantum simulators for Tomonaga-Luttinger liquids. *J. Phys.: Condens. Matter*, **25**, 014004.

Part V

Condensates in Astrophysics and Cosmology

Editorial Notes

One of the most fascinating connections in the study of Bose-Einstein condensation (BEC) is the possibility of using condensed matter and atomic systems to answer questions on astrophysical scales. Astrophysics and cosmology have a long history of using ideas from condensed matter. One well-known example is the use of spontaneous symmetry-breaking concepts from studies of ferromagnetism in understanding the Higgs boson and the early universe. In another example, Bardeen-Cooper-Schrieer (BCS) fermion pairing has been proposed for the state of some neutron stars, as reviewed here in Chapter 29.

There are two general ways in which BEC studies connect to astrophysics and cosmology. One way is to use laboratory condensates to simulate astrophysical effects. Chapter 30 reviews recent work on using Helium-3 to simulate cosmological broken symmetries, with the related Kibble-Zurek mechanism discussed for ultracold atoms in Chapters 6–7. Both cold atom[1] and polariton[2] systems have also been proposed to simulate Hawking radiation from black holes.

Another connection is proposals that bosonic particles actually have undergone Bose-Einstein condensation on the grand cosmological scale. Gerry Brown proposed universal kaon BEC in the 1995 green book.[3] The concept of the chiral quark–antiquark condensate, for which Nambu received the Nobel Prize in 2008, has had major impact in the field of nuclear physics, but is not reviewed in this book.[4] More speculatively, the appearance of Bose-Einstein condensation through

[1] Steinhauer, J. (2014), Observation of self-amplifying Hawking radiation in an analogue black-hole laser, *Nat. Phys.*, **10**, 864; Boron, D., et al. (2015), Quantum signature of analog Hawking radiation in momentum space, *Phys. Rev. Lett.*, **115**, 025301.

[2] Nguyen, H. S., et al. (2015) Acoustic black hole in a stationary hydrodynamic flow of microcavity polaritons, *Phys. Rev. Lett.*, **114**, 036402.

[3] Brown G., in *Bose-Einstein Condensation*, A. Griffin, D.W. Snoke, and S. Stringari, eds. (Cambridge University Press, 1995).

[4] Braun-Munzinger, P. and Wambach, J. (2009), Colloquium: phase diagram of strongly interacting matter, *Rev. Mod. Phys.*, 81, 1031.

the gravitationally self-induced thermalization of cold axions and the implications to cold dark matter are discussed in Chapter 31, and our final chapter makes an interesting claim that Bose-Einstein condensation of gravitons could explain the anticipated behavior of black holes and form a new way to understand quantum gravity (Chapter 32).

29

Bose-Einstein Condensates in Neutron Stars

CHRISTOPHER J. PETHICK

Niels Bohr International Academy, Niels Bohr Institute, University of Copenhagen, Denmark
NORDITA, KTH Royal Institute of Technology and Stockholm University, Stockholm, Sweden

THOMAS SCHÄFER

Department of Physics, North Carolina State University, Raleigh, USA

ACHIM SCHWENK

Institut für Kernphysik, Technische Universität Darmstadt, Germany
ExtreMe Matter Institute, GSI Helmholtzzentrum für Schwerionenforschung GmbH, Darmstadt, Germany

In the two decades since the appearance of the book *Bose-Einstein Condensation* in 1995, there have been a number of developments in our understanding of dense matter. After a brief overview of neutron star structure and the Bose-Einstein condensed phases that have been proposed, we describe selected topics, including neutron and proton pairing gaps; the physics of the inner crust of neutron stars, where a neutron fluid penetrates a lattice of nuclei, meson condensates; and pairing in dense quark matter. Especial emphasis is placed on basic physical effects and on connections to the physics of cold atomic gases.

29.1 Introduction

Neutron stars contain strongly interacting matter under extreme conditions. The high densities, up to $\sim 10^{15}\,\mathrm{g\,cm^{-3}}$ in the interior, combined with the physics of the strong interaction, which involves a rich spectrum of particles and attractive interaction channels, can lead to the realisation of various condensates in neutron stars. In this chapter, we give an overview of these possibilities, ranging from pairing of nucleons, to condensates of mesons, to quark pairing. After a description of the structure of neutron stars, we provide an introduction to the various condensates, focussing on new developments since the book *Bose-Einstein Condensation* in 1995 [1].

We first discuss superfluid phases of neutrons and protons. These are rather well established, and the present challenge is to make reliable calculations of pairing gaps and critical temperatures. This area has benefitted significantly from the connections to cold atomic Fermi gases with resonantly tuned strong interactions. If hyperons are present in neutron stars, additional paired phases of hyperons are possible. At asymptotically high densities, the interaction between quarks becomes weak and the pairing of quarks is well established theoretically in this limit. We discuss how the nature of the gluon-exchange interaction gives rise to special features for quark pairing. At intermediate densities, the condensation of pions and kaons is possible. Whether or not these phases exist in nature is an open question because of the difficulty of making reliable calculations of the effects of strong correlations: this is true irrespective of whether one approaches the problem from low densities, using hadron degrees of freedom, or from high densities, where quark degrees of freedom are the natural choice. Finally, we discuss briefly some observational consequences of condensation in neutron stars, including the cooling of neutron stars, and the impact on rotational behaviour and on phenomena involving the neutron star crust.

29.2 Neutron Star Structure

Matter in a neutron star ranges in density from typical terrestrial values at the surface to greater than nuclear density at the centre. In all but the outermost parts of the star, the thermal energy $k_B T$, where k_B is the Boltzmann constant and T the temperature, is low compared with characteristic excitation energies of the system: the star is thus a low temperature system in which quantum effects play an important role. Despite their name, neutron stars contain ingredients other than neutrons. At the surface of the star, matter in its ground state consists of $^{56}_{26}$Fe nuclei, with an equal number of electrons to ensure charge neutrality: thus, in the nucleus the number of neutrons, 30, is only slightly greater than that of protons, 26. The Fermi energy of the electrons increases rapidly with depth, and it becomes energetically favourable for electrons to be captured by protons, thereby producing nuclei that are more neutron-rich.[1]

At a density of around 4×10^{11} g cm^{-3}, around one-thousandth of nuclear density, the highest occupied neutron levels are no longer bound, a situation referred to as ‘neutron drip’. As a consequence, at higher densities the lattice of nuclei is permeated by a neutron fluid. With further increase in density, the density of this neutron fluid increases, and nuclei become even more neutron rich and occupy an increasing fraction of space. Nuclei merge to form a uniform fluid of neutrons and protons at a density of around one-half of nuclear density, and the fraction of protons is $\sim$ 5%. At densities just below that for the transition to a uniform

[1] For a review of the physics of the outer parts of neutron stars, see Ref. [2].

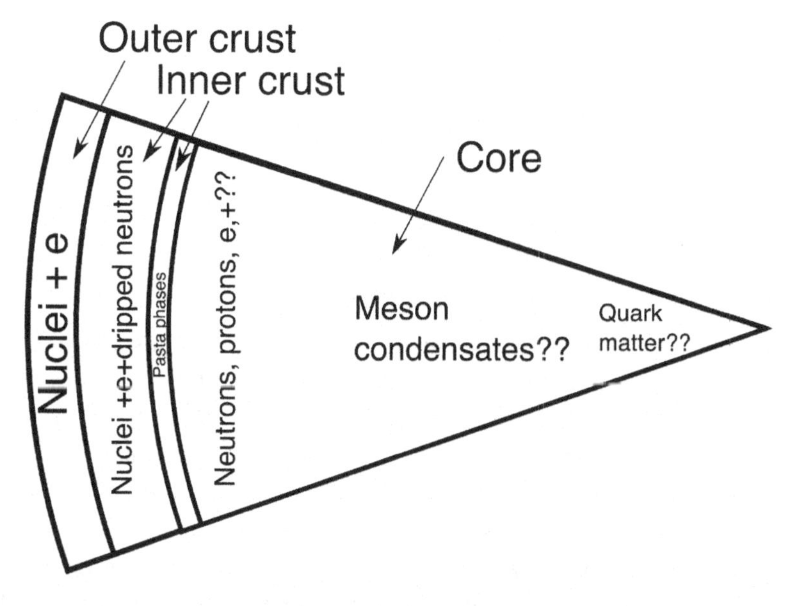

Figure 29.1 Schematic picture of the phases encountered in a neutron star.

medium, nuclei can form highly nonspherical shapes such as rods or sheets in what are referred to as 'pasta phases' because of their resemblance to spaghetti and lasagna. At higher densities, other constituents can appear: among these are muons (which are present when the electron chemical potential exceeds the muon restmass energy), hyperons and possible phases with deconfined quarks. A schematic view of a slice of a neutron star is shown in Fig. 29.1. The figure is not to scale. For neutron stars in the mass range that can be observed, the radius is $\sim$ 12 km [3], the outer crust is some hundreds of metres thick and the inner crust about half a kilometre.

29.3 Condensates in Neutron Stars

We now give an overview of the various condensates that have been proposed. These may be classified according to the baryon number, B, of the condensed boson.[2] At the lower densities encountered in the inner crust and the outer part of core of the star, condensates of neutron pairs and of proton pairs, which have $B = 2$, have been studied extensively. These are analogous to condensates in conventional metallic superconductors and, in the case of pairs with nonzero orbital angular momentum, the superfluid phases of liquid ^{3}He. At supranuclear densities,

[2] Irrespective of whether the condensed entity is a pair of fermionic excitations, as in a superconductor, or a bosonic excitation, such as a meson, we shall refer to it simply as a 'boson'.

condensates of pions and of kaons, which have $B = 0$, have been proposed. At even higher densities, one expects matter to consist of deconfined quarks, and interactions between these can lead to condensates of pairs of quarks, which have $B = 2/3$.

29.3.1 Pairing of Nucleons

Neutron Pairing

Even before the first identification of neutron stars in the cosmos and shortly after the formulation of the Bardeen-Cooper-Schrieffer (BCS) theory of superconductivity [4], Migdal in a side remark in a paper on superfluidity and the moment of inertia of atomic nuclei commented that, in matter in the interior of neutron stars, neutrons would pair with a transition temperature of order 1 MeV (10^{10} K) and that this would lead to interesting cosmological phenomena [5]. Estimates of transition temperatures for neutrons were made by Ginzburg and Kirzhnits [6], who also pointed out that the heat capacity of matter at temperatures significantly below the transition temperature would be reduced, thereby increasing the cooling rate of the star.

Initially, pairing in the singlet, S-wave state ($^1S_0$) was considered. However, on the basis of calculations of the pairing gap which used interactions deduced from nucleon–nucleon scattering data, Hoffberg et al. showed that, at densities in excess of roughly nuclear density, pairing in a spin-triplet state with unit orbital angular momentum would lead to a lower energy [7]. Explicitly, the state was found to be $^3P_2$, in which the orbital and spin angular momenta are aligned.[3] This takes advantage of the fact that the nuclear spin-orbit interaction is attractive (in contrast to atomic physics, where it is repulsive), thereby favouring parallel spin and orbital angular momenta. The state is closely related to the superfluid states of liquid ^{3}He, which also have unit orbital angular momentum, but with the important difference that for ^{3}He the spin-orbit coupling is much weaker, since it is due to the dipole–dipole interaction between the nuclear magnetic moments of the atoms.

Proton Pairing

In the outer core, the ratio of protons to neutrons is of order 5%. In free space, the interaction of two protons is closely equal to that between two neutrons. This is due to the fact that the effects of electromagnetic interactions are small compared with the nuclear interactions, which are approximately invariant under rotations

[3] As a consequence of the tensor character of nucleon–nucleon interactions, the state contains an admixture of the $^3F_2$ state, but for simplicity we shall refer to the state as $^3P_2$.

in isospin space. If the interaction between two protons in matter were the same as that in free space, this would imply that protons would be paired, with a gap that depended on the proton density in the same way as the neutron pairing gap depended on the neutron Fermi momentum. In other words, the proton gap as a function of the total mass density would have the same basic form as that for neutrons, but shifted to $\sim$ 20 times higher densities. However, this picture is over-simplified, because the interaction between two protons is modified by the presence of the much denser neutron medium. In addition, at these densities the contributions from many-body interactions are expected to be important. A reliable calculation of the proton-pairing gap taking into account both these effects is an important open problem.

29.3.2 Meson Condensates

Pion Condensation

The composition of matter in neutron stars was initially discussed in terms of models which treated the particles as being independent. In particular, Bahcall and Wolf [8] pointed out that pions would be degenerate if their density were high enough. If pions are treated as free particles, negative pions would appear in matter when the electron chemical potential became equal to the pion rest mass. A macroscopic number of pions would then appear in the lowest energy state, forming a coherent state of the pion field. Because of interactions, the picture is more complicated, as described in detail by Baym and Campbell [9]. One of the key findings is that the most energetically favourable pion state is one with nonzero momentum. This is due to the fact that the pion is a pseudoscalar Goldstone boson, and therefore the matrix element for absorption of a pion by a nucleon is proportional to $\boldsymbol{\sigma} \cdot \mathbf{q}$, where $\boldsymbol{\sigma}$ is the nucleon spin operator and $\mathbf{q}$ the pion momentum. For example, the interaction mixes a negative pion with proton-hole–neutron-particle states, thereby decreasing the energy of the pion. Neutron and proton states with momenta differing by the pion momentum are mixed by interaction with the pion field, and the elementary fermionic excitations are linear combinations of neutrons and protons. It is difficult to make reliable predictions of the threshold density for pion condensation because central, tensor and spin–orbit correlations as well as many-body forces are important at supranuclear densities.

Kaon Condensation

Because the mass of the strange quark is greater than that of up and down quarks, strange particles are not present in low-density matter in equilibrium. The mass of the strange quark is roughly 100 MeV, and therefore, when chemical potentials of constituents of matter (relative to their rest masses) are of order 100 MeV and greater, it is relevant to ask whether strange particles could be present. In the normal state, there is a possibility of Σ^- and Λ^0 hyperons being present. Kaplan and Nelson [10] proposed that, because of the attractive interaction between kaons and nucleons predicted by chiral theories, condensates of kaons could appear in neutron stars. In Ref. [10], the calculations were made in the mean-field approximation. Subsequently, it was demonstrated that correlation effects would reduce the attraction between kaons and nucleons, thereby increasing the threshold density [11]. Estimating the threshold density for kaon condensation is challenging because of the paucity of experimental information of interactions of strange particles with other particles, in addition to the difficulties described for pion condensation.

Pairing of Quarks

Up to this point, we have discussed possible Bose-Einstein condensed states of dense matter on the basis of a description in terms of hadronic degrees of freedom. In the regime of very high density, it is more appropriate to use a Fermi gas of quarks as a starting point [12, 13]. It is difficult to quantify the meaning of 'high density' in this context. In the case of high-temperature and zero baryon density, numerical simulations of quantum chromodynamics (QCD) on a space-time lattice show that a transition from hadronic matter to a quark gluon plasma takes place at a temperature of $T_c \simeq 170$ MeV. The transition is a smooth crossover, and the equation of state of the plasma at temperatures $T \gtrsim 1.5T_c$ can be described in terms of quark and gluon quasiparticles. At nonzero baryon density, lattice simulations cannot be carried out because of the fermion sign problem, and as a result there is no reliable information on the phase diagram at high baryon density. Using the mean thermal momentum at T_c as a guide for the Fermi momentum in quark matter near the critical quark chemical potential, we get $\mu_c \simeq 500$ MeV, corresponding to a baryon density $\rho \simeq 10\rho_0$, where $\rho_0 \approx 0.16$ fm^{-3} is the nuclear saturation density. This is quite large, but there are indications that the transition from quark matter to nuclear matter is smooth [14], so that calculations in the high-density limit may provide useful constraints on the behaviour of matter at moderate density.

The presence of a Fermi surface combined with the attractive interaction between quarks implies that the BCS mechanism will lead to Cooper pairing and superfluidity even if the gluon-mediated interaction is weak. This was proposed by Ivanenko and Kurdgelaidze [15] even before the development of QCD. Subsequently, following the understanding of asymptotic freedom, which implies that interactions between quarks become weak at high densities, pairing between quarks of different flavours was considered in Refs. [16, 17], see [18] for a review. Interest in pairing of quarks revived about two decades ago with the appreciation that large pairing gaps can arise in states in which the flavours of quarks are correlated with their colours.

29.4 Recent Developments

In this section, we describe some of the developments during the past two decades. As we shall show, there are a number of points of contact between the physics of dense matter and that of ultracold atomic gases.

29.4.1 Uniform Neutron Matter

As described earlier, at densities above that for neutron drip, matter consists of a lattice of nuclei immersed in an electron gas with an interstitial fluid of neutrons, whose density increases from very low values just above neutron drip to ones approaching nuclear density at the inner edge of the crust. For most of this density range, the nuclei occupy a small fraction of space and to a very good approximation the neutron fluid may be treated as being homogeneous. This neutron fluid cannot be studied directly in the lab, and therefore its properties must be determined theoretically. The interaction between two neutrons in an S-wave state at low energy is described by the scattering length, which is $a_s \approx -18.5$ fm, and by an effective range $r_e \approx 2.7$ fm, which is relevant for the inner crust. Since the interaction is attractive, a low-density neutron gas will be paired in the $^1S_0$ state. In what we shall refer to as the BCS approximation, one assumes that the interaction between two neutrons is the same as the interaction in free space, and the effect of the medium on the interaction and on the single-particle properties is neglected. One then finds that the neutron pairing gap in the dilute limit is given by

$$\Delta_{\rm BCS} = \frac{8}{e^2} E_F \exp\left(-\frac{\pi}{2k_F|a_s|}\right), \tag{29.1}$$

where $E_F = \hbar^2 k_F^2/2m_n$ is the Fermi energy, k_F being the Fermi wavenumber and m_n the neutron mass.

It came as something of a surprise to find that this is a poor approximation even in the limit of low densities. This was well understood long ago by Gor'kov and Melik-Barkhudarov [19] in a paper that was largely unnoticed until attention was drawn to it after the experimental realisation of degenerate atomic Fermi gases [20]. They showed that in the low-density limit the gap is given by

$$\Delta = \left(\frac{2}{e}\right)^{7/3} E_F \exp\left(-\frac{\pi}{2k_F|a_s|}\right), \tag{29.2}$$

which is $(4e)^{-1/3} \approx 0.45$ of its value in the BCS approximation. The suppression is readily understood in terms of the modification of the interaction between two neutrons by the presence of other neutrons. At the lowest densities, the excitations in the medium are particle–hole pairs, which correspond to either density fluctuations or spin fluctuations. Exchange of density fluctuations leads to an attractive interaction, as is well known in the case of conventional superconductors, where the density fluctuation is a lattice phonon. However, exchange of spin fluctuations, i.e., particle–hole pairs with spin 1, gives a repulsive interaction. For exchange of spin fluctuations with spin-projection $m_S = 0$, the interaction between a neutron with spin up and one with spin down is positive, because the two neutrons exchanging the spin fluctuation have opposite spins and therefore couple to the spin fluctuation with opposite signs. For excitations with $m_S = \pm 1$, the interaction is again repulsive because, although the interaction is intrinsically attractive, it corresponds to an exchange interaction in the pairing channel since it reverses the roles of the spin-up and spin-down neutrons undergoing pairing, thereby introducing an additional minus sign.

At higher densities, other many-body effects must be taken into account, and a variety of methods have been used to calculate the properties of neutron matter. The best present-day techniques are a family of Quantum Monte Carlo methods, which are accurate enough at lower densities that it is possible to calculate pairing gaps from the differences between the energies of systems with odd and even particle numbers (For a review of these developments, see [21]). Thanks to these methods, coupled with the analytical results in the low-density limit, the neutron gap is now well understood at densities less than about one-tenth of nuclear density, where the effects of the two-body interactions beyond the S-wave contribution and of three-body interactions may be neglected. The current challenge is to calculate neutron pairing gaps at higher densities at which correlations become stronger and three-body interactions play an important role. (At saturation density, the leading four-body forces in chiral effective field theory have been shown to be very small, an order of magnitude smaller than 3N interactions.)

To calculate gaps for non-S-wave pairing is more difficult, because these depend not on the pairing interaction averaged over angles between the initial and final momenta of the fermions undergoing scattering, as in the case of S-wave pairing, but on deviations of the pairing interaction from this average value. For $^3P_2$ pairing, the role of correlations, in particular from induced spin–orbit interactions, has only been explored at low order in a many-body expansion [22]; these calculations led to very small $^3P_2$ gaps, but the possibilities that the $^3P_2$ gap vanishes or that pairing is stronger in some other channel are not excluded.

29.4.2 Superfluid Neutrons in the Inner Crust

In the inner crust, the neutron superfluid permeates a lattice of nuclei. Low-frequency, long-wavelength phenomena may be described by a two-fluid model, the two fluids corresponding, loosely speaking, to the superfluid neutrons and the nuclei [23]. For a spatially uniform neutron fluid, at zero temperature Galilean invariance leads to the conclusion that the neutron superfluid density is equal to the total neutron density. However, for neutrons in a lattice of nuclei, the neutron superfluid density is less than the total neutron density because of what in the language of quantum liquids is referred to as 'backflow' and in the neutron star literature as 'entrainment'. The two terms reflect different aspects of the problem, the first emphasising the disturbance of the neutron current density caused by the nuclei, the second the fact that part of the neutron density is locked to the nuclei and does not contribute to the superfluid flow.

The neutron superfluid density is an important quantity in models of a number of phenomena observed in pulsars. One is sudden speed-ups (glitches) of the rotation frequency of the pulsar, which in some models is attributed to superfluid neutrons weakly coupled to the crust and other normal parts of the star (for reviews of glitch phenomena, see [24]): here the moment of inertia of the superfluid neutrons, which is directly related to the neutron superfluid density, plays a key role (see, e.g., [25, 26]). Another is in explaining quasiperiodic oscillations observed in the X-ray afterglows of giant flares from highly magnetised neutron stars, a possible model for which is oscillations of the crust of the neutron star [27, 28]. The frequency of these modes is sensitive to the neutron superfluid density in the crust.

To the extent that the lattice is rigid, the system thus resembles a superfluid atomic Fermi gas in a three-dimensional optical lattice. An important difference, however, is that the number of neutrons per unit cell can be as high as $\sim 10^3$. As a consequence of the large number of neutrons per unit cell, a correspondingly large number of bands in the neutron band structure must be taken into account: this represents a considerable challenge. If the effects of neutron pairing are weak, the neutron superfluid density is simply related to the response of a current in the normal state to a vector potential. Chamel has performed mean-field (Hartree-Fock) calculations of the band structure of neutrons in the normal state and finds that the neutron superfluid density can be a factor of 10 or more smaller than the density of neutrons between nuclei, which one might expect to be a reasonable first estimate of the superfluid density [29].

An important open problem is to make improved calculations of the neutron superfluid density that take into account both band structure and pairing. As an indication of the need to include both effects simultaneously, one may mention the fact that in the part of the inner crust where the neutron gap attains its largest values, the coherence length, which is the dimension of a neutron pair, is less than the lattice spacing. The corresponding problem for fermionic atoms with resonant interactions in a one-dimensional optical lattice has been addressed in Ref. [30], and recently Watanabe has extended these calculations to atoms with interactions that are not resonant [31]. What is needed is an extension of this work to three-dimensional lattices and to higher numbers of atoms per unit cell.

29.4.3 A Dilute Solution of Protons in Neutrons

In the outer core of a neutron star, matter is expected to be a uniform fluid of neutrons with a $\sim 5\%$ admixture of protons (together with an equal admixture of electrons to maintain charge neutrality). Since the protons are a minority component and the coupling of neutrons and protons is strong, interactions between protons in the medium are modified by the presence of the neutrons. Dilute mixtures of atoms have been studied extensively, as have dilute solutions of ^{3}He in liquid ^{4}He, where solutions with a concentration of a few per cent have received particular attention.

To illustrate how insights from cold gases may be exploited, let us consider the interaction of two protons with opposite spins induced by the presence of the nuclear medium.[4] Since the neutron medium has both spin and density degrees of freedom, the induced interaction has two contributions, one from exchange of density fluctuations (as in dilute solutions of ^{3}He) and the other from exchange of spin fluctuations (see, e.g., Ref. [34]). A new feature of the proton solutions compared with the other dilute solutions is that tensor forces are important at the relevant densities, and the theory needs to be developed to include them. In addition, the effects of three-body interactions, which are relatively unimportant in atomic gases, play an important role at densities comparable to that of nuclei and above.

29.4.4 Atomic Analogue of Pion Condensation

The production of cold gases of atoms with large magnetic moments has stimulated interest in creating atomic systems in which there is a ‘magnetic field condensation’ akin to the condensation of the pion field. In this state, the atoms exhibit a static spin-density wave, which is accompanied by a spin polarisation of the nucleons. As an example, we mention the creation of cold, trapped dysprosium gases, which have magnetic moments roughly ten times those of alkali atoms [35, 36]. (Dipolar condensates are briefly considered in Chapter 18.) Viewed from the perspective of the particles, the similarity between the two situations is clear: the dipole–dipole interaction between atoms and the one-pion-exchange (OPE) interaction between nucleons both have a tensor structure.

An important difference between the two cases is the sign of the interaction: the electromagnetic dipole–dipole is negative for dipoles with the same orientation

[4] The analogous problem of the interaction of minority fermions in the presence of a majority fermion component has previously been considered in the context of cold atomic gases [32, 33].

arranged head-to-tail, but positive for like dipoles lying side-by-side. While the tensor force between a neutron and a proton in the isospin singlet channel has the same sign as for magnetic diples, the opposite is the case for the spins of neutrons interacting via the one-pion-exchange interaction. For a sufficiently strong dipolar interaction, uniform matter is unstable to formation of a spin-density wave. For dipolar atoms in a one-dimensional structure, the local spin polarisation in the energetically favoured state is perpendicular to the direction of the modulation [37], while for neutrons interacting via the OPE interaction the polarisation is in the same direction as the modulation, as indicated schematically in Fig. 29.2. If the dipolar interaction is weak compared with the central part of the interaction, a ferromagnetic state is predicted to be favourable.

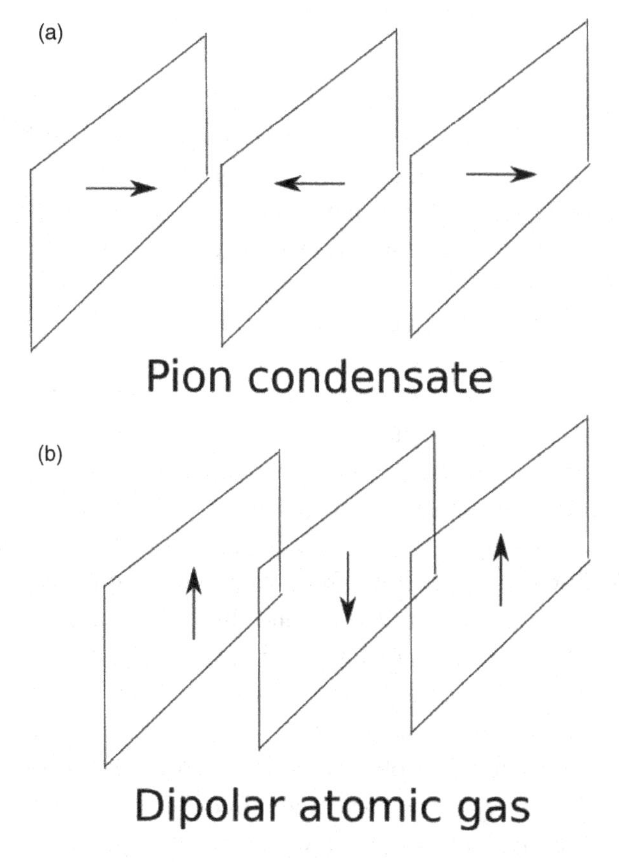

Figure 29.2 Schematic picture of the neutron spin-density wave in a neutral pion condensed state (a) and in an atomic gas with dipolar interactions (b). The magnetisation density is uniform on the planes indicated and varies smoothly in between. Due to nonlinear effects, the spin-density wave will be accompanied by a density wave with a wave vector equal to twice that of the spin-density wave. For this reason, the phases are referred to as smectic because of their resemblance to smectic liquid crystals.

The magnetic dipolar interaction differs from the OPE interaction in that it is long range. As a consequence, in finite geometry, such as a trapped atomic cloud, the equilibrium state will be influenced by the energy of the magnetic field outside the cloud. As an example, consider matter which in bulk is predicted to be ferromagnetic: in a cloud, the direction of the magnetisation will vary in direction in order to reduce the magnetic field energy outside the cloud, an effect analogous to formation of domains in a solid ferromagnet. However, because of the absence of the lattice, the thickness of the domain walls is comparable to the size of a domain. Further work is required to determine in detail the structure of dipolar atomic gases, as a function of the strengths of the central and dipolar parts of the interaction, the strength of the magnetic field and the finite extent of the atomic cloud.

29.4.5 Quark Matter

Normal State

Even in the normal state, quark matter is an interesting system because it is an example of a marginal Fermi liquid: the velocity of a quasiparticle with momentum $\mathbf{p}$ close in magnitude to the Fermi momentum, p_F, tends to zero as $p \to p_F$, whereas in a normal Fermi liquid the velocity tends to a constant. The reason for this unusual behaviour is that, because gluons are massless, the gluon-exchange interaction is long range. The interaction is made up of colour-electric (longitudinal) and colour-magnetic (transverse) contributions in essentially the same way as for the electric and magnetic contributions to the interaction between two electrical charges. The coupling of gluons to quarks is governed by a coupling constant g that depends on the momentum transfer. The sign of g, like that of the charge of the electron, is a matter of convention, and we take it to be positive. The dimensionless quantity $g^2/\hbar c$ is the QCD analogue of the fine structure constant $\alpha = e^2/\hbar c$ in QED. Asymptotic freedom implies that the typical interaction between quarks decreases as the density and the Fermi momentum increase. If one neglects the effects of the medium on the gluons, the quark self-energy due to emission and absorption of a gluon diverges. In a medium, the electric part of the interaction is cut off at wavenumbers less than the screening wavenumber $k_D \sim g p_F$ by Debye screening.[5] However, because quarks do not possess a magnetic charge, magnetic fields are not screened at zero frequency, but at nonzero frequency ω, magnetic interactions are cut off at wavenumbers less than $(g^2|\omega|p_F^2)^{1/3}$ by Landau damping. In more technical terms, the transverse gluon propagator has the form [38]

[5] To avoid making formulae cumbersome, in the remainder of this section we use units in which c, $\hbar$ and the Boltzmann constant k_B are equal to unity.

$$D_{ij}(\omega, \mathbf{k}) = \frac{\delta_{ij} - \hat{k}_i \hat{k}_j}{\omega^2 - k^2 + i(\pi/4)k_D^2 \omega/k}, \tag{29.3}$$

for $|\omega| \leq k$. In the physics of normal metals, the analogous phenomenon for electrodynamics is known as the anomalous skin effect: when the mean free path of an electron becomes larger than the wavelength of an applied magnetic field, the length relevant in determining the response is the wavelength rather than the mean free path.

So far, we have concentrated on the interaction between quarks and have assumed that near the Fermi surface in the normal state the spectrum is linear in $p - p_F$. However, from a one-loop calculation using Eq. (29.3) for the one-gluon transverse propagator, one finds that the quark propagator behaves as

$$S(\nu, \mathbf{p}) = \frac{Z_F}{\nu - v_F(p - p_F)}, \tag{29.4}$$

with $Z_F \sim v_F \sim [g^2 \log(k_D/|\nu|)]^{-1}$. Thus the residue of the quasiparticle pole in the propagator and the quasiparticle velocity vanish on the would-be Fermi surface.[6] One might suspect that this breakdown of Fermi liquid theory would severely affect estimates of gaps in superfluid phases, but this turns out not to be the case, as we shall describe below.

Order Parameter

We turn now to the pairing interaction between quarks. Since quarks are nearly massless, the system is relativistic and $p_F \simeq \mu_q$, where $\mu_q \simeq \mu_B/3$, is the quark chemical potential and μ_B the baryon chemical potential. In relativistic theory, fermions of a particular species are described by four-component (Dirac) fields rather than the two-component spinors corresponding to the two spin components in nonrelativistic theory. The fields describing quarks, $q^a_{\alpha f}$, are thus labelled by a Dirac index α, in addition to a colour index a (red, blue or green) and a flavour index f (up, down or strange).[7] The interaction between quarks is mediated by gluon fields A^c_μ, which couple to the colour charges of the quarks. Here, $c = 1, \ldots, 8$ labels traceless hermitean 3×3 matrices that act on the colour indices of the quarks. The one-gluon exchange interaction is attractive in the colour-antisymmetric quark–quark channel, and therefore, on the basis of the analogy with the nonrelativistic problem, one would expect quark matter to be unstable to formation of a state with quark pairs (diquarks). These pairs may be viewed as building blocks in the formation of

[6] The analogous effect for the electron gas, in that case due to exchange of transverse *photons*, was investigated long ago in Ref. [39]. However, in a nonrelativistic electron gas, the effects of transverse photons are of order v_F^2/c^2 smaller than those due to the Coulomb interaction and are not observable experimentally.

[7] Quarks of other flavours have much higher masses and would appear only at densities considerably higher than those anticipated to occur in neutron stars.

baryons: diquarks $D_a = \epsilon_{abc} q^b q^c$ can bind with quarks q^a to form colour neutral baryons $B = D_a q^a$.

The order parameter is of the form

$$\langle q^a_{\alpha f} (C\Gamma)^{\alpha\beta} q^b_{\beta g} \rangle = \phi^{ab}_{fg} , \tag{29.5}$$

where C is the charge conjugation matrix and Γ is a Dirac matrix.[8] There are many possible channels, and the ground state will depend on the flavour composition and the strength of the interaction. At very high density, matter is approximately flavour symmetric, and asymptotic freedom implies that the ground state can be determined in weak coupling QCD. One finds [40, 41]

$$\langle q^a_{\alpha f} (C\gamma_5)^{\alpha\beta} q^b_{\beta g} \rangle = \phi \left(\delta^a_f \delta^b_g - \delta^a_g \delta^b_f \right) , \tag{29.6}$$

which is called the colour–flavour locked (CFL) state.[9] The order parameter is antisymmetric in colour, as required by the interaction, and antisymmetric in both flavour and spin. This implies that the quark pairs form spin singlets. As the density is lowered, the larger mass of the strange quark becomes more important, flavour symmetry is broken and states with a smaller spin–flavour symmetry may appear.

The colour–flavour locked state spontaneously breaks the chiral symmetry of the QCD Lagrangian and exhibits low-energy excitations with the quantum numbers of pions and kaons. This can be seen as follows: The relation $C\gamma_5 = C(P_L - P_R)$, where $P_{L,R}$ are projectors on left- and right-handedness, implies a fixed phase relation between the left- and right-handed components of the diquark condensate. The colour and flavour orientations of these condensates can be characterised by the matrices

$$X^a_f = \epsilon^{abc} \epsilon_{fgh} \langle (q_L)^b_g (q_L)^c_h \rangle, \qquad Y^a_f = \epsilon^{abc} \epsilon_{fgh} \langle (q_R)^b_g (q_R)^c_h \rangle , \tag{29.7}$$

where the ϵ-tensors take into account the antisymmetry of the CFL state under exchange of colour and of flavour. The quantities X^a_f and Y^a_f depend on the gauge choice, but one can define a gauge invariant order parameter for chiral symmetry breaking by the relation $\Sigma^g_f = X^a_f (Y^\dagger)^g_a$.

The CFL ground state corresponds to $\Sigma \sim \mathbf{1}$, and low-energy modes with the quantum numbers of pions and kaons are described by oscillations of Σ in the pion direction, $\Sigma \sim \lambda^{1,2,3}$, or the kaon direction, $\Sigma \sim \lambda^{4-7}$. Here λ^a are the Gell-Mann matrices, a generalisation of the isospin matrices to flavour $SU(3)$.

[8] This expression may be regarded as the generalisation to relativistic particles of the Nambu formalism for superconductivity. There one works with (two-component) Pauli spinors and the order parameter is give in terms of spinor field operators ψ_σ by $\langle \psi_\sigma \sigma_i \psi_{\sigma'} \rangle$, where σ_i is a Pauli matrix. For example, with $\Gamma = \gamma_5$ the order parameter (29.5) corresponds to the relativistic way of writing down singlet pairs, since the nonrelativistic limit of $C\gamma_5$ is σ_2.

[9] This is the simplest choice for the order parameter and corresponds to a particular gauge choice. A more general form may be obtained by performing a unitary transformation of the colour variables, which will leave unaltered variables that can be measured experimentally.

In response to quark masses and lepton chemical potentials, the CFL ground state can become polarised in the pion or kaon directions in flavour space. A nonzero strange quark mass, for example, favours a ground state in which there are fewer strange quarks than up and down quarks. This can be realised by a neutral kaon condensate $\Sigma \sim \exp(i\alpha\lambda^6)$ [42]. This is a homogeneous condensate, analogous to the kaon condensate discussed from the point of view of hadronic matter in Section 29.3.2. For larger strange quark masses or lower densities, interactions between kaons and gapless fermion modes can lead to the formation of a standing wave kaon condensate [43]. This phase is analogous to the pion condensed state discussed above and to standing wave ground states in Bose-Einstein condensed atomic gases; see [44, 45]. Fermion quasiparticles in these systems are quarks surrounded by a diquark polarisation cloud [46], similar to the state $B = D_a q^a$ described above. As the strength of the interaction increases, these states can continuously evolve into tightly bound baryons. A similar crossover can be studied in cold atomic gases of three fermionic species [47].

Gaps

Superfluid gaps in quark matter have a very different dependence on coupling compared with dilute atomic gases. Since the gluon exchange interaction is of order g^2, on the basis of the calculations for a dilute atomic gas, Eq. (29.2), one might have expected $\Delta \sim \mu_q e^{-\mathrm{const.}/g^2}$. In fact, for weak coupling one finds [48, 49, 50, 51, 41]

$$\Delta \simeq C_1 \frac{\mu_q}{g^5} \exp\left(-C_2/g\right), \tag{29.8}$$

where

$$C_1 = \frac{512\pi^4}{2^{1/3}} \left(\frac{2}{3}\right)^{5/2} e^{-(\pi^2+4)/8}, \quad \text{and} \quad C_2 = \frac{3\pi^2}{\sqrt{2}}. \tag{29.9}$$

The reason for the difference lies in the long-range character of the gluon exchange interaction, as discussed above. Transverse gluons do not upset the basic mechanism of the BCS instability, but they do modify the magnitude of the gap. The gap equation has two logarithmic infrared divergences, one related to the BCS mechanism and one caused by the unscreened transverse gluon exchange. The coefficient of the $1/g$ term in the exponent is determined by transverse gluon exchange, and the exponential term in C_1 is related to the wave function renormalisation Z_F discussed above. As one sees, the fact that, in the normal state, quark matter is a marginal Fermi liquid has only a modest effect on the gap. The pre-exponential factor is set by electric gluon exchanges and is sensitive to the symmetries of the order parameter. For $\mu_q \simeq 500$ MeV, the gap is of order 10–20 MeV.

Superfluidity and Superconductivity

In the CFL phase, the densities of u-, s- and d quarks are equal and, consequently, in bulk matter there is no need for other particles, such as electrons, to ensure electrical neutrality. The CFL phase is a superfluid but an electrical insulator. Heuristically, this may be understood as being a consequence of the fact that the sum of the charges of the three flavours of quarks vanishes.[10]

When the flavour symmetry is broken, the net charge density of the quarks is nonzero, and this is compensated by a background of electrons. In this case, matter becomes a charged superfluid and is an electrical superconductor.

29.5 Observational Considerations

We now comment briefly on some observational consequences of Bose-Einstein condensation in neutron stars. Models of glitches and X-ray flares that involve superfluid flow have been mentioned in Section 29.4.2. A more extensive account of observational effects may be found in Ref. [52].

29.5.1 Neutron Star Structure

The mass–radius relation for neutron stars depends on the equation of state of matter, which is affected by Bose-Einstein condensation. In the case of pairing of nucleons and quarks, the predicted gaps are small compared with Fermi energies and, consequently, pairing has only a small effect on structure. For meson condensates, the picture could be very different because, if it were energetically favourable to create a pion or kaon condensate, the energy would be reduced compared with that of the system with no condensate, and matter will become softer. This would lead to a lower maximum mass of a neutron star. The observation within the past few years of two neutron stars with masses approximately twice that of the sun with small error bars [53, 54] is most simply understood in terms of an equation of state that is similar to what is expected for models based on nucleon degrees of freedom, without significant softening due to formation of a meson condensate. However, models of dense matter based on quark degrees of freedom can, with appropriate choice of parameters, give equations of state stiff enough to account for observed neutron star masses.

29.5.2 Neutron Star Cooling

For the first 10^5–10^6 years after formation, a neutron star cools primarily by neutrino emission from its core. In favourable cases, emission of X-rays from

[10] This argument is an oversimplification, as is explained in Section II A.3 of Ref. [18].

the surface of the star can be detected and the temperature of the core deduced. Since the rate of neutrino emission in dense matter depends on the microscopic properties of the matter, observations of cooling therefore have potential for providing information about the stellar interior (for a review, see [55]). The path from the microscopic properties of dense matter to observed cooling curves has many steps, but the past two decades have seen a steady increase in sophistication of the modelling and the precision of observations.

As an illustration of the potential such observations have for pinning down the properties of dense matter, we give a recent example. Pairing of nucleons reduces neutrino emission rates at temperatures well below the transition temperature because of the reduction in the number of thermal excitations. However, just below the transition temperature, emission of neutrino–antineutrino pairs can occur by annihilation of two thermal excitations in the superfluid. In the normal state, the process is forbidden by conservation of baryon number, but in the superfluid, excitations are linear combinations of normal-state particles and holes and the process is allowed. This process leads to accelerated cooling for a limited temperature range. The reported detection of changes over a period of 10 years in the surface temperature of the neutron star produced in the Cassiopeia A supernova approximately 330 years ago has been interpreted as being due to neutrons in the core undergoing a transition to a state with $^3P_2$ pairing [56, 57]. While the analysis of the observations has been called into question, this case brings out the promise that such observations have for shedding light on pairing in neutron star cores.

29.6 Concluding Remarks

The study of dense matter in neutron stars has profited from work on terrestrial quantum liquids and ultracold gases. In addition, the studies of dense matter have provided inspiration for experiments in the laboratory. In the future, one may expect this synergy to continue.

Within the next few years, one may expect significant improvements in calculations of properties of matter at densities of order nuclear matter density by the application of chiral effective field theories [58]. Chiral effective field theory, which incorporates the symmetries of quantum chromodynamics, provides a systematic framework for treating interactions between nucleons and, when extended to strange particles, hyperons. Using a momentum expansion scheme, the theory includes the long-range interactions due to the exchange of pions (and kaons) explicitly and general contact interactions for the shorter-range parts. Moreover, it makes predictions for consistent three- and higher-body interactions.

There are also aspects of superfluidity in neutron stars that are relatively little studied, among which one may mention superconductivity of protons in the pasta phases. The appreciation of the fact that defects can occur in these phases and thereby lead to multiply connected structures [59] implies that the proton superfluid may exhibit topologically nontrivial structures (flux lines) that are trapped by the defects.

Acknowledgements: We are grateful to Dmitry Kobyakov for helpful conversations and to Gentaro Watanabe for communicating to us his results for the superfluid density of an atomic Fermi gas in a one-dimensional optical lattice. This work was supported in part by the NewCompStar network, COST Action MP1304, the US Department of Energy grant DE-FG02-03ER41260 and by the ERC Grant No. 307986 STRONGINT.

References

[1] Griffin, A., Snoke, D. W., and Stringari, S. (eds). 1995. *Bose-Einstein Condensation*. Cambridge: Cambridge University Press.

[2] Pethick, C. J., and Ravenhall, D. G. 1995. Matter at large neutron excess and the physics of neutron star crusts. *Annu. Rev. Nucl. Part. Sci.*, **45**, 429–484.

[3] Hebeler, K., Lattimer, J. M., Pethick, C. J., and Schwenk, A. 2013. Equation of state and neutron star properties constrained by nuclear physics and observation. *Ap. J.*, **773**, 11.

[4] Bardeen, J., Cooper, L. N., and Schrieffer, J. R. 1957. Theory of superconductivity. *Phys. Rev.*, **108**, 1175–1204.

[5] Migdal, A. B. 1960. Superfluidity and the moments of inertia of nuclei. *Sov. Phys. JETP*, **10**, 176.

[6] Ginzburg, V. L., and Kirzhnits, D. A. 1965. On the superfluidity of neutron stars. *Sov. Phys. JETP*, **20**, 1346–1348.

[7] Hoffberg, M., Glassgold, A. E., Richardson, R. W., and Ruderman, M. 1970. Anisotropic superfluidity in neutron star matter. *Phys. Rev. Lett.*, **24**, 775–777.

[8] Bahcall, J. N., and Wolf, R. A. 1965. Neutron stars. I. Properties at absolute zero temperature. *Phys. Rev.*, **140**, B1445–1451.

[9] Baym, G., and Campbell, D. K. 1978. Chiral symmetry and pion condensation. Pages 1031–1094 of: Rho, M., and Wilkinson, D. H. (eds), *Mesons in Nuclei*, vol. III. Amsterdam: North Holland.

[10] Kaplan, D. B., and Nelson, A. E. 1986. Strange goings on in dense nucleonic matter. *Phys. Lett. B*, **175**, 57–63.

[11] Pandharipande, V. R., Pethick, C. J., and Thorsson, V. 1995. Kaon energies in dense matter. *Phys. Rev. Lett.*, **75**, 4567–4570.

[12] Collins, J. C., and Perry, M. J. 1975. Superdense matter: neutrons or asymptotically free quarks? *Phys. Rev. Lett.*, **34**, 1353–1356.

[13] Baym, G., and Chin, S. A. 1976. Can a neutron star be a giant MIT bag? *Phys. Lett. B*, **62**, 241–244.

[14] Schäfer, T., and Wilczek, F. 1999. Continuity of quark and hadron matter. *Phys. Rev. Lett.*, **82**, 3956–3959.

[15] Ivanenko, D., and Kurdgelaidze, D. F. 1969. Remarks on quark stars. *Nuovo Cim. Lett.*, **2**, 13–16.

[16] Barrois, B. C. 1977. Superconducting quark matter. *Nucl. Phys. B*, **129**, 390–396.

[17] Bailin, D., and Love, A. 1979. Superfluid quark matter. *J. Phys. A Math. Gen.*, **12**, L283–L289.
[18] Alford, M. G., Schmitt, A., Rajagopal, K., and Schäfer, T. 2008. Color superconductivity in dense quark matter. *Rev. Mod. Phys.*, **80**, 1455–1515.
[19] Gor'kov, L. P., and Melik-Barkhudarov, T. K. 1961. Contribution to the theory of superfluidity in an imperfect Fermi gas. *Sov. Phys. JETP*, **13**, 1018–1022.
[20] Heiselberg, H., Pethick, C. J., Smith, H., and Viverit, L. 2000. Influence of induced interactions on the superfluid transition in dilute Fermi gases. *Phys. Rev. Lett.*, **85**, 2418–2421.
[21] Gezerlis, A., Pethick, C. J., and Schwenk, A. 2014. Pairing and superfluidity of nucleons in neutron stars. Pages 580–615 of: Bennemann, K. H., and Ketterson, J. B. (eds), *Novel Superfluids*, vol. 2. Oxford: Oxford University Press.
[22] Schwenk, A., and Friman, B. 2004. Polarization contributions to the spin-dependence of the effective interaction in neutron matter. *Phys. Rev. Lett.*, **92**, 082501.
[23] Pethick, C. J., Chamel, N., and Reddy, S. 2010. Superfluid dynamics in neutron star crusts. *Prog. Theor. Phys. Suppl.*, **186**, 9–16.
[24] Haskell, B., and Melatos, A. 2015. Models of pulsar glitches. *Int. J. Mod. Phys. D*, **24**, 1530008.
[25] Andersson, N., Glampedakis, K., Ho, W. C. G., and Espinoza, C. M. 2012. Pulsar glitches: the crust is not enough. *Phys. Rev. Lett.*, **109**, 241103.
[26] Chamel, N. 2013. Crustal entrainment and pulsar glitches. *Phys. Rev. Lett.*, **110**, 011101.
[27] Duncan, R. C. 1998. Global seismic oscillations in soft gamma repeaters. *Ap. J. Lett.*, **498**, L45–L49.
[28] Steiner, A. W., and Watts, A. L. 2009. Constraints on neutron star crusts from oscillations in giant flares. *Phys. Rev. Lett.*, **103**, 181101.
[29] Chamel, N. 2012. Neutron conduction in the inner crust of a neutron star in the framework of the band theory of solids. *Phys. Rev. C*, **85**, 035801.
[30] Watanabe, G., Orso, G., Dalfovo, F., Pitaevskii, L. P., and Stringari, S. 2008. Equation of state and effective mass of the unitary Fermi gas in a one-dimensional periodic potential. *Phys. Rev. A*, **78**, 063619.
[31] Watanabe, G. private communication.
[32] Mora, C., and Chevy, F. 2010. Normal phase of an imbalanced Fermi gas. *Phys. Rev. Lett.*, **104**, 230402.
[33] Yu, Z., Zöllner, S., and Pethick, C. J. 2010. Comment on 'Normal phase of an imbalanced Fermi gas'. *Phys. Rev. Lett.*, **105**, 188901.
[34] Baldo, M., and Schulze, H.-J. 2007. Proton pairing in neutron stars. *Phys. Rev. C*, **75**, 025802.
[35] Lu, M., Youn, S. H., and Lev, B. L. 2010. Trapping ultracold dysprosium: a highly magnetic gas for dipolar physics. *Phys. Rev. Lett.*, **104**, 063001.
[36] Lu, M., Burdick, N. Q., and Lev, B. L. 2012. Quantum degenerate dipolar Fermi gas. *Phys. Rev. Lett.*, **108**, 215301.
[37] Maeda, K., Hatsuda, T., and Baym, G. 2013. Antiferrosmectic ground state of two-component dipolar Fermi gases: an analog of meson condensation in nuclear matter. *Phys. Rev. A*, **87**, 021604.
[38] Baym, G., Monien, H., Pethick, C. J., and Ravenhall, D. G. 1990. Transverse interactions and transport in relativistic quark-gluon and electromagnetic plasmas. *Phys. Rev. Lett.*, **64**, 1867–1870.
[39] Holstein, T., Norton, R. E., and Pincus, P. 1973. de Haas-van Alphen effect and the specific heat of an electron gas. *Phys. Rev. B*, **8**, 2649–2656.

[40] Alford, M. G., Rajagopal, K., and Wilczek, F. 1999. Color flavor locking and chiral symmetry breaking in high density QCD. *Nucl. Phys. B*, **537**, 443–458.
[41] Schäfer, T. 2000. Patterns of symmetry breaking in QCD at high baryon density. *Nucl. Phys. B*, **575**, 269–284.
[42] Bedaque, P. F., and Schäfer, T. 2002. High-density quark matter under stress. *Nucl. Phys. A*, **697**, 802–822.
[43] Schäfer, T. 2006. Meson supercurrent state in high-density QCD. *Phys. Rev. Lett.*, **96**, 012305.
[44] Son, D. T., and Stephanov, M. A. 2006. Phase diagram of cold polarized Fermi gas. *Phys. Rev. A*, **74**, 013614.
[45] Radzihovsky, L., and Sheehy, D. E. 2010. Imbalanced Feshbach-resonant Fermi gases. *Rep. Prog. Phys.*, **73**, 076501.
[46] Kryjevski, A., and Schäfer, T. 2005. An effective theory for baryons in the CFL phase. *Phys. Lett. B*, **606**, 52–58.
[47] Rapp, R., Zarand, G., Honerkamp, C., and Hofstetter, W. 2007. Color superfluidity and 'baryon' formation in ultracold fermions. *Phys. Rev. Lett.*, **98**, 160405.
[48] Son, D. T. 1999. Superconductivity by long-range color magnetic interaction in high-density quark matter. *Phys. Rev. D*, **59**, 094019.
[49] Schäfer, T., and Wilczek, F. 1999. Superconductivity from perturbative one-gluon exchange in high density quark matter. *Phys. Rev. D*, **60**, 114033.
[50] Pisarski, R. D., and Rischke, D. H. 2000. Color superconductivity in weak coupling. *Phys. Rev. D*, **61**, 074017.
[51] Brown, W. E., Liu, J. T., and Ren, H.-c. 2000. Perturbative nature of color superconductivity. *Phys. Rev. D*, **61**, 114012.
[52] Page, D., Lattimer, J. M., Prakash, M., and Steiner, A. W. 2014. Stellar superfluids. Pages 505–579 of: Bennemann, K. H., and Ketterson, J. B. (eds), *Novel Superfluids*, vol. 2. Oxford: Oxford University Press.
[53] Demorest, P. B., Pennucci, T., Ransom, S. M., Roberts, M. S. E., and Hessels, J. W. T. 2010. A two-solar-mass neutron star measured using Shapiro delay. *Nature*, **467**, 1081–1083.
[54] Antoniadis, J., Freire, P. C. C., Wex, N., Tauris, T. M., Lynch, R. S., van Kerkwijk, M. H., Kramer, M., Bassa, C., Dhillon, Vik, S., Driebe, T., Hessels, J. W. T., Kaspi, V. M., Kondratiev, V. I., Langer, N., Marsh, T. R., McLaughlin, M. A., Pennucci, T. T., Ransom, S. M., Stairs, I. H., van Leeuwen, J., Verbiest, J. P. W., and Whelan, D. G. 2013. A massive pulsar in a compact relativistic binary. *Science*, **340**, 1233232.
[55] Yakovlev, D. G., and Pethick, C. J. 2004. Neutron star cooling. *Annu. Rev. Astron. Astrophys.*, **42**, 169–210.
[56] Page, D., Prakash, M., Lattimer, J. M., and Steiner, A. W. 2011. Rapid cooling of the neutron star in Cassiopeia A triggered by neutron superfluidity in dense matter. *Phys. Rev. Lett.*, **106**, 081101.
[57] Shternin, P. S., Yakovlev, D. G., Heinke, C. O., Ho, W. C. G., and Patnaude, D. J. 2011. Cooling neutron star in the Cassiopeia A supernova remnant: evidence for superfluidity in the core. *Mon. Not. Roy. Astron. Soc.*, L108–L112.
[58] Epelbaum, E., Hammer, H.-W., and Meissner, U.-G. 2009. Modern theory of nuclear forces. *Rev. Mod. Phys.*, **81**, 1773–1825.
[59] Schneider, A. S., Berry, D. K., Briggs, C. M., Caplan, M. E., and Horowitz, C. J. 2014. Nuclear 'waffles'. *Phys. Rev. C*, **90**, 055805.

30

A Simulated Cosmological Metric: The Superfluid ^{3}He Condensate

GEORGE R. PICKETT

Department of Physics, Lancaster University, UK

Since superfluid ^{3}He in the zero temperature limit is a pure condensate, all the constituent particles obey the same wavefunction, equivalent to a global set of field equations acting on all. Serendipitously, the symmetries broken to create the superfluid are also very similar to those broken by the metric of the universe soon after the Big Bang. The superfluid thus provides a powerful 'tabletop' medium for simulating cosmological processes. We explain here the analogies and describe our exploiting them in experiments to simulate cosmic string formation and brane annihilation in the early universe, yielding insight into what are otherwise experimentally inaccessible processes.

30.1 Introduction

The superfluid transition in liquid ^{3}He can be regarded as a Bose-Einstein condensation of the fermionic Cooper pairs (or a variation thereof) resulting in a dense condensate, at the other end of the scale from the tenuous cold-gas equivalents. While the cold gases provide interesting laboratories for looking at problems of wider relevance, their limited size and densities can be a disadvantage. The superfluid ^{3}He condensate, while less straightforward to achieve, has a number of beneficial properties which make it ideal for many experiments.

The analogies between superfluid dynamics and the behaviour of pure Euler liquids have long been apparent. However, the superfluid case is complicated, at medium temperatures ($0 < T < T_c$), by the presence of 'normal' fluid, i.e., the residual gas of those unpaired particles not contributing to the condensate. Superfluid ^{3}He is simpler in this regard than superfluid ^{4}He in that the condensate (or superfluid fraction) can be identified with the Cooper pairs and the normal fluid fraction with the remaining unpaired ^{3}He atoms. In the case of our experiments, we avoid this latter complication by working at the zero-temperature limit, where the normal fraction is negligible. In this region, the superfluid flow does indeed correspond to that of an Euler fluid and is irrotational. However, we can simulate rotation by introducing vortices, or line defects, along the cores of which the condensation is suppressed. Around these vortices, circulation is allowed, since the local liquid flow remains irrotational. In the superfluid, there is an added constraint that since the flow is generated by a gradient in the phase of the wavefunction, ϕ, the phase change around any loop, $\Delta\phi$, is restricted to modulo 2π and thus any vortices must be quantised.

However, in the present context we are more interested in analogue properties. Coherent condensates have delocalised component particles and the system can be described by a single wavefunction governing the whole liquid. In other words, the system is determined not by a local description of each particle but by a global set of field equations. The important point here is that while this is extremely unusual in 'material' systems, there are many others where a general field-equation description is the rule. In the context of this chapter, we are interested in the extensions to cosmology. The fact that our superfluid ^{3}He condensate (or vacuum, in the appropriate language) is governed by our field equations immediately gives us access to cosmological equivalents governed by very similar field equations. Consequently, we can utilise our condensate as an equivalent of the cosmological metric.

To see how this analogy operates, following the work of Volovik [1], we use the broken symmetry description of the condensate. (For issues related to symmetry breaking in condensates, see Chapter 5.) We begin by considering the evolution of the universe after the Big Bang. We cannot repeat the Big Bang experiment; we can only construct a model and compare the outcomes with what we observe today. What we can currently observe, however, is the behaviour of the four fundamental forces, in order of decreasing magnitude: the strong force, the weak force, electromagnetism and gravity. Neglecting for the moment gravity, which does not fit comfortably in this scheme and is in any case not easy to manipulate, we note that the other three converge in magnitude if measured at higher and higher energies (or equivalently shorter and shorter timescales). This leads to the conclusion that at some early point in the evolution of the universe these forces were undifferentiated. We already have the theory that unifies the weak force and electromagnetism, thus we may surmise that there was initially a single undifferentiated 'force'. As the energy scale of the universe fell, at some point there must have been a phase transition when the strong force became differentiated, followed by another which led to the separation of the weak force and electromagnetism. The differentiation of the strong force is associated with the breaking of symmetry SU(3), the weak force with SU(2) and electromagnetism with choosing a gauge which breaks U(1). What the symmetry of the initial Grand Unified Theory state was is more problematic. Anyway, for present purposes it does not matter. We can take it as SU(5) and thus the whole cascade of transitions at each force differentiation can be represented by

SU(5)→SU(3)×SU(2)×U(1). To give a picture of what these symmetries mean, we note that breaking U(1) is equivalent to choosing a point around a circle, or in other words a direction in two-space. SU(2), governing the weak force, at its simplest, toggles (for example) an up-quark into a down-quark while simultaneously toggling an electron into a neutrino, in other words acting as a reversing switch. We have not been able to find a simple analogy for SU(3). Anyway, in brief, the broken symmetries of the universe (neglecting gravity) can be expressed in the most simple form by SU(3)×SU(2)×U(1).

Now let us look at the helium superfluids, beginning with superfluid ^{4}He. The constituent particles (^{4}He atoms) are spherically symmetric, with no angular momentum, either in the nucleus or in the surrounding electron cloud, with no significant internal structure other than size, and are labelled only with a simple mass. Thus the superfluid properties involve only mass flow. When the system passes through the superfluid transition, the liquid must choose a value for the quantum mechanical phase ϕ, which may take any value from 0 to 2π. In other words, the liquid is forced to choose a direction in 2-space and, in the language above, that means breaking $U(1)$.

Superfluid ^{3}He is more complex. The component Cooper pairs consist of two ^{3}He atoms each with a nuclear spin of 1/2, and thus the pair must have a total nuclear spin S of either 1 or 0. The p-wave state, with $S = 1$, is preferred (since it minimises the probability of the two ^{3}He atoms occupying the same position). To preserve symmetry, the angular momentum then must be odd and, in the ground state, the pair angular momentum is in fact $L = 1$. The pairs thus behave rather as loose dimers with the two component atoms remotely orbiting each other.

The Cooper pairs thus have very much a structure and in consequence the superfluid can exist in several distinct phases (in the phase-diagram sense). However, at its most general, the condensate wavefunction has to choose a phase, i.e., break U(1); the nuclear spin has to point somewhere, i.e., choose a direction in three-space, i.e., break the relevant symmetry SO(3); and the angular momentum of the mutually orbiting pair has also to choose a direction breaking a further SO(3). Thus when liquid ^{3}He passes through the superfluid transition, the wavefunction breaks the symmetries SO(3)×SO(3)×U(1). Now this is clearly not the same as the broken symmetries of the universe, SU(3)×SU(2)×U(1), but the parallels are striking. The comparison is illustrated in Fig. 30.1, where the evolution of the universe in the early stages is presented as a cooling process (indicated by the equivalent temperature) through the various symmetry-breaking transitions, alongside the similar processes occurring when the two heliums, ^{4}He and ^{3}He, cool through their superfluid transitions, presented in the same way.

Since superfluid ^{3}He is a complex subject, we outline here only enough background for the following discussion. Liquid ^{3}He is very close to being

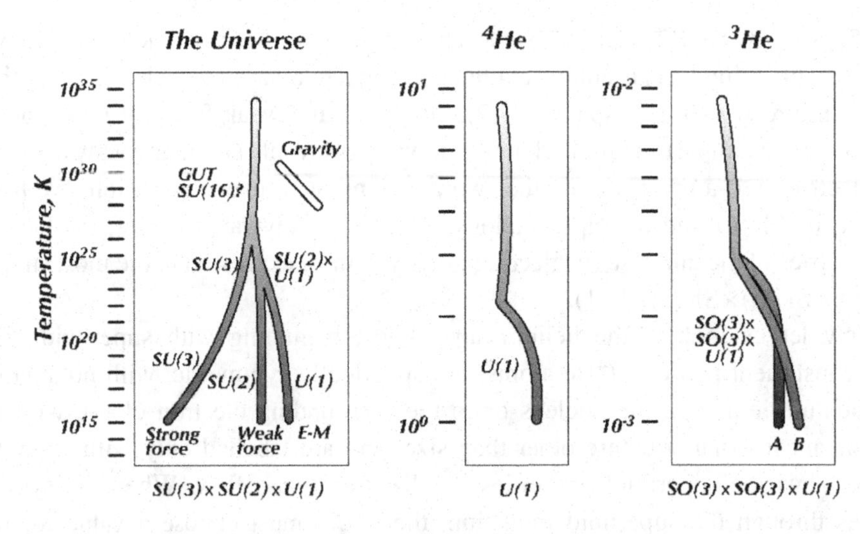

Figure 30.1 A comparison of the symmetry-breaking processes undergone by the universe in the early stages, as a function of temperature, with the similar processes as ^{4}He and ^{3}He cool through the superfluid transition. The processes are similar; only the temperature scales are somewhat different.

ferromagnetic at low temperatures, and at the transition temperature, spin fluctuations in the liquid bind two ^{3}He atoms together to form the Cooper pair. The transition only occurs at milliKelvin temperatures and is pressure dependent (since the pairing action is stronger at higher density), ranging from 2.5 mK at 34 bar to only 0.94 mK at zero pressure. It is interesting to note that while the pairs are bound by the indirect attraction of the nuclear spins of the two component atoms, those familiar with nuclear magnetism will know that nuclear–nuclear spin interactions have strengths in temperature units of the order of microKelvin, not milliKelvin. This is an immediate indication that we are dealing with a coherent condensate rather than an ensemble of independent pairs. Since all the Cooper pairs are in the same state, in a very real sense we are not just coupling spin 1 with spin 2 in each pair individually, but rather coupling coherently all the spin 1s with the all spin 2s of all the pairs within a coherence length.

Since the pairs have structure, this manifests itself in the properties of the condensate, and the liquid exhibits various directionalities. Examples include the directions of the spin and angular momentum vectors, which are global properties since all pairs are in the same state. The superfluid thus behaves in many ways as a liquid crystal, but here the directionalities are more fundamental and precise. The most commonly occurring superfluid phases are the A phase and the B phase. These have very different gap structures, as shown in Fig. 30.2, which will be important later, but without going into the details we note that the B phase has an isotropic

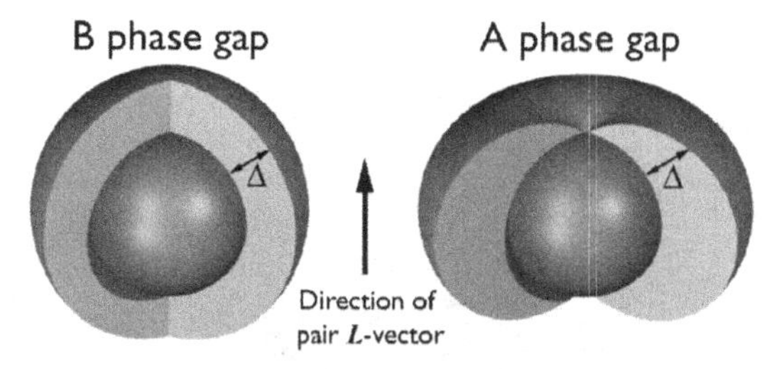

Figure 30.2 The energy gap structure of the two common phases of superfluid ^{3}He. In the B phase, the gap is isotropic and has the same size in all directions. In the A phase, the gap has zero value nodes at the 'poles' which are aligned along thc L-vcctor.

gap, equal all around the Fermi surface (but with different pair spin composition depending on the direction), whereas the A phase has nodes where the gap falls to zero at the 'poles' aligning with the direction of the L-vector.

30.2 The Quasiparticle Blackbody Radiator

The device which we use to observe the processes at the $T \approx 0$ limit is the quasiparticle blackbody radiator (BBR) [2], developed by Shaun Fisher when he was a graduate student, shown in Fig. 30.3. This is precisely what it says: a small container of a few millimetres linear dimension immersed in the superfluid, with a small hole in one wall connecting the liquid inside with the surrounding bulk liquid. This is our radiator, the exact analogy of the classical blackbody radiator, but one in which the active medium is the quasipartile/quasihole gas of the superfluid rather than a gas of photons. Inside the container are two vibrating wire resonators (VWRs), as can be seen in the figure. These devices consist of an approximately semicircular loop of superconducting wire, with the 'legs' anchored. The loop is able to flap perpendicular to its plane. In an applied small vertical magnetic field, the Lorentz force on an AC current driven through the wire at the mechanical resonance drives the device into oscillation normal to the loop plane. Since the wire is superconducting, no ohmic voltages are generated and the voltage measured between the legs simply represents the rate of flux cutting by the moving wire and is thus proportional to the wire velocity. This yields a Lorentzian resonance with width proportional to the damping of the surrounding medium, in this case the excitation gas.

One of the resonators is used as a thermometer and the other as a heater. The damping of a moving wire in superfluid ^{3}He does not seem, at first sight, the most sensible way to detect the vanishingly tenuous gas of normal excitations in the liquid close to absolute zero. However, since the excitation dispersion curve has

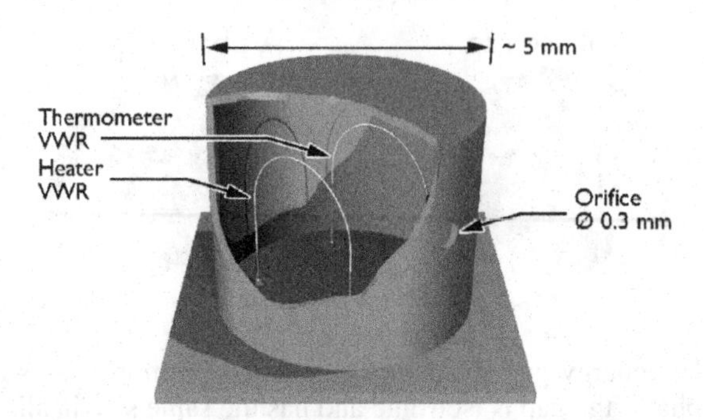

Figure 30.3 A typical blackbody radiator. This one is of copper. The cylindrical box of around 3 mm × 3 mm has a Ø0.3 mm hole to make contact to the outside bulk liquid. Inside the box are two vibrating wire resonators (VWRs), one acting as a thermometer and the other acting as a heater.

the Bardeen-Cooper-Schrieer (BCS) form, the damping applied by this tenuous gas is many orders of magnitude larger than that from a similar Newtonian gas. The reasons for this high damping can be followed in [3], but, put in very simple terms, since the gap has the BCS form, the excitations near the minima (the only states occupied at low temperatures) have tiny energies but still have momenta of p_F, i.e., very much larger than the momenta of a gas of classical particles with the same energy. As a further bonus, since the density of the excitation gas falls very rapidly with temperature, (following the gap Boltzmann factor $\exp(-\Delta/kT)$), the damping on the wire also falls similarly, providing an extremely precise thermometer. A second wire resonator in the radiator can be driven above the critical velocity for pair breaking to act as a heater. Thus, by applying a known heating pulse to the heater resonator and observing the temperature rise in the box, as indicated by the thermometer wire, we can calibrate the BBR. Such devices are extremely sensitive, being able to detect heat inputs below the fW level, as shown in Fig. 30.4.

30.3 'Cosmological' Experiments in the Laboratory

30.3.1 The Big Bang and Cosmic Strings

One of the first experiments suggested by the superfluid/universe analogies was the search for topological defects remaining in the liquid after a rapid cooling passage through the superfluid transition. The basic idea is that as the system approaches the transition from above, thermal fluctuations ensure that the cooler parts of the liquid cross the transition independently. Each independent region makes a random choice of phase ϕ. These regions then grow and coalesce, the result being that after the transition the superfluid consists of a patchwork of regions with different phases, in other words, a 'phase glass'. The jumble of phase gradients at the boundaries

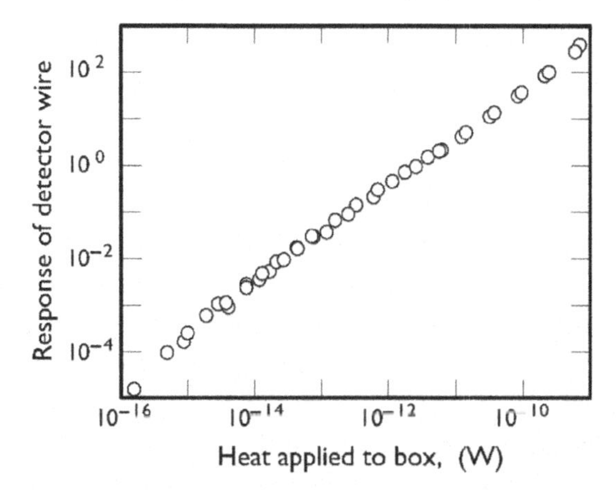

Figure 30.4 A calibration of a quasiparticle blackbody radiator. Note that the response is more or less linear over more than 6 orders of magnitude in power, and sensitive to power inputs as low as 100s of attowatts.

subsequently anneal away in an attempt to produce a gradientless configuration. Of course, the topology does not always allow this, and stable topological defects can be left in the structure.

The relaxation time of the thermal fluctuations experiences a critical slowing approaching a transition, in principle reaching infinity at the actual point of transition. Thus, as the system cools towards the transition, at some point the cooling rate overtakes the ability of the fluctuations to follow and the existing pattern of fluctuations remains frozen until the transition is passed. This means that the size of the domains passing independently through the transition depends on the cooling rate.

What is described above is the Kibble mechanism first introduced by Tom Kibble to explain the formation of topological defects in the 'metric' of the early universe [4]. Kibble's original ideas have been extended by Żurek [5] to condensed-matter systems, such as discussed here, where the process is generally referred to as the Kibble-Żurek Mechanism.

The first experiments were tried on liquid ^{4}He. Unfortunately, there is no gentle way to quench liquid ^{4}He rapidly through the transition other than by a violent decompression to take the system suddenly across the transition, which is too violent for precise results [6].

Fortunately, ^{3}He, on the other hand, can be cooled through the transition in a controlled and very localised way by bombardment with neutrons. Since the ^{3}He nucleus is an alpha particle minus one neutron, it has a very large capture cross section for thermal neutrons. However, a stable alpha particle does not survive, but instead a triton and proton are created releasing 764 keV of energy according to the process $n + {}^3He^{++} = {}^3H^+ + p^+ + 764keV$. The two output particles fly apart, carrying the excess energy which is rapidly transferred to the surrounding atoms, heating a very localised region of normal fluid to well above the superfluid transition. This hot region, of micron dimensions, subsequently recools very rapidly as the quasiparticles disperse into the surrounding cold superfluid, producing a very rapid phase transition. This is our thermal quench.

Cooling through the phase transition can in principle lead to the creation of defects of zero, one and two dimensions, that is, monopoles, strings and branes. Despite searches in the universe and in condensed matter systems for monopoles and branes, in fact it seems that linear defects are the most likely outcomes. These form after the initial phase 'glass' anneals to leave a path through the medium around which the phase changes by 2π. In the superfluid, this would represent a vortex and in the early universe a cosmic string. The linear core of such a defect has a heavily distorted order parameter where the condensation is suppressed. The core material thus reflects the structure of the previous higher-temperature less-ordered phase.

The Kibble-Żurek process is depicted in Fig. 30.5, for the case of our superfluid ^{3}He experiment. Part A of the figure shows the initial energy deposition following the neutron capture. This produces a localised hot spot of normal fluid (B), analogous to early moment of the Big Bang when all forces were unified. As the system rapidly cools again, thermal fluctuations mean that various protodomains form in the system (C). Since there is no time for the ordering information to be transmitted from one domain to the next, the regions are casually disconnected and

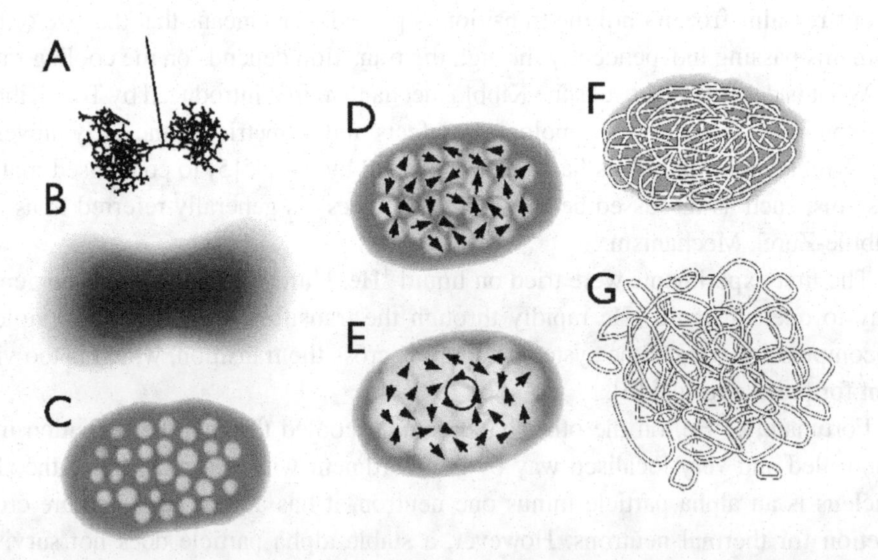

Figure 30.5 Schematic of the Kibble-Żurek mechanism, applied in this case to vortex creation via a rapid neutron-induced phase transition in ^{3}He-B. See text.

must independently choose an order parameter (D). As the system cools further, the protodomains grow and coalesce, leaving a phase 'glass' (E).

A subsequent relaxation to a uniform state may not always be possible. For example, regions may form in which the phase evolves by 2π around some central core. This corresponds to the formation of a line defect (vortex/string). Many such defects may form in random locations producing a tangle (F), which then subsequently evolves/decays on much longer time scales (G).

One such experiment was carried out in Grenoble [7] using a blackbody radiator, with the superfluid ^{3}He inside irradiated with neutrons. (See also Chapters 6 and 7 for related experiments with dilute ultracold atoms and their interpretation.) Measurements of the resultant neutron events observed are shown in Fig. 30.6. The damping measured in the BBR (effectively the temperature) is shown as a function of time during neutron irradiation with a neutron source placed near the cryostat. Any event releasing energy in the BBR will lead to a rapid increase in the temperature followed by a slow decay to equilibrium. In the figure, the most prominent temperature jumps are those associated with events depositing the equivalent of 600 to 700 keV into the BBR. These are the capture processes we are looking for. Further details can be found in Ref. [7].

The energy of each of the observed events can be measured and then plotted as a histogram of event-count number versus energy. The resultant spectrum is dominated by a clear peak associated with the neutron capture processes, with the vast

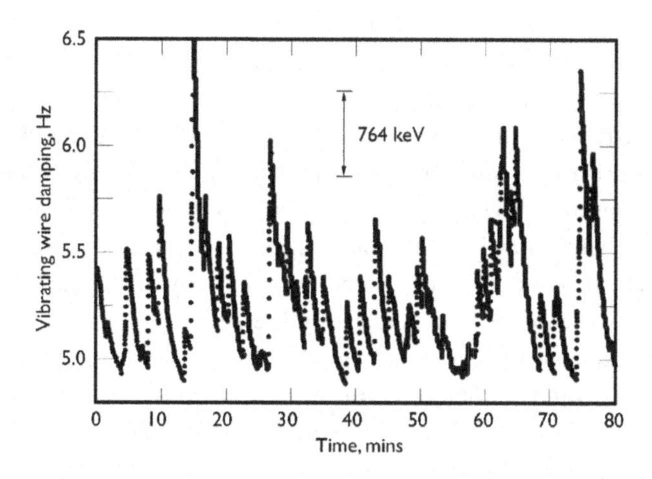

Figure 30.6 Measurements of events in a blackbody radiator in the presence of a neutron source. There are many events with different energy releases, but those corresponding to around 600 to 700 keV, the capture processes, dominate the picture. Adapted with with kind permission from Springer Science+Business Media B.V. from Bäuerle, C., et al. (1998) Superfluid ^{3}He Simulation of Cosmic String Creation in the Early Universe, *J. Low Temp. Phys.*, **110**, 13 [8].

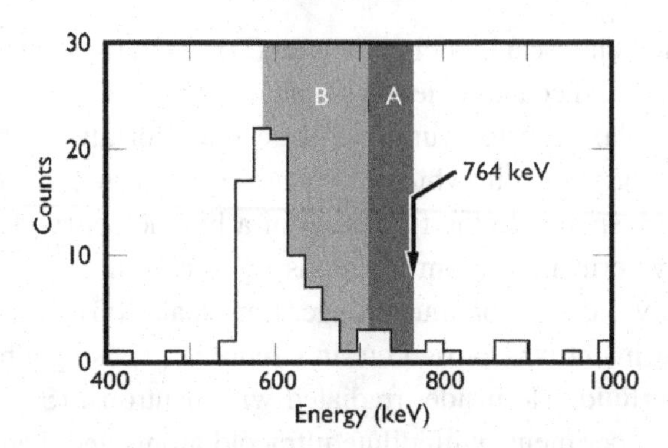

Figure 30.7 The measured neutron absorption peak at 19.4 bar. The peak 'should' occur at 764 keV, but is lower. The grey regions A and B denote the 'missing energy'. Bar A denotes what we estimate is lost to ultraviolet productions, and the remainder, B, must represent the energy used for vortex production, via the Kibble-Żurek mechanism.

majority of background events occurring at much lower energies. A spectrum taken in this way is shown in Fig. 30.7. Significantly, while we know that the neutron capture process is exothermic and associated with an energy release of 764 keV, we see in the figure that the neutron-capture peak occurs at a significantly lower energy. Part of the discrepancy can be attributed to the emission of uv scintillation photons (in other words, some of the energy is converted to uv). However, we know the value of these losses from other experiments. The remaining energy deficit is pressure dependent and is in excellent quantitative agreement with the amount of energy which we would expect to be stored in the tangle of vortices predicted by the Kibble-Żurek mechanism [5].

Simultaneously with the above, a related experiment was taking place in Helsinki, measuring the same phenomenon but by an entirely different method [9]. These measurements were made in a rotating cryostat. Now, as mentioned in the introduction, superflow is irrotational. Thus if we take a container of superfluid and bring it into rotation, the liquid remains stationary. This remains the case so long as we prevent vortices from nucleating (i.e., either by not rotating too fast or by having a container with very smooth walls). Thus we can set up the odd situation where the normal fluid is rotating with the container and the superfluid remains nonrotational 'in the frame of the fixed stars'. This difference in the velocities of the two fluids increases the energy of the system enormously and thus there is a high cost of maintaining the superfluid in the nonrotating state. However, if at this point we irradiate with neutrons, we again go through the Kibble-Żurek process and form a microscopic bundle of vortex loops. In this case, however, these are snatched by the normal flow and are stretched, creating an avalanche which rapidly fills the cell with a vortex lattice to simulate an average rotation matching that of the normal fluid. The resulting vortex lattice can be detected by nuclear magnetic resonance (NMR) (since the flow distorts the orientation of the pair nuclear spins). Thus the experiment confirms that vorticity is formed by neutron irradiation. The two experiments are complementary; the Grenoble measurement detects the energy lost to the vortex tangle but does not actually detect the vortices, while the Helsinki measurement detects the vortices but not the energy.

There are some interesting philosophical questions arising from these experiments involving just how random the choice of phase can be when the system crosses the transition. Are the independent regions condensing really completely independent of all influence of the others? For example, if we imagine just three regions arranged in an equilateral triangle which just happen to 'randomly' choose phases of 0, $2\pi/3$ and $4\pi/3$, then they would most simply anneal to give a phase change of $\phi = 2\pi$ around the loop with a vortex down the centre. That of course now has an angular momentum, which came from where?

30.3.2 Branes

The universe, as we currently observe it, suffers from the 'flatness' problem, the fact that the current universe, despite lacking any causal connection, looks the same in all directions. The currently accepted solution to this problem is that at some time in the past all regions visible to us have been in causal connection, but a subsequent period of very sudden expansion has separated them. This is inflation. In the braneworld scenario of cosmology, it has been suggested that epochs of inflation may have been triggered and halted by the consequences of brane collisions. Branes are defects of lower dimensionality in a higher dimensional matrix. For example, a two-dimensional surface would represent a 'brane' in a three-dimensional world. It has been suggested, for example, that our universe may exist as a three-dimensional brane in a four-dimensional matrix. (For example if the three larger forces, strong, weak and E-M, were confined to our three-dimensional [3D] brane but the flux associated with gravity could 'leak' into the fourth dimension, that might explain gravity's weakness.) However, these considerations are so far removed from everyday reality that it is not easy to bring physical insight to bear on the problem.

One way of trying to inject some insight into this otherwise remote subject is to devise an analogue experiment on a similar system in the laboratory. This we can also achieve in the ^{3}He condensate as it provides an excellent brane analogue in the form of the phase boundary between the two common phases of superfluid ^{3}He.

Our interest is not in triggering the analogue of inflation, but rather in looking for the traces of topological defects which might be left after a brane collision/annihilation. These might possibly still be observable in the universe today if there was indeed an epoch dominated by brane dynamics in the early universe.

As pointed out above, the superfluid ^{3}He condensate exhibits directionalities in the liquid. All the nuclear spins must point in the same direction (within some healing length) and similarly for the orbital moments. Thus the condensate has a very rich anisotropic structure. This is known as the 'texture', a term borrowed from the liquid-crystal field.

By creating a brane/antibrane pair in superfluid ^{3}He and then subsequently annihilating the two, we can indeed investigate whether defects are indeed created in the superfluid texture (the superfluid analogue of space-time).

Earlier in the chapter, we introduced the idea of the broken symmetries in the superfluid ^{3}He condensate. That is somewhat of a simplification, since the superfluid breaks some further subsymmetries to exist in several distinct phases. The two stablest phases are the A and B phases mentioned earlier. Under the zero pressure and low temperature conditions of our experiments, the B phase is stable up to a magnetic field of 340 mT, above which there is a transition to the A phase. Thus, by applying a field gradient, we can stabilise and manipulate an A-B interface. Since the order parameter transforms continuously between the two phases, the interface must itself be coherent. It is worth emphasising that this is a coherent two-dimensional structure sandwiched between two different coherent phases on the two sides. This represents the most coherent two-dimensional structure we know of. This will play the role of our analogue cosmological two-brane in the three-dimensional bulk matrix.

To make such an investigation, we set up an experiment inside a blackbody radiator [10]. Using a profiled magnetic field, we can introduce a slab of A phase across a B-phase background. This creates an A-B and a B-A interface, our brane and antibrane. Decreasing the field reduces the thickness of the A phase slab, bringing the A-B/B-A interfaces closer and finally forcing them into a simulated brane/antibrane annihilation as the thin sliver of A phase finally disappears. After this 'annihilation', we examine the texture of the bulk B phase, which remains for evidence of defect creation.

The experimental cell is shown in Fig. 30.8. It consists of a vertical cylinder of superfluid, with an orifice at the top, connecting with the bulk fluid outside. At the top and bottom of the cylinder are vibrating wire resonators to act as excitation generators and detectors. By running a resonator at the base above the pair-breaking velocity, we generate a flux of excitations travelling from the generating wire up the cylinder and out through the orifice at the top. This allows us to measure the impedance of the cylinder to excitation flow. The flow is an awkward mixture of ballistic transport and scattering off the walls, so we simply measure the ratio of the density of excitations at the top and bottom and use that as our measure of the 'impedance'. This flow acts as the probe for defects in the texture. To understand the properties of the liquid which allow this, we need to refer again to the structure

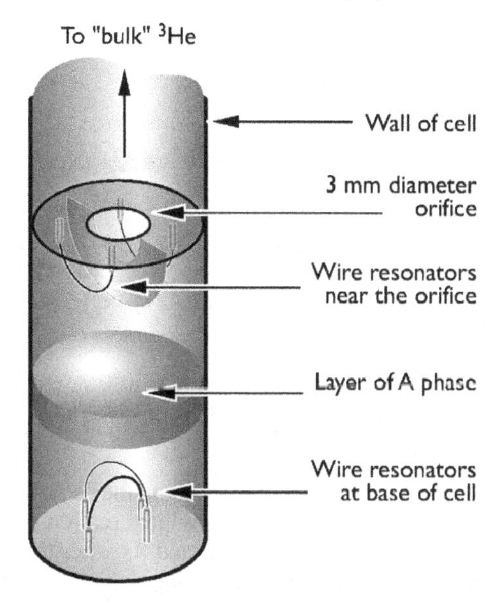

Figure 30.8 The experimental cell for studying 'brane annihilation. An excitation beam up the cylinder is generated by a 'heater' VWR at the base of the cell and the impedance superfluid measured as the slab of A phase halfway up the cylinder is 'annihilated'; see text. Adapted with permission from Bradley, D. I., et al. (2008), Relic topological defects from brane annihilation simulated in superfluid ^{3}He, *Nat. Phys.*, **4**, 46 [9].

of the energy gap as shown in Fig. 30.2. At its simplest, when a magnetic field is applied to the B phase, the gap becomes distorted, with the equatorial gap growing slightly and the polar gap decreasing. If our sudden annihilation leads to relic defects in the texture, then the direction of the L-vector in the region of the defect will be disturbed. Since the gap distortion, although generated by the applied magnetic field, is aligned along the L-vector, that means that excitations moving through the liquid no longer see a smooth landscape but rather a mountainous terrain as the gap varies from place to place depending on the direction of L.

To make the measurement, we increase the field over a flat region halfway up the cylinder. (That is allowed by Maxwell's equations with a reversed Helmholtz pair.) At some point, this field will exceed the 0.34 T needed to stabilise the A phase, and a layer of A phase will form. We first measure the impedance of the cylinder when the field is just below that required to stabilise the A phase. We then increase the field just enough to create the A phase layer and measure the impedance again. Finally, we rapidly reduce the field to its original value, annihilating the two phase interfaces, and measure the impedance one final time.

The field configuration for the final measurement is the same for that of the first. If defects are indeed created, then they will mess up the texture, and the impedance

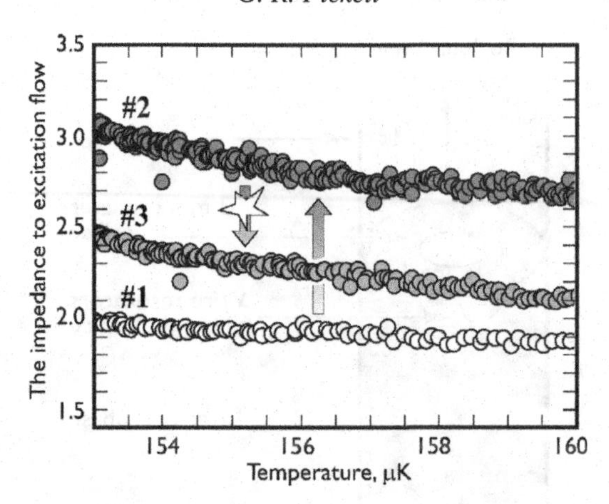

Figure 30.9 Measurements of the 'impedance' along the cell of the previous figure. #1: Measured at a field in the centre just below the creation of the A-phase slab, #2: Measured with the A=phase slab in place, and #3: after the annihilation. Note the data of #1 and [#3 are not the same indicating defects in the texture after the loss of the A-phase slab, see text.

measured finally will be greater than that measured initially. The crux question is, therefore, are these two measurements the same or different?

The results for a set of measurements is shown in Fig. 30.9. The data of #1 are taken with the field just below the threshold for A phase to appear. We then ramp up the field just enough to create the A-phase slab and take another set of data, #2. We see that the impedance is higher as the excitation flux has to negotiate two additional phase boundaries in its path. We then suddenly return the field to its initial value. The two boundaries/branes annihilate (with a 'bang'), leaving B phase but with defects where the annihilation took place. The measurement taken afterwards, #3, shows a higher value than the measurement of #1, showing that defects are indeed formed by the process. One should note the temperatures of the experiment, $\sim$100 μK. It is paradoxical that this simulation of something which may have happened in the early universe at 10^{30} degrees is only possible at temperatures of 10^{-4} K.

Of course, in the superfluid experiment, the curvature of the colliding phase boundaries will provide very sharp 'edges' which will rip through the liquid under the action of surface tension. This process may well dominate the creation of defects. However, similar quantum fluctuations in branes should have the same effect, and treating brane collision as an interaction of two perfect planes is not realistic. In these and similar ways, the experiment can inform the more remote theory.

Acknowledgements: The author would like to acknowledge the contributions of the Lancaster Ultralow Temperature Group to a large fraction of the work discussed in this chapter, and would like to dedicate the work to Shaun Fisher, whose role was paramount but sadly curtailed by his recent and very untimely death.

References

[1] G. E. Volovik, *Exotic Properties of Superfluid ^{3}He* (World Scientific, Singapore, 1992).
[2] S. N. Fisher, A. M. Guénault, C. J. Kennedy, and G. R. Pickett, Blackbody source and detector of ballistic quasiparticles in He^{3}B: emission angle from a wire moving at supercritical velocity, *Phys. Rev. Lett.*, 69 (1992): 1073–6.
[3] S. N. Fisher, A. M. Guénault, C. J. Kennedy, and G. R. Pickett, Beyond the two-fluid model: transition from linear behavior to a velocity-independent force on a moving object in ^{3}He B, *Phys. Rev. Lett.*, 63 (1989): 2566–9.
[4] T. W. B. Kibble, Topology of cosmic domains and strings, *J. Phys.*, A9 (1976): 1387–98.
[5] W. H. Zurek, Cosmological experiments in superfluid-helium, *Nature*, 317 (1985): 505–8.
[6] P. C. Hendry, N. S. Lawson, R. A. M. Lee, and others, Generation of defects in superfluid ^{4}He as an analogue of the formation of cosmic strings, *Nature*, 368 (1994): 315–7.
[7] C. Bäuerle, Yu. M. Bunkov, S. N. Fisher, and others, Laboratory simulation of cosmic string formation in the early Universe using superfluid ^{3}He, *Nature*, 382 (1996): 332–4.
[8] C. Bäuerle, Yu. M. Bunkov, S. N. Fisher, H. Godfrin, and G. R. Pickett, Superfluid 3He simulation of cosmic string creation in the early universe, *Journal of Low Temperature Physics*, 110 (1998): 13–22.
[9] V. M. H. Ruutu, V. B. Eltsov, A. J. Gill, and others, Vortex formation in neutron-irradiated superfluid ^{3}He as an analogue of cosmological defect formation, *Nature*, 382 (1996): 334–6.
[10] D. I. Bradley, S. N. Fisher, A.M. Guénault, and others, Relic topological defects from brane annihilation simulated in superfluid ^{3}He, *Nature Physics*, 4 (2008): 46–9.

31

Cosmic Axion Bose-Einstein Condensation

NILANJAN BANIK AND PIERRE SIKIVIE
Department of Physics, University of Florida, Gainesville, USA

QCD axions are a well-motivated candidate for cold dark matter. Cold axions are produced in the early universe by vacuum realignment, axion string decay, and axion domain wall decay. We show that cold axions thermalize via their gravitational self-interactions and form a Bose-Einstein condensate. As a result, axion dark matter behaves differently from the other proposed forms of dark matter. The differences are observable.

31.1 QCD Axions

The theory of strong interactions, called quantum chromodynamics, or QCD for short, has in its Lagrangian density a "θ-term" [1, 2, 3, 4]

$$\mathcal{L}_\theta = \theta \frac{g_s^2}{32\pi^2} \tilde{G}^{a\mu\nu} G^a_{\mu\nu}, \tag{31.1}$$

where θ is an angle between 0 and 2π, g_s is the coupling constant for strong interactions, and $G^a_{\mu\nu}$ is the gluon field tensor. The θ-term is a 4-divergence and therefore has no effects in perturbation theory. However, it can be shown to have nonperturbative effects, and these are important at low energies/long distances. Since $\mathcal{L}_\theta$ is P and CP odd, QCD violates those discrete symmetries when $\theta \neq 0$. The strong interactions are observed to be P and CP symmetric, and therefore θ must be small. The experimental upper bound on the electric dipole moment of the neutron implies $\theta \lesssim 0.7 \times 10^{-11}$ [5, 6]. In the Standard Model of particle physics, there is no reason for θ to be small; it is expected to be of order one. That θ is less than 10^{-11} is a puzzle, referred to as the strong CP problem.

Peccei and Quinn proposed [7, 8] solving the strong CP problem by introducing a global $U(1)_{\rm PQ}$ symmetry which is spontaneously broken. When some conditions are met, the parameter θ is promoted to a dynamical field $\frac{\phi(x)}{f_a}$, where f_a is the energy scale at which $U(1)_{\rm PQ}$ is spontaneously broken, and $\phi(x)$ the associated

Nambu-Goldstone boson field. The theory now depends on the expectation value of $\phi(x)$. The latter minimizes the QCD effective potential. It can be shown that the minima of the QCD effective potential occur where $\theta = 0$ [9]. The strong CP problem is thus solved if there is a Peccei-Quinn symmetry.

Axions are the quanta of the field $\phi(x)$ [10, 11]. Axions acquire mass due to the nonperturbative effects that make QCD depend on θ. The axion mass is given by

$$m_a \simeq 10^{-6}\text{eV}\left(\frac{10^{12}\text{ GeV}}{f_a}\right) \tag{31.2}$$

when the temperature is zero.

31.2 Production of Cold Axions

The equation of motion for $\phi(x)$ is

$$D_\mu D^\mu \phi(x) + V_a'(\phi(x)) = 0, \tag{31.3}$$

where V_a' is the derivative of the effective potential with respect to the axion field and D_μ is the covariant derivative with respect to space-time coordinates. The effective potential may be written

$$V_a = m_a(t)^2 f_a^2 \left[1 - \cos\left(\frac{\phi(x)}{f_a}\right)\right]. \tag{31.4}$$

The axion mass is temperature and hence time dependent. It reaches its zero-temperature value, Eq. (31.2), at temperatures well below 1 GeV. At temperatures much larger than 1 GeV, m_a is practically zero. The axion field starts to oscillate [12, 13, 14] at a time t_1 after the Big Bang given by

$$m(t_1) \cdot t_1 = 1. \tag{31.5}$$

Throughout we use units in which $\hbar = c = 1$. t_1 is approximately $2 \times 10^{-7}\text{s}\left(\frac{f_a}{10^{12}\text{ GeV}}\right)^{1/3}$. The temperature of the primordial plasma at that time is $T_1 \simeq 1\text{ GeV}\left(\frac{10^{12}\text{ GeV}}{f_a}\right)^{1/6}$. The $\phi(x)$ oscillations describe a population of axions called "of vacuum realignment." Their momenta are of order t_1^{-1} at time t_1, and are red-shifted by the expansion of the universe after t_1:

$$\delta p(t) \sim \frac{1}{t_1}\frac{R(t_1)}{R(t)} \tag{31.6}$$

where $R(t)$ is the scale factor. As a result, the axions are nonrelativistic soon after t_1, and today they are extremely cold. The fact that they are naturally abundant, weakly coupled, and very cold, and that they solve the strong CP problem as well, makes axions an attractive candidate for the dark matter of the universe.

The number of axions produced depends on various circumstances, in particular whether inflation occurred before or after the phase transition in which $U(1)_{PQ}$ is spontaneously broken, hereafter called the PQ phase transition. For a review, see Ref. [15]. If inflation occurs afterward, it homogenizes the axion field within the observable universe. The initial value of the axion field may then be accidentally close to the CP conserving value, in which case the cold axion population from vacuum realignment is suppressed. If inflation occurs before the PQ phase transition, there is always a vacuum realignment contribution (because the axion field has random unrelated values in different QCD horizons) and there are additional contributions from axion string decay and axion domain wall decay. The number density of cold axions is

$$n(t) \simeq \frac{4 \times 10^{47}}{\text{cm}^3} X \left(\frac{f_a}{10^{12}\text{GeV}} \right)^{5/3} \left(\frac{R(t_1)}{R(t)} \right)^3 \tag{31.7}$$

where X is a fudge factor. If inflation occurs before the PQ phase transition, X is of order 2 or 20 depending on whose estimate of the string decay contribution one believes. If inflation occurs after the PQ phase transition, X is of order $\frac{1}{2}\sin^2\alpha_1$, where $\alpha_1 = \phi(t_1)/f_a$ is the initial misalignment angle.

Cold axions are effectively stable because their lifetime is vastly longer than the age of the universe. The number of axions is effectively conserved. The phase-space density of cold axions implied by Eqs. (31.6) and (31.7) is [16]

$$\mathcal{N} \sim n \frac{(2\pi)^3}{\frac{4\pi}{3}(m\delta v)^3} \sim 10^{61} X \left(\frac{f_a}{10^{12}\text{ GeV}} \right)^{8/3}. \tag{31.8}$$

$\mathcal{N}$ is the average occupation number of those axion states that are occupied. Because their phase-space density is huge and their number is conserved, cold axions may form a Bose-Einstein condensate (BEC). The remaining necessary and sufficient condition for the axions to form a BEC is that they thermalize. Assuming thermal equilibrium, the critical temperature is [16, 17]

$$T_c(t) = \left(\frac{\pi^2 n(t)}{\zeta(3)} \right)^{1/3} \simeq 300 \text{ GeV } X^{1/3} \left(\frac{f_a}{10^{12}\text{ GeV}} \right)^{5/9} \frac{R(t_1)}{R(t)}. \tag{31.9}$$

The critical temperature is enormous because the cosmic axion density is so very high. The formula given in Eq. (31.9) differs from the one for atoms because, in thermal equilibrium, most of the noncondensate axions would be relativistic.

The question is whether the axions thermalize. This is not at all obvious since axions are extremely weakly coupled. Note that for Bose-Einstein condensation to occur, it is not necessary that full thermal equilibrium be reached. It is sufficient that the rate of condensation into the lowest energy available state be larger than the inverse age of the universe. Whether this happens is the issue which we address next.

31.3 Axion–Axion Interactions

Axions interact by $\lambda\phi^4$ self-interactions and by gravitational self-interactions. In this section, we discuss these two processes in detail and calculate the corresponding relaxation rates [17]. Let us introduce a cubic box of volume $V = L^3$, with periodic boundary conditions at the surface. The axion field and its canonically conjugate field $\pi(\mathbf{x}, t)$ may be written as

$$\phi(\mathbf{x}, t) = \sum_{\mathbf{n}} (a_{\mathbf{n}}(t)\Phi_{\mathbf{n}}(\mathbf{x}) + a_{\mathbf{n}}^{\dagger}(t)\Phi_{\mathbf{n}}^{*}(\mathbf{x})) \tag{31.10}$$

$$\pi(\mathbf{x}, t) = \sum_{\mathbf{n}} (-i\omega_{\mathbf{n}})(a_{\mathbf{n}}(t)\Phi_{\mathbf{n}}(\mathbf{x}) - a_{\mathbf{n}}^{\dagger}(t)\Phi_{\mathbf{n}}^{*}(\mathbf{x})) \tag{31.11}$$

inside the box, where

$$\Phi_{\mathbf{n}}(\mathbf{x}) = \frac{e^{i\mathbf{p}_{\mathbf{n}}\cdot\mathbf{x}}}{\sqrt{2\omega_{\mathbf{n}}V}}\ , \tag{31.12}$$

$\mathbf{p}_{\mathbf{n}} = 2\pi\mathbf{n}/L$, $\mathbf{n} = (n_1, n_2, n_3)$, where n_1, n_2, n_3 are integers, and $\omega_{\mathbf{n}} = \sqrt{p_{\mathbf{n}}^2 + m^2}$. The creation and annihilation operators satisfy canonical equal-time commutation relations

$$[a_{\mathbf{n}}(t), a_{\mathbf{n}'}^{\dagger}(t)] = \delta_{\mathbf{n},\mathbf{n}'}, \qquad [a_{\mathbf{n}}(t), a_{\mathbf{n}'}(t)] = 0. \tag{31.13}$$

The Hamiltonian, including $\lambda\phi^4$ self-interactions, is

$$H = \sum_{\mathbf{n}} \omega_{\mathbf{n}} a_{\mathbf{n}}^{\dagger} a_{\mathbf{n}} \ + \sum_{\mathbf{n}_1,\mathbf{n}_2,\mathbf{n}_3,\mathbf{n}_4} \frac{1}{4} \Lambda_{s\ \mathbf{n}_1,\mathbf{n}_2}^{\mathbf{n}_3\mathbf{n}_4} a_{\mathbf{n}_1}^{\dagger} a_{\mathbf{n}_2}^{\dagger} a_{\mathbf{n}_3} a_{\mathbf{n}_4}, \tag{31.14}$$

where

$$\Lambda_{s\ \mathbf{n}_1,\mathbf{n}_2}^{\mathbf{n}_3\mathbf{n}_4} = \frac{-\lambda}{4m^2V}\delta_{\mathbf{n}_1+\mathbf{n}_2,\mathbf{n}_3+\mathbf{n}_4}. \tag{31.15}$$

The Kronecker-delta ensures 3-momentum conservation. When deriving Eq. (31.14), axion number violating terms such as $aaaa$, $a^{\dagger}a^{\dagger}a^{\dagger}a^{\dagger}$, $a^{\dagger}aaa$, and $a^{\dagger}a^{\dagger}a^{\dagger}a$ are neglected. Indeed, in lowest order they allow only processes that are forbidden by energy-momentum conservation. In higher orders, they do lead to axion number violating processes but only on times scales that are vastly longer than the age of the universe.

The gravitational self-interactions of the axion fluid are described by Newtonian gravity since we only consider interactions on subhorizon scales. The interaction Hamiltonian is

$$H_g = -\frac{G}{2}\int d^3x\, d^3x' \,\frac{\rho(\mathbf{x},t)\rho(\mathbf{x}',t)}{|\mathbf{x}-\mathbf{x}'|}, \tag{31.16}$$

where $\rho(\mathbf{x},t) = \frac{1}{2}(\pi^2 + m^2\phi^2)$ is the axion energy density. In terms of creation and annihilation operators [17]

$$H_g = \sum_{\mathbf{n}_1,\mathbf{n}_2,\mathbf{n}_3,\mathbf{n}_4} \frac{1}{4}\Lambda_g{}^{\mathbf{n}_3\mathbf{n}_4}_{\mathbf{n}_1,\mathbf{n}_2} a^\dagger_{\mathbf{n}_1} a^\dagger_{\mathbf{n}_2} a_{\mathbf{n}_3} a_{\mathbf{n}_4}, \tag{31.17}$$

where

$$\Lambda_g{}^{\mathbf{n}_3\mathbf{n}_4}_{\mathbf{n}_1,\mathbf{n}_2} = -\frac{4\pi G m^2}{V}\delta_{\mathbf{n}_1+\mathbf{n}_2,\mathbf{n}_3+\mathbf{n}_4}\left(\frac{1}{|\mathbf{p}_{\mathbf{n}_1}-\mathbf{p}_{\mathbf{n}_3}|^2} + \frac{1}{|\mathbf{p}_{\mathbf{n}_1}-\mathbf{p}_{\mathbf{n}_4}|^2}\right). \tag{31.18}$$

H_g must be added to the RHS of Eq. (31.14).

In summary, we have found that the axion fluid is described by a set of coupled quantum harmonic oscillators. We now estimate the resulting relaxation rates. There are two different regimes of relaxation depending on the relative values of the relaxation rate Γ and the energy dispersion $\delta\omega$. The condition $\Gamma << \delta\omega$ defines the "particle kinetic regime," whereas $\Gamma >> \delta\omega$ defines the "condensed regime." Most physical systems relax in the particle kinetic regime. Axions, on the other hand, relax in the condensed regime.

31.3.1 Particle Kinetic Regime

When $\Gamma << \delta\omega$, the rate of change of the occupation numbers $\mathcal{N}_i$ $(i = 1,2,..M)$ of M coupled oscillators is given by

$$\langle\dot{\mathcal{N}_l}\rangle = \sum_{i,j,k=1}^{M} \frac{1}{2}|\Lambda^{kl}_{ij}|^2[\mathcal{N}_i\mathcal{N}_j(\mathcal{N}_l+1)(\mathcal{N}_k+1) - (\mathcal{N}_i+1)(\mathcal{N}_j+1)\mathcal{N}_l\mathcal{N}_k]2\pi\delta(\Omega^{lk}_{ij}) + \mathcal{O}(\Lambda^3), \tag{31.19}$$

where $\Omega^{kl}_{ij} = \omega_k + \omega_l - \omega_i - \omega_j$, and the Λ^{kl}_{ij} are the relevant couplings, such as are given in Eqs. (31.15) and (31.18) for axions. If we substitute the couplings due to $\lambda\phi^4$ interactions, Eq. (31.15), and replace the sums over modes by integrals over momenta, we obtain [18, 17]

$$\langle\dot{\mathcal{N}_1}\rangle = \frac{1}{2\omega_1}\int \frac{d^3p_2}{(2\pi)^3 2\omega_2}\frac{d^3p_3}{(2\pi)^3 2\omega_3}\frac{d^3p_4}{(2\pi)^3 2\omega_4}\lambda^2(2\pi)^4\delta^4(p_1+p_2-p_3-p_4) \times \frac{1}{2}[(\mathcal{N}_1+1)(\mathcal{N}_2+1)\mathcal{N}_3\mathcal{N}_4 - \mathcal{N}_1\mathcal{N}_2(\mathcal{N}_3+1)(\mathcal{N}_4+1)], \tag{31.20}$$

where $\mathcal{N}_1 \equiv \mathcal{N}_{\mathbf{p}_1}$ and so forth. When the states are not highly occupied ($\mathcal{N} \lesssim 1$), Eq. (31.20) implies the standard formula for the relaxation rate

$$\Gamma \sim \frac{\dot{\mathcal{N}}}{\mathcal{N}} \sim n\sigma\delta v, \tag{31.21}$$

where $\sigma = \lambda^2/64\pi m^2$ is the scattering cross section due to $\lambda\phi^4$ interactions, n is the particle density, and δv is the velocity dispersion. On the other hand, when the states are highly occupied ($\mathcal{N} >> 1$), Eq. (31.20) implies

$$\Gamma \sim n\sigma\delta v\mathcal{N}. \tag{31.22}$$

The relaxation rate is enhanced by the degeneracy factor, which is huge ($\mathcal{N} \sim 10^{61}$) in the axion case. The process of Bose-Einstein condensation occurs as a result of scatterings $a(\mathbf{p}_1) + a(\mathbf{p}_2) \leftrightarrow a(\mathbf{p}_3) + a(\mathbf{p}_4)$ in which $\mathcal{N}_1$, $\mathcal{N}_2$ and $\mathcal{N}_3$ are of order the large degeneracy factor $\mathcal{N}$ whereas $\mathcal{N}_4 << \mathcal{N}$. Eq. (31.20) implies that, as a result of such scatterings, the occupation number of the lowest available energy state grows exponentially with the rate given in Eq. (31.22) [18, 16, 19].

In contrast to $\lambda\phi^4$ interactions, gravitational interactions are long range. The cross section for gravitational scattering is infinite due to the contribution from very small angle (forward) scattering. But forward scattering does not contribute to relaxation, whereas scattering through large angles does contribute. (The issue does not arise in the case of $\lambda\phi^4$ interactions, for which there is no peak in the differential cross section for forward scattering, and scattering is generically through large angles.) The upshot is that Eqs. (31.21) and (31.22) are still valid for estimating the relaxation rate by gravitational interactions in the particle kinetic regime provided one uses for σ the cross section for large angle scattering. That cross section is finite and equals

$$\sigma_g \sim \frac{4G^2m^2}{(\delta v)^4} \tag{31.23}$$

in order of magnitude.

31.3.2 Condensed Regime

When $\Gamma >> \delta\omega$, one cannot use Eq. (31.19) because the derivation of that equation involves an averaging over time that is valid only when $\Gamma << \delta\omega$. Instead we will use the equations

$$i\dot{a}_l(t) = \omega_l a_l(t) + \sum_{i,j,k=1}^{M} \frac{1}{2}\Lambda_{kl}^{ij} a_k^\dagger a_i a_j, \tag{31.24}$$

which follow directly from the Hamiltonian, Eq. (31.14). It is convenient to define $c_l(t) \equiv a_l(t)e^{i\omega_l t}$, in terms of which Eq. (31.24) becomes

$$\dot{c}_l(t) = -i \sum_{i,j,k=1}^{M} \frac{1}{2}\Lambda_{kl}^{ij} c_k^\dagger c_i c_j e^{i\Omega_{ij}^{kl} t} \tag{31.25}$$

where $\Omega_{ij}^{kl} \equiv \omega_k + \omega_l - \omega_i - \omega_j$, as before. Further, because the occupation numbers of the occupied states are huge, we write c_l as a sum of a classical part C_l and a quantum part d_l

$$c_l(t) = C_l(t) + d_l(t). \tag{31.26}$$

The C_l are c-number functions of order $\sqrt{\mathcal{N}_l}$ describing the bulk of the axion fluid. They satisfy the equations of motion

$$\dot{C}_l(t) = -i \sum_{i,j,k=1}^{M} \frac{1}{2} \Lambda_{kl}^{ij} C_k^* C_i C_j e^{i\Omega_{ij}^{kl} t}. \tag{31.27}$$

The d_l and $d_l^\dagger$ are annihilation and creation operators satisfying canonical commutation relations. Quantum statistics plays the essential role in determining the *outcome* of relaxation to be the Bose-Einstein distribution. However, we may use classical physics to estimate the *rate* of relaxation. The relaxation rate is the inverse time scale over which $C_l(t)$ changes by an amount of order $C_l(t)$.

The sum in Eq. (31.27) is dominated by those states that are highly occupied. Let K be the number of such states. Using the fact that in the condensed regime $\Omega_{ij}^{kl} t << 1$, we may rewrite Eq. (31.27) as

$$\dot{C}_l(t) \sim -i \sum_{i,j,k=1}^{K} \frac{1}{2} \Lambda_{kl}^{ij} C_k^* C_i C_j. \tag{31.28}$$

If we substitute Eq. (31.15) for $\lambda\phi^4$ interactions, we get

$$\dot{C}_{\mathbf{p}_1}(t) \sim i \frac{\lambda}{4m^2 V} \sum_{\mathbf{p}_2, \mathbf{p}_3} \frac{1}{2} C_{\mathbf{p}_2}^* C_{\mathbf{p}_3} C_{\mathbf{p}_4}, \tag{31.29}$$

where $\mathbf{p}_4 = \mathbf{p}_1 + \mathbf{p}_2 - \mathbf{p}_3$ and the sum is restricted to the highly occupied states. The sum is similar to a random walk with each step of order $\sim \mathcal{N}^{3/2}$ and the number of steps of order K^2. Hence

$$\dot{C}_{\mathbf{p}} \sim \frac{\lambda}{4m^2 V} K \mathcal{N}^{3/2} \sim \frac{\lambda}{4m^2 V} N \mathcal{N}^{1/2} \tag{31.30}$$

where we used $K \sim N/\mathcal{N}$. Since $C_l \sim \sqrt{\mathcal{N}}$, the relaxation rate due to $\lambda\phi^4$ interactions in the condensed regime is [16, 17]

$$\Gamma_\lambda \sim \frac{1}{4} n \lambda m^{-2} \tag{31.31}$$

where $n = N/V$ is the number density of the particles in highly occupied states. Likewise, the relaxation rate for gravitational scattering is found to be

$$\Gamma_g \sim 4\pi G n m^2 \ell^2 \tag{31.32}$$

where $\ell = 1/\delta p$ is the correlation length of the particles.

The expressions estimating the relaxation rates in the condensed regime, Eqs. (31.31) and (31.32), are very different from the expression, Eq. (31.22), in the particle kinetic regime. In particular, in the condensed regime, the relaxation rate is first order in the coupling, whereas it is second order in the particle kinetic regime. But the expressions are compatible. At the boundary between the two regimes, where $\delta\omega \sim \Gamma$, the two estimates agree. At that boundary, up to factors of order 2 or so,

$$\delta v \mathcal{N} \sim \delta v \frac{n}{(\delta p)^3} \sim \frac{n}{m^2 \delta\omega} \sim \frac{n}{m^2 \Gamma} \ . \tag{31.33}$$

Substituting this into Eq. (31.22) yields Eq. (31.31). This is similarly the case for the relaxation rate due to gravitational self-interactions.

31.4 Axion BEC

For a system of particles to form a BEC, four conditions must be satisfied:

I. The particles must be identical bosons.
II. Their number must be conserved.
III. They must be degenerate, i.e., the average occupation number $\mathcal{N}$ of the states that they occupy should be order 1 or larger.
IV. They must thermalize.

When the four conditions are satisfied, a macroscopically large fraction of the particles go to the lowest energy available state. It may be useful to clarify the notion of *lowest energy available state* [20]. Thermalization involves interactions. By lowest energy available state, we mean the lowest energy state that can be reached by the thermalizing interactions. In general, the system has states of yet lower energy. For example, and at the risk of stating the obvious, when a beaker of superfluid ^{4}He is sitting on a table, the condensed atoms are in their lowest energy available state. This is not their absolute lowest energy state since the energy of the condensed atoms can be lowered by placing the beaker on the floor. In the case of atoms, it is relatively clear what state the atoms condense into when BEC occurs. The case of axions is more confusing because the thermalizing interactions, both gravity and the $\lambda\phi^4$ self-interactions, are attractive and therefore cause the system to be unstable. When the system is unstable, the restriction to the lowest energy *available* state is especially crucial. (See also Chapter 7 for a discussion of issues of condensate formation.)

We saw in the first two sections that, for cold dark matter axions, the first three conditions for BEC are manifestly satisfied. In this section, we show that the fourth condition is satisfied as well [16, 17]. Cold axions will thermalize if their relaxation time τ is shorter than the age t of the universe, or equivalently if their relaxation rate $\Gamma \equiv 1/\tau$ is greater than the Hubble expansion rate $H \sim 1/t$.

The cold axion energy dispersion is

$$\delta\omega(t) \simeq \frac{(\delta p(t))^2}{2m(t)} . \tag{31.34}$$

In view of Eqs. (31.5) and (31.6), $\delta\omega(t_1) \sim 1/t_1$. If axions thermalize at time t_1, we have $\Gamma(t_1) > 1/t_1$ and therefore the thermalization is in the condensed regime or at the border between the particle kinetic and condensed regimes. After time t_1, $\delta\omega(t) < 1/t$ since $m(t)$ increases sharply for a period after t_1 whereas $(\delta p(t))^2 \propto R(t)^{-2} \propto 1/t$, since $R(t) \propto \sqrt{t}$ in the radiation dominated era. So after t_1, axions can only thermalize in the condensed regime.

To see whether the axions thermalize by $\lambda\phi^4$ self-interactions at time t_1, we may use either Eq. (31.31) or (31.22). Both estimates yield $\Gamma_\lambda(t_1) \sim H(t_1)$, indicating that the axions thermalize at time t_1 by $\lambda\phi^4$ self-interactions but only barely so. After t_1, we must use Eq. (31.31). It informs us that $\Gamma_\lambda(t)/H(t) \propto R(t)^{-3}t \propto t^{-\frac{1}{2}}$, i.e., that even if axions thermalize at time t_1, they stop doing so shortly thereafter. Nothing much changes as a result of this brief epoch of thermalization since in either case, whether it occurs or not, the correlation length $\ell(t) \equiv 1/\delta p(t) \sim t_1 R(t)/R(t_1)$.

To see whether the axions thermalize by gravitational self-interactions, we use Eq. (31.32). It implies

$$\Gamma_g(t)/H(t) \sim 8\pi G n m^2 \ell^2 t \sim 5 \cdot 10^{-7} \frac{R(t_1)}{R(t)} \frac{t}{t_1} X \left(\frac{f_a}{10^{12}\text{GeV}} \right)^{\frac{2}{3}} \tag{31.35}$$

once the axion mass has reached its zero temperature value, shortly after t_1. Gravitational self-interactions are too slow to cause thermalization of cold axions near the QCD phase transition but, because $\Gamma_g/H \propto R^{-1}(t)t \propto R(t)$, they do cause the cold axions to thermalize later on. The right-hand side (RHS) of Eq. (31.35) reaches one at a time t_{BEC} when the photon temperature is of order

$$T_{\text{BEC}} \sim 500 \text{ eV } X \left(\frac{f_a}{10^{12} \text{ GeV}} \right)^{\frac{1}{2}} . \tag{31.36}$$

The axions thermalize then and form a BEC as a result of their gravitational self-interactions. The whole idea may seem far-fetched because we are used to thinking that gravitational interactions among particles are negligible. The axion case is special, however, because almost all particles are in a small number of states with very long de Broglie wavelength, and gravity is long range.

Systems dominated by gravitational self-interactions are inherently unstable. In this regard, the axion BEC differs from the BECs that occur in superfluid ^{4}He and dilute gases. The axion fluid is subject to the Jeans gravitational instability, and this is so whether the axion fluid is a BEC or not [16]. The Jeans instability causes density perturbations to grow at a rate of order of the Hubble rate $H(t)$, i.e., on a time scale of order the age of the universe at the moment under consideration. Each mode of the axion fluid is Jeans unstable. We showed, however, that after $t_{\rm BEC}$, the thermalization rate is faster than the Hubble rate. The rate at which quanta of the axion field jump between modes is faster than the rate at which the Jeans instability develops. So the modes are essentially frozen on the time scale over which the axions thermalize.

Finally, we comment on a misapprehension that appears in the literature. The axions do not condense in the lowest momentum mode $\mathbf{p} = 0$. Condensation into the $\mathbf{p} = 0$ state would mean that the fluid becomes homogeneous and at rest. Of course, this is not what happens in the axion case since the axion fluid is Jeans unstable. Despite a common misconception, it is not a rule of BEC that the particles condense into the $\mathbf{p} = 0$ state. The rule instead is that they condense into the lowest energy available state, as defined earlier. Only in empty space, and only if the total linear momentum and the total angular momentum of the particles are zero, is the lowest energy state a state of zero momentum. It should be obvious that the particles do not condense in the $\mathbf{p} = 0$ state if they are moving or rotating. Nonetheless, Bose-Einstein condensation occurs.

31.5 Observational Implications

For a long time, it was thought that axions and the other proposed forms of cold dark matter behave in the same way on astronomical scales and are therefore indistinguishable by observation. Axion BEC changed that. On time scales longer than their thermalization time scale τ, axions almost all go to the lowest energy state available to them. The other dark matter candidates, such as weakly interacting masssive particles (WIMPs) and sterile neutrinos, do not do this. It was shown in Ref. [16] that, on all scales of observational interest, density perturbations in axion BEC behave in exactly the same way as those in ordinary cold dark matter provided the density perturbations are within the horizon and in the linear regime. On the other hand, when density perturbations enter the horizon, or in second order of perturbation theory, axions generally behave differently from ordinary cold dark matter because the axions rethermalize so that the state most axions are in tracks the lowest energy available state.

A distinction between axions and the other forms of cold dark matter arises in second order of perturbation theory, in the context of the tidal torquing of galactic halos. Tidal torquing is the mechanism by which galaxies acquire angular momentum. Before they fall onto a galactic halo, the axions thermalize sufficiently fast that the axions that are about to fall into a particular galactic gravitational potential well go to their lowest energy available state consistent with the total angular momentum they acquired from nearby protogalaxies through tidal torquing [20]. That state is a state of net overall rotation, more precisely a state of rigid rotation on the turnaround sphere. In contrast, ordinary cold dark matter falls into a galactic gravitational potential well with an irrotational velocity field [21]. The inner caustics are different in the two cases. In the case of net overall rotation, the inner caustics are rings [22] whose cross section is a section of the elliptic umbilic D_{-4} catastrophe [23], called caustic rings for short. If the velocity field of the infalling particles is irrotational, the inner caustics have a "tent-like" structure, which is described in detail in Ref. [21] and which is quite distinct from caustic rings. Evidence was found for caustic rings. A summary of the evidence is given in Ref. [24]. Furthermore, it was shown in Ref. [25] that the assumption that the dark matter is axions explains not only the existence of caustic rings but also their detailed properties, in particular the pattern of caustic ring radii and their overall size.

Vortices appear in the axion BEC as it is spun up by tidal torquing. The vortices in the axion BEC are attractive, unlike those in superfluid ^{4}He and dilute gases. Hence a large fraction of the vortices in the axion BEC join into a single big vortex along the rotation axis of the galaxy [20]. Baryons and ordinary cold dark matter particles that may be present, such as WIMPs and/or sterile neutrinos, are entrained by the axion BEC and acquire the same velocity distribution. The resulting baryonic angular momentum distribution gives a good qualitative fit [20] to the angular momentum distributions observed in dwarf galaxies [26]. This resolves a long-standing problem with ordinary cold dark matter called the "galactic angular momentum problem" [27, 28]. A minimum fraction of cold dark matter must be axions to explain the data. That fraction is of order 35% [20].

References

[1] 't Hooft, G. 1976a. Symmetry breaking through Bell-Jackiw anomalies. *Phys. Rev. Lett.*, **37**, 8–11.

[2] 't Hooft, G. 1976b. Computation of the quantum effects due to a four-dimensional pseudoparticle. *Phys. Rev.*, **D 14**, 3432–3450.

[3] Jackiw, R., and Rebbi, C. 1976. Vacuum periodicity in a Yang-Mills quantum theory. *Phys. Rev. Lett.*, **37**, 172–175.

[4] Callan, C. G., Jr., Dashen, R. F., and Gross, D. J. 1976. The structure of the gauge theory vacuum. *Phys. Lett.*, **B 63**, 334–340.

[5] Baker, C. A., Doyle, D. D., Geltenbort, P., Green, K., van der Grinten, M. G. D., et al. 2006. An improved experimental limit on the electric dipole moment of the neutron. *Phys. Rev. Lett.*, **97**, 131801.

[6] Kim, J. E., and Carosi, G. 2010. Axions and the strong CP problem. *Rev. Mod. Phys.*, **82**, 557–602.
[7] Peccei, R. D., and Quinn, H. R. 1977. CP conservation in the presence of instantons. *Phys. Rev. Lett.*, **38**, 1440–1443.
[8] Peccei, R. D., and Quinn, H. R. 1977. Constraints imposed by CP conservation in the presence of instantons. *Phys. Rev.*, **D 16**, 1791–1797.
[9] Vafa, C., and Witten, E. 1984. Parity conservation in QCD. *Phys. Rev. Lett.*, **53**, 535.
[10] Weinberg, S. 1978. A new light boson? *Phys. Rev. Lett.*, **40**, 223–226.
[11] Wilczek, F. 1978. Problem of strong p and t invariance in the presence of instantons. *Phys. Rev. Lett.*, **40**, 279–282.
[12] Preskill, J., Wise, M. B., and Wilczek, F. 1983. Cosmology of the invisible axion. *Phys. Lett.*, **B 120**, 127–132.
[13] Abbott, L. F., and Sikivie, P. 1983. A cosmological bound on the invisible axion. *Phys. Lett.*, **B 120**, 133–136.
[14] Dine, M., and Fischler, W. 1983. The not so harmless axion. *Phys. Lett.*, **B 120**, 137–141.
[15] Sikivie, P. 2008. Axion cosmology. *Lect. Notes Phys.*, **741**, 19–50.
[16] Sikivie, P., and Yang, Q. 2009. Bose-Einstein condensation of dark matter axions. *Phys. Rev. Lett.*, **103**, 111301.
[17] Erken, O., Sikivie, P., Tam, H., and Yang, Q. 2012. Cosmic axion thermalization. *Phys. Rev.*, **D 85**, 063520.
[18] Semikoz, D. V., and Tkachev, I. I. 1997. Condensation of bosons in kinetic regime. *Phys. Rev.*, **D 55**, 489–502.
[19] Berges, J., and Jaeckel, J. 2014. Far from equilibrium dynamics of Bose-Einstein condensation for axion dark matter.
[20] Banik, N., and Sikivie, P. 2013. Axions and the galactic angular momentum distribution. *Phys. Rev.*, **D 88**, 123517.
[21] Natarajan, A., and Sikivie, P. 2006. The inner caustics of cold dark matter halos. *Phys. Rev.*, **D 73**, 023510.
[22] Sikivie, P. 1998. Caustic rings of dark matter. *Phys. Lett.*, **B 432**, 139–144.
[23] Sikivie, P. 1999. The caustic ring singularity. *Phys. Rev.*, **D 60**, 063501.
[24] Duffy, L. D., and Sikivie, P. 2008. The caustic ring model of the Milky Way halo. *Phys. Rev.*, **D 78**, 063508.
[25] Sikivie, P. 2011. The emerging case for axion dark matter. *Phys. Lett.*, **B 695**, 22–25.
[26] van den Bosch, F. C., Burkert, A., and Swaters, R. A. 2001. The angular momentum content of dwarf galaxies: new challenges for the theory of galaxy formation. *Mon. Not. Roy. Astron. Soc.*, **326**, 1205.
[27] Navarro, J. F., and Steinmetz, M. 2000. The core density of dark matter halos: a critical challenge to the lambda-cdm paradigm? *Astrophys. J.*, **528**, 607–611.
[28] Burkert, A., and D'Onghia, E. 2004. Galaxy formation and the cosmological angular momentum problem. *Astrophys. Space Sci. Libr.*, **319**, 341.

32

Graviton BECs: A New Approach to Quantum Gravity

GIA DVALI

Arnold Sommerfeld Center for Theoretical Physics, Department für Physik,
Ludwig-Maximilians-Universität München, Germany
Max-Planck-Institut für Physik, München, Germany
Center for Cosmology and Particle Physics, Department of Physics,
New York University, NY, USA

CESAR GOMEZ

Arnold Sommerfeld Center for Theoretical Physics, Department für Physik,
Ludwig-Maximilians-Universität München, Germany
Instituto de Física Teórica, Universidad Autónoma de Madrid, Spain

We outline an alternative view to quantum gravity, based on the idea that a black hole can be understood as a Bose-Einstein condensate of gravitons (quanta of gravitational energy) at the critical point of the quantum phase transition, with black hole radiation and evaporation being a manifestation of how the graviton many-body system maintains itself at criticality. Within this approach, the de Sitter invariance is quantum mechanically broken as a $(1/N)$ effect, where N is the number of gravitons.

32.1 Introduction

Although classical General Relativity (GR) has had enormously interesting experimental predictions, going from the deflection of light to the expansion of the universe, we are still missing an efficient way to consume the marriage between GR and quantum mechanics.

We can easily identify the two main reasons that make this desired marriage extremely difficult. Quantum mechanics, as well as Quantum Field Theory (QFT), is based on notions such as particle and interaction that are defined relative to an absolute space-time. They can be extended to curved backgrounds, but when the geometry starts to affect concepts such as global time, paradoxes unavoidably appear. Moreover, the main message of GR, namely that geometry itself is a dynamical notion, creates its own set of conceptual problems. Should we think of geometry as on an equal footing as we think of other fields such as Yang Mills that we quantize on a privileged absolute Minkowski space-time? And, if that is the right way to proceed, how should we deal with the problems of renormalizability and unitarity that this naive approach immediately creates?

As it is well known the root of these perturbative problems lies in the dimensions of the coupling (the Newton constant) defining the strength of the gravitational interaction. In Wilsonian terminology, the gravitational interaction is defined by an irrelevant operator that flows nicely into the infrared (IR) but goes out of control in the ultraviolet (UV) whenever we go beyond the Planck scale, a length scale that acquires the Wilsonian meaning of a natural cutoff.

In quantum mechanics, in order to resolve small distances we need to design processes involving high momentum transfer. This view is based on a classical space-time, relative to which we can give sense to the notion of short distances. However, once we turn on gravity we reach a limit to this naive way of thinking. In fact, for ultraplanckian energies we unavoidably need to take into account the quantum nature of space-time itself. This manifests, among other things, into the contamination of the resolution process by the gravitons sourced by the energy used in such a process.

The clue to understanding the marriage of quantum mechanics and gravity lies in identifying the nature of these gravitons. If the energy used to resolve shorter and shorter distances is concentrated in few very hard gravitons, we reach a strong coupling problem and we cannot say anything. However, it could happen that this energy is distributed into a large number N of weakly coupled soft gravitons. If that is the case, we transform a strong coupling problem into weakly coupled large N physics. The would-be strong coupling effects become translated into a large-N effect of very soft quanta. However, this effect will not solve the problem if the probability of creating so many soft gravitons is quantum mechanically much smaller that the one of creating few hard gravitons.

At first sight, creating many soft gravitons comes with a penalty of the order of $\alpha(N)^N$, for $\alpha(N)$ the typical weak coupling among the N generated gravitons. In order to compensate this natural weak coupling suppression, we need to inject new physics directly related to the collective properties of the so-generated system of soft gravitons. To unveil these collective effects and to describe how they avoid the large weak coupling suppression will be the main target of our story.

Before entering into details, we should say few words to frame our approach into the standard discussion on quantum gravity. Quantum mechanically dynamical fluctuations of the metric can be understood as creation of gravitons; however, this simpleminded quantum mechanical view seems to enter into conflict with the Holy principles of GR, such as general covariance. As it is well known, general covariance creates problems with the notion of locality and leads to a canonical quantization of GR based on the Wheeler–de Witt equation that does not involve time. In other words, dynamical fluctuations of the metric prevent the existence of a standard notion of time evolution. However, we can try to push this puzzle into one of a different nature, namely how a natural notion of *physical clock* emerges from the quantum version of the metric fluctuations once they are understood as collections of gravitons. In more precise terms, we shall not focus on the invariance with respect to changes of formal clocks, understood as tools to define coordinates, but instead on the quantum generation of a physical clock. We shall say few words regarding this issue at the very end of this chapter.

The main ideas reported in this chapter have been presented in detail in Refs. [1, 2, 3, 4, 5, 6, 7], to which the interested reader is referred for further analysis.

32.1.1 Classicalization and Self-Completeness

When the previous intuitive picture is combined with quantum mechanics, something quite extraordinary happens. Indeed, if when we increase the center of mass energy we unavoidably source a large number of soft quanta, we are forced to conclude that flowing into the deep UV pushes us into a large N regime, where the quantum $\hbar$ effects become effectively $1/N$ effects. In other words, we are pushed into a regime that in the $N = \infty$ limit we could call classical. This form of UV flow is what we have baptized as *classicalization*.

Classical GR contains itself the germ of this classicalization phenomena, something that we have known for many years. In fact, in gravity, any probe of energy E has an effective size determined by the gravitational radius $r_g(E) = EL_P^2$, where L_P denotes the Planck length (see Eq. (32.1)). When we increase the energy, we reach a regime where this gravitational size is bigger than the quantum wavelength $r_g(E) > \hbar/E$. This happens at the threshold for black hole formation.

Since gravity associates with any amount of energy E a universal size $r_g(E)$, we can immediately associate with E a unique given action, namely $Er_g(E)$ (we shall use $c = 1$ in what follows). Thus, quantum mechanics, in the form of the old fashion Bohr-Sommerfeld rule, allows us to estimate the associated number N of quanta (gravitons) as simply $N(E) = Er_g(E)/\hbar$.

The meaning of this quantum mechanical estimate is to inform us how many gravitons we need to account for the gravitational self-energy of an arbitrary system with total energy E. In fact, the gravitational self-energy of an amount of energy E located in a region of size L is $O(Mr_g/L = N(E)\hbar/L)$, i.e, it can be encoded into a set of $N(E)$ quanta with the typical wavelength determined by the physical size of the system L. These gravitons interact with a gravitational strength $\alpha = L_P^2/L^2$. For those distributions of energy localized in a region of size L, much larger than $r_g(E)$, the gravitational self-energy is almost a negligible effect compared to the total energy E.

In order to measure the strength of quantum large-N effects, we define the coupling $\lambda \equiv \alpha N$. In the regime where L is larger than $r_g(E)$, we get $\lambda < 1$. The black hole formation takes place when $L = r_g(E)$, i.e, at $\lambda = 1$.

Hence if we try to resolve distances L smaller than $r_g(E)$, we can only do it encoding the gravitational self-energy in terms of a number n smaller than $N(E)$ of hard gravitons. In that case, we could use these hard gravitons as probes of shorter and shorter distances. This possibility creates, however, a strong coupling problem.

Therefore, the possibility of working out quantum gravity without entering into the strong coupling regime requires two basic ingredients:

- To impose a $\lambda = 1$ barrier, or in other words, to impose a lower bound, equal to $N(E)$, on the number of quanta (gravitons) accounting for the gravitational self energy.
- To unveil the special features of the $\lambda = 1$ point, as a quantum critical point. In particular, to understand how collective critical phenomena can avoid the quantum suppression to create a large number of weakly coupled soft quanta.

Whenever this can be done, without invoking new degrees of freedom, we shall say that the theory is self-complete.

32.1.2 The Meaning of Quantum Gravity Effects: 1/N Versus e^{-N} and the Semiclassical Limit

As already discussed, in the presence of gravity, a given amount of mass M defines a gravitational size $r_g = 2G_N M$, where G_N denotes Newton's gravitational constant. Once we consider the problem quantum mechanically, i.e., with $\hbar$ nonzero, the system is characterized by an extra parameter $N(M)$. In terms of $\hbar$, we can define the Planck length and mass as

$$L_P^2 = \hbar G_N \qquad \text{and} \qquad M_P L_P = \hbar \, . \tag{32.1}$$

How does N enter into the definition of the classical limit? The answer is that we should take the limit $\hbar = 0$ and therefore $N(M) = \infty$, keeping both G_N and M finite, and consequently also doing the same for r_g.

Let us now consider the *semiclassical limit*, in which we are interested in calculating quantum mechanical effects on a classical background sourced by some external mass upon ignoring back reaction effects. In the spirit of GR, what we shall do is to work out the quantum field theory in a given geometrical background. Since we are not going to take into account back reaction effects, the only thing we need to use is the geometrical features of the space-time background, such as the size of the horizon given by r_g, while we can happily ignore any other extragravitational effect. Thus, for all practical purposes, in this semiclassical limit, we take $M = \infty$, $G_N = 0$, but we keep r_g finite, i.e., we reduce all the gravitational effects to the effect of an external geometry, in this simple case characterized by the horizon size r_g. Moreover, in this semiclassical limit, we work out quantum effects and therefore we keep $\hbar$ different from zero. What about N?

Since we ignore all gravitational effects except the ones derived from working in a given geometry, we are setting $L_P = 0$ and therefore $N = \infty$, while keeping $r_g = \sqrt{N} L_P$ finite. The key point, of the so-defined semiclassical limit, is that it is defined as a $N = \infty$ limit and therefore it is blind to any finite N effect.

Therefore, what is the properly *quantum gravity* regime? This is identified as the regime where not only $\hbar$ is different from zero but also N is kept finite. In this purely quantum regime, we should distinguish between two types of different quantum effects, namely those that go as $1/N$ (perturbative) and those that contribute as e^{-N} (nonperturbative).

The difference between both types of effects is easy to figure out. Although both effects require to work with finite N, their nature is very different. While $1/N$ effects depend on the nature of the quanta and on their interaction Hamiltonian, e^{-N} effects depend on quantum processes involving the system as a whole. In what follows, we shall be mostly interested in $1/N$ effects.

32.1.3 Hints from Perturbation Theory and Asymptotic Series

In quantum mechanics, as well as in quantum field theory, we are used to perturbative expansions in the coupling $F(g) = \sum F_n g^n$ with zero radius of convergence. These series are asymptotic and generically start to diverge for $n \sim a/g$, where a is some numerical coefficient. The origin of this divergence is simply related with the growth of the coefficients F_n as a factorial $n!$. The reason for these factorials is related with the number of Feynman diagrams contributing to a given order of the perturbative expansion. The standard way to treat this problem is defining the sum by the optimal truncation at $n \sim a/g$. This truncation leads to an error that is of the order $e^{-a/g}$, i.e, *nonperturbative*, but exponentially small. The so-called tran-series requires to complete the perturbative expansion with an additional series in powers of $e^{-a/g}$, the instanton contributions.

An interesting case of this phenomena takes place when we consider processes $2 \to N$ in some quantum field theories. For instance, in $\lambda\phi^4$ theory, the amplitude $2 \to N$ at weak coupling λ should be expected to be very small. This is intuitively obvious thinking in the time reversal process of N particles producing just two outgoing particles. However, the naive perturbative computation leads to a puzzle since due to the growth of the number of tree-level diagrams contributing to the $2 \to N$ amplitude, we can find that for $N \gtrsim 1/\lambda$ the amplitude starts to diverge. Does that imply that our intuitive expectation based on the time reversal process is wrong and that the process where a very large number N of particles produces just two is dominant, instead of being exponentially suppressed? Most likely the answer is no.

In fact, what accounts for the regime where the perturbative computation breaks down are nonperturbative effects that are again exponentially small, as should be expected.

With these ideas in mind, let us address a similar question in the case of gravity. While in a $\lambda\phi^4$ theory the amplitudes for $2 \rightarrow n$ for $n < N$ were perfectly fine, and the problem starts for large number of particles, the situation in quantum gravity is just the opposite. Indeed, if we consider just standard quantum gravity amplitudes at tree level, the amplitudes $2 \rightarrow n$ for small n and for center of mass energy s larger than M_P^2 violate perturbative unitarity. In a certain sense, we can think of them as the analog of the amplitudes $2 \rightarrow N$ in $\lambda\phi^4$ theory for $N > 1/\lambda$. However, here the intuitive argument based on the time reversal process is not working. So, what happens?

Before answering this, we need to see how, *for a fixed value* of s, the amplitude behaves when we increase n. Interestingly enough, and contrary to what happens in the $\lambda\phi^4$ theory, the amplitudes start to smooth down and they do that *perturbatively*, i.e, this is the regime where perturbation theory is working nicely.

The value of N analog to $N \sim 1/\lambda$ for the $\lambda\phi^4$ theory now depends on the center of mass energy s as $N(s) \sim s/M_P^2$ and the corresponding amplitude as $e^{-N(s)}$. This suppression is the one that we intuitively expect, and it is analog to the one we obtain in $\lambda\phi^4$ theory. However, there exist two crucial differences between those two cases. In the case of gravity, this exponential suppression *is not a nonperturbative effect* and moreover the regime that calls for nonperturbative input is the one of amplitudes $2 \rightarrow n$ with small number n of hard outgoing gravitons.

This is the first main lesson we learn from this exercise. For quantum gravity and within the perturbative scheme, what looks ill defined are trans-Planckian processes with a small number of outgoing gravitons while processes with large number of outgoing soft gravitons are perturbatively perfectly well defined. In this precise sense, gravity works in the opposite way as theories like $\lambda\phi^4$.

If we shall follow the same philosophy as in the $\lambda\phi^4$ theory, we should truncate the amplitudes for number of outgoing gravitons smaller than $N(s)$. However, we certainly need something else, since the remaining contributions are exponentially suppressed.

The simplest way to understand the perturbative exponential suppression of amplitudes $2 \rightarrow N(s)$ is thinking that the $N(s)$ outgoing gravitons are forming some sort of *coherent state*. This assumption immediately accounts for the appareance of the exponential suppression. However, the system of a large number of outgoing gravitons can have some collective properties, completely hidden from the perturbative point of view, that can compensate the perturbative exponential suppression. Indeed, we can think that the collective system can have a large *nonperturbative* degeneration. By that we mean a large number of gapless low lying collective modes. This is indeed what we shall find in our proposed Bose-Einstein condensation (BEC) model of black holes.

Why do we think here in terms of black holes? The reason is because what we expect to happen in ultraplanckian scattering is black hole formation due to the concentration of energy in a small region of space. Thus the obvious guess, triggered by the perturbative analysis, is that the $N(s)$ gravitons, at which the amplitude starts to behave nicely perturbatively, should combine into a coherent quantum state with special collective features. That should be the germ for the quantum BEC model of black holes.

32.2 The BEC Model of Black Holes

Everybody knows that when a certain amount of mass is localized in a region of size smaller or equal to the corresponding gravitational radius, a black hole is formed. By that we refer to the geometry created by such distribution of mass, a geometry that is characterized by having a horizon that hides inside a singularity of spacelike type.

In this background, geometry quantum field theory can be defined, and it presents some very interesting features, first unveiled by Hawking. The most spectacular quantum effect in this geometry is particle creation around the horizon with a thermal spectrum. Heuristically, the simplest way to understand this thermal radiation is observing that the timelike Killing vector, relative to which the asymptotic observer defines energy, becomes, inside the black hole, spacelike. This opens the possibility to use quantum uncertainty, in the form of vacuum polarization in the near-horizon region, to induce real particle production. In a nutshell, one of the particles in the pair becomes negative energy with respect to the spacelike Killing inside the black hole, while the other positive energy particle is radiated. The negative energy particle falling in the interior effectively reduces the mass of the black hole and eventually can lead to total evaporation. In this picture, the black hole radiation is a pure vacuum effect triggered by the horizon geometry.

This way to model the evaporation process leads to a serious problem known as *information paradox*. Indeed, in the radiation process, the state of the emitted quanta – once we trace over the quanta falling in the interior – is characterized by an entanglement entropy of the order ln2 for each emission. If the process continues and if we assume that the same vacuum mechanism underlines the subsequent processes of emission, then the final result, once the black hole is completely evaporated, will be a mixed state for the radiated quanta. This will imply that the black hole formation–evaporation process is not satisfying standard quantum unitarity, i.e., a pure state eventually evolves into a mixed state. Therefore, in order to make the final state pure, we need some purification mechanism. This purification involves, necessarily, some deviations from exact thermality during the evaporation process.

An important and hot topic of discussion lies in identifying the effects responsible from such deviations from exact thermality. If we assume unitarity, these effects should be important enough to purify the final state. What makes the problem an interesting puzzle is when we compare it with the standard evaporation of a hot body. In this case, we have as an important input the number N of constituents of the system. If we assume that during the evaporation the system is effectively divided into two systems, one with $n(t)$ radiated quanta and the other with the $N - n(t)$ remaining constituents, for t the evaporation time, it is obvious that the entanglement entropy, defined for the radiated quanta after tracing over the remaining black hole constituents, starts to decrease at the moment n is of the order $N/2$. This moment defines Page's time. From this moment on, the radiation needs to deviate from thermality and the black hole should be able to deliver information. The puzzle appears because at this time the black hole can be arbitrarily large if we have started with N large enough. Thus, the geometry in the near-horizon region can be as smooth and weak as we wish, and therefore it looks, at first sight, that nothing could be wrong with Hawking's semiclassical picture. This is in essence the information paradox.

The lesson we can extract from Page's argument is simply that the information paradox appears because we are implicitly assuming $N = \infty$, i.e, we are working in the semiclassical limit as it was defined above. The purification mechanism, implicit in Page's argument, not only requires us to keep N finite but also to reduce from N to $N - 1$ in each emission process.

Once we realize this point, we need to ask ourselves how N enters into the computation of the black hole radiation.

Recall that for the black hole, N has the simple interpretation of being the average number of quanta of typical wavelength R_g we need to use to account, quantum mechanically, for the gravitational self-energy. In a zero-order approximation, the simplest quantum mechanical model we can associate with a black hole is a many-body system of N attractive bosons with wavelength of the order $\sqrt{N}L_P$. The strength of the interaction is defined gravitationally and becomes simply $1/N$. Thus, the key question we need to answer is: *what is so special about this many-body system in order to explain some of the peculiar and mysterious features of black hole geometry?*

The first guess about what is the most natural candidate for the ground state of this system of gravitons is, of course, a weakly interacting Bose-Einstein condensate. To avoid difficulties, we can think in a very large black hole with N much larger than one. The typical one-particle spectrum for this system will have a gap of the order $1/\sqrt{N}$ while the typical quantity representing the effective quantum effects is $1/N$. In normal conditions, we should expect that for very large N, quantum fluctuations of the system will induce a very small deviation from an exact BEC ground state. In that case, the one-particle density matrix is a matrix with only one nonvanishing eigenvalue in the diagonal, and the corresponding von Neuman entropy will vanish. If that will be the case, we should have few hopes to model black hole geometry in terms of such a BEC. In particular, such a BEC appears as a very bad candidate to store information in the form of von Neumann entropy for the one-particle reduced density matrix. Thus, what is the new ingredient, in this many-body model, that can create hopes to have grasped the essential microscopic dynamics of black holes?

The answer is in essence quite simple and somehow expected. This many-body system, for these particular values of the number of bodies and interaction strength, is *at a quantum critical point*. What is the meaning of this criticality?

It means that the low-lying collective excitations are gapless for these values. In other words, a large number of quantum states of order N are populating a range of energies of order $1/N$. This is extremely interesting for an obvious reason. Quantum effects of order $1/N$ are populating the first N low-lying states of the spectrum. The first important consequence is that now quantum effects are making N diagonal elements of the one-particle reduced density matrix nonvanishing, creating a large amount of von Neumann entropy. This phenomenon happens suddenly, as it is typical of a quantum phase transition, and it represents the main difference between the attractive gas of gravitons accounting for the gravitational self-energy of a normal system (for instance, a neutron star) and a black hole.

32.2.1 Hawking Radiation

But, what about Hawking radiation? Once Bekenstein assigned an entropy to the black hole and once we associate the black hole formation with a thermalization process, the thermodynamic consistency of the whole picture requires us to identify the mechanism underlying thermal radiation and the subsequent process of black hole evaporation. As already stressed, it is very important to distinguish between the emission process and its back reaction, i.e., the evaporation process itself. While the first process survives in the semiclassical limit of infinite N, the second can only be understood if we keep N finite and we keep track of changes of N by one unit, i.e., if we track $1/N$ effects.

Let us start focusing our attention on the emission process. How can we understand such emission in the BEC graviton model?

The underlying physics is quite simple. In the BEC graviton model, the black hole is in a first approximation a bound state of gravitons. Since these gravitons are interacting, there would be scattering processes among them that eventually can push one of the constituent gravitons away from the condensate, i.e., with an energy bigger than the escape energy. Since in the limit of very large N we can treat the system in the mean-field approximation, we should think the former processes in the infinite N limit as a transfer of momentum of one graviton with the collective system defined by the rest of gravitons. In the large N limit, we can easily estimate the rate of emission that turns out to be of one graviton in a time of order $\sqrt{N}$ in Planck units. Thus, this mean-field emission rate can be expressed as

$$\frac{dN}{dt} = \frac{1}{\sqrt{N} L_P}. \tag{32.2}$$

Let us now see what happens after this emission takes place if we keep N finite. The remaining $N - 1$ gravitons are slightly away (by $1/N$ effects) from the critical point, and they need some time to reorganize themselves as a new $(N - 1)$ black hole. Two potential sources of quantum corrections immediately enter into the game. If the next emission takes place after the thermalization of the remaining gravitons, we can again proceed as before to define the following emission, and this second emission process will not have memory of the first one. However, if the thermalization time is larger than the typical time of a new rescattering, then the new emission will differ from the former one, tracking extra memory of the whole evaporation process.

In order to grasp the key of the BEC approach to black holes, it is crucial to understand the special role of gravitons in the whole picture. Let us briefly elaborate this point. Naively, we could think that an attractive gas of gravitons has the same fate as any other system where attraction, at some point, dominates over pressure, namely to collapse. Thus, somebody can ask, what is the new ingredient we are adding using gravitons instead of any other form of gravitationally interacting matter? Of course, although we can certainly form a black hole putting together bottles of wine, it will be very silly to claim that the main constituents of the black hole are bottles of wine! Thus, what is so special about gravitons? The key ingredient is that the gravitons accounting for *the gravitational self interaction* reaccommodate, at each step, their wavelength to keep the *criticality conditions*. It is this criticality which is essentially new and does not exist for any other form of collapsing matter. The nuclei composing a star don't undergo any phase transition when the collapse starts, and they will continue collapsing. However, for the self-sourced gravitons, the situation is radically different, and they reach, at the moment of black hole formation, a critical point. In the BEC approach, the black hole radiation and the consequent evaporation reflects how the many-body system of gravitons keeps itself at criticality.

32.2.2 Quantum Break Time and Scrambling

The notion of black holes as scramblers was first introduced when it was realized that perturbed black holes should thermalize in a time $t \geq R \log S_{BH}$ for S_{BH} the black hole entropy and R the black hole radius. It was then suggested that black holes may saturate this bound, a property that has become known as fast scrambling. The associated time scale is now known as scrambling time.

Consider a quantum mechanical system whose Hilbert space is a direct product $\mathcal{H} = \mathcal{H}_A \otimes \mathcal{H}_B$ in a state described by the density matrix ρ. The conventional measure of entanglement between the subsystems is the von Neumann entropy of the reduced density matrix. A system is called a scrambler if it dynamically thermalizes in the sense that, if prepared in an atypical state, it evolves toward typicality. That is, even for an initial state that has little or no entanglement between subsystems, the time evolution is such that the reduced density matrices are finally close to thermal density matrices. The scrambling time is simply the characteristic time scale associated to this process. It can be described as the time it takes for a perturbed system, one that is described by a product state, to evolve back into a strongly entangled state. It can also be interpreted as the time necessary to distribute any information entering the system among all its constituents.

The quantum meaning of the scrambling time becomes more transparent if we rewrite it as $t_{\text{scrambling}} \sim R \log(S/\hbar)$, with S now denoting the action of the black hole. This is the typical expression for the *quantum break time* provided the system is near an instability, where quantum break time denotes the mean time scale for the breakdown of the classical (mean-field) description. Hence, we shall identify as a necessary condition for a system to behave as a fast scrambler to have a quantum break time scaling logarithmically with the number of constituents.

In the context of quantum chaos, it has long been known that under certain conditions, the classical description breaks down much quicker than the naively expected polynomial quantum break time. Specifically, in the vicinity of an instability for the classical description, i.e., positive local Lyapunov exponent λ, the quantum break time usually goes as $t_{\text{break}} \sim \lambda^{-1} \log \frac{S}{\hbar}$. This exactly resembles the logarithmic scaling of the scrambling time. In fact, the black hole scrambling time coincides with the typical quantum break time if the microscopic description of the black hole contains an instability characterized by a Lyapunov exponent $\lambda \sim 1/R$. The BEC model contains such an instability which survives in the semiclassical limit ($L_P = 0$, $N = \infty$, with $\sqrt{N}L_P$ fixed). The characteristic time scale is given by $R = \sqrt{N}L_P$, which classically becomes the black hole radius. Hence, we expect the Lyapunov exponent to be set by $1/R$.

The relation between scrambling and quantum break time is even stronger if the classical limit of the relevant system not only contains a local instability but also exhibits classical chaos. For such systems, it has been claimed – and checked to some extent – that the time scale of thermalization is of the same order as t_{break}.

A necessary condition for having a quantum break time t_b scaling like $\log N$ for some initial many body state Ψ_0 is the exponential growth with time of small fluctuations $\delta\Psi(t)$, where $\Psi = \Psi_0 + \delta\Psi$. In linear approximation, the equation controlling $\delta\Psi$ is the Bogoliubov–De Gennes equation. As discussed above, a significant departure from the mean-field approximation as well as generation of entanglement for the reduced one-particle density matrix requires a growth in time of the depleted, i.e., of the noncondensed, particles. Nicely enough, the equations controlling the growth of depleted particles are the same as the ones controlling the small fluctuations of the Gross-Pitaevskii equation, and therefore we can translate the problem of finding a time t_b scaling like $\log N$ into the simpler problem of the *stability of the Gross-Pitaevskii equation.*

We can understand the short break time more concretely if we think about the difference between the exact evolution and the mean-field evolution as the addition of a small perturbation to the exact Hamiltonian. Since an unstable system is exponentially sensitive to perturbations of the Hamiltonian, then the time for the evolution of states to differ substantially is very short. The instability is controlled by the Lyapunov exponent λ, while the pre-exponential factor will depend on the size of the perturbation. The quantum break time is the time when this becomes important, so we can naturally expect it to scale like $t_b \sim \lambda^{-1} \log N$.

In summary, the BEC model of black holes as an attractive Bose condensate at a critical point explains their nature as fast scramblers. In the $N = \infty$ limit, the quantum break time becomes infinity, and that is the reason why this effect cannot be seen in the semiclassical approximation. Contrary to Hawking temperature, the scrambling time is an intrinsic quantum property of black holes and therefore it can be explained only on the basis of some microscopic model such as the BEC model.

32.3 BECs and the Quantum Consistency of the Cosmological Constant

Already in 1977, Gibbons and Hawking (GH) noticed the important similarities between black holes and the cosmological constant as it is represented in eternal de Sitter space-time. In both cases, we have horizon and near-horizon particle creation and temperature. The notion of cosmological horizon is, however, observer dependent, and the thermal radiation created by the near-horizon geometry is of the same type as Unruh effect in Rindler space-time. Indeed, a geodesic observer in de Sitter will detect a thermal bath at GH temperature $T = 1/R_H$. However, the GH temperature is de Sitter invariant, and contrary to the case of Hawking radiation for the black hole, the GH particle creation in de Sitter space is not leading to any obvious instability of de Sitter.

In brief, in eternal de Sitter, the GH radiation should be exactly thermal as it is exactly thermal Unruh effect in Rindler space-time. This is in sharp contrast with what happens in the case of black holes, where Hawking thermality is not expected to be exactly thermal if the whole evaporation process satisfies unitarity. Heuristically, it is easy to understand why the very notion of eternal de Sitter implies exact thermality for the GH radiation. Imagine that it will not be the case. Hence, we could imagine that the observer cannot only detect the radiation but also resolve small correlations among the radiated quanta. If that could be the case, by this process of resolving the radiation, the observer will be effectively gaining information about what is going on behind her cosmological horizon. In practice, this resolution of the GH radiation will be equivalent to purify the GH thermal density matrix. However, such a possibility enters immediately in conflict with the notion of cosmological constant as a classical external source, that is, creating the cosmological horizon and therefore preventing the observer from having any information about what happens behind.

Thus, the very notion of cosmological constant requires us, in order to be consistent, to assume that GH radiation should be exactly thermal. This is what we easily assume for the Unruh radiation in Rindler. However, there exists a very important difference between Rindler and de Sitter, and that difference is related again with the protagonist of this lecture, namely N. In the case of Rindler, we cannot associate with the space a finite entropy, and therefore we are working with $N = \infty$, while in the case of de Sitter we can associate with the space-time a finite value of N, namely the GH entropy defined as the area of the cosmological horizon in Planck units.

Once we realize that de Sitter comes with a finite value of N, the first query is, of course, what is the meaning of this N, and why the GH radiation is protected with respect to $1/N$ or e^{-N} effects?

What could be the microscopic meaning of N in the case of de Sitter? A natural guess, after our discussion of black holes, is to imagine that N has a similar microscopic meaning for de Sitter space, namely, to account for the number of gravitons per Hubble patch self-sourced by the external cosmological constant. These gravitons can be imagined as having a well-defined frequency determined by the Hubble radius and an infinite wave length. Moreover, we can imagine that these gravitons, as it happens in the case of the black hole, are at a critical point. From our point of view, these are reasonable assumptions, once we realize that the cosmological constant (Λ), as an external source, comes with some gravitational self-interaction characterized by a length scale (the analog of the black hole gravitational radius) that is the Hubble radius

$$R_H = \frac{1}{\sqrt{\Lambda}} \,. \tag{32.3}$$

The quantum state representing this many-body system of gravitons is pure. This is far from surprising since what this state is describing is the many-body system of gravitons self-sourced by the cosmological constant.

If we accept this basic assumption, we can use the dynamics among these gravitons to understand both the GH radiation as well as to check under what conditions this radiation could be exactly thermal and what could be the meaning of this exact thermality.

If N is finite, rescattering among these gravitons will produce depletion phenomena similarly to what we have described in the case of Hawking radiation. The observer will see the spontaneous creation of quanta with frequencies of the order of $\hbar/R_H$, i.e, a very IR phenomenon. However, these quanta cannot, as it is the case of black holes, induce the effective reduction of the cosmological constant. The process of depletion is pure quantum noise of the system of N gravitons and not variation of the external classical dark energy density. This reflects the well-known fact about the difficulty to define energy in de Sitter. The quantum many-body effects lead to a deviation of the system from a mean-field approximation, or – in more precise words – they turn the would-be classical system of N gravitons representing the cosmological constant gravitational self-energy, into a quantum system. This process of loss of quantum coherence of the system of gravitons defines a quantum time and a quantum clock. In this clock, the life of the cosmological constant, as a well-defined classical concept, is determined by the quantum break time of the gravitational system. This time can be roughly estimated to be of $O(M_P^5/\Lambda^{3/2})$. After such a time, the notion of classical cosmological constant becomes quantum mechanically inconsistent.

In summary, within the BEC approach to cosmology, we find that de Sitter invariance is quantum mechanically broken and that the breaking is a $1/N$ effect with

$$N = \frac{R_H^2}{L_P^2} \,. \tag{32.4}$$

References

[1] G. Dvali and C. Gomez: Black hole's quantum N-portrait, *Fortsch. Phys.* 61 (2013) 742–767

[2] G. Dvali and C. Gomez: Black hole's 1/N hair, *Phys. Lett.* B 719 (2013) 419–423

[3] G. Dvali and C. Gomez: Black holes as critical point of quantum phase transition, *Eur. Phys. J.* C 74 (2014) 2752

[4] G. Dvali, D. Flassig, C. Gomez, A. Pritzel, and N. Wintergerst, Scrambling in the black hole portrait, *Phys. Rev.* D 88 (2013), 124041

[5] G. Dvali and C. Gomez: Quantum compositeness of gravity: black holes, AdS and inflation, *JCAP* (2014) 01, 023
[6] G. Dvali, C. Gomez, R. S. Isermann, D. Lst, and S. Stieberger Black hole formation and classicalization in ultra-Planckian 2N scattering, *Nucl. Phys.* B 893 (2015) 187–235
[7] G. Dvali and C. Gomez: Quantum exclusion of positive cosmological constant? *Ann. Phys.* (Berlin) **528**, 68–73 (2016)

Universal Bose-Einstein Condensation Workshop

Universal Themes of Bose-Einstein Condensation
Lorentz Center, Leiden, The Netherlands, 11–15 March 2013
List of Registered Participants

- A. Joy Allen (*Joint Quantum Centre Durham-Newcastle, Newcastle University, U.K.*)
- Nils Andersson (*University of Southampton, U.K.*)
- Jildou Baarsma (*Utrecht University, Netherlands*)
- Gordon Baym (*Univesity of Illinois at Urbana-Champaign, U.S.*)
- Natalia Berloff (*University of Cambridge, U.K.*)
- Keith Burnett (*Sheffield Unversity, U.K.*) *[Co-organiser]*
- Iacopo Carusotto (*Italian Institute of Optics–National Research Council of Italy [INO-CNR] Bose-Einstein Condensation [BEC] Center, University of Trento, Italy*)
- Cheng Chin (*University of Chicago, U.S.*)
- Nigel Cooper (*University of Cambridge, U.K.*)
- Matthew Davis (*University of Queensland, Australia*)
- Brian Demarco (*Univesity of Illinois at Urbana-Champaign, U.S.*)
- Sergej Demokritov (*University of Münster, Germany*)
- Hui Deng (*University of Michigan, U.S.*)
- Benoit Deveaud-Plédran (*École polytechnique fédérale de Lausanne [EPFL], Lausanne, Switzerland*)
- Rembert Duine (*Utrecht University, Netherlands*)
- Gia Dvali (*Ludwig-Maximilians-University, Munich, Germany*)
- Paul Eastham (*Trinity College Dublin, Ireland*)
- Ramy El-Ganainy (*Max-Planck Institute for the Physics of Complex Systems, Dresden, Germany*)
- Tilman Esslinger (*Swiss Federal Institute of Technology [ETH] Zurich, Switzerland*)
- Jonathan Fellows (*University of Warwick, U.K.*)
- Jason Fleischer (*Princeton University, U.S.*)
- Jozsef Fortagh (*University of Tübingen, Germany*)
- Thierry Giamarchi (*University of Geneva, Switzerland*)
- Zoran Hadzibabic (*University of Cambridge, U.K.*)
- Hartmut Haug (*Goethe University Frankfurt, Germany*)
- Markus Karl (*Heidelberg University, Germany*)
- Jonathan Keeling (*University of St Andrews, U.K.*)
- Wolfgang Ketterle (*Massachusetts Institute of Technology, U.S.*)
- Na Young Kim (*Stanford University, U.S.*)
- Sergei Klimin (*University of Antwerpen, Belgium*)
- Lucia Komendova (*University of Antwerpen, Belgium*)
- Anthony Leggett (*Univesity of Illinois at Urbana-Champaign, U.S.*)
- Peter Littlewood (*Argonne National Laboratory & University of Chicago, US*) *[Co-organiser]*
- Ke Liu (*Leiden University, Netherlands*)
- Francesca Maria Marchetti (*Universidad Autonoma de Madrid, Spain*)
- Peter Mason (*Joint Quantum Centre Durham-Newcastle, Durham University, U.K.*)
- Meera Parish (*University College London, U.K.*)

- George Pickett (*Lancaster University, U.K.*)
- Nick P. Proukakis (*Joint Quantum Centre Durham-Newcastle, Newcastle University, U.K.*) *[Co-organiser]*
- Guillaume Salomon (*Institut d'Optique, Palaiseau cedex, France*)
- Laurent Sanchez-Palencia (*Institut d'Optique, Palaiseau cedex, France*)
- Joerg Schmiedmayer (*Atominstitut, Vienna University of Technology, Austria*)
- Pierre Sikivie (*University of Florida, U.S.*)
- Maurice Skolnick (*Sheffield University, U.K.*)
- David Snoke (*University of Pittsburgh, U.S.*) *[Co-organiser]*
- Ian Spielman (*Joint Quantum Institute, Gaithersburg, U.S.*)
- Henk Stoof (*Utrecht University, Netherlands*) *[Co-organiser]*
- Sandro Stringari (*INO-CNR BEC Center, University of Trento, Italy*)
- Jacques Tempere (*Antwerpen, Belgium*)
- Hugo Teras (*Institut Pascal, Clermont Université, France*)
- Peter van der Straten (*Utrecht University, Netherlands*)
- N.J. (Klaasjan) Van Druten (*University of Amsterdam, Netherlands*)
- Wim Vassen (*University of Amsterdam, Netherlands*)
- Martin Weitz (*University of Bonn, Germany*)
- Angela White (*Joint Quantum Centre Durham-Newcastle, Newcastle University, U.K.*)
- Tod Wright (*University of Queensland, Brisbane, Australia*)

We gratefully acknowledge the financial and organisational support of the Lorentz Center (in particular, Corrie Kuster, Mieke Schutte and conference photographer Auke Planjer), and further support by Europhysics Letters. *Details and presentation slides can be found at www.lorentzcenter.nl/lc/web/2013/546/info.php3?wsid=546.*

Figure 1 *Conference photo of Universal Themes of BEC Workshop.*

Contributors

- **Allen, A. Joy** *(Joint Quantum Centre (JQC) Durham-Newcastle, School of Mathematics and Statistics, Newcastle University, Newcastle upon Tyne, NE1 7RU, UK)*
- **Altman, Ehud** *(Department of Physics, University of California, Berkeley, California 94720, USA, and Department of Condensed Matter Physics, Weizmann Institute of Science, Rehovet, Israel)*
- **Banik, Nilanjan** *(Department of Physics, University of Florida, Gainesville, FL 32611, USA)*
- **Barenghi, Carlo F.** *(Joint Quantum Centre (JQC) Durham-Newcastle, School of Mathematics and Statistics, Newcastle University, Newcastle upon Tyne, NE1 7RU, UK)*
- **Baumberg, Jeremy J.** *(Cavendish Laboratory, University of Cambridge, Cambridge CB3 0HE, UK)*
- **Bender, Scott A.** *(Department of Physics and Astronomy, University of California, Los Angeles, CA 90095, USA)*
- **Berloff, Natalia G.** *(Skolkovo Institute of Science and Technology Novaya St., 100, Karakorum Building, 4th floor Skolkovo 143025 Russian Federation, and DAMTP, Centre for Mathematical Sciences, Wilberforce Road, Cambridge CB3 0WA, UK)*
- **Bloch, Immanuel** *(Max-Planck Institute of Quantum Optics, Hans-Kopfermann Str. 1, 85748 Garching, Germany, and Ludwig-Maximilians University, Schellingstrasse 4, 80799 Munich, Germany)*
- **Brataas, Arne** *(Department of Physics, Norwegian University of Science and Technology, NO-7491 Trondheim, Norway)*
- **Carusotto, Iacopo** *(INO-CNR BEC Center and Dipartimento di Fisica, Università di Trento, Via Sommarive 14, I-38123 Povo, Italy)*
- **Chen, Hua** *(Department of Physics, University of Texas at Austin, USA)*

- **Chen, Leiming** *(College of Science, China University of Mining and Technology, Xuzhou, Jiangsu 221116, Peoples Republic of China)*
- **Chin, Cheng** *(James Franck Institute, Enrico Fermi Institute, University of Chicago, Chicago, IL 60637, USA)*
- **Chiocchetta, Alessio** *(SISSA – International School for Advanced Studies and INFN, via Bonomea 265, 34136 Trieste, Italy)*
- **Christmann, Gabriel** *(Foundation for Research and Technology–Hellas, Institute of Electronic Structure and Laser, P.O. Box 1527, 71110 Heraklion, Crete, Greece)*
- **Cooper, Nigel R.** *(T.C.M. Group, Cavendish Laboratory, J.J. Thomson Avenue, Cambridge CB3 0HE, UK)*
- **Daley, Andrew J.** *(Department of Physics and Scottish Universities Physics Alliance, University of Strathclyde, Glasgow G4 0NG, Scotland, UK)*
- **Dalibard, Jean** *(Laboratoire Kastler Brossel, Collège de France, CNRS, ENS-PSL Research University, UPMC-Sorbonne Universités, 11 place Marcelin Berthelot, 75005, Paris, France)*
- **Davis, Matthew J.** *(School of Mathematics and Physics, University of Queensland, St. Lucia QLD 4072, Australia and JILA, 440 UCB, University of Colorado, Boulder, Colorado 80309, USA)*
- **Demokritov, Sergej** *(Institute for Applied Physics and Center for Nonlinear Science, University of Muenster, Corrensstr. 2–4, 48149 Muenster, Germany, and Institute of Metal Physics, Ural Division of RAS, Yekaterinburg 620041, Russia)*
- **Diehl, Sebastian** *(Institut für Theoretische Physik, Universität zu Köln, D-50937 Cologne, Germany)*
- **Dominici, Lorenzo** *(Istituto di Nanotecnologia (NANOTEC) — Consiglio Nazionale delle Ricerche (CNR), Via Arnesano, 73100 Lecce, Italy)*
- **Duine, Rembert A.** *(Institute for Theoretical Physics and Center for Extreme Matter and Emergent Phenomena, Leuvenlaan 4, 3584 CE, Utrecht, The Netherlands)*
- **Dvali, Gia** *(Arnold Sommerfeld Center for Theoretical Physics, Department für Physik, Ludwig-Maximilians-Universität München, Theresienstr. 37, 80333 München, Germany, and Max-Planck-Institut für Physik, Föhringer Ring 6, 80805 München, Germany, and Center for Cosmology and Particle Physics, Department of Physics, New York University, 4 Washington Place, New York, NY 10003, USA)*
- **Eastham, Paul R.** *(School of Physics and CRANN, Trinity College Dublin, Dublin 2, Ireland)*
- **Edelman, Alex** *(James Franck Institute and Department of Physics, University of Chicago, Chicago, IL 60637, USA)*

- **Gambassi, Andrea** *(SISSA – International School for Advanced Studies and INFN, via Bonomea 265, 34136 Trieste, Italy)*
- **Gardiner, Simon A.** *(Joint Quantum Centre (JQC) Durham-Newcastle, Department of Physics, Durham University, Durham DH1 3LE, UK)*
- **Gasenzer, Thomas** *(Kirchhoff-Institut für Physik, Universität Heidelberg, Im Neuenheimer Feld 227, 69120 Heidelberg, Germany, and ExtreMe Matter Institute EMMI, GSI Helmholtzzentrum für Schwerionenforschung, 64291 Darmstadt, Germany)*
- **Giamarchi, Thierry** *(Department of Quantum Matter Physics, University of Geneva, Geneva, Switzerland)*
- **Goldman, Nathan** *(Center for Nonlinear Phenomena and Complex Systems, Université Libre de Bruxelles, CP 231, Campus Plaine, B-1050 Brussels, Belgium)*
- **Gomez, Cesar** *(Arnold Sommerfeld Center for Theoretical Physics, Department für Physik, Ludwig-Maximilians-Universität München, Theresienstr. 37, 80333 München, Germany, and Instituto de Física Teórica UAM-CSIC, C-XVI, Universidad Autónoma de Madrid, Cantoblanco, 28049 Madrid, Spain)*
- **Greytak, Thomas** *(Department of Physics, Massachusetts Institute of Technology, Cambridge, MA, USA)*
- **Guda, Kurumuthy** *(Department of Physics and Astronomy, University of Sheffield, Sheffield, S3 7RH, UK)*
- **Keeling, Jonathan** *(SUPA, School of Physics and Astronomy, University of St. Andrews, St. Andrews KY16 9SS UK)*
- **Ketterle, Wolfgang** *(Department of Physics, Massachusetts Institute of Technology, Cambridge, MA, USA)*
- **Kim, Na Young** *(Edward L. Ginzton Laboratory, Stanford University, 348 Via Pueblo Mall, Stanford, CA 94305, USA); present address: Institute for Quantum Computing, Department of Electrical and Computer Engineering, University of Waterloo, 200 University Avenue West, Waterloo, ON, N2L 3G1, Canada)*
- **Klaers, Jan** *(Institut für Angewandte Physik, Universität Bonn, Wegelerstr. 8, 53115 Bonn, Germany; present address: Institute of Quantum Electronics, ETH Zürich, Auguste-Piccard-Hof 1, 8093 Zürich, Switzerland)*
- **Kleppner, Daniel** *(Department of Physics, Massachusetts Institute of Technology, Cambridge, MA, USA)*
- **Kollath, Corinna** *(HISKP, University of Bonn, Nussallee 14-16, 53115 Bonn, Germany)*
- **Krizhanovskii, Dmitry N.** *(Department of Physics and Astronomy, University of Sheffield, Sheffield, S3 7RH, UK)*

- **Langen, Tim** *(Vienna Center for Quantum Science and Technology, Atominstitut, TU Wien, Stadionallee 2, 1020 Wien, Austria; present address: JILA, NIST, and Department of Physics, University of Colorado, Boulder, CO 80309, USA)*
- **LeBlanc, Lindsay J.** *(Department of Physics, University of Alberta, Edmonton AB, T6G 2E1, Canada)*
- **Littlewood, Peter B.** *(James Franck Institute and Department of Physics, University of Chicago, Chicago, IL 60637, USA, and Argonne National Laboratory, Lemont, IL, 60439, USA)*
- **MacDonald, Allan H.** *(Department of Physics, University of Texas at Austin, USA)*
- **Malpuech, Guillaume** *(Institut Pascal, PHOTON-N2, Clermont Université, Blaise Pascal University, CNRS,24 Avenue des Landais, 63177 Aubière Cedex, France)*
- **Nitsche, Wolfgang H.** *(Edward L. Ginzton Laboratory, Stanford University, 348 Via Pueblo Mall, Stanford, CA 94305, USA; present address: Halliburton, 3000 North Sam Houston Parkway East, Houston, TX 77032, USA)*
- **Parker, Nick G.** *(Joint Quantum Centre [JQC] Durham-Newcastle, School of Mathematics and Statistics, Newcastle University, Newcastle upon Tyne, NE1 7RU, UK)*
- **Pethick, Christopher J.** *(Niels Bohr International Academy, Niels Bohr Institute, University of Copenhagen, Blegdamsvej 17, DK-2100 Copenhagen Ø, Denmark, and NORDITA, KTH Royal Institute of Technology and Stockholm University, Roslagstullsbacken 23, SE-10691 Stockholm, Sweden)*
- **Pickett, George R.** *(Department of Physics, Lancaster University, Lancaster, LA1 4YB, UK)*
- **Pitaevskii, Lev** *(INO-CNR BEC Center and Dipartimento di Fisica, Università di Trento, Via Sommarive 14, I-38123 Povo, Italy, and Kapitza Institute for Physical Problems, RAS, ul Kosygina 2, 119334 Moscow, Russia)*
- **Proukakis, Nick P.** *(Joint Quantum Centre (JQC) Durham-Newcastle, School of Mathematics and Statistics, Newcastle University, Newcastle upon Tyne, NE1 7RU, UK)*
- **Rosenow, Bernd** *(Institut für Theoretische Physik, Universität Leipzig, D-04103, Leipzig, Germany)*
- **Rüegg, Christian** *(Paul Scherrer Institute, Laboratory for Neutron Scattering and Imaging, Switzerland, and University of Geneva, Department of Quantum Matter Physics, Switzerland)*
- **Salman, Hayder** *(School of Mathematics, University of East Anglia Norwich, NR4 7TJ, UK)*

- **Sanvitto, Daniele** *(Istituto di Nanotecnologia (NANOTEC) — Consiglio Nazionale delle Ricerche (CNR), Via Arnesano, 73100 Lecce, Italy)*
- **Savvidis, Pavlos G.** *(Foundation for Research and Technology–Hellas, Institute of Electronic Structure and Laser, P.O. Box 1527, 71110 Heraklion, Crete, Greece, and Department of Materials Science and Technology, University of Crete, P.O. Box 2208, 71003 Heraklion, Crete, Greece, and Cavendish Laboratory, University of Cambridge, Cambridge CB3 0HE, UK)*
- **Schäfer, Thomas** *(Department of Physics, North Carolina State University, Raleigh, NC 27695-8202, USA)*
- **Schmiedmayer, Jörg** *(Vienna Center for Quantum Science and Technology, Atominstitut, TU Wien, Stadionallee 2, 1020 Wien, Austria)*
- **Schwenk, Achim** *(Institut für Kernphysik, Technische Universität Darmstadt, D-64289 Darmstadt, Germany, and ExtreMe Matter Institute EMMI, GSI Helmholtzzentrum für Schwerionenforschung GmbH, D-64291 Darmstadt, Germany)*
- **Sich, Maksym** *(Department of Physics and Astronomy, University of Sheffield, Sheffield, S3 7RH, UK)*
- **Sieberer, Lukas M.** *(Department of Physics, University of California, Berkeley, California 94720, USA, and Department of Condensed Matter Physics, Weizmann Institute of Science, Rehovet, Israel, and Institute for Theoretical Physics, University of Innsbruck, Austria)*
- **Sikivie, Pierre** *(Department of Physics, University of Florida, Gainesville, FL 32611, USA)*
- **Skolnick, Maurice S.** *(Department of Physics and Astronomy, University of Sheffield, Sheffield, S3 7RH, UK)*
- **Smith, Robert P.** *(Cavendish Laboratory, University of Cambridge, J. J. Thomson Avenue, Cambridge, CB3 0HE, UK)*
- **Snoke, David W.** *(Department of Physics and Astronomy, University of Pittsburgh, Pittsburgh, PA 15260, USA)*
- **Solnyshkov, Dmitry D.** *(Institut Pascal, PHOTON-N2, Clermont Université, Blaise Pascal University, CNRS,24 Avenue des Landais, 63177 Aubière Cedex, France)*
- **Spielman, Ian B.** *(Joint Quantum Institute, National Institute of Standards and Technology, and University of Maryland, Gaithersburg MD 20899, USA)*
- **Stringari, Sandro** *(INO-CNR BEC Center and Dipartimento di Fisica, Università di Trento, Via Sommarive 14, I-38123 Povo, Italy)*
- **Terças, Hugo** *(Institute for Theoretical Physics, University of Innsbruck, 6020 Innsbruck, Austria; present address: Physics of Information and Quantum Technologies Group, Instituto de Telecomunicações, 1049-001 Lisboa, Portugal)*

- **Toner, John** *(Department of Physics and Institute of Theoretical Science, University of Oregon, Eugene, OR 97403, USA)*
- **Tserkovnyak, Yaroslav** *(Department of Physics and Astronomy, University of California, Los Angeles, CA 90095, USA)*
- **Ueda, Masahito** *(Department of Physics, University of Tokyo, 7-3-1 Hongo, Bunkyo-ku, Tokyo 113-0033, Japan)*
- **Weitz, Martin** *(Institut für Angewandte Physik, Universität Bonn, Wegelerstr. 8, 53115 Bonn, Germany)*
- **Wright, Tod M.** *(School of Mathematics and Physics, University of Queensland, St. Lucia QLD 4072, Australia)*
- **Yamamoto, Yoshihisa** *(Edward L. Ginzton Laboratory, Stanford University, 348 Via Pueblo Mall, Stanford, CA 94305, USA, and IMPACT Project, Japan Science and Technology Agency, Chiyoda-ku, Tokyo 102-0076, Japan)*

Index

Printed in the United States
by Baker & Taylor Publisher Services